UNITED STATES

DEPARTMENT OF AGRICULTURE

NATIONAL AGRICULTURAL STATISTICS SERVICE

AGRICULTURAL

STATISTICS

2016

ISBN: 978-1-59888-960-4

Agricultural Statistics 2016

Agricultural Statistics, 2016 was prepared under the direction of Jackie Ross, Agricultural Statistics Board, National Agricultural Statistics Service. Carolyne Foster, Phoebe Hilliard, Sherrie Pendarvis, and Tanya Ray were responsible for coordination and technical editorial work.

The USDA and NASS invite you to explore their information via their respective web sites: **http://www.usda.gov/** and **http://www.nass.usda.gov/.**

For information on NASS products you may call the **Agricultural Statistics Hotline, 1–800–727–9540 or send e-mail to nass@nass.usda.gov.**

We gratefully acknowledge the cooperation of the contributors to this publication. Source notes below each table credit the various Government agencies which collaborated in providing information.

CONTENTS

Introduction

Agricultural Statistics is published each year to meet the diverse need for a reliable reference book on agricultural production, supplies, consumption, facilities, costs, and returns. Its tables of annual data cover a wide variety of facts in forms suited to most common use.

Inquiries concerning more current or more detailed data, past and prospective revisions, or the statistical methodology used should be addressed directly to the agency credited with preparing the table. Most of the data were prepared or compiled in the U.S. Department of Agriculture.

The historical series in this volume are generally of data from 2006 and later.

Foreign agricultural trade statistics include Government as well as non-Government shipments of merchandise from the United States and Territories to foreign countries. They do not include U.S. shipments to the U.S. Armed Forces abroad for their own use or shipments between the States and U.S. Territories. The world summaries of production and trade of major farm products are prepared by the U.S. Department of Agriculture from reports of the U.S. Department of Commerce, official statistics of foreign governments, other foreign source materials, reports of U.S. Agricultural Attache and Foreign Service Officers, and the result of office research.

Statistics presented in many of the tables represent actual counts of the items covered. Most of the statistics relating to foreign trade and to Government programs, such as numbers and amounts of loans made to farmers, and amounts of loans made by the Commodity Credit Corporation, etc., are data of this type. A large number of other tables, however, contain data that are estimates made by the Department of Agriculture.

The estimates for crops, livestock, and poultry made by the U.S. Department of Agriculture are prepared mainly to give timely current State and national totals and averages. They are based on data obtained by sample surveys of farmers and of people who do business with farmers. The survey data are supplemented by information from the Censuses of Agriculture taken every five years and check data from various sources. Being estimates, they are subject to revision as more data become available from commercial or Government sources. Unless otherwise indicated, the totals for the United States shown in the various tables on area, production, numbers, price, value, supplies, and disposition are based on official Department estimates. They exclude States for which no official estimates are compiled.

DEFINITIONS

"Value of production" as applied to crops in the various tables, is derived by multiplying production by the estimated season average price received by farmers for that portion of the commodity actually sold. In the case of fruits and vegetables, quantities not harvested because of low prices or other economic factors are not included in value of production. The word "Value" is used in the inventory tables on livestock and poultry to mean value of the number of head on the inventory date. It is derived by multiplying the number of head by an estimated value per head as of the date.

The word "Year" (alone) in a column heading means calendar year unless otherwise indicated. "Ton" when used in this book without qualifications means a short ton of 2,000 pounds.

WEIGHTS, MEASURES, AND CONVERSION FACTORS

The following table on weights, measures, and conversion factors covers the most important agricultural products, or the products for which such information is most frequently asked of the U.S. Department of Agriculture. It does not cover all farm products nor all containers for any one product.

The information has been assembled from State schedules of legal weights, various sources within the U.S. Department of Agriculture, and other Government agencies. For most products, particularly fruits and vegetables, there is a considerable variation in weight per unit of volume due to differences in variety or size of commodity, condition and tightness of pack, degree to which the container is heaped, etc. Effort has been made to select the most representative and fairest average for each product. For those commodities which develop considerable shrinkage, the point of origin weight or weight at harvest has been used.

The approximate or average weights as given in this table do not necessarily have official standing as a basis for packing or as grounds for settling disputes. Not all of them are recognized as legal weight. The table was prepared chiefly for use of workers in the U.S. Department of Agriculture who have need of conversion factors in statistical computations.

WEIGHTS, MEASURES, AND CONVERSION FACTORS
(See explanatory text just preceding this table)

WEIGHTS AND MEASURES

Commodity	Unit[1]	Approximate net weight U.S.	Metric	Commodity	Unit[1]	Approximate net weight U.S.	Metric
		Pounds	Kilograms			Pounds	Kilograms
Alfalfa seed	Bushel	60	27.2	Celery	Crate[8]	60	27.2
Apples	....do	48	21.8	Cherries	Lug (Camp-		
Do	Loose pack	38–42	17.2–19.1		bell)[9]	16	7.3
Do	Tray pack	40–45	18.1–20.4	Do	Lug	20	9.1
Do	Cell pack	37–41	16.8–18.6	Clover seed	Bushel	60	27.2
Apricots	Lug (brent-			Coffee	Bag	132.3	60
	wood)[2]	24	10.9	Corn:			
Western	4–basket crate[3]	26	11.8	Ear, husked ...	Bushel[10]	70	31.8
Artichokes:				Shelled	....do	56	25.4
Globe	Ctn, by count			Meal	....do	50	22.7
	and loose			Oil	Gallon[7]	7.7	3.5
	pack	20–25	9.1–11.3	Syrup	....do	11.72	5.3
Jerusalem	Bushel	50	22.7	Sweet	Wirebound		
Asparagus	Crate (NJ)	30	13.6		crate	50	22.7
Avocados	Lug[4]	12–15	5.4–6.8	Do	Ctn, packed 5		
Bananas	Fiber folding				oz. ears	50	22.7
	box[5].	40	18.1	Do	WDB crate,		
Barley	Bushel	48	21.8		4½–5 oz.		
Beans:					(from FL &		
Lima, dry	....do	56	25.4		NJ)	42	19.1
Other, dry	....do	60	27.2	Cotton	Bale[11], gross ...	500	227
	Sack	100	45.4	Do	Bale[11], net	480	218
Lima	Bushel	28–32	12.7–14.5	Cottonseed	Bushel[12]	32	14.5
unshelled				Cottonseed oil ...	Gallon[7]	7.7	3.5
Snap	....do	28–32	12.7–14.5	Cowpeas	Bushel	60	27.2
Beets:				Cranberries	Barrel	100	45.4
Topped	Sack	25	11.3	Do	¼–bbl. box[13] ...	25	11.3
Bunched	½ crate 2 dz-			Cream, 40–per-			
	bchs	36–40	16.3–18.1	cent butterfat	Gallon	8.38	3.80
Berries				Cucumbers	Bushel	48	21.8
frozen pack:				Dewberries	24–qt. crate	36	16.3
Without sugar	50–gal. barrel ...	380	172	Eggplant	Bushel	33	15.0
3 + 1 pack	....do	425	193	Eggs, average			
2 + 1 pack	....do	450	204	size	Case, 30 dozen	47.0	21.3
Blackberries	12, ½-pint bas-	6	2.7	Escarole	Bushel	25	11.3
	ket.			Figs, fresh	Box single		
Bluegrass seed	Bushel	14–30	6.4–13.6		layer[14]	6	2.7
Broccoli	Wirebound	20–25	9.1–11.3	Flaxseed	Bushel	56	25.4
	crate.			Flour, various	Bag	100	45.4
Broomcorn (6	Bale	333	151	Do	Ctn or Crate,		
bales per ton)				Garlic	Bulk	30	13.6
Broomcorn seed	Bushel	44–50	20.0–22.7		Ctn of 12 tubes		
Brussels sprouts	Ctn, loose pack	25	11.3		or 12 film bag		
Buckwheat	Bushel	48	21.8		pkgs 12		
Butter	Block	55,68	25,30.9		cloves each ..	10	4.5
Cabbage	Open mesh bag	50	22.7	Grapefruit:			
Do	Flat crate (1¾			Florida and			
	bu)	50–60	22.7–27.2	Texas	½–box mesh		
Do	Ctn, place pack	53	24.0		bag	40	18.1
Cantaloups	Crate[6]	40	18.1	Florida	1⅗ bu. box	85	38.6
Carrots	Film plastic			Texas	1⅖ bu. box	80	36.3
	Bags, mesh			California and			
	sacks & car-			Arizona	Box[15,16]	67	30.4
	tons holding			Grapes:			
	48 1 lb. film			Eastern	12–qt. basket ...	20	9.1
Without tops ..	bags	55	24.9	Western	Lug	28	12.7
Castor beans	Burlap sack	74–80	33.6–36.3	Do	4–basket		
Castor oil	Bushel	41	18.6		crate[17]	20	9.1
Cauliflower	Gallon[7]	8	3.6	Hempseed	Bushel	44	20.0
Do	W.G.A. crate	50–60	22.7–27.2	Hickory nuts	....do	50	22.7
	Fiberboard box			Honey	Gallon	11.84	5.4
	wrapper			Honeydew			
	leaves			melons	⅔ Ctn	28–32	12.7–14.5
	removed film-			Hops	Bale, gross	200	90.7
	wrapped, 2						
	layers	23–35	10.4–15.9				

See footnotes on page ix.

WEIGHTS AND MEASURES—Continued

Commodity	Unit [1]	Approximate net weight U.S. (Pounds)	Approximate net weight Metric (Kilograms)	Commodity	Unit [1]	Approximate net weight U.S. (Pounds)	Approximate net weight Metric (Kilograms)
Horseradish roots	Bushel	35	15.9	Do	Std box, 4/5 bu	45–48	20.4–21.8
Do	Sack	50	22.7	Do	Ctn, Tight-fill pack	36–37	16.3–16.7
Hungarian millet seed	Bushel	48–50	21.8–22.7	Peas: Green,			
Kale	Ctn or crate	25	11.3	unshelled	Bushel	28–30	12.7–13.6
Kapok seed	...do	35–40	15.9–18.1	Dry	...do	60	27.2
Lard	Tierce	375	170	Peppers, green	...do	25–30	11.3–13.6
Lemons: California and				Do	1½ bu carton	28	12.7
Arizona	Box [18]	76	34.5	Perilla seed	Bushel	37–40	16.8–18.1
Do	Carton	38	17.2	Pineapples	Carton	40	18.1
Lentils	Bushel	60	27.2	Plums and			
Lettuce, iceberg	carton packed 24.	43–52	19.5–23.6	prunes:	Ctn & lugs	28	12.7
Lettuce, hot-				Do	½-bu. basket	30	13.6
house	24-qt. basket	10	4.5	Popcorn: On ear	Bushel [10]	70	31.8
Limes (Florida)	Box	88	39.9	Shelled	...do	56	25.4
Linseed oil	Gallon [7]	7.7	3.5	Poppy seed	...do	46	20.9
Malt	Bushel	34	15.4	Potatoes	Bushel	60	27.2
Maple syrup	Gallon	11.02	5.0	Do	Barrel	165	74.8
Meadow fescue seed	Bushel	24	10.9	Do	Box	50	22.7
Milk	Gallon	8.6	3.9	Do	...do	100	45.4
Millet	Bushel	48–60	21.8–27.2	Quinces	Bushel	48	21.8
Molasses: edible	Gallon	11.74	5.3	Rapeseed	...do	50–60	22.7–27.2
inedible	...do	11.74	5.3	Raspberries	½-pint baskets	6	2.7
Mustard seed	Bushel	58–60	26.3–27.2	Redtop seed	Bushel	50–60	22.7–27.2
Oats	...do	32	14.5	Refiners' syrup	Gallon	11.45	5.2
Olives	Lug	25–30	11.3–13.6	Rice: Rough	Bushel	45	20.4
Olive oil	Gallon	7 7.6	3.4	Do	Bag	100	45.4
Onions, dry	Sack	50	22.7	Do	Barrel	162	73.5
Onions, green bunched	Ctn, 24-dz bchs	10–16	4.5–7.3	Milled	Pocket or bag	100	45.4
Oranges:				Rosin	Drum, net	520	236
Florida	Box	90	40.8	Rutabagas	Bushel	56	25.4
Texas	Box	85	38.5	Rye	...do	56	25.4
California and Arizona	Box [15]	75	34.0	Sesame seed	...do	46	20.9
Do	Carton	38	17.2	Shallots	Crate (4–7 doz. bunches)	20–35	9.1–15.9
Orchardgrass seed	Bushel	14	6.4	Sorgo: Seed	Bushel	50	22.7
Palm oil	Gallon [7]	7.7	3.5	Syrup	Gallon	11.55	5.2
Parsnips	Bushel	50	22.7	Sorghum grain [19]	Bushel	56	25.4
Peaches	...do	48	21.8	Soybeans	...do	60	27.2
Do	2 layer ctn or lug	22	10.0	Soybean oil	Gallon [7]	7.7	3.5
Do	¾-Bu, Ctn/crate	38	17.2	Spelt	Bushel	40	18.1
Peanut oil	Gallon [7]	7.7	3.5	Spinach	...do	18–20	8.2–9.1
Peanuts, unshelled: Virginia type	Bushel	17	7.7	Strawberries	24-qt. crate	36	16.3
Runners, South-east-				Do	12-pt. crate	9–11	4.1–5.0
ern	...do	21	9.5	Sudangrass seed	Bushel	40	18.1
Spanish: South-				Sugarcane: Syrup (sulfured or			
eastern	...do	25	11.3	un-sulfured)	Gallon	11.45	5.2
South-				Sunflower seed	Bushel	24–32	10.9–14.5
western	...do	25	11.3	Sweet potatoes	Bushel [20]	55	24.9
Pears:				Do	Crate	50	22.7
California	Bushel	48	21.8	Tangerines: Florida	Box	95	43.1
Other	...do	50	22.7	Arizona	Box	75	34.0
				California	Box	75	34.0

See footnotes on page ix.

WEIGHTS AND MEASURES—Continued

Commodity	Unit [1]	Approximate net weight		Commodity	Unit [1]	Approximate net weight	
		U.S.	Metric			U.S.	Metric
		Pounds	*Kilograms*			*Pounds*	*Kilograms*
Timothy seed	Bushel	45	20.4	Turnips:			
Tobacco:				Without tops ..	Mesh sack	50	22.7
Maryland	Hogshead	775	352	Bunched	Crate [6]	70–80	31.8–36.3
Flue-cured	.....do	950	431	Turpentine	Gallon	7.23	3.3
Burley	.....do	975	442	Velvetbeans			
Dark air-cured	.....do	1,150	522	(hulled)	Bushel	60	27.2
Virginia fire-				Vetch seed	.....do	60	27.2
cured	.....do	1,350	612	Walnuts	Sacks	50	22.7
Kentucky and				Water 60° F	Gallon	8.33	3.8
Tennessee				Watermelons	Melons of aver-		
fire-cured	.....do	1,500	680		age or me-		
Cigar-leaf	Case	250–365	113–166		dium size	25	11.3
Do	Bale	150–175	68.0–79.4	Wheat	Bushel	60	27.2
Tomatoes	Crate	60	27.2	Various com-			
Do	Lug box	32	14.5	modities	Short ton	2,000	907
Do	2-layer flat	21	9.5	Do	Long ton	2,240	1,016
Tomatoes, hot-				Do	Metric ton	2,204.6	1,000
house	12-qt. basket	20	9.1				
Tung oil	Gallon [7]	7.8	3.5				

See footnotes on page ix.

To Convert From Avoirdupois Pounds

To	Multiply by
Kilograms	0.45359237
Metric tons	0.00045359237

Conversion Factors

1 Metric ton=2,204.622 pounds
1 Kilogram=2.2046 pounds
1 Acre=0.4047 hectares
1 Hectare=2.47 acres
1 Square mile=640 acres=259 hectares
1 Gallon=3.7853 liters

CONVERSION FACTORS

Commodity	Unit	Approximate equivalent
Apples	1 pound dried	7 pounds fresh; beginning 1943, 8 pounds fresh
Do	1 pound chops	5 pounds fresh
Do	1 case canned [21]	1.4 bushels fresh
Applesauce	do [21]	1.2 bushels fresh
Apricots	1 pound dried	6 pounds fresh
Barley flour	100 pounds	4.59 bushels barley
Beans, lima	1 pound shelled	2 pounds unshelled
Beans, snap or wax	1 case canned [22]	0.008 ton fresh
Buckwheat flour	100 pounds	3.47 bushels buckwheat
Calves	1 pound live weight	0.611 pound dressed weight (1999 average)
Cattle	do	0.607 pound dressed weight (1999 average)
Cane syrup	1 gallon	5 pounds sugar
Cherries, tart	1 case canned [21]	0.023 ton fresh
Chickens	1 pound live weight	0.72 pound ready-to-cook weight
Corn, shelled	1 bushel (56 lbs.)	2 bushels (70 pounds) of husked ear corn
Corn, sweet	1 case canned [22]	0.030 ton fresh
Cornmeal:		
Degermed	100 pounds	3.16 bushels corn, beginning 1946
Nondegermed	do	2 bushels corn, beginning 1946
Cotton	1 pound ginned	3.26 pounds seed cotton, including trash [23]
Cottonseed meal	1 pound	2.10 pounds cottonseed
Cottonseed oil	do	5.88 pounds cottonseed
Dairy products:		
Butter	do	21.1 pounds milk
Cheese	do	10 pounds milk
Condensed milk, whole	do	2.3 pounds milk
Dry cream	do	19 pounds milk
Dry milk, whole	do	7.6 pounds milk
Evaporated milk, whole	do	2.14 pounds milk
Malted milk	do	2.6 pounds milk
Nonfat dry milk	do	11 pounds liquid skim milk
Ice cream [24]	1 gallon	15 pounds milk
Ice cream [24] (eliminating fat from butter and concentrated milk).	do	12 pounds milk
Eggs	1 case	47 pounds
Eggs, shell	do	41.2 pounds frozen or liquid whole eggs
Do	do	10.3 pounds dried whole eggs
Figs	1 pound dried	3 pounds fresh in California; 4 pounds fresh elsewhere
Flaxseed	1 bushel	About 2½ gallons oil
Grapefruit, Florida	1 case canned [22]	0.64 box fresh fruit
Hogs	1 pound live weight	0.737 pound dressed weight, excluding lard (1999 average)
Linseed meal	1 pound	1.51 pounds flaxseed
Linseed oil	do	2.77 pounds flaxseed
Malt	1 bushel (34 lbs.)	1 bushel barley (48 lbs.)
Maple syrup	1 gallon	8 pounds maple sugar
Nuts:		
Almonds, imported	1 pound shelled	3½ pounds unshelled
Almonds, California	do	2.22 pounds unshelled through 1949; 2 pounds thereafter
Brazil	do	2 pounds unshelled
Cashews	do	4.55 pounds unshelled
Chestnuts	do	1.19 pounds unshelled
Filberts	do	2.22 pounds unshelled through 1949; 2.5 pounds thereafter
Pecans:		
Seedling	do	2.78 pounds unshelled
Improved	do	2.50 pounds unshelled
Pignolias	do	1.3 pounds unshelled
Pistachios	do	2 pounds unshelled
Walnuts:		
Black	do	5.88 pounds unshelled
Persian (English)	do	2.67 pounds unshelled
Oatmeal	100 pounds	7.6 bushels oats, beginning 1943
Oranges, Florida	1 case canned juice [22]	0.53 box fresh
Peaches, California, freestone	1 pound dried	5⅓ pounds fresh through 1918; 6 pounds fresh for 1919–28; and 6½ pounds fresh from 1929 to date
Peaches, California, clingstone	do	7½ pounds fresh
Peaches, clingstone	1 case canned [21]	1 bushel fresh
Do	do	0.0230 ton fresh
Peanuts	1 pound shelled	1½ pounds unshelled
Pears	1 pound dried	6½ pounds fresh
Pears, Bartlett	1 case canned [22]	1.1 bushels fresh
Do	do	0.026 ton fresh

See footnotes on page ix.

CONVERSION FACTORS—Continued

Commodity	Unit	Approximate equivalent
Peas, green	1 pound shelled	2½ pounds unshelled
Do	1 case canned [22]	0.009 ton fresh (shelled)
Prunes	1 pound dried	2.7 pounds fresh in California; 3 to 4 pounds fresh elsewhere
Raisins	1 pound	4.3 pounds fresh grapes
Rice, milled (excluding brewers)	100 pounds	152 pounds rough or unhulled rice
Rye flour	....do	2.23 bushels rye, beginning 1947
Sheep and lambs	1 pound live weight	0.504 pound dressed weight (1999 average)
Soybean meal	1 pound	1.27 pounds soybeans
Soybean oil	....do	5.49 pounds soybeans
Sugar	1 ton raw	0.9346 ton refined
Tobacco	1 pound farm-sales weight ..	Various weights of stemmed and unstemmed, according to aging and the type of tobacco (See circular 435, U.S. Dept. of Agr.)
Tomatoes	1 case canned [22]	0.018 ton fresh
Turkeys	1 pound live weight	0.80 pound ready-to-cook weight
Wheat flour	100 pounds	2.30 bushels wheat [25]
Wool, domestic apparel shorn	1 pound greasy	0.48 pounds scoured
Wool, domestic apparel pulled	....do	0.73 pound scoured

[1] Standard bushel used in the United States contains 2,150.42 cubic inches; the gallon, 231 cubic inches; the cranberry barrel, 5,826 cubic inches; and the standard fruit and vegetable barrel, 7,056 cubic inches. Such large-sized products as apples and potatoes sometimes are sold on the basis of a heaped bushel, which would exceed somewhat the 2,150.42 cubic inches of a bushel basket level full. This also applies to such products as sweetpotatoes, peaches, green beans, green peas, spinach, etc.

[2] Approximate inside dimensions, 4⅝ by 12½ by 16⅛ inches.

[3] Approximate inside dimensions, 4½ by 16 by 16⅛ inches.

[4] Approximate dimensions, 4½ by 13½ by 16⅛ inches.

[5] Approximate inside dimensions, 13 by 12 by 32 inches.

[6] Approximate inside dimensions, 13 by 18 by 21⅝ inches.

[7] This is the weight commonly used in trade practices, the actual weight varying according to temperature conditions.

[8] Approximate inside dimensions, 9¾ by 16 by 20 inches.

[9] Approximate inside dimensions, 4⅛ by 11½ by 14 inches.

[10] The standard weight of 70 pounds is usually recognized as being about 2 measured bushels of corn, husked, on the ear, because it required 70 pounds to yield 1 bushel, or 56 pounds, of shelled corn.

[11] For statistical purposes the bale of cotton is 500 pounds or 480 pounds net weight. Prior to Aug. 1, 1946, the net weight was estimated at 478 pounds. Actual bale weights vary considerably, and the customary average weights of bales of foreign cotton differ from that of the American square bale.

[12] This is the average weight of cottonseed, although the legal weight in some States varies from this figure of 32 pounds.

[13] Approximate inside dimensions, 9¼ by 10½ by 15 inches.

[14] Approximate inside dimensions, 1¾ by 11 by 16⅛ inches.

[15] Approximate inside dimensions, 11½ by 11½ by 24 inches.

[16] Beginning with the 1993-94 season, net weights for California Desert Valley and Arizona grapefruit were increased from 64 to 67 pounds, equal to the California other area net weight, making a 67 pound net weight apply to all of California.

[17] Approximate inside dimensions, 4¾ by 16 by 16⅛ inches.

[18] Approximate inside dimensions, 9⅞ by 13 by 25 inches.6 by 16 by 16⅛ inches.

[19] Includes both sorghum grain (kafir, milo, hegari, etc.) and sweet sorghum varieties.

[20] This average of 55 pounds indicates the usual weight of sweetpotatoes when harvested. Much weight is lost in curing or drying and the net weight when sold in terminal markets may be below 55 pounds.

[21] Case of 24 No. 2½ cans.

[22] Case of 24 No. 303 cans.

[23] Varies widely by method of harvesting.

[24] The milk equivalent of ice cream per gallon is 15 pounds. Reports from plants indicate about 81 percent of the butterfat in ice cream is from milk and cream, the remainder being from butter and concentrated milk. Thus the milk equivalent of the milk and cream in a gallon of ice cream is about 12 pounds.

[25] This is equivalent to 4.51 bushels of wheat per barrel (196 pounds) of flour and has been used in conversions, beginning July 1, 1957. Because of changes in milling processes, the following factors per barrel of flour have been used for earlier periods: 1790–1879, 5 bushels; 1880–1908, 4.75 bushels, 1909–17, 4.7 bushels; 1918 and 1919, 4.5 bushels; 1920, 4.6 bushels; 1921–44, 4.7 bushels; July 1944–Feb. 1946, 4.57 bushels; March 1946–Oct. 1946, average was about 4.31 bushels; and Nov. 1946–June 1957, 4.57 bushels.

CHAPTER I
STATISTICS OF GRAIN AND FEED

This chapter contains tables for wheat, rye, rice, corn, oats, barley, sorghum grain, and feedstuffs. Estimates are given of area, production, disposition, supply and disappearance, prices, value of production, stocks, foreign production and trade, price-support operations, animal units fed, and feed consumed by livestock and poultry.

Table 1-1.—Total grain: Supply and disappearance, United States, 2005–2014 [1]

Year [2]	Supply				Disappearance			Ending stocks
	Beginning stocks	Production	Imports	Total	Domestic use	Exports	Total disappearance	
	Million metric tons	Million metric tons	Million metric tons	Million metric tons	Million metric tons	Million metric tons	Million metric tons	Million metric tons
2005	74.7	363.1	4.8	442.6	280.2	90.7	370.9	71.7
2006	71.7	335.5	6.5	413.7	277.8	86.0	363.8	49.9
2007	49.9	412.0	7.3	469.1	307.2	107.6	414.8	54.3
2008	54.3	400.4	7.1	461.9	314.4	81.6	396.0	65.9
2009	66.3	418.0	6.4	490.8	331.1	83.4	414.5	76.3
2010	76.3	399.6	6.0	481.9	333.1	90.8	423.9	58.0
2011	58.0	385.1	6.8	449.9	326.3	74.2	400.0	49.9
2012	49.9	355.7	11.0	416.6	318.7	53.1	371.8	44.7
2013	44.7	433.9	9.0	487.1	345.3	90.2	435.6	51.8
2014 [3]	51.5	442.4	8.4	503.3	348.8	84.8	433.2	....

[1] Aggregate data on corn, sorghum, barley, oats, wheat, rye, and rice. [2] The marketing year for corn and sorghum begins September 1; for oats, barley, wheat, and rye, June 1; and for rice, August 1. [3] Estimate.
ERS, Market and Trade Economics Division, (202) 694–5313.

Table 1-2.—Wheat: Area, yield, production, and value, United States, 2006–2015

Year	Area		Yield per harvested acre	Production	Marketing year average price per bushel received by farmers [2]	Value of production [2]
	Planted [1]	Harvested				
	1,000 acres	1,000 acres	Bushels	1,000 bushels	Dollars	1,000 dollars
2006	57,334	46,800	38.6	1,808,416	4.26	7,694,734
2007	60,460	50,999	40.2	2,051,088	6.48	13,289,326
2008	63,617	56,036	44.8	2,511,896	6.78	16,701,285
2009	59,017	49,841	44.3	2,208,918	4.87	10,607,218
2010	52,620	46,883	46.1	2,163,023	5.70	12,579,125
2011	54,277	45,687	43.6	1,993,111	7.24	14,269,225
2012	55,294	48,758	46.2	2,252,307	7.77	17,383,149
2013	56,236	45,332	47.1	2,134,979	6.87	14,604,442
2014	56,841	46,385	43.7	2,026,310	5.99	11,914,954
2015	54,644	47,094	43.6	2,051,752	5.00	10,203,360

[1] Includes area seeded in preceding fall for winter wheat. [2] Includes allowance for loans outstanding and purchases by the Government valued at the average loan and purchase rate, by States, where applicable.
NASS, Crops Branch, (202) 720–2127.

Table 1-3.—Wheat, by type: Area, yield, production, and value, United States, 2006–2015

Year	Area		Yield per harvested acre	Production	Marketing year average price per bushel received by farmers [2]	Value of production [2]
	Planted [1]	Harvested				
	1,000 acres	*1,000 acres*	*Bushels*	*1,000 bushels*	*Dollars*	*1,000 dollars*

Winter wheat

Year	Planted [1]	Harvested	Yield per harvested acre	Production	Price	Value of production [2]
	1,000 acres	*1,000 acres*	*Bushels*	*1,000 bushels*	*Dollars*	*1,000 dollars*
2006	40,565	31,107	41.6	1,294,461	4.17	5,367,806
2007	45,012	35,938	41.7	1,499,241	6.13	9,077,574
2008	46,781	40,000	47.1	1,885,575	6.57	12,054,269
2009	43,287	34,550	44.0	1,521,077	4.71	7,070,719
2010	36,576	31,219	46.5	1,452,313	5.37	7,835,595
2011	40,596	32,378	46.1	1,493,130	6.81	10,154,257
2012	40,897	34,609	47.1	1,630,387	7.55	12,245,482
2013	43,230	32,650	47.3	1,542,902	6.89	10,590,949
2014	42,409	32,299	42.6	1,377,216	5.92	8,036,108
2015	39,461	32,257	42.5	1,370,188	4.90	6,716,666

Durum wheat

Year	Planted [1]	Harvested	Yield per harvested acre	Production	Price	Value of production [2]
	1,000 acres	*1,000 acres*	*Bushels*	*1,000 bushels*	*Dollars*	*1,000 dollars*
2006	1,870	1,815	29.5	53,475	4.43	243,992
2007	2,156	2,119	34.1	72,224	9.92	692,512
2008	2,721	2,574	31.3	80,467	9.26	704,365
2009	2,512	2,386	44.0	104,930	5.47	569,360
2010	2,503	2,462	41.2	101,482	5.98	633,469
2011	1,337	1,280	36.8	47,043	9.68	445,186
2012	2,138	2,122	38.4	81,501	8.18	682,317
2013	1,400	1,338	43.3	57,976	7.46	432,733
2014	1,407	1,346	40.2	54,056	8.81	482,417
2015	1,936	1,896	43.5	82,484	7.80	617,694

Other spring wheat [3]

Year	Planted [1]	Harvested	Yield per harvested acre	Production	Price	Value of production [2]
	1,000 acres	*1,000 acres*	*Bushels*	*1,000 bushels*	*Dollars*	*1,000 dollars*
2006	14,899	13,878	33.2	460,480	4.46	2,082,936
2007	13,292	12,942	37.1	479,623	7.16	3,519,240
2008	14,115	13,462	40.5	545,854	7.31	3,942,651
2009	13,218	12,905	45.2	582,911	5.23	2,967,139
2010	13,541	13,202	46.1	609,228	6.49	4,110,061
2011	12,344	12,029	37.7	452,938	8.24	3,669,782
2012	12,259	12,027	44.9	540,419	8.24	4,455,350
2013	11,606	11,344	47.1	534,101	6.73	3,580,760
2014	13,025	12,740	46.7	595,038	5.75	3,396,429
2015	13,247	12,941	46.3	599,080	4.85	2,869,000

[1] Seeded in preceding fall for winter wheat.   [2] Obtained by weighting State prices by quantity sold.   [3] Includes small quantities of Durum wheat grown in other States.
NASS, Crops Branch, (202) 720–2127.

Table 1-4.—Wheat: Stocks on and off farms, United States, 2006–2015

Year beginning September	All wheat							
	On farms				Off farms [1]			
	Sept. 1	Dec. 1	Mar. 1	Jun. 1	Sept. 1	Dec. 1	Mar. 1	Jun. 1
	1,000 bushels	1,000 bushels	1,000 bushels	1,000 bushels	1,000 bushels	1,000 bushels	1,000 bushels	1,000 bushels
2006	572,020	403,250	192,450	73,190	1,178,525	911,408	664,278	382,963
2007	495,000	289,540	91,990	25,635	1,221,927	842,398	617,280	280,183
2008	635,700	454,000	280,400	140,745	1,222,183	968,089	759,664	515,760
2009	836,000	558,800	348,250	209,900	1,373,338	1,222,891	1,008,107	765,737
2010	812,100	550,000	288,010	130,915	1,637,517	1,382,946	1,137,292	732,083
2011	633,000	405,200	217,100	112,030	1,513,669	1,257,318	982,245	630,590
2012	572,900	399,500	236,970	120,150	1,542,209	1,271,079	997,860	597,739
2013	555,000	398,400	237,530	96,995	1,314,637	1,076,451	819,435	493,288
2014	713,450	472,800	278,710	155,170	1,193,770	1,056,830	861,697	597,224
2015	650,200	503,450	319,800	197,210	1,446,889	1,242,799	1,052,355	784,091

Year beginning September	Durum wheat [2]							
	On farms				Off farms [1]			
	Sept. 1	Dec. 1	Mar. 1	Jun. 1	Sept. 1	Dec. 1	Mar. 1	Jun. 1
	1,000 bushels	1,000 bushels	1,000 bushels	1,000 bushels	1,000 bushels	1,000 bushels	1,000 bushels	1,000 bushels
2006	31,500	25,900	17,100	8,950	31,524	25,447	21,736	12,430
2007	34,700	17,600	8,100	2,350	35,764	22,170	17,058	5,938
2008	36,200	26,100	18,700	13,300	22,595	18,405	13,571	11,774
2009	74,100	50,600	34,300	23,900	27,686	25,181	21,216	10,749
2010	71,200	46,600	35,700	22,100	28,931	21,742	20,720	13,366
2011	34,900	24,500	17,900	15,200	28,828	23,507	17,899	10,270
2012	43,600	36,700	21,400	13,600	24,842	24,306	21,088	9,450
2013	42,900	32,800	20,700	12,800	23,465	21,175	17,430	8,724
2014	38,700	23,900	16,200	10,250	19,121	20,147	21,454	15,406
2015	44,900	35,700	17,700	12,190	29,146	24,650	24,840	15,661

[1] Includes stocks at mills, elevators, warehouses, terminals, and processors.　[2] Included in all wheat.
NASS, Crops Branch, (202) 720–2127.

Table 1-5.—Wheat: Supply and disappearance, by class, United States, 2011–2015 [1]

Item	Year beginning June				
	2011	2012	2013	2014	2015
	Million bushels	Million bushels	Million bushels	Million bushels	Million bushels
All wheat:					
Stocks, June 1	862	743	718	590	752
Production	1,999	2,252	2,135	2,026	2,062
Supply [2]	2,974	3,119	3,025	2,768	2,927
Exports [3]	1,051	1,012	1,176	864	775
Domestic disappearance	1,180	1,389	1,259	1,151	1,177
Stocks, May 31	743	718	590	752	976
Hard red winter:					
Stocks, June 1	386	317	343	237	294
Production	780	998	747	739	830
Supply [2]	1,166	1,333	1,109	985	1,130
Exports [3]	397	382	446	272	226
Domestic disappearance	452	608	426	419	458
Stocks, May 31	317	343	237	294	446
Soft red winter:					
Stocks, June 1	171	185	124	113	154
Production	458	413	568	455	359
Supply [2]	661	616	713	581	531
Exports [3]	165	194	283	134	120
Domestic disappearance	310	297	318	293	255
Stocks, May 31	185	124	113	154	157
Hard red spring:					
Stocks, June 1	185	151	165	169	212
Production	398	503	491	556	568
Supply [2]	618	698	733	792	828
Exports [3]	243	233	246	274	252
Domestic disappearance	224	301	318	306	304
Stocks, May 31	151	165	169	212	272
Durum:					
Stocks, June 1	35	25	23	22	26
Production	50	82	58	54	84
Supply [2]	122	145	129	126	143
Exports [3]	27	29	32	37	29
Domestic disappearance	70	93	76	64	86
Stocks, May 31	25	23	22	26	28
White:					
Stocks, June 1	85	64	63	50	67
Production	314	257	271	224	221
Supply [2]	406	328	341	283	294
Exports [3]	219	175	170	147	147
Domestic disappearance	124	90	121	69	73
Stocks, May 31	64	63	50	67	74

[1] Data except production are approximations. [2] Total supply includes imports. [3] Import and exports include flour and products in wheat equivalent.
ERS, Market and Trade Economics Division, (202) 694–5285.

Table 1-6.—Wheat: Supply and disappearance, United States, 2006–2015

Year be-ginning June	Supply				Disappearance						Ending stocks May 31
	Begin-ning stocks	Produc-tion	Im-ports [1]	Total	Domestic use				Ex-ports [1]	Total dis-appear-ance	
					Food	Seed	Feed [2]	Total			
	Million bushels	*Million bushels*	*Million bushels*	*Million bushels*	*Million bushels*	*Million bushels*	*Million bushels*	*Million bushels*	*Million bushels*	*Million bushels*	*Million bushels*
2006	571	1,808	122	2,501	938	82	117	1,137	908	2,045	456
2007	456	2,051	113	2,620	948	88	16	1,051	1,263	2,314	306
2008	306	2,499	127	2,932	927	78	255	1,260	1,015	2,275	657
2009	657	2,218	119	2,993	919	69	150	1,138	879	2,018	976
2010	976	2,207	97	3,279	926	71	129	1,126	1,291	2,417	862
2011	862	1,999	112	2,974	941	76	162	1,180	1,051	2,231	743
2012	743	2,252	124	3,119	951	73	365	1,389	1,012	2,401	718
2013	718	2,135	172	3,025	955	76	228	1,259	1,176	2,435	590
2014	590	2,026	151	2,768	958	79	114	1,151	864	2,015	752
2015	752	2,062	113	2,927	957	67	152	1,177	775	1,952	976

[1] Imports and exports include flour and other products expressed in wheat equivalent. [2] Approximates feed and residual use and includes negligible quantities used for distilled spirits.
Totals may not add due to independent rounding.
ERS, Market and Trade Economics Division, (202) 694–5296.

Table 1-7.—All Wheat: Area, yield, and production, by State and United States, 2013–2015

State	Area planted [1]			Area harvested		
	2013	2014	2015	2013	2014	2015
	1,000 acres	*1,000 acres*	*1,000 acres*	*1,000 acres*	*1,000 acres*	*1,000 acres*
Alabama	310	255	260	285	225	220
Arizona	87	85	150	84	83	142
Arkansas	680	465	350	610	395	240
California	690	530	465	394	220	210
Colorado	2,310	2,759	2,408	1,639	2,358	2,147
Delaware	85	80	70	78	75	65
Florida	25	15	25	19	10	15
Georgia	430	300	215	360	230	145
Idaho	1,321	1,271	1,200	1,261	1,196	1,135
Illinois	880	740	540	840	670	520
Indiana	460	390	290	435	335	260
Iowa	30	26	20	21	15	15
Kansas	9,500	9,600	9,200	8,450	8,800	8,700
Kentucky	700	630	560	610	510	440
Louisiana	265	160	110	255	150	92
Maryland	345	340	355	260	250	270
Michigan	620	550	510	590	470	475
Minnesota	1,227	1,262	1,532	1,184	1,212	1,473
Mississippi	400	230	150	385	215	120
Missouri	1,080	880	760	985	740	610
Montana	5,400	5,985	5,520	5,165	5,650	5,265
Nebraska	1,470	1,550	1,490	1,140	1,450	1,210
Nevada	31	21	12	15	10	8
New Jersey	34	33	27	29	25	20
New Mexico	440	380	385	100	105	190
New York	125	120	120	115	95	110
North Carolina	990	830	650	925	770	570
North Dakota	6,105	7,960	7,990	6,025	7,490	7,915
Ohio	660	620	520	640	545	480
Oklahoma	5,600	5,300	5,300	3,400	2,800	3,800
Oregon	880	830	835	868	818	828
Pennsylvania	185	185	195	155	150	175
South Carolina	280	230	170	265	220	160
South Dakota	2,494	2,514	2,756	1,839	2,364	2,236
Tennessee	640	530	455	575	475	395
Texas	6,300	6,000	6,000	2,350	2,250	3,550
Utah	138	130	125	124	117	119
Virginia	335	290	260	290	260	210
Washington	2,210	2,320	2,280	2,175	2,250	2,215
West Virginia	9	10	9	7	7	4
Wisconsin	315	295	230	265	250	210
Wyoming	150	140	145	120	125	130
United States	56,236	56,841	54,644	45,332	46,385	47,094

See footnote(s) at end of table.

GRAIN AND FEED

Table 1-7.—All Wheat: Area, yield, and production, by State and United States, 2013–2015—Continued

State	Yield per harvested acre			Production		
	2013	2014	2015	2013	2014	2015
	Bushels	*Bushels*	*Bushels*	*1,000 bushels*	*1,000 bushels*	*1,000 bushels*
Alabama	69.0	69.0	68.0	19,665	15,525	14,960
Arizona	99.4	110.1	101.0	8,348	9,136	14,346
Arkansas	62.0	63.0	56.0	37,820	24,885	13,440
California	82.5	83.4	79.4	32,500	18,350	16,680
Colorado	25.3	38.1	37.1	41,488	89,812	79,635
Delaware	64.0	72.0	65.0	4,992	5,400	4,225
Florida	59.0	39.0	43.0	1,121	390	645
Georgia	60.0	49.0	43.0	21,600	11,270	6,235
Idaho	82.2	78.4	77.4	103,592	93,717	87,850
Illinois	67.0	67.0	65.0	56,280	44,890	33,800
Indiana	73.0	76.0	68.0	31,755	25,460	17,680
Iowa	52.0	49.0	52.0	1,092	735	780
Kansas	38.0	28.0	37.0	321,100	246,400	321,900
Kentucky	75.0	71.0	73.0	45,750	36,210	32,120
Louisiana	58.0	62.0	39.0	14,790	9,300	3,588
Maryland	67.0	70.0	64.0	17,420	17,500	17,280
Michigan	75.0	74.0	81.0	44,250	34,780	38,475
Minnesota	56.7	54.8	59.9	67,152	66,468	88,294
Mississippi	58.0	58.0	48.0	22,330	12,470	5,760
Missouri	57.0	58.0	53.0	56,145	42,920	32,330
Montana	39.0	37.1	35.2	201,635	209,470	185,415
Nebraska	35.0	49.0	38.0	39,900	71,050	45,980
Nevada	87.0	105.0	81.3	1,305	1,050	650
New Jersey	54.0	53.0	50.0	1,566	1,325	1,000
New Mexico	44.0	28.0	25.0	4,400	2,940	4,750
New York	68.0	63.0	63.0	7,820	5,985	6,930
North Carolina	57.0	58.0	53.0	52,725	44,660	30,210
North Dakota	45.4	46.3	46.7	273,343	347,068	370,023
Ohio	70.0	74.0	67.0	44,800	40,330	32,160
Oklahoma	31.0	17.0	26.0	105,400	47,600	98,800
Oregon	62.1	54.3	47.3	53,904	44,444	39,195
Pennsylvania	68.0	65.0	65.0	10,540	9,750	11,375
South Carolina	54.0	52.0	46.0	14,310	11,440	7,360
South Dakota	42.2	55.5	46.2	77,558	131,260	103,406
Tennessee	71.0	66.0	68.0	40,825	31,350	26,860
Texas	29.0	30.0	30.0	68,150	67,500	106,500
Utah	44.2	50.3	48.5	5,484	5,882	5,775
Virginia	62.0	68.0	66.0	17,980	17,680	13,860
Washington	66.9	48.2	50.4	145,530	108,460	111,540
West Virginia	52.0	64.0	60.0	364	448	240
Wisconsin	58.0	65.0	74.0	15,370	16,250	15,540
Wyoming	24.0	38.0	32.0	2,880	4,750	4,160
United States	47.1	43.7	43.6	2,134,979	2,026,310	2,051,752

[1] Includes area planted preceding fall.
NASS, Crops Branch, (202) 720–2127.

Table 1-8.—Wheat, by type: Area, yield, and production, by State and United States, 2013–2015

State	Area planted [1]			Area harvested		
	2013	2014	2015	2013	2014	2015
	Winter wheat					
	1,000 acres	*1,000 acres*	*1,000 acres*	*1,000 acres*	*1,000 acres*	*1,000 acres*
Alabama	310	255	260	285	225	220
Arizona	12	8	5	10	7	2
Arkansas	680	465	350	610	395	240
California	620	490	400	345	190	150
Colorado	2,300	2,750	2,400	1,630	2,350	2,140
Delaware	85	80	70	78	75	65
Florida	25	15	25	19	10	15
Georgia	430	300	215	360	230	145
Idaho	780	780	750	740	730	700
Illinois	880	740	540	840	670	520
Indiana	460	390	290	435	335	260
Iowa	30	26	20	21	15	15
Kansas	9,500	9,600	9,200	8,450	8,800	8,700
Kentucky	700	630	560	610	510	440
Louisiana	265	160	110	255	150	92
Maryland	345	340	355	260	250	270
Michigan	620	550	510	590	470	475
Minnesota	27	42	52	24	32	43
Mississippi	400	230	150	385	215	120
Missouri	1,080	880	760	985	740	610
Montana	2,000	2,500	2,350	1,900	2,240	2,220
Nebraska	1,470	1,550	1,490	1,140	1,450	1,210
Nevada	23	15	8	12	9	6
New Jersey	34	33	27	29	25	20
New Mexico	440	380	385	100	105	190
New York	125	120	120	115	95	110
North Carolina	990	830	650	925	770	570
North Dakota	215	870	200	200	555	190
Ohio	660	620	520	640	545	480
Oklahoma	5,600	5,300	5,300	3,400	2,800	3,800
Oregon	790	750	740	780	740	735
Pennsylvania	185	185	195	155	150	175
South Carolina	280	230	170	265	220	160
South Dakota	1,300	1,210	1,420	670	1,080	970
Tennessee	640	530	455	575	475	395
Texas	6,300	6,000	6,000	2,350	2,250	3,550
Utah	120	120	115	110	109	110
Virginia	335	290	260	290	260	210
Washington	1,700	1,700	1,650	1,670	1,640	1,590
West Virginia	9	10	9	7	7	4
Wisconsin	315	295	230	265	250	210
Wyoming	150	140	145	120	125	130
United States	43,230	42,409	39,461	32,650	32,299	32,257
	Other spring wheat					
Colorado	10	9	8	9	8	7
Idaho	530	480	440	510	455	425
Minnesota	1,200	1,220	1,480	1,160	1,180	1,430
Montana	2,950	3,050	2,550	2,830	2,980	2,440
Nevada	8	6	4	3	1	2
North Dakota	5,100	6,250	6,700	5,060	6,140	6,650
Oregon	90	80	95	88	78	93
South Dakota	1,190	1,300	1,330	1,165	1,280	1,260
Utah	18	10	10	14	8	9
Washington	510	620	630	505	610	625
United States	11,606	13,025	13,247	11,344	12,740	12,941
	Durum wheat					
Arizona	75	77	145	74	76	140
California	70	40	65	49	30	60
Idaho	11	11	10	11	11	10
Montana	450	435	620	435	430	605
North Dakota	790	840	1,090	765	795	1,075
South Dakota	4	4	6	4	4	6
United States	1,400	1,407	1,936	1,338	1,346	1,896

See footnote(s) at end of table.

Table 1-8.—Wheat, by type: Area, yield, and production, by State and United States, 2013-2015—Continued

State	Yield per harvested acre			Production		
	2013	2014	2015	2013	2014	2015
	Bushels	*Bushels*	*Bushels*	*1,000 bushels*	*1,000 bushels*	*1,000 bushels*
Winter wheat						
Alabama	69.0	69.0	68.0	19,665	15,525	14,960
Arizona	80.0	100.0	103.0	800	700	206
Arkansas	62.0	63.0	56.0	37,820	24,885	13,440
California	80.0	80.0	70.0	27,600	15,200	10,500
Colorado	25.0	38.0	37.0	40,750	89,300	79,180
Delaware	64.0	72.0	65.0	4,992	5,400	4,225
Florida	59.0	39.0	43.0	1,121	390	645
Georgia	60.0	49.0	43.0	21,600	11,270	6,235
Idaho	86.0	80.0	82.0	63,640	58,400	57,400
Illinois	67.0	67.0	65.0	56,280	44,890	33,800
Indiana	73.0	76.0	68.0	31,755	25,460	17,680
Iowa	52.0	49.0	52.0	1,092	735	780
Kansas	38.0	28.0	37.0	321,100	246,400	321,900
Kentucky	75.0	71.0	73.0	45,750	36,210	32,120
Louisiana	58.0	62.0	39.0	14,790	9,300	3,588
Maryland	67.0	70.0	64.0	17,420	17,500	17,280
Michigan	75.0	74.0	81.0	44,250	34,780	38,475
Minnesota	43.0	49.0	58.0	1,032	1,568	2,494
Mississippi	58.0	58.0	48.0	22,330	12,470	5,760
Missouri	57.0	58.0	53.0	56,145	42,920	32,330
Montana	43.0	41.0	41.0	81,700	91,840	91,020
Nebraska	35.0	49.0	38.0	39,900	71,050	45,980
Nevada	90.0	110.0	90.0	1,080	990	540
New Jersey	54.0	53.0	50.0	1,566	1,325	1,000
New Mexico	44.0	28.0	25.0	4,400	2,940	4,750
New York	68.0	63.0	63.0	7,820	5,985	6,930
North Carolina	57.0	58.0	53.0	52,725	44,660	30,210
North Dakota	43.0	49.0	44.0	8,600	27,195	8,360
Ohio	70.0	74.0	67.0	44,800	40,330	32,160
Oklahoma	31.0	17.0	26.0	105,400	47,600	98,800
Oregon	62.0	55.0	47.0	48,360	40,700	34,545
Pennsylvania	68.0	65.0	65.0	10,540	9,750	11,375
South Carolina	54.0	52.0	46.0	14,310	11,440	7,360
South Dakota	39.0	55.0	44.0	26,130	59,400	42,680
Tennessee	71.0	66.0	68.0	40,825	31,350	26,860
Texas	29.0	30.0	30.0	68,150	67,500	106,500
Utah	44.0	50.0	48.0	4,840	5,450	5,280
Virginia	62.0	68.0	66.0	17,980	17,680	13,860
Washington	69.0	52.0	56.0	115,230	85,280	89,040
West Virginia	52.0	64.0	60.0	364	448	240
Wisconsin	58.0	65.0	74.0	15,370	16,250	15,540
Wyoming	24.0	38.0	32.0	2,880	4,750	4,160
United States	47.3	42.6	42.5	1,542,902	1,377,216	1,370,188
Other spring wheat						
Colorado	82.0	64.0	65.0	738	512	455
Idaho	77.0	76.0	70.0	39,270	34,580	29,750
Minnesota	57.0	55.0	60.0	66,120	64,900	85,800
Montana	37.0	35.0	31.0	104,710	104,300	75,640
Nevada	75.0	60.0	55.0	225	60	110
North Dakota	46.5	47.5	48.0	235,290	291,650	319,200
Oregon	63.0	48.0	50.0	5,544	3,744	4,650
South Dakota	44.0	56.0	48.0	51,260	71,680	60,480
Utah	46.0	54.0	55.0	644	432	495
Washington	60.0	38.0	36.0	30,300	23,180	22,500
United States	47.1	46.7	46.3	534,101	595,038	599,080
Durum wheat						
Arizona	102.0	111.0	101.0	7,548	8,436	14,140
California	100.0	105.0	103.0	4,900	3,150	6,180
Idaho	62.0	67.0	70.0	682	737	700
Montana	35.0	31.0	31.0	15,225	13,330	18,755
North Dakota	38.5	35.5	39.5	29,453	28,223	42,463
South Dakota	42.0	45.0	41.0	168	180	246
United States	43.3	40.2	43.5	57,976	54,056	82,484

¹ Includes area planted preceding fall.
NASS, Crops Branch, (202) 720-2127.

Table 1-9.—Wheat: Support operations, United States, 2006–2015

Marketing year beginning June 1	Income support payment rates per bushel [1]	Program price levels per bushel		Put under loan		Acquired by CCC under loan program [5]	Owned by CCC at end of marketing year
		Loan [2]	Target/Reference [3]	Quantity	Percentage of production [4]		
	Dollars	Dollars	Dollars	Million bushels	Percent	Million bushels	Million bushels
2006/2007 ...	0.52/0.00	2.75	3.92	94	5.2	0	41
2007/2008 ...	0.52/0.00	2.75	3.92	36	1.8	0	0
2008/2009 ...	0.52/0.00	2.75	3.92	84	3.4	0	0
2009/2010 ...	0.52/0.00	2.75	3.92	103	4.6	0	0
2010/2011 ...	0.52/0.00	2.94	4.17	67	3.0	0	0
2011/2012 ...	0.52/0.00	2.94	4.17	36	1.8	0	0
2012/2013 ...	0.52/0.00	2.94	4.17	28	1.2	0	0
2013/2014 ...	0.52/0.00	2.94	4.17	25	1.2	0	0
2014/2015 ...	0.00	2.94	5.50	43	2.1	0	0
2015/2016 ...	0.61	2.94	5.50	81	3.9	0	0

[1] The first entry is the direct payment rate and the second entry is the counter-cyclical payment rate for 2004/2005-2013/2014. For 2009/2010-2013/2014, producers who participate in the Average Crop Revenue (ACRE) program get a 20 percent reduction in their direct payment, not calculated in this table. For 2014/2015 and after, the entry is the price loss coverage payment rate. Agricultural Risk Coverage (ARC) is also available, but payment rates are established at the county or farm levels.   [2] For 2009/2010-2013/2014, producers who participate in the ACRE program get a 30 percent reduction in their loan rate, not calculated in this table.   [3] Target applies to 2006/2007-2013/2014 and Reference applies beginning with 2014/2015.   [4] Percent of production is on a grain basis.   [5] Acquisitions as of July 31, 2016
FSA, Food Grains, (202) 720–7787.

Table 1-10.—Wheat: Marketing year average price and value, by State and United States, 2013–2015

State	Marketing year average price per bushel			Value of production		
	2013	2014	2015	2013	2014	2015
	Dollars	Dollars	Dollars	1,000 dollars	1,000 dollars	1,000 dollars
Alabama	6.85	5.95	5.15	134,705	92,374	77,044
Arizona	8.74	8.36	9.25	71,970	75,600	131,928
Arkansas	7.04	5.62	5.20	266,253	139,854	69,888
California	7.99	7.50	7.90	258,563	138,392	122,295
Colorado	6.97	5.88	4.45	289,581	528,491	354,126
Delaware	6.80	4.90	4.50	33,946	26,460	19,013
Florida	5.75	5.10	4.15	6,446	1,989	2,677
Georgia	6.50	5.35	4.60	140,400	60,295	28,681
Idaho	7.05	6.16	5.45	731,620	582,170	478,800
Illinois	6.52	4.98	3.85	366,946	223,552	130,130
Indiana	6.42	5.22	4.90	203,867	132,901	86,632
Iowa	6.35	5.81	3.70	6,934	4,270	2,886
Kansas	6.99	6.07	4.85	2,244,489	1,495,648	1,561,215
Kentucky	6.57	5.62	5.35	300,578	203,500	171,842
Louisiana	7.10	6.10	4.50	105,009	56,730	16,146
Maryland	6.85	4.90	4.50	119,327	85,750	77,760
Michigan	6.71	5.74	5.65	296,918	199,637	217,384
Minnesota	6.68	5.48	4.75	448,576	363,900	422,065
Mississippi	6.92	5.75	5.15	154,524	71,703	29,664
Missouri	6.55	5.36	4.15	367,750	230,051	134,170
Montana	6.87	6.23	5.00	1,386,401	1,298,295	938,817
Nebraska	6.95	5.77	4.65	277,305	409,959	213,807
Nevada	6.92	6.20	5.45	9,022	6,426	3,548
New Jersey	6.60	4.80	4.75	10,336	6,360	4,750
New Mexico	6.80	6.32	4.85	29,920	18,581	23,038
New York	6.60	5.61	5.35	51,612	33,576	37,076
North Carolina	6.28	5.16	4.85	331,113	230,446	146,519
North Dakota	6.62	5.74	4.85	1,795,566	1,956,030	1,792,247
Ohio	6.54	5.60	4.60	292,992	225,848	147,936
Oklahoma	6.99	6.34	4.95	736,746	301,784	489,060
Oregon	7.03	6.78	5.55	378,452	300,311	217,433
Pennsylvania	6.83	6.06	5.25	71,988	59,085	59,719
South Carolina	6.00	4.80	4.55	85,860	54,912	33,488
South Dakota	6.84	5.54	4.75	530,269	727,517	484,148
Tennessee	7.00	5.63	5.30	285,775	176,501	142,358
Texas	7.11	6.40	4.80	484,547	432,000	511,200
Utah	7.94	7.07	5.40	42,893	41,040	29,832
Virginia	6.91	5.74	5.45	124,242	101,483	75,537
Washington	6.95	6.55	5.60	1,014,032	714,858	629,124
West Virginia	6.75	5.70	5.40	2,457	2,554	1,296
Wisconsin	6.12	4.75	4.45	94,064	77,188	69,153
Wyoming	7.10	5.67	4.55	20,448	26,933	18,928
United States	6.87	5.99	5.00	14,604,442	11,914,954	10,203,360

NASS, Crops Branch, (202) 720–2127.

Table 1-11.—International Wheat: Area, yield, and production in specified countries, 2013/2014–2015/2016

Country	Area			Yield per hectare			Production		
	2013/ 2014	2014/ 2015	2015/ 2016	2013/ 2014	2014/ 2015	2015/ 2016	2013/ 2014	2014/ 2015	2015/ 2016
	1,000 hec- tares	*1,000 hec- tares*	*1,000 hec- tares*	*Metric tons*	*Metric tons*	*Metric tons*	*1,000 metric tons*	*1,000 metric tons*	*1,000 metric tons*
Australia	12,613	12,155	12,750	2.01	1.90	1.92	25,303	23,076	24,500
Canada	10,442	9,480	9,600	3.59	3.10	2.88	37,530	29,420	27,600
China	24,117	24,069	24,140	5.06	5.24	5.39	121,930	126,208	130,190
European Union ...	25,884	26,783	26,846	5.59	5.85	5.90	144,585	156,656	158,457
India	30,003	30,473	30,600	3.12	3.15	2.83	93,506	95,850	86,530
Iran	6,400	6,800	6,800	2.27	1.91	2.06	14,500	13,000	14,000
Pakistan	8,660	9,199	9,180	2.80	2.82	2.78	24,211	25,979	25,478
Russia	23,399	23,636	25,600	2.23	2.50	2.38	52,091	59,080	61,000
Turkey	7,700	7,710	7,860	2.44	1.98	2.48	18,750	15,250	19,500
Ukraine	6,566	6,300	7,100	3.39	3.93	3.84	22,278	24,750	27,250
Others	45,711	45,501	44,744	2.24	2.22	2.28	102,567	101,055	101,971
Total foreign	201,495	202,106	205,220	3.26	3.32	3.30	657,251	670,324	676,476
United States	18,345	18,771	19,058	3.17	2.94	2.93	58,105	55,147	55,840
Total	219,840	220,877	224,278	3.25	3.28	3.27	715,356	725,471	732,316

FAS, Office of Global Analysis, (202) 720-6301. Prepared or estimated on the basis of official USDA production, supply, and distribution statistics from foreign governments.

Table 1-12.—Wheat and flour: United States imports, 2006–2015

Year beginning June	All wheat grain	All wheat flour [1]	All wheat products [2]	Total all wheat [3]
	1,000 bushels	*1,000 bushels* [4]	*1,000 bushels* [4]	*1,000 bushels*
2006	92,928	11,853	17,089	121,870
2007	85,806	10,710	16,115	112,631
2008	101,964	9,785	15,221	126.970
2009	93,003	9,720	15,868	118,591
2010	69,053	11,283	16,582	96,918
2011	83,336	10,666	18,068	112,069
2012	96,103	10,151	18,063	124,317
2013	141,665	11,535	19,267	172,467
2014	116,973	14,331	19,963	151,267
2015	76,561	15,827	20,523	112,912

[1] Includes meal, groats, and durum.   [2] Includes bulgur, couscous, and selected categories of pasta.   [3] Totals may not add due to rounding.   [4] Expressed in grain-equivalent bushels.
ERS, Market and Trade Economics Division, (202) 694–5285.

Table 1-13.—Wheat: International trade, 2013/2014–2015/2016

Country	2013/2014	2014/2015	2015/2016
	1,000 metric tons	*1,000 metric tons*	*1,000 metric tons*
Principle exporting countries:			
Argentina	2,250	5,301	7,000
Australia	18,615	16,605	17,000
Brazil	80	1,691	1,300
Canada	23,270	24,116	22,000
European Union	32,032	35,418	32,500
Kazakhstan	8,100	5,539	6,500
Mexico	1,322	1,104	1,100
Russia	18,568	22,800	23,000
Turkey	4,441	4,062	4,700
Ukraine	9,755	11,269	15,500
Others	15,476	12,986	11,042
Total Foreign	133,909	140,891	141,642
United States	32,001	23,249	21,092
Total	165,910	164,140	162,734
Principle importing countries:			
Algeria	7,484	7,257	7,700
Brazil	7,066	5,374	6,500
Egypt	10,170	11,063	11,000
European Union	3,976	5,975	6,300
Indonesia	7,392	7,478	8,100
Japan	6,123	5,878	5,700
Korea, South	4,288	3,942	4,500
Mexico	4,636	4,446	4,400
Nigeria	4,580	4,244	4,400
Philippines	3,482	5,099	4,300
Others	94,463	93,932	94,176
Total Foreign	153,660	154,688	157,076
United States	4,710	4,068	3,266
Total	158,370	158,756	160,342

FAS, Office of Global Analysis, (202) 720-6301. Prepared or estimated on the basis of official USDA production, supply, and distribution statistics from foreign governments.

Table 1-14.—Wheat and flour: United States exports by country of destination, 2014–2016

Country of destination	Year		
	2014	2015	2016
	Metric tons	*Metric tons*	*Metric tons*
Wheat:			
Mexico	2,928,478	2,609,799	2,763,704
Japan	2,975,729	2,765,612	2,721,679
Philippines	2,350,577	2,021,012	2,613,306
Brazil	2,494,135	441,088	1,254,817
Taiwan	1,043,554	1,048,476	1,094,375
Korea, South	1,419,804	1,124,398	1,088,612
Nigeria	2,224,855	1,899,653	1,073,620
Indonesia	922,669	428,307	961,075
China	574,676	548,765	900,689
Colombia	738,091	684,416	837,779
Thailand	495,779	754,849	653,207
Morocco	9,669	20,132	539,763
Guatemala	463,559	547,491	510,442
Ethiopia(*)	177,779	248,800	492,817
Yemen(*)	322,882	389,013	463,365
Peru	570,375	375,799	434,567
Chile	339,066	260,143	427,191
Italy(*)	497,580	792,226	420,629
Venezuela	574,102	274,546	409,530
Rest of World	4,291,430	4,014,762	4,102,997
World Total	25,414,789	21,249,287	23,764,164
Wheat flour:			
Mexico	110,368	156,323	206,844
Canada	98,097	111,512	122,196
Kenya	22,373	11,230	5,387
Dominican Republic	3,266	2,359	2,478
Netherlands Antilles(*)	3,305	1,695	1,995
Korea, South	4,630	3,432	1,772
Sweden	1,903	1,473	1,723
Tanzania	0	2,400	1,300
Djibouti	800	1,558	1,295
Bahamas, The	1,591	1,511	1,256
Malaysia	629	674	1,043
Leeward-Windward Islands(*)	1,299	868	989
Venezuela	0	0	715
United Arab Emirates	569	1,006	594
Trinidad and Tobago	501	385	574
China	270	601	442
Brazil	160	237	432
Panama	153	50	401
Japan	40	19	364
Rest of World	8,707	11,692	4,709
World Total	258,660	309,025	356,509

(*) Denotes a country that is a summarization of its component countries.
FAS, Office of Global Analysis, (202) 720-6301. Prepared or estimated on the basis of official USDA production, supply, and distribution, supply, and distribution statistics from foreign governments.

**Table 1-15.—Rye: Area, yield, production, disposition, and value,
United States, 2006–2015**

Year	Area		Yield per harvested acre	Production	Marketing year average price per bushel received by farmers	Value of production
	Planted [1]	Harvested				
	1,000 acres	*1,000 acres*	*Bushels*	*1,000 bushels*	*Dollars*	*1,000 dollars*
2006	1,396	274	26.3	7,193	3.32	23,895
2007	1,334	252	25.0	6,311	5.01	31,604
2008	1,345	286	29.1	8,315	6.35	52,803
2009	1,256	251	27.1	6,791	5.06	34,355
2010	1,256	270	27.7	7,480	5.18	38,767
2011	1,227	239	25.3	6,051	7.81	47,250
2012	1,271	250	26.2	6,542	7.69	50,304
2013	1,451	278	27.4	7,626	7.95	60,598
2014	1,434	258	27.9	7,189	7.74	55,639
2015	1,569	360	31.9	11,496	6.52	75,004

[1] Area planted in preceding fall.
NASS, Crops Branch, (202) 720–2127.

Table 1-16.—Rye: Supply and disappearance, United States, 2006–2015

Year beginning June	Supply				Disappearance							Ending stocks May 31
	Beginning stocks	Production	Imports	Total	Domestic use					Exports	Total disappearance [2]	
					Food	Seed	Industry [1]	Feed	Total			
	1,000 bushels	*1,000 bushels*	*1,000 bushels*	*1,000 bushels*	*1,000 bushels*	*1,000 bushels*	*1,000 bushels*	*1,000 bushels*	*1,000 bushels*	*1,000 bushels*	*1,000 bushels*	*1,000 bushels*
2006 ..	706	7,193	5,899	13,798	3,300	3,000	3,000	3,947	13,247	70	13,317	481
2007 ..	481	6,311	7,064	13,856	3,300	3,000	3,000	3,909	13,209	251	13,460	396
2008 ..	396	7,979	3,953	12,328	3,300	3,000	3,000	2,203	11,503	316	11,819	509
2009 ..	509	6,993	4,251	11,753	3,300	3,000	3,000	1,448	10,748	73	10,821	932
2010 ..	932	7,480	5,552	13,964	3,300	3,000	3,000	3,714	13,014	149	13,163	801
2011 ..	801	6,051	5,994	12,846	3,310	3,000	3,010	2,917	12,237	157	12,394	452
2012 ..	452	6,542	8,966	15,960	3,400	3,000	3,020	5,829	15,249	310	15,559	401
2013 ..	401	7,626	9,213	17,240	3,430	3,000	3,030	7,227	16,687	268	16,955	285
2014 ..	285	7,189	9,320	16,794	3,460	3,000	3,040	6,465	15,965	240	16,205	589
2015 ..	589	11,616	8,758	20,963	3,630	3,000	3,050	10,647	20,327	181	20,508	455

[1] Includes commercial adhesives, packaging materials, thatching, mattresses, hats, and paper.   [2] Totals may not add due to independent rounding.
ERS, Market and Trade Economics Division, (202) 694–5302.

Table 1-17.—Rye: Area, yield, and production, by State and United States, 2013–2015

State	Area planted [1]			Area harvested		
	2013	2014	2015	2013	2014	2015
	1,000 acres	*1,000 acres*	*1,000 acres*	*1,000 acres*	*1,000 acres*	*1,000 acres*
Georgia	190	170	210	40	20	30
Oklahoma	260	240	240	80	55	80
Other States [2]	1,001	1,024	1,119	158	183	250
United States	1,451	1,434	1,569	278	258	360

State	Yield per harvested acre			Production		
	2013	2014	2015	2013	2014	2015
	Bushels	*Bushels*	*Bushels*	*1,000 bushels*	*1,000 bushels*	*1,000 bushels*
Georgia	27.0	27.0	14.0	1,080	540	420
Oklahoma	20.0	9.0	24.0	1,600	495	1,920
Other States [2]	31.3	33.6	36.6	4,946	6,154	9,156
United States	27.4	27.9	31.9	7,626	7,189	11,496

[1] Includes area planted preceding fall.　[2] Other States include Illinois, Kansas, Michigan, Minnesota, Nebraska, New York, North Carolina, North Dakota, Pennsylvania, South Carolina, South Dakota, Texas, and Wisconsin.
NASS, Crops Branch, (202) 720–2127.

Table 1-18.—Rye: Marketing year average price and value, by State and United States, 2013–2015

State	Marketing year average price per bushel			Value of production		
	2013	2014	2015	2013	2014	2015
	Dollars	*Dollars*	*Dollars*	*1,000 dollars*	*1,000 dollars*	*1,000 dollars*
Georgia	8.35	7.75	6.65	9,018	4,185	2,793
Oklahoma	8.40	11.80	8.60	13,440	5,841	16,512
Other States [2]	7.71	7.41	6.08	38,140	45,613	55,699
United States	7.95	7.74	6.52	60,598	55,639	75,004

[1] Other States include Illinois, Kansas, Michigan, Minnesota, Nebraska, New York, North Carolina, North Dakota, Pennsylvania, South Carolina, South Dakota, Texas, and Wisconsin.
NASS, Crops Branch, (202) 720-2127.

Table 1-19.—International Rye: Area, yield, and production in specified countries, 2013/2014–2015/2016

Country	Area 2013/ 2014	2014/ 2015	2015/ 2016	Yield per hectare 2013/ 2014	2014/ 2015	2015/ 2016	Production 2013/ 2014	2014/ 2015	2015/ 2016
	1,000 hectares	1,000 hectares	1,000 hectares	Metric tons	Metric tons	Metric tons	1,000 metric tons	1,000 metric tons	1,000 metric tons
Argentina	35	51	32	1.49	1.90	1.91	52	97	61
Australia	35	35	36	0.57	0.57	0.61	20	20	22
Belarus	322	320	325	2.01	2.71	2.46	648	867	800
Canada	87	88	95	2.56	2.48	2.37	223	218	225
European Union	2,577	2,114	2,014	3.94	4.19	3.95	10,151	8,858	7,960
Kazakhstan	39	43	38	1.10	1.42	0.97	43	61	37
Norway	3	3	3	3.67	4.00	4.00	11	12	12
Russia	1,777	1,853	1,250	1.89	1.77	1.68	3,360	3,279	2,100
Turkey	140	140	140	2.50	2.50	2.50	350	350	350
Ukraine	279	185	150	2.29	2.57	2.60	638	475	390
Others	38	38	37	1.37	1.34	1.35	52	51	50
Total Foreign	5,332	4,870	4,120	2.92	2.93	2.91	15,548	14,288	12,007
United States	113	104	146	1.72	1.76	2.00	194	183	292
Total	5,445	4,974	4,266	2.89	2.91	2.88	15,742	14,471	12,299

FAS, Office of Global Analysis, (202) 720-6301. Prepared or estimated on the basis of official USDA production, supply, and distribution, supply, and distribution statistics from foreign governments.

Table 1-20.—Rye:[1] International trade, 2013/2014–2015/2016 [2]

Country	2013/2014	2014/2015	2015/2016
	1,000 metric tons	1,000 metric tons	1,000 metric tons
Principle exporting countries:			
Belarus	5	5	5
Canada	118	86	100
European Union	169	184	150
Russia	73	114	80
Ukraine	51	22	20
Others			
Total Foreign	416	411	355
United States	7	6	5
Total	423	417	360
Principle importing countries:			
European Union	77	102	70
Israel	33	25	25
Japan	37	22	25
Korea, South	8	4	10
Norway	17	8	15
Russia	5	5	5
Switzerland	6	6	5
Others	1	5	
Total Foreign	184	177	155
United States	234	237	229
Total	418	414	384

[1] Flour and products reported in terms of grain equivalent. [2] Year beginning July 1.
FAS, Office of Global Analysis, (202) 720-6301. Prepared or estimated on the basis of official USDA production, supply, and distribution statics from foreign governments.

Table 1-21.—Rice, rough: Area, yield, production, and value, United States, 2006–2015 [1]

Year	Area planted	Area harvested	Yield per acre	Production	Marketing year average price per cwt. received by farmers	Value of production
	1,000 acres	*1,000 acres*	*Pounds*	*1,000 cwt*	*Dollars*	*1,000 dollars*
2006	2,838	2,821	6,898	194,585	9.96	1,990,783
2007	2,761	2,748	7,219	198,388	12.80	2,600,871
2008	2,995	2,976	6,846	203,733	16.80	3,603,460
2009	3,135	3,103	7,085	219,850	14.40	3,209,236
2010	3,636	3,615	6,725	243,104	12.70	3,183,213
2011	2,689	2,617	7,067	184,941	14.50	2,737,423
2012	2,700	2,679	7,463	199,939	15.10	3,067,365
2013	2,490	2,469	7,694	189,953	16.30	3,181,993
2014	2,954	2,933	7,576	222,215	13.40	3,075,618
2015	2,469	2,575	7,470	192,343	12.90	2,579,001

[1] Sweet rice yield and production as short grain beginning in 2003.
NASS, Crops Branch, (202) 720–2127.

Table 1-22.—Rice, rough: Stocks on and off farms, United States, 2006–2015

Year beginning previous December	On farms			Off farms [1]		
	Dec. 1	Mar. 1	Aug. 1	Dec. 1	Mar. 1	Aug. 1
	1,000 cwt	*1,000 cwt*	*1,000 cwt*	*1,000 cwt*	*1,000 cwt*	*1,000 cwt*
2006	58,630	30,865	1,553	101,518	80,416	35,825
2007	52,420	28,015	1,220	97,706	76,145	33,713
2008	48,250	22,923	395	102,815	81,623	23,981
2009	47,530	21,286	876	91,071	70,042	23,787
2010	51,880	23,465	1,200	104,726	80,516	29,176
2011	63,617	33,895	3,772	109,869	86,720	37,917
2012	41,180	23,825	910	105,689	82,666	35,052
2013	39,385	18,430	640	108,650	85,839	29,880
2014	36,935	14,995	305	96,636	73,959	26,299
2015 [2]	47,865	25,810	1,480	110,061	86,658	41,293

[1] Stocks at mills and in attached warehouses, in warehouses not attached to mills, and in ports or in transit.   [2] Preliminary.
NASS, Crops Branch, (202) 720–2127.

Table 1-23.—Rice, by length of grain: Harvest, yield, and production, United States, 2006–2015

Year	Area harvested			Yield per acre			Production		
	Long grain	Medium grain	Short grain	Long grain	Medium grain	Short grain	Long grain	Medium grain	Short grain
	1,000 acres	*1,000 acres*	*1,000 acres*	*Pounds*	*Pounds*	*Pounds*	*1,000 cwt*	*1,000 cwt*	*1,000 cwt*
2006	2,186	574	61	6,727	7,631	6,098	147,063	43,802	3,720
2007	2,052	630	66	6,980	8,105	6,197	143,235	51,063	4,090
2008	2,350	575	51	6,522	8,203	6,490	153,257	47,166	3,310
2009	2,265	786	52	6,743	8,052	7,373	152,725	63,291	3,834
2010	2,826	746	43	6,486	7,660	6,195	183,296	57,144	2,664
2011	1,739	834	44	6,691	7,861	6,880	116,352	65,562	3,027
2012	1,979	643	57	7,291	8,059	6,737	144,280	51,819	3,840
2013	1,767	655	47	7,464	8,384	6,685	131,896	54,915	3,142
2014	2,196	700	37	7,407	8,175	6,292	162,665	57,222	2,328
2015	1,843	695	37	7,218	8,155	7,119	133,032	56,677	2,634

NASS, Crops Branch, (202) 720-2127.

Table 1-24.—Rice, rough, by length of grain: Stocks in all positions, United States, 2007–2016

Year beginning previous December	Dec. 1	Mar. 1	Jun. 1 [1]	Aug. 1	Oct. 1 [2]
			Long grain		
	1,000 cwt	*1,000 cwt*	*1,000 cwt*	*1,000 cwt*	*1,000 cwt*
2007	109,301	76,127		25,738	77
2008	103,620	69,207	35,580	16,101	([4])
2009	96,994	64,226	34,293	17,698	372
2010	103,430	67,910	37,041	19,532	([4])
2011	123,521	86,864	54,636	32,847	([4])
2012	93,027	68,469	37,552	22,034	([4])
2013	99,111	69,313	(NA)	19,439	([4])
2014	85,149	56,125	31,622	14,010	([4])
2015	106,786	71,243	42,295	23,917	([4])
2016 [3]	91,814	66,307	36,305		
			Medium grain		
	1,000 cwt	*1,000 cwt*	*1,000 cwt*	*1,000 cwt*	*1,000 cwt*
2007	37,225	25,857		8,372	2,506
2008	43,520	32,584	16,284	7,196	1,084
2009	37,989	24,755	12,722	6,093	938
2010	49,264	33,395	18,344	9,527	3,053
2011	46,853	31,548	15,327	7,870	1,366
2012	50,935	35,978	20,272	12,983	1,870
2013	45,349	32,558	(NA)	10,057	2,029
2014	45,086	30,481	17,294	11,631	6,691
2015	48,483	39,505	27,988	18,169	7,892
2016 [3]	50,582	40,718	26,960		
			Short grain		
	1,000 cwt	*1,000 cwt*	*1,000 cwt*	*1,000 cwt*	*1,000 cwt*
2007	3,600	2,176		823	412
2008	3,925	2,755	1,554	1,079	([4])
2009	3,618	2,347	1,301	872	522
2010	3,912	2,676	2,023	1,317	([4])
2011	3,112	2,203	1,410	972	([4])
2012	2,907	2,044	1,282	945	([4])
2013	3,575	2,398	(NA)	1,024	([4])
2014	3,336	2,348	1,435	963	([4])
2015	2,657	1,720	1,057	687	([4])
2016 [3]	2,653	1,737	999		

(NA) Not available. [1] Estimates began in 2008. [2] California only. [3] Preliminary. [4] Not published to avoid disclosing individual reports.
NASS, Crops Branch, (202) 720–2127.

Table 1-25.—Rough and milled rice (rough equivalent): Supply and disappearance, United States, 2004–2013 [1]

Year beginning August	Supply				Disappearance					Ending stocks July 31
	Beginning stocks	Production	Imports [2]	Total	Food, industrial, & residual [3]	Seed	Total	Exports [2]	Total disappearance	
	Million cwt	*Million cwt*	*Million cwt*	*Million cwt*	*Million cwt*	*Million cwt*	*Million cwt*	*Million cwt*	*Million cwt*	*Million cwt*
2004	23.7	232.4	13.2	269.2	118.5	4.2	122.7	108.8	231.5	37.7
2005	37.7	222.8	17.1	277.7	116.3	3.5	119.8	114.9	234.7	43.0
2006	43.0	194.6	20.6	258.2	124.7	3.4	128.1	90.8	218.8	39.3
2007	39.3	198.4	23.9	261.6	123.2	3.7	126.8	105.3	232.1	29.5
2008	29.5	203.7	19.2	252.4	123.8	3.9	127.6	94.4	222.0	30.4
2009	30.4	219.9	19.0	269.3	119.9	4.5	124.4	108.3	232.8	36.5
2010	36.5	243.1	18.3	297.9	133.6	3.3	136.9	112.6	249.5	48.5
2011	48.5	184.9	19.4	252.8	107.5	3.3	110.8	100.9	211.7	41.1
2012	41.1	199.5	21.1	261.7	115.6	3.1	118.6	106.6	225.3	36.4
2013	36.4	189.9	23.1	249.4	121.3	3.6	124.9	92.7	217.6	31.8

[1] Consolidated supply and disappearance of rough and milled rice. Milled rice data converted to a rough basis using annually derived extraction rates as factors. [2] Trade data from Bureau of the Census. [3] The residual includes unaccounted losses in transporting, processing, and marketing.
ERS, Market and Trade Economics Division, (202) 694–5292.

Table 1-26.—Rice, by length of grain: Area, yield, and production, by State and United States, 2013–2015

State	Area harvested			Yield per acre			Production		
	2013	2014	2015	2013	2014	2015	2013	2014	2015
Long grain									
	1,000 acres	1,000 acres	1,000 acres	Pounds	Pounds	Pounds	1,000 cwt	1,000 cwt	1,000 cwt
Arkansas	950	1,265	1,045	7,560	7,570	7,380	71,820	95,761	77,121
California	6	4	7	5,700	7,300	6,700	342	292	469
Louisiana	392	393	351	7,330	7,150	6,990	28,734	28,100	24,535
Mississippi	124	189	149	7,400	7,420	7,110	9,176	14,024	10,594
Missouri	154	207	167	7,030	6,830	7,040	10,826	14,138	11,757
Texas	141	138	124	7,800	7,500	6,900	10,998	10,350	8,556
United States	1,767	2,196	1,843	7,464	7,407	7,218	131,896	162,665	133,032
Medium grain									
	1,000 acres	1,000 acres	1,000 acres	Pounds	Pounds	Pounds	1,000 cwt	1,000 cwt	1,000 cwt
Arkansas	119	214	240	7,570	7,540	7,150	9,008	16,136	17,160
California	510	402	378	8,670	8,800	9,100	44,217	35,376	34,398
Louisiana	21	69	64	6,670	7,020	6,650	1,401	4,844	4,256
Mississippi	-	1	-	-	7,200	-	-	72	-
Missouri	2	6	7	7,080	6,700	6,500	142	402	455
Texas	3	8	6	4,900	4,900	6,800	147	392	408
United States	655	700	695	8,384	8,175	8,155	54,915	57,222	56,677
Short grain									
	1,000 acres	1,000 acres	1,000 acres	Pounds	Pounds	Pounds	1,000 cwt	1,000 cwt	1,000 cwt
Arkansas	1	1	1	6,000	6,000	6,000	60	60	60
California	46	36	36	6,700	6,300	7,150	3,082	2,268	2,574
United States	47	37	37	6,685	6,292	7,119	3,142	2,328	2,634

NASS, Crops Branch, (202) 720–2127.

Table 1-27.—Rice: Area, yield, and production, by State and United States, 2013–2015 [1]

State	Area planted			Area harvested		
	2013	2014	2015	2013	2014	2015
	1,000 acres	1,000 acres	1,000 acres	1,000 acres	1,000 acres	1,000 acres
Arkansas	1,076	1,486	1,306	1,070	1,480	1,286
California	567	445	423	562	442	421
Louisiana	418	466	420	413	462	415
Mississippi	125	191	150	124	190	149
Missouri	159	216	182	156	213	174
Texas	145	150	133	144	146	130
United States	2,490	2,954	2,614	2,469	2,933	2,575

State	Yield per harvested acre			Production		
	2013	2014	2015	2013	2014	2015
	Pounds	Pounds	Pounds	1,000 cwt	1,000 cwt	1,000 cwt
Arkansas	7,560	7,560	7,340	80,888	111,957	94,341
California	8,480	8,580	8,890	47,641	37,936	37,441
Louisiana	7,300	7,130	6,940	30,135	32,944	28,791
Mississippi	7,400	7,420	7,110	9,176	14,096	10,594
Missouri	7,030	6,830	7,020	10,968	14,540	12,212
Texas	7,740	7,360	6,900	11,145	10,742	8,964
United States	7,694	7,576	7,470	189,953	222,215	192,343

[1] Sweet rice acreage included with short grain.
NASS, Crops Branch, (202) 720–2127.

Table 1-28.—Rice: Marketing year average price and value, by State and United States, 2013–2015

State	Marketing year average price per cwt.			Value of production		
	2013	2014	2015	2013	2014	2015
	Dollars	*Dollars*	*Dollars*	*1,000 dollars*	*1,000 dollars*	*1,000 dollars*
Arkansas	15.20	12.00	11.60	1,229,498	1,343,484	1,094,356
California	20.90	21.80	20.10	995,697	827,005	752,564
Louisiana	15.50	12.90	11.90	467,093	424,978	342,613
Mississippi	15.40	12.20	11.70	141,310	171,971	123,950
Missouri	15.10	11.00	12.20	165,617	159,940	148,986
Texas	16.40	13.80	13.00	182,778	148,240	116,532
United States	16.30	13.40	12.90	3,181,993	3,075,618	2,579,001

NASS, Crops Branch, (202) 720–2127.

Table 1-29.—Rice, milled, by length of grain: Stocks in all positions, United States, 2007–2016

Year beginning previous December	Whole kernels (head rice)				
	Dec. 1	Mar. 1	Jun. 1 [1]	Aug. 1	Oct. 1 [2]
	Long grain				
	1,000 cwt	*1,000 cwt*	*1,000 cwt*	*1,000 cwt*	*1,000 cwt*
2007	2,803	2,454	....	1,989	[4]
2008	2,638	2,546	2,015	2,065	[4]
2009	2,504	2,300	3,251	1,658	[4]
2010	2,022	2,370	2,043	2,511	[4]
2011	2,665	3,038	3,291	1,924	[4]
2012	2,699	1,856	1,906	1,550	1
2013	2,574	3,144	(NA)	1,704	[4]
2014	3,348	2,379	[4]	1,556	[4]
2015	2,308	1,952	2,523	1,792	[4]
2016 [3]	2,716	2,616	3,463	....	....
	Medium grain				
	1,000 cwt	*1,000 cwt*	*1,000 cwt*	*1,000 cwt*	*1,000 cwt*
2007	653	792	....	536	[4]
2008	958	1,735	850	508	[4]
2009	1,531	978	823	689	[4]
2010	1,496	1,472	1,188	799	454
2011	1,743	1,803	981	845	399
2012	1,406	935	1,041	496	281
2013	1,638	772	(NA)	726	[4]
2014	1,285	1,584	1,601	472	603
2015	1,694	1,139	1,662	878	1,430
2016 [3]	1,279	1,459	1,030	....	....
	Short grain				
	1,000 cwt	*1,000 cwt*	*1,000 cwt*	*1,000 cwt*	*1,000 cwt*
2007	55	98	....	48	[4]
2008	92	69	78	59	[4]
2009	80	69	57	51	36
2010	73	74	47	60	[4]
2011	97	62	53	45	[4]
2012	199	49	60	38	69
2013	99	64	(NA)	61	28
2014	58	48	[4]	47	[4]
2015	51	59	54	37	[4]
2016 [3]	75	62	41	....	....

See footnote(s) at end of table.

**Table 1-29.—Rice, milled, by length of grain: Stocks in all positions,
United States, 2007–2016—Continued**

Year beginning previous December	Broken kernels[5]				
	Dec. 1	Mar. 1	Jun. 1[1]	Aug. 1	Oct. 1[2]
	Second heads				
	1,000 cwt	1,000 cwt	1,000 cwt	1,000 cwt	1,000 cwt
2007	240	562	....	307	([4])
2008	853	852	906	488	([4])
2009	661	794	828	1,465	([4])
2010	1,374	707	961	670	([4])
2011	888	634	817	992	([4])
2012	1,037	1,021	778	870	([4])
2013	1,231	1,225	(NA)	1,104	([4])
2014	1,243	1,064	1,159	797	([4])
2015	727	1,663	1,227	1,120	([4])
2016 [3]	1,385	1,550	1,634	....	....
	Screenings				
	1,000 cwt	1,000 cwt	1,000 cwt	1,000 cwt	1,000 cwt
2007	90	([4])	....	81	....
2008	195	163	145	206	....
2009	42	64	61	3	....
2010	52	20	34	93	....
2011	110	91	153	75	....
2012	192	82	145	51	....
2013	136	271	(NA)	153	....
2014	96	118	72	92	....
2015	118	115	153	88	....
2016[3]	56	61	70	....	....
	Brewers				
	1,000 cwt	1,000 cwt	1,000 cwt	1,000 cwt	1,000 cwt
2007	163	([4])	....	150	([4])
2008	533	239	379	249	([4])
2009	437	527	704	211	([4])
2010	662	252	247	247	([4])
2011	799	647	673	786	([4])
2012	710	551	532	573	([4])
2013	1,241	1,189	(NA)	384	([4])
2014	575	651	589	747	([4])
2015	748	941	1,153	141	([4])
2016[3]	734	387	678	....	....

(NA) Not available.   [1] Estimates began in 2008.   [2] California only.   [3] Preliminary.   [4] Not published to avoid disclosing individual operations.   [5] Screenings included in second heads in California.
NASS, Crops Branch, (202) 720–2127.

Table 1-30.—Rice, rough: Support operations, United States, 2012–2015

Marketing year beginning August 1	Income support payment rates per cwt [1]	Program price levels per cwt		Put under loan		Acquired by CCC under loan program [4]	Owned by CCC at end of marketing year
		Loan [2]	Target/ Reference [3]	Quantity	Percentage of production		
	Dollars	*Dollars*	*Dollars*	*1,000 cwt*	*Percent*	*1,000 cwt*	*1,000 cwt*
Long grain							
2012/2013	0.52/0.00	6.50	10.50	45.4	31.47	0	0
2013/2014	0.52/0.00	6.50	10.50	19.0	14.37	0	0
2014/2015	2.10	6.50	14.00	24.8	15.25	0	0
2015/2016	3.0	6.50	14.00	18.3	13.79	0	0
Medium grain [5]							
2012/2013	0.52/0.00	6.50	10.50	20.5	36.80	0	0
2013/2014	0.52/0.00	6.50	10.50	17.1	29.50	0	0
2014/2015	0.00	6.50	14.00	17.4	29.29	0	0
2015/2016	2.70	6.50	14.00	19.7	33.29	0	0
Temperate Japonica [6]							
2012/2013	(NA)	(NA)	(NA)	(NA)	(NA)	(NA)	(NA)
2013/2014	(NA)	(NA)	(NA)	(NA)	(NA)	(NA)	(NA)
2014/2015	0	6.50	16.40	14.8	39.4	0	0
2015/2016	0	6.50	16.40	16.2	43.8	0	0

(NA) Not applicable. [1] The first entry is the direct payment rate and the second entry is the counter-cyclical payment rate for 2012/2013-2013/14. For 2009/2010-2013/2014, producers who participate in the Average Crop Revenue (ACRE) program get a 20 percent reduction in their direct payment, not calculated in this table. For 2014/2015 and after, the entry is the price loss coverage payment rate. Agricultural Risk Coverage (ARC) is also available, but payment rates are established at the county or farm levels. For 2015/2016, projected payment rate is based on August 2016 WASDE MY price. [2] For 2009/2010-2013/2014, producers who participate in the ACRE program get a 30 percent reduction in their loan rate, not calculated in this table. [3] Target applies to 2012/2013-2013/2014 and Reference applies beginning with 2014/2015. [4] Acquisitions as of July 31, 2016. [5] Medium Grain includes all short grain rice and medium grain rice for the marketing assistance loan program. The quantity under loan for 2014/15 includes the medium and short grain rice in California. The medium and short grain rice in California is also shown in the Temperate Japonica section below. [6] Temperate Japonica Rice is medium and short grain rice in California starting with the 2014/15 crop. Prior to the 2014/15 crop, temperate japonica rice was classified as medium grain rice for program purposes.
FSA, Food Grains, (202) 720-7787.

Table 1-31.—Rice: United States exports (milled basis), by country of destination, 2014–2016 [1]

Country of destination	Year		
	2014	2015	2016
	Metric tons	*Metric tons*	*Metric tons*
Mexico	738,921	821,334	856,419
Haiti	361,578	403,328	418,747
Japan	288,507	334,837	346,760
Venezuela	230,691	208,787	292,041
Honduras	160,544	196,137	270,925
El Salvador	77,609	49,933	264,739
Canada	253,089	228,144	221,923
Korea, South	35,141	156,193	163,586
Guatemala	90,939	103,297	148,708
Colombia	96,415	303,454	140,333
Saudi Arabia	101,319	115,900	123,471
Costa Rica	70,110	77,062	115,755
Jordan	89,042	90,205	102,180
Iraq	140,562	93,915	93,080
Libya	21,162	80,918	92,833
Turkey	244,502	107,233	78,121
Taiwan	50,979	57,014	60,224
Panama	24,809	110,833	45,907
United Kingdom	21,595	30,153	30,309
Rest of World	268,308	293,148	293,548
World Total [2]	3,365,821	3,861,824	4,159,607

(*) Denotes a country that is a summarization of its component countries. [1] Year beginning Jan 1. [2] Includes countries not shown.
FAS, Grain and Feed Division, (202) 720–6219. www.fas.usda.gov/grain/default.html.

Table 1-32.—International Rice, milled: Area, yield, and production in specified countries, 2013/2014–2015/2016

Country	Area			Yield per hectare			Production		
	2013/ 2014	2014/ 2015	2015/ 2016	2013/ 2014	2014/ 2015	2015/ 2016	2013/2014	2014/2015	2015/2016
	1,000 hectares	1,000 hectares	1,000 hectares	Metric tons	Metric tons	Metric tons	1,000 metric tons	1,000 metric tons	1,000 metric tons
Bangladesh	11,750	11,790	12,000	4.39	4.39	4.33	34,390	34,500	34,600
Brazil	2,400	2,295	2,150	5.09	5.42	5.41	8,300	8,465	7,905
Burma	7,050	7,030	6,800	2.65	2.80	2.80	11,957	12,600	12,200
China	30,312	30,310	30,210	6.72	6.81	6.89	142,530	144,560	145,770
India	44,136	43,740	42,750	3.62	3.62	3.61	106,646	105,480	103,000
Indonesia	12,100	11,830	11,660	4.72	4.73	4.77	36,300	35,560	35,300
Japan	1,621	1,608	1,586	6.72	6.71	6.63	7,931	7,849	7,653
Philippines	4,800	4,705	4,650	3.92	4.02	3.93	11,858	11,915	11,500
Thailand	10,920	10,270	9,460	2.84	2.77	2.55	20,460	18,750	15,900
Vietnam	7,788	7,792	7,660	5.79	5.80	5.89	28,161	28,234	28,200
Others	27,857	28,128	27,742	2.29	2.27	2.27	63,773	63,739	62,956
Total Foreign	160,734	159,498	156,668	2.94	2.96	2.97	472,306	471,652	464,984
United States	999	1,187	1,042	8.62	8.49	8.37	6,117	7,106	6,107
Total	161,733	160,685	157,710	2.96	2.98	2.99	478,423	478,758	471,091

FAS, Office of Global Analysis, (202) 720-6301. Prepared or estimated on the basis of official USDA production, supply, and distribution, supply, and and distribution statistics from foreign governments.

Table 1-33.—Rice, milled equivalent: International trade, 2013/2014–2015/2016

Country	2013/2014	2014/2015	2015/2016
	1,000 metric tons	1,000 metric tons	1,000 metric tons
Principle exporting countries:			
Argentina	465	325	520
Brazil	819	930	830
Burma	1,688	1,750	1,800
Cambodia	1,000	1,150	900
Guyana	346	502	536
India	10,149	11,871	8,600
Pakistan	3,200	4,000	4,600
Thailand	10,969	9,779	10,000
Uruguay	890	800	950
Vietnam	6,325	6,606	7,000
Others	2,902	2,697	2,746
Total Foreign	38,753	40,410	38,482
United States	3,005	3,207	3,175
Total	41,758	43,617	41,657
Principle importing countries:			
China	4,000	4,700	5,000
European Union	1,530	1,703	1,550
Indonesia	1,225	1,198	2,000
Iran	1,600	1,350	1,450
Iraq	950	1,100	1,200
Malaysia	989	1,000	1,020
Nigeria	2,800	3,500	2,500
Philippines	1,200	1,800	2,000
Saudi Arabia	1,410	1,420	1,450
South Africa	1,144	980	1,080
Others	20,770	21,448	20,062
Total Foreign	37,618	40,199	39,312
United States	734	783	762
Total	38,352	40,982	40,074

FAS, Office of Global Analysis, (202) 720-6301. Prepared or estimated on the basis of official USDA production, supply, and distribution statistics from foreign governments.

Table 1-34.—Food grains: Average price, selected markets and grades, 2006–2015 [1]

Calendar year [2]	Kansas City Wheat, No. 1 Hard Winter, Ordinary Protein (rail)	Wheat, No. 1 Hard Winter, 13% protein (rail)	Wheat, No. 2 Soft Red Winter (rail)	Minneapolis (rail) Wheat, No. 1 Hard Amber Durum (milling) (rail)	Wheat, No. 1 Dark Northern Spring (rail), 14% protein	Rye, No. 2, 20 day delivery (truck)	Portland Wheat No. 1 Soft White	St. Louis Wheat, No. 2 Soft Red Winter (truck)
	Dollars per bushel	Dollars per bushel	Dollars per bushel	Dollars per bushel	Dollars per bushel	Dollars per bushel	Dollars per bushel	Dollars per bushel
2006	5.11	5.21	4.27	(NA)	5.19	3.25	4.07	3.47
2007	6.85	7.06	6.27	11.33	7.01	6.24	7.29	5.96
2008	8.92	9.82	7.72	23.25	11.16	7.12	7.93	6.32
2009	5.80	6.29	5.06	....	7.21	4.35	5.28	5.55
2010	5.90	6.58	6.03	8.75	7.79	4.58	5.45	5.36
2011	8.44	9.17	7.87	....	11.36	....	7.08	7.49
2012	8.47	8.96	7.25	....	9.69	....	7.74	7.56
2013	7.22	8.60	6.67	(NA)	8.08	(NA)	7.67	6.97
2014	7.89	7.95	5.72	(NA)	8.73	5.75	7.09	6.45
2015	6.05	6.16	4.43	(NA)	7.00	5.75	5.88	4.81

Calendar year [2]	Chicago Wheat, No. 2 Soft Red Winter (rail)	Denver Wheat, No. 1 Hard Winter (truck red)	S.W. Louisiana Milled Rice Medium	Long	Arkansas Milled Rice Medium	Long	Texas Milled Rice Long
	Dollars per bushel	Dollars per bushel	Dollars per cwt	Dollars per cwt	Dollars per cwt	Dollars per cwt	Dollars per cwt
2006	3.58	4.47	22.50	71.46	21.56	17.82	19.38
2007	5.85	6.05	23.44	19.28	22.81	19.50	21.58
2008	6.75	7.85	36.49	34.97	38.85	35.93	36.41
2009	4.43	5.04	39.56	24.91	40.57	26.46	27.88
2010	5.36	5.00	30.80	23.12	29.68	23.91	26.46
2011	6.83	7.36	33.47	26.20	34.99	27.10	29.25
2012	7.44	7.44	28.89	25.77	27.91	26.18	27.18
2013	6.84	7.27	30.56	28.48	28.68	28.52	29.61
2014	5.89	6.38	30.76	27.04	31.30	27.62	29.15
2015	5.08	4.77	27.88	23.30	27.41	23.35	25.31

(NA) Not available. [1] Simple average of daily prices. [2] For wheat and rye, crop year begins in June. For rice, crop year begins in August.
AMS, Livestock and Grain Market News branch, (202) 720–6231.

Table 1-35.—Corn: Area, yield, production, and value, United States, 2006–2015

Year	Area planted, all purposes	Corn for grain Area harvested	Yield per harvested acre	Production	Marketing year average price per bushel	Value of production	Corn for silage Area harvested	Yield per harvested acre	Production
	1,000 acres	1,000 acres	Bushels	1,000 bushels	Dollars	1,000 dollars	1,000 acres	Tons	1,000 tons
2006	78,327	70,638	149.1	10,531,123	3.04	32,083,011	6,487	16.2	105,294
2007	93,527	86,520	150.7	13,073,875	4.20	54,666,959	6,060	17.5	106,229
2008	85,982	78,570	153.3	12,043,203	4.06	49,104,148	5,971	18.7	111,792
2009	86,382	79,490	164.4	13,067,156	3.55	46,640,936	5,605	19.3	108,221
2010	88,192	81,446	152.6	12,425,330	5.18	64,529,628	5,567	19.3	107,314
2011	91,936	83,879	146.8	12,313,956	6.22	76,651,242	5,935	18.4	109,091
2012	97,291	87,365	123.1	10,755,111	6.89	74,155,299	7,419	15.7	116,148
2013	95,365	87,451	158.1	13,828,964	4.46	61,927,548	6,281	18.8	118,296
2014	90,597	83,136	171.0	14,215,532	3.70	52,951,760	6,371	20.1	128,048
2015	87,999	80,749	168.4	13,601,198	3.60	49,038,819	6,221	20.4	126,894

NASS, Crops Branch, (202) 720–2127.

GRAIN AND FEED

Table 1-36.—Corn: Area, yield, and production, by State and United States, 2013–2015

State	Area planted for all purposes			Corn for grain		
				Area harvested		
	2013	2014	2015	2013	2014	2015
	1,000 acres	*1,000 acres*	*1,000 acres*	*1,000 acres*	*1,000 acres*	*1,000 acres*
Alabama	320	300	260	295	285	245
Arizona	85	75	70	51	28	34
Arkansas	880	540	460	870	530	445
California	600	520	430	180	95	60
Colorado	1,220	1,150	1,100	980	1,010	950
Connecticut	27	26	26	(NA)	(NA)	(NA)
Delaware	180	175	170	174	168	164
Florida	115	75	80	78	40	50
Georgia	510	350	330	465	310	285
Idaho	350	320	280	115	80	70
Illinois	12,000	11,900	11,700	11,800	11,750	11,500
Indiana	6,000	5,900	5,650	5,830	5,770	5,480
Iowa	13,600	13,700	13,500	13,050	13,300	13,050
Kansas	4,300	4,050	4,150	4,000	3,800	3,920
Kentucky	1,530	1,520	1,400	1,430	1,430	1,310
Louisiana	680	400	400	670	390	390
Maine	31	31	31	(NA)	(NA)	(NA)
Maryland	480	500	440	420	430	380
Massachusetts	16	16	16	(NA)	(NA)	(NA)
Michigan	2,600	2,550	2,350	2,230	2,210	2,070
Minnesota	8,600	8,200	8,100	8,140	7,550	7,600
Mississippi	860	510	510	830	485	490
Missouri	3,350	3,500	3,250	3,200	3,380	3,080
Montana	120	130	105	75	75	50
Nebraska	9,950	9,300	9,400	9,550	8,950	9,150
Nevada	7	4	2	(NA)	(NA)	(NA)
New Hampshire	14	15	15	(NA)	(NA)	(NA)
New Jersey	90	85	80	80	79	72
New Mexico	120	125	125	38	48	40
New York	1,200	1,140	1,080	690	680	590
North Carolina	930	840	790	860	780	730
North Dakota	3,850	2,800	2,750	3,600	2,530	2,560
Ohio	3,900	3,700	3,550	3,730	3,470	3,260
Oklahoma	370	320	310	310	290	280
Oregon	80	80	65	36	39	30
Pennsylvania	1,480	1,460	1,340	1,090	1,030	940
Rhode Island	2	2	2	(NA)	(NA)	(NA)
South Carolina	350	295	295	335	280	260
South Dakota	6,200	5,800	5,400	5,860	5,320	5,030
Tennessee	890	920	780	810	840	730
Texas	2,350	2,250	2,300	1,950	1,990	1,970
Utah	83	75	60	31	28	15
Vermont	92	92	92	(NA)	(NA)	(NA)
Virginia	510	500	450	360	350	300
Washington	190	215	170	105	110	75
West Virginia	53	51	50	36	36	35
Wisconsin	4,100	4,000	4,000	3,030	3,110	3,000
Wyoming	100	90	85	67	60	59
United States	95,365	90,597	87,999	87,451	83,136	80,749

See footnote(s) at end of table.

Table 1-36.—Corn: Area, yield, and production, by State and United States, 2013–2015—Continued

State	Corn for grain					
	Yield per harvested acre			Production		
	2013	2014	2015	2013	2014	2015
	1,000 acres	*1,000 acres*	*1,000 acres*	*1,000 acres*	*1,000 acres*	*1,000 acres*
Alabama	147.0	159.0	147.0	43,365	45,315	36,015
Arizona	177.0	210.0	210.0	9,027	5,880	7,140
Arkansas	186.0	187.0	181.0	161,820	99,110	80,545
California	191.0	165.0	157.0	34,380	15,675	9,420
Colorado	131.0	146.0	142.0	128,380	147,460	134,900
Connecticut	(NA)	(NA)	(NA)	(NA)	(NA)	(NA)
Delaware	166.0	200.0	192.0	28,884	33,600	31,488
Florida	133.0	135.0	141.0	10,374	5,400	7,050
Georgia	175.0	170.0	171.0	81,375	52,700	48,735
Idaho	181.0	200.0	207.0	20,815	16,000	14,490
Illinois	178.0	200.0	175.0	2,100,400	2,350,000	2,012,500
Indiana	177.0	188.0	150.0	1,031,910	1,084,760	822,000
Iowa	164.0	178.0	192.0	2,140,200	2,367,400	2,505,600
Kansas	126.0	149.0	148.0	504,000	566,200	580,160
Kentucky	170.0	158.0	172.0	243,100	225,940	225,320
Louisiana	173.0	183.0	171.0	115,910	71,370	66,690
Maine	(NA)	(NA)	(NA)	(NA)	(NA)	(NA)
Maryland	158.0	175.0	164.0	66,360	75,250	62,320
Massachusetts	(NA)	(NA)	(NA)	(NA)	(NA)	(NA)
Michigan	155.0	161.0	162.0	345,650	355,810	335,340
Minnesota	159.0	156.0	188.0	1,294,260	1,177,800	1,428,800
Mississippi	176.0	185.0	175.0	146,080	89,725	85,750
Missouri	136.0	186.0	142.0	435,200	628,680	437,360
Montana	115.0	100.0	110.0	8,625	7,500	5,500
Nebraska	169.0	179.0	185.0	1,613,950	1,602,050	1,692,750
Nevada	(NA)	(NA)	(NA)	(NA)	(NA)	(NA)
New Hampshire	(NA)	(NA)	(NA)	(NA)	(NA)	(NA)
New Jersey	139.0	157.0	147.0	11,120	12,403	10,584
New Mexico	190.0	195.0	180.0	7,220	9,360	7,200
New York	137.0	148.0	143.0	94,530	100,640	84,370
North Carolina	142.0	132.0	113.0	122,120	102,960	82,490
North Dakota	110.0	124.0	128.0	396,000	313,720	327,680
Ohio	174.0	176.0	153.0	649,020	610,720	498,780
Oklahoma	145.0	147.0	129.0	44,950	42,630	36,120
Oregon	188.0	190.0	188.0	6,768	7,410	5,640
Pennsylvania	146.0	154.0	147.0	159,140	158,620	138,180
Rhode Island	(NA)	(NA)	(NA)	(NA)	(NA)	(NA)
South Carolina	129.0	117.0	93.0	43,215	32,760	24,180
South Dakota	137.0	148.0	159.0	802,820	787,360	799,770
Tennessee	156.0	168.0	160.0	126,360	141,120	116,800
Texas	136.0	148.0	135.0	265,200	294,520	265,950
Utah	170.0	160.0	173.0	5,270	4,480	2,595
Vermont	(NA)	(NA)	(NA)	(NA)	(NA)	(NA)
Virginia	154.0	145.0	161.0	55,440	50,750	48,300
Washington	215.0	215.0	215.0	22,575	23,650	16,125
West Virginia	147.0	149.0	148.0	5,292	5,364	5,180
Wisconsin	145.0	156.0	164.0	439,350	485,160	492,000
Wyoming	127.0	138.0	159.0	8,509	8,280	9,381
United States	158.1	171.0	168.4	13,828,964	14,215,532	13,601,198

(NA) Not available.
NASS, Crops Branch, (202) 720–2127.

GRAIN AND FEED

Table 1-37.—Corn: Supply and disappearance, United States, 2005–2014

Year beginning September 1	Supply				Disappearance					Ending stocks Aug. 31
	Beginning stocks	Produc-tion	Imports	Total [1]	Domestic use			Exports	Total dis-appear-ance	
					Feed and residual	Food, seed, and industrial	Total			
	Million bushels	Million bushels	Million bushels	Million bushels	Million bushels	Million bushels	Million bushels	Million bushels	Million bushels	Million bushels
2005	2,114	11,112	9	13,235	6,115	3,019	9,134	2,134	11,268	1,967
2006	1,967	10,531	12	12,510	5,540	3,541	9,081	2,125	11,207	1,304
2007	1,304	13,038	20	14,362	5,858	4,442	10,300	2,437	12,737	1,624
2008	1,624	12,092	14	13,729	5,133	5,025	10,207	1,849	12,056	1,673
2009	1,673	13,067	8	14,749	5,101	5,961	11,062	1,979	13,041	1,708
2010	1,708	12,425	28	14,161	4,777	6,426	11,202	1,831	13,033	1,128
2011	1,128	12,314	29	13,471	4,519	6,424	10,943	1,539	12,482	989
2012	989	10,755	160	11,904	4,315	6,038	10,353	730	11,083	821
2013	821	13,829	36	14,686	5,040	6,493	11,534	1,920	13,454	1,232
2014	1,232	14,216	32	15,479	5,315	6,568	11,883	1,864	13,748	1,731

Latest data may be preliminary or projected. [1] Total may not add due to rounding.
ERS, Market and Trade Economics Division, (202) 694–5313.

Table 1-38.—Corn: Stocks on and off farms, United States, 2007–2016

Year beginning previous December	On farms				Off farms [1]			
	Dec. 1	Mar. 1	Jun. 1	Sep. 1 [2]	Dec. 1	Mar. 1	Jun. 1	Sep. 1 [2]
	1,000 bushels	1,000 bushels	1,000 bushels	1,000 bushels	1,000 bushels	1,000 bushels	1,000 bushels	1,000 bushels
2007	5,627,000	3,330,000	1,826,600	460,100	3,305,707	2,738,250	1,706,843	843,547
2008	6,530,000	3,780,000	1,970,900	499,950	3,748,085	3,078,722	2,057,117	1,124,200
2009	6,482,000	4,085,000	2,205,400	607,500	3,590,106	2,869,145	2,056,027	1,065,811
2010	7,405,000	4,548,000	2,131,400	485,100	3,497,460	3,145,787	2,178,671	1,222,687
2011	6,302,000	3,384,000	1,681,500	314,950	3,754,769	3,139,228	1,988,832	812,695
2012	6,175,000	3,192,000	1,482,000	313,700	3,471,823	2,831,356	1,666,204	675,327
2013	4,586,000	2,669,200	1,260,100	275,000	3,446,732	2,730,726	1,506,144	546,185
2014	6,380,000	3,860,500	1,863,200	462,000	4,072,532	3,147,623	1,988,516	769,904
2015	7,087,000	4,380,000	2,275,000	593,000	4,124,380	3,369,806	2,177,988	1,138,164
2016 [3]	6,829,000	....	....	....	4,382,605	....	....	....

[1] Includes stocks at mills, elevators, warehouses, terminals, and processors. [2] Old crop only. [3] Preliminary.
NASS, Crops Branch, (202) 720–2127.

Table 1-39.—Corn: Utilization for silage, by State and United States, 2013–2015

State	Silage								
	Area harvested			Yield per acre			Production		
	2013	2014	2015	2013	2014	2015	2013	2014	2015
	1,000 acres	*1,000 acres*	*1,000 acres*	*Tons*	*Tons*	*Tons*	*1,000 tons*	*1,000 tons*	*1,000 tons*
Alabama	9	9	9	17.0	17.0	15.0	153	153	135
Arizona	33	46	35	30.0	29.0	31.0	990	1,334	1,085
Arkansas	2	2	2	18.0	18.0	15.0	36	36	30
California	415	420	365	26.5	26.0	25.5	10,998	10,920	9,308
Colorado	100	110	120	23.0	25.0	25.5	2,300	2,750	3,060
Connecticut	23	22	21	19.0	20.0	18.5	437	440	389
Delaware	5	5	4	17.0	24.0	20.0	85	120	80
Florida	35	30	25	18.0	13.0	17.0	630	390	425
Georgia	35	35	40	20.0	19.5	22.0	700	683	880
Idaho	230	235	205	26.0	28.0	29.0	5,980	6,580	5,945
Illinois	100	80	90	19.0	22.0	20.0	1,900	1,760	1,800
Indiana	130	100	90	23.0	22.0	17.0	2,990	2,200	1,530
Iowa	390	310	340	19.0	20.0	24.0	7,410	6,200	8,160
Kansas	150	150	170	13.0	14.0	18.5	1,950	2,100	3,145
Kentucky	80	75	70	21.0	21.0	20.0	1,680	1,575	1,400
Louisiana	3	2	1	18.0	18.0	14.0	54	36	14
Maine	27	27	27	17.5	18.5	18.5	473	500	500
Maryland	55	60	45	21.0	22.0	22.0	1,155	1,320	990
Massachusetts	13	13	13	18.0	20.0	19.0	234	260	247
Michigan	350	320	260	17.5	20.5	19.0	6,125	6,560	4,940
Minnesota	380	500	450	16.5	18.0	21.5	6,270	9,000	9,675
Mississippi	10	10	10	16.0	14.0	16.0	160	140	160
Missouri	80	80	100	14.0	18.0	14.0	1,120	1,440	1,400
Montana	41	51	50	23.0	22.0	23.0	943	1,122	1,150
Nebraska	260	260	220	16.0	21.0	20.0	4,160	5,460	4,400
Nevada	6	3	2	24.0	20.0	24.0	144	60	48
New Hampshire	13	14	14	20.0	21.0	20.0	260	294	280
New Jersey	9	5	7	20.0	20.0	21.0	180	100	147
New Mexico	79	73	83	25.0	26.0	25.0	1,975	1,898	2,075
New York	500	450	480	17.0	18.0	17.0	8,500	8,100	8,160
North Carolina	45	50	50	17.0	19.0	16.0	765	950	800
North Dakota	140	230	150	12.0	14.5	14.0	1,680	3,335	2,100
Ohio	130	190	240	19.5	20.5	20.0	2,535	3,895	4,800
Oklahoma	37	20	15	21.0	17.0	17.0	777	340	255
Oregon	43	40	34	27.0	25.0	24.0	1,161	1,000	816
Pennsylvania	380	410	390	20.5	20.0	20.0	7,790	8,200	7,800
Rhode Island	2	2	2	20.5	20.0	17.0	41	40	34
South Carolina	10	11	13	18.0	14.0	14.0	180	154	182
South Dakota	280	400	330	13.0	15.5	16.0	3,640	6,200	5,280
Tennessee	60	60	40	19.0	21.0	18.0	1,140	1,260	720
Texas	220	210	250	19.0	22.0	21.0	4,180	4,620	5,250
Utah	49	45	42	23.0	22.0	23.0	1,127	990	966
Vermont	85	85	88	15.0	18.0	17.0	1,275	1,530	1,496
Virginia	125	125	125	20.0	19.5	21.0	2,500	2,438	2,625
Washington	85	105	95	27.0	28.0	26.0	2,295	2,940	2,470
West Virginia	16	14	14	19.0	18.0	18.0	304	252	252
Wisconsin	980	850	970	16.5	18.5	19.5	16,170	15,725	18,915
Wyoming	31	27	25	24.0	24.0	23.0	744	648	575
United States	6,281	6,371	6,221	18.8	20.1	20.4	118,296	128,048	126,894

NASS, Crops Branch, (202) 720–2127.

GRAIN AND FEED

Table 1-40.—Corn for grain: Marketing year average price and value, by State and United States, 2013–2015

State	Marketing year average price per bushel			Value of production		
	2013	2014	2015	2013	2014	2015
	Dollars	Dollars	Dollars	1,000 dollars	1,000 dollars	1,000 dollars
Alabama	4.71	3.75	3.85	204,249	169,931	138,658
Arizona	5.89	5.36	5.10	53,169	31,517	36,414
Arkansas	5.12	4.13	4.10	828,518	409,324	330,235
California	5.33	4.81	4.50	183,245	75,397	42,390
Colorado	4.61	3.95	3.70	591,832	582,467	499,130
Delaware	4.94	3.83	3.80	142,687	128,688	119,654
Florida	4.51	3.65	3.80	46,787	19,710	26,790
Georgia	5.17	4.17	4.00	420,709	219,759	194,940
Idaho	4.91	4.16	4.70	102,202	66,560	68,103
Illinois	4.52	3.71	3.65	9,493,808	8,718,500	7,345,625
Indiana	4.47	3.75	3.85	4,612,638	4,067,850	3,164,700
Iowa	4.49	3.71	3.50	9,609,498	8,783,054	8,769,600
Kansas	4.49	3.78	3.75	2,262,960	2,140,236	2,175,600
Kentucky	4.67	3.94	3.80	1,135,277	890,204	856,216
Louisiana	5.10	4.15	4.00	591,141	296,186	266,760
Maryland	4.84	3.79	3.80	321,182	285,198	236,816
Michigan	4.18	3.65	3.50	1,444,817	1,298,707	1,173,690
Minnesota	4.30	3.58	3.40	5,565,318	4,216,524	4,857,920
Mississippi	5.05	4.24	4.00	737,704	380,434	343,000
Missouri	4.57	3.54	3.65	1,988,864	2,225,527	1,596,364
Montana	4.27	3.77	4.05	36,829	28,275	22,275
Nebraska	4.47	3.77	3.60	7,214,357	6,039,729	6,093,900
New Jersey	4.66	3.80	3.75	51,819	47,131	39,690
New Mexico	5.18	4.35	4.60	37,400	40,716	33,120
New York	4.52	4.11	4.25	427,276	413,630	358,573
North Carolina	4.96	4.19	4.30	605,715	431,402	354,707
North Dakota	3.91	3.34	3.20	1,548,360	1,047,825	1,048,576
Ohio	4.41	3.78	3.80	2,862,178	2,308,522	1,895,364
Oklahoma	5.09	4.11	4.00	228,796	175,209	144,480
Oregon	5.44	4.45	4.20	36,818	32,975	23,688
Pennsylvania	4.47	3.90	3.80	711,356	618,618	525,084
South Carolina	4.67	3.90	4.00	201,814	127,764	96,720
South Dakota	4.05	3.34	3.25	3,251,421	2,629,782	2,599,253
Tennessee	4.87	3.81	3.85	615,373	537,667	449,680
Texas	5.14	4.42	4.15	1,363,128	1,301,778	1,103,693
Utah	5.47	4.13	4.70	28,827	18,502	12,197
Virginia	4.80	3.90	4.00	266,112	197,925	193,200
Washington	5.29	4.90	4.65	119,422	115,885	74,981
West Virginia	4.70	3.85	3.95	24,872	20,651	20,461
Wisconsin	4.38	3.67	3.40	1,924,353	1,780,537	1,672,800
Wyoming	4.08	3.80	3.60	34,717	31,464	33,772
United States	4.46	3.70	3.60	61,927,548	52,951,760	49,038,819

NASS, Crops Branch, (202) 720–2127.

Table 1-41.—International Corn: Area, yield, and production in specified countries, 2013/2014–2015/2016

Country	Area			Yield per hectare			Production		
	2013/ 2014	2014/ 2015	2015/ 2016	2013/ 2014	2014/ 2015	2015/ 2016	2013/2014	2014/2015	2015/2016
	1,000 hectares	1,000 hectares	1,000 hectares	Metric tons	Metric tons	Metric tons	1,000 metric tons	1,000 metric tons	1,000 metric tons
Argentina	3,400	3,300	3,400	7.65	8.18	7.94	26,000	27,000	27,000
Brazil	15,800	15,750	16,200	5.06	5.40	5.19	80,000	85,000	84,000
Canada	1,480	1,227	1,310	9.59	9.36	10.38	14,194	11,487	13,600
China	36,318	37,123	38,120	6.02	5.81	5.89	218,490	215,646	224,580
European Union	9,664	9,530	9,287	6.69	7.95	6.22	64,635	75,793	57,751
India	9,066	9,090	9,000	2.68	2.66	2.33	24,259	24,170	21,000
Indonesia	3,120	3,100	3,180	2.92	2.90	2.96	9,100	9,000	9,400
Mexico	7,052	7,325	7,000	3.24	3.48	3.43	22,880	25,480	24,000
Russia	2,322	2,596	2,650	5.01	4.36	4.91	11,635	11,325	13,000
Ukraine	4,825	4,625	4,080	6.40	6.15	5.71	30,900	28,450	23,300
Others	52,753	51,317	50,234	2.62	2.64	2.53	138,015	135,242	127,018
Total Foreign ...	145,800	144,983	144,461	4.39	4.47	4.32	640,108	648,593	624,649
United States ...	35,390	33,644	32,678	9.93	10.73	10.57	351,272	361,091	345,486
Total	181,190	178,627	177,139	5.47	5.65	5.48	991,380	1,009,684	970,135

FAS, Office of Global Analysis, (202) 720-6301. Prepared or estimated on the basis of offical USDA production, supply, and distribution statistics from foreign governments.

Table 1-42.—Corn: International trade, 2013/2014–2015/2016

Country	2013/2014	2014/2015	2015/2016
	1,000 metric tons	1,000 metric tons	1,000 metric tons
Principle exporting countries:			
Argentina	17,102	18,500	17,000
Brazil	20,967	34,450	28,000
Burma	700	850	900
Canada	1,949	423	1,000
European Union	2,405	4,026	1,500
Mexico	501	784	1,000
Paraguay	2,372	3,287	2,300
Russia	4,192	3,213	3,800
Serbia	1,736	2,964	1,700
Ukraine	20,004	19,661	15,500
Others	10,393	5,692	5,515
Total Foreign	82,321	93,850	78,215
United States	48,783	47,359	41,912
Total	131,104	141,209	120,127
Principle importing countries:			
Algeria	3,739	4,381	4,200
Colombia	4,436	4,496	4,500
Egypt	8,726	7,826	8,250
European Union	16,015	8,646	16,000
Iran	5,500	6,200	5,000
Japan	15,121	14,656	14,700
Korea, South	10,406	10,168	10,000
Malaysia	3,476	3,221	3,500
Mexico	10,954	11,269	11,500
Taiwan	4,189	3,821	4,100
Others	40,542	47,839	46,125
Total Foreign	123,104	122,523	127,875
United States	909	804	1,270
Total	124,013	123,327	129,145

FAS, Office of Global Analysis, (202) 720-6301. Prepared or estimated on the basis of official USDA production, supply, and distribution statistics from foreign governments.

GRAIN AND FEED

Table 1-43.—Corn: Support operations, United States, 2006/2007–2015/2016

Marketing year beginning September 1	Income support payment rates per bushel[1]	Program price levels per bushel		Put under loan		Acquired by CCC under loan program[5]	Owned by CCC at end of marketing year
		Loan[2]	Target/ eference[3]	Quantity	Percentage of production[4]		
	Dollars	Dollars	Dollars	Million bushels	Percent	Million bushels	Million bushels
2006/2007	0.28/0.00	1.95	2.63	1,108	10.5	0	0
2007/2008	0.28/0.00	1.95	2.63	1,218	9.3	0	0
2008/2009	0.28/0.00	1.95	2.63	1,074	8.9	0	0
2009/2010	0.28/0.00	1.95	2.63	934	7.1	0	0
2010/2011	0.28/0.00	1.95	2.63	801	6.4	0	0
2011/2012	0.28/0.00	1.95	2.63	574	4.6	0	0
2012/2013	0.28/0.00	1.95	2.63	368	3.4	0	0
2013/2014	0.28/0.00	1.95	2.63	460	3.3	0	0
2014/2015	0.00	1.95	3.70	574	4.0	0	0
2015/2016	0.10	1.95	3.70	746	5.9	0	0

[1] The first entry is the direct payment rate and the second entry is the counter-cyclical payment rate for 2004/2005-2013/2014. For 2009/2010-2013/2014, producers who participate in the Average Crop Revenue (ACRE) program get a 20 percent reduction in their direct payment, not calculated in this table. For 2014/2015 and after, the entry is the price loss coverage payment rate. For 2015/2016, projected payment rate based on August 2016 WASDE MY price. Agricultural Risk Coverage (ARC) is also available, but payment rates are established at the county or farm levels. [2] For 2009/2010-2013/2014, producers who participate in the ACRE program get a 30 percent reduction in their loan rate, not calculated in this table. [3] Target applies to 2003/2004-2013/2014 and Reference applies beginning with 2014/2015. [4] Percent of production is on a grain basis. [5] Acquisitions as of July 30, 2016.
FSA, Feed Grains and Oilseeds, (202) 720–2711.

Table 1-44.—Corn: United States exports, specified by country of destination, 2014–2016[1]

Country of destination	2014	2015	2016
	Metric tons	Metric tons	Metric tons
Mexico	10,341,603	11,811,586	13,922,240
Japan	12,574,819	11,049,438	11,941,197
Korea, South	4,614,158	3,450,215	4,839,045
Colombia	4,324,495	4,354,592	4,527,583
Peru	2,392,749	1,742,869	2,690,128
Taiwan	1,688,757	1,815,770	2,672,708
Saudi Arabia	1,072,800	1,150,986	1,953,149
Venezuela	1,194,507	528,763	1,116,316
Egypt	2,981,207	967,744	962,188
Guatemala	788,152	811,608	954,811
Canada	907,037	1,262,005	878,346
Morocco	232,542	268,286	772,927
El Salvador	443,012	516,799	711,938
Chile	14,098	72,940	690,064
Algeria	76,021	238,846	678,575
Costa Rica	744,513	614,784	674,514
Honduras	426,279	423,230	587,091
Vietnam	391,065	3,967	540,372
Dominican Republic	659,544	489,391	515,511
Rest of World	3,619,528	2,866,987	4,347,925
World Total	49,486,886	44,440,806	55,976,628

(*) Denotes a country that is a summarization of its component countries. [1] Compiled from U.S. Census data. Excludes seed, popcorn.
FAS, Office of Global Analysis, (202) 720-6301.

Table 1-45.—Oats: Area, yield, production, and value, United States, 2006–2015

Year	Area		Yield per harvested acre	Production	Marketing year average price per bushel received by farmers	Value of production
	Planted [1]	Harvested				
	1,000 acres	1,000 acres	Bushels	1,000 bushels	Dollars	1,000 dollars
2006	4,166	1,564	59.8	93,522	1.87	180,899
2007	3,763	1,504	60.1	90,430	2.63	247,644
2008	3,260	1,414	63.7	90,051	3.15	273,843
2009	3,349	1,340	67.9	91,043	2.02	200,997
2010	3,113	1,267	64.6	81,856	2.52	218,536
2011	2,349	875	57.3	50,126	3.49	175,523
2012	2,700	1,005	61.2	61,486	3.89	243,018
2013	2,980	1,009	64.1	64,642	3.75	239,807
2014	2,753	1,035	67.9	70,232	3.21	241,472
2015	3,088	1,276	70.2	89,535	2.12	220,398

[1] Relates to the total area of oats sown for all purposes, including oats sown in the preceding fall.
NASS, Crops Branch, (202) 720–2127.

Table 1-46.—Oats: Stocks on and off farms, United States, 2006–2015

Year beginning September	On farms				Off farms [1]			
	Sep. 1	Dec. 1	Mar. 1	Jun. 1	Sep. 1	Dec. 1	Mar. 1	Jun. 1
2006	60,800	53,000	33,900	18,400	39,284	45,889	37,158	32,198
2007	53,650	43,100	31,000	16,100	34,710	51,331	47,988	50,674
2008	52,800	42,600	30,200	17,480	66,296	72,322	65,250	66,619
2009	54,500	43,000	30,900	17,600	73,875	67,629	67,091	62,716
2010	45,850	34,100	26,950	14,580	70,722	66,911	59,361	53,049
2011	30,300	24,600	19,550	11,070	47,391	54,244	55,044	43,869
2012	34,100	26,100	18,900	11,380	50,872	47,051	33,726	24,957
2013	37,150	25,650	19,800	9,710	26,339	22,394	15,323	15,029
2014	41,400	31,300	20,810	15,120	32,910	35,670	38,609	38,625
2015	47,800	36,750	26,800	18,350	46,066	46,098	48,494	38,456

[1] Includes stocks at mills, elevators, warehouses, terminals, and processors.
NASS, Crops Branch, (202) 720–2127.

Table 1-47.—Oats: Supply and disappearance, United States, 2005–2014

Year beginning June 1	Supply				Disappearance					Ending stocks May 31
					Domestic use				Total disappearance	
	Beginning stocks	Production	Imports	Total	Feed and residual	Food, seed and industrial	Total	Exports		
	Million bushels	Million bushels	Million bushels	Million bushels	Million bushels	Million bushels	Million bushels	Million bushels	Million bushels	Million bushels
2005	58	115	91	264	136	74	209	2	211	53
2006	53	94	106	252	125	74	199	3	202	51
2007	51	90	123	264	120	74	195	3	198	67
2008	67	89	115	270	108	75	183	3	186	84
2009	84	91	95	270	113	74	188	2	190	80
2010	80	82	85	247	103	74	177	3	180	68
2011	68	50	94	212	78	76	154	2	157	55
2012	55	61	93	209	96	76	172	1	173	36
2013	36	65	97	198	98	73	172	2	173	25
2014	25	70	107	202	70	77	147	2	149	54

Latest data may be preliminary or projected. Totals may not add due to rounding.
ERS, Market and Trade Economics Division, (202) 694–5313.

Table 1-48.—Oats: Area, yield, and production, by State and United States, 2013–2015

State	Area planted [1]			Area harvested		
	2013	2014	2015	2013	2014	2015
	1,000 acres	*1,000 acres*	*1,000 acres*	*1,000 acres*	*1,000 acres*	*1,000 acres*
Alabama	60	50	55	20	15	20
Arkansas	11	12	11	7	8	8
California	150	120	120	15	10	10
Colorado	55	45	45	12	9	10
Georgia	50	60	65	18	20	25
Idaho	70	70	75	15	15	15
Illinois	40	35	40	25	25	25
Indiana	20	20	15	10	10	5
Iowa	220	145	125	60	55	57
Kansas	100	85	95	20	15	40
Maine	28	32	30	26	31	29
Michigan	50	55	75	30	40	50
Minnesota	240	230	280	105	125	160
Missouri	30	25	30	14	13	14
Montana	50	45	50	22	16	22
Nebraska	150	110	135	25	30	40
New York	75	55	70	46	40	40
North Carolina	35	33	35	13	17	16
North Dakota	225	235	275	135	105	140
Ohio	50	50	70	25	35	40
Oklahoma	60	60	40	7	10	7
Oregon	30	30	35	13	18	11
Pennsylvania	95	90	95	50	60	65
South Carolina	20	21	24	9	10	9
South Dakota	260	250	325	120	100	145
Texas	450	450	520	40	45	55
Utah	40	20	20	5	3	2
Virginia	10	10	12	2	3	4
Washington	20	25	18	5	5	5
Wisconsin	255	255	280	105	140	195
Wyoming	31	30	23	10	7	12
United States	2,980	2,753	3,088	1,009	1,035	1,276

State	Yield per harvested acre			Production		
	2013	2014	2015	2013	2014	2015
	Bushels	*Bushels*	*Bushels*	*1,000 bushels*	*1,000 bushels*	*1,000 bushels*
Alabama	60.0	65.0	50.0	1,200	975	1,000
Arkansas	73.0	90.0	60.0	511	720	480
California	80.0	100.0	60.0	1,200	1,000	600
Colorado	65.0	60.0	80.0	780	540	800
Georgia	60.0	54.0	45.0	1,080	1,080	1,125
Idaho	73.0	82.0	86.0	1,095	1,230	1,290
Illinois	69.0	80.0	77.0	1,725	2,000	1,925
Indiana	71.0	74.0	59.0	710	740	295
Iowa	66.0	64.0	73.0	3,960	3,520	4,161
Kansas	42.0	56.0	65.0	840	840	2,600
Maine	67.0	70.0	80.0	1,742	2,170	2,320
Michigan	62.0	69.0	67.0	1,860	2,760	3,350
Minnesota	57.0	63.0	78.0	5,985	7,875	12,480
Missouri	53.0	65.0	65.0	742	845	910
Montana	54.0	69.0	53.0	1,188	1,104	1,166
Nebraska	65.0	80.0	67.0	1,625	2,400	2,680
New York	67.0	63.0	58.0	3,082	2,520	2,320
North Carolina	70.0	67.0	66.0	910	1,139	1,056
North Dakota	62.0	73.0	74.0	8,370	7,665	10,360
Ohio	63.0	63.0	63.0	1,575	2,205	2,520
Oklahoma	38.0	38.0	39.0	266	380	273
Oregon	100.0	85.0	88.0	1,300	1,530	968
Pennsylvania	62.0	58.0	55.0	3,100	3,480	3,575
South Carolina	59.0	62.0	58.0	531	620	522
South Dakota	77.0	93.0	87.0	9,240	9,300	12,615
Texas	46.0	38.0	48.0	1,840	1,710	2,640
Utah	62.0	69.0	85.0	310	207	170
Virginia	70.0	78.0	76.0	140	234	304
Washington	68.0	70.0	54.0	340	350	270
Wisconsin	65.0	62.0	72.0	6,825	8,680	14,040
Wyoming	57.0	59.0	60.0	570	413	720
United States	64.1	67.9	70.2	64,642	70,232	89,535

[1] Relates to the total area of oats sown for all purposes, including oats sown in the preceding fall.
NASS, Crops Branch, (202) 720–2127.

Table 1-49.—Oats: Support operations, United States, 2006/2007–2015/2016

Marketing Year beginning June 1	Income support payment rates per bushel [1]	Program price levels per bushel		Put under loan		Acquired by CCC under loan program [5]	Owned by CCC at end of marketing year
		Loan [2]	Target [3]	Quantity	Percentage of production [4]		
	Dollars	*Dollars*	*Dollars*	*Million bushels*	*Percent*	*Million bushels*	*Million bushels*
2006/2007	0.02/0.00	1.33	1.44	1.7	1.8	0	0
2007/2008	0.02/0.00	1.33	1.44	1.2	1.3	0	0
2008/2009	0.02/0.00	1.33	1.44	1.1	1.3	0	0
2009/2010	0.02/0.00	1.33	1.44	1.1	1.2	0	0
2010/2011	0.02/0.00	1.39	1.79	0.7	0.9	0	0
2011/2012	0.02/0.00	1.39	1.79	0.2	0.4	0	0
2012/2013	0.02/0.00	1.39	1.79	0.3	0.5	0	0
2013/2014	0.02/0.00	1.39	1.79	0.3	0.5	0	0
2014/2015	0.00	1.39	2.40	0.5	0.7	0	0
2015/2016	0.28	1.39	2.40	1.1	1.3	0	0

[1] The first entry is the direct payment rate and the second entry is the counter-cyclical payment rate for 2004/2005-2013/2014. For 2009/2010-2013/2014, producers who participate in the Average Crop Revenue (ACRE) program get a 20 percent reduction in their direct payment, not calculated in this table. For 2014/2015 and after, the entry is the price loss coverage payment rate. Agricultural Risk Coverage (ARC) is also available, but payment rates are established at the county or farm levels. [2] For 2009/2010-2013/2014, producers who participate in the ACRE program get a 30 percent reduction in their loan rate, not calculated in this table. [3] Target applies to 2003/2004-2013/2014 and Reference applies beginning with 2014/2015. [4] Percent of production is on a grain basis. [5] Acquisitions as of July 30, 2016.
FSA, Feed Grains and Oilseeds, (202) 720–2711.

Table 1-50.—Oats: Marketing year average price and value of production, by State and United States, 2013–2015

State	Marketing year average price per bushel			Value of production		
	2013	2014	2015	2013	2014	2015
	Dollars	*Dollars*	*Dollars*	*1,000 dollars*	*1,000 dollars*	*1,000 dollars*
Alabama	3.90	5.00	4.50	4,680	4,875	4,500
Arkansas	3.40	3.95	2.95	1,737	2,844	1,416
California	3.38	3.20	3.20	4,056	3,200	1,920
Colorado	4.91	3.62	3.60	3,830	1,955	2,880
Georgia	3.80	4.50	4.10	4,104	4,860	4,613
Idaho	3.80	2.64	2.95	4,161	3,247	3,806
Illinois	4.07	4.58	2.90	7,021	9,160	5,583
Indiana	4.25	4.55	3.15	3,018	3,367	929
Iowa	3.93	3.60	2.65	15,563	12,672	11,027
Kansas	4.25	3.61	2.40	3,570	3,032	6,240
Maine	2.45	2.35	1.60	4,432	5,100	3,712
Michigan	3.68	3.40	2.00	6,845	9,384	6,700
Minnesota	3.78	3.22	2.10	22,623	25,358	26,208
Missouri	3.85	4.35	2.70	2,857	3,676	2,457
Montana	2.87	3.30	3.50	3,410	3,643	4,081
Nebraska	4.18	3.01	2.30	6,793	7,224	6,164
New York	3.60	3.66	2.45	11,095	9,223	5,684
North Carolina	3.30	3.10	2.20	3,003	3,531	2,323
North Dakota	3.14	2.42	2.10	26,282	18,549	21,756
Ohio	3.75	4.65	3.40	5,906	10,253	8,568
Oklahoma	3.80	4.25	2.35	1,011	1,615	642
Oregon	4.61	4.53	5.05	5,993	6,931	4,888
Pennsylvania	3.91	3.79	3.40	12,121	13,189	12,155
South Carolina	3.60	3.60	3.20	1,912	2,232	1,670
South Dakota	3.67	3.01	2.30	33,911	27,993	29,015
Texas	3.63	5.93	4.10	6,679	10,140	10,824
Utah	4.42	3.75	3.60	1,370	776	612
Virginia	3.10	2.90	2.00	434	679	608
Washington	4.00	2.55	2.55	1,360	893	689
Wisconsin	4.07	3.52	1.90	27,778	30,554	26,676
Wyoming	3.95	3.19	2.85	2,252	1,317	2,052
United States	3.75	3.21	2.12	239,807	241,472	220,398

NASS, Crops Branch, (202) 720–2127.

GRAIN AND FEED

Table 1-51.—International Oats: Area, yield, and production in specified countries, 2013/2014–2015/2016

Country	Area			Yield per hectare			Production		
	2013/ 2014	2014/ 2015	2015/ 2016	2013/ 2014	2014/ 2015	2015/ 2016	2013/ 2014	2014/ 2015	2015/ 2016
	1,000 hec-tares	1,000 hec-tares	1,000 hec-tares	Metric tons	Metric tons	Metric tons	1,000 metric tons	1,000 metric tons	1,000 metric tons
Argentina	215	245	235	2.07	2.14	2.06	445	525	485
Australia	715	683	860	1.76	1.59	1.51	1,255	1,087	1,300
Belarus	133	151	140	2.65	3.46	2.86	352	522	400
Brazil	170	150	190	2.24	2.00	1.84	380	300	350
Canada	1,113	928	1,050	3.51	3.21	3.27	3,906	2,979	3,430
Chile	136	90	85	4.49	4.68	5.53	610	421	470
China	200	200	200	2.90	3.00	3.00	580	600	600
European Union	2,652	2,542	2,627	3.16	3.09	2.91	8,380	7,845	7,640
Russia	3,007	3,077	2,835	1.64	1.71	1.60	4,932	5,267	4,550
Ukraine	241	245	215	1.94	2.49	2.28	467	610	490
Others	687	651	689	1.83	1.82	1.80	1,257	1,187	1,239
Total Foreign	9,269	8,962	9,126	2.43	2.38	2.30	22,564	21,343	20,954
United States	408	419	516	2.30	2.43	2.52	938	1,019	1,300
Total	9,677	9,381	9,642	2.43	2.38	2.31	23,502	22,362	22,254

FAS, Office of Global Analysis, (202) 720-6301. Prepared or estimated on the basis of official USDA production, supply, and distribution, supply, and and distribution statistics from foreign governments.

Table 1-52. Oats: International trade, 2013/2014–2015/2016

Country	2013/2014	2014/2015	2015/2016
	1,000 metric tons	1,000 metric tons	1,000 metric tons
Principle exporting countries:			
Argentina	2	2	5
Australia	260	250	350
Brazil	6	9	5
Canada	1,662	1,689	1,600
Chile	56	85	75
European Union	316	219	230
Kazakhstan	6	2	5
Russia	6	9	20
Ukraine	6	39	70
Others	1	1	
Total Foreign	2,321	2,305	2,360
United States	23	26	29
Total	2,344	2,331	2,389
Principle importing countries:			
Albania	5	5	5
Algeria	8	16	30
Canada	24	13	10
China	116	162	250
Ecuador	26	21	20
Japan	46	47	50
Mexico	97	84	100
Norway	39	3	40
South Africa	19	50	25
Switzerland	52	52	50
Others	19	28	30
Total Foreign	451	481	610
United States	1,674	1,852	1,637
Total	2,125	2,333	2,247

FAS, Office of Global Analysis, (202) 720-6301. Prepared or estimated on the basis of official USDA production, supply, and distribution statistics from foreign governments.

Table 1-53.—Barley: Area, yield, production, and value, United States, 2006–2015

Year	Area		Yield per harvested acre	Production	Marketing year average price per bushel received by farmers	Value of production
	Planted [1]	Harvested				
	1,000 acres	*1,000 acres*	*Bushels*	*1,000 bushels*	*Dollars*	*1,000 dollars*
2006	3,452	2,951	61.1	180,165	2.85	498,691
2007	4,018	3,502	60.0	210,110	4.02	834,954
2008	4,239	3,775	63.3	239,072	5.37	1,253,452
2009	3,568	3,114	72.8	226,603	4.66	968,676
2010	2,872	2,465	73.1	180,241	3.86	691,573
2011	2,564	2,241	69.1	154,788	5.35	808,741
2012	3,660	3,274	66.9	218,990	6.43	1,371,381
2013	3,528	3,040	71.3	216,745	6.06	1,265,798
2014	3,031	2,497	72.7	181,542	5.30	914,955
2015	3,558	3,109	68.9	214,297	5.50	1,117,323

[1] Barley sown for all purposes, including barley sown in the preceding fall.
NASS, Crops Branch, (202) 720–2127.

Table 1-54.—Barley: Stocks on and off farms, United States, 2006–2015

Year beginning September	On farms				Off farms [1]			
	Sep. 1	Dec. 1	Mar. 1	June 1	Sep. 1	Dec. 1	Mar. 1	June 1
	1,000 bushels	*1,000 bushels*	*1,000 bushels*	*1,000 bushels*	*1,000 bushels*	*1,000 bushels*	*1,000 bushels*	*1,000 bushels*
2006	112,850	83,650	38,310	14,580	99,939	89,171	78,756	54,300
2007	105,600	62,050	28,270	9,950	83,095	73,728	82,154	58,273
2008	127,750	77,050	44,310	27,010	81,669	95,766	84,791	61,723
2009	154,050	114,630	67,370	40,440	85,414	91,759	89,985	75,059
2010	125,070	91,660	57,700	26,040	98,818	88,720	80,424	63,311
2011	93,050	55,320	26,480	9,670	82,007	82,999	67,248	50,317
2012	112,550	72,580	35,180	15,840	85,226	85,473	81,897	64,557
2013	105,620	81,340	43,830	19,110	90,470	88,063	77,734	63,145
2014	97,820	74,510	41,990	20,940	81,997	81,625	76,247	57,639
2015	135,840	96,650	....	....	83,132	84,474	....	....

[1] Includes stocks at mills, elevators, warehouses, terminals, and processors.
NASS, Crops Branch, (202) 720–2127.

Table 1-55.—Barley: Supply and disappearance, United States, 2005–2014

Year beginning June 1	Supply				Disappearance					Ending stocks May 31
					Domestic use					
	Beginning stocks	Produc-tion	Imports	Total	Feed and residual	Food, seed, and in-dustrial	Total	Exports	Total dis-appear-ance	Total
	Million bushels	*Million bushels*	*Million bushels*	*Million bushels*	*Million bushels*	*Million bushels*	*Million bushels*	*Million bushels*	*Million bushels*	*Million bushels*
2005	128	212	5	346	48	162	210	28	238	108
2006	108	180	12	300	49	162	211	20	231	69
2007	69	210	29	308	30	169	199	41	240	68
2008	68	240	29	337	67	169	236	13	249	89
2009	89	227	17	332	47	164	211	6	216	115
2010	115	180	9	305	50	159	208	9	216	89
2011	89	155	16	260	37	155	192	9	200	60
2012	60	219	23	302	66	147	213	9	222	80
2013	80	217	19	316	66	153	219	14	234	82
2014	82	182	24	287	43	151	194	14	209	79

Includes quantity under loan and farmer–owned reserve.   Latest data may be preliminary or projected.   Totals may not add due to independent rounding.
ERS, Market and Trade Economics Division, (202) 694–5313.

Table 1-56.—Barley: Area, yield, and production, by State and United States, 2013–2015

State	Area planted [1]			Area harvested		
	2013	2014	2015	2013	2014	2015
	1,000 acres	*1,000 acres*	*1,000 acres*	*1,000 acres*	*1,000 acres*	*1,000 acres*
Arizona	75	36	17	69	32	16
California	95	80	70	42	25	25
Colorado	63	57	65	58	54	63
Delaware	43	41	32	33	31	22
Idaho	650	600	580	620	550	550
Kansas	17	16	13	11	10	8
Maine	20	13	13	17	12	12
Maryland	75	70	50	52	45	35
Michigan	10	9	11	9	8	6
Minnesota	90	75	135	75	60	120
Montana	990	920	970	830	770	850
New York	11	12	11	8	8	9
North Carolina	19	20	19	14	15	14
North Dakota	760	620	1,120	720	535	1,050
Oregon	63	50	49	50	38	37
Pennsylvania	75	70	55	60	50	40
South Dakota	37	28	37	19	17	19
Utah	40	32	27	30	20	16
Virginia	72	56	46	44	28	16
Washington	205	115	110	195	105	100
Wisconsin	33	26	28	16	16	15
Wyoming	85	85	100	68	68	86
United States	3,528	3,031	3,558	3,040	2,497	3,109

State	Yield per harvested acre			Production		
	2013	2014	2015	2013	2014	2015
	Bushels	*Bushels*	*Bushels*	*1,000 bushels*	*1,000 bushels*	*1,000 bushels*
Arizona	118.0	125.0	120.0	8,142	4,000	1,920
California	75.0	73.0	55.0	3,150	1,825	1,375
Colorado	133.0	124.0	130.0	7,714	6,696	8,190
Delaware	78.0	86.0	80.0	2,574	2,666	1,760
Idaho	93.0	94.0	97.0	57,660	51,700	53,350
Kansas	47.0	35.0	39.0	517	350	312
Maine	53.0	68.0	85.0	901	816	1,020
Maryland	85.0	77.0	69.0	4,420	3,465	2,415
Michigan	52.0	53.0	56.0	468	424	336
Minnesota	69.0	52.0	77.0	5,175	3,120	9,240
Montana	52.0	58.0	52.0	43,160	44,660	44,200
New York	52.0	47.0	45.0	416	376	405
North Carolina	67.0	71.0	72.0	938	1,065	1,008
North Dakota	64.0	67.0	64.0	46,080	35,845	67,200
Oregon	70.0	50.0	52.0	3,500	1,900	1,924
Pennsylvania	68.0	71.0	65.0	4,080	3,550	2,600
South Dakota	54.0	52.0	37.0	1,026	884	703
Utah	78.0	83.0	84.0	2,340	1,660	1,344
Virginia	82.0	79.0	75.0	3,608	2,212	1,200
Washington	72.0	60.0	48.0	14,040	6,300	4,800
Wisconsin	49.0	47.0	55.0	784	752	825
Wyoming	89.0	107.0	95.0	6,052	7,276	8,170
United States	71.3	72.7	68.9	216,745	181,542	214,297

[1] Includes area planted in the preceding fall.
NASS, Crops Branch, (202) 720–2127.

Table 1-57.—Barley: Marketing year average price and value, by State and United States, 2013-2015

State	Marketing year average price per bushel			Value of production		
	2013	2014	2015	2013	2014	2015
	Dollars	*Dollars*	*Dollars*	*1,000 dollars*	*1,000 dollars*	*1,000 dollars*
Arizona	5.13	4.44	4.20	41,768	17,760	8,064
California	5.81	5.24	4.80	18,302	9,563	6,600
Colorado	5.73	5.78	6.20	44,201	38,703	50,778
Delaware	4.75	3.65	2.75	12,227	9,731	4,840
Idaho	6.42	5.11	5.75	370,177	264,187	306,763
Kansas	4.52	5.25	2.55	2,337	1,838	796
Maine	4.15	3.55	3.75	3,739	2,897	3,825
Maryland	4.13	3.60	2.70	18,255	12,474	6,521
Michigan	4.75	3.89	3.80	2,223	1,649	1,277
Minnesota	6.22	5.38	5.35	32,189	16,786	49,434
Montana	6.32	5.33	5.85	272,771	238,038	258,570
New York	4.61	3.70	4.00	1,918	1,391	1,620
North Carolina	4.00	3.80	3.40	3,752	4,047	3,427
North Dakota	6.09	5.30	4.80	280,627	189,979	322,560
Oregon	4.02	3.56	3.15	14,070	6,764	6,061
Pennsylvania	3.89	3.70	2.60	15,871	13,135	6,760
South Dakota	4.25	3.50	3.75	4,361	3,094	2,636
Utah	4.17	3.13	2.80	9,758	5,196	3,763
Virginia	3.80	3.60	3.20	13,710	7,963	3,840
Washington	4.12	3.54	3.35	57,845	22,302	16,080
Wisconsin	6.49	5.25	5.45	5,088	3,948	4,496
Wyoming	6.71	5.98	5.95	46,609	43,510	48,612
United States	6.06	5.30	5.50	1,265,798	914,955	1,117,323

NASS, Crops Branch, (202) 720-2127.

Table 1-58.—International Barley: Area, yield, and production in specified countries, 2013/2014-2015/2016

Country	Area			Yield per hectare			Production		
	2013/ 2014	2014/ 2015	2015/ 2016	2013/ 2014	2014/ 2015	2015/ 2016	2013/ 2014	2014/ 2015	2015/ 2016
	1,000 hectares	*1,000 hectares*	*1,000 hectares*	*Metric tons*	*Metric tons*	*Metric tons*	*1,000 metric tons*	*1,000 metric tons*	*1,000 metric tons*
Argentina	1,270	900	1,000	3.74	3.22	3.60	4,750	2,900	3,600
Australia	3,814	3,836	4,100	2.41	2.09	2.12	9,174	8,014	8,700
Canada	2,652	2,136	2,350	3.86	3.33	3.50	10,237	7,119	8,225
European Union	12,408	12,394	12,327	4.81	4.88	4.95	59,674	60,460	60,962
Iran	1,635	1,575	1,600	1.71	2.03	2.06	2,800	3,200	3,300
Kazakhstan	1,837	1,909	2,038	1.38	1.26	1.31	2,539	2,412	2,675
Morocco	1,690	1,440	1,600	1.60	1.18	2.19	2,700	1,700	3,500
Russia	8,024	8,803	8,050	1.92	2.27	2.12	15,389	20,026	17,100
Turkey	3,330	3,400	3,400	2.19	1.18	2.18	7,300	4,000	7,400
Ukraine	3,233	3,200	3,000	2.34	2.95	2.92	7,561	9,450	8,750
Others	9,385	8,932	9,242	1.87	1.98	1.87	17,514	17,712	17,262
Total Foreign	49,278	48,525	48,707	2.83	2.82	2.90	139,638	136,993	141,474
United States	1,230	1,011	1,258	3.84	3.91	3.71	4,719	3,953	4,666
Total	50,508	49,536	49,965	2.86	2.85	2.92	144,357	140,946	146,140

FAS, Office of Global Analysis, (202) 720-6301. Prepared or estimated on the basis of official USDA production, supply, and distribution statistics from foreign governments.

Table 1-59.—Barley: International trade, 2013/2014–2015/2016

Country	2013/2014	2014/2015	2015/2016
	1,000 metric tons	1,000 metric tons	1,000 metric tons
Principle exporting countries:			
Argentina	2,891	1,552	2,100
Australia	6,216	5,219	6,000
Canada	1,559	1,516	1,350
European Union	5,741	9,547	8,600
India	441	431	200
Kazakhstan	416	483	750
Moldova	10	62	50
Russia	2,681	5,336	3,700
Serbia	28	7	25
Ukraine	2,476	4,456	4,500
Others	54	83	50
Total Foreign	22,513	28,692	27,325
United States	311	312	305
Total	22,824	29,004	27,630
Principle importing countries:			
Algeria	511	876	650
Brazil	318	484	400
China	4,891	9,859	7,500
Iran	900	2,200	1,300
Japan	1,294	1,097	1,300
Jordan	1,009	890	700
Kuwait	436	412	400
Libya	681	1,001	900
Saudi Arabia	9,000	8,200	8,500
Tunisia	646	483	450
Others	2,994	3,942	3,210
Total Foreign	22,680	29,444	25,310
United States	408	513	435
Total	23,088	29,957	25,745

FAS, Office of Global Analysis, (202) 720-6301. Prepared or estimated on the basis of official USDA production, supply, and distribution statistics from foreign governments.

Table 1-60.—Barley: Support operations, United States, 2006/2007–2015/2016

Marketing year beginning June 1	Income support payment rates per bushel [1]	Program price levels per bushel		Put under loan		Acquired by CCC under loan program [5]	Owned by CCC at end of marketing year
		Loan [2]	Target [3]	Quantity	Percentage of production [4]		
	Dollars	Dollars	Dollars	Million bushels	Percent	Million bushels	Million bushels
2006/2007	0.24/0.00	1.85	2.24	9.3	5.1	0	0
2007/2008	0.24/0.00	1.85	2.24	4.4	2.1	0	0
2008/2009	0.24/0.00	1.85	2.24	6.7	2.8	0	0
2009/2010	0.24/0.00	1.85	2.24	12.7	5.6	0	0
2010/2011	0.24/0.00	1.95	2.63	6.4	3.6	0	0
2011/2012	0.24/0.00	1.95	2.63	2.8	1.8	0	0
2012/2013	0.24/0.00	1.95	2.63	3.1	1.4	0	0
2013/2014	0.24/0.00	1.95	2.63	4.2	1.9	0	0
2014/2015	0.00	1.95	4.95	3.9	2.1	0	0
2015/2016	0.00	1.95	4.95	8.3	3.9	0	0

[1] The first entry is the direct payment rate and the second entry is the counter-cyclical payment rate for 2004/2005-2013/2014. For 2009/2010-2013/2014, producers who participate in the Average Crop Revenue (ACRE) program get a 20 percent reduction in their direct payment, not calculated in this table. For 2014/2015 and after, the entry is the price loss coverage payment rate. Agricultural Risk Coverage (ARC) is also available, but payment rates are established at the county or farm levels. [2] For 2009/2010-2013/2014, producers who participate in the ACRE program get a 30 percent reduction in their loan rate, not calculated in this table. [3] Target applies to 2003/2004-2013/2014 and Reference applies beginning with 2014/2015. [4] Percent of production is on a grain basis. [5] Acquisitions as of July 30, 2016.
FSA, Feed Grains and Oilseeds, (202) 720–2711.

Table 1-61.—Grains and grain products: Total and per capita civilian consumption as food, United States, 2003–2013

Calendar year [1]	Wheat			Rye		Rice (milled)	
	Total consumed [2]	Per capita consumption of food products		Total consumed [2]	Per capita consumption of rye flour	Total consumed [4]	Per capita consumption
		Flour [3]	Non-milled product				
	Million bushels	Pounds	Pounds	Million bushels	Pounds	Million cwt.	Pounds
2004	905	135	2.4	3.3	0.5	56.6	19.4
2005	917	134	2.4	3.3	0.5	57.1	19.4
2006	938	136	2.4	3.3	0.5	60.4	20.3
2007	948	138	2.4	3.3	0.5	61.0	20.3
2008	927	137	2.4	3.3	0.5	62.2	20.5
2009	920	135	2.3	3.3	0.5	62.6	20.4
2010	928	135	2.3	3.3	0.5	63.1	20.4
2011	925	133	2.3	3.3	0.5	(NA)	(NA)
2012	952	134	2.3	3.4	0.5	(NA)	(NA)
2013	945	135	2.3	3.4	0.3	....	....

Calendar year [1]	Corn						Oats		Barley	
	Total consumed [5]	Per capita consumption of food products					Total consumed [6]	Per capita consumption of oat food products	Total consumed [7]	Per capita consumption of food products [8]
		Flour and meal	Hominy and grits	Syrup	Dextrose	Starch				
	Million bushels	Pounds	Pounds	Pounds	Pounds	Pounds	Million bushels	Pounds	Million bushels	Pounds
2003	986	18.3	7.4	76.2	3.1	4.6	62.4	4.7	6.5	0.7
2004	973	18.6	7.8	75.6	3.3	4.5	63.0	4.7	6.6	0.7
2005	989	18.8	8.1	74.5	3.2	4.5	62.9	4.6	6.7	0.7
2006	980	19.0	8.5	72.1	3.1	4.4	64.5	4.7	6.6	0.7
2007	958	19.1	8.9	70.0	3.0	4.4	66.0	4.7	6.7	0.7
2008	....	19.3	9.3	66.6	2.8	4.4	67.6	4.8	6.8	0.7
2009	....	19.3	9.3	....	2.7	4.4	74.5	5.3	6.9	0.7
2010	....	19.3	9.3	....	2.9	4.5	74.0	5.2	6.8	0.7
2011	....	19.9	9.6	....	2.9	4.6	76.0	5.3	6.8	0.7
2012	....	19.8	9.5	....	2.7	4.6	76.0	5.2	6.5	0.6

Estimates of corn syrup and sugar are unofficial estimates; industry data were not reported after April 1968.
(NA) Not available. [1] Data are in marketing year; for corn, September 1-August 31; for oats and barley, June 1-May 31; and rice, August 1-July 31. Wheat, rye, syrup, and sugar are in calendar year. [2] Excludes quantities used in alcoholic beverages. [3] Includes white, whole wheat, and semolina flour. [4] Does not include shipments to U.S. territories. Excludes rice used in alcoholic beverages. Includes imports and rice used in processed foods and pet foods. [5] Includes an allowance for the quantity used as hominy and grits. This series is not adjusted for trade. [6] Oats used in oatmeal, prepared breakfast foods, infant foods, and food products. [7] Malt for food, breakfast food uses, pearl barley, and flour. [8] Malt equivalent of barley food products.
ERS, Market & Trade Economics Division, (202) 694-5290. All figures are estimates based on data from private industry sources, the U.S. Department of Commerce, the Internal Revenue Service, and other Government agencies.

Table 1-62.—Sorghum: Area, yield, production, and value, United States, 2006–2015

Year	Area planted for all purposes [1]	Sorghum for grain [2]					Sorghum for silage		
		Area harvested	Yield per harvested acre	Production	Marketing year average price per cwt [3]	Value of production [3]	Area harvested	Yield per harvested acre	Production
	1,000 acres	1,000 acres	Bushels	1,000 bushels	Dollars	1,000 dollars	1,000 acres	Tons	1,000 tons
2006 ..	6,522	4,937	56.1	276,824	5.88	883,204	347	13.3	4,612
2007 ..	7,712	6,792	73.2	497,445	7.28	1,925,312	392	13.4	5,246
2008 ..	8,404	7,312	65.1	475,855	5.72	1,641,963	412	13.9	5,710
2009 ..	6,599	5,502	69.4	381,605	5.75	1,202,331	255	14.5	3,693
2010 ..	5,369	4,806	71.9	345,464	8.96	1,616,773	272	12.5	3,412
2011 ..	5,451	3,945	54.0	212,993	10.70	1,259,635	226	10.4	2,353
2012 ..	6,259	4,995	49.6	247,742	11.30	1,606,237	353	11.9	4,196
2013 ..	8,076	6,585	59.6	392,331	7.64	1,716,927	380	14.3	5,420
2014 ..	7,138	6,401	67.6	432,575	7.20	1,721,330	315	13.1	4,123
2015 ..	8,459	7,851	76.0	596,751	5.90	2,080,568	306	14.6	4,475

[1] Grain and sweet sorghum for all uses, including syrup.   [2] Includes both grain sorghum for grain, and sweet sorghum for grain or seed.   [3] Based on the reported price of grain sorghum.
NASS, Crops Branch, (202) 720–2127.

Table 1-63.—Sorghum grain: Stocks on and off farms, United States, 2007–2016

Year beginning previous Dec.	On farms				Off farms [1]			
	Dec. 1	Mar. 1	Jun. 1	Sep. 1	Dec. 1	Mar. 1	Jun. 1	Sep. 1
	1,000 bushels	1,000 bushels	1,000 bushels	1,000 bushels	1,000 bushels	1,000 bushels	1,000 bushels	1,000 bushels
2007	38,100	17,100	5,380	2,150	174,094	125,122	69,490	29,903
2008	51,400	26,100	7,000	3,550	239,850	159,808	94,019	49,200
2009	54,400	32,200	12,000	4,400	243,290	173,650	90,215	50,312
2010	48,000	23,680	10,700	4,500	202,759	151,873	77,162	36,740
2011	30,500	13,020	3,140	2,030	207,168	158,027	76,894	25,420
2012	27,850	12,800	4,120	1,160	123,101	95,266	54,405	21,792
2013	17,600	10,850	2,710	602	122,247	80,685	38,398	14,551
2014	32,950	15,950	4,500	1,945	198,441	159,784	87,924	32,087
2015	30,500	9,000	2,960	1,895	192,094	110,856	31,329	16,514
2016 [2]	51,500	....	....	....	270,720	....	....	....

[1] Includes stocks at mills, elevators, warehouses, terminals, and processors.   [2] Preliminary.
NASS, Crops Branch, (202) 720–2127.

Table 1-64.—Sorghum: Supply and disappearance, United States, 2005–2014

Year beginning September 1	Supply			Disappearance						Ending stocks Aug. 31
	Beginning stocks	Production	Total	Domestic use			Exports	Total disappearance		
				Feed and residual	Food, seed and industrial	Total				
	Million bushels	Million bushels	Million bushels	Million bushels	Million bushels	Million bushels	Million bushels	Million bushels		Million bushels
2005	57	393	450	140	50	190	194	384		66
2006	66	277	343	113	45	158	153	311		32
2007	32	497	530	165	35	200	277	477		53
2008	53	476	529	233	94	331	143	474		55
2009	55	382	436	141	90	231	164	395		41
2010	41	345	387	123	85	208	152	359		27
2011	27	213	241	69	85	154	63	218		23
2012	23	248	280	93	95	189	76	265		15
2013	15	392	408	93	70	162	211	374		34
2014	34	433	467	80	15	96	353	449		18

Includes quantity under loan and farmer–owned reserve.   Latest data may be preliminary or projected.   Totals may not add due to independent rounding.
ERS, Market and Trade Economics Division, (202) 694–5313.

Table 1-65.—Sorghum: Area, yield, and production, by State and United States, 2013–2015

State	Area planted for all purposes			Sorghum for grain		
				Area harvested		
	2013	2014	2015	2013	2014	2015
	1,000 acres	1,000 acres	1,000 acres	1,000 acres	1,000 acres	1,000 acres
Arizona	33	25	24	17	8	4
Arkansas	130	170	450	125	165	440
Colorado	400	345	440	240	280	400
Georgia	55	40	50	40	23	34
Illinois	23	23	38	20	21	34
Kansas	3,150	2,850	3,400	2,850	2,700	3,200
Louisiana	115	100	77	113	96	74
Mississippi	65	110	120	62	105	115
Missouri	70	85	155	60	73	140
Nebraska	250	210	270	145	160	240
New Mexico	125	110	125	68	60	90
Oklahoma	320	370	440	270	310	410
South Dakota	340	200	270	275	150	220
Texas	3,000	2,500	2,600	2,300	2,250	2,450
United States	8,076	7,138	8,459	6,585	6,401	7,851

State	Yield per harvested acre			Production		
	2013	2014	2015	2013	2014	2015
	Bushels	Bushels	Bushels	1,000 bushels	1,000 bushels	1,000 bushels
Arizona	75.0	100.0	92.0	1,275	800	368
Arkansas	102.0	97.0	98.0	12,750	16,005	43,120
Colorado	24.0	30.0	55.0	5,760	8,400	22,000
Georgia	50.0	41.0	48.0	2,000	943	1,632
Illinois	94.0	106.0	94.0	1,880	2,226	3,196
Kansas	59.0	74.0	88.0	168,150	199,800	281,600
Louisiana	107.0	93.0	85.0	12,091	8,928	6,290
Mississippi	94.0	80.0	79.0	5,828	8,400	9,085
Missouri	82.0	101.0	94.0	4,920	7,373	13,160
Nebraska	67.0	82.0	96.0	9,715	13,120	23,040
New Mexico	34.0	42.0	47.0	2,312	2,520	4,230
Oklahoma	55.0	56.0	52.0	14,850	17,360	21,320
South Dakota	80.0	63.0	83.0	22,000	9,450	18,260
Texas	56.0	61.0	61.0	128,800	137,250	149,450
United States	59.6	67.6	76.0	392,331	432,575	596,751

NASS, Crops Branch, (202) 720–2127.

Table 1-66.—Sorghum: Utilization for silage, by State and United States, 2013–2015

State	Silage								
	Area harvested			Yield per acre			Production		
	2013	2014	2015	2013	2014	2015	2013	2014	2015
	1,000 acres	1,000 acres	1,000 acres	Tons	Tons	Tons	1,000 tons	1,000 tons	1,000 tons
Arizona	15	17	20	22.0	23.0	22.0	330	391	440
Arkansas	1	2	2	17.0	17.0	9.0	17	34	18
Colorado	30	10	10	13.0	11.0	14.0	390	110	140
Georgia	10	14	12	10.0	11.0	12.0	100	154	144
Illinois	2	1	2	16.0	18.0	15.0	32	18	30
Kansas	110	70	105	14.0	11.0	15.0	1,540	770	1,575
Louisiana	1	1	1	14.0	13.0	11.0	14	13	11
Mississippi	2	2	2	14.0	12.0	8.0	28	24	16
Missouri	8	10	10	17.0	17.0	19.0	136	170	190
Nebraska	30	20	10	10.0	12.0	12.5	300	240	125
New Mexico	16	33	29	13.0	13.0	12.0	208	429	348
Oklahoma	10	15	15	20.0	10.0	12.0	200	150	180
South Dakota	25	20	18	13.0	11.0	13.5	325	220	243
Texas	120	100	70	15.0	14.0	14.5	1,800	1,400	1,015
United States	380	315	306	14.3	13.1	14.6	5,420	4,123	4,475

NASS, Crops Branch, (202) 720–2127.

Table 1-67.—Sorghum grain: Marketing year average price and value of production, by State and United States, 2013–2015

State	Marketing year average price per cwt			Value of production		
	2013	2014	2015	2013	2014	2015
	Dollars	Dollars	Dollars	1,000 dollars	1,000 dollars	1,000 dollars
Arizona	10.10	9.34	9.20	7,211	4,184	1,896
Arkansas	8.85	7.07	7.75	63,189	63,367	187,141
Colorado	7.66	6.92	5.65	24,708	32,552	69,608
Georgia	9.52	8.52	8.35	10,662	4,499	7,631
Illinois	7.42	5.97	6.20	7,812	7,442	11,097
Kansas	7.46	7.10	5.45	702,463	794,405	859,443
Louisiana	8.70	7.76	7.90	58,907	38,798	27,827
Mississippi	8.62	7.05	8.10	28,133	33,163	41,210
Missouri	8.05	6.35	6.70	22,179	26,218	49,376
Nebraska	7.38	7.17	5.95	40,150	52,679	76,769
New Mexico	5.76	6.64	6.75	7,458	9,370	15,989
Oklahoma	7.49	6.94	6.10	62,287	67,468	72,829
South Dakota	6.57	5.95	6.00	80,942	31,487	61,354
Texas	8.33	7.23	7.15	600,826	555,698	598,398
United States	7.64	7.20	5.90	1,716,927	1,721,330	2,080,568

NASS, Crops Branch, (202) 720–2127.

Table 1-68.—Sorghum grain: Support operations, United States, 2006/2007–2015/2016

Marketing year beginning September 1	Income support payment rates per cwt[1]	Program price levels per cwt		Put under support		Acquired by CCC under loan program[5]	Owned by CCC at end of marketing year
		Loan[2]	Target[3]	Quantity	Percentage of production[4]		
	Dollars	Dollars	Dollars	Million cwt	Percent	Millions cwt	Million cwt
2006/2007	0.63/0.00	3.48	4.59	1.9	1.2	0	0
2007/2008	0.63/0.00	3.48	4.59	1.8	0.6	0	0
2008/2009	0.63/0.00	3.48	4.59	4.5	1.7	0	0
2009/2010	0.63/0.00	3.48	4.59	1.8	0.8	0	0
2010/2011	0.63/0.00	3.48	4.70	0.5	0.3	0	0
2011/2012	0.63/0.00	3.48	4.70	0.2	0.2	0	0
2012/2013	0.63/0.00	3.48	4.70	0.2	0.1	0	0
2013/2014	0.63/0.00	3.48	4.70	0.3	0.1	0	0
2014/2015	0.00	3.48	7.05	0.4	0.2	0	0
2015/2016	0.65	3.48	7.05	1.4	0.4	0	0

[1] The first entry is the direct payment rate and the second entry is the counter-cyclical payment rate for 2004/2005-2013/2014. For 2009/2010-2013/2014, producers who participate in the Average Crop Revenue (ACRE) program get a 20 percent reduction in their direct payment, not calculated in this table. For 2014/2015 and after, the entry is the price loss coverage payment rate. For 2015/2016, projected based on August 2016 WASDE MY price. Agricultural Risk Coverage (ARC) is also available, but payment rates are established at the county or farm levels. [2] For 2009/2010-2013/2014, producers who participate in the ACRE program get a 30 percent reduction in their loan rate, not calculated in this table. [3] Target applies to 2003/2004-2013/2014 and Reference applies beginning with 2014/2015. [4] Percent of production is on a grain basis. [5] Acquisitions as of July 30, 2016.
FSA, Feed Grains and Oilseeds, (202) 720–2711.

Table 1-69.—International Sorghum: Area, yield, and production in specified countries, 2013/2014–2015/2016

Country	Area			Yield per hectare			Production		
	2013/ 2014	2014/ 2015	2015/ 2016	2013/ 2014	2014/ 2015	2015/ 2016	2013/ 2014	2014/ 2015	2015/ 2016
	1,000 hec- tares	*1,000 hec- tares*	*1,000 hec- tares*	*Metric tons*	*Metric tons*	*Metric tons*	*1,000 metric tons*	*1,000 metric tons*	*1,000 metric tons*
Argentina	1,000	770	870	4.40	4.55	4.48	4,400	3,500	3,900
Australia	532	651	700	2.41	3.23	3.14	1,282	2,104	2,200
Brazil	730	730	700	2.59	2.74	2.86	1,890	2,000	2,000
Burkina	1,800	1,800	1,800	1.04	1.02	1.06	1,880	1,836	1,900
China	582	619	610	4.97	4.66	4.75	2,892	2,885	2,900
Ethiopia	1,818	1,800	1,800	2.26	2.22	2.11	4,114	4,000	3,800
India	5,793	5,500	6,000	0.96	0.99	0.92	5,542	5,450	5,500
Mexico	2,073	1,715	1,600	4.10	3.66	3.56	8,500	6,270	5,700
Nigeria	5,000	5,500	5,300	1.32	1.22	1.16	6,592	6,700	6,150
Sudan	4,356	8,626	8,000	0.52	0.73	0.69	2,249	6,281	5,500
Others	12,631	12,546	12,459	0.95	0.95	0.96	11,947	11,874	12,000
Total Foreign	36,315	40,257	39,839	1.41	1.31	1.29	51,288	52,900	51,550
United States	2,665	2,590	3,177	3.74	4.24	4.77	9,966	10,988	15,158
Total	38,980	42,847	43,016	1.57	1.49	1.55	61,254	63,888	66,708

FAS, Office of Global Analysis, (202) 720-6301. Prepared or estimated on the basis of official USDA production, supply, and distribution statistics from foreign governments.

Table 1-70.—Sorghum: International trade, 2013/2014–2015/2016

Country	2013/2014	2014/2015	2015/2016
	1,000 metric tons	*1,000 metric tons*	*1,000 metric tons*
Principle exporting countries:			
Argentina	1,279	1,000	1,000
Australia	385	1,650	1,000
Brazil	11	13	30
China	11	9	25
Ethiopia	75	75	25
India	87	121	100
Kenya	56	40	30
Nigeria	50	100	50
South Africa	27	25	25
Ukraine	229	156	100
Others	111	125	90
Total Foreign	2,321	3,314	2,475
United States	5,362	8,965	8,255
Total	7,683	12,279	10,730
Principle importing countries:			
Chile	184	150	100
China	4,161	10,162	7,000
Colombia	104		50
Ethiopia	50	75	150
European Union	193	134	80
Japan	1,003	903	950
Kenya	92	98	100
Mexico	162	29	700
Pakistan	15	32	130
Sudan	75	100	75
Others	575	625	385
Total Foreign	6,614	12,308	9,720
United States	2	10	127
Total	6,616	12,318	9,847

FAS, Office of Global Analysis, (202) 720-6301. Prepared or estimated on the basis of official USDA production, supply, and distribution statistics from foreign governments.

Table 1-71.—International Mixed grain: Area, yield, and production in specified countries, 2013/2014–2015/2016

Country	Area 2013/ 2014	Area 2014/ 2015	Area 2015/ 2016	Yield per hectare 2013/ 2014	Yield per hectare 2014/ 2015	Yield per hectare 2015/ 2016	Production 2013/ 2014	Production 2014/ 2015	Production 2015/ 2016
	1,000 hectares	1,000 hectares	1,000 hectares	metric tons	metric tons	metric tons	1,000 metric tons	1,000 metric tons	1,000 metric tons
Bangladesh	35	35	35	0.71	0.71	0.71	25	25	25
Canada	66	63	62	2.80	3.00	2.90	185	189	180
European Union ...	3,953	4,035	4,144	3.89	4.17	3.83	15,373	16,808	15,869
Switzerland	9	10	10	5.67	6.00	6.00	51	60	60
Turkey	100	100	100	1.15	1.15	1.15	115	115	115
Total Foreign	4,163	4,243	4,351	3.78	4.05	3.73	15,749	17,197	16,249
Total	4,163	4,243	4,351	3.78	4.05	3.73	15,749	17,197	16,249

FAS, Office of Global Analysis, (202) 720-6301. Prepared or estimated on the basis of official USDA production, supply, and distribution statistics from foreign governments.

Table 1-72.—Commercial feeds: Disappearance for feed, United States, 2005–2014

Year beginning October	Oilseed cake and meal Soybean	Oilseed cake and meal Cotton-seed	Oilseed cake and meal Linseed	Oilseed cake and meal Peanut[1]	Oilseed cake and meal Sun-flower	Oilseed cake and meal Total	Animal protein Tankage and meat meal	Animal protein Fish meal	Animal protein Dried milk[2]	Animal protein Total
	1,000 tons	1,000 tons	1,000 tons	1,000 tons	1,000 tons	1,000 tons	1,000 tons	1,000 tons	1,000 tons	1,000 tons
2005	33,195	3,355	269	117	298	37,234	2,254	199	269	2,722
2006	34,355	3,049	275	119	356	38,154	2,375	215	292	2,882
2007	33,232	2,589	210	116	343	36,490	2,398	213	250	2,861
2008	30,752	1,807	129	102	357	33,147	2,271	223	250	2,744
2009	30,639	1,755	210	91	415	33,110	2,343	200	250	2,793
2010	30,301	2,452	209	112	281	33,354	2,350	200	250	2,800
2011	31,541	2,406	194	109	189	34,439	2,444	198	463	3,105
2012	28,999	2,351	199	123	231	31,903	2,445	198	368	3,012
2013	29,547	1,797	152	130	257	31,883	2,445	198	219	2,862
2014	32,300	2,180	212	138	222	35,053	2,535	198	248	2,982

Year beginning October	Mill products[3] Wheat millfeeds	Mill products[3] Gluten feed and meal[4]	Mill products[3] Rice millfeeds	Mill products[3] Alfalfa meal	Mill products[3] Total	Total commercial feeds
	1,000 tons	1,000 tons	1,000 tons	1,000 tons	1,000 tons	1,000 tons
2005	6,753	3,514	641	0	10,908	50,865
2006	6,873	4,624	545	0	12,042	53,078
2007	6,776	4,560	568	0	11,904	51,256
2008	6,464	5,167	570	0	12,201	48,092
2009	6,400	5,075	575	0	12,050	47,953
2010	6,500	5,050	575	0	12,125	48,279
2011	6,168	5,772	580	0	12,521	50,064
2012	6,278	5,707	535	0	12,520	47,435
2013	6,271	5,649	525	0	12,445	47,190
2014	9,101	5,849	525	0	15,475	53,509

[1] Year beginning August 1.   [2] Includes dried skim milk, and whey for feed, but does not include any milk products fed on farms.   [3] Other mill products that are not listed include screenings, hominy, and oats feed etc., for which no statistics are available.   [4] Adjusted for export data.

ERS, Market and Trade Economics Division, (202) 694-5313.

Table 1-73.—High-protein feeds: Quantity for feeding, high-protein animal units, quantity per animal unit, and prices, United States, 2005–2014

Year beginning October	Quantity for feeding [1]						High-protein animal units	Quantity per animal unit	High protein feed prices
	Oilseed meal			Animal protein	Grain protein [3]	Total			
	Soybean meal	Other oilseed meals [2]	Total						
	1,000 tons	1,000 tons	1,000 tons	1,000 tons	1,000 tons	1,000 tons	Million units	Pounds	Index numbers 1992=100
2005	36,515	3,724	40,239	3,047	2,092	45,378	72	1,267	88
2006	37,791	3,497	41,288	3,219	2,753	47,260	72	1,317	105
2007	36,555	3,002	39,557	3,232	2,715	45,505	71	1,273	170
2008	33,827	2,207	36,034	3,092	3,077	42,203	71	1,191	168
2009	33,703	2,263	35,966	3,149	3,022	42,137	70	1,196	150
2010	33,331	2,814	36,145	3,158	3,007	42,310	69	1,220	175
2011	34,695	2,675	37,370	3,334	3,437	44,142	68	1,298	200
2012	31,899	2,680	34,579	3,307	3,398	41,284	68	1,221	235
2013	32,502	2,156	34,658	3,260	3,364	41,282	67	1,237	247
2014	35,530	2,538	38,068	3,379	3,483	44,930	68	1,324	189

Latest data may be preliminary or projected. [1] In terms of 44 percent protein soybean meal equivalent. [2] Includes cottonseed, linseed, peanut meal, and sunflower meal. [3] Beginning 1974, adjusted for exports of corn gluten feed and meal.
ERS, Market and Trade Economics Division, (202) 694–5313.

Table 1-74.—Feed concentrates: Fed to livestock and poultry, 2005–2014

Year beginning October	Feed grains				Wheat [2]	Rye [2]	By-product feeds [3]	Total concentrates	Grain consuming animal units	Concentrates fed per grain-consuming animal unit
	Corn [1]	Sorghum [1]	Oats [2] and barley [2]	Total						
	Million tons	Million tons	Million tons	Million tons	Million tons	Million tons	Million tons	Million tons	Millions	Tons
2005	171.2	3.9	3.9	179.0	3.0	0.1	58.8	240.9	91.5	2.63
2006	155.1	3.2	4.0	162.3	5.1	0.1	60.6	228.1	92.7	2.46
2007	164.0	4.6	3.4	172.0	4.6	0.1	59.0	235.7	95.1	2.48
2008	145.1	6.6	3.4	155.1	3.7	0.1	55.2	214.1	92.7	2.31
2009	133.7	3.9	3.4	141.1	4.5	0.1	54.6	200.3	91.6	2.19
2010	126.5	3.4	3.6	133.6	4.6	0.1	54.6	192.9	92.4	2.09
2011	126.5	1.9	2.6	131.1	11.5	0.1	56.6	199.2	93.1	2.14
2012	120.8	2.6	3.4	126.9	11.4	0.1	53.8	192.1	92.3	2.08
2013	141.1	2.6	2.7	146.4	1.8	0.1	53.9	202.2	91.0	2.22
2014	148.8	2.3	1.4	152.5	4.8	0.1	60.7	218.0	93.2	2.34

Latest data may be preliminary or projected. [1] Marketing year beginning Sept. 1. [2] Marketing year beginning June 1. [3] Oilseed meals, animal protein feeds, mill byproducts, and mineral supplements.
ERS, Market and Trade Economics Division, (202) 694–5313.

Table 1-75.—Feed: Consumed per head and per unit of production, by class of livestock or poultry, with quantity expressed in equivalent feeding value of corn, 2005–2014

Year beginning October	Dairy cattle			Beef cattle				Sheep and lambs	
	Milk cows		Other dairy cattle per head	Cattle on feed per head Jan. 1[1]	Other beef cattle per head	All beef cattle per head	Cattle and calves per 100 pounds produced[2]	Per head	Per 100 pounds produced[3]
	Per head	Per 100 pounds milk produced							
	Pounds	Pounds	Pounds	Pounds	Pounds	Pounds	Pounds	Pounds	Pounds
2005	13,125	68	6,565	9,978	5,321	6,129	1,301	1,279	1,679
2006	12,775	64	6,506	9,466	5,303	6,037	1,267	1,272	1,691
2007	12,779	64	6,507	9,471	5,303	6,053	1,273	1,272	1,720
2008	12,394	61	6,442	8,909	5,283	5,905	1,215	1,265	1,744
2009	12,226	59	6,409	8,615	5,274	5,844	1,217	1,262	1,683
2010	11,658	55	6,295	7,773	5,244	5,695	1,159	1,252	1,693
2011	11,730	55	6,329	7,836	5,258	5,737	1,136	1,256	1,662
2012	11,721	54	6,341	8,127	5,248	5,765	1,146	1,253	1,650
2013	11,788	54	6,341	8,373	5,255	5,790	1,123	1,256	1,615
2014	12,075	55	6,429	8,410	5,273	5,801	1,136	1,265	1,554

Year beginning October	Poultry								Hogs per 100 pounds produced	Horses and mules two years and over per head
	Hens and pullets		Chickens raised		Broilers produced		Turkeys raised			
	Per head Jan. 1	Per 100 eggs	Per head	Per 100 pounds live weight	Per head	Per 100 pounds produced	Per head	Per 100 pounds produced		
	Pounds	Pounds	Pounds	Pounds	Pounds	Pounds	Pounds	Pounds	Pounds	Pounds
2005	131	50	31	933	12	217	98	357	584	3,875
2006	124	47	30	982	11	202	92	340	552	3,806
2007	124	47	30	1,030	11	200	92	334	564	3,807
2008	116	43	28	984	10	176	87	278	504	3,732
2009	113	42	27	1,068	10	178	84	288	484	9,944
2010	101	37	24	992	9	154	76	264	461	10,018
2011	103	38	25	952	9	155	76	263	464	10,034
2012	100	37	23	935	9	153	79	255	431	9,745
2013	101	37	24	976	9	154	80	259	447	10,271
2014	108	39	26	1,053	10	168	82	274	493	10,489

[1] Feed consumed by all cattle divided by the number on feed Jan. 1. [2] Feed for all cattle, except milk cows, divided by the net live-weight production of cattle and calves. It includes the growth on dairy heifers and calves as well as all beef cattle. [3] Including wool produced.
ERS, Market and Trade Economics Division, (202) 694–5313.

Table 1-76.—Feed: Consumed by livestock and poultry, by type of feed, with quantity expressed in equivalent feeding value of corn, 2005–2014

Year beginning October	Concentrates	Harvested roughage	Pasture	Total
	Million tons	Million tons	Million tons	Million tons
2005	266	87	157	509
2006	254	82	162	498
2007	261	80	163	504
2008	237	83	158	478
2009	227	83	153	463
2010	203	81	150	435
2011	210	76	153	440
2012	206	76	151	433
2013	206	78	153	437
2014	227	95	140	463

Latest data may be preliminary or projected.
ERS, Market and Trade Economics Division, (202) 694–5313.

Table 1-77.—Animal units fed: Grain-consuming, roughage-consuming, and grain-and-roughage-consuming, United States, 2005–2014 [1]

Year beginning October	Grain-consuming [2]	Roughage-consuming [3]	Grain and roughage-consuming [4]
	1,000 units	1,000 units	1,000 units
2005	91,490	71,647	78,731
2006	92,749	71,753	79,289
2007	95,118	71,479	80,042
2008	92,708	70,887	78,782
2009	91,636	70,386	78,033
2010	92,851	69,233	77,786
2011	93,232	68,023	77,237
2012	92,267	67,593	76,601
2013	90,420	66,756	75,314
2014	92,576	67,409	76,371

Latest data may be preliminary or projected. [1] Index series based on average feeding rates for years 1969–71. In calculations for the feeding years 1969 to date, cattle numbers used are the new categories shown in the Livestock and Poultry Inventory, published by NASS, USDA. [2] Livestock and poultry numbers weighted by all concentrates consumed. [3] Livestock and poultry numbers weighted by all roughage (including pasture) consumed. [4] Livestock and poultry numbers weighted by all feed (including pasture) fed to livestock.
ERS, Market and Trade Economics Division, (202) 694–5313.

Table 1-78.—Feed grains: Average price, selected markets and grades, 2006–2015 [1]

Calendar year	Kansas City			Minneapolis			
	Corn, No. 2 Yellow (truck)	Corn, No. 2 White (rail)	Sorghum, No. 2 Yellow (truck)	Corn, No. 2 Yellow	Barley, No. 3 or Better malting	Duluth Barley, No. 2 Feed	Oats, No. 2 White
	Dollars per bushel	Dollars per bushel	Dollars per cwt.	Dollars per bushel	Dollars per bushel	Dollars per bushel	Dollars per bushel
2006	2.42	2.03	4.27	2.24	3.20		2.24
2007	4.61	4.43	6.05	3.38	2.02	3.95	2.98
2008	5.12	5.32	8.41	4.76	6.81		3.91
2009	3.60	3.90	5.57	3.46	4.26		2.21
2010	4.08	4.27	6.94	3.65	4.70	(NA)	2.74
2011	6.81	7.11	11.73	6.58	6.96		3.72
2012	7.17	7.29	11.81	6.87	7.06		3.67
2013	6.28	6.35	10.40	5.97	6.35	4.47	3.90
2014	4.07	4.30	7.30	3.83	6.23	(NA)	4.02
2015	3.76	3.82	6.97	3.45	6.23	(NA)	2.84

Calendar year	Omaha: Corn, No. 2 Yellow (truck)	Chicago: Corn, No. 2 Yellow	Texas High Plains: Sorghum, No. 2 Yellow	Memphis Corn, No. 2 Yellow	St. Louis: Corn, No. 2 Yellow (truck)
	Dollars per bushel	Dollars per bushel	Dollars per cwt	Dollars per bushel	Dollars per cwt
2006	2.31	2.43	5.06	2.66	2.34
2007	3.54	3.67	7.10	3.71	3.74
2008	5.04	5.12	9.53	5.07	5.11
2009	3.56	3.76	6.52	3.69	3.78
2010	4.03	4.16	7.81	4.25	4.26
2011	6.72	6.78	12.64	6.85	7.08
2012	7.06	6.86	13.15	6.95	7.08
2013	6.23	5.81	10.51	6.07	6.25
2014	3.99	4.03	7.26	4.33	4.84
2015	3.66	3.68	6.72	3.87	3.87

(NA) Not available. [1] Simple average of daily prices.
AMS, Livestock and Grain Market News Branch, (202) 720–6231.

Table 1-79.—Feedstuffs: Average price per ton bulk, in wholesale lots, at leading markets, 2004–2013

Year beginning October	Soybean meal 44% protein Decatur	Soybean meal 48% protein Decatur	Cottonseed meal 41% protein Kansas City	Cottonseed meal 41% protein Memphis	Linseed meal 34% protein Minneapolis	Meat meal 50% protein Kansas City	Fish meal 60% protein Gulf Coast	Wheat bran Kansas City	Wheat middlings Minneapolis
	Dollars per ton	Dollars per ton	Dollars per ton	Dollars per ton	Dollars per ton	Dollars per ton	Dollars per ton	Dollars per ton	Dollars per ton
2004	(1)	237.30	193.58	167.68	148.09	190.63	524.97	67.82	64.19
2005	(1)	188.17	156.59	128.89	115.70	169.19	....	54.34	44.53
2006	(1)	175.60	171.84	141.87	116.12	151.43	707.27	72.68	61.12
2007	(1)	230.39	187.53	166.49	148.36	225.96	850.53	87.31	87.20
2008	(1)	331.09	298.72	265.82	227.05	326.48	866.06	134.33	136.20
2009	(1)	347.73	293.25	265.21	231.77	334.69	861.06	90.17	89.12
2010	(1)	313.47	249.72	212.57	210.80	299.05	1,265.78	94.62	94.44
2011	....	333.73	310.43	275.43	242.60	342.84	1,198.89	176.20	172.74
2012	....	439.77	336.79	303.37	282.67	391.49	1,168.09	175.06	192.91
2013	(NA)	464.61	430.80	333.42	(NA)	467.00	(NA)	129.57	182.40

Year beginning October	Wheat shorts or middlings Kansas City	Wheat millrun Portland	Gluten feed 21% protein Illinois Points	Hominy feed Midwest	Distillers' dried grains Lawrenceburg	Brewers' dried grains Columbus	Alfalfa meal Dehydrated, 17% protein Kansas City	Alfalfa meal Sun-cured Kansas City	Blackstrap molasses New Orleans
	Dollars per ton	Dollars per ton	Dollars per ton	Dollars per ton	Dollars per ton	Dollars per ton	Dollars per ton	Dollars per ton	Dollars per ton
2004	67.82	85.00	68.83	77.02	106.04	(1)	121.35	109.26	57.28
2005	54.23	74.72	68.17	50.50	75.47	(1)	135.83	110.57	(NA)
2006	72.53	84.51	69.51	59.84	89.04	(1)	174.13	161.77	(NA)
2007	....	129.30	81.34	108.64	113.38	(1)	206.53	179.50	(NA)
2008	134.31	185.85	....	153.50	....	(1)	236.28	205.77	(NA)
2009	90.24	120.48	....	100.53	114.23	(1)	224.93	189.19	(NA)
2010	94.62	124.06	....	115.61	122.77	(1)	210.36	178.16	(NA)
2011	176.20	187.71	....	206.85	202.42	....	254.72	243.29	....
2012	174.78	219.71	....	....	200.00	....	347.23	....	....
2013	129.99	228.57	166.35	202.42	236.07	(NA)	339.28	275.70	(NA)

(NA) Not available.   ¹ Discontinued.
AMS, Livestock and Grain Market News Branch, (202) 720–6231.

Table 1-80.—International Millet: Area, yield, and production in specified countries, 2013/2014–2015/2016

Country	Area			Yield per hectare			Production		
	2013/ 2014	2014/ 2015	2015/ 2016	2013/ 2014	2014/ 2015	2015/ 2016	2013/2014	2014/2015	2015/2016
	1,000 hectares	1,000 hectares	1,000 hectares	metric tons	metric tons	metric tons	1,000 metric tons	1,000 metric tons	1,000 meric tons
Burkina	1,327	1,200	1,200	0.81	0.87	0.92	1,079	1,038	1,100
Chad	800	1,000	1,000	0.73	0.68	0.70	582	683	700
China	716	772	750	2.44	2.34	2.40	1,746	1,809	1,800
Ethiopa	455	440	440	1.87	1.70	1.59	850	750	700
India	9,687	8,903	8,950	1.20	1.31	1.19	11,663	11,630	10,680
Mali	1,437	1,600	1,600	0.80	1.00	1.00	1,152	1,600	1,600
Niger	7,100	7,000	7,000	0.42	0.41	0.46	2,995	2,900	3,200
Nigeria	4,000	4,000	4,000	1.25	1.20	1.20	5,000	4,800	4,800
Senegal	714	850	850	0.72	0.48	0.76	515	409	650
Sudan	1,501	2,884	2,500	0.24	0.38	0.28	359	1,085	700
Others	3,898	3,919	3,994	0.79	0.82	0.86	3,090	3,231	3,440
Total	31,635	32,568	32,284	0.92	0.92	0.91	29,031	29,935	29,370

FAS, Office of Global Analysis, (202) 720-6301. Prepared or estimated on the basis of official USDA production, supply, and distribution statistics from foreign governments.

Table 1-81.—Proso millet: Area, yield, production, and value, United States, 2006–2015

Year	Area		Yield per harvested acre	Production	Marketing year average price per bushel received by farmers	Value of production
	Planted	Harvested				
	1,000 acres	*1,000 acres*	*Bushels*	*1,000 dollars*	*Dollars*	*1,000 dollars*
2006	580	475	21.5	10,195	4.09	41,748
2007	570	520	32.5	16,900	4.67	78,975
2008	520	460	32.3	14,880	3.23	48,017
2009	350	265	33.5	8,875	2.87	25,460
2010	390	363	31.8	11,535	4.54	52,419
2011	370	338	27.1	9,149	6.01	54,974
2012	335	205	15.0	3,075	15.10	46,548
2013	720	638	28.9	18,436	4.61	84,915
2014	505	430	31.4	13,483	3.35	45,104
2015	445	418	33.9	14,159	2.85	40,420

NASS, Crops Branch, (202) 720–2127.

Table 1-82.—Proso millet: Area, yield, and production, by State and United States, 2013–2015

State	Area planted			Area harvested		
	2013	2014	2015	2013	2014	2015
	1,000 acres	*1,000 acres*	*1,000 acres*	*1,000 acres*	*1,000 acres*	*1,000 acres*
Colorado	370	310	270	330	250	260
Nebraska	160	120	105	143	111	97
South Dakota	190	75	70	165	69	61
United States	720	505	445	638	430	418

State	Yield per acre			Production		
	2013	2014	2015	2013	2014	2015
	Bushels	*Bushels*	*Bushels*	*1,000 bushels*	*1,000 bushels*	*1,000 bushels*
Colorado	25.0	32.5	34.5	8,250	8,125	8,970
Nebraska	32.0	29.0	34.0	4,576	3,219	3,298
South Dakota	34.0	31.0	31.1	5,610	2,139	1,891
United States	28.9	31.4	33.9	18,436	13,483	14,159

NASS, Crops Branch, (202) 720–2127.

Table 1-83.—Proso millet: Marketing year average price and value, by State and United States, 2013–2015

State	Marketing year average price per bushel			Value of production		
	2013	2014	2015	2013	2014	2015
	Dollars	*Dollars*	*Dollars*	*1,000 dollars*	*1,000 dollars*	*1,000 dollars*
Colorado	4.31	3.30	2.88	35,558	26,813	25,834
Nebraska	4.73	3.25	2.76	21,644	10,462	9,102
South Dakota	4.94	3.66	2.90	27,713	7,829	5,484
United States	4.61	3.35	2.85	84,915	45,104	40,420

NASS, Crops Branch, (202) 720–2127.

CHAPTER II
STATISTICS OF COTTON, TOBACCO, SUGAR CROPS, AND HONEY

In addition to tables on cotton, tobacco, sugar, and honey, this chapter includes tables on fibers other than cotton and syrups. Cottonseed data, however, are in the following chapter on oilseeds, fats, and oils.

Table 2-1.—Cotton: Area, yield, production, market year average price, and value, United States, 2006–2015

Year	Area		Yield per harvested acre	Production	Marketing year average price per pound received by farmers	Value of production
	Planted	Harvested				
	1,000 acres	1,000 acres	Pounds	1,000 bales [1]	Dollars	1,000 dollars
2006	15,274.0	12,731.5	814	21,587.8	0.484	5,013,238
2007	10,827.2	10,489.1	879	19,206.9	0.613	5,652,907
2008	9,471.0	7,568.7	813	12,825.4	0.491	3,023,632
2009	9,149.5	7,533.7	776	12,183.0	0.648	3,786,674
2010	10,974.2	10,698.7	812	18,101.8	0.846	7,346,868
2011	14,735.4	9,460.9	790	15,573.2	0.935	6,985,976
2012	12,264.4	9,321.8	892	17,313.8	0.757	6,291,813
2013	10,407.0	7,544.4	821	12,909.2	0.838	5,191,505
2014	11,037.4	9,346.8	838	16,319.4	0.657	5,147,241
2015	8,580.5	8,074.9	769	12,888.0	0.622	3,862,448

[1] Value based on 480-pound net weight bale.
NASS, Crops Branch, (202) 720–2127.

Table 2-2.—Cotton: Marketing year average price per pound, and value, by State and United States, 2013–2015

State	Marketing year average price per pound			Value of production		
	2013	2014	2015	2013	2014	2015
	Dollars	Dollars	Dollars	1,000 dollars	1,000 dollars	1,000 dollars
Upland:						
Alabama	0.819	0.603	0.621	231,941	189,004	163,944
Arizona	0.780	0.641	0.632	179,712	150,763	86,458
Arkansas	0.794	0.644	0.597	274,406	243,277	136,116
California	0.905	0.853	0.620	144,655	87,620	49,104
Florida	0.820	0.667	0.600	68,880	61,471	43,200
Georgia	0.825	0.654	0.629	918,720	806,774	694,416
Kansas	0.774	0.559	0.599	15,232	12,879	7,763
Louisiana	0.784	0.626	0.575	122,680	121,394	52,440
Mississippi	0.815	0.658	0.599	281,273	340,476	192,638
Missouri	0.838	0.650	0.585	199,511	177,840	113,724
New Mexico	0.771	0.655	0.650	22,205	20,122	16,536
North Carolina	0.815	0.668	0.572	299,659	319,037	143,320
Oklahoma	0.781	0.553	0.599	57,732	71,403	106,382
South Carolina	0.819	0.655	0.589	141,523	166,003	42,408
Tennessee	0.799	0.626	0.604	158,777	148,437	87,556
Texas	0.746	0.587	0.583	1,493,194	1,739,868	1,609,080
Virginia	0.802	0.664	0.586	58,129	70,756	40,504
United States	0.779	0.613	0.593	4,668,229	4,727,124	3,585,589
American-Pima:						
Arizona	(D)	(D)	(D)	(D)	(D)	(D)
California	1.720	1.570	1.340	503,616	376,800	231,552
New Mexico	(D)	(D)	(D)	(D)	(D)	(D)
Texas	1.710	1.300	1.250	12,312	17,472	16,800
United States	1.720	1.550	1.326	523,276	420,117	276,859
All:						
United States	0.838	0.657	0.622	5,191,505	5,147,241	3,862,448

(D) Withheld to avoid disclosing data for individual operations.
NASS, Crops Branch, (202) 720–2127.

Table 2-3.—Cotton: Area, yield, production, and type by State and United States, 2013–2015

State	Area planted			Area harvested		
	2013	2014	2015	2013	2014	2015
Upland:	*1,000 acres*	*1,000 acres*	*1,000 acres*	*1,000 acres*	*1,000 acres*	*1,000 acres*
Alabama	365.0	350.0	315.0	359.0	348.0	307.0
Arizona	160.0	150.0	89.0	159.0	149.0	88.0
Arkansas	310.0	335.0	210.0	305.0	330.0	207.0
California	93.0	57.0	47.0	92.0	56.0	46.0
Florida	131.0	107.0	85.0	127.0	105.0	83.0
Georgia	1,370.0	1,380.0	1,130.0	1,340.0	1,370.0	1,120.0
Kansas	27.0	31.0	16.0	26.0	29.0	16.0
Louisiana	130.0	170.0	115.0	128.0	168.0	112.0
Mississippi	290.0	425.0	320.0	287.0	420.0	315.0
Missouri	255.0	250.0	185.0	246.0	245.0	175.0
New Mexico	39.0	43.0	35.0	31.0	33.0	31.0
North Carolina	465.0	465.0	385.0	460.0	460.0	355.0
Oklahoma	185.0	240.0	215.0	125.0	210.0	205.0
South Carolina ...	258.0	280.0	235.0	250.0	278.0	136.0
Tennessee	250.0	275.0	155.0	233.0	270.0	140.0
Texas	5,800.0	6,200.0	4,800.0	3,100.0	4,600.0	4,500.0
Virginia	78.0	87.0	85.0	77.0	86.0	84.0
United States	10,206.0	10,845.0	8,422.0	7,345.0	9,157.0	7,920.0
American Pima:						
Arizona	1.5	15.0	17.5	1.5	14.5	17.0
California	187.0	155.0	117.0	186.0	154.0	116.0
New Mexico	3.5	5.4	7.0	3.4	5.3	6.9
Texas	9.0	17.0	17.0	8.5	16.0	15.0
United States	201.0	192.4	158.5	199.4	189.8	154.9
All:						
United States	10,407.0	11,037.4	8,580.5	7,544.4	9,346.8	8,076.9

State	Yield per harvested acre			Production [1]		
	2013	2014	2015	2013	2014	2015
Upland:	*Pounds*	*Pounds*	*Pounds*	*1,000 bales[2]*	*1,000 bales[2]*	*1,000 bales[2]*
Alabama	789	901	866.0	590.0	653.0	554.0
Arizona	1,449	1,579	1,511.0	480.0	490.0	277.0
Arkansas	1,133	1,145	1,092.0	720.0	787.0	471.0
California	1,737	1,834	1,722.0	333.0	214.0	165.0
Florida	661	878	885.0	175.0	192.0	153.0
Georgia	831	900	966.0	2,320.0	2,570.0	2,255.0
Kansas	757	794	1,050.0	41.0	48.0	35.0
Louisiana	1,223	1,154	810.0	326.0	404.0	189.0
Mississippi	1,203	1,232	1,024.0	719.0	1,078.0	672.0
Missouri	968	1,117	1,097.0	496.0	570.0	400.0
New Mexico	929	931	929.0	60.0	64.0	60.0
North Carolina	799	1,038	713.0	766.0	995.0	527.0
Oklahoma	591	615	876.0	154.0	269.0	374.0
South Carolina ...	691	912	547.0	360.0	528.0	155.0
Tennessee	853	878	1,046.0	414.0	494.0	305.0
Texas	646	644	610.0	4,170.0	6,175.0	5,720.0
Virginia	941	1,239	817	151.0	222.0	143.0
United States	802	826	755	12,275.0	15,753.0	12,455.0
American Pima:						
Arizona	1,024	993	875	3.2	30.0	31.0
California	1,574	1,558	1,494	610.0	500.0	361.0
New Mexico	847	761	904	6.0	8.4	13.0
Texas	847	840	896	15.0	28.0	28.0
United States	1,527	1,432	1,342	634.2	566.4	433.0
All:						
United States	821	838	766	12,909.2	16,319.4	12,888.0

[1] Production ginned and to be ginned.   [2] 480-pound net weight bales.
NASS, Crops Branch, (202) 720–2127.

Table 2-4.—Cotton, American Upland: Support operations, United States, 2008/2009–2015/2016

Marketing Year beginning August 1	Income support payment rates per pound [1]	Program price levels per pound		Put under Loan		Acquired by CCC under loan program	Owned by CCC at end of marketing year
		Loan [2]	Target [3]	Quantity	Percentage of production		
	Cents	Cents	Cents	1,000 bales	Percent	1,000 bales	1,000 bales
2008/2009	6.67/12.58	52.00	71.25	10,002	80.7	4	0
2009/2010	6.67/1.68	52.00	71.25	8,271	70.2	0	0
2010/2011	6.67/0	52.00	71.25	11,398	64.8	0	0
2011/2012	6.67/0	52.00	71.25	7,262	49.3	1	0
2012/2013	6.67/0	52.00	71.25	8,330	50.4	0	0
2013/2014	6.67/0	52.00	71.25	3,981	32.4	0	0
2014/2015	(NA)	52.00	(NA)	7,625	48.4	0	0
2015/2016 [4]	(NA)	52.00	(NA)	6,758	54.3	0	0

(NA) Not available. [1] The first entry is the direct payment rate and the second entry is the counter-cyclical payment rate for 2004/2005-2013/2014. For 2009/2010-2013/2014, producers who participate in the Average Crop Revenue (ACRE) program get a 20 percent reduction in their direct payment, not calculated in this table. The Agricultural Act of 2014 excludes upland cotton as a covered commodity for the Agricultural Risk Coverage and Price Loss Coverage Programs. [2] For 2009/2010-2013/2014, producers who participate in the ACRE program get a 30 percent reduction in their loan rate, not calculated in this table. [3] Target applies to 2003/2004-2013/2014 and Reference applies beginning with 2014/2015. [4] Aquisitions as of July 31, 2016.
FSA, Fibers, Peanuts, Tobacco, (202) 401–0062.

Table 2-5.—International Cotton: Area, yield, and production in specified countries, 2013/2014–2015/2016

Country	Area			Yield per hectare			Production		
	2013/ 2014	2014/ 2015	2015/ 2016	2013/ 2014	2014/ 2015	2015/ 2016	2013/ 2014	2014/ 2015	2015/ 2016
	1,000 hectares	1,000 hectares	1,000 hectares	Kilograms	Kilograms	Kilograms	1,000 480 lb. bales	1,000 480 lb. bales	1,000 480 lb. bales
Argentina	436	205	285	2,047.00	2,443.00	1,910.00	4,100	2,300	2,500
Brazil	1,120	1,020	950	1,555.00	1,494.00	1,536.00	8,000	7,000	6,700
Burkina	647	660	625	421.00	445.00	418.00	1,250	1,350	1,200
China	4,800	4,400	3,400	1,486.00	1,484.00	1,524.00	32,750	30,000	23,800
India	11,700	12,700	11,800	577.00	506.00	494.00	31,000	29,500	26,800
Mali	450	538	580	411.00	433.00	413.00	850	1,070	1,100
Pakistan	2,900	2,950	2,800	713.00	782.00	544.00	9,500	10,600	7,000
Turkey	330	430	370	1,517.00	1,620.00	1,559.00	2,300	3,200	2,650
Turkmenistan	575	545	500	587.00	609.00	566.00	1,550	1,525	1,300
Uzbekistan ...	1,300	1,285	1,285	687.00	661.00	627.00	4,100	3,900	3,700
Others	5,480	5,506	5,062	2.21	2.25	2.08	12,097	12,386	10,528
Total foreign	29,738	30,239	27,657	3.61	3.40	3.16	107,497	102,831	87,278
United States	3,053	3,783	3,261	921.00	939.00	864.00	12,909	16,319	12,943
Total	32,791	34,022	30,918	3.67	3.50	3.24	120,406	119,150	100,221

FAS, Office of Global Analysis, (202) 720–6301. Prepared or estimated on the basis of official USDA production, supply, and distribution statistics from foreign governments.

Table 2-6.—Cotton: Supply and distribution, United States, 2004–2013

Year beginning August 1	Supply			Distribution				
	Beginning of season total [2]	Ginnings in season [1]	Total supply [2]	Consumption [1]			Exports	Carryover, end of season [2]
				Upland	American Pima	Total		
	1,000 bales	*1,000 bales*	*1,000 bales*	*1,000 bales*	*1,000 bales*	*1,000 bales*	*1,000 bales*	*1,000 bales*
2004	3,381	22,576	25,957	5,968	60	6,028	13,593	5,411
2005	5,368	23,253	28,576	5,604	49	5,653	17,038	5,877
2006	5,878	20,998	26,872	(D)	(D)	4,745	12,631	9,221
2007	9,223	18,713	27,929	(D)	(D)	4,499	13,237	9,699
2008	9,699	12,462	22,154	(D)	(D)	3,439	12,875	6,135
2009	6,136	11,832	17,963	(D)	(D)	3,336	11,687	2,852
2010	2,850	17,643	20,786	(D)	(D)	3,490	13,958	2,940
2011	2,600	15,153	18,190	(D)	(D)	3,300	11,710	3,350
2012	3,350	16,834	20,680	(D)	(D)	3,500	13,030	3,900
2013 [3]	3,900	12,521	16,421	(D)	(D)	3,600	10,400	2,800

(D) Withheld to avoid disclosing data for individual companies. [1] Ginnings during the 12 months, Aug. 1–July 31. Includes an allowance for "city crop" which consists of rebaled samples and pickings from cotton damaged by fire and weather. [2] May include small volume of foreign growths. [3] Preliminary.
AMS, Cotton and Tobacco Program, (901) 384–3016. Compiled from reports of the Bureau of the Census. Bureau of the Census discontinued the report containing this information.

Table 2-7.—Cotton, American Upland: Percentage distribution of fiber strength, United States, 2009–2013

Fiber strength [1]	Year				
	2009	2010	2011	2012	2013
17 and below	·	·	*	-	-
18	·	·	*	*	-
19	·	·	*	·	·
20	·	·	*	·	*
21	·	·	*	*	*
22	*	*	0.1	*	*
23	0.1	*	0.2	0.1	*
24	0.4	0.2	0.3	0.2	0.1
25	1.4	0.6	0.9	0.8	0.4
26	4.9	2.1	2.8	3.1	1.7
27	11.8	5.9	6.9	8.2	5.4
28	19.7	12.5	13.0	13.7	11.8
29	21.9	18.7	17.6	17.2	17.0
30	17.8	20.6	18.6	18.5	19.0
31	11.6	17.1	16.1	16.1	17.7
32	5.9	11.5	10.9	11.1	13.4
33	2.4	6.2	6.5	6.1	7.7
34	1.1	2.9	3.7	3.0	3.7
35	0.6	1.2	1.6	1.3	1.5
36 and above	0.3	0.5	0.7	0.6	0.5
Average	29.1	30.0	30.0	29.9	30.3

[1] Fiber strength expressed in terms of ⅛″ gage (grams per tex). *Less than 0.05 percent.
AMS, Cotton and Tobacco Program, (901) 384–3016.

Table 2-8.—Cotton, American Upland: Estimated percentage of the crop forward contracted by growers, by States, 2006–2013

State	Crop of—							
	2006	2007	2008	2009	2010	2011	2012	2013
	Percent	*Percent*	*Percent*	*Percent*	*Percent*	*Percent*	*Percent*	*Percent*
Alabama	15	2	10	3	15	41	32	30
Arizona	-	4	6	-	16	1	5	3
Arkansas	12	3	5	-	17	63	33	18
California	*	*	-	-	14	7	8	4
Florida	-	*	16	10	30	18	8	15
Georgia	2	*	12	16	29	20	18	40
Louisiana	22	39	68	2	64	57	49	15
Mississippi	14	14	6	13	6	54	21	11
Missouri	3	-	19	*	9	78	35	29
New Mexico	-	-	-	-	-	-	-	1
North Carolina	9	1	10	4	29	52	14	33
Oklahoma	-	-	-	-	-	-	-	-
South Carolina	6	6	10	2	22	30	7	34
Tennessee	3	-	1	*	10	69	38	20
Texas	3	3	11	3	22	33	5	5
United States	7	4	12	4	19	38	14	17

*Less than 0.5 percent.
AMS, Cotton and Tobacco Program, (901) 384–3016.

Table 2-9.—Cotton, American Upland: Carryover and crop, running bales, by grade groupings, United States, 2003–2012

Year beginning August 1	White color grades					Light spotted color grades				Other color grades[1]	All grades[2]
	21 and higher	31	41	51	61 and 71	22 and higher	32	42	52 and lower		
	1,000 bales	1,000 bales	1,000 bales	1,000 bales	1,000 bales	1,000 bales	1,000 bales	1,000 bales	1,000 bales	1,000 bales	1,000 bales
Carryover:											
2003	596	988	1,804	502	8	37	193	475	251	115	4,972
2004	435	1,573	1,106	54	1	22	47	51	7	19	3,314
2005	975	1,042	1,609	530	18	42	154	505	186	339	5,402
2006	1,642	2,178	1,466	90	1	68	92	146	62	66	5,810
2007	1,874	3,909	2,611	132	3	59	133	209	127	39	9,096
2008	2,373	4,149	2,466	123	4	77	137	200	126	38	9,692
2009	852	1,999	2,536	236	0	45	64	84	55	3	5,874
2010	336	408	1,547	234	0	42	17	83	98	71	2,838
2011	654	842	666	228	9	21	73	138	222	41	2,899
2012	340	595	1,633	202	0	44	18	97	81	72	3,081
Crop:											
2003	3,971	7,755	4,423	193	2	156	278	319	67	124	17,290
2004	4,063	5,228	7,079	1,955	45	180	605	1,328	567	782	21,832
2005	7,698	8,029	4,297	541	5	303	591	699	312	164	22,638
2006	3,785	8,145	6,842	397	4	146	296	425	198	24	20,262
2007	6,376	3,794	4,788	592	4	188	238	1,184	745	16	17,925
2008	2,160	4,557	4,285	391	2	169	171	161	166	12	12,075
2009	2,696	3,419	3,665	776	21	55	97	241	419	30	11,419
2010	7,099	5,693	2,668	382	7	221	264	402	278	124	17,138
2011	3,475	5,029	4,190	321	4	348	199	416	155	118	14,255
2012	4,075	4,805	5,545	285	1	421	310	374	80	157	16,053

[1] Includes all color grades of Spotted, Tinged, Yellow Stained, and Below Grade.　[2] Bales classed as reported by AMS, Cotton and Tobacco Program.
AMS, Cotton and Tobacco Program, (901) 384–3016.

Table 2-10.—Cotton, American Upland: Carryover and crop, running bales, by staple groupings, United States, 2003-2012

Year beginning August 1	Staple										All staples[1]
	26 and shorter	28	29	30	31	32	33	34	35	36 and longer	
	1,000 bales	1,000 bales	1,000 bales	1,000 bales	1,000 bales	1,000 bales	1,000 bales	1,000 bales	1,000 bales	1,000 bales	1,000 bales
Carryover:											
2003	-	(2)	15	35	69	214	708	1,495	1,357	1,079	4,972
2004	-	1	3	14	33	142	389	1,189	869	674	3,314
2005	-	1	4	17	77	213	543	1,128	1,615	1,803	5,402
2006	-	-	(2)	4	32	173	510	1,582	1,849	1,659	5,810
2007	-	(2)	(2)	5	62	382	924	1,873	2,236	3,613	9,096
2008	-	(2)	(2)	6	62	368	892	1,827	2,312	4,225	9,692
2009	-	-	-	3	11	61	337	816	1,423	3,224	5,874
2010	-	-	(2)	-	(2)	63	48	255	825	1,647	2,838
2011	-	(2)	(2)	8	44	176	335	611	765	958	2,899
2012	-	-	(2)	-	(2)	63	47	298	907	1,766	3,081
Crop:											
2003	(2)	1	10	57	202	624	2,205	4,873	4,805	4,512	17,290
2004	(2)	1	9	56	196	723	2,175	4,630	6,543	7,499	21,832
2005	(2)	(2)	1	16	127	650	2,460	5,892	7,261	6,232	22,638
2006	(2)	1	7	29	136	588	1,764	3,735	5,181	8,821	20,262
2007	-	(2)	2	14	113	524	1,574	3,293	4,376	8,030	17,925
2008	(2)	(2)	1	7	41	195	685	1,675	2,541	6,930	12,075
2009	(2)	(2)	2	11	39	120	488	1,828	3,461	5,468	11,419
2010	-	(2)	1	8	61	317	1,128	2,788	4,371	8,465	17,138
2011	(2)	1	7	32	123	386	947	1,969	3,329	7,461	14,255
2012	(2)	1	4	22	90	285	675	1,574	2,651	9,783	16,052

[1] Carryover as reported by the Bureau of the Census, Crop as reported by AMS, Cotton and Tobacco Program.　[2] Less than 500 bales.
AMS, Cotton and Tobacco Program, (901) 384–3016.

Table 2-11.—Cotton, American Pima: Carryover and crop, running bales, by grade and staple, United States, 2007-2012

Year beginning August 1	Grade					Staple				All grades and staples [1]
	01 and 02	03	04	05	06 and 07	42 and shorter	44	46	48 and longer	
	1,000 bales	*1,000 bales*	*1,000 bales*	*1,000 bales*	*1,000 bales*	*1,000 bales*	*1,000 bales*	*1,000 bales*	*1,000 bales*	*1,000 bales*
Carryover:										
2007	76.0	45.1	1.7	1.9	0.1	0.0	51.4	56.1	18.1	125.6
2008	65.5	75.0	0.8	2.9	1.6	0.0	31.7	91.7	22.7	146.2
2009	243.5	15.3	2.1	0.3	0.5	0.1	25.4	128.2	107.8	261.5
2010	14.2	0.5	0.1	0.0	0.0	0.0	1.6	1.9	11.3	14.8
2011	8.5	6.9	20.3	2.5	2.3	0.0	3.8	14.4	23.1	41.3
2012	256.0	8.9	2.4	0.4	0.4	0.0	28.1	35.6	204.3	268.0
Crop:										
2007	784.1	29.0	7.3	1.9	0.3	0.4	51.8	400.7	369.9	822.7
2008	391.1	18.5	1.8	0.3	1.0	0.0	10.5	126.7	275.5	412.7
2009	324.1	51.9	7.7	0.9	0.0	0.0	10.8	118.2	255.5	384.6
2010	411.5	57.9	10.8	3.0	0.4	0.1	8.5	96.8	377.9	483.3
2011	755.3	49.2	11.9	6.4	0.9	0.1	13.6	171.4	638.7	823.8
2012	716.0	31.7	2.3	0.5	0.2	0.3	25.2	212.8	512.5	750.8

[1] Carryover as reported by the Bureau of the Census; crop as reported by AMS, Cotton and Tobacco Program.
AMS, Cotton and Tobacco Program, (901) 384–3016.

Table 2-12.—Cotton, Upland: Average staple length of Upland cotton classed, by State and United States, 2007–2013

State	Average staple length (32ds of an inch) [1]						
	2007	2008	2009	2010	2011	2012	2013
Alabama	33.8	34.3	34.8	34.3	35.6	35.8	35.5
Arizona	35.7	36.3	36.2	36.4	36.0	36.0	36.2
Arkansas	35.0	36.1	35.6	35.5	36.1	36.2	36.8
California	37.2	38.1	38.0	38.2	37.7	37.4	37.5
Florida	34.4	35.0	34.8	35.1	36.1	36.2	35.8
Georgia	34.4	34.5	34.9	34.9	35.9	36.0	36.0
Kansas	35.1	35.7	35.6	35.2	34.9	34.6	34.4
Louisiana	34.8	34.5	35.1	35.0	34.9	35.4	35.6
Mississippi	34.6	35.9	35.5	34.9	35.8	36.1	36.3
Missouri	34.8	36.0	35.7	35.6	36.3	36.1	36.1
New Mexico	37.0	37.2	36.5	36.5	36.5	36.3	36.7
North Carolina	33.9	34.8	35.0	34.8	34.8	35.9	36.1
Oklahoma	35.4	36.0	35.5	35.4	35.2	34.9	35.2
South Carolina	33.6	35.2	35.1	35.1	35.5	36.1	36.2
Tennessee	33.3	35.1	35.0	34.7	35.4	35.1	35.7
Texas	36.0	36.3	35.6	35.7	34.5	35.1	35.4
Virginia	34.0	34.5	35.4	34.3	34.8	36.0	36.3
Other States	(NA)	(NA)	(NA)	(NA)	(NA)	(NA)	(NA)
United States	35.3	35.7	35.5	35.5	35.5	35.7	35.8

(NA) Not available.   [1] Average calculated on numerical equivalents of the staple-length designations. For example, 7/8-inch = 28, 29/32-inch = 29, etc.
AMS, Cotton and Tobacco Program, (901) 384–3016.

Table 2-13.—Cotton: United States exports by country of destination, 2014–2016

Country of destination	Year		
	2014	2015	2016
	Metric tons	*Metric tons*	*Metric tons*
Cotton linters:			
Spain	6,945	2,162	4,233
China	8,802	11,305	3,626
Japan	6,203	6,330	2,010
Germany(*)	1,490	669	955
France(*)	155	197	149
Mexico	7	128	140
United Kingdom	86	0	79
Canada	90	53	53
Sweden	12	12	23
Vietnam	7	12	20
South Africa	0	0	16
Chile	0	5	10
Venezuela	0	0	8
Guatemala	37	14	8
United Arab Emirates	0	0	6
Malaysia	0	0	5
Peru	0	0	3
Argentina	0	0	2
Australia(*)	47	0	0
Barbados	0	10	0
Belgium-Luxembourg(*)	0	137	0
Brazil	0	16	0
Costa Rica	0	0	0
Egypt	0	0	0
Czech Republic	0	7	0
Indonesia	0	0	0
Israel(*)	2	0	0
Jamaica	0	10	0
Kuwait	0	0	0
Rest of World	68	102	0
World Total	23,949	21,170	11,348
Cotton < 1:			
Taiwan	2,890	3,893	12,121
Thailand	3,455	3,695	7,064
Vietnam	775	628	4,390
Japan	2,072	2,726	2,524
China	3,495	2,189	2,312
Turkey	2,565	3,818	1,707
Indonesia	342	0	1,127
Mexico	5,815	1,021	937
Gabon	0	0	559
Korea, South	2,643	83	255
Italy(*)	0	1	196
Switzerland(*)	0	0	173
Brazil	0	0	134
Austria	0	97	99
Dominican Republic	2	0	94
Panama	39	179	58
Pakistan	0	0	39
India	217	0	28
Colombia	20	0	23
Germany(*)	1	0	20
Nicaragua	0	0	18
Morocco	0	0	12
New Zealand(*)	0	0	11
Chile	0	0	11
Spain	0	4	6
Ecuador	51	0	5
Belgium-Luxembourg(*)	98	0	5
Australia(*)	1	0	3
Saudi Arabia	0	0	3
Rest of World	1,956	194	4
World Total	26,436	18,527	33,937

See footnote(s) at end of table.

Table 2-13.—Cotton: United States exports by country of destination, 2014–2016—Continued

Country of destination	Year		
	2014	2015	2016
	Metric tons	Metric tons	Metric tons
Cotton > 1 < 1 1/8:			
China	160,499	222,099	295,002
Turkey	182,387	205,645	195,301
Mexico	203,900	145,821	122,538
Vietnam	99,770	110,575	89,481
Indonesia	188,477	173,830	83,817
Pakistan	33,763	39,873	39,492
Taiwan	26,949	32,784	35,763
Thailand	32,419	34,640	33,512
Korea, South	21,176	22,804	27,690
Bangladesh	24,076	20,729	22,105
El Salvador	12,634	12,973	14,945
Guatemala	18,962	17,488	14,167
Colombia	4,773	10,315	12,531
Japan	27,688	32,733	11,957
Peru	2,014	15,327	11,502
Morocco	8,108	0	10,003
Malaysia	18,548	17,786	9,732
Ecuador	12,321	14,967	8,279
Brazil	7,461	7,202	6,859
Philippines	5,373	4,983	6,628
India	371	2,890	2,839
Hong Kong	458	853	1,794
Egypt	78	243	986
Singapore	494	447	808
Venezuela	138	319	695
Germany(*)	219	180	138
Bahrain	199	1,692	102
Tunisia	343	0	98
Belgium-Luxembourg(*)	100	114	68
Rest of World	18,957	13,745	212
World Total	1,112,655	1,163,056	1,059,044
Pima >= 1 3/8:			
India	5,783	12,362	37,926
China	27,568	12,870	25,759
Vietnam	1,810	1,276	3,589
Egypt	1,536	1,647	3,125
Turkey	9,291	1,725	3,075
Pakistan	3,565	5,225	2,972
Peru	2,444	3,630	2,366
Indonesia	6,409	480	2,031
Japan	3,167	1,566	1,276
Germany(*)	2,141	1,680	1,247
Thailand	3,466	4,166	943
Korea, South	8,670	61	453
Mexico	346	78	217
Austria	38	0	199
Colombia	78	136	174
Brazil	38	1	137
Taiwan	882	801	134
Honduras	150	99	118
Philippines	0	193	97
Italy(*)	196	135	97
Guatemala	58	21	20
United Kingdom	0	20	19
Belgium-Luxembourg(*)	0	0	2
Australia(*)	0	0	0
Bangladesh	1,049	0	0
Ecuador	0	0	0
Hong Kong	0	0	0
Iraq	0	0	0
Mauritius	0	0	0
Rest of World	282	271	0
World Total	78,966	48,443	85,977

See footnote(s) at and of table.

Table 2-13.—Cotton: United States exports by country of destination, 2014–2016—Continued

Country of destination	Year		
	2014	2015	2016
	Metric tons	Metric tons	Metric tons
Cotton Other > 1 1/8:			
Vietnam	55,660	207,948	233,928
Turkey	209,393	150,081	204,449
China	298,488	288,477	194,003
Indonesia	60,586	78,209	129,044
Pakistan	11,719	46,682	104,339
Korea, South	63,341	118,195	91,233
India	19,309	25,444	61,783
Thailand	50,243	58,645	57,677
Taiwan	30,008	35,333	32,249
Bangladesh	21,659	17,301	30,581
Peru	47,185	34,106	26,826
Brazil	2,993	20	20,163
Mexico	21,321	9,575	20,082
Malaysia	1,768	15,222	15,697
Egypt	4,959	18,185	12,340
Costa Rica	1,231	8,497	11,248
Colombia	4,413	4,505	8,496
Japan	8,538	7,015	7,668
Guatemala	2,948	3,543	6,462
Ecuador	4,879	4,759	6,202
El Salvador	1,026	2,636	3,818
Honduras	3,104	2,008	2,409
Bahrain	258	1,686	2,291
Germany(*)	1,296	1,685	1,347
Nicaragua	1	0	1,310
Philippines	942	3,796	1,057
Italy(*)	1,120	3,976	874
Venezuela	1,600	1,358	676
Singapore	730	0	383
Rest of World	17,995	17,250	1,400
World Total	948,713	1,166,137	1,290,035

(*) Denotes a country that is a summarization of its component countries.
Users should use cautious interpretation on quantity reports using mixed units of measure.   Quantity line items will only include statistics on the units of measure that are equal to, or are able to be converted to, the assigned unit of measure of the grouped commodities.
FAS, Office of Global Analysis, (202) 720–6301.

Table 2-14.—Cotton excluding linters: United States exports and imports for consumption, by country of origin, 2013–2016

Country of origin	Year beginning August			
	2013	2014	2015	2016
	Metric tons	*Metric tons*	*Metric tons*	*Metric tons*
Exports:				
Vietnam	214,697	218,744	431,952	536,908
Turkey	428,171	425,149	301,445	331,768
China	1,076,103	518,029	477,366	305,891
Indonesia	138,703	167,108	189,264	221,683
Mexico	218,857	209,870	216,318	216,537
Pakistan	96,283	39,359	72,636	129,455
India	31,783	30,083	48,122	112,268
Korea, South	86,220	102,342	151,073	103,898
Thailand	86,572	89,582	101,146	99,196
Taiwan	68,190	67,543	79,901	83,997
Bangladesh	50,465	45,382	40,105	58,271
Guatemala	21,766	29,955	36,469	42,245
Peru	59,968	62,082	52,703	37,471
Brazil	13,750	11,139	21	30,437
Malaysia	15,101	3,902	30,624	27,199
Japan	26,275	26,410	24,279	26,414
Colombia	28,611	23,474	22,129	22,861
Egypt	13,782	10,033	21,312	15,465
El Salvador	25,873	19,574	20,422	13,550
Ecuador	20,473	10,303	9,742	12,835
Costa Rica	77	1,231	8,527	11,251
Philippines	7,156	8,402	11,191	8,012
Bahrain	2,393	716	2,539	4,086
Venezuela	1,253	2,151	4,248	3,515
Honduras	3,135	3,453	3,838	2,629
Germany(*)	5,905	3,856	3,431	2,613
Italy(*)	1,587	1,810	4,560	1,974
Nicaragua	87	176	1,464	1,332
Spain	1,673	5,593	7,432	992
Rest of World	45,328	29,324	21,906	4,239
World Total	2,790,237	2,166,771	2,396,162	2,468,993
Imports:				
Brazil	0	0	0	5,086
India	21	207	1,062	1,407
Egypt	1,096	1,440	954	733
Kazakhstan	0	0	0	35
Tajikistan	0	0	0	20
Germany(*)	0	21	14	15
France(*)	0	0	33	1
Taiwan	0	0	0	1
Turkey)	965	1,171	537	1
Mexico	14	0	1	0
China	1	0	0	0
Japan	-	0	-	0
South Africa	0	0	0	-
United Kingdom	0	0	0	-
Canada	0	0	0	0
Hong Kong	0	-	0	0
Italy(*)	0	0	0	0
Kyrgyzstan	20	0	0	0
Pakistan	37	56	59	0
Sweden	0	0	-	0
Zambia	94	0	0	0
World Total	2,248	2,894	2,660	7,298

(*) Denotes a country that is a summarization of its component countries.
All zeroes for a data item may show that statistics exist in the other import type.
Users should use cautious interpretation on quantity reports using mixed units of measure.   Quantity line items will only include statistics on the units of measure that are equal to, or are able to be converted to, the assigned unit of measure of the grouped commodities.
FAS, Office of Global Analysis, (202) 720-6301. Data Source: Department of Commerce, U.S. Census Bureau, Foreign Trade Statistics.

Table 2-15.—Cotton: International trade, 2013/2014–2015/2016 [1]

Country	2013/2014	2014/2015	2015/2016
	1,000 bales	*1,000 bales*	*1,000 bales* [2]
Principle exporting countries:			
Australia	4,852	2,393	2,750
Benin	525	675	675
Brazil	2,230	3,910	4,200
Burkina	1,250	1,125	1,300
Cote d'Ivoire	725	850	800
Greece	1,285	1,168	900
India	9,261	4,199	5,500
Mali	800	750	1,200
Turkmenistan	1,625	1,500	1,000
Uzbekistan	2,700	2,450	2,300
Others	5,030	5,110	4,776
Total foreign	30,283	24,130	25,401
United States	10,530	11,246	9,500
Total	40,813	35,376	34,901
Principle importing countries:			
Bangladesh	5,300	5,400	5,750
China	14,122	8,284	5,000
India	675	1,226	900
Indonesia	2,989	3,345	3,100
Korea, South	1,286	1,321	1,300
Mexico	1,040	830	975
Pakistan	1,200	835	2,300
Thailand	1,546	1,475	1,425
Turkey	4,246	3,675	3,800
Vietnam	3,200	4,300	5,200
Others	5,662	5,009	5,159
Total foreign	41,266	35,700	34,909
United States	13	12	10
Total	41,279	35,712	34,919

[1] Marketing year beginning Aug. 1. [2] 480-pound net weight.
FAS, Office of Global Analysis, (202) 720–6301. Prepared or estimated on the basis of official USDA production, supply, and distribution statistics from foreign governments.

Table 2-16.—Cotton, American Upland: High, low, and season average spot prices for the base quality in the designated markets, cents per pound, 2005–2013

Season beginning August 1	Grade 41 Staple 34 [1]		
	Average	High	Low
	Cents	Cents	Cents
2005	48.96	53.25	43.46
2006	48.67	60.67	42.84
2007	61.50	79.16	50.34
2008	47.87	62.69	36.28
2009	67.76	78.90	50.98
2010	137.88	209.60	80.18
2011	85.81	110.87	60.71
2012	75.24	87.55	65.11
2013	80.28	89.14	63.53

[1] Prices are for mixed lots, net weight, compressed, FOB car/truck.
AMS, Cotton and Tobacco Program, (901) 384–3016.

Table 2-17.—Cotton, American Upland: Percentage distribution of mike readings, by specified groups, United States, 2004–2013

Year beginning August 1	Mike groups						
	26 and below	27 to 29	30 to 32	33 to 34	35 to 49	50 to 52	53 and above
	Percent						
2004	0.4	1.5	3.4	3.7	83.8	6.4	0.8
2005	*	1.5	4.0	4.4	82.0	6.5	5.8
2006	1.1	1.8	2.7	2.3	79.2	10.8	1.8
2007	0.1	0.6	1.8	2.8	87.4	6.5	0.8
2008	0.5	1.1	2.8	3.9	77.2	9.7	1.4
2009	1.9	2.6	3.9	3.7	81.0	6.0	0.9
2010	0.1	0.4	1.3	1.8	75.6	16.9	3.8
2011	0.1	0.3	0.8	1.1	80.9	14.0	2.9
2012	0.4	0.9	1.8	1.9	75.0	15.7	4.3
2013	0.4	1.3	3.1	3.2	78.2	12.2	1.6

(*) Less than 0.05 percent.
AMS, Cotton and Tobacco Program, (901) 384–3016.

Table 2-18.—Cotton, American Upland: Average spot prices for specified grades of staple 34 in the designated markets for mixed lots, net weight, compressed, FOB car/truck, cents per pound, 2004–2013

Year beginning August 1	White				Light Spotted			Spotted	
	Color 31 Leaf 3	Color 41 Leaf 4	Color 51 Leaf 5	Color 61 Leaf 6	Color 32 Leaf 3	Color 42 Leaf 4	Color 52 Leaf 5	Color 33 Leaf 3	Color 43 Leaf 4
	Cents								
2004	48.40	45.61	41.59	39.11	45.70	43.30	40.38	42.51	40.75
2005	51.33	48.96	44.84	42.34	48.72	46.42	43.41	45.98	44.05
2006	50.83	48.67	44.56	42.12	48.39	46.25	43.24	45.75	43.81
2007	63.46	61.50	57.35	54.95	61.16	59.01	56.09	58.58	56.59
2008	49.83	47.87	43.73	41.33	47.55	45.37	42.48	44.95	42.87
2009	69.97	67.76	63.47	61.21	67.44	65.26	62.37	64.84	62.76
2010	139.98	137.88	133.45	131.28	137.56	135.38	132.49	134.96	132.88
2011	87.64	85.81	81.16	79.29	85.32	81.52	80.16	82.84	79.32
2012	76.87	75.24	70.56	68.76	74.67	70.81	69.46	72.16	68.67
2013	81.68	80.28	75.64	73.83	79.71	77.39	74.49	77.20	75.17

AMS, Cotton and Tobacco Program, (901) 384–3016.

Table 2-19.—Cotton, American Upland: Average spot prices for specified staple lengths of Color 41 Leaf 4 in the designated markets for mixed lots, net weight, compressed, FOB car/truck, cents per pound, 2004–2013

Year beginning August 1	Staple							
	28	29	30	31	32	33	34	35
	Cents							
2004	41.54	41.54	42.13	43.28	43.32	44.07	45.61	47.02
2005	44.26	44.26	44.96	46.13	46.14	46.84	48.96	50.36
2006	43.92	43.92	44.67	45.79	45.89	46.53	48.67	49.97
2007	56.75	56.75	57.50	58.62	58.50	59.27	61.50	62.69
2008	43.12	43.12	43.87	44.99	44.82	45.62	47.87	49.04
2009	62.83	62.83	63.58	64.71	64.71	65.51	67.76	68.98
2010	132.76	132.76	133.51	134.63	134.83	135.63	137.88	139.09
2011	79.54	79.54	79.54	79.54	82.20	82.66	85.81	86.90
2012	69.18	69.18	69.18	69.18	71.52	72.03	75.24	76.52
2013	75.17	75.17	75.17	75.17	77.05	77.44	80.28	81.58

AMS, Cotton and Tobacco Program, (901) 384–3016.

Table 2-20.—Cotton, American Upland: Season average spot prices for the base quality, by designated markets, cents per pound, 2008–2013 [1]

Market	Color 41, Leaf 4, Staple 34 [2]					
	2008	2009	2010	2011	2012	2013
	Cents					
Southeast	48.97	70.13	139.70	87.75	77.50	82.74
North Delta	47.99	69.30	139.02	86.90	76.56	81.85
South Delta	47.99	69.30	139.02	86.95	76.56	81.85
East TX–OK	47.08	66.57	135.51	83.30	73.78	79.27
West Texas	46.93	66.38	135.10	83.31	73.68	79.13
Desert SW	47.03	65.40	137.91	85.73	73.86	78.23
SJ Valley	49.10	67.20	138.91	86.74	74.74	78.88
Average	47.87	67.76	137.88	85.81	75.21	80.28

[1] Year beginning August 1.   [2] Prices are for mixed lots, net weight, compressed, FOB car/truck.
AMS, Cotton and Tobacco Program, (901) 384–3016.

Table 2-21.—Sugarbeets: Area, yield, production, and value, United States, 2006–2015 [1]

Year	Area		Yield per harvested acre	Production	Marketing year average price per ton received by farmers	Value of production
	Planted	Harvested				
	1,000 acres	1,000 acres	tons	1,000 tons		1,000 dollars
2006	1,366.2	1,303.6	26.1	34,064	44.20	1,193,151
2007	1,268.8	1,246.8	25.5	31,834	42.00	1,337,173
2008	1,090.7	1,004.5	26.8	26,881	48.10	1,294,144
2009	1,185.8	1,148.5	25.9	29,783	51.50	1,532,634
2010	1,171.9	1,156.1	27.7	32,034	66.90	2,142,162
2011	1,232.8	1,213.2	23.8	28,896	69.40	2,004,116
2012	1,230.1	1,204.1	29.3	35,224	66.40	2,338,789
2013	1,198.0	1,154.0	28.4	32,789	46.60	1,536,821
2014	1,162.5	1,146.3	27.3	31,285	46.00	1,437,575
2015	1,158.8	1,144.3	30.8	35,278	....	....

[1] Relates to year of intended harvest for fall planted beets in central California and to year of planting for overwintered beets in central and southern California.
NASS, Crops Branch, (202) 720–2127.

Table 2-22.—Sugarbeets: Area, yield, and production, by State and
United States, 2013–2015

State	Area planted			Area harvested			Yield per harvested acre			Production		
	2013	2014	2015	2013	2014	2015	2013	2014	2015	2013	2014	2015
	1,000 acres	1,000 acres	1,000 acres	1,000 acres	1,000 acres	1,000 acres	Tons	Tons	Tons	1,000 tons	1,000 tons	1,000 tons
California [1] ...	24.4	24.3	25.0	24.3	22.5	25.0	43.4	42.6	44.2	1,055	959	1,105
Colorado	26.8	29.6	27.5	25.7	29.3	27.3	33.5	31.3	35.1	861	917	958
Idaho	175.0	170.0	171.0	174.0	169.0	169.0	36.2	37.3	38.1	6,299	6,304	6,439
Michigan	154.0	151.0	152.0	153.0	150.0	151.0	26.2	29.3	31.7	4,009	4,395	4,787
Minnesota	462.0	440.0	443.0	426.0	434.0	435.0	26.0	22.5	28.0	11,076	9,765	12,180
Montana	43.4	45.1	44.1	42.8	44.4	43.7	29.2	32.3	32.8	1,250	1,434	1,433
Nebraska	46.0	49.1	47.5	44.2	45.9	46.8	29.7	29.1	28.4	1,313	1,336	1,329
North Dakota	227.0	215.0	208.0	225.0	214.0	206.0	25.3	23.8	27.9	5,693	5,093	5,747
Oregon	9.4	7.5	9.2	9.3	7.2	9.1	38.4	34.5	39.3	357	248	358
Wyoming	30.0	30.9	31.5	29.7	30.0	31.4	29.5	27.8	30.0	876	834	942
United States	1,198.0	1,162.5	1,158.8	1,154.0	1,146.3	1,144.3	28.4	27.3	30.8	32,789	31,285	35,278

[1] Relates to year of intended harvest for fall planted beets in central California and to year of planting for overwintered beets in central and southern California.
NASS, Crops Branch, (202) 720–2127.

Table 2-23.—Sugarbeets: Production and value, by State and
United States, 2013–2014

State	Production		Marketing year average price per ton received by farmers		Value of production	
	2013	2014	2013	2014	2013	2014
	1,000 tons	1,000 tons	Dollars	Dollars	1,000 dollars	1,000 dollars
California [1]	1,055	959	43.00	43.00	45,365	41,237
Colorado	861	917	35.60	46.70	30,652	42,824
Idaho	6,299	6,304	40.00	45.00	251,960	283,680
Michigan	4,009	4,395	53.60	49.10	214,882	215,795
Minnesota	11,076	9,765	52.60	45.10	582,598	440,402
Montana	1,250	1,434	39.40	49.70	49,250	71,270
Nebraska	1,313	1,336	37.80	48.70	49,631	65,063
North Dakota	5,693	5,093	44.90	44.20	255,616	225,111
Oregon	357	248	40.00	45.00	14,280	11,160
Wyoming	876	834	37.20	49.20	32,587	41,033
United States	32,789	31,285	46.60	46.00	1,526,821	1,437,575

[1] Relates to year of intended harvest for fall planted beets in central California and to year of planting for overwintered beets in central and southern California.
NASS, Crops Branch, (202) 720–2127.

Table 2-24.—Sugarcane for sugar and seed: Area, yield, production, and value, United States, 2006–2015

Year	Area harvested			Yield per acre [1]			Production [1]		
	For sugar	For seed	Total	For sugar	For seed	For sugar and seed	For sugar	For seed	Total
	1,000 acres	*1,000 acres*	*1,000 acres*	*Tons*	*Tons*	*Tons*	*1,000 tons*	*1,000 tons*	*1,000 tons*
2006	846.6	51.1	897.7	33.0	31.4	32.9	27,962	1,602	29,564
2007	827.9	51.7	879.6	34.2	32.8	34.1	28,273	1,696	29,969
2008	821.6	46.4	868.0	31.8	31.7	31.8	26,131	1,472	27,603
2009	817.0	56.9	873.9	34.9	34.1	34.8	28,494	1,938	30,432
2010	825.3	52.2	877.5	31.1	32.5	31.2	25,663	1,697	27,360
2011	827.1	45.5	872.6	33.5	32.7	33.5	27,738	1,486	29,224
2012	854.9	47.5	902.4	35.7	36.4	35.7	30,500	1,727	32,227
2013	859.6	51.2	910.8	33.8	33.9	33.8	29,023	1,738	30,761
2014	823.7	44.8	868.5	35.1	34.1	35.0	28,895	1,529	30,424
2015	847.5	45.2	892.7	36.5	35.0	36.5	30,967	1,582	32,549

Year	Marketing year average price per ton received by farmers	Value of production	
		Of cane used for sugar	Of cane used for sugar and seed [2]
	Dollars	*1,000 dollars*	*1,000 dollars*
2005	28.40	701,920	754,529
2006	30.40	849,157	897,601
2007	29.40	831,218	880,616
2008	29.50	771,134	814,479
2009	34.80	991,424	1,056,613
2010	41.70	1,069,537	1,140,636
2011	47.20	1,308,951	1,379,498
2012	41.90	1,276,631	1,348,361
2013	31.40	910,377	962,807
2014	34.70	1,002,138	1,054,657

[1] Net tons.   [2] Price per ton of cane for sugar used in evaluating value of production for seed.
NASS, Crops Branch, (202) 720–2127.

Table 2-25.—Sugarcane for sugar and seed: Production and value, by State and United States, 2013–2014

State	Sugarcane for sugar						Sugar and seed: Value of production [1]	
	Production [2]		Price per ton		Value of production			
	2013	2014	2013	2014	2013	2014	2013	2014
	1,000 tons	*1,000 tons*	*Dollars*	*Dollars*	*1,000 dollars*	*1,000 dollars*	*1,000 dollars*	*1,000 dollars*
Florida	13,720	15,053	35.10	36.80	481,572	553,950	505,440	579,158
Hawaii	1,352	1,261	57.50	43.10	77,740	54,349	80,328	56,289
Louisiana	12,505	11,387	25.90	33.60	323,880	382,603	349,158	407,400
Texas	1,446	1,194	18.80	9.41	27,185	11,236	27,881	11,810
United States	29,023	28,895	31.40	34.70	910,377	1,002,138	962,807	1,054,657

[1] Price per ton of cane for sugar used in evaluating value of production for seed.   [2] Net tons.
NASS, Crops Branch, (202) 720–2127.

Table 2-26.—Sugarcane for sugar and seed: Area, yield, and production, by State and United States, 2013–2015

State	Area harvested			Yield per acre [1]			Cane production [1]		
	2013	2014	2015	2013	2014	2015	2013	2014	2015
For sugar:	1,000 acres	1,000 acres	1,000 acres	Tons	Tons	Tons	1,000 tons	1,000 tons	1,000 tons
Florida	400.0	392.0	409.0	34.3	38.4	39.8	13,720	15,053	16,278
Hawaii	15.5	14.2	16.5	87.2	88.8	86.2	1,352	1,261	1,422
Louisiana	410.0	386.0	385.0	30.5	29.5	31.0	12,505	11,387	11,935
Texas	34.1	31.5	37.0	42.4	37.9	36.0	1,446	1,194	1,332
United States	859.6	823.7	847.5	33.8	35.1	36.5	29,023	28,895	30,967
For seed:									
Florida	16.0	16.0	16.0	42.5	42.8	43.2	680	685	691
Hawaii	2.2	2.2	2.2	20.5	20.4	20.0	45	45	44
Louisiana	32.0	25.0	25.0	30.5	29.5	31.0	976	738	775
Texas	1.0	1.6	2.0	37.0	37.9	36.0	37	61	72
United States	51.2	44.8	45.2	33.9	34.1	35.0	1,738	1,529	1,582
For sugar and seed:									
Florida	416.0	408.0	425.0	34.6	38.6	39.9	14,400	15,738	16,969
Hawaii	17.7	16.4	18.7	78.9	79.6	78.4	1,397	1,306	1,466
Louisiana	442.0	411.0	410.0	30.5	29.5	31.0	13,481	12,125	12,710
Texas	35.1	33.1	39.0	42.3	37.9	36.0	1,483	1,255	1,404
United States	910.8	868.5	892.7	33.8	35.0	36.5	30,761	30,424	32,549

[1] Net tons.
NASS, Crops Branch, (202) 720–2127.

Table 2-27.—Sugar, cane (raw value [1]): Refiners' raw stocks, receipts, meltings, continental United States, 2005–2014

Year	Jan. 1 stocks	Receipts [2]	Meltings
	1,000 tons	1,000 tons	1,000 tons
2005	244	5,215	5,270
2006	217	5,543	5,405
2007	358	5,388	5,464
2008	304	5,634	5,329
2009	468	5,459	5,577
2010	346	5,753	5,843
2011	257	6,062	5,534
2012	498	6,172	5,805
2013	574	6,025	5,938
2014	592	6,173	6,135

[1] Raw value is the equivalent in terms of 96° sugar.   [2] Receipts include refiners' total offshore raw sugar receipts in continental U.S. ports, whether entered through the customs or held pending availability of quota and raw cane sugar produced from sugarcane in the continental United States.
FSA, Dairy and Sweeteners Analysis, (202) 720–3451.

Table 2-28.—Sugar, cane and beet: Domestic marketings, by source of supply, continental United States, 2012–2014

Area of supply	2012	2013	2014
	1,000 tons	1,000 tons	1,000 tons
Domestic areas:			
Mainland (beet)	4,816	6,037	4,998
Mainland and Hawaii (cane)	6,168	6,390	6,591
Total domestic areas	10,984	12,427	11,589

FSA, Dairy and Sweeteners Analysis Division, (202) 720–3451. Source: U.S. Census.

Table 2-29.—Sugar, cane and beet (refined): Stocks, production and receipts, and deliveries, continental United States, 2005–2014

Item and year	Cane sugar refineries	Beet sugar factories	Importers of direct consumption sugar	Mainland cane sugar mills [1]	Total
Jan. 1 Stocks [2]	*1,000 tons*	*1,000 tons*	*1,000 tons*	*1,000 tons*	*1,000 tons*
2005	368	1,782	0	4	2,154
2006	328	1,429	0	7	1,764
2007	452	1,792	0	3	2,247
2008	400	1,806	0	4	2,210
2009	440	1,464	0	5	1,909
2010	484	1,456	0	8	1,948
2011	466	1,691	0	4	2,161
2012	315	1,597	0	6	1,918
2013	388	2,013	0	3	2,404
2014	572	1,603	0	5	2,180
Production and Receipts					
2005	5,112	4,690	197	19	10,018
2006	5,741	4,758	576	16	11,091
2007	5,525	5,219	733	21	11,498
2008	5,460	4,937	2,961	28	13,386
2009	5,867	4,434	2,848	34	13,183
2010	5,753	4,883	3,396	23	14,055
2011	6,062	4,666	3,891	19	14,638
2012	6,360	5,237	3,431	16	15,044
2013	6,579	5,712	3,319	17	15,627
2014	6,572	5,014	3,326	10	14,922
Deliveries [3]					
2005	5,453	5,012	197	17	10,679
2006	5,587	4,419	576	19	10,601
2007	5,520	5,206	733	20	11,479
2008	5,397	5,258	2,961	27	13,643
2009	5,768	4,441	2,848	32	13,089
2010	6,025	4,631	3,396	20	14,072
2011	6,043	4,755	3,891	17	14,706
2012	6,168	4,816	3,431	18	14,433
2013	6,390	6,037	3,319	15	15,761
2014	6,591	4,998	3,326	11	14,926

[1] Sugar for human consumption only.   [2] Stocks include sugar in bond and in Customs custody and control.   [3] Consists of all refined sugar.
FSA, Dairy and Sweeteners Analysis, (202) 720–3451.

Table 2-30.—Sugar (raw and refined): Average price per pound at specified markets, 2006–2015

Year	Cane sugar Raw, 96 centrifugal		Refined beet: Mid-west	Retail price, granulated: United States
	Intercontinental contract No. 11, average of nearby futures, world price	Intercontinental contract No. 16, average of nearby futures, U.S. price		
	Cents	*Cents*	*Cents*	*Cents*
2006	14.65	22.14	33.10	49.58
2007	9.91	20.99	25.06	51.48
2008	12.11	21.30	32.54	52.91
2009	17.91	24.93	38.10	57.03
2010	22.49	35.97	53.23	62.86
2011	27.22	38.12	56.22	68.30
2012	21.69	28.90	43.38	69.41
2013	17.46	20.46	27.22	64.32
2014	16.34	24.15	32.86	60.90
2015	13.14	24.88	33.88	65.36

ERS, Market and Trade Economics Division, (202) 694–5247. Compiled from the following sources: (New York) Coffee, Sugar & Cocoa Exchange; the U.S. Department of Labor, Bureau of Labor Statistics; Milling and Baking News, New York Board of Trade; Intercontinental Exchange,.

Table 2-31.—Sugar, centrifugal: International trade, 2012/2013–2014/2015

Country	2012/2013	2013/2014	2014/2015
	1,000 Metric tons, raw value		
Principle exporting countries:			
Australia	3,100	3,242	3,561
Brazil	27,650	26,200	23,950
Colombia	542	900	845
Cuba	775	992	880
European Union	1,662	1,552	1,610
Guatemala	1,911	2,100	2,200
India	1,261	2,806	2,425
Mexico	2,091	2,661	1,545
South Africa	356	868	772
Thailand	6,693	7,200	8,000
Others	9,236	8,851	8,177
Total Foreign	55,277	57,372	53,965
United States	249	278	168
Total	55,526	57,650	54,133
Principle importing countries:			
Algeria	2,014	1,854	1,844
Bangladesh	1,547	2,085	1,982
China	3,802	4,275	5,058
European Union	3,790	3,262	2,600
India	3,570	3,570	3,050
Indonesia	1,806	1,909	1,881
Iran	1,966	1,897	2,063
Korea, South	1,450	1,470	1,465
Malaysia	1,403	1,320	1,431
United Arab Emirates	2,583	2,108	2,366
Others	25,134	24,120	23,422
Total Foreign	49,065	47,870	47,162
United States	2,925	3,395	3,238
Total	51,990	51,265	50,400

FAS, Office of Global Analysis, (202) 720–6301.   Prepared or estimated on the basis of official USDA production, supply, and distribution statistics from foreign governments.

Table 2-32.—U.S. Sugar (raw value): Supply and use, includes Puerto Rico, 2006–2015

Fiscal Year	Production	Beginning stocks	Imports	Exports	Total deliveries, domestic use
	1,000 short tons	1,000 short tons	1,000 short tons	1,000 short tons	1,000 short tons
2006	8,445	1,698	2,080	422	10,704
2007	8,152	1,799	2,620	203	10,607
2008	7,531	1,664	3,082	136	11,152
2009	7,963	1,534	3,320	211	11,422
2010	7,831	1,498	3,738	248	11,313
2011	8,488	1,378	3,632	269	11,776
2012	8,981	1,979	3,224	274	12,246
2013	8,462	2,158	3,742	306	12,019
2014	8,656	1,810	3,553	185	12,155
2015	8,925	1,815	3,375	75	12,235

ERS, Market and Trade Economics Division, (202) 694–5247. Source: ERS Sugar and Sweetener Outlook

Table 2-33.—Honey: United States exports and imports for consumption, by country of origin, 2014–2016

Country of origin	2014	2015	2016
	Metric tons	Metric tons	Metric tons
Exports:			
Canada	888	1,089	769
Philippines	269	495	741
China	317	425	641
Korea, South	213	560	488
Kuwait	416	311	418
Yemen(*)	421	206	402
United Arab Emirates	358	272	397
Japan	537	320	230
Hong Kong	59	75	165
Singapore	69	65	104
Taiwan	223	67	83
Australia(*)	20	12	78
Pakistan	59	19	74
Mexico	1	42	66
Bahrain	30	25	62
Barbados	40	56	41
Bahamas, The	43	30	40
Brazil	43	4	38
Panama	25	29	35
Bermuda	35	43	29
Netherlands Antilles(*)	26	30	23
Saudi Arabia	35	95	20
Indonesia	517	11	19
Venezuela	0	0	19
France(*)	0	0	12
Trinidad and Tobago	0	0	8
Bangladesh	0	1	7
Cayman Islands	2	3	5
Cambodia	0	0	5
Rest of World	315	822	26
World Total	4,961	5,107	5,045
Imports:			
Vietnam	47,107	36,973	38,514
Argentina	36,888	27,081	34,708
India	20,290	36,123	29,364
Brazil	19,249	15,440	19,062
Canada	5,612	8,234	13,508
Ukraine	8,876	11,411	11,086
Mexico	7,254	5,364	4,557
Thailand	3,458	10,753	4,238
New Zealand(*)	1,625	1,992	1,858
Turkey	2,581	5,196	1,852
Uruguay	5,362	7,243	1,767
Taiwan	2,524	4,443	1,662
Germany(*)	348	577	904
Dominican Republic	899	1,070	485
France(*)	146	419	455
Burma	459	1,093	428
Spain	147	195	363
Sierra Leone	0	0	208
Bulgaria	411	173	192
Russia	83	115	143
Australia(*)	66	66	129
United Kingdom	8	23	120
Hungary	48	105	118
Greece	81	90	108
China	0	98	105
Italy(*)	144	112	100
Austria	77	90	85
Poland	64	72	58
Romania	8	19	55
Rest of World	1,962	676	284
World Total	165,777	175,244	166,515

(*) Denotes a country that is a summarization of its component countries.
FAS, Office of Global Analysis, (202) 720-6301.   Data Source: U.S. Census Bureau Trade Data.

Table 2-34.—Honey: Number of colonies, yield, production, stocks, price, and value, United States, 2006–2015[1]

State	Honey producing colonies[2]	Yield per colony	Production[3]	Stocks Dec 15[4]	Average price per pound[5]	Value of production
	1,000	Pounds	1,000 pounds	1,000 pounds	Cents	1,000 dollars
2006	2,394	64.7	154,910	60,484	100.5	155,685
2007	2,443	60.7	148,341	52,635	107.7	159,763
2008	2,342	69.9	163,789	51,159	142.1	232,744
2009	2,498	58.6	146,416	37,516	147.3	215,671
2010	2,692	65.6	176,462	45,018	161.9	285,692
2011	2,491	59.6	148,357	36,761	176.5	261,850
2012	2,539	56.0	142,296	31,829	199.2	283,454
2013	2,640	56.6	149,499	38,160	214.1	320,077
2014	2,740	65.1	178,270	41,192	217.3	387,381
2015	2,660	58.9	156,544	42,203	209.0	327,177

[1] For producers with 5 or more colonies that also qualify as a farm. Colonies which produced honey in more than one State were counted in each State.   [2] Honey producing colonies are the maximum number of colonies from which honey was taken during the year. It is possible to take honey from colonies which did not survive the entire year.   [3] Due to rounding, total colonies multiplied by total yield may not exactly equal production.   [4] Stocks held by producers.   [5] Average price per pound based on expanded sales.
NASS, Livestock Branch, (202) 720–3570.

Table 2-35.—Honey: Number of colonies, yield, production, stocks, price and value, by State and United States, 2015[1]

State	Honey producing colonies[2]	Yield per colony	Production	Stocks Dec 15[3]	Average price per pound[4]	Value of production[5]
	1,000	Pounds	1,000 pounds	1,000 pounds	Cents	1,000 dollars
Alabama	7	47	329	13	383	1,260
Arizona	26	49	1,274	306	217	2,765
Arkansas	24	72	1,728	121	202	3,491
California	275	30	8,250	1,485	204	16,830
Colorado	29	51	1,479	399	218	3,224
Florida	220	54	11,880	832	197	23,404
Georgia	69	40	2,760	221	242	6,679
Hawaii	14	102	1,428	71	195	2,785
Idaho	89	32	2,848	1,082	192	5,468
Illinois	8	51	408	155	432	1,763
Indiana	6	53	318	165	338	1,075
Iowa	36	50	1,800	990	233	4,194
Kansas	8	36	288	107	352	1,014
Kentucky	5	46	230	55	386	888
Louisiana	44	99	4,356	348	193	8,407
Maine	10	47	470	47	551	2,590
Michigan	90	58	5,220	1,984	243	12,685
Minnesota	122	68	8,296	2,157	183	15,182
Mississippi	15	83	1,245	87	264	3,287
Missouri	10	52	520	52	350	1,820
Montana	146	83	12,118	3,757	194	23,509
Nebraska	57	48	2,736	1,450	202	5,527
New Jersey	12	27	324	207	420	1,361
New York	58	62	3,596	899	294	10,572
North Carolina ...	12	45	540	103	452	2,441
North Dakota	490	74	36,260	9,428	180	65,268
Ohio	17	50	850	357	408	3,468
Oregon	71	38	2,698	809	252	6,799
Pennsylvania	17	53	901	225	363	3,271
South Carolina ...	14	67	938	38	409	3,836
South Dakota	290	66	19,140	9,379	179	34,261
Tennessee	7	59	413	78	407	1,681
Texas	126	66	8,316	1,164	209	17,380
Utah	27	42	1,134	147	193	2,189
Vermont	5	52	260	62	423	1,100
Virginia	6	38	228	50	567	1,293
Washington	73	44	3,212	1,221	164	5,268
West Virginia	5	35	175	32	444	777
Wisconsin	52	67	3,484	1,603	241	8,396
Wyoming	38	77	2,926	146	190	5,559
Other States[6][7]	30	39	1,168	371	503	5,875
United States[7][8]	2,660	58.9	156,544	42,203	209.0	327,177

[1] For producers with 5 or more colonies that also qualify as a farm.   Colonies which produced honey in more than one State were counted in each State.   [2] Honey producing colonies are the maximum number of colonies from which honey was taken during the year.   It is possible to take honey from colonies which did not survive the entire year.   [3] Stocks held by producers.   [4] Average price per pound based on expanded sales.   [5] Value of production is equal to production multiplied by average price per pound.   [6] Alaska, Connecticut, Delaware, Maryland, Massachusetts, Nevada, New Hampshire, New Mexico, Oklahoma, and Rhode Island not published separately to avoid disclosing data for individual operations.   [7] Due to rounding, total colonies multiplied by total yield may not exactly equal production.   [8] United States value of production will not equal summation of States.
NASS, Livestock Branch, (202) 720–3570.

Table 2-36.—U.S. per capita caloric sweeteners estimated deliveries for domestic food and beverage, use by calendar year 2006–2015 [1] [2]

Calendar year	U.S. population[3] (July 1)	Refined sugar[4]	Corn Sweetener				Pure honey	Edible syrups	Total caloric sweeteners
			HFCS	Glucose syrup	Dextrose	Total			
									Millions
2006	298.8	62.2	57.8	13.7	3.1	74.7	1.2	0.7	138.7
2007	301.7	61.2	55.9	13.7	3.0	72.6	0.9	0.6	135.3
2008	304.5	65.1	52.7	13.4	2.8	68.8	1.0	0.6	135.5
2009	307.2	63.4	49.7	13.0	2.7	65.3	0.9	0.6	130.3
2010	309.3	66.0	48.4	12.6	2.9	64.0	1.0	0.7	131.7
2011	311.7	65.9	46.7	12.2	2.9	61.8	1.1	0.7	129.5
2012	314.1	66.6	45.8	12.5	2.7	61.0	1.1	0.7	129.4
2013	316.4	68.0	43.7	12.0	2.6	58.4	1.1	0.7	128.1
2014	318.9	68.4	43.6	12.2	3.0	58.8	1.2	0.8	129.2
2015	321.4	69.1	42.5	12.3	3.0	57.7	1.3	0.9	128.9

[1] Per capita deliveries of sweeteners by U.S. processors and refiners and direct-consumption imports to food manufacturers, retailers, and other end users represent the per capita supply of caloric sweeteners. The data exclude deliveries to manufacturers of alcoholic beverages. Actual human intake of caloric sweeteners is lower because of uneaten food, spoilage, and other losses. See Table 51 of the Sugar and Sweeteners Yearbook series for estimated intake of sugar.   [2] Totals may not add due to rounding.   [3] Source: U.S. Census Bureau.   [4] Based on U.S. sugar deliveries for domestic food and beverage use.
ERS, Market and Trade Economics Division (202) 694–5247. Source: ERS Sugars and Sweeteners Outlook.

Table 2-37.—Tobacco: Area, yield, production, price, and value, United States, 2006–2015

Year	Area harvested	Yield per acre	Production[1]	Marketing year average price per pound received by farmers	Value of production
	Acres	*Pounds*	*1,000 pounds*	*Dollars*	*1,000 dollars*
2006	339,000	2,147	727,897	1.665	1,211,885
2007	356,000	2,213	787,653	1.693	1,329,235
2008	354,490	2,258	800,504	1.859	1,488,069
2009	354,040	2,323	822,581	1.837	1,511,196
2010	337,500	2,128	718,190	1.782	1,279,920
2011	325,040	1,841	598,252	1.857	1,110,737
2012	336,245	2,268	762,709	2.027	1,557,604
2013	355,675	2,034	723,579	2.177	1,574,982
2014	378,360	2,316	876,415	2.094	1,835,308
2015	328,650	2,178	715,946	2.003	1,434,099

[1] Production figures are on farm-sales-weight basis.
NASS, Crops Branch, (202) 720–2127.

Table 2-38.—Tobacco: Area, yield, and production, by State and United States, 2013–2015

State	Area harvested			Yield per harvested acre			Production		
	2013	2014	2015	2013	2014	2015	2013	2014	2015
	Acres	*Acres*	*Acres*	*Pounds*	*Pounds*	*Pounds*	*1,000 pounds*	*1,000 pounds*	*1,000 pounds*
Connecticut	(D)	(D)	(D)	(D)	(D)	(D)	(D)	(D)	(D)
Georgia	12,800	15,000	13,500	1,750	2,300	2,400	22,400	34,500	32,400
Kentucky	87,200	91,700	72,900	2,147	2,337	2,055	187,240	214,280	149,830
Massachusetts	(D)	(D)	(D)	(D)	(D)	(D)	(D)	(D)	(D)
North Carolina	181,900	193,400	173,000	1,994	2,347	2,198	362,660	453,860	380,250
Ohio	2,100	2,000	1,900	2,200	2,150	1,900	4,620	4,300	3,610
Pennsylvania ...	8,900	9,100	7,900	2,389	2,445	2,290	21,260	22,250	18,090
South Carolina	14,500	15,800	13,000	1,700	2,100	2,000	24,650	33,180	26,000
Tennessee	21,400	24,250	20,900	2,083	2,151	2,333	44,570	52,155	48,770
Virginia	24,250	24,330	23,050	2,170	2,370	2,275	52,613	57,651	52,430
Other States[1] ..	2,625	2,780	2,500	1,358	1,525	1,826	3,566	4,239	4,239
United States ...	355,675	378,360	328,650	2,034	2,316	2,178	723,579	876,415	715,946

[1] Includes data withheld above.
NASS, Crops Branch, (202) 720–2127.

Table 2-39.—Tobacco: Stocks owned by dealers and manufacturers, by types, United States, 2007–2013 (farm-sales-weight basis) [1]

Type and year	Jan. 1	Apr. 1	July 1	Oct. 1
	1,000 pounds	*1,000 pounds*	*1,000 pounds*	*1,000 pounds*
Flue-cured, types 11–14:				
2007	671,018	570,171	493,248	578,776
2008	581,279	483,696	396,757	452,740
2009	564,889	436,658	360,324	448,901
2010	618,862	449,823	369,054	369,054
2011	561,073	454,548	381,925	363,831
2012	515,681	376,519	319,719	308,391
2013	472,916	372,058	260,393	353,534
Virginia fire-cured, type 21:				
2007	3,167	3,668	3,288	2,717
2008	3,131	3,154	2,833	2,579
2009	3,384	2,894	2,785	2,696
2010	886	1,092	802	802
2011	4,417	4,020	3,729	3,405
2012	3,983	3,848	591	723
2013	829	1,019	975	616
Kentucky and Tennessee fire-cured, types 22–23:				
2007	103,320	117,804	108,637	100,535
2008	111,458	121,405	112,796	103,306
2009	125,167	140,069	136,463	126,011
2010	96,965	102,349	91,278	91,278
2011	161,124	163,421	153,526	144,198
2012	164,412	170,764	104,149	98,740
2013	109,023	113,561	106,232	21,126
Burley, type 31:				
2007	422,568	426,348	361,305	296,177
2008	321,546	337,271	282,561	256,163
2009	283,223	297,075	265,545	239,152
2010	274,244	279,984	237,656	237,656
2011	255,695	275,863	248,528	208,171
2012	239,932	254,687	207,965	170,638
2013	174,645	227,961	194,321	164,151
Maryland, type 32:				
2007	375	1,190	372	1,028
2008	249	971	930	410
2009	116	970	43	30
2010	24	2,048	132	132
2011	20	2,449	87	85
2012	75	2,108	1,851	899
2013	569	700	326	13
One Sucker and Green River, types 35–36:[2]				
2007	39,818	44,456	40,765	35,775
2008	43,183	45,956	43,018	39,047
2009	49,492	53,357	53,517	51,812
2010	38,844	43,502	37,822	37,822
2011	61,528	63,485	59,590	54,906
2012	59,641	62,751	41,870	38,813
2013	39,261	44,603	40,580	7,075
Virginia sun-cured, type 37:				
2007	17	8	0	0
2008	0	5	5	5
2009	5	5	0	13
2010	13	12	12	12
2011	45	45	35	45
2012	45	45	45	45
2013	45	45	66	45
Pennsylvania seedleaf, type 41:				
2007	9,891	10,221	7,899	6,909
2008	6,375	9,953	9,210	7,932
2009	7,666	11,350	10,850	10,620
2010	7,262	13,550	13,547	13,547
2011	13,995	17,941	15,861	15,861
2012	15,467	20,723	14,305	16,499
2013	17,080	20,249	17,608	16,393

See footnote(s) at end of table.

Table 2-39.—Tobacco: Stocks owned by dealers and manufacturers, by types, United States, 2007–2013 (farm-sales-weight basis) [1]—Continued

Type and year	Jan. 1	Apr. 1	July 1	Oct. 1
	1,000 pounds	1,000 pounds	1,000 pounds	1,000 pounds
Connecticut Valley, types 51–52:				
2007	1,713	1,790	1,950	1,762
2008	1,730	1,398	1,837	533
2009	1,554	1,286	1,409	1,671
2010	1,062	953	1,178	943
2011	1,176	593	978	526
2012	149	229	41	155
2013	166	319	320	252
Wisconsin binder, types 54–55:				
2007	6,707	6,564	5,675	4,930
2008	4,826	5,378	4,497	3,777
2009	3,647	4,500	3,805	3,201
2010	2,798	4,492	4,335	4,612
2011	4,778	5,528	5,100	4,866
2012	4,287	5,205	4,008	3,809
2013	3,318	4,055	3,528	3,197
Cigar Wrapper, type 61:				
2007	727	966	511	1,149
2008	768	810	591	779
2009	611	278	239	742
2010	327	222	231	688
2011	1,071	478	356	759
2012	269	316	82	494
2013	354	404	329	290
Perique, type 72:				
2007	27	29	28	43
2008	43	42	22	36
2009	39	36	19	127
2010	105	93	117	165
2011	412	217	200	253
2012	246	216	321	374
2013	356	335	485	458
Other miscellaneous domestic, type 73:				
2007	3,558	1,851	2,661	2,781
2008	1,730	3,101	3,195	3,979
2009	5,351	3,998	2,871	2,024
2010	1,546	2,280	2,359	7,437
2011	8,351	8,611	8,670	8,598
2012	8,846	8,566	8,417	9,116
2013	9,340	9,379	8,956	1,743
Foreign-growth cigar leaf, types 81–89:				
2007	91,323	84,390	82,627	79,698
2008	84,538	85,535	81,340	81,468
2009	99,181	103,158	93,970	91,668
2010	85,370	98,216	77,507	78,171
2011	87,334	76,810	71,096	67,092
2012	71,631	68,683	55,380	83,975
2013	89,706	87,402	79,635	48,256
Foreign-growth cigarette and smoking, types 91–99:				
2007	766,925	753,161	757,311	721,959
2008	711,251	719,283	711,278	670,380
2009	655,356	668,814	621,702	623,288
2010	621,793	609,022	588,708	561,716
2011	587,319	578,762	546,270	530,319
2012	519,810	530,565	500,967	512,542
2013	501,423	492,380	329,528	496,577

[1] Stocks shown have been converted to a farm-sales-weight basis—the equivalent of weight at the time of sale-thereby making these data of leaf-tobacco stocks comparable with the leaf-tobacco production.   [2] One Sucker and Green leaf combined.

AMS Cotton and Tobacco Program, (901) 384–3016.

Table 2-40.—Tobacco products: Cigars, cigarettes, chewing and smoking tobacco, and snuff, manufactured in the United States, 2004–2013

Year	Cigars		Cigarettes		Chewing tobacco			
	Large	Small	Large [1]	Small	Firm	Moist	Twist	Looseleaf
	Millions	Millions	Millions	Millions	1,000 pounds	1,000 pounds	1,000 pounds	1,000 pounds
2004	4,341.7	3,359.8	0.0	492,749.4	1,403	271	651	37,012
2005	3,674.2	4,665.1	0.0	498,974.7	1,173	230	601	37,226
2006	4,256.2	5,291.3	0.0	483,678.0	1,098	199	551	36,406
2007	4,797.3	5,870.4	0.0	449,728.5	1,009	176	538	35,066
2008	4,984.4	6,478.0	0.0	396,115.4	909	144	500	30,935
2009	8,231.5	2,729.0	0.0	338,107.6	756	114	470	27,973
2010	10,407.5	1,461.3	0.0	326,653.6	699	86	390	25,974
2011	10,313.9	1,153.5	0.0	318,863.8	674	57	295	23,056
2012	9,910.7	1,258.1	0.0	303,752.3	537	37	320	22,109
2013	7,820.1	1,129.7	0.0	292,720.1	491	29	289	20,524
	Taxable removals and domestic invoices [2]							
2004	4,319.2	2,701.6	0.0	374,977.6	1,325	245	656	35,721
2005	4,441.0	3,772.1	0.0	363,260.2	1,166	201	614	35,701
2006	4,499.5	4,233.7	0.0	364,177.7	1,050	174	561	35,486
2007	4,658.7	4,791.3	0.0	347,960.2	978	150	539	32,721
2008	4,771.1	5,440.1	0.0	334,942.7	881	133	512	30,103
2009	7,944.2	2,150.6	0.0	308,117.2	736	72	457	27,002
2010	9,901.8	924.1	0.0	292,714.3	658	68	392	31,429
2011	9,887.0	753.9	0.0	286,315.5	638	42	292	28,606
2012	9,408.6	704.9	0.0	279,830.8	510	23	323	27,148
2013	7,788.4	622.4	0.0	266,205.8	486	19	291	25,639
	Tax-free removals and exports							
2004	114.5	658.6	0.0	111,202.4	28	19	0	55
2005	98.2	689.7	0.0	124,117.2	18	19	0	56
2006	100.0	830.7	0.0	116,649.0	21	20	0	59
2007	115.0	1,024.9	0.0	94,935.3	20	18	0	60
2008	152.6	857.0	0.0	61,698.4	19	19	0	96
2009	110.2	674.0	0.0	33,210.5	18	13	0	134
2010	291.6	528.7	0.0	30,369.1	15	13	0	134
2011	257.0	427.2	0.0	28,341.1	13	9	0	150
2012	401.2	492.6	0.0	21,815.9	15	10	0	118
2013	594.3	265.2	0.0	22,544.3	1	6	0	150

Year	Smoking tobacco			Snuff	Total chewing, smoking, and snuff
	Pipe	Granulated	Cigarette cut		
	1,000 pounds	1,000 pounds	1,000 pounds	1,000 pounds	1,000 pounds
2004	4,512	0	11,626	79,333	134,808
2005	4,280	0	13,109	81,951	138,570
2006	4,067	0	12,388	86,041	140,750
2007	4,117	0	12,164	90,153	143,223
2008	3,442	0	13,707	94,416	144,053
2009	5,102	0	8,394	95,528	138,337
2010	7,948	0	4,909	99,733	139,739
2011	18,484	0	4,239	106,194	152,999
2012	15,809	0	4,659	108,947	152,418
2013	8,436	0	2,933	113,503	146,205
	Taxable removals and domestic invoices [2]				
2004	3,773	0	11,675	74,718	128,113
2005	3,483	0	12,873	79,060	133,098
2006	3,149	0	12,311	83,618	136,349
2007	3,138	0	12,132	88,255	137,913
2008	2,949	0	13,735	93,112	141,425
2009	4,549	0	8,284	93,080	135,491
2010	7,807	0	5,063	97,290	138,164
2011	9,010	0	4,131	102,137	141,713
2012	8,873	0	3,185	106,111	146,173
2013	8,436	0	2,933	111,078	148,882
	Tax-free removals and exports				
2004	652	0	0	726	1,480
2005	446	0	0	785	1,324
2006	747	0	0	749	1,596
2007	942	0	0	740	1,780
2008	381	0	0	797	1,312
2009	239	0	0	745	1,156
2010	241	0	4	788	1,176
2011	507	0	11	845	1,520
2012	298	0	8	879	1,328
2013	303	0	5	839	1,304

[1] Weighing more than three pounds per thousand.   [2] Includes cigars and cigarettes imported or brought into the United States and Puerto Rico.
AMS Cotton and Tobacco Program, (901) 384–3016.

CHAPTER III

STATISTICS OF OILSEEDS, FATS, AND OILS

This chapter includes information on cottonseed, flaxseed, olive oil, peanuts, soybeans, and fats and oils. Most butter statistics are included in the chapter on dairy and poultry statistics. Lard data are mostly in the chapter on livestock.

Table 3-1.—Cottonseed: All cotton harvested area and cottonseed production, farm disposition, marketing year average price per ton received by farmers, and value, United States, 2006–2015

Year	Harvested area of all cotton	Cottonseed				
		Production [2]	Farm disposition		Marketing year average price	Value of production
			Sales to oil mills	Other [1]		
	1,000 acres	*1,000 tons*	*1,000 tons*	*1,000 tons*	*Dollars/tons*	*1,000 dollars*
2006	12,731.5	7,347.9	3,608.3	3,739.6	111.00	814,151
2007	10,489.1	6,588.7	3,635.1	2,953.6	162.00	1,069,849
2008	7,568.7	4,300.3	2,526.5	1,773.8	223.00	962,708
2009	7,533.7	4,148.8	2,277.9	1,870.9	158.00	670,027
2010	10,698.7	6,096.1	3,252.9	2,843.2	161.00	988,656
2011	9,460.9	5,370.0	2,695.0	2,675.0	260.00	1,413,343
2012	9,321.8	5,666.0	2,985.0	2,681.0	252.00	1,456,245
2013	7,544.4	4,203.0	2,070.0	2,133.0	246.00	1,054,003
2014	9,346.8	5,125.0	2,459.0	2,666.0	194.00	1,015,607
2015	8,074.9	4,043.0	1,916.0	2,127.0	227.00	942,234

(NA) Not available. [1] Includes planting seed, feed, exports, inter-farm sales, shrinkage, losses, and other uses. [2] 2015 production estimates based on 3-year average lint-seed ratio.
NASS, Crops Branch, (202) 720–2127.

Table 3-2.—Cottonseed: Production and farm disposition, by State and United States, 2014 and 2015

State	Production		Farm disposition				Seed for planting [2]	
	2014	2015	Sales to oil mills		Other [1]		2014	2015
			2014	2015	2014	2015		
	1,000 tons	*1,000 tons*	*1,000 tons*	*1,000 tons*	*1,000 tons*	*1,000 tons*	*1,000 tons*	*1,000 tons*
Alabama	193.0	162.0	40.0	22.0	153.0	140.0	1.5	1.8
Arizona	180.0	98.0	-	-	180.0	98.0	0.9	0.9
Arkansas	275.0	156.0	191.0	106.0	84.0	50.0	1.4	1.8
California	267.0	199.0	48.0	31.0	219.0	168.0	1.2	1.6
Florida	53.0	41.0	41.0	31.0	12.0	10.0	0.5	0.4
Georgia	740.0	615.0	304.0	266.0	436.0	349.0	5.3	5.6
Kansas	15.0	11.0	-	-	15.0	11.0	0.1	0.1
Louisiana	136.0	61.0	106.0	47.0	30.0	14.0	0.8	1.0
Mississippi	333.0	215.0	216.0	122.0	117.0	93.0	2.5	2.9
Missouri	200.0	154.0	140.0	102.0	60.0	52.0	1.1	1.5
New Mexico	24.0	24.0	-	-	24.0	24.0	0.3	0.3
North Carolina	302.0	156.0	50.0	28.0	252.0	128.0	2.4	1.9
Oklahoma	87.0	121.0	63.0	84.0	24.0	37.0	1.4	1.5
South Carolina	156.0	43.0	72.0	17.0	84.0	26.0	1.1	1.1
Tennessee	152.0	105.0	132.0	89.0	20.0	16.0	1.1	1.5
Texas	1,946.0	1,844.0	1,046.0	964.0	900.0	880.0	33.7	29.3
Virginia	66.0	38.0	10.0	7.0	56.0	31.0	0.5	0.5
United States	5,125.0	4,043.0	2,459.0	1,916.0	2,666.0	2,127.0	55.8	53.7

-Represents zero. [1] Includes planting seed, feed, exports, inter-farm sales, shrinkage, losses, and other uses. [2] Included in "other" farm disposition. Seed for planting is produced in crop year shown, but used in the following year.
NASS, Crops Branch, (202) 720–2127.

Table 3-3.—Cottonseed: Marketing year average price per ton and value of production, by State and United States, crop of 2013–2015

State	Marketing year average price per ton			Value of production		
	2013	2014	2015	2013	2014	2015
	Dollars	Dollars	Dollars	1,000 dollars	1,000 dollars	1,000 dollars
Alabama	218.00	180.00	203.00	35,970	35,100	32,683
Arizona	269.00	263.00	323.00	43,847	45,408	35,207
Arkansas	259.00	221.00	266.00	65,268	63,648	44,156
California	372.00	328.00	284.00	132,060	90,528	55,380
Florida	197.00	170.00	208.00	7,486	6,800	9,152
Georgia	202.00	173.00	195.00	141,602	130,442	132,015
Kansas	(D)	(D)	(D)	(D)	(D)	(D)
Louisiana	245.00	195.00	239.00	28,910	27,105	15,535
Mississippi	243.00	195.00	265.00	53,460	59,670	57,240
Missouri	253.00	210.00	269.00	51,865	43,680	39,812
New Mexico	298.00	240.00	236.00	4,172	3,600	5,428
North Carolina	227.00	198.00	228.00	57,885	62,964	36,936
Oklahoma	226.00	176.00	215.00	10,170	14,080	26,660
South Carolina	203.00	193.00	200.00	21,924	27,599	9,200
Tennessee	255.00	209.00	272.00	35,445	32,604	26,112
Texas	254.00	181.00	222.00	347,472	354,579	414,918
Virginia	(D)	(D)	(D)	(D)	(D)	(D)
United States	246.00	194.00	227.00	1,054,003	1,015,607	942,234

(D) Withheld to avoid disclosing data for individual operations.
NASS, Crops Branch, (202) 720–2127.

Table 3-4.—Cottonseed: Crushings, output of products and product prices, United States, 2006–2015

Year beginning August	Quantity crushed	Cottonseed products and prices			
		Oil		Cake and meal	
		Quantity	Price [1]	Quantity	Price [2]
	1,000 tons	Million pounds	Cents per pound	1,000 tibs	Dollars per ton
2006	2,680	849	35.7	1,241	150.36
2007	2,706	856	73.6	1,262	253.81
2008	2,240	669	37.1	938	255.23
2009	1,901	617	40.3	883	220.90
2010	2,563	835	54.5	1,163	273.84
2011	2,400	755	53.2	1,090	275.13
2012	2,500	800	48.6	1,125	331.52
2013	2,000	630	60.7	900	377.71
2014	1,900	610	45.7	855	304.27
2015 [3]	1,500	480	46.5-49.5	675	240.00-270.00

[1] Tanks, f.o.b. Valley Points.   [2] 41 percent protein, solvent, Memphis.   [3] Forecast.
ERS, Field Crops Branch, (202) 694–5300. Compiled from annual reports of the U.S. Department of Commerce.

Table 3-5.—International meal, cottonseed: Production in specified countries, 2013/2014–2015/2016

Country	2013/2014	2014/2015	2015/2016
	1,000 metric tons	1,000 metric tons	1,000 metric tons
Australia	330	294	313
Brazil	1,130	990	950
Burma	168	169	160
China	4,432	4,159	3,306
India	4,269	4,315	4,057
Mexico	160	206	195
Pakistan	1,700	1,860	1,302
Turkey	330	455	380
Turkmenistan	227	222	196
Uzbekistan	660	655	655
Others	1,412	1,341	1,243
Total Foreign	14,818	14,666	12,757
United States	816	776	612
Total	15,634	15,442	13,369

FAS, Office of Global Analysis, (202) 720-6301.   Prepared or estimated on the basis of official USDA production, supply, and distribution statistics from foreign governments.

Table 3-6.—Cottonseed, Cottonseed oil, and cottonseed cake and meal: United States exports by country of destination, 2014–2016

Country of destination	2014	2015	2016
	Metric tons	*Metric tons*	*Metric tons*
Cottonseed:			
Mexico	61,222	27,875	99,730
Saudi Arabia	38,918	36,769	45,071
Korea, South	66,492	43,023	39,267
Japan	28,228	22,767	13,337
Canada	1,446	786	742
United Arab Emirates	1,244	1,132	695
Oman	230	0	576
Morocco	48	0	400
Jordan	557	0	381
Costa Rica	0	3	42
Colombia	1	1	17
Guatemala	0	0	7
Trinidad and Tobago	0	5	7
Panama	0	2	5
El Salvador	0	1	0
Honduras	11	0	0
Australia(*)	1,444	20	0
Bahrain	0	19	0
China	102	0	0
Dominican Republic	0	2	0
Hong Kong	8	6	0
Lebanon	25	0	0
Netherlands Antilles(*)	0	1	0
Nicaragua	0	1	0
Qatar	0	115	0
World Total	199,975	132,529	200,278
Cottonseed oil:			
Mexico	36,433	24,767	19,599
South Africa	0	0	3,929
Canada	11,935	12,316	3,474
Australia(*)	181	1,293	1,872
Korea, South	0	529	512
Japan	2,118	913	474
Guatemala	0	134	311
Malaysia	523	4,224	241
Portugal	107	0	127
India	60	21	101
Israel(*)	0	0	73
United Kingdom	0	17	64
Brazil	69	81	37
United Arab Emirates	0	0	27
Costa Rica	0	0	19
Panama	41	5	7
Bahamas, The	0	0	4
Nicaragua	0	0	2
Singapore	71	1	1
Chile	2	0	1
Netherlands	926	467	1
Turkey	316	68	1
Colombia	15	2	0
Argentina	525	0	0
France(*)	0	4	0
Rest of World	13,369	148	0
World Total	66,689	44,990	30,877
Cottonseed cake & meal:			
Mexico	52,664	43,823	70,234
Germany(*)	3,019	1,993	2,543
Canada	3,738	5,229	2,532
Korea, South	4,342	2,763	1,651
United Kingdom	2,288	2,150	1,642
Brazil	1,046	2,195	1,376
Ecuador	2,331	344	1,344
Italy(*)	947	843	1,332
Austria	1,095	1,011	912
Hungary	88	98	198
Portugal	285	117	189
Japan	564	167	94
Spain	4	11	9
France(*)	1	0	4
Colombia	4	2	2
Bermuda	0	1	1
Belgium-Luxembourg(*)	5	0	0
Dominican Republic	0	0	0
India	0	0	0
Slovakia	0	0	0
Netherlands	1	0	0
Suriname	0	7	0
Taiwan	1	0	0
World Total	72,423	60,753	84,063

(*) Denotes a country that is a summarization of its component countries.
FAS, Office of Global Analysis, (202) 720-6301. Data Source: U.S. Census Bureau Trade Data

Table 3-7.—International oilseed, cottonseed: Area and production in specified countries, 2013/2014–2015/2016

Country	Area			Production		
	2013/2014	2014/2015	2015/2016	2013/2014	2014/2015	2015/2016
	1,000 hectares	*1,000 hectares*	*1,000 hectares*	*1,000 metric tons*	*1,000 metric tons*	*1,000 metric tons*
Australia	436	205	285	1,881	711	800
Brazil	1,120	1,020	925	2,671	2,360	2,266
Burma	300	300	285	368	370	350
China	4,800	4,400	3,400	12,835	11,757	9,330
European Union	311	355	315	480	523	383
India	11,700	12,700	11,800	12,950	12,500	11,400
Pakistan	2,900	2,950	2,800	4,100	4,600	3,000
Turkey	330	430	370	740	1,030	853
Turkmenistan	575	545	500	607	598	509
Uzbekistan	1,300	1,285	1,285	1,607	1,528	1,450
Others	5,072	5,039	4,785	3,631	3,718	3,230
Total Foreign	28,844	29,229	26,750	41,870	39,695	33,571
United States	3,053	3,783	3,261	3,813	4,649	3,768
Total	31,897	33,012	30,011	45,683	44,344	37,339

FAS, Office of Global Analysis, (202) 720–6301. Prepared or estimated on the basis of official USDA production, supply, and distribution statistics from foreign governments.

Table 3-8.—International oil, cottonseed: Production in specified countries, 2013/2014–2015/2016 [1]

Country	2013/2014	2014/2015	2015/2016
	1,000 metric tons	*1,000 metric tons*	*1,000 metric tons*
Australia	113	100	109
Brazil	384	336	320
China	60	61	58
European Union	1,487	1,396	1,110
India	1,305	1,320	1,244
Mexico	61	79	74
Pakistan	560	615	430
Turkey	130	185	155
Turkmenistan	85	83	74
Uzbekistan	233	230	230
Others	463	442	409
Total Foreign	4,881	4,847	4,213
United States	286	276	218
Total	5,167	5,123	4,431

[1] Year beginning July 1.
FAS, Office of Global Analysis, (202) 720–6301. Prepared or estimated on the basis of official USDA production, supply, and distribution statistics from foreign governments.

Table 3-9.—Flaxseed: Area, yield, production, disposition, and value, United States, 2006–2015

Year	Area planted	Area harvested	Yield per harvested acre	Production	Marketing year average price per bushel received by farmers	Value of production
	1,000 acres	*1,000 acres*	*Bushels*	*1,000 bushels*	*Dollars*	*1,000 dollars*
2006	813	767	14.4	11,019	5.80	63,961
2007	354	349	16.9	5,896	13.00	76,521
2008	354	340	16.8	5,716	12.70	72,773
2009	317	314	23.6	7,423	8.15	60,373
2010	421	418	21.7	9,056	12.20	110,251
2011	178	173	16.1	2,791	13.90	38,570
2012	349	336	17.3	5,798	13.80	79,919
2013	181	172	19.5	3,356	13.80	46,325
2014	311	302	21.1	6,368	11.80	75,077
2015	463	456	22.1	10,095	8.85	89,869

NASS, Crops Branch, (202) 720–2127.

Table 3-10.—Flaxseed: Area, yield, and production, by State and United States, 2013–2015

State	Area planted			Area harvested		
	2013	2014	2015	2013	2014	2015
	1,000 acres	*1,000 acres*	*1,000 acres*	*1,000 acres*	*1,000 acres*	*1,000 acres*
Minnesota	4	2	3	4	2	3
Montana	20	28	31	16	25	30
North Dakota	150	275	410	146	270	405
South Dakota	7	6	19	6	5	18
United States	181	311	463	172	302	456

State	Yield per harvested acre			Production		
	2013	2014	2015	2013	2014	2015
	Bushels	*Bushels*	*Bushels*	*1,000 bushels*	*1,000 bushels*	*1,000 bushels*
Minnesota	19.0	24.0	14.0	76	48	42
Montana	15.0	17.0	15.0	240	425	450
North Dakota	20.0	21.5	23.0	2,920	5,805	9,315
South Dakota	20.0	18.0	16.0	120	90	288
United States	19.5	21.1	22.1	3,356	6,368	10,095

NASS, Crops Branch, (202) 720–2127.

Table 3-11.—Flaxseed: Marketing year average price and value of production, by State and United States, 2013–2015

State	Marketing year average price per bushel			Value of production		
	2013	2014	2015	2013	2014	2015
	Dollars	*Dollars*	*Dollars*	*1,000 dollars*	*1,000 dollars*	*1,000 dollars*
Minnesota	13.80	11.00	8.50	1,049	528	357
Montana	14.00	11.80	9.80	3,360	5,015	4,410
North Dakota	13.80	11.80	8.85	40,296	68,499	82,438
South Dakota	13.50	11.50	9.25	1,620	1,035	2,664
United States	13.80	11.80	8.85	46,325	75,077	89,869

NASS, Crops Branch, (202) 720–2127.

Table 3-12.—Flaxseed: Support operations, United States, 2006/2007–2015/2016

Marketin year beginning June 1	Income support payment rates per bushels [1]	Program price levels per bushel		Put under loan		Acquired by CCC under loan program [4]	Owned by CCC at end of marketing year
		Loan [2]	Target/ Reference [3]	Quantity	Percentage of production		
	Dollars	Dollars	Dollars	1,000 bushels	Percent	1,000 bushels	1,000 bushels
2006/2007	0.45/0.00	5.21	5.66	598.2	5.4	0	0
2007/2008	0.45/0.00	5.21	5.66	131.0	2.2	0	0
2008/2009	0.45/0.00	5.21	5.66	141.1	2.5	0	0
2009/2010	0.45/0.00	5.21	5.66	80.4	1.1	0	0
2010/2011	0.45/0.00	5.65	7.10	69.6	0.8	0	0
2011/2012	0.45/0.00	5.65	7.10	17.9	0.6	0	0
2012/2013	0.45/0.00	5.65	7.10	26.8	0.5	0	0
2013/2014	0.45/0.00	5.65	7.10	10.7	0.3	0	0
2014/2015	0.00	5.65	11.28	37.5	0.6	0	0
2015/2016	2.28	5.65	11.28	269.6	2.7	0	0

[1] The first entry is the direct payment rate and the second entry is the counter-cyclical payment rate for 2004/2005-2013/2014. For 2009/2010-2013/2014, producers who participate in the Average Crop Revenue (ACRE) program get a 20 percent reduction in their direct payment, not calculated in this table. For 2014/2015 and after, the entry is the price loss coverage payment rate. For 2015/2016, projected based on August 2016 WASDE MY price. Agricultural Risk Coverage (ARC) is also available, but payment rates are established at the county or farm levels.   [2] For 2009/2010-2013/2014, producers who participate in the ACRE program get a 30 percent reduction in their loan rate, not calculated in this table.   [3] Target applies to 2003/2004-2013/2014 and Reference applies beginning with 2014/2015.   [4] Acquisitions as of July 30, 2016.
FSA, Feed Grains and Oilseeds, (202) 720–2711.

Table 3-13.—Flaxseed: Supply and disappearance, United States, 2006–2015

Year beginning June	Supply				Disappearance			
	Stocks June 1	Production	Imports	Total	Total used for seed	Exports	Crushings [1]	Total domestic disappearance [2]
	1,000 bushels	1,000 bushels	1,000 bushels	1,000 bushels	1,000 bushels	1,000 bushels	1,000 bushels	1,000 bushels
2006	3,535	11,019	5,464	20,018	287	1,788	14,900	15,786
2007	2,444	5,896	8,019	16,359	287	2,221	11,700	12,627
2008	1,512	5,716	4,794	12,022	257	432	8,150	9,038
2009	2,552	7,423	6,283	16,258	341	1,752	12,000	12,949
2010	1,557	9,056	6,040	16,653	144	2,130	11,635	12,352
2011	2,170	2,791	8,286	13,247	279	654	10,500	11,473
2012	1,120	5,798	6,928	13,846	147	1,020	11,000	11,902
2013	924	3,356	6,759	11,039	252	599	8,700	9,677
2014	763	6,368	7,351	14,482	340	528	11,850	13,146
2015	808	10,095	4,351	15,254	284	600	12,700	12,938

[1] From domestic and imported seed.   [2] Total supply minus exports and stocks June 1 of following year.
ERS, Field Crops Branch, (202) 694–5300.

Table 3-14.—Flaxseed and linseed oil and meal: Average price Minneapolis, 2006–2015

Year	Average price received by farmers per bushel	Minneapolis	
		Oil, per pound [1]	Meal, per ton [2]
	Dollars	Cents	Dollars
2006	5.80	53.99	124.61
2007	13.00	44.37	191.54
2008	12.70	70.31	227.66
2009	8.15	86.52	217.24
2010	12.20	67.49	223.23
2011	13.90	(NA)	238.35
2012	13.80	(NA)	320.13
2013	13.80	(NA)	359.42
2014	12.00	(NA)	263.90
2015	11.80	(NA)	175.00-205.00

(NA) Not available.   [1] Raw oil in tank cars.   [2] Bulk carlots, 34 percent protein.
ERS, Field Crops Branch, (202) 694–5300.

Table 3-15.—Flaxseed and products: Flaxseed crushed; production, imports, and exports of linseed oil, cake, and meal; and June 1 stocks of oil, United States, 2006–2015

Year beginning June	Total flaxseed crushed	Linseed oil			Linseed cake and meal		
		Stocks June 1	Production	Exports	Production	Imports for consumption	Exports
	1,000 bushels	Million pounds	Million pounds	Million pounds	1,000 tons	1,000 tons	1,000 tons
2006	14,900	29	291	76	268	17	10
2007	11,700	51	228	74	211	9	10
2008	8,150	26	159	66	147	10	28
2009	12,000	73	234	103	216	3	10
2010	11,635	37	227	101	209	7	7
2011	10,500	38	205	89	189	8	3
2012	11,000	35	215	94	198	6	5
2013	8,700	35	170	58	157	1	6
2014	11,850	35	231	52	213	3	4
2015	12,700	35	248	30	229	6	4

ERS, Field Crops Branch, (202) 694–5300.

Table 3-16.—Peanuts: Area, yield, production, disposition, marketing year average price per pound received by farmers, and value, United States, 2006–2015

Year	Area planted	Peanuts for nuts				
		Area harvested	Yield per acre	Production [1]	Marketing year average price	Value of production
	1,000 acres	1,000 acres	Pounds	1,000 pounds	Cents	1,000 dollars
2006	1,243.0	1,210.0	2,863	3,464,250	17.7	612,798
2007	1,230.0	1,195.0	3,073	3,672,250	20.5	758,626
2008	1,534.0	1,507.0	3,426	5,162,400	23.0	1,193,617
2009	1,116.0	1,079.0	3,421	3,691,650	21.7	793,147
2010	1,288.0	1,255.0	3,312	4,156,840	22.5	938,611
2011	1,140.6	1,080.6	3,386	3,658,590	31.8	1,168,587
2012	1,638.0	1,604.0	4,211	6,753,880	30.1	2,026,326
2013	1,067.0	1,043.0	4,001	4,173,170	24.9	1,055,095
2014	1,353.5	1,322.5	3,923	5,188,665	22.0	1,158,251
2015	1,625.0	1,567.0	3,963	6,210,590	19.0	1,186,903

[1] Estimates comprised of quota and non-quota peanuts.
NASS, Crops Branch, (202) 720–2127.

Table 3-17.—Peanuts, farmer stock: Stocks, production, and quantity milled, United States, 2006–2015

Year beginning August	Stocks Aug. 1 [1]	Production harvested for nuts [1]	Imports	Total supply	Milled [1][2]
	1,000 pounds	1,000 pounds	1,000 pounds	1,000 pounds	1,000 pounds
2006	1,402,614	3,464,250	48	4,866,912	3,914,354
2007	730,134	3,672,250	0	4,402,384	3,783,154
2008	346,948	5,162,400	194	5,509,542	3,901,712
2009	1,359,950	3,691,650	1,243	5,052,843	3,930,088
2010	991,394	4,156,840	163	5,148,397	3,976,460
2011	769,016	3,658,590	1,299	4,428,905	3,949,494
2012	272,838	6,753,880	3,117	7,029,835	4,979,182
2013	1,924,996	4,173,170	169	6,098,335	4,955,484
2014	661,774	5,188,665	7	6,248,196	4,775,454
2015	1,445,310	6,001,357	287	7,446,954	4,998,028

[1] Net weight basis.   [2] Includes peanuts milled for seed.
NASS, Crops Branch, (202) 720–2127, ERS, and Foreign trade from the Bureau of the Census.

Table 3-18.—Peanuts: Crushings, and oil and meal stocks, production, and foreign trade, United States, 2006–2015

Year beginning August	Peanuts crushed (shelled basis)	Peanut oil				Peanut cake and meal	
		Stocks Aug. 1 [1]	Production of crude	Imports	Exports [2]	Stocks Aug. 1 [3]	Production
	1,000 pounds	1,000 pounds	1,000 pounds	1,000 pounds	1,000 pounds	1,000 pounds	1,000 pounds
2006	385,375	11,730	166,450	104,623	11,009	4,908	223,537
2007	372,980	19,824	158,144	75,697	12,979	5,651	211,733
2008	334,296	6,024	142,666	54,155	9,311	4,949	190,748
2009	326,779	4,491	139,903	73,184	10,764	3,792	185,452
2010	441,017	6,888	190,110	60,012	15,923	7,045	250,043
2011	453,835	6,836	188,479	28,290	15,646	9,959	250,037
2012	493,205	2,478	210,702	10,360	12,598	10,446	270,328
2013	497,272	(NA)	209,808	55,265	8,953	(NA)	268,554
2014	506,677	(NA)	214,041	30,041	18,788	(NA)	278,380
2015	531,770	(NA)	226,219	93,430	10,253	3,197	291,193

(NA) Not available.   [1] Crude plus refined.   [2] Reported as edible peanut oil and crude peanut oil; in this tabulation added without converting.   [3] Holding at producing mills only.
NASS, Crops Branch, (202) 720–2127, ERS, and Foreign trade from the Bureau of the Census.

Table 3-19.—Cleaned peanuts (roasting stock): Supply and disposition, United States, 2006–2015

Year beginning August	Supply				Disposition		
	Stocks Aug. 1	Production	Imports	Total	Exports	Domestic disappearance	
						Total	Per capita
	1,000 pounds	*1,000 pounds*	*1,000 pounds*	*1,000 pounds*	*1,000 pounds*	*1,000 pounds*	*Pounds*
2006	56,993	221,618	48	278,659	19,600	216,956	0.72
2007	42,103	257,386	0	299,489	56,323	185,371	0.61
2008	57,795	282,284	194	340,273	67,091	212,695	0.69
2009	60,487	257,414	1,243	319,144	55,430	210,785	0.69
2010	52,929	268,956	163	322,048	78,693	184,385	0.60
2011	58,970	226,481	1,299	286,750	66,819	179,965	0.58
2012	39,966	323,731	3,117	366,814	194,821	100,098	0.32
2013	71,895	323,607	169	395,671	143,875	178,643	0.56
2014	73,153	248,170	7	321,330	168,932	152,398	0.48
2015	40,335	251,484	287	292,106	418,888	(126,782)	(489.06)

NASS, Crops Branch, (202) 720–2127, ERS, and Foreign trade from the Bureau of the Census.

Table 3-20.—Shelled peanuts (all grades): Supply, exports, and quantity crushed, United States, 2006–2015

Year beginning August	Supply						Exports	Crushed
	Stocks Aug. 1		Production		Imports	Total		
	Edible	Oil stock	Edible	Oil stock				
	1,000 pounds	*1,000 pounds*	*1,000 pounds*	*1,000 pounds*	*1,000 pounds*	*1,000 pounds*	*1,000 pounds*	*1,000 pounds*
2006	510,097	21,499	2,415,495	347,243	42,888	3,337,222	437,663	385,375
2007	528,918	33,401	2,291,603	319,186	53,471	3,226,579	520,508	372,980
2008	431,593	39,508	2,442,345	253,778	61,199	3,228,423	494,738	334,296
2009	511,261	22,320	2,547,434	280,888	49,454	3,411,357	402,732	326,779
2010	554,295	35,498	2,450,639	357,130	46,031	3,443,593	395,554	441,017
2011	473,878	43,380	2,399,094	345,565	163,852	3,425,769	352,994	453,835
2012	466,310	52,883	3,125,786	351,284	72,871	4,069,134	741,103	493,205
2013	547,965	33,883	3,098,392	373,008	60,738	4,113,986	708,141	497,272
2014	519,824	25,364	2,997,078	391,728	64,345	3,998,339	676,509	506,677
2015	431,674	31,012	3,141,099	442,650	62,604	4,109,039	838,677	531,770

NASS, Crops Branch, (202) 720–2127, ERS, and Foreign trade from the U.S. Bureau of the Census.

Table 3-21.—Peanuts: Shelled (raw basis) by types, used in primary products and apparent disappearance of peanuts, United States, 2006–2015

Type and year beginning August	Edible grades used in products					Apparent disappear-ance [2]
	Peanut butter [1]	Peanut snack	Peanut candy	Other products	Total	
	1,000 pounds	*1,000 pounds*	*1,000 pounds*	*1,000 pounds*	*1,000 pounds*	*1,000 pounds*
Virginia and Valencia:						
2006	113,689	75,858	29,542	1,103	220,196	
2007	125,497	71,059	27,909	979	225,445	
2008	110,737	52,925	26,342	1,766	191,770	
2009	(3)	50,812	17,361	(3)	198,770	
2010	(3)	62,708	16,070	(3)	211,194	
2011	(3)	78,333	17,856	(3)	203,958	
2012	82,981	83,722	17,731	8,452	192,888	
2013	86,759	85,298	17,109	13,373	202,536	
2014	102,340	91,909	12,079	26,440	232,768	
2015	108,156	89,766	14,319	26,258	238,494	
Runner:						
2006	869,014	328,167	329,806	8,263	1,535,250	
2007	878,026	344,551	279,564	9,666	1,511,807	
2008	981,546	303,730	276,212	8,043	1,569,531	
2009	1,056,699	290,358	286,277	13,120	1,646,454	
2010	1,076,521	319,529	365,260	13,036	1,774,346	
2011	1,091,541	303,631	360,797	14,840	1,770,809	
2012	1,143,108	309,860	347,428	12,195	1,812,591	
2013	1,128,206	337,934	(3)	(3)	1,844,490	
2014	1,196,277	329,930	348,367	26,737	1,901,311	
2015	1,186,810	408,726	346,831	35,128	1,977,495	
Spanish:						
2006	(3)	11,104	14,335	(3)	36,211	
2007	(3)	9,556	12,994	(3)	31,321	
2008	(3)	10,823	13,721	(3)	34,990	
2009	(3)	11,793	11,957	(3)	31,269	
2010	(3)	12,940	14,122	(3)	35,207	
2011	(3)	8,104	16,025	(3)	27,390	
2012	(3)	6,847	16,755	(3)	25,389	
2013	(3)	6,564	15,996	(3)	(3)	
2014	(3)	6,638	15,410	(3)	27,188	
2015	(3)	7,200	16,355	(3)	28,225	
All types:						
2006	993,445	415,131	373,684	9,397	1,791,657	2,732,015
2007	1,012,263	425,166	320,467	10,676	1,768,572	2,702,007
2008	1,102,698	367,478	316,275	9,840	1,796,291	2,633,643
2009	1,191,821	352,963	315,595	15,840	1,876,219	2,682,110
2010	1,213,229	395,177	395,452	16,890	2,020,748	2,880,304
2011	1,197,748	390,068	394,678	19,661	2,020,748	2,742,724
2012	1,227,859	400,429	381,914	20,664	2,030,866	3,417,047
2013	1,218,170	429,796	395,726	29,103	2,072,795	3,508,060
2014	1,303,755	428,477	375,856	53,179	2,161,267	3,471,308
2015	1,299,634	505,692	377,505	61,388	2,244,219	3,513,946

[1] Excludes peanut butter made by manufacturers for own use in candy. Includes peanut butter used in spreads, sandwiches, and cookies. [2] Apparent disappearance represents stocks beginning of year plus production, minus stocks at end of year. Includes edible grades and oilstocks. [3] Not published to avoid disclosure of individual operations.
NASS, Crops Branch, (202) 720–2127, and ERS.

Table 3-22.—Peanuts: Area, yield, and production, by State and United States, 2013–2015

State	Area planted			Peanuts for nuts		
	2013	2014	2015	Area harvested		
				2013	2014	2015
	1,000 acres	*1,000 acres*	*1,000 acres*	*1,000 acres*	*1,000 acres*	*1,000 acres*
Alabama	140.0	175.0	200.0	138.0	173.0	197.0
Florida	140.0	175.0	190.0	131.0	167.0	180.0
Georgia	430.0	600.0	785.0	426.0	589.0	777.0
Mississippi	34.0	32.0	44.0	33.0	31.0	42.0
New Mexico	7.0	4.5	5.0	7.0	4.5	5.0
North Carolina	82.0	94.0	90.0	81.0	93.0	88.0
Oklahoma	17.0	12.0	10.0	16.0	11.0	9.0
South Carolina ...	81.0	112.0	112.0	78.0	108.0	82.0
Texas	120.0	130.0	170.0	117.0	127.0	168.0
Virginia	16.0	19.0	19.0	16.0	19.0	19.0
United States	1,067.0	1,353.5	1,625.0	1,043.0	1,322.5	1,567.0

State	Peanuts for nuts					
	Yield per harvested acre			Production		
	2013	2014	2015	2013	2014	2015
	Pounds	*Pounds*	*Pounds*	*1,000 pounds*	*1,000 pounds*	*1,000 pounds*
Alabama	3,550	3,150	3,350	489,900	544,950	659,950
Florida	3,950	4,000	3,650	517,450	668,000	657,000
Georgia	4,430	4,135	4,470	1,887,180	2,435,515	3,473,190
Mississippi	3,700	4,000	3,600	122,100	124,000	151,200
New Mexico	3,100	3,500	3,000	21,700	15,750	15,000
North Carolina	3,900	4,320	3,400	315,900	401,760	299,200
Oklahoma	3,700	4,000	3,500	59,200	44,000	31,500
South Carolina ...	3,500	3,800	3,200	273,000	410,400	262,400
Texas	3,620	3,620	3,500	423,540	459,740	588,000
Virginia	3,950	4,450	3,850	63,200	84,550	73,150
United States	4,001	3,923	3,963	4,173,170	5,188,665	6,210,590

NASS, Crops Branch, (202) 720–2127.

Table 3-23.—Peanuts: Marketing year average price, and value of production, by State and United States, 2013–2015

State	Marketing year average price per pound			Value of production		
	2013	2014	2015	2013	2014	2015
	Dollars	*Dollars*	*Dollars*	*1,000 dollars*	*1,000 dollars*	*1,000 dollars*
Alabama	0.226	0.199	0.178	110,717	108,445	106,912
Florida	0.242	0.215	0.188	125,223	143,620	123,516
Georgia	0.240	0.206	0.192	452,923	501,716	656,433
Mississippi	0.243	0.208	0.174	29,670	25,792	23,436
New Mexico	0.334	0.303	0.215	7,248	4,772	3,240
North Carolina	0.275	0.262	0.221	86,873	105,261	70,910
Oklahoma	0.292	0.261	0.213	17,286	11,484	7,308
South Carolina	0.264	0.245	0.190	72,072	100,548	52,854
Texas	0.321	0.291	0.214	135,956	133,784	126,420
Virginia	0.271	0.270	0.214	17,127	22,829	15,874
United States	0.249	0.220	0.193	1,055,095	1,158,251	1,186,903

NASS, Crops Branch, (202) 720–2127.

Table 3-24.—Peanuts, farmer's stock: Support operations, United States, 2010/2011–2015/2016

Marketing year beginning August 1	Income support payment rates per short ton [1]	Program price levels per short ton		Put under support		Acquired by CCC under loan program	Owned by CCC at end of marketing year
		Loan [2]	Target Reference [3]	Quantity	Percentage of production		
	Dollars	Dollars	Dollars	1,000 short tons	Percent	1,000 short tons	1,000 short tons
2010/2011	36.00/9.00	355.00	495.00	1,811	87.1	0.0	0.0
2011/2012	36.00/0.00	355.00	495.00	1,402	76.6	0.0	0.0
2012/2013	36.00/0.00	355.00	495.00	2,640	78.2	55.9	0.0
2013/2014	36.00/0.00	355.00	495.00	1,452	69.6	6.1	2.5
2014/2015	95.00	355.00	535.00	2,123	81.8	173.2	6.8
2015/2016 [4]	155.00	355.00	535.00	2,233	71.9	0	0.0

[1] The first entry is the direct payment rate and the second entry is the counter-cyclical payment rate for 2004/2005-2013/2014. The 2014 Act eliminated direct payments and replaced the counter-cyclical rate with a new reference price. For 2009/2010-2013/2014, producers who participate in the Average Crop Revenue (ACRE) program get a 20 percent reduction in their direct payment, not calculated in this table. For 2014/2015 and after, the entry is the price loss coverage payment rate. For 2015/2016, projected based on August 2016 WASDE MY price. Agricultural Risk Coverage (ARC) is also available, but payment rates are established at the county or farm levels. [2] For 2009/2010-2013/2014, producers who participate in the ACRE program get a 30 percent reduction in their loan rate, not calculated in this table. [3] Target applies to 2003/2004-2013/2014 and Reference applies beginning with 2014/2015. [4] Acquisitions as of July 31, 2016.
FSA, Fibers, Peanuts, Tobacco, (202) 401–0062.

Table 3-25.—International oilseed, peanut: Area and production in specified countries, 2013/2014–2015/2016

Country	Area			Production		
	2013/2014	2014/2015	2015/2016	2013/2014	2014/2015	2015/2016
	1,000 hectares	1,000 hectares	1,000 hectares	1,000 metric tons	1,000 metric tons	1,000 metric tons
Argentina	378	341	329	997	1,188	1,070
Burma	885	885	885	1,375	1,375	1,375
Cameroon	463	470	400	636	640	550
China	4,633	4,604	4,600	16,972	16,482	16,700
India	5,400	4,600	4,500	5,650	4,900	4,700
Indonesia	655	630	615	1,160	1,150	1,130
Nigeria	2,500	2,500	2,500	3,000	3,000	3,000
Senegal	917	879	770	677	669	725
Sudan	2,162	1,254	2,184	1,767	963 871	
Tanzania	944	840	840	900	800	800
Others	6,052	6,116	6,084	6,089	5,898	5,886
Total Foreign	24,989	23,119	23,707	39,223	37,065	37,807
United States	422	535	634	1,893	2,354	2,817
Total	25,411	23,654	24,341	41,116	39,419	40,624

FAS, Office of Global Analysis, (202) 720–6301. Prepared or estimated on the basis of official USDA production, supply, and distribution statistics from foreign governments.

Table 3-26.—International meal, peanut: Production in specified countries, 2013/2014–2015/2016

Country	2013/2014	2014/2015	2015/2016
	1,000 metric tons	1,000 metric tons	1,000 metric tons
Argentina	90	110	112
Burkina	94	90	90
Burma	325	325	325
Cameroon	57	57	57
China	3,470	3,344	3,462
India	1,410	1,300	1,253
Nigeria	240	240	240
Senegal	120	120	130
Sudan	222	222	260
Tanzania	160	160	160
Others	539	560	545
Total Foreign	6,727	6,528	6,634
United States	122	126	146
Total	6,849	6,654	6,780

FAS, Office of Global Analysis, (202) 720–6301. Prepared or estimated on the basis of official USDA production, supply, and distribution statistics from foreign governments.

Table 3-27.—Soybeans: Area, yield, production, and value, United States, 2006–2015

Year	Area planted	Soybeans for beans				
		Area harvested	Yield per acre	Production	Marketing year average price per bushel received by farmers	Value of production
	1,000 acres	*1,000 acres*	*Bushels*	*1,000 bushels*	*Dollars*	*1,000 dollars*
2006	75,522	74,602	42.9	3,196,726	6.43	20,468,267
2007	64,741	64,146	41.7	2,677,117	10.10	26,974,406
2008	75,718	74,681	39.7	2,967,007	9.97	29,458,225
2009	77,451	76,372	44.0	3,360,931	9.59	32,163,204
2010	77,404	76,610	43.5	3,331,306	11.30	37,571,277
2011	75,046	73,776	42.0	3,097,179	12.50	38,542,177
2012	77,198	76,144	40.0	3,042,044	14.40	43,723,144
2013	76,840	76,253	44.0	3,357,984	13.00	43,582,901
2014	83,276	82,591	47.5	3,927,090	10.10	39,474,861
2015	82,650	81,814	48.0	3,929,160	8.80	34,535,320

NASS, Crops Branch, (202) 720–2127.

Table 3-28.—Soybeans: Stocks on and off farms, United States, 2006–2015

Year beginning previous December	On farms				Off farms [1]			
	Dec. 1	Mar. 1	June 1	Sep. 1 [2]	Dec. 1	Mar. 1	June 1	Sep. 1 [2]
	1,000 bushels	*1,000 bushels*	*1,000 bushels*	*1,000 bushels*	*1,000 bushels*	*1,000 bushels*	*1,000 bushels*	*1,000 bushels*
2006	1,345,000	872,000	495,500	176,300	1,156,426	797,206	495,199	273,026
2007	1,461,000	910,000	500,000	143,000	1,240,366	876,887	592,185	430,810
2008	1,128,500	593,000	226,600	47,000	1,231,860	840,982	449,543	158,034
2009	1,189,000	656,500	226,300	35,100	1,086,432	645,289	369,859	103,098
2010	1,229,500	609,200	232,600	35,400	1,109,050	660,868	338,523	115,485
2011	1,091,000	505,000	217,700	48,500	1,187,084	743,800	401,583	166,513
2012	1,139,000	555,000	179,000	38,250	1,230,885	819,488	488,465	131,120
2013	910,000	456,700	171,100	39,550	1,056,161	541,320	263,564	101,007
2014	955,000	381,900	109,100	21,325	1,198,621	611,928	295,945	70,666
2015 [3]	1,218,000	609,200	246,300	49,700	1,309,744	717,399	380,768	140,910

[1] Includes stocks at mills, elevators, warehouses, terminals, and processors.   [2] Old crop only.   [3] Preliminary.
NASS, Crops Branch, (202) 720–2127.

Table 3-29.—Soybeans, soybean meal, and oil: Average price at specified markets, 2006–2015

Year [1]	Soybeans per bushel: No. 1 Yellow Chicago	Soybean oil per pound crude, tanks, f.o.b. Decatur	Soybean meal per short ton: 48 percent protein Decatur
	Dollars	*Cents*	*Dollars*
2006	6.92	31.02	205.44
2007	12.22	52.03	335.94
2008	10.09	32.16	331.17
2009	9.65	35.95	311.27
2010	12.97	53.20	345.52
2011	13.49	51.90	393.53
2012	14.86	47.13	468.11
2013	13.51	38.23	489.94
2014	9.84	31.60	368.49
2015 [2]	9.45	29.75	322.50

[1] Year beginning September for soybeans and October for oil and meal.   [2] Preliminary.
ERS, Field Crops Branch, (202) 694–5300.

Table 3-30.—Soybeans: Supply and disappearance, United States, 2006–2015

Year beginning September	Supply				
	Stocks by position			Production	Total [1]
	Farm	Terminal market, interior mill, elevator, and warehouse	Total		
	1,000 bushels	1,000 bushels	1,000 bushels	1,000 bushels	1,000 bushels
2006	176,300	273,026	449,326	3,196,726	3,655,086
2007	143,000	430,810	573,810	2,677,117	3,260,798
2008	47,000	158,034	205,034	2,967,007	3,185,304
2009	35,100	103,098	138,198	3,360,931	3,513,717
2010	35,400	115,485	150,885	3,331,306	3,496,640
2011	48,500	166,513	215,013	3,097,179	3,328,324
2012	38,250	131,120	169,370	3,042,044	3,251,930
2013	39,550	101,007	140,557	3,357,984	3,570,254
2014	21,325	70,666	91,991	3,927,090	4,052,308
2015 [2]	49,700	140,910	190,610	3,929,160	4,144,500

Year beginning September	Disappearance			
	Crushed [3]	Seed, feed and residual	Exports	Total
	1,000 bushels	1,000 bushels	1,000 bushels	1,000 bushels
2006	1,807,706	157,074	1,116,496	3,081,276
2007	1,803,407	93,527	1,158,829	3,055,764
2008	1,661,922	105,890	1,279,294	3,047,106
2009	1,751,686	112,098	1,499,048	3,362,832
2010	1,648,043	128,607	1,504,978	3,281,627
2011	1,703,019	89,601	1,366,335	3,158,954
2012	1,688,903	94,944	1,327,526	3,111,373
2013	1,733,888	106,550	1,637,826	3,478,263
2014	1,873,494	144,829	1,843,376	3,861,698
2015 [2]	1,900,000	109,600	1,940,000	3,949,500

[1] Includes imports.　[2] Preliminary.　[3] Reported by the U.S. Department of Commerce.
ERS, Field Crops Branch, (202) 694–5300.

Table 3-31.—Soybeans: Support operations, United States, 2006/2007–2015/2016

Marketing year beginning September 1	Income support payment rates per bushels [1]	Program price levels per bushel		Put under loan		Acquired by CCC under loan program [4]	Owned by CCC at end of marketing year
		Loan [2]	Target/ Reference [3]	Quantity	Percentage of production		
	Dollars	Dollars	Dollars	Million bushels	Percent	Million bushels	Million bushels
2006/2007	0.44/0.00	5.00	5.80	397	12.5	0	0
2007/2008	0.44/0.00	5.00	5.80	181	6.8	0	0
2008/2009	0.44/0.00	5.00	5.80	189	6.4	0	0
2009/2010	0.44/0.00	5.00	5.80	123	3.7	0	0
2010/2011	0.44/0.00	5.00	6.00	108	3.3	0	0
2011/2012	0.44/0.00	5.00	6.00	98	3.2	0	0
2012/2013	0.44/0.00	5.00	6.00	58	1.9	0	0
2013/2014	0.44/0.00	5.00	6.00	49	1.5	0	0
2014/2015	0.00	5.00	8.40	80	2.0	0	0
2015/2016	0.00	5.00	8.40	124	3.2	0	0

[1] The first entry is the direct payment rate and the second entry is the counter-cyclical payment rate for 2004/2005-2013/2014. For 2009/2010-2013/2014, producers who participate in the Average Crop Revenue (ACRE) program get a 20 percent reduction in their direct payment, not calculated in this table. For 2014/2015 and after, the entry is the price loss coverage payment rate. For 2015/2016, projected based on August 2016 WASDE MY price. Agricultural Risk Coverage (ARC) is also available, but payment rates are established at the county or farm levels.　[2] For 2009/2010-2013/2014, producers who participate in the ACRE program get a 30 percent reduction in their loan rate, not calculated in this table.　[3] Target applies to 2003/2004-2013/2014 and Reference applies beginning with 2014/2015.　[4] Acquisitions as of July 30, 2016.
FSA, Feed Grains and Oilseeds, (202) 720–2711.

Table 3-32.—Soybeans for beans: Area, yield, and production, by State and United States, 2013–2015

State	Area planted			Area harvested		
	2013	2014	2015	2013	2014	2015
	1,000 acres	*1,000 acres*	*1,000 acres*	*1,000 acres*	*1,000 acres*	*1,000 acres*
Alabama	440	480	500	430	470	490
Arkansas	3,270	3,230	3,200	3,240	3,200	3,170
Delaware	165	185	175	163	183	173
Florida	32	39	33	30	37	31
Georgia	235	300	325	230	290	315
Illinois	9,500	9,800	9,800	9,480	9,770	9,720
Indiana	5,200	5,450	5,550	5,190	5,440	5,500
Iowa	9,300	9,850	9,850	9,250	9,770	9,800
Kansas	3,600	4,000	3,900	3,540	3,960	3,860
Kentucky	1,670	1,760	1,840	1,660	1,750	1,810
Louisiana	1,130	1,410	1,430	1,120	1,395	1,395
Maryland	485	510	520	480	505	515
Michigan	1,930	2,050	2,030	1,920	2,040	2,020
Minnesota	6,700	7,350	7,600	6,620	7,270	7,550
Mississippi	2,010	2,210	2,300	1,990	2,190	2,270
Missouri	5,650	5,650	4,550	5,610	5,590	4,480
Nebraska	4,800	5,400	5,300	4,770	5,330	5,270
New Jersey	90	105	105	88	103	103
New York	280	330	305	278	327	301
North Carolina	1,480	1,750	1,820	1,450	1,730	1,790
North Dakota	4,650	5,900	5,750	4,630	5,870	5,720
Ohio	4,500	4,700	4,750	4,490	4,690	4,740
Oklahoma	345	375	395	335	365	375
Pennsylvania	560	570	580	555	565	575
South Carolina	320	450	475	310	440	370
South Dakota	4,600	5,150	5,150	4,580	5,110	5,120
Tennessee	1,580	1,640	1,750	1,550	1,610	1,720
Texas	105	155	130	92	135	115
Virginia	610	650	630	600	640	620
West Virginia	23	27	27	22	26	26
Wisconsin	1,580	1,800	1,880	1,550	1,790	1,870
United States	76,840	83,276	82,650	76,253	82,591	81,814

State	Yield per harvested acre			Production		
	2013	2014	2015	2013	2014	2015
	Bushels	*Bushels*	*Bushels*	*1,000 bushels*	*1,000 bushels*	*1,000 bushels*
Alabama	43.5	40.0	41.0	18,705	18,800	20,090
Arkansas	43.5	49.5	49.0	140,940	158,400	155,330
Delaware	40.5	47.5	40.0	6,602	8,693	6,920
Florida	41.0	43.0	38.0	1,230	1,591	1,178
Georgia	40.5	40.0	43.0	9,315	11,600	13,545
Illinois	50.0	56.0	56.0	474,000	547,120	544,320
Indiana	51.5	55.5	50.0	267,285	301,920	275,000
Iowa	45.5	51.0	56.5	420,875	498,270	553,700
Kansas	37.0	35.5	38.5	130,980	140,580	148,610
Kentucky	50.0	47.5	49.0	83,000	83,125	88,690
Louisiana	48.5	56.5	41.0	54,320	78,818	57,195
Maryland	39.5	46.0	40.0	18,960	23,230	20,600
Michigan	44.5	42.5	49.0	85,440	86,700	98,980
Minnesota	42.0	41.5	50.0	278,040	301,705	377,500
Mississippi	46.0	52.0	46.0	91,540	113,880	104,420
Missouri	36.0	46.5	40.5	201,960	259,935	181,440
Nebraska	53.5	54.0	58.0	255,195	287,820	305,660
New Jersey	39.5	44.0	32.0	3,476	4,532	3,296
New York	48.0	44.5	43.0	13,344	14,552	12,943
North Carolina	33.5	40.0	32.0	48,575	69,200	57,280
North Dakota	30.5	34.5	32.5	141,215	202,515	185,900
Ohio	49.5	52.5	50.0	222,255	246,225	237,000
Oklahoma	30.5	28.0	31.0	10,218	10,220	11,625
Pennsylvania	49.0	49.0	44.0	27,195	27,685	25,300
South Carolina	28.5	35.0	26.5	8,835	15,400	9,805
South Dakota	40.5	45.0	46.0	185,490	229,950	235,520
Tennessee	46.5	46.0	46.0	72,075	74,060	79,120
Texas	25.5	38.5	26.0	2,346	5,198	2,990
Virginia	38.5	39.5	34.5	23,100	25,280	21,390
West Virginia	46.5	51.0	48.0	1,023	1,326	1,248
Wisconsin	39.0	44.0	49.5	60,450	78,760	92,565
United States	44.0	47.5	48.0	3,357,984	3,927,090	3,929,160

NASS, Crops Branch, (202) 720–2127.

Table 3-33.—Soybeans: Crushings, and oil and meal stocks, production, and foreign trade, United States, 2006–2015

Year beginning October	Soybeans crushed	Soybean oil			Soybean cake and meal		
		Stocks Oct. 1	Production	Exports	Stocks Oct. 1	Production	Exports
	1,000 bushels	*Million pounds*	*Million pounds*	*Million pounds*	*1,000 tons*	*1,000 tons*	*1,000 tons*
2006	1,807,706	3,010	20,489	1,877	314	43,032	8,804
2007	1,803,407	3,085	20,580	2,911	343	42,284	9,242
2008	1,661,922	2,485	18,745	2,193	294	39,102	8,497
2009	1,751,686	2,861	19,615	3,359	235	41,707	11,159
2010	1,648,043	3,406	18,888	3,233	302	39,251	9,081
2011	1,703,019	2,675	19,740	1,464	350	41,036	9,750
2012	1,688,903	2,590	19,820	2,163	300	39,875	11,146
2013	1,733,888	1,655	20,130	1,877	275	40,685	11,546
2014	1,873,494	1,165	21,399	2,014	250	45,062	13,150
2015 [1]	1,900,000	1,855	22,080	2,400	260	44,940	11,800

[1] Preliminary.
ERS, Field Crops Branch, (202) 694–5300. Data from the U.S. Department of Commerce.

Table 3-34.—Soybeans for beans: Marketing year average price and value, by State and United States, 2013–2015

State	Marketing year average price per bushel			Value of production		
	2013	2014	2015	2013	2014	2015
	Dollars	*Dollars*	*Dollars*	*1,000 dollars*	*1,000 dollars*	*1,000 dollars*
Alabama	12.90	10.00	8.80	241,295	188,000	176,792
Arkansas	13.10	10.60	9.45	1,846,314	1,679,040	1,467,869
Delaware	12.40	9.60	8.40	81,865	83,453	58,128
Florida	11.80	8.90	8.00	14,514	14,160	9,424
Georgia	13.30	10.50	8.95	123,890	121,800	121,228
Illinois	13.20	10.20	9.00	6,256,800	5,580,624	4,898,880
Indiana	13.20	10.20	8.85	3,528,162	3,079,584	2,433,750
Iowa	13.10	9.96	8.65	5,513,463	4,962,769	4,789,505
Kansas	12.80	9.63	8.35	1,676,544	1,353,785	1,240,894
Kentucky	13.10	10.50	9.20	1,087,300	872,813	815,948
Louisiana	13.40	10.90	9.75	727,888	859,116	557,651
Maryland	12.40	9.82	8.40	235,104	228,119	173,040
Michigan	12.90	10.10	8.60	1,102,176	875,670	851,228
Minnesota	12.90	9.96	8.60	3,586,716	3,004,982	3,246,500
Mississippi	13.20	11.00	9.85	1,208,328	1,252,680	1,028,537
Missouri	13.10	9.99	9.00	2,645,676	2,596,751	1,632,960
Nebraska	12.70	9.73	8.55	3,240,977	2,800,489	2,613,393
New Jersey	12.40	9.85	8.45	43,102	44,640	27,851
New York	12.80	9.69	8.75	170,803	141,009	113,251
North Carolina	13.10	10.20	8.50	636,333	705,840	486,880
North Dakota	12.40	9.49	8.40	1,751,066	1,921,867	1,561,560
Ohio	13.00	10.30	8.85	2,889,315	2,536,118	2,097,450
Oklahoma	12.90	9.90	8.70	131,812	101,178	101,138
Pennsylvania	13.00	9.80	8.30	353,535	271,313	209,990
South Carolina	13.30	10.30	8.60	117,506	158,620	90,558
South Dakota	12.50	9.37	8.40	2,318,625	2,154,632	1,978,368
Tennessee	13.00	10.60	9.30	936,975	785,036	735,816
Texas	12.60	9.55	8.40	29,560	49,641	25,116
Virginia	13.00	9.90	9.05	300,300	250,272	193,580
West Virginia	12.90	10.00	9.00	13,197	13,260	11,232
Wisconsin	12.80	10.00	8.50	773,760	787,600	786,803
United States	13.00	10.10	8.80	43,582,901	39,474,861	34,535,320

NASS, Crops Branch, (202) 720–2127.

Table 3-35.—International oilseed, soybean: Area and production in specified countries, 2013/2014–2015/2016

Country	Area			Production		
	2013/2014	2014/2015	2015/2016	2013/2014	2014/2015	2015/2016
	1,000 hectares	*1,000 hectares*	*1,000 hectares*	*1,000 metric tons*	*1,000 metric tons*	*1,000 metric tons*
Argentina	19,400	19,300	20,000	53,500	61,400	58,500
Bolivia	1,000	1,082	1,280	2,400	2,650	3,100
Brazil	30,100	32,100	33,300	86,700	96,200	100,000
Canada	1,860	2,235	2,200	5,359	6,049	6,235
China	6,791	6,800	6,550	11,950	12,150	12,000
India	12,200	10,908	11,650	9,500	8,700	8,000
Paraguay	3,255	3,240	3,400	8,190	8,100	8,800
Russia	1,202	1,907	2,030	1,517	2,362	2,640
Ukraine	1,351	1,800	2,125	2,774	3,900	3,925
Uruguay	1,312	1,333	1,120	3,300	3,109	3,110
Others	3,675	4,011	3,974	6,166	7,067	6,962
Total Foreign	82,146	84,716	87,629	191,356	211,687	213,272
United States	30,858	33,423	33,109	91,389	106,878	106,934
Total	113,004	118,139	120,738	282,745	318,565	320,206

FAS, Office of Global Analysis, (202) 720–6301. Prepared or estimated on the basis of official USDA production, supply, and distribution statistics from foreign governments.

Table 3-36.—Soybeans: United States exports by country of destination, 2014–2016

Country of destination	2014	2015	2016
	Metric tons	*Metric tons*	*Metric tons*
China	30,827,882	27,317,811	35,848,523
Mexico	3,550,945	3,581,452	3,639,647
Indonesia	1,953,744	1,886,972	2,574,699
Japan	1,835,802	2,395,121	2,360,086
Netherlands	946,600	1,119,010	1,961,397
Taiwan	1,444,639	1,345,043	1,518,637
Germany(*)	1,151,046	2,191,796	1,193,065
Thailand	480,009	551,315	908,865
Spain	1,123,706	1,041,898	895,232
Vietnam	681,726	650,162	859,957
Bangladesh	327,405	768,685	602,408
Korea, South	693,909	496,118	520,364
Colombia	314,770	517,011	491,904
Malaysia	300,095	320,499	389,446
Tunisia	236,679	152,036	362,771
Egypt	640,505	418,100	324,560
Costa Rica	134,431	269,409	307,252
Canada	282,236	210,992	304,089
Pakistan	0	314,363	300,653
France(*)	122,599	104,165	272,466
Philippines	104,975	108,783	265,699
United Kingdom	255,842	200,185	229,897
Saudi Arabia	207,080	280,840	214,841
Italy(*)	101	50,090	201,452
Turkey	785,281	458,642	157,369
Russia	329,945	510,507	155,547
Iran	0	62,999	142,598
Peru	51,219	162,789	141,731
Israel(*)	124,974	73	74,141
Rest of World	658,599	672,237	431,559
World Total	49,566,745	48,159,102	57,650,854

(*) Denotes a country that is a summarization of its component countries.
FAS, Office of Global Analysis, (202) 720–6301. Data Source: U.S. Census Bureau Trade Data.

Table 3-37.—Soybean oil: United States exports by country of destination, 2014–2016

Country of destination	2014	2015	2016
	Metric tons	Metric tons	Metric tons
Mexico	197,882	231,865	257,374
China	150,112	17,259	143,269
Dominican Republic	141,046	111,377	108,001
Korea, South	36,384	48,930	82,948
Colombia	48,890	32,468	75,087
Venezuela	4,992	81,919	50,090
Guatemala	41,481	42,839	42,977
Peru	64,869	116,053	41,248
Jamaica	18,012	24,301	31,328
Iran	0	0	23,054
Morocco	44,402	81,474	19,005
Egypt	9,001	6,014	18,499
Canada	29,251	26,942	16,369
Cuba	0	0	15,200
Algeria	0	10,409	14,700
Nicaragua	45,761	21,713	12,610
Trinidad and Tobago	8,883	10,480	11,014
Pakistan	0	13,101	7,253
Panama	6,163	3,954	6,308
Senegal	4,180	10,000	4,759
Japan	4,180	2,584	3,151
United Arab Emirates	2,549	5,958	3,065
Libya	0	0	3,000
Haiti	295	535	2,754
Bahamas, The	2,613	2,221	2,148
Lebanon	1,892	1,657	1,670
Saudi Arabia	943	6,102	1,494
Kuwait	562	820	897
Hong Kong	6,575	962	865
Rest of World	26,217	50,119	6,581
World Total	897,133	962,055	1,006,717

(*) Denotes a country that is a summarization of its component countries.   Users should use cautious interpretation on quantity reports using mixed units of measure.   Quantity line items will only include statistics on the units of measure that are equal to, or are able to be converted to, the assigned unit of measure of the grouped commodities.
FAS, Office of Global Analysis, (202) 720–6301.

Table 3-38.—Soybean cake and meal: United States exports by country of destination, 2014–2016

Country of destination	2014	2015	2016
	Metric tons	Metric tons	Metric tons
Mexico	1,602,069	2,014,094	2,170,142
Philippines	1,104,343	1,509,177	1,846,745
Canada	987,980	820,230	769,452
Colombia	400,886	759,258	720,861
Dominican Republic	374,441	472,833	505,907
Thailand	552,392	802,493	402,102
Guatemala	306,873	362,685	356,692
Ecuador	364,017	371,322	350,931
Venezuela	842,618	540,926	311,944
Peru	12,090	203,017	291,024
Honduras	192,404	223,776	233,462
Vietnam	368,539	319,107	211,016
Panama	151,185	192,767	210,521
El Salvador	162,775	180,595	199,060
Japan	212,173	174,980	160,271
Morocco	135,914	139,029	157,119
Sri Lanka	60,534	128,180	149,028
Bangladesh	44,548	55,759	144,476
Saudi Arabia	122,691	72,711	126,687
Burma	8,310	57,077	125,763
Egypt	97,321	166,257	121,052
Jamaica	103,681	109,510	117,451
Nicaragua	74,550	123,399	116,756
Cuba	129,437	127,940	97,599
Turkey	154,321	30,150	58,002
Belgium-Luxembourg(*)	31,407	40,302	57,919
Costa Rica	46,456	69,102	56,049
Cambodia	1,977	23,386	46,601
Trinidad and Tobago	54,271	57,954	44,728
Rest of World	1,682,374	1,350,580	444,166
World Total	10,382,577	11,498,593	10,603,523

(*) Denotes a country that is a summarization of its component countries.   Users should use cautious interpretation on quantity reports using mixed units of measure.   Quantity line items will only include statistics on the units of measure that are equal to, or are able to be converted to, the assigned unit of measure of the grouped commodities.
FAS, Office of Global Analysis, (202) 720–6301. Data Source: Department of Commerce, U.S. Census Bureau, Foreign Trade Statistics.

Table 3-39.—International oil, soybean: Production in specified countries, 2013/2014–2015/2016

Country	2013/2014	2014/2015	2015/2016
	1,000 metric tons	*1,000 metric tons*	*1,000 metric tons*
Argentina	6,785	7,687	8,710
Bolivia	400	445	475
Brazil	7,070	7,660	7,680
China	12,335	13,347	14,655
European Union	2,553	2,698	2,717
India	1,478	1,245	1,150
Mexico	720	745	745
Paraguay	640	697	783
Russia	609	654	717
Taiwan	355	393	402
Others	2,928	3,668	4,114
Total Foreign	35,873	39,239	42,148
United States	9,131	9,706	9,857
Total	45,004	48,945	52,005

FAS, Office of Global Analysis, (202) 720–6301.   Prepared or estimated on the basis of official USDA production, supply, and distribution statistics from foreign governments.

Table 3-40.—International trade: Meal, soybean in specified countries, 2013/2014–2015/2016

Country	2013/2014	2014/2015	2015/2016
	1,000 metric tons	*1,000 metric tons*	*1,000 metric tons*
Principle exporting countries:			
Argentina	24,972	28,545	32,800
Bolivia	1,608	1,625	1,650
Brazil	13,948	14,390	15,600
Canada	241	212	275
China	2,017	1,595	1,850
European Union	296	362	400
India	2,742	1,072	150
Paraguay	2,504	2,450	2,980
Russia	507	505	450
Ukraine	80	216	250
Others	769	675	699
Total Foreign	49,684	51,647	57,104
United States	10,474	11,929	10,160
Total	60,158	63,576	67,264
Principle importing countries:			
European Union	18,135	19,260	20,700
Indonesia	3,983	3,850	4,500
Iran	2,683	1,948	2,100
Japan	1,976	1,699	1,850
Korea, South	1,825	1,751	1,900
Malaysia	1,397	1,465	1,500
Mexico	1,410	1,795	2,000
Philippines	2,337	2,200	2,400
Thailand	2,665	3,017	3,050
Vietnam	3,342	4,276	4,600
Others	17,825	18,635	19,964
Total Foreign	57,578	59,896	64,564
United States	347	302	340
Total	57,925	60,198	64,904

FAS, Office of Global Analysis, (202) 720–6301.   Prepared or estimated on the basis of official USDA production, supply, and distribution statistics from foreign governments.

Table 3-41.—International trade: Oil, soybean in specified countries, 2013/2014–2015/2016

Country	2013/2014	2014/2015	2015/2016
Principle exporting countries:	*1,000 metric tons*	*1,000 metric tons*	*1,000 metric tons*
Argentina	4,087	5,093	5,925
Bolivia	371	350	370
Brazil	1,378	1,510	1,390
Canada	92	118	170
European Union	766	1,010	1,000
Malaysia	157	170	176
Paraguay	650	660	735
Russia	332	423	445
Ukraine	118	136	160
Vietnam	91	104	100
Others	531	526	524
Total Foreign	8,573	10,100	10,995
United States	851	914	953
Total	9,424	11,014	11,948
Principle importing countries:			
Algeria	629	620	640
Bangladesh	443	508	520
China	1,353	773	850
Colombia	288	315	330
Egypt	230	480	400
India	1,830	2,799	3,700
Iran	551	421	450
Morocco	444	430	460
Peru	355	355	380
Venezuela	403	473	440
Others	2,665	2,772 952	
Total Foreign	9,191	9,946	11,122
United States	75	120	136
Total	9,266	10,066	11,258

FAS, Office of Global Analysis, (202) 720–6301.   Prepared or estimated on the basis of official USDA production, supply, and distribution statistics from foreign governments.

Table 3-42.—International trade: Oil, soybean local in specified countries, 2013/2014–2015/2016

Country	2013/2014	2014/2015	2015/2016
Principle exporting countries:	*1,000 metric tons*	*1,000 metric tons*	*1,000 metric tons*
Argentina	7,190	8,590	8,540
Brazil	7,340	7,740	7,680
Total Foreign	14,530	16,330	16,220
Total	14,530	16,330	16,220

FAS, Office of Global Analysis, (202) 720–6301.   Prepared or estimated on the basis of official USDA production, supply, and distribution statistics from foreign governments.

Table 3-43.—International trade: Oilseed, soybean in specified countries, 2013/2014–2015/2016

Country	2013/2014	2014/2015	2015/2016
Principle exporting countries:	1,000 metric tons	1,000 metric tons	1,000 metric tons
Argentina	7,842	10,573	11,800
Bolivia	141	25	150
Brazil	46,829	50,612	58,000
Canada	3,469	3,853	4,200
China	215	143	200
India	183	234	200
Paraguay	4,800	4,375	4,600
Russia	24	312	350
Ukraine	1,261	2,422	2,200
Uruguay	3,195	2,850	2,850
Others	168	312	351
Total Foreign	68,127	75,711	84,901
United States	44,574	50,169	45,994
Total	112,701	125,880	130,895
Principle importing countries:			
China	70,364	78,350	82,000
Egypt	1,694	1,947	2,050
European Union	13,293	13,388	13,200
Indonesia	2,241	2,000	2,300
Japan	2,894	3,004	2,900
Mexico	3,842	3,819	3,850
Russia	2,048	1,986	2,250
Taiwan	2,335	2,520	2,550
Thailand	1,798	2,411	2,350
Turkey	1,608	2,197	2,400
Others	7,710	9,619	11,497
Total Foreign	109,827	121,241	127,347
United States	1,954	904	816
Total	111,781	122,145	128,163

FAS, Office of Global Analysis, (202) 720–6301. Prepared or estimated on the basis of official USDA production, supply, and distribution statistics from foreign governments.

Table 3-44.—International trade: Oilseed, Soybeans local in specified countries, 2013/2014–2015/2016

Country	2013/2014	2014/2015	2015/2016
Principle exporting countries:	1,000 metric tons	1,000 metric tons	1,000 metric tons
Argentina	7,433	11,800	11,400
Brazil	45,747	54,635	56,650
Total Foreign	53,180	66,435	68,050
Total	53,180	66,435	68,050
Principle importing countries:			
Argentina	2	2	2
Brazil	579	325	300
Total Foreign	581	327	302
Total	581	327	302

FAS, Office of Global Analysis, (202) 720–6301. Prepared or estimated on the basis of official USDA production, supply, and distribution statistics from foreign governments.

Table 3-45.—Sunflower, all: Area, yield, production, and value, United States, 2006–2015

Year	Area planted	Area harvested	Yield per harvested acre	Production	Price per cwt	Value of production
	1,000 acres	*1,000 acres*	*Pounds*	*1,000 pounds*	*Dollars*	*1,000 dollars*
2006	1,950.0	1,770.0	1,211	2,143,613	14.50	308,832
2007	2,070.0	2,012.0	1,426	2,868,870	21.70	614,736
2008	2,516.5	2,396.0	1,429	3,422,840	21.80	704,105
2009	2,030.0	1,953.5	1,554	3,036,460	15.10	458,959
2010	1,951.5	1,873.8	1,460	2,735,570	23.30	633,778
2011	1,543.0	1,457.8	1,398	2,038,275	29.10	589,282
2012	1,920.0	1,840.0	1,487	2,736,060	25.40	699,970
2013	1,575.5	1,464.6	1,380	2,021,765	23.50	443,296
2014	1,560.8	1,510.1	1,469	2,219,050	21.60	497,775
2015	1,859.1	1,799.4	1,625	2,923,730	22.50	559,257

NASS, Crops Branch, (202) 720–2127.

Table 3-46.—Sunflower, oil varieties: Area, yield, production, and value, United States, 2006–2015

Year	Area planted	Area harvested	Yield per harvested acre	Production	Price per cwt.	Value of production
	1,000 acres	*1,000 acres*	*Pounds*	*1,000 pounds*	*Dollars*	*1,000 dollars*
2006	1,658.0	1,514.0	1,181	1,787,966	14.10	249,848
2007	1,765.5	1,719.0	1,445	2,483,585	21.40	527,925
2008	2,163.0	2,062.0	1,452	2,993,510	19.50	572,979
2009	1,698.0	1,653.0	1,563	2,584,010	13.80	359,331
2010	1,463.0	1,422.5	1,458	2,074,500	22.60	457,135
2011	1,289.5	1,233.4	1,397	1,722,675	28.00	480,412
2012	1,658.0	1,589.8	1,484	2,359,775	24.40	580,404
2013	1,279.0	1,200.9	1,363	1,637,205	19.30	322,605
2014	1,174.0	1,139.5	1,460	1,664,090	19.30	323,469
2015	1,550.5	1,510.0	1,579	2,383,870	18.00	431,325

NASS, Crops Branch, (202) 720–2127.

Table 3-47.—Sunflower, non-oil varieties: Area, yield, production, and value, United States, 2006–2015

Year	Area planted	Area harvested	Yield per harvested acre	Production	Price per cwt.	Value of production
	1,000 acres	*1,000 acres*	*Pounds*	*1,000 pounds*	*Dollars*	*1,000 dollars*
2006	292.0	256.0	1,389	355,647	16.80	58,984
2007	304.5	293.0	1,315	385,285	22.90	86,811
2008	353.5	334.0	1,285	429,330	31.30	131,126
2009	332.0	300.5	1,506	452,450	22.10	99,628
2010	488.5	451.3	1,465	661,070	26.60	176,643
2011	253.5	224.4	1,406	315,600	33.80	108,870
2012	262.0	250.2	1,504	376.285	29.60	119,566
2013	296.5	263.7	1,458	384,560	31.30	120,691
2014	391.3	370.6	1,497	554,960	31.50	174,306
2015	308.6	289.4	1,865	539,860	23.00	127,932

NASS, Crops Branch, (202) 720–2127.

Table 3-48.—Sunflower, all: Area, yield, production, and value, by type, State and United States, 2014–2015

Type and State	Area planted		Area harvested		Yield per harvested acre	
	2014	2015	2014	2015	2014	2015
Oil:	*1,000 acres*	*1,000 acres*	*1,000 acres*	*1,000 acres*	*Pounds*	*Pounds*
California	44.0	33.0	44.0	33.0	1,300	1,300
Colorado	35.0	60.0	32.0	57.0	1,400	1,200
Kansas	45.0	57.0	42.0	53.0	1,370	1,520
Minnesota	47.0	77.0	45.0	75.0	1,450	1,650
Nebraska	27.0	29.0	25.0	27.0	1,160	1,580
North Dakota	520.0	620.0	510.0	605.0	1,340	1,470
Oklahoma	3.0	3.5	1.5	3.0	1,400	1,600
South Dakota	410.0	580.0	400.0	570.0	1,670	1,840
Texas	43.0	91.0	40.0	87.0	1,420	950
United States	1,174.0	1,550.5	1,139.5	1,510.0	1,460	1,579
Non-oil:						
California	3.5	1.4	3.5	1.4	1,350	1,300
Colorado	11.5	13	11	12	1,900	1,400
Kansas	18	27	17	25	2,000	2,200
Minnesota	15	24	14.5	23.5	1,560	1,800
Nebraska	11	20	10.5	17.5	1,750	2,100
North Dakota	145	100	139	97	1,180	1,850
Oklahoma	1.3	2.2	1.1	2	1,000	900
South Dakota	125	99	122	92	1,710	1,970
Texas	61	22	52	19	1,550	1,300
United States	391.3	308.6	370.6	289.4	1,497	1,865
Total:						
California	47.5	34.4	47.5	34.4	1,304	1,300
Colorado	46.5	73.0	43.0	69.0	1,528	1,235
Kansas	63.0	84.0	59.0	78.0	1,552	1,738
Minnesota	62.0	101.0	59.5	98.5	1,477	1,686
Nebraska	38.0	49.0	35.5	44.5	1,335	1,784
North Dakota	665.0	720.0	649.0	702.0	1,306	1,523
Oklahoma	4.3	5.7	2.6	5.0	1,231	1,320
South Dakota	535.0	679.0	522.0	662.0	1,679	1,858
Texas	104.0	113.0	92.0	106.0	1,493	1,013
United States	1,565.3	1,859.1	1,510.1	1,799.4	1,469	1,625

Type and State	Production		Marketing year average price per cwt		Value of production	
	2014	2015 [1]	2014	2015 [1]	2014	2015 [1]
Oil:	*1,000 lbs*	*1,000 lbs*	*Dollars*	*Dollars*	*1,000 dollars*	*1,000 dollars*
California	57,200	42,900	25.00	25.00	14,300	14,300
Colorado	44,800	68,400	(D)	(D)	(D)	(D)
Kansas	57,540	80,560	(D)	(D)	(D)	(D)
Minnesota	65,250	123,750	(D)	19.60	(D)	13,181
Nebraska	29,000	42,660	17.10	17.00	4,959	4,509
North Dakota	683,400	889,350	19.50	18.90	133,263	135,997
Oklahoma	2,100	4,800	18.70	17.60	393	393
South Dakota	668,000	1,048,800	18.80	17.10	125,584	124,248
Texas	56,800	82,650	(D)	18.00	(D)	(D)
Other States [1]	....	....	20.00	17.40	....	30,898
United States	1,664,090	2,383,870	19.30	18.00	323,469	323,526
Non-oil:						
California	4,725	1,820	28.00	34.00	1,323	619
Colorado	20,900	16,800	(D)	(D)	(D)	(D)
Kansas	34,000	55,000	(D)	(D)	(D)	(D)
Minnesota	22,620	42,300	(D)	22.60	(D)	9,560
Nebraska	18,375	36,750	33.00	27.80	6,064	10,217
North Dakota	164,020	179,450	31.50	29.00	51,666	52,041
Oklahoma	1,100	1,800	34.00	25.00	374	450
South Dakota	208,620	181,240	31.60	16.80	65,924	30,448
Texas	80,600	24,700	(D)	25.00	(D)	6,175
Other States [1]	....	....	31.90	....	....	....
United States	554,960	539,860	31.50	23.00	174,306	127,932
Total:						
California	75,150	61,250	27.10	25.20	20,351	15,434
Colorado	45,600	63,300	25.60	31.00	11,022	14,712
Kansas	82,000	91,540	22.30	26.00	18,896	22,660
Minnesota	69,250	87,870	22.50	20.40	15,752	21,392
Nebraska	32,975	45,055	23.60	23.50	7,779	10,573
North Dakota	600,560	848,600	23.20	22.40	132,052	190,843
Oklahoma	5,180	3,200	23.90	24.00	1,238	767
South Dakota	996,800	876,620	19.60	19.80	209,341	192,675
Texas	114,250	137,400	23.50	25.90	26,865	35,540
Other States [1]	....	....	(X)	(X)	(X)	(X)
United States	2,021,765	2,214,835	21.40	21.50	443,296	504,596

(D) Withheld to avoid disclosing data for individual operations.   (X) Not applicable.   [1] Includes data withheld above.
NASS, Crops Branch, (202) 720–2127.

Table 3-49.—International Oilseed, sunflowerseed: Area and production in specified countries, 2013/2014–2015/2016

Country	Area			Production		
	2013/2014	2014/2015	2015/2016	2013/2014	2014/2015	2015/2016
	1,000 hectares	*1,000 hectares*	*1,000 hectares*	*1,000 metric tons*	*1,000 metric tons*	*1,000 metric tons*
Argentina	1,300	1,440	1,300	2,000	3,160	2,600
Burma	570	540	540	360	450	450
China	930	949	920	2,424	2,492	2,300
European Union	4,615	4,279	4,147	9,052	8,932	7,750
India	750	550	550	670	420	460
Pakistan	822	769	701	573	513	534
Russia	6,795	6,371	6,350	9,842	8,374	9,095
South Africa	600	576	615	832	661	625
Turkey	690	530	490	1,400	1,200	1,000
Ukraine	5,300	5,300	5,235	11,600	10,200	11,300
Others	1,171	1,272	1,273	1,952	2,206	1,978
Total Foreign	23,543	22,576	22,121	40,705	38,608	38,092
United States	593	611	728	917	1,007	1,326
Total	24,136	23,187	22,849	41,622	39,615	39,418

FAS, Office of Global Analysis, (202) 720–6301.   Prepared or estimated on the basis of official USDA production, supply, and distribution statistics from foreign governments.

Table 3-50—Sunflowerseed: United States exports by country of destination, 2014–2016

Country	2014	2015	2016
	Metric tons	*Metric tons*	*Metric tons*
Spain	24,354	22,130	21,378
Mexico	14,096	15,806	19,420
Canada	24,264	21,662	18,342
Iraq	528	5,808	5,435
Turkey	11,624	5,567	4,911
Jordan	5,594	2,524	3,066
Israel(*)	4,580	2,162	2,493
Korea, South	2,095	2,889	2,381
Egypt	2,054	1,978	2,039
Greece	1,881	1,920	1,610
France(*)	3,546	2,434	1,454
Romania	2,936	780	1,188
Japan	1,112	1,297	1,104
Kuwait	4,738	506	649
Indonesia	10	3,024	513
Germany(*)	767	285	504
Morocco	888	451	454
United Kingdom	695	497	444
Colombia	34	37	390
Ecuador	630	535	376
United Arab Emirates	629	487	297
Lebanon	624	264	231
Taiwan	417	81	180
Guatemala	80	417	135
Chile	126	93	131
Poland	428	342	120
Singapore	128	54	110
Burma	0	0	83
Qatar	160	98	60
Rest of World	2,933	3,465	361
World Total	111,948	97,594	89,858

(*) Denotes a country that is a summarization of its component countries.
FAS, Office of Global Analysis, (202) 720–6301. Data Source: U.S. Census Bureau Trade Data.

Table 3-51.—Sunflowerseed oil: United States exports by country of destination, 2014–2016

Country	2014	2015	2016
	Metric tons	Metric tons	Metric tons
Canada	17,609	18,473	18,946
Mexico	4,661	4,734	9,528
Japan	3,942	3,327	3,431
Taiwan	63	587	835
Honduras	5	306	702
Vietnam	1,140	258	664
Colombia	1	11	534
China	418	254	505
Guyana	0	22	357
Netherlands	2,531	198	279
Latvia	0	0	236
Korea, South	633	8	235
Ecuador	0	0	88
India	8	0	42
Costa Rica	518	149	41
Thailand	13	28	25
Jordan	0	0	22
Indonesia	4	61	20
Venezuela	0	0	15
United Kingdom	4,094	0	15
Panama	0	0	12
Netherlands Antilles(*)	0	0	8
Malaysia	0	0	6
Singapore	48	33	5
Brazil	0	0	4
Barbados	2	0	3
South Africa	0	0	2
Uruguay	0	0	2
Ireland	0	0	2
Rest of World	3,166	554	2
World Total	38,855	29,003	36,565

(*) Denotes a country that is a summarization of its component countries.
FAS, Office of Global Analysis, (202) 720–6301.

Table 3-52.—International oil, sunflowerseed: Production in specified countries, 2013/2014–2015/2016

Country	2013/2014	2014/2015	2015/2016
	1,000 metric tons	1,000 metric tons	1,000 metric tons
Argentina	934	1,135	1,140
Bolivia	113	123	142
Burma	110	150	150
China	481	466	420
European Union	3,170	3,180	2,980
Russia	3,593	3,366	3,530
Serbia	123	161	161
South Africa	347	285	258
Turkey	845	725	595
Ukraine	4,750	4,325	4,663
Others	853	787	783
Total foreign	15,319	14,703	14,822
United States	197	166	258
Total	15,516	14,869	15,080

FAS, Office of Global Analysis, (202) 720–6301.   Prepared or estimated on the basis of official USDA production, supply, and distribution statistics from foreign governments.

Table 3-53.—Sunflowerseed cake and meal: United States exports by country of destination, 2011–2013

Country	2011	2012	2013
	Metric tons	Metric tons	Metric tons
Israel(*)	0	0	14,057
Canada	2,137	3,652	4,823
Dominican Republic	0	0	38
Mexico	825	38	18
Korea, South	0	2	0
World Total	2,962	3,693	18,936

(*) Denotes a country that is a summarization of its component countries.
FAS, Office of Global Analysis, (202) 720–6301.

Table 3-54.—Mint oil: Production and value, by State and United States, 2013–2015

State	Production			Price per pound			Value of production		
	2013	2014	2015	2013	2014	2015	2013	2014	2015
	1,000 pounds	1,000 pounds	1,000 pounds	Dollars	Dollars	Dollars	1,000 dollars	1,000 dollars	1,000 dollars
Peppermint:									
California	153	168	156	27.70	26.30	24.80	4,238	4,418	3,869
Idaho	1,550	1,544	1,596	22.30	21.50	21.40	34,565	33,196	34,154
Indiana	425	510	400	27.60	26.00	24.00	11,730	13,260	9,600
Michigan	42	(D)	(D)	27.60	(D)	(D)	1,159	(D)	(D)
Oregon	1,849	1,800	1,995	24.30	24.80	23.00	44,931	44,640	45,885
Washington	1,925	1,500	1,540	23.60	21.00	19.90	45,430	31,500	30,646
Wisconsin	171	(D)	(D)	24.60	(D)	(D)	4,207	(D)	(D)
Other States	-	197	195	(NA)	22.10	22.10	(NA)	4,350	4,307
United States	6,115	5,719	5,882	23.90	23.00	21.80	146,260	131,364	128,461
Spearmint:									
Idaho	125	156	189	20.00	21.00	19.20	2,500	3,276	3,629
Indiana	263	226	189	20.10	26.00	20.00	5,286	5,876	3,780
Michigan	119	(D)	(D)	18.00	(D)	(D)	2,142	(D)	(D)
Oregon	265	325	338	20.80	20.90	19.70	5,512	6,793	6,659
Washington	2,133	1,938	2,218	18.50	18.80	18.00	39,402	36,411	39,974
Native	1,365	1,190	1,378	17.50	17.90	17.00	23,888	21,301	23,426
Scotch	768	748	840	20.20	20.20	19.70	15,514	15,110	16,548
Wisconsin	21	(D)	(D)	21.00	(D)	(D)	441	(D)	(D)
Other States	-	139	136	(NA)	18.90	17.50	(NA)	2,621	2,380
United States	2,926	2,784	3,070	18.90	19.80	20.30	55,283	54,977	56,422

- Represents zero.   (D) Withheld to avoid disclosing data for individual operations.   (NA) Not available.
NASS, Crops Branch, (202) 720–2127.

Table 3-55.—Peppermint oil: Area, yield, production, and value, United States, 2006–2015

Year	Area harvested	Yield per harvested acre	Production	Price per pound	Value of production
	1,000 acres	Pounds	1,000 pounds	Dollars	1,000 dollars
2006	77.7	91	7,105	12.70	89,911
2007	63.6	89	5,636	13.60	76,866
2008	60.0	92	5,499	15.90	87,450
2009	69.8	91	6,379	20.10	128,118
2010	72.8	89	6,495	20.30	131,805
2011	75.5	89	6,700	23.10	154,626
2012	75.7	87	6,549	24.10	158,121
2013	68.5	89	6,115	23.90	146,260
2014	63.5	90	5,719	23.00	131,364
2015	65.2	90	5,882	21.80	128,461

NASS, Crops Branch (202), 720–2127.

Table 3-56.—Spearmint oil: Area, yield, production, and value, United States, 2006–2015

Year	Area harvested	Yield per harvested acre	Production	Price per pound	Value of production
	1,000 acres	Pounds	1,000 pounds	Dollars	1,000 dollars
2006	18.5	110	2,038	11.30	23,044
2007	19.8	126	2,493	12.60	31,495
2008	20.4	118	2,399	14.90	35,765
2009	20.5	132	2,698	16.50	44,597
2010	18.6	125	2,318	16.20	37,553
2011	17.3	132	2,286	18.60	42,438
2012	20.0	119	2,386	19.50	46,521
2013	24.5	119	2,926	18.90	55.283
2014	24.4	114	2,784	19.80	54,977
2015	27.2	113	3,070	20.30	56,422

NASS, Crops Branch, (202) 720–2127.

Table 3-57.—International oil, olive: Production in specified countries, 2013/2014–2015/2016

Country	2013/2014	2014/2015	2015/2016
	1,000 metric tons	1,000 metric tons	1,000 metric tons
Algeria	57	70	73
Australia	14	20	20
European Union	2,475	1,550	2,085
Jordan	25	23	29
Lebanon	18	21	20
Libya	15	15	15
Morocco	120	120	130
Syria	145	105	180
Tunisia	70	340	150
Turkey	140	170	143
Others	16	21	17
Total foreign	3,095	2,455	2,862
United States	5	5	5
Total	3,100	2,460	2,867

FAS, Office of Global Analysis, (202) 720–6301.   Prepared or estimated on the basis of official USDA production, supply, and distribution statistics from foreign governments.

Table 3-58.—Fats and oils: Wholesale price per pound, 2011–2015[1]

Item and market	2011	2012	2013	2014	2015
	Cents	Cents	Cents	Cents	Cents
Castor oil, No. 1, Brazilian, tanks, imported, New York	(NA)	(NA)	(NA)	(NA)	(NA)
Coconut oil, crude, tanks, f.o.b. New York	80.60	54.40	44.26	60.21	52.27
Corn oil, crude, tank cars, f.o.b. Decatur	61.37	55.74	43.13	38.36	38.24
Cottonseed oil, crude, tank cars, f.o.b. Valley	54.60	53.00	46.60	60.33	47.01
Linseed oil, raw, tank cars, Minneapolis	(NA)	(NA)	(NA)	(NA)	(NA)
Palm oil, U.S. ports, refined	55.98	48.13	41.13	39.86	32.10
Canola oil, Midwest	59.10	57.56	52.81	42.46	36.70
Soybean oil, crude, tank cars, f.o.b. Decatur	54.05	51.02	44.98	36.83	30.20
Sunflower oil, crude, Minneapolis	93.83	77.98	62.79	59.77	68.03
Tallow, edible, number 1 delivered Chicago	53.40	47.99	42.52	38.91	28.42
Tung oil, imported, drums, f.o.b. New York	(NA)	(NA)	(NA)	(NA)	(NA)

(NA) Not available.   [1] All prices are calendar year basis.
ERS, Market and Trade Economics Division, Field Crops Branch, (202) 694–5300. Compiled from the Chemical Marketing Reporter, Milling and Baking News, and the Agricultural Marketing Service.

CHAPTER IV
STATISTICS OF VEGETABLES AND MELONS

This chapter contains statistics on potatoes, sweet potatoes, and commercial vegetables and melons.

For potatoes and sweet potatoes, the estimates of area, production, value, and farm disposition pertain to the total crop and include quantities produced both for sale and for use on farms where grown. Potato statistics are shown on a within-year seasonal grouping of winter, spring, summer, and fall crops, by States. Some States have production in more than one seasonal group.

Commercial vegetables for fresh market include 24 principal vegetable and melon crops in the major producing States. These estimates relate to crops which are grown primarily for sale and do not include vegetables and melons produced in farm and non-farm gardens. The bulk of the production of the principal vegetable and melon crops is for consumption in the fresh state. The commercial estimates of these crops include local market production from areas near consuming centers as well as production from well recognized commercial areas which specialize in producing supplies for shipment to distant markets. For fresh market vegetables and melons, value per unit and total value are on a f.o.b. basis.

For processing vegetables, the estimates of area, production, and value for each of 10 crops relate to production used by commercial canners, freezers, and other processors, except dehydrators. These estimates include raw products grown by processors themselves and those grown under contract or purchased on the open market. This production and the actual area harvested are not duplicated in the fresh market estimates for the same commodities. The production of those vegetables used for processing for which regular processing estimates are not made is included in the fresh market estimates. The processed segment of production for asparagus, broccoli, and cauliflower, combined with fresh market production during the year, is published separately at the end of the season. For processed vegetables, value per unit and total value are at processing plant door.

STATISTICS OF VEGETABLES AND MELONS

Table 4-1.—Vegetables, commercial: Area, production, and value of principal crops, United States, 2006-2015

Year	Area [1]	
	Fresh market [2]	Processing [3]
	Acres	Acres
2006	1,829,840	1,253,350
2007	1,784,290	1,249,230
2008	1,710,884	1,224,410
2009	1,681,774	1,264,849
2010	1,685,155	1,142,000
2011	1,624,535	1,057,480
2012	1,626,510	1,135,605
2013	1,695,810	1,053,160
2014	1,633,690	1,089,920
2015	1,624,710	....

Year	Production [4]	
	Fresh market [2]	Processing [3]
	1,000 cwt	Tons
2006	460,812	15,910,370
2007	459,421	17,799,410
2008	445,551	17,473,670
2009	437,980	19,518,030
2010	438,324	17,599,990
2011	426,656	17,001,540
2012	429,988	18,225,775
2013	418,009	17,168,580
2014	409,104	....
2015	400,042	15,696,690

Year	Value [5]	
	Fresh market [2]	Processing [3]
	1,000 dollars	1,000 dollars
2006	10,150,783	1,343,800
2007	10,047,825	1,609,544
2008	10,297,645	1,896,796
2009	10,757,613	2,102,214
2010	10,777,069	1,659,332
2011	10,532,246	1,771,192
2012	9,797,128	2,002,481
2013	11,488,248	2,092,316
2014	10,729,140	2,171,607
2015	11,899,035	....

[1] Area for fresh market is area for harvest, including any partially harvested or not harvested because of low prices or other economic factors. Area for processing is area harvested.   [2] Area, production, and farm value of the following crops for which regular seasonal estimates are prepared in major producing States: Artichokes, asparagus, snap beans, broccoli, cabbage, cantaloups, carrots, cauliflower, celery, sweet corn, cucumbers, garlic, honeydew melons, head lettuce, leaf lettuce, romaine lettuce, onions, bell peppers, chile peppers, pumpkins, spinach, squash, tomatoes, and watermelons.   [3] Area, production, and farm value of the following 8 crops in all States: Lima beans, snap beans, carrots, sweet corn, cucumbers (pickles), green peas, spinach, and tomatoes. Production of other vegetables processed included in fresh market series of estimates.   [4] Production for fresh market excludes some quantities not marketed because of low prices or other economic factors.   [5] Value for all fresh market vegetables.   For processing vegetables, value at processing plant door.
NASS, Crops Branch, (202) 720–2127.

Table 4-2.—Vegetables, commercial: Area of principal crops, by State and United States, 2013–2015 [1]

State	Fresh market [2]			Processing [3]		
	2013	2014	2015	2013	2014	2015
	Acres	Acres	Acres	Acres	Acres	Acres
Alabama	5,100	5,000	4,900			
Arizona	105,900	110,000	112,800			
Arkansas	2,600	2,500	2,300			
California	766,500	727,900	711,900	279,500	308,460	313,550
Colorado	13,400	12,100	11,750			
Connecticut	3,600	4,000	3,600			
Delaware	6,100	6,000	5,900	26,150	27,800	28,000
Florida	161,600	150,200	154,500			
Georgia	87,900	87,200	90,100	3,600	2,050	2,750
Idaho	9,000	6,900	8,000			
Illinois	25,000	25,900	20,800	29,200	30,100	36,100
Indiana	15,300	15,520	14,400			
Maine	1,500	1,600	1,400			
Maryland	9,040	9,180	8,670	10,440	12,990	11,490
Massachusetts	4,700	4,300	3,300			
Michigan	49,400	48,500	47,900			
Minnesota				183,710	178,030	158,000
Mississippi	2,000	2,100	1,800			
Missouri	3,000	2,700	2,600		(D)	(D)
Nevada	4,050	4,150	4,200			
New Hampshire	1,400	1,400	1,300			
New Jersey	23,000	22,900	22,600	4,100	4,200	3,800
New Mexico	14,700	12,800	13,300			
New York	63,430	58,470	56,210	30,050	39,320	39,510
North Carolina	28,700	30,500	28,300			
Ohio	30,640	30,800	29,100	12,100	10,900	9,700
Oklahoma	2,200	2,600	4,000			
Oregon	27,550	27,750	29,400	56,650	53,000	51,800
Pennsylvania	22,280	21,860	18,730			
Rhode Island	650	690	(D)			
South Carolina	15,000	15,500	14,500			
Tennessee	8,500	10,130	10,990			
Texas	54,850	52,230	49,050	8,800	8,500	8,270
Utah	1,400	1,400	1,300			
Vermont	830	910	(D)			
Virginia	8,490	8,370	8,360	(D)	(D)	
Washington	39,600	39,700	54,100	117,090	115,900	127,930
Wisconsin	9,100	8,100	8,000	175,600	177,700	172,220
Other States [4]	(X)	(X)	1,290	116,170	128,820	112,630
United States	1,628,010	1,571,860	1,561,350	1,053,160	1,097,770	1,075,750

(D) Withheld to avoid disclosing data for individual operations. [1] Area for fresh market and for processing is area harvested. [2] Area of the following crops for which regular seasonal estimates are prepared in major producing States: Artichokes, asparagus, snap beans, broccoli, cabbage, cantaloups, carrots, cauliflower, celery, sweet corn, cucumbers, garlic, honeydew melons, head lettuce, leaf lettuce, romaine lettuce, onions, bell peppers, Chile pepper, spinach, tomatoes, and watermelons. [3] Includes Lima beans, snap beans, carrots, sweet corn, cucumbers (pickles), green peas, spinach, and tomatoes. Other vegetables processed (dual purpose) included in fresh market series of estimates. [4] Other States include Alabama, Arkansas, Florida, Iowa, Idaho, Indiana, Massachusetts, Michigan, North Carolina, Pennsylvania, and South Carolina.

NASS, Crops Branch, (202) 720–2127.

Table 4-3.—Vegetables, commercial: Production of principal crops, by State and United States, 2013–2015

State	Fresh market [1]			Processing [2]		
	2013	2014	2015	2013	2014	2015
	1,000 cwt	*1,000 cwt*	*1,000 cwt*	*Tons*	*Tons*	*Tons*
Alabama	746	957	860			
Arizona	29,829	31,816	29,510			
Arkansas	486	473	490			
California	214,294	209,678	207,069	12,258,850	14,150,110	14,483,930
Colorado	5,234	4,601	4,462			
Connecticut	234	260	209			
Delaware	1,218	1,190	1,150	88,490	104,400	99,190
Florida	31,727	29,317	31,177			
Georgia	18,780	16,258	18,540	16,020	11,020	15,390
Idaho	7,380	5,658	5,760			
Illinois	5,927	8,227	3,759	103,430	148,420	153,950
Indiana	3,377	3,952	3,018			
Maine	83	112	85			
Maryland	1,379	1,484	1,398	49,500	55,620	50,780
Massachusetts	291	284	251			
Michigan	8,015	8,153	8,054			
Minnesota				915,250	858,650	905,855
Mississippi	400	378	315			
Missouri	843	837	572		(D)	(D)
Nevada	2,161	1,364	1,369			
New Hampshire	97	91	88			
New Jersey	3,863	3,977	3,789	19,220	19,490	16,710
New Mexico	3,923	4,234	5,142			
New York	11,585	10,097	9,658	129,650	175,150	174,320
North Carolina	5,604	4,284	5,303			
Ohio	4,403	4,599	4,192	205,500	194,950	133,690
Oklahoma	242	364	540			
Oregon	14,922	15,750	15,555	332,070	299,650	316,040
Pennsylvania	2,436	2,612	2,168			
Rhode Island	36	46	(D)			
South Carolina	4,103	3,547	4,173			
Tennessee	1,397	1,677	1,489			
Texas	13,119	11,337	9,821	75,540	75,270	76,080
Utah	763	742	715			
Vermont	46	46	(D)			
Virginia	1,440	1,315	1,252	(D)	(D)	
Washington	15,524	17,479	20,579	900,650	904,550	982,660
Wisconsin	2,102	1,908	2,018	1,117,870	1,068,500	1,035,130
Other States [3]		(X)	60	956,540	1,182,880	927,870
United States	418,009	409,104	404,590	17,168,580	19,248,660	19,371,595

(D) Withheld to avoid disclosing data for individual operations.   [1] Production of the following crops for which regular seasonal estimates are prepared in major producing States: Artichokes, asparagus, snap beans, broccoli, cabbage, cantaloups, carrots, cauliflower, celery, sweet corn, cucumbers, garlic, honeydew melons, head lettuce, leaf lettuce, romaine lettuce, onions, bell peppers, Chile peppers, spinach, squash, tomatoes, and watermelons.   [2] Includes Lima beans, snap beans, carrots, sweet corn, cucumbers (pickles), green peas, spinach, and tomatoes. Other vegetables processed (dual purpose) included in fresh market series of estimates.   [3] Other States include Alabama, Arkansas, Florida, Iowa, Idaho, Indiana, Massachusetts, Michigan, North Carolina, Pennsylvania, and South Carolina.
NASS, Crops Branch, (202) 720–2127.

Table 4-4.—Vegetables, commercial: Value of principal crops, by State and United States, 2013–2015

State	Fresh market [1]			Processing [2]		
	2013	2014	2015	2013	2014	2015
	1,000 dollars	*1,000 dollars*	*1,000 dollars*	*1,000 dollars*	*1,000 dollars*	*1,000 dollars*
Alabama	15,998	20,111	27,376			
Arizona	1,071,047	685,608	1,040,269			
Arkansas	11,232	13,646	13,970			
California	6,188,023	6,328,525	7,262,285	1,134,826	1,434,537	1,382,592
Colorado	100,126	67,591	77,462			
Connecticut	9,126	9,620	10,450			
Delaware	19,218	19,754	18,451	21,562	29,255	27,293
Florida	1,253,028	1,044,176	1,102,153			
Georgia	460,639	405,264	470,261	5,200	4,354	6,673
Idaho	71,478	44,083	49,803			
Illinois	59,810	64,205	21,798	25,441	22,608	19,587
Indiana	68,366	53,230	53,844			
Maine	4,233	5,040	3,825			
Maryland	23,992	26,323	24,036	9,739	12,764	11,662
Massachusetts	15,132	14,200	10,040			
Michigan	187,289	173,711	209,590			
Minnesota				183,765	134,945	125,256
Mississippi	4,200	4,423	3,780			
Missouri	8,683	7,868	4,976	(D)	(D)	(D)
Nevada	69,213	45,012	45,177			
New Hampshire	5,723	5,005	4,840			
New Jersey	124,187	132,080	156,510	4,526	4,765	3,745
New Mexico	90,397	95,917	147,714			
New York	344,310	289,250	276,984	32,070	53,502	34,841
North Carolina	140,146	112,377	142,874			
Ohio	139,070	127,182	135,446	34,811	29,379	19,289
Oklahoma	3,328	4,295	7,290			
Oregon	160,064	132,338	177,341	54,157	53,692	57,835
Pennsylvania	74,462	80,583	81,045			
Rhode Island	2,160	2,438	(D)			
South Carolina	73,396	78,887	91,541			
Tennessee	53,758	76,184	76,550			
Texas	241,054	213,212	162,677	14,749	16,306	12,405
Utah	8,373	7,835	8,385			
Vermont	2,760	3,128	(D)			
Virginia	53,774	43,572	52,668	(D)	(D)	(D)
Washington	291,188	259,064	435,325	132,317	115,646	129,801
Wisconsin	39,265	33,403	40,818	255,146	157,920	157,759
Other States [3]			3,880	184,007	256,746	207,998
United States	11,488,248	10,729,140	12,451,434	2,092,316	2,326,419	2,196,736

(D) Withheld to avoid disclosing data for individual operations.   [1] Value of the following crops for which regular seasonal estimates are prepared in major producing States: Artichokes, asparagus, snap beans, broccoli, cabbage, cantaloups, carrots, cauliflower, celery, sweet corn, cucumbers, garlic, honeydew melons, head lettuce, leaf lettuce, romaine lettuce, onions, bell peppers, Chile peppers, spinach, tomatoes, and watermelons.   [2] Includes Lima beans, snap beans, carrots, sweet corn, cucumbers (pickles), green peas, spinach, and tomatoes. Other vegetables processed (dual purpose) included in fresh market series of estimates.   [3] Other States include Alabama, Arkansas, Florida, Iowa, Idaho, Indiana, Massachusetts, Michigan, North Carolina, Pennsylvania, and South Carolina.

NASS, Crops Branch, (202) 720–2127.

Table 4-5.— Artichokes for fresh market and processing, commercial crop: Area, yield, production, value, and total value, United States, 2006-2015

Year	Area harvested	Yield per acre	Production	Value	
				Per cwt	Total
	Acres	Cwt	1,000 cwt	Dollars	1,000 dollars
2006	8,700	135	1,175	42.00	49,350
2007	9,600	110	1,056	55.00	58,080
2008	8,800	130	1,144	47.80	54,683
2009	8,600	125	1,075	56.20	60,415
2010	7,200	120	864	50.20	43,373
2011	7,400	135	999	51.10	51,049
2012	7,300	145	1,059	54.30	57,504
2013	7,100	135	959	61.00	58,499
2014	7,300	130	949	57.60	54,662
2015	6,500	135	878	83.00	72,874

NASS, Crops Branch, (202) 720–2127.

Table 4-6.—Artichokes for fresh market and processing: Area, production, and value per hundredweight, California, 2013–2015

Crop	Area harvested			Production			Value per unit		
	2013	2014	2015	2013	2014	2015	2013	2014	2015
	Acres	Acres	Acres	1,000 cwt	1,000 cwt	1,000 cwt	Dollars per cwt	Dollars per cwt	Dollars per cwt
California	7,300	7,100	6,500	959	949	878	61.00	57.60	83.00

NASS, Crops Branch, (202) 720–2127.

Table 4-7.—Asparagus, commercial crop: Area, yield, production, value per hundredweight and per ton, and total value, United States, 2010-2015

Year	Total crop				
	Area for harvest	Yield per acre	Production	Value [1]	
				Per cwt	Total
	Acres	Cwt	1,000 cwt	Dollars	1,000 dollars
2010	28,000	29	799	114.00	90,777
2011	26,800	31	820	112.00	91,896
2012	25,300	30	761	110.00	83,662
2013	24,500	31	762	121.00	92,448
2014	23,800	31	743	99.30	73,782
2015	23,500	29	685	110.00	75,648

Year	Fresh market			Processing		
	Production	Value [1]		Production	Value [2]	
		Per cwt	Total		Per ton	Total
	1,000 cwt	Dollars	1,000 dollars	Tons	Dollars	1,000 dollars
2010	679	122.00	82,597	6,000	1,360.00	8,180
2011	650	122.00	79,047	8,500	1,510.00	12,849
2012	620	117.00	72,592	7,050	1,570.00	11,070
2013	620	131.00	81,332	7,100	1,570.00	11,116
2014	568	106.00	60,432	8,750	1,530.00	13,350
2015	530	119.00	62,870	7,750	1,650.00	12,778

[1] Price and value at point of first sale. [2] Price and value at processing plant door.
NASS, Crops Branch, (202) 720–2127.

Table 4-8.—Asparagus for Fresh Market and Processing, commercial crop: Area planted and harvested, yield, production, price and value - by State and United States, 2013–2015

State	Area planted			Area harvested		
	2013	2014	2015	2013	2014	2015
	Acres	Acres	Acres	Acres	Acres	Acres
California	12,000	11,500	11,300	368	341	290
Michigan	10,300	10,300	10,000	206	220	228
Washington	4,500	4,000	4,000	188	182	167
United States	26,800	25,800	25,300	762	743	685

State	Yield per acre			Production		
	2013	2014	2015	2013	2014	2015
	Cwt	Cwt	Cwt	1,000 cwt	1,000 cwt	1,000 cwt
California	32	31	27	368	341	290
Michigan	23	24	26	206	220	228
Washington	47	48	44	188	182	167
United States	31	31	29	762	743	685

State	Price per cwt			Value of Production		
	2013	2014	2015	2013	2014	2015
	Dollars	Dollars	Dollars	1,000 dollars	1,000 dollars	1,000 dollars
California	152.00	116.00	139.00	55,936	39,556	40,310
Michigan	90.50	93.20	86.60	18,640	20,505	19,754
Washington	95.10	75.40	93.30	17,872	13,721	15,584
United States	121.00	99.30	110.00	92,448	73,782	75,648

NASS, Crops Branch, (202) 720–2127.

Table 4-9.—Asparagus for Fresh Market and Processing: Production, price and value by utilization - States and United States, 2013–2015

Utilization and State	Production		
	2013	2014	2015
	1,000 cwt	1,000 cwt	1,000 cwt
Fresh market			
California [1]	368	341	290
Other States [2]	252	227	240
United States	620	568	530
	tons	tons	tons
Processing			
Other States [2]	7,100	8,750	7,750
United States	7,100	8,750	7,750
Canning	2,200	2,000	2,300
Freezing	4,900	6,750	5,450

Utilization and State	Price per unit			Value of production		
	2013	2014	2015	2013	2014	2015
	Dollars per cwt	Dollars per cwt	Dollars per cwt	1,000 dollars	1,000 dollars	1,000 dollars
Fresh market						
California [1]	152.00	116.00	139.00	55,936	39,556	40,310
Other States [2]	101.00	92.00	94.00	25,396	20,876	22,560
United States	131.00	106.00	119.00	81,332	60,432	62,870
	Dollars per ton	Dollars per ton	Dollars per ton	1,000 dollars	1,000 dollars	1,000 dollars
Processing						
Other States [2]	1,570.00	1,530.00	1,650.00	11,116	13,350	12,778
United States	1,570.00	1,530.00	1,650.00	11,116	13,350	12,778
Canning	1,720.00	1,680.00	1,620.00	3,784	3,360	3,726
Freezing	1,500.00	1,480.00	1,660.00	7,332	9,990	9,052

[1] Includes a small amount of processing asparagus.   [2] Other States include Michigan and Washington.
NASS, Crops Branch, (202) 720–2127.

Table 4-10.—Lima beans for processing, commercial crop: Area, yield, production, value per ton, and total value, United States, 2006-2015

Year	Area harvested	Yield per acre	Production	Value [1]	
				Per ton	Total
	Acres	Tons	Tons	Dollars	1,000 dollars
2006	43,050	1.31	56,330	398.00	22,444
2007	39,330	1.35	53,100	423.00	22,450
2008	38,270	1.28	49,150	500.00	24,584
2009	34,740	1.38	48,030	519.00	24,945
2010	42,430	1.47	62,230	473.00	29,456
2011	30,960	1.42	45,780	525.00	23,700
2012	32,900	1.85	61,015	524.00	31,950
2013	28,930	1.68	48,620	582.00	28,312
2014	27,760	1.82	50,400	580.00	29,236
2015	29,200	1.72	50,255	596.00	29,955

[1] Price and value at processing plant door.
NASS, Crops Branch, (202) 720-2127.

Table 4-11.—Snap beans for fresh market, commercial crop: Area, yield, production, value hundredweight, and total value, United States, 2006-2015

Year	Area harvested	Yield per acre	Production	Value [1]	
				Per cwt	Total
	Acres	Cwt	1,000 cwt	Dollars	1,000 dollars
2006	93,900	66	6,213	50.00	310,420
2007	96,400	67	6,502	61.20	397,611
2008	90,600	65	5,849	53.30	312,010
2009	88,300	55	4,823	53.50	257,990
2010	85,800	56	4,770	60.40	287,885
2011	83,500	55	4,591	59.50	273,201
2012	76,700	57	4,340	62.70	271,922
2013	73,100	58	4,252	68.90	293,073
2014	68,600	54	3,720	60.20	224,052
2015	71,170	56	3,952	60.00	236,993

[1] Price and value at point of first sale.
NASS, Crops Branch, (202) 720-2127.

Table 4-12.—Snap beans for fresh market: Area, production, and value per hundredweight, by State and United States, 2013–2015

State	Area harvested			Production			Value per unit		
	2013	2014	2015	2013	2014	2015	2013	2014	2015
	Acres	*Acres*	*Acres*	*1,000 cwt*	*1,000 cwt*	*1,000 cwt*	*Dollars per cwt*	*Dollars per cwt*	*Dollars per cwt*
California	5,700	4,200	4,400	570	462	528	63.20	78.10	70.90
Florida	28,800	26,600	27,500	1,728	1,330	1,238	80.00	58.20	61.60
Georgia	11,200	9,800	10,300	616	392	515	42.60	39.10	38.50
Maryland	1,100	1,000	870	63	58	47	34.00	39.00	41.00
Michigan	2,600	2,600	2,900	125	182	212	57.00	44.90	54.20
New Jersey	2,500	2,600	2,400	80	86	77	50.80	38.20	59.80
New York	10,200	9,800	10,700	561	617	696	89.80	84.50	84.00
North Carolina	4,100	4,300	2,800	160	172	115	67.00	38.00	31.00
South Carolina	600	600	400	30	35	11	78.00	48.70	75.00
Tennessee	4,500	5,400	7,000	243	308	448	50.00	58.00	41.00
Virginia	1,800	1,700	1,900	76	78	65	48.00	42.00	65.00
United States	73,100	68,600	71,170	4,252	3,720	3,952	68.90	60.20	60.00

NASS, Crops Branch, (202) 720–2127.

Table 4-13.—Snap beans for processing, commercial crop: Area, yield, production, value per ton, and total value, United States, 2006-2015

Year	Area harvested	Yield per acre	Production	Value [1]	
				Per ton	Total
	Acres	*Tons*	*Tons*	*Dollars*	*1,000 dollars*
2006	203,240	3.87	785,950	157.00	123,218
2007	198,770	3.79	753,730	168.00	126,620
2008	196,600	4.08	802,060	219.00	175,697
2009	196,179	4.16	816,440	191.00	156,092
2010	191,660	3.98	762,080	192.00	146,477
2011	158,550	4.20	665,870	235.00	156,563
2012	166,355	4.39	730,680	271.00	197,811
2013	149,770	4.45	666,780	320.00	213,250
2014	159,070	4.28	681,240	250.00	170,066
2015	158,920	4.81	764,900	233.00	178,578

[1] Price and value at processing plant door.
NASS, Crops Branch, (202) 720–2127.

Table 4-14.—Snap beans for processing, commercial crop: Area, production, and value per ton, by State and United States, 2013–2015

State	Area harvested			Production			Value per unit		
	2013	2014	2015	2013	2014	2015	2013	2014	2015
	Acres	*Acres*	*Acres*	*Tons*	*Tons*	*Tons*	*Dollars per ton*	*Dollars per ton*	*Dollars per ton*
Illinois	(1)	(1)	(1)	(1)	(1)	(1)	(1)	(1)	(1)
Indiana	5,000	4,700	3,600	14,520	14,100	13,620	247.00	210.00	230.00
Michigan	17,800	(1)	16,000	77,100	(1)	71,200	240.00	(1)	230.00
Minnesota	(1)	20,420	(1)	(1)	(1)	(1)	(1)	(1)	(1)
New York	18,900	(1)	(1)	62,270	81,680	(1)	264.00	338.00	(1)
Oregon	10,550	8,500	(1)	66,850	34,850	(1)	204.00	400.00	(1)
Pennsylvania	7,900	11,400	5,300	21,630	34,310	23,330	284.00	303.00	317.00
Wisconsin	59,800	64,100	65,300	298,570	312,280	329,530	340.00	185.00	202.00
Other States [2]	29,820	49,950	68,720	125,840	204,020	327,220	425.00	282.00	260.00
United States	149,770	159,070	158,920	666,780	681,240	764,900	320.00	250.00	233.00

[1] Withheld to avoid disclosing data for individual operations.   [2] Includes data withheld above and/or data for States not listed in this table.
NASS, Crops Branch, (202) 720–2127.

Table 4-15.—Broccoli, commercial crop: Area, yield, production, value per hundredweight and per ton, and total value, United States, 2006-2015 [1]

Year	Area for harvest	Yield per acre	Production	Value [2] Per cwt	Value [2] Total	Production	Value [2] Per cwt	Value [2] Total	Production	Value [3] Per ton	Value [3] Total
				Total crop			Fresh market			Processing	
	Acres	Cwt	1,000 cwt	Dollars	1,000 dollars	1,000 cwt	Dollars	1,000 dollars	Tons	Dollars	1,000 dollars
2006 ...	130,900	145	19,040	33.30	634,394	18,538	33.70	624,827	25,110	381.00	9,567
2007 ...	129,900	148	19,188	36.20	694,922	18,287	36.70	671,681	45,040	516.00	23,241
2008 ...	126,900	158	20,086	35.90	721,307	19,412	36.20	701,884	33,720	576.00	19,423
2009 ...	126,000	158	19,890	39.90	794,124	19,410	39.80	773,124	24,000	875.00	21,000
2010 ...	130,200	148	19,289	37.70	727,463	18,879	37.60	709,833	20,500	860.00	17,630
2011 ...	123,200	147	18,149	35.20	639,120	17,756	35.40	628,792	19,673	525.00	10,328
2012 ...	126,400	162	20,472	33.60	687,727	20,076	33.80	678,619	19,800	460.00	9,108
2013 ...	131,500	162	21,360	42.80	913,520	20,664	43.20	893,336	34,800	580.00	20,184
2014 ...	126,400	163	20,600	40.00	824,961	19,910	40.70	809,608	34,500	445.00	15,353
2015 ...	123,400	157	19,408	47.40	918,981	18,624	48.10	896,323	39,200	578.00	22,658

[1] Sprouting broccoli only. Does not include broccoli rabe nor heading (cauliflower) broccoli.   [2] Price and value at point of first sale.   [3] Price and value at processing plant door.
NASS, Crops Branch, (202) 720–2127.

Table 4-16.—Broccoli, commercial crop: Area, production, and value per hundredweight, and per ton, by State and United States, 2013–2015 [1]

State	Area harvested 2013	2014	2015	Production 2013	2014	2015	Value per unit 2013	2014	2015
	Acres	Acres	Acres	1,000 cwt	1,000 cwt	1,000 cwt	Dollars per cwt	Dollars per cwt	Dollars per cwt
Arizona	7,500	6,400	8,400	900	800	1,008	57.10	37.40	52.10
California	124,000	120,000	115,000	20,460	19,800	18,400	42.10	40.20	47.10
United States ...	131,500	126,400	123,400	21,360	20,600	19,408	42.80	40.00	47.40

State	Fresh market Production 2013	2014	2015	Value per unit 2013	2014	2015	Processing Production 2013	2014	2015	Value per unit 2013	2014	2015
	1,000 cwt	1,000 cwt	1,000 cwt	Dollars per cwt	Dollars per cwt	Dollars per cwt	Tons	Tons	Tons	Dollars per ton	Dollars per ton	Dollars per ton
Arizona	900	800	1,008	57.10	37.40	52.10	....	....	....	....	....	....
California	19,764	19,110	17,616	42.60	40.80	47.90	34,800	34,500	39,200	580.00	445.00	578.00
United States ...	20,664	19,910	18,624	43.20	40.70	48.10	34,800	34,500	39,200	580.00	445.00	578.00

[1] Sprouting broccoli only. Does not include broccoli rabe nor heading (cauliflower) broccoli.
NASS, Crops Branch, (202) 720–2127.

Table 4-17.—Cabbage for fresh market, commercial crop: Area, yield, production, value and total value, United States, 2006-2015

Year	Area harvested	Yield per acre	Production	Value Per cwt	Value Total
	Acres	Cwt	1,000 cwt	Dollars	1,000 dollars
2006	69,250	338	23,411	14.10	324,365
2007	69,050	346	23,886	16.40	386,373
2008	65,460	373	24,396	14.70	353,793
2009	63,000	345	21,761	15.60	333,537
2010	66,260	350	23,210	17.40	397,314
2011	58,760	343	20,134	17.60	347,381
2012	56,800	348	19,752	18.50	359,094
2013	59,480	366	21,750	20.10	429,419
2014	59,650	354	21,141	19.70	416,303
2015	55,920	360	20,113	19.20	386,085

NASS, Crops Branch, (202) 720–2127.

Table 4-18.—Cabbage for fresh market: Area, production, and value per hundredweight, by State and United States, 2013–2015

State	Area harvested 2013	Area harvested 2014	Area harvested 2015	Production 2013	Production 2014	Production 2015 [1]	Value per unit 2013	Value per unit 2014	Value per unit 2015
	Acres	Acres	Acres	1,000 cwt	1,000 cwt	1,000 cwt	Dollars per cwt	Dollars per cwt	Dollars per cwt
Arizona	2,800	3,700	2,600	1,316	1,906	819	29.00	21.70	23.00
California	14,000	13,500	13,800	5,670	5,670	5,865	21.70	24.50	26.80
Colorado	1,200	(2)	1,500	564	(2)	653	16.40	(D)	16.90
Florida	8,300	8,800	8,200	2,739	2,992	2,706	22.70	16.70	12.50
Georgia	5,100	4,700	3,700	1,734	1,598	1,258	15.00	12.00	10.50
Michigan	3,000	3,100	3,300	690	837	957	16.00	16.00	17.00
New Jersey	1,500	1,500	1,600	570	533	624	17.00	17.80	18.40
New York	8,800	8,300	8,100	3,960	3,320	3,240	19.70	21.80	18.40
North Carolina	2,900	2,800	2,400	783	644	528	19.00	15.00	13.00
Ohio	1,300	1,400	1,100	436	490	407	16.00	13.00	10.00
Pennsylvania	980	960	830	234	212	208	16.40	19.80	21.30
Texas	6,100	6,200	5,500	2,074	1,550	1,815	17.90	21.50	17.00
Virginia	400	(2)	(2)	112	(2)	(2)	14.00	(2)	(2)
Wisconsin	3,100	2,900	(2)	868	725	(2)	16.40	14.20	(2)
Other States		1,790	3,290		664	1,033		11.70	17.80
United States	59,480	59,650	55,920	21,750	21,141	20,113	20.10	19.70	19.20

[1] Includes some quantities of fall storage in New York harvested but not sold because of shrinkage and loss. [2] Not published to avoid disclosure of individual operations.
NASS, Crops Branch, (202) 720–2127.

Table 4-19.—Cantaloupes for fresh market, commercial crop: Area, yield, production, value, and total value, United States, 2006-2015

Year	Area harvested	Yield per acre	Production	Value	
				Per cwt	Total
	Acres	*Cwt*	*1,000 cwt*	*Dollars*	*1,000 dollars*
2006	79,300	246	19,498	17.20	335,526
2007	73,820	277	20,426	14.80	302,485
2008	71,730	269	19,294	18.50	356,781
2009	73,930	258	19,059	18.10	344,122
2010	73,330	256	18,808	16.20	304,896
2011	70,400	266	18,692	18.40	344,832
2012	63,430	263	16,706	19.10	319,348
2013	68,600	265	18,173	17.60	320,029
2014	58,200	234	13,612	22.10	300,634
2015	51,600	260	13,402	19.50	261,521

NASS, Crops Branch, (202) 720–2127.

Table 4-20.—Cantaloupes for fresh market: Area, production, and value per hundredweight, by State and United States, 2013–2015

State	Area harvested			Production			Value per unit		
	2013	2014	2015	2013	2014	2015	2013	2014	2015
	Acres	*Acres*	*Acres*	*1,000 cwt*	*1,000 cwt*	*1,000 cwt*	*Dollars per cwt*	*Dollars per cwt*	*Dollars per cwt*
Arizona	15,800	16,000	14,500	3,634	3,840	3,553	20.00	24.30	22.90
California	42,500	31,000	27,500	12,750	8,060	8,250	16.00	21.00	17.40
Colorado	600	400	500	120	82	123	40.00	37.90	31.00
Georgia	2,600	3,300	2,800	390	396	504	20.50	15.00	17.30
Indiana	2,000	1,800	1,700	460	396	272	25.00	15.10	28.00
Maryland	600	600	500	43	77	60	27.00	27.00	28.00
Pennsylvania	1,200	1,200	1,200	184	144	154	25.20	28.10	31.20
South Carolina	1,400	1,600	1,100	364	352	270	17.00	24.70	11.50
Texas	1,900	2,300	1,800	228	265	216	31.00	31.00	31.80
United States	68,600	58,200	51,600	18,173	13,612	13,402	17.60	22.10	19.50

NASS, Crops Branch, (202) 720–2127.

Table 4-21.—Carrots for fresh market, commercial crop: Area, production, and value per hundredweight, by State and United States, 2013-2015

State	Area harvested			Production			Value per unit		
	2013	2014	2015	2013	2014	2015	2013	2014	2015
	Acres	Acres	Acres	1,000 cwt	1,000 cwt	1,000 cwt	Dollars per cwt	Dollars per cwt	Dollars per cwt
California	62,500	65,500	63,000	20,000	20,960	19,530	29.60	28.20	32.70
Michigan	1,500	1,600	1,500	435	496	360	16.60	15.80	20.00
Texas	1,600	1,400	(D)	336	378	(D)	22.00	20.00	(D)
Other States [1]	5,800	5,700	7,050	3,484	3,545	4,392	24.70	23.20	26.40
United States	71,400	74,200	71,550	24,255	25,379	24,282	28.60	27.10	31.40

[1] Other States include Colorado, Georgia, and Washington.
NASS, Crops Branch, (202) 720-2127.

Table 4-22.—Carrots for processing, commercial crop: Area, production, and value per ton, by State and United States, 2013-2015

State	Area harvested			Production			Value per unit		
	2013	2014	2015	2013	2014	2015	2013	2014	2015
	Acres	Acres	Acres	Tons	Tons	Tons	Dollars per tons	Dollars per tons	Dollars per tons
Washington	4,200	(1)	(1)	127,680	(1)	(1)	104.00	(1)	(1)
Wisconsin	4,100	3,700	3,800	116,850	107,100	97,730	106.00	89.30	98.80
Other States [1]	5,010	7,730	7,220	107,420	214,720	196,270	162.00	130.00	135.00
United States	13,310	11,430	11,020	351,950	321,820	294,000	122.00	116.00	123.00

[1] Includes data withheld above and/or data for States not listed in this table.
NASS, Crops Branch, (202) 720-2127.

Table 4-23.—Cauliflower, commercial crop: Area, yield, production, value per hundredweight and per ton, and total value, United States, 2006-2015 [1]

Year	Total crop					Fresh market				Processing		
	Area for harvest	Yield per acre	Production	Value [2]		Production	Value [2]		Production	Value [3]		
				Per cwt.	Total		Per cwt.	Total		Per ton	Total	
	Acres	Cwt	1,000 cwt	Dollars	1,000 dollars	1,000 cwt	Dollars	1,000 dollars	Tons	Dollars	1,000 dollars	
2006	39,350	176	6,965	31.40	219,008	6,678	32.30	215,607	14,350	237.00	3,401	
2007	37,820	181	6,828	34.20	233,413	6,616	34.40	227,689	10,600	540.00	5,724	
2008	36,700	181	6,648	40.40	268,531	6,485	40.70	263,912	8,160	566.00	4,619	
2009	38,600	186	7,167	44.00	315,551	7,000	44.30	310,290	8,350	630.00	5,261	
2010	38,460	184	7,087	41.70	295,186	6,972	41.80	291,647	5,755	615.00	3,539	
2011	36,430	183	6,649	46.00	305,539	6,399	46.80	299,164	12,500	510.00	6,375	
2012	36,070	185	6,690	35.80	239,203	6,542	35.90	235,059	7,400	560.00	4,144	
2013	36,830	181	6,662	44.30	295,450	6,572	44.50	292,615	4,500	630.00	2,835	
2014	35,070	181	6,361	49.80	316,937	6,286	50.10	314,649	3,750	610.00	2,288	
2015	38,110	176	6,721	55.20	371,108	6,673	55.40	369,956	2,400	480.00	1,152	

[1] Includes heading (cauliflower) broccoli.   [2] Price and value at point of first sale.   [3] Price and value at processing plant door.   [4] Not published to avoid disclosure of individual operations.
NASS, Crops Branch, (202) 720-2127.

Table 4-24.—Cauliflower, commercial crop: Area, production, and value per hundredweight and per ton, by State and United States, 2013–2015 [1]

State	Area harvested			Production			Value per unit		
	2013	2014	2015	2013	2014	2015	2013	2014	2015
	Acres	Acres	Acres	1,000 cwt	1,000 cwt	1,000 cwt	Dollars per cwt	Dollars per cwt	Dollars per cwt
Arizona	3,800	3,600	4,500	741	720	700	56.10	54.90	77.30
California	32,600	31,300	33,500	5,868	5,580	5,976	42.70	49.20	52.70
New York	430	470	500	53	61	45	59.80	43.00	50.00
United States	36,830	35,400	35,070	6,662	6,361	6,721	44.30	49.80	55.20

State	Fresh market						Processing					
	Production			Value per unit			Production			Value per unit		
	2013	2014	2015	2013	2014	2015	2013	2014	2015	2013	2014	2015
	1,000 cwt	1,000 cwt	1,000 cwt	Dollars per cwt	Dollars per cwt	Dollars per cwt	Tons	Tons	Tons	Dollars per ton	Dollars per ton	Dollars per ton
Arizona	741	720	700	56.10	54.90	77.30	....	....	....	....	....	....
California	5,778	5,505	5,928	42.90	49.50	52.90	4,500	3,750	2,400	630.00	610.00 0.00	....
New York	53	61	45	59.80	43.00	50.00	....	....	....	....	....	....
United States	6,572	6,286	6,673	44.50	50.10	55.40	4,500	3,750	2,400	630.00	610.00	480.00

[1] Includes heading (cauliflower) broccoli.
NASS, Crops Branch, (202) 720–2127.

Table 4-25.—Celery, commercial crop: Area, production, and value per hundredweight, by State and United States, 2013–2015 [1]

State	Area harvested			Production			Value per unit		
	2013	2014	2015	2013	2014	2015	2013	2014	2015
	Acres	Acres	Acres	1,000 cwt	1,000 cwt	1,000 cwt	Dollars per cwt	Doll ars per cwt	Dollars per cwt
California	27,000	27,200	28,500	16,968	17,408	17,100	25.80	17.00	25.60
Michigan	1,800	1,700	1,600	1,035	985	930	19.70	19.10	20.30
United States	28,800	28,900	30,100	18,003	18,393	18,030	25.40	17.10	25.30

[1] Mostly for fresh market use, but includes some quantities used for processing.
NASS, Crops Branch, (202) 720–2127.

Table 4-26.—Celery, commercial crop: Area, yield, production, value per hundredweight, and total value, United States, 2006–2015 [1]

Year	Area for harvest	Yield per acre	Production	Value [2]	
				Per cwt	Total
	Acres	Cwt	1,000 cwt	Dollars	1,000 dollars
2006	27,700	694	19,230	18.20	350,454
2007	28,400	705	20,011	20.40	408,001
2008	28,300	708	20,025	18.50	369,684
2009	28,500	704	20,074	20.10	404,039
2010	28,000	712	19,923	18.60	371,153
2011	28,200	687	19,362	19.70	381,780
2012	29,000	681	19,752	18.20	358,988
2013	28,800	625	18,003	25.40	457,765
2014	28,900	636	18,393	17.10	314,134
2015	30,100	599	18,030	25.30	456,338

[1] Mostly for fresh market use, but includes quantities used for processing. [2] Price and value at point of first sale.
NASS, Crops Branch, (202) 720–2127.

Table 4-27.—Corn, sweet, commercial crop: Area, production, and value per hundredweight and per ton, by State and United States, 2013–2015

Utilization and State	Area harvested			Production			Value per unit		
	2013	2014	2015	2013	2014	2015	2013	2014	2015
Fresh market:	Acres	Acres	Acres	1,000 cwt	1,000 cwt	1,000 cwt	Dollars per cwt	Dollars per cwt	Dollars per cwt
Alabama	1,100	1,100	1,000	72	81	55	26.00	21.60	26.50
California	36,000	27,800	29,900	6,120	4,726	5,681	25.70	34.00	28.00
Colorado	4,100	4,000	3,400	656	680	612	18.70	16.20	15.20
Connecticut	3,600	4,000	3,600	234	260	209	39.00	37.00	50.00
Delaware	3,400	3,500	3,600	354	357	389	25.00	25.00	22.00
Florida	39,500	34,000	36,900	5,530	4,590	5,166	27.50	28.20	30.00
Georgia	23,000	22,400	25,700	4,025	2,912	3,598	25.60	25.50	21.40
Illinois	4,900	6,100	6,100	451	769	580	28.00	22.20	16.80
Indiana	5,300	5,300	5,100	350	440	281	42.60	29.90	43.00
Maine	1,500	1,600	1,400	83	112	85	51.00	45.00	45.00
Maryland	3,700	3,900	3,600	196	230	216	38.00	35.00	28.00
Massachusetts ..	4,700	4,300	3,300	291	284	251	52.00	50.00	40.00
Minnesota	9,000	9,200	9,000	900	828	927	26.00	24.70	26.00
New Hampshire	1,400	1,400	1,300	97	91	88	59.00	55.00	55.00
New Jersey	6,000	6,000	5,800	480	540	522	29.30	31.20	32.20
New York	22,600	18,100	16,700	2,486	1,774	1,403	31.40	23.90	22.40
North Carolina ..	4,600	4,600	4,600	423	308	391	22.50	27.00	31.00
Ohio	15,500	15,100	14,500	1,550	1,314	1,088	25.60	29.90	28.00
Oregon	3,900	4,900	6,900	534	720	1,242	20.00	24.00	9.40
Pennsylvania	12,000	11,300	9,300	852	791	623	31.40	36.30	38.50
Rhode Island	(D)	690	(D)	(D)	46	(D)	(D)	53.00	(D)
Texas	2,100	2,950	4,200	158	280	332	21.80	22.00	21.50
Vermont	(D)	910	(D)	(D)	46	(D)	(D)	68.00	(D)
Virginia	2,800	2,900	2,900	305	261	235	27.00	23.00	31.00
Washington	12,000	11,500	25,300	1,908	1,817	3,441	37.00	27.00	6.30
Wisconsin	4,200	3,500	3,700	424	392	477	29.60	25.60	29.30
Other States	1,480	(X)	1,290	82	(X)	60	60.00	(X)	65.00
United States	228,380	211,050	229,090	28,561	24,649	27,952	28.20	28.70	24.00
Processing:	Acres	Acres	Acres	Tons	Tons	Tons	Dollars per ton	Dollars per ton	Dollars per ton
Illinois	(D)	(D)	(D)	(D)	(D)	(D)	(D)	(D)	(D)
Minnesota	118,700	109,400	103,200	799,720	749,440	761,320	150.00	113.00	96.20
Oregon	(D)	23,200	21,200	(D)	220,480	198,610	(D)	126.00	140.00
Washington	64,400	69,400	78,800	641,320	692,600	722,240	121.00	108.00	106.00
Wisconsin	66,500	65,800	60,700	583,960	542,160	490,230	150.00	96.20	87.90
Other States [1] ...	65,350	44,480	43,600	526,830	363,140	315,710	137.00	138.00	111.00
United States	314,950	312,280	307,500	2,551,830	2,567,820	2,488,110	140.00	113.00	103.00

(D) Withheld to avoid disclosing data for individual operations. (X) Not applicable. [1] Includes data withheld above and/or data for States not listed in this table.
NASS, Crops Branch, (202) 720–2127.

Table 4-28.—Corn, sweet, commercial crop: Area, yield, production, value per hundredweight and per ton, and total value, United States, 2006-2015

Year	Fresh market					Processing				
	Area harvested	Yield per acre	Production	Value [1]		Area harvested	Yield per acre	Production	Value [2]	
				Per cwt	Total				Per ton	Total
	Acres	Cwt	1,000 cwt	Dollars	1,000 dollars	Acres	Tons	Tons	Dollars	1,000 dollars
2006 ...	218,300	118	25,745	23.00	590,859	384,700	8.02	3,085,550	66.80	205,965
2007 ...	234,000	122	28,504	22.70	646,374	367,600	7.88	2,897,430	81.80	236,908
2008 ...	230,280	124	28,501	26.00	739,650	360,600	7.85	2,832,490	120.00	340,486
2009 ...	230,950	122	28,077	29.50	827,941	379,500	8.52	3,234,080	104.00	335,519
2010 ...	235,800	118	27,875	25.90	721,980	335,200	8.04	2,694,210	85.30	229,786
2011 ...	224,050	118	26,336	26.70	702,217	331,650	8.04	2,664,820	116.00	309,622
2012 ...	217,350	128	27,910	26.30	735,104	360,480	8.21	2,959,350	127.00	376,694
2013 ...	228,380	125	28,561	28.20	806,488	314,950	8.10	2,551,830	140.00	357,804
2014 ...	211,050	117	24,649	28.70	708,332	312,280	8.22	2,567,820	113.00	289,573
2015 ...	229,090	122	27,952	24.00	671,933	307,500	8.09	2,488,110	103.00	255,480

[1] Price and value at point of first sale. [2] Price and value at processing plant door.
NASS, Crops Branch, (202) 720–2127.

Table 4-29.—Cucumbers for fresh market: Area, production, and value per hundredweight, by State and United States, 2013-2015

State	Area harvested			Production			Value per unit		
	2013	2014	2015	2013	2014	2015	2013	2014	2015
	Acres	*Acres*	*Acres*	*1,000 cwt*	*1,000 cwt*	*1,000 cwt*	*Dollars per cwt*	*Dollars per cwt*	*Dollars per cwt*
California	3,800	3,800	3,500	760	684	718	25.90	24.20	29.60
Florida	10,700	9,400	10,600	2,408	2,444	1,696	31.70	26.70	28.20
Georgia	6,500	6,500	6,900	1,528	1,008	1,725	28.90	23.20	23.90
Maryland	300	340	460	17	27	31	25.00	28.00	23.00
Michigan	3,500	3,400	3,700	665	680	592	21.00	20.40	22.80
New Jersey	3,200	3,100	3,200	576	667	688	22.00	22.40	23.90
New York	1,700	1,700	1,900	238	272	304	38.50	37.10	56.90
North Carolina	4,600	4,800	4,800	966	552	696	17.00	19.00	18.00
South Carolina	1,600	1,800	1,100	176	252	127	27.40	24.90	21.80
Texas	2,000	2,400	1,500	230	283	116	22.50	18.60	18.50
Virginia	460	330	320	39	26	32	27.50	44.50	40.00
United States	38,360	37,570	37,980	7,603	6,895	6,725	26.80	24.40	26.30

NASS, Crops Branch, (202) 720-2127.

Table 4-30.—Cucumbers (for pickles), commercial crop: Area, yield, production, value per ton, total value, and pickle stocks, United States, 2006-2015

Year	Processing					Pickle stocks on hand Dec. 1 [2] [3]
	Area harvested	Yield per acre	Production	Value [1]		
				Per ton	Total	
	Acres	*Tons*	*Tons*	*Dollars*	*1,000 dollars*	*Tons*
2006 ...	103,000	4.90	505,190	305.00	153,968	444,306
2007 ...	101,500	5.33	541,230	325.00	175,822	376,732
2008 ...	96,600	5.87	567,100	316.00	178,998	447,969
2009 ...	97,500	5.63	548,640	328.00	179,836	189,525
2010 ...	87,900	6.27	551,370	337.00	185,928	183,465
2011 ...	82,630	5.83	482,030	360.00	173,425	259,515
2012 ...	84,560	5.68	480,060	342.00	164,151	275,003
2013 ...	82,100	5.76	473,140	313.00	148,101	212,820
2014 ...	84,430	6.36	536,910	378.00	202,705	83,160
2015 ...	85,110	6.27	533,460	324.00	172,715	55,048

[1] Price and value at processing plant door. [2] Stocks in hands of original salters of both salt and dill pickles, sold and unsold, in tanks and barrels, on Dec. 1. [3] Includes stocks of fresh-pack pickles.
NASS, Crops Branch, (202) 720-2127.

Table 4-31.—Cucumbers (for pickles), commercial crop: Area, production, and value per ton, by State and United States, 2013-2015

State	Area harvested			Production			Value per unit		
	2013	2014	2015	2013	2014	2015	2013	2014	2015
	Acres	*Acres*	*Acres*	*Tons*	*Tons*	*Tons*	*Dollars per ton*	*Dollars per ton*	*Dollars per ton*
Florida	(D)	19,200	18,010	(D)	150,910	148,040	(D)	555.00	435.00
Indiana	1,900	2,300	(D)	5,700	11,500	(D)	350.00	320.00	(D)
Michigan	28,000	27,400	29,000	162,400	167,140	171,100	230.00	210.00	220.00
North Carolina	(D)	(D)	(D)	(D)	(D)	(D)	(D)	(D)	(D)
Ohio	7,000	6,500	5,000	52,500	40,950	27,000	325.00	300.00	260.00
South Carolina	(D)	(D)	(D)	(D)	(D)	(D)	(D)	(D)	(D)
Texas	4,000	4,200	3,950	18,800	19,570	21,800	430.00	540.00	320.00
Wisconsin	5,900	5,400	5,400	38,410	33,800	32,890	288.00	297.00	290.00
Other States [1]	35,300	19,430	23,750	195,330	113,040	132,630	371.00	418.00	355.00
United States	82,100	84,430	85,110	473,140	536,910	533,460	313.00	378.00	324.00

(D) Withheld to avoid disclosing data for individual operations. [1] Other States include Alabama, California Delaware, Georgia, and Maryland.
NASS, Crops Branch, (202) 720-2127.

Table 4-32.—Cucumbers for fresh market, commercial crop: Area, yield, production, value, and total value, United States, 2006-2015

Year	Area harvested	Yield per acre	Production	Value	
				Per cwt	Total
	Acres	Cwt	1,000 cwt	Dollars	1,000 dollars
2006	50,740	179	9,079	25.30	229,775
2007	50,960	190	9,700	24.60	238,925
2008	46,600	189	8,830	24.80	218,779
2009	45,420	204	9,269	25.60	237,147
2010	43,410	192	8,351	22.90	190,886
2011	39,000	184	7,194	26.70	191,814
2012	38,300	214	8,182	25.00	204,464
2013	38,360	198	7,603	26.80	203,894
2014	37,570	184	6,895	24.40	168,038
2015	37,980	177	6,725	26.30	176,983

NASS, Crops Branch, (202) 720–2127.

Table 4-33.—Garlic for fresh market and processing, commercial crop: Area, yield, production, value, and total value, United States, 2006-2015

Year	Area harvested	Yield per acre	Production	Value	
				Per cwt	Total
	Acres	Cwt	1,000 cwt	Dollars	1,000 dollars
2006	26,120	165	4,312	29.50	127,067
2007	24,810	165	4,104	41.20	169,218
2008	25,440	168	4,282	43.60	186,807
2009	22,230	174	3,878	49.70	192,872
2010	22,850	164	3,752	71.10	266,884
2011	25,150	167	4,204	68.20	286,820
2012	25,950	166	4,319	52.60	227,090
2013	23,900	162	3,867	60.00	231,977
2014	23,800	163	3,868	68.80	266,265
2015	23,600	162	3,825	73.00	279,294

NASS, Crops Branch, (202) 720–2127.

Table 4-34.—Garlic for fresh market and processing: Area, production, and value per hundredweight, by State and United States, 2013–2015

State	Area harvested			Production			Value per unit		
	2013	2014	2015	2013	2014	2015	2013	2014	2015
	Acres	Acres	Acres	1,000 cwt	1,000 cwt	1,000 cwt	Dollars per cwt	Dollars per cwt	Dollars per cwt
California	23,000	23,000	22,500	3,795	3,795	3,713	60.30	69.30	71.00
Nevada	550	550	(D)	61	61	(D)	33.00	33.00	(D)
Oregon	350	250	(D)	11	12	(D)	100.00	90.00	(D)
United States	23,900	23,800	23,600	3,867	3,868	3,825	60.00	68.80	73.00

NASS, Crops Branch, (202) 720–2127.

Table 4-35.—Honeydew melons, commercial crop: Area, yield, production, value per hundredweight, and total value, United States, 2006–2015

Year	Area harvested	Yield per acre	Production	Value [1]	
				Per cwt	Total
	Acres	Cwt	1,000 cwt	Dollars	1,000 dollars
2006	18,300	231	4,221	18.20	76,943
2007	17,550	236	4,144	17.70	73,517
2008	17,200	215	3,690	17.80	65,636
2009	15,100	242	3,657	15.60	57,093
2010	16,850	220	3,704	15.50	57,418
2011	15,400	236	3,628	22.10	80,008
2012	13,600	242	3,286	20.70	67,973
2013	14,450	249	3,605	20.90	75,389
2014	14,450	259	3,739	23.70	88,649
2015	13,600	261	3,553	22.70	80,548

[1] Price and value at point of first sale.
NASS, Crops Branch, (202) 720–2127.

Table 4-36.—Honeydew melons, commercial crop: Area, production, and value per hundredweight, by State and United States, 2013–2015

State	Area harvested			Production			Value per unit		
	2013	2014	2015	2013	2014	2015	2013	2014	2015
	Acres	Acres	Acres	1,000 cwt	1,000 cwt	1,000 cwt	Dollars per cwt	Dollars per cwt	Dollars per cwt
Arizona	3,300	3,500	2,300	693	805	564	21.70	32.60	24.10
California	10,500	10,500	10,700	2,730	2,835	2,889	20.20	21.00	22.20
Texas	650	450	600	182	99	100	28.60	29.00	28.20
United States	14,450	14,450	13,600	3,605	3,739	3,553	20.90	23.70	22.70

NASS, Crops Branch, (202) 720–2127.

Table 4-37.—Head lettuce, commercial crop: Area, production, and value per hundredweight, by State and United States, 2013–2015

State	Area harvested			Production			Value per unit		
	2013	2014	2015	2013	2014	2015	2013	2014	2015
	Acres	Acres	Acres	1,000 cwt	1,000 cwt	1,000 cwt	Dollars per cwt	Dollars per cwt	Dollars per cwt
Arizona	33,000	34,500	32,500	11,550	12,248	10,238	31.80	12.80	24.80
California	96,000	91,000	86,500	33,600	33,670	32,870	24.90	28.60	30.30
United States	129,000	125,500	119,000	45,150	45,918	43,108	26.70	24.40	29.00

NASS, Crops Branch, (202) 720–2127.

Table 4-38.—Head lettuce, commercial crop: Area, yield, production, value per hundredweight, and total value, United States, 2006–2015

Year	Area harvested	Yield per acre	Production	Value [1]	
				Per cwt	Total
	Acres	Cwt	1,000 cwt	Dollars	1,000 dollars
2006	178,800	350	62,494	16.90	1,054,941
2007	161,800	355	57,474	21.70	1,247,941
2008	148,700	356	52,952	20.10	1,063,132
2009	135,000	372	50,180	22.40	1,121,724
2010	132,000	380	50,120	21.10	1,057,504
2011	130,500	381	49,665	23.00	1,142,267
2012	141,600	360	50,976	17.70	901,238
2013	129,000	350	45,150	26.70	1,203,930
2014	125,500	366	45,918	24.40	1,119,736
2015	119,000	362	43,108	29.00	1,249,863

[1] Price and value at point of first sale.
NASS, Crops Branch, (202) 720–2127.

Table 4-39.—Leaf lettuce for fresh market, commercial crop: Area, yield, production, value per hundredweight, and total value, United States, 2006–2015

Year	Area harvested	Yield per acre	Production	Value[1]	
				Per cwt	Total
	Acres	*Cwt*	*1,000 cwt*	*Dollars*	*1,000 dollars*
2006	55,900	238	13,317	34.80	463,859
2007	54,600	224	12,240	30.50	373,692
2008	52,300	244	12,781	32.20	411,719
2009	49,100	241	11,845	38.70	458,765
2010	51,200	254	13,004	38.40	499,538
2011	47,400	252	11,939	34.60	413,484
2012	52,500	244	12,825	35.30	452,858
2013	55,000	230	12,650	38.30	484,036
2014	54,100	240	12,984	37.90	491,602
2015	47,800	258	12,335	57.50	709,037

[1] Price and value at point of first sale.
NASS, Crops Branch, (202) 720–2127.

Table 4-40.—Leaf lettuce for fresh market: Area, production, and value per hundredweight, by State and United States, 2013–2015

State	Area harvested			Production			Value per unit		
	2013	2014	2015	2013	2014	2015	2013	2014	2015
	Acres	*Acres*	*Acres*	*1,000 cwt*	*1,000 cwt*	*1,000 cwt*	*Dollars per cwt*	*Dollars per cwt*	*Dollars per cwt*
Arizona	7,500	8,100	9,300	1,725	1,944	2,325	54.50	31.40	63.00
California	47,500	46,000	38,500	10,925	11,040	10,010	35.70	39.00	56.20
United States	55,000	54,100	47,800	12,650	12,984	12,335	38.30	37.90	57.50

NASS, Crops Branch, (202) 720–2127.

Table 4-41.—Romaine lettuce for fresh market, commercial crop: Area, yield, production, value per hundredweight, and total value, United States, 2006–2015

Year	Area harvested	Yield per acre	Production	Value[1]	
				Per cwt	Total
	Acres	*Cwt*	*1,000 cwt*	*Dollars*	*1,000 dollars*
2006	86,400	307	26,500	22.40	593,866
2007	82,400	320	26,409	24.80	655,533
2008	77,400	294	22,774	21.00	479,006
2009	76,100	294	22,355	27.40	612,716
2010	79,300	345	27,389	23.90	655,659
2011	81,900	344	28,142	31.50	886,342
2012	88,900	318	28,256	22.40	633,489
2013	91,100	292	26,620	33.10	880,373
2014	85,400	289	24,679	31.00	763,887
2015	84,400	301	25,421	39.40	1,001,841

[1] Price and value at point of first sale.
NASS, Crops Branch, (202) 720–2127.

Table 4-42.—Romaine lettuce for fresh market: Area, production, and value per hundredweight, by State and United States, 2013–2015

State	Area harvested			Production			Value per unit		
	2013	2014	2015	2013	2014	2015	2013	2014	2015
	Acres	*Acres*	*Acres*	*1,000 cwt*	*1,000 cwt*	*1,000 cwt*	*Dollars per cwt*	*Dollars per cwt*	*Dollars per cwt*
Arizona	20,100	21,900	20,900	6,030	6,899	6,688	48.00	23.10	45.60
California	71,000	63,500	63,500	20,590	17,780	18,733	28.70	34.00	37.20
United States	91,100	85,400	84,400	26,620	24,679	25,421	33.10	31.00	39.40

NASS, Crops Branch, (202) 720–2127.

Table 4-43.—Onions, commercial crop: Area, yield, production, shrinkage and loss, value per hundredweight, and total value, United States, 2006–2015 [1]

Year	Area harvested	Yield per acre	Production [2]	Shrinkage and loss	Value [3]	
					Per cwt	Total
	Acres	*Cwt*	*1,000 cwt*	*1,000 cwt*	*Dollars*	*1,000 dollars*
2006	163,780	446	73,066	5,863	16.10	1,084,099
2007	160,080	497	79,638	6,274	11.10	816,061
2008	153,490	489	75,120	5,072	11.90	834,386
2009	151,060	500	75,599	5,170	15.00	1,054,227
2010	149,270	493	73,599	6,112	15.60	1,049,704
2011	147,630	502	74,097	5,882	10.90	742,236
2012	146,870	487	71,495	4,923	14.20	942,340
2013	143,340	486	69,652	4,858	15.00	969,170
2014	139,850	499	69,815	4,147	13.60	892,325
2015	132,900	507	67,380	4,718	17.60	993,360

[1] Mostly for fresh market use, but includes some quantities used for processing. [2] Includes storage crop onions harvested but not sold because of shrinkage and loss. [3] Price and value at point of first sale.
NASS, Crops Branch, (202) 720–2127.

Table 4-44.—Onions for fresh market: Area harvested, production, shrinkage and loss, and value per hundredweight, by State and United States, 2013–2015

Season and State	Area harvested			Production			Shrinkage and loss			Value per unit		
	2013	2014	2015	2013	2014	2015	2013	2014	2015	2013	2014	2015
	Acres	*Acres*	*Acres*	*1,000 cwt*	*1,000 cwt*	*1,000 cwt*	*1,000 cwt*	*1,000 cwt*	*1,000 cwt*	*Dollars per cwt*	*Dollars per cwt*	*Dollars per cwt*
Spring [1]:												
California	6,800	6,800	6,700	2,720	2,992	3,015	....	....	....	13.20	12.30	17.70
Georgia	11,100	11,000	11,200	2,997	2,585	2,688	....	....	....	44.10	41.90	45.30
Texas	9,700	9,000	4,000	3,492	2,340	800	....	....	....	23.80	24.00	24.60
United States	27,600	26,800	21,900	9,209	7,917	6,503	....	....	....	27.30	25.40	30.00
Summer non-storage [1]:												
California	7,800	7,800	7,500	3,820	3,744	3,750	....	....	....	6.40	7.00	7.00
Nevada	(D)	(D)	(D)	(D)	(D)	(D)	....	....	....	(D)	(D)	(D)
New Mexico	6,100	5,100	5,100	2,623	3,060	3,264	....	....	....	15.60	18.70	28.00
Texas	(D)	(D)	(D)	(D)	(D)	(D)	....	....	....	(D)	(D)	(D)
Washington [2]	2,000	2,000	2,000	900	800	760	....	....	....	27.40	34.60	45.80
Other States [3]	4,000	4,250	3,900	2,280	1,622	1,393	....	....	....	32.50	34.10	33.40
United States	19,900	19,150	18,500	9,625	9,226	9,167	....	....	....	17.10	18.00	21.70
Summer storage:												
California [4]	29,300	30,000	30,000	11,700	13,129	12,900	250	250	250	9.10	13.30	11.70
Colorado	4,000	3,600	3,200	1,700	1,800	1,472	190	198	110	18.90	19.00	20.90
Idaho	9,000	6,900	8,000	7,380	5,658	5,760	1,110	410	829	11.40	8.40	10.10
Michigan	2,700	2,500	2,400	810	925	792	113	65	79	15.90	12.90	14.10
New York	6,500	8,000	7,500	2,015	2,360	2,588	40	151	285	16.00	15.30	17.60
Oregon												
Malheur	10,900	9,300	9,200	7,848	7,440	6,992	1,020	670	909	11.30	8.40	10.10
Other	10,000	10,300	9,300	6,100	6,798	6,045	854	1,240	1,209	12.60	8.90	13.20
Washington	20,000	20,000	19,900	11,600	13,000	13,731	1,160	1,040	947	13.50	8.90	13.90
Wisconsin	1,800	1,700	1,500	810	791	645	49	44	5	16.40	17.50	16.60
Other States [3]	1,640	1,600	1,500	857	771	785	72	79	95	13.10	11.70	13.10
United States	95,840	93,900	92,500	50,820	52,672	51,710	4,858	4,858	4,718	12.10	10.80	12.80
Total summer	115,740	113,050	111,000	60,443	61,898	60,877	....	....	....	12.90	12.00	14.20
Total, spring and summer	143,340	139,850	132,900	69,652	69,815	67,380	....	....	....	15.00	13.60	17.60

(D) Withheld to avoid disclosing data for individual operations. (NA) Not available. [1] Primarily fresh market. [2] Includes Walla Walla and other non-storage onions. [3] Other States include Nevada and Texas for Summer Non-storage Onions and Ohio and Utah for Summer Storage Onions. [4] Primarily dehydrated processing.
NASS, Crops Branch, (202) 720–2127.

Table 4-45.—Onions (fresh market): Foreign trade, United States, 2006–2015[1]

Year beginning July	Imports	Domestic exports
	1,000 cwt	*1,000 cwt*
2006	8,656	6,236
2007	7,337	5,368
2008	6,639	6,130
2009	8,177	6,101
2010	8,783	7,478
2011	8,337	6,479
2012	9,306	6,223
2013	10,652	7,226
2014	10,436	6,187
2015	11,539	6,488

[1] Includes bulb onions, onion sets, and pearl onions.
ERS, Market and Trade Economics Division, (202) 694–5326. Compiled from reports of the U.S. Department of Commerce.

Table 4-46.—Peas, green (processing), commercial crop: Area, yield, production, value per ton, and total value, United States, 2006–2015

Year	Area harvested	Yield per acre	Production	Value[1] Per ton	Value[1] Total
	Acres	*Tons*	*Tons*	*Dollars*	*1,000 dollars*
2006	195,900	2.00	392,420	247.00	96,778
2007	202,000	2.07	419,080	259.00	108,702
2008	209,700	1.96	411,780	360.00	148,052
2009	205,400	2.15	441,680	319.00	140,707
2010	172,600	2.00	345,640	287.00	99,216
2011	159,100	1.85	294.920	399.00	117,682
2012	194,500	2.09	407,250	434.00	176,818
2013	178,300	2.00	356,050	427.00	151,928
2014	187,600	1.93	362,860	367.00	133,261
2015	166,200	2.47	411,320	320.00	131,568

[1] Price and value at processing plant door.
NASS, Crops Branch, (202) 720–2127.

Table 4-47.—Peas, green (processing), commercial crop: Area, production, and value per ton, by State and United States, 2013–2015[1]

State	Area harvested 2013	Area harvested 2014	Area harvested 2015	Production 2013	Production 2014	Production 2015	Value per unit 2013	Value per unit 2014	Value per unit 2015
	Acres	*Acres*	*Acres*	*Tons*	*Tons*	*Tons*	*Dollars per ton*	*Dollars per ton*	*Dollars per ton*
Delaware	5,000	5,000	(D)	10,800	11,300	(D)	263.00	340.00	(D)
Minnesota	60,800	64,000	48,500	87,630	78,150	112,450	640.00	567.00	402
Oregon	21,100	19,600	(D)	40,300	41,430	(D)	265.00	253.00	294
Washington	44,500	40,500	42,600	122,380	118,440	147,210	287.00	249.00	262
Wisconsin	36,000	36,600	34,300	75,550	70,640	81,120	521.00	378.00	334
Other States[2]	10,900	21,900	40,800	19,390	42,900	70,540	403.00	433.00	(D)
United States	178,300	187,600	166,200	356,050	362,860	411,320	427.00	367.00	320

[1] Shelled basis; 2½ pounds of peas in the shell produce approximately 1 pound of shelled peas.   [2] Other States include Illinois, Maryland, New Jersey, and New York.
NASS, Crops Branch, (202) 720–2127.

Table 4-48.—Chile peppers for fresh market and processing: Area, production, and value per hundredweight, by State and United States, 2013–2015 [1] [2]

State	Area harvested			Production			Value per unit		
	2013	2014	2015	2013	2014	2015	2013	2014	2015
	Acres	*Acres*	*Acres*	*1,000 cwt*	*1,000 cwt*	*1,000 cwt*	*Dollars per cwt*	*Dollars per cwt*	*Dollars per cwt*
Arizona	1,300	1,400	1,300	72	80	81	86.50	33.80	48.30
California	6,900	6,900	6,400	2,640	3,186	2,424	38.70	51.80	31.50
New Mexico	8,600	7,700	7,700	1,300	1,174	1,334	38.10	33.00	30.80
Texas	3,200	3,100	2,700	198	185	195	51.50	52.40	73.50
United States	20,000	19,100	18,100	4,210	4,625	4,034	39.90	46.70	33.60

[1] Chile peppers are defined as all peppers excluding bell peppers.   [2] Estimates include both fresh and dry product combined.
NASS, Crops Branch, (202) 720–2127.

Table 4-49.—Chile peppers for fresh market and processing, commercial crop: Area, yield, production, value, and total value, United States, 2006–2015 [1]

Year	Area harvested	Yield per acre	Production	Value	
				Per cwt	Total
	Acres	*Cwt*	*1,000 cwt*	*Dollars*	*1,000 dollars*
2006	28,200	169	4,779	21.90	104,775
2007	24,900	156	3,877	29.90	115,745
2008	25,200	168	4,235	26.60	112,753
2009	28,500	177	5,042	27.80	140,405
2010	23,000	202	4,644	29.80	138,586
2011	22,100	178	3,928	35.00	137,657
2012	20,400	212	4,315	36.20	156,039
2013	20,000	211	4,210	39.90	168,146
2014	19,100	242	4,625	46.70	216,107
2015	18,100	223	4,034	33.60	135,743

[1] Chile peppers are defined as all peppers excluding bell peppers.   Estimates include both fresh and dry product combined.
NASS, Crops Branch, (202) 720–2127.

Table 4-50.—Bell peppers for fresh market and processing: Area, production, and value per hundredweight, by State and United States, 2013–2015

State	Area harvested			Production			Value per unit		
	2013	2014	2015	2013	2014	2015	2013	2014	2015
	Acres	*Acres*	*Acres*	*1,000 cwt*	*1,000 cwt*	*1,000 cwt*	*Dollars per cwt*	*Dollars per cwt*	*Dollars per cwt*
California	20,000	19,400	19,400	8,465	9,432	8,831	41.70	36.10	50.00
Florida	12,300	11,900	12,200	3,075	3,094	4,392	46.00	53.10	50.20
Georgia	3,500	3,800	3,600	630	798	1,008	43.60	51.70	48.10
Michigan	1,500	1,400	1,400	375	364	392	37.00	37.00	38.00
New Jersey	3,100	2,800	2,500	977	952	763	28.30	31.00	45.00
North Carolina	1,700	2,100	2,200	425	441	517	37.00	44.00	42.00
Ohio	2,700	2,600	2,500	486	455	575	43.00	38.00	43.00
United States	44,800	44,000	43,800	14,433	15,536	16,478	41.60	40.30	48.90

NASS, Crops Branch, (202) 720–2127.

Table 4-51.—Bell peppers for fresh market and processing, commercial crop: Area, yield, production, value, hundredweight, and total value, United States, 2006-2015

Year	Area harvested	Yield per acre	Production	Value	
				Per cwt	Total
	Acres	Cwt	1,000 cwt	Dollars	1,000 dollars
2006	53,100	296	15,710	33.70	528,652
2007	54,000	298	16,100	33.10	532,799
2008	50,800	312	15,867	40.10	635,662
2009	49,400	331	16,354	34.10	557,749
2010	49,000	309	15,161	39.10	592,266
2011	46,400	351	16,275	37.80	615,740
2012	45,500	345	15,687	32.80	514,000
2013	44,800	322	14,433	41.60	599,932
2014	44,000	353	15,536	40.30	626,089
2015	43,800	376	16,478	48.90	806,115

NASS, Crops Branch, (202) 720–2127.

Table 4-52.—Potatoes: Area, yield, production, season average price, and value, United States, 2006–2015

Year	Area planted	Area harvested	Yield per harvested acre	Production	Season average price per cwt received by farmers [1]	Value of production
	1,000 acres	1,000 acres	Cwt	1,000 cwt	Dollars	1,000 dollars
2006 ...	1,139.4	1,120.2	393	440,698	7.31	3,208,632
2007 ...	1,141.9	1,122.2	396	444,875	7.51	3,339,710
2008 ...	1,059.6	1,046.9	396	415,055	9.09	3,770,462
2009 ...	1,071.2	1,044.0	414	432,601	8.25	3,557,574
2010 ...	1,026.7	1,008.9	401	404,549	9.20	3,724,821
2011 ...	1,100.7	1,078.5	399	430,037	9.41	4,044,869
2012 ...	1,154.8	1,138.5	408	464,970	8.63	4,017,245
2013 ...	1,063.9	1,050.9	414	434,652	9.75	4,237,284
2014 ...	1,062.6	1,051.1	421	442,170	8.88	3,928,211
2015 ...	1,066.1	1,054.4	418	441,205	8.76	3,865,538

[1] 2002-2007 obtained by weighting State prices by quantity sold. 2008 obtained by weighting State prices by production.
NASS, Crops Branch, (202) 720–2127.

Table 4-53.—Potatoes: Production, seed used, and disposition, United States, 2006–2015

Year	Production	Total used for seed	Used on farms where produced		Sold
			For seed, feed, and household use	Shrinkage and loss	
	1,000 cwt	1,000 cwt	1,000 cwt	1,000 cwt	1,000 cwt
2006 ...	440,698	26,374	4,750	29,639	406,309
2007 ...	444,875	24,476	4,105	29,561	411,209
2008 ...	415,055	24,595	4,139	26,438	384,478
2009 ...	432,601	24,027	4,535	29,135	398,931
2010 ...	404,549	25,100	4,227	24,996	375,326
2011 ...	430,037	26,527	4,146	27,789	398,102
2012 ...	464,970	25,656	4,850	28,505	431,615
2013 ...	434,652	25,249	4,323	26,211	404,118
2014 ...	442,170	26,259	4,192	26,762	411,216
2015 ...	441,205	25,715	4,631	26,509	410,065

NASS, Crops Branch, (202) 720–2127.

Table 4-54.—Potatoes: Area, production, and marketing year price per hundredweight received by farmers, by State and United States, 2013–2015

Season and State	Area harvested			Yield			Production		
	2013	2014	2015	2013	2014	2015	2013	2014	2015
	1,000 acres	*1,000 acres*	*1,000 acres*	*Cwt*	*Cwt*	*Cwt*	*1,000 cwt*	*1,000 cwt*	*1,000 cwt*
Spring:									
Arizona	3.4	3.5	3.5	280	310	290	952	1,085	1,015
California	26.5	24.8	26.7	410	470	385	10,865	11,656	10,280
Florida	29.5	29.3	29.6	240	240	230	7,080	7,032	6,808
North Carolina	13.5	13.5	12.7	240	210	210	3,240	2,835	2,667
United States	72.9	71.1	72.5	304	318	286	22,137	22,608	20,770
Summer:									
Delaware	1.4	1.2	(D)	280	290	(D)	392	348	(D)
Illinois	6.7	6.4	6.9	370	415	380	2,479	2,656	2,622
Kansas	4.3	4.1	3.6	350	340	335	1,505	1,394	1,206
Maryland	2.1	2.3	2.4	310	380	330	651	874	792
Missouri	9.0	7.9	8.1	300	270	305	2,700	2,133	2,471
New Jersey	2.4	1.9	(D)	230	225	(D)	552	428	(D)
Texas	17.7	20.6	18.2	460	335	375	8,142	6,901	6,825
Virginia	3.9	4.5	4.7	210	250	220	819	1,125	1,034
Other States	(NA)	(NA)	3.2	(NA)	(NA)	245	(NA)	(NA)	784
United States	47.5	48.9	47.1	363	324	334	17,240	15,859	15,734
Fall:									
California	7.3	8.3	8.4	480	470	420	3,504	3,901	3,528
Colorado	54.6	59.8	57.4	372	388	393	20,304	23,196	22,575
San Luis Valley	49.6	53.9	51.8	365	380	385	18,104	20,482	19,943
All other areas	5.0	5.9	5.6	440	460	470	2,200	2,714	2,632
Idaho	316.0	320.0	322.0	415	415	405	131,131	132,880	130,400
10 S.W. counties	17.0	16.0	16.0	520	515	500	8,840	8,240	8,000
All other counties	299.0	304.0	306.0	409	410	400	122,291	124,640	122,400
Maine	54.0	50.5	50.5	290	290	320	15,660	14,645	16,160
Massachusetts	3.9	3.6	3.6	260	285	305	1,014	1,026	1,098
Michigan	44.0	42.5	45.0	360	370	390	15,840	15,725	17,550
Minnesota	45.0	41.0	40.5	385	400	400	17,325	16,400	16,200
Montana	11.1	11.3	10.9	310	320	325	3,441	3,616	3,543
Nebraska	18.3	16.9	15.3	460	470	450	8,418	7,943	6,885
Nevada	(D)	(D)	(D)	(D)	(D)	(D)	(D)	(D)	(D)
New Mexico	(D)	(D)	(D)	(D)	(D)	(D)	(D)	(D)	(D)
New York	17.1	15.8	14.8	290	275	280	4,959	4,345	4,144
North Dakota	78.0	77.0	80.0	290	310	345	22,620	23,870	27,600
Ohio	1.8	1.5	1.2	280	280	230	504	420	276
Oregon	39.6	38.9	38.9	545	580	560	21,582	22,562	21,784
Pennsylvania	6.6	5.2	5.3	290	275	280	1,914	1,430	1,484
Rhode Island	0.5	0.5	0.6	260	245	135	130	123	81
Washington	160.0	165.0	170.0	600	615	590	96,000	101,475	100,300
Wisconsin	62.0	64.0	62.5	420	410	445	26,040	26,240	27,813
Other States [1]	10.7	9.3	7.9	457	420	415	4,889	3,906	3,280
United States	930.5	931.1	934.8	425	434	433	395,275	403,703	404,701
All Potatoes:									
United States	1,050.9	1,051.1	1,054.4	414	421	418	434,652	442,170	441,205

(D) Withheld to avoid disclosing data for individual operations. (NA) Not available.   [1] Includes data withheld above.
NASS, Crops Branch, (202) 720–2127.

Table 4-55.—Fall potatoes: Total stocks held by growers, local dealers, and processors - 13 Fall States: 2014 and 2015 [1]

State	Crop of 2014			
	Dec. 1	Feb. 1	Apr. 1	June 1
	1,000 cwt	*1,000 cwt*	*1,000 cwt*	*1,000 cwt*
California	2,500	1,600	700	(D)
Colorado	17,200	13,000	8,500	4,000
Idaho	95,000	74,000	49,000	23,500
Maine	11,600	8,500	5,500	2,200
Michigan	9,400	5,000	2,200	(D)
Minnesota	10,000	7,800	5,000	2,800
Montana	3,500	3,400	2,200	(D)
Nebraska	4,800	3,300	2,100	700
New York	2,100	1,400	500	(D)
North Dakota	16,900	12,200	7,100	2,600
Oregon	17,700	13,200	8,200	2,700
Washington	57,000	44,000	29,000	12,500
Wisconsin	18,000	13,400	8,700	4,500
Other States [2]	-	-	-	835
United States	265,700	200,800	128,700	56,335
Klamath Basin [3]	5,000	3,000	1,500	(D)

State	Crop of 2015			
	Dec. 1	Feb. 1	Apr. 1	June 1
	1,000 cwt	*1,000 cwt*	*1,000 cwt*	*1,000 cwt*
California	2,400	1,600	1,200	290
Colorado	17,200	13,000	8,500	4,000
Idaho	90,000	71,000	46,000	21,500
Maine	12,500	8,900	5,800	2,300
Michigan	10,900	6,700	3,500	(D)
Minnesota	10,400	7,800	4,700	1,400
Montana	3,400	3,300	1,900	(D)
Nebraska	4,400	3,100	2,200	(D)
New York	2,200	1,500	600	(D)
North Dakota	19,500	14,500	9,200	3,600
Oregon	17,000	12,000	7,500	3,000
Washington	56,000	42,000	27,000	11,500
Wisconsin	17,500	13,400	7,800	2,500
Other States [2]	-	-	-	1,620
United States	263,600	198,500	125,900	51,710
Klamath Basin [3]	5,100	3,000	1,650	(D)

- Represents zero.   (D) Withheld to avoid disclosing data for individual operations.   (NA) Not available.   [1] Stocks are defined as the quantity (whether sold or not) remaining in storage for all purposes and uses, including seed potatoes that are not yet moved, and shrinkage, waste, and other losses that occur after the date of each estimate.   [2] Includes data withheld above.   [3] Includes potato stocks in California and Klamath County, Oregon.
NASS, Crops Branch, (202) 720–2127.

Table 4-56.—Fall potatoes: Production and total stocks held by growers and local dealers, 13 Major States, 2006–2015

Crop year	Production	Total stocks						
		Dec. 1	Following year					
			Jan. 1	Feb. 1	Mar. 1	Apr. 1	May 1	June 1
	1,000 cwt	*1,000 cwt*	*1,000 cwt*	*1,000 cwt*	*1,000 cwt*	*1,000 cwt*	*1,000 cwt*	*1,000 cwt*
2006	389,527	258,900	225,800	192,200	159,500	120,900	79,050	44,460
2007	397,753	265,500	232,300	199,300	163,400	125,500	83,960	50,420
2008	369,866	243,700	213,200	183,900	152,700	115,800	78,100	45,300
2009	383,962	265,800	234,300	203,500	169,700	128,700	89,610	55,120
2010	357,467	240,200	209,400	180,300	148,500	111,000	72,000	41,320
2011	382,318	253,000	(NA)	187,500	(NA)	115,650	(NA)	43,340
2012	410,367	271,500	(NA)	204,600	(NA)	(NA)	(NA)	(NA)
2013	386,824	(NA)	(NA)	(NA)	(NA)	119,050	(NA)	46,885
2014	396,798	265,700	(NA)	200,800	(NA)	128,700	(NA)	56,335
2015	398,482	263,600	(NA)	198,500	(NA)	125,900	(NA)	51,710

(NA) Not available.
NASS, Crops Branch, (202) 720–2127.

Table 4-57.—Potatoes: Utilization and processing, United States, 2014 and 2015

Utilization - 30 Program States	2014	2015
	1,000 cwt	*1,000 cwt*
Sales		
Table stock	107,344	110,960
Processing	280,330	272,538
Other sales		
Livestock feed	768	919
Seed	22,774	25,648
Other Sales Total	23,542	26,567
Total sales	411,216	410,065
Non-sales		
Seed used on farms where grown	3,343	3,765
Household use and used for feed on farms where grown	849	866
Shrinkage and loss	26,762	26,509
Total non-sales	30,954	31,140
Total production	442,170	441,205

Processing - United States	2014	2015
	1,000 cwt	*1,000 cwt*
Processing		
Chips and shoestring	73,960	58,122
All Dehydrated (including starch and flour)	48,707	47,893
Frozen french fries	152,832	152,329
Other frozen products	9,208	13,573
Canned potatoes	435	478
Other canned products (hash, stews, soups)	886	730
Other (including fresh pack, potato salad, vodka, etc)	6,907	6,377
Total	292,935	279,502

NASS, Crops Branch, (202) 720–2127.

Table 4-58.—Potatoes: Production, seed used, and farm disposition, by seasonal groups, crop of 2014

Season and State	Production	Total used for seed	Used on farms where produced		Sold
			For seed, feed, and household use	Shrinkage and loss	
	1,000 cwt	*1,000 cwt*	*1,000 cwt*	*1,000 cwt*	*1,000 cwt*
Spring					
Arizona	1,085	(D)	(D)	(D)	(D)
California	11,656	1,036	8	172	11,476
Florida	7,032	897	-	531	6,501
North Carolina	2,835	(D)	(D)	(D)	(D)
Other States [1]	(NA)	350	6	6	3,908
United States	22,608	2,283	14	709	21,885
Summer					
Delaware	348	20	1	1	346
Illinois	2,656	150	26	-	2,630
Kansas	1,394	107	-	50	1,344
Maryland	874	38	1	1	872
Missouri	2,133	204	-	-	2,133
New Jersey	428	40	2	1	425
Texas	6,901	630	50	51	6,800
Virginia	1,125	97	2	8	1,115
United States	15,859	1,286	82	112	15,665
Fall					
California	3,901	226	2	39	3,860
Colorado	23,196	1,537	961	2,330	19,905
Idaho	132,880	7,703	890	8,630	123,360
Maine	14,645	1,071	327	1,818	12,500
Massachusetts	1,026	83	12	5	1,009
Michigan	15,725	1,058	315	450	14,960
Minnesota	16,400	984	136	659	15,605
Montana	3,616	276	226	200	3,190
Nebraska	7,943	432	162	544	7,237
Nevada	(D)	(D)	(D)	(D)	(D)
New Mexico	(D)	(D)	(D)	(D)	(D)
New York	4,345	346	82	240	4,023
North Dakota	23,870	1,782	360	2,200	21,310
Ohio	420	32	5	5	410
Oregon	22,562	952	70	1,375	21,117
Pennsylvania	1,430	120	13	14	1,403
Rhode Island	123	11	-	1	122
Washington	101,475	4,318	260	6,340	94,875
Wisconsin	26,240	1,501	275	846	25,119
Other States [1]	3,906	258	-	245	3,661
United States	403,703	22,690	4,096	25,941	373,666
All potatoes					
United States	442,170	26,259	4,192	26,762	411,216

- Represents zero.   (D) Withheld to avoid disclosing data for individual operations.   (NA) Not applicable.   [1] Includes data withheld above.
NASS, Crops Branch, (202) 720-2127.

Table 4-59.—Potatoes: Production, seed used, and farm disposition, by seasonal groups, crop of 2015

Season and State	Production	Total used for seed	Used on farms where produced		Sold
			For seed, feed, and household use	Shrinkage and loss	
	1,000 cwt	*1,000 cwt*	*1,000 cwt*	*1,000 cwt*	*1,000 cwt*
Spring					
Arizona	1,015	(D)	(D)	(D)	(D)
California	10,280	754	8	935	9,337
Florida	6,808	658	1	-	6,807
North Carolina	2,667	(D)	(D)	(D)	(D)
Other States [1]	(NA)	422	5	33	3,644
United States	20,770	1,834	14	968	19,788
Summer					
Delaware	(D)	(D)	(D)	(D)	(D)
Illinois	2,622	158	-	12	2,610
Kansas	1,206	105	-	46	1,160
Maryland	792	36	1	1	790
Missouri	2,471	194	-	10	2,461
New Jersey	(D)	(D)	(D)	(D)	(D)
Texas	6,825	584	-	-	6,825
Virginia	1,034	70	2	3	1,029
Other States	784	55	-	2	782
United States	15,734	1,202	3	74	15,657
Fall					
California	3,528	154	2	141	3,385
Colorado	22,575	1,415	930	2,030	19,615
Idaho	130,400	8,060	1,089	8,420	120,891
Maine	16,160	1,029	322	1,615	14,223
Massachusetts	1,098	88	16	9	1,073
Michigan	17,550	1,152	335	540	16,675
Minnesota	16,200	840	104	826	15,270
Montana	3,543	294	208	140	3,195
Nebraska	6,885	479	251	529	6,105
Nevada	(D)	(D)	(D)	(D)	(D)
New Mexico	(D)	(D)	(D)	(D)	(D)
New York	4,144	288	46	290	3,808
North Dakota	27,600	1,782	456	2,234	24,910
Ohio	276	29	5	5	266
Oregon	21,784	1,053	207	1,305	20,272
Pennsylvania	1,484	122	28	22	1,434
Rhode Island	81	12	2	1	78
Washington	100,300	4,323	268	6,220	93,812
Wisconsin	27,813	1,361	222	1,004	26,587
Other States [1]	3,280	198	123	136	3,021
United States	404,701	22,679	4,614	25,467	374,620
All potatoes					
United States	441,205	25,715	4,631	26,509	410,065

- Represents zero.   (D) Withheld to avoid disclosing data for individual operations.   [1] Includes data withheld above.
NASS, Crops Branch, (202) 720–2127.

Table 4-60.—Potatoes, fresh & seed: United States exports by country of destination and imports by country of origin, 2011–2013

Country	2011	2012	2013
	Metric tons	Metric tons	Metric tons
Imports			
Canada	491,495	352,959	390,719
Belgium-Luxembourg(*)	0	0	23
Ecuador	0	0	3
Peru	25	9	1
Dominican Republic	5	0	0
Germany(*)	45	0	0
India	0	1	0
Mexico	19	0	0
Leeward-Windward Islands(*)	0	1	0
World Total	491,588	352,970	390,746
Exports			
Canada	284,785	240,858	270,969
Mexico	73,570	79,948	82,954
Taiwan	9,300	18,704	18,158
Japan	9,138	14,665	17,312
Korea, South	16,654	18,661	16,291
Malaysia	12,596	17,181	16,105
Philippines	4,431	5,289	10,136
Indonesia	74	3,024	8,323
Singapore	6,136	5,875	6,086
Thailand	8,767	11,533	4,877
Dominican Republic	2,107	5,105	4,676
Hong Kong	3,798	4,030	3,922
Nicaragua	1,139	1,541	2,856
Bahamas, The	1,207	1,883	2,467
Panama	820	984	1,621
Russia	2,284	4,987	1,579
Guatemala	1,809	2,678	1,472
Vietnam	344	2,397	1,284
Brazil	176	0	1,106
Costa Rica	2,695	830	1,069
Leeward-Windward Islands(*)	769	707	999
Trinidad and Tobago	267	731	860
United Arab Emirates	82	164	722
El Salvador	778	565	504
Cayman Islands	125	241	347
Poland	0	0	270
Sri Lanka	197	150	256
Jamaica	105	46	174
Congo (Kinshasa)	0	100	160
Belize	168	153	135
Rest of World	4,726	4,437	1,122
World Total	449,047	447,468	478,809

(*) Denotes a country that is a summarization of its component countries.
FAS, Office of Global Analysis, (202) 720-6301.

Table 4-61.—Potatoes (fresh market): Foreign trade, United States, 2006–2015 [1]

Year beginning July	Imports for consumption	Domestic exports
	1,000 cwt.	1,000 cwt.
2006/2007	10,681	10,681
2007/2008	10,852	10,852
2008/2009	10,550	10,550
2009/2010	9,050	9,050
2010/2011	10,384	10,384
2011/2012	9,041	9,041
2012/2013	7,375	7,375
2013/2014	10,833	10,833
2014/2015	8,514	8,514
2015/2016	10,105	10,105

[1] Includes seed.
ERS, (202) 694–5253. Based on data from U.S. Department of Commerce, U.S. Census Bureau.

Table 4-62.—Pumpkins for fresh market and processing: Area, production, and value per hundredweight, by State and United States, 2013-2015

State	Area harvested			Production			Value per unit		
	2013	2014	2015	2013	2014	2015	2013	2014	2015
	Acres	Acres	Acres	1,000 cwt	1,000 cwt	1,000 cwt	Dollars per cwt	Dollars per cwt	Dollars per cwt
California	5,900	5,800	6,100	1,947	1,798	1,464	15.60	16.10	17.80
Illinois	20,100	19,800	14,700	5,476	7,458	3,179	8.62	6.32	3.79
Michigan	6,600	6,100	5,000	978	970	770	11.50	11.80	15.70
New York	6,000	5,200	4,400	960	690	489	31.40	29.70	22.60
Ohio	6,100	6,300	5,500	1,004	1,053	842	15.40	17.80	17.10
Pennsylvania	5,900	5,900	5,200	856	1,050	794	16.50	15.70	18.40
United States	50,300	49,100	40,900	11,221	13,019	7,538	13.20	11.00	12.00

NASS, Crops Branch, (202) 720-2127.

Table 4-63.—Pumpkins for fresh market and processing, commercial crop: Area, yield, production, value, hundredweight, and total value, United States, 2006-2015

Year	Area harvested	Yield per acre	Production	Value	
				Per cwt	Total
	Acres	Cwt	1,000 cwt	Dollars	1,000 dollars
2006	43,700	240	10,484	9.98	104,623
2007	45,900	250	11,458	10.80	123,519
2008	43,400	246	10,663	12.90	137,072
2009	44,100	211	9,313	11.00	102,730
2010	48,500	222	10,748	11.00	117,791
2011	46,900	228	10,705	10.60	113,085
2012	50,500	238	12,036	12.10	146,033
2013	50,300	223	11,221	13.20	148,517
2014	49,100	265	13,019	11.00	143,158
2015	40,900	184	7,538	12.00	90,214

NASS, Crops Branch, (202) 720-2127.

Table 4-64.—Spinach for fresh market: Area, production, and value per hundredweight, by State and United States, 2013-2015

State	Area harvested			Production			Value per unit		
	2013	2014	2015	2013	2014	2015	2013	2014	2015
	Acres	*Acres*	*Acres*	*1,000 cwt*	*1,000 cwt*	*1,000 cwt*	*Dollars per cwt*	*Dollars per cwt*	*Dollars per cwt*
Arizona	7,200	8,000	10,300	1,368	1,240	1,288	48.30	39.90	42.90
California	24,300	26,000	26,700	3,645	4,160	4,272	41.60	45.50	44.40
New Jersey	1,300	1,300	1,400	254	221	210	46.60	37.70	49.30
Texas	1,300	1,500	2,100	260	225	263	38.50	44.00	21.00
Other States[1]	840	740	690	92	73	43	35.80	37.20	37.20
United States	34,940	37,540	41,190	5,619	5,919	6,076	43.20	43.90	43.20

[1] Other States include Colorado and Maryland.
NASS, Crops Branch, (202) 720-2127.

Table 4-65.—Spinach for fresh market, commercial crop: Area, yield, production, value per hundredweight, and total value, United States, 2006-2015

Year	Area harvested	Yield per acre	Production	Value[1]	
				Per cwt	Total
	Acres	*Cwt*	*1,000 cwt*	*Dollars*	*1,000 dollars*
2006	36,500	166	6,045	29.90	180,774
2007	31,900	159	5,079	32.30	163,952
2008	34,334	161	5,519	34.00	187,497
2009	34,934	192	6,699	39.50	264,564
2010	30,735	184	5,648	42.70	241,050
2011	31,435	191	6,010	40.40	242,823
2012	34,440	156	5,360	42.20	226,232
2013	34,940	161	5,619	43.20	242,846
2014	37,540	158	5,919	43.90	259,700
2015	41,190	148	6,076	43.90	262,407

[1] Price and value at point of first sale.
NASS, Crops Branch, (202) 720-2127.

Table 4-66.—Spinach for processing, commercial crop: Area, yield, production, value per ton, and total value, United States, 2006-2015

Year	Area harvested	Yield per acre	Production	Value[1]	
				Per ton	Total
	Acres	*Tons*	*Tons*	*Dollars*	*1,000 dollars*
2006	9,400	7.40	69,560	127.00	8,809
2007	11,400	8.58	97,800	104.00	10,123
2008	10,200	10.15	103,540	124.00	12,831
2009	10,100	9.47	95,660	127.00	12,144
2010	10,200	8.34	85,050	143.00	12,153
2011	7,500	10.69	80,180	131.00	10,465
2012	7,400	10.18	75,320	137.00	10,320
2013	8,800	10.08	88,710	129.00	11,451
2014	9,100	9.92	90,310	125.00	11,312
2015	7,200	10.44	75,200	138.00	10,394

[1] Price and value at processing plant door.
NASS, Crops Branch, (202) 720-2127.

Table 4-67.—Sweet Potatoes: Area, yield, production, season average price per hundredweight received by farmers, and value, United States, 2006–2015

Year	Area harvested	Yield per acre	Production	Market year average price [1]	Value of production
	1,000 acres	Cwt	1,000 cwt	Dollars	1,000 dollars
2006	87.3	188	16,401	18.20	298,388
2007	97.4	186	18,070	18.30	330,060
2008	97.3	190	18,443	21.20	390,572
2009	96.9	201	19,469	21.80	423,677
2010	116.9	204	23,845	19.80	472,218
2011	129.7	208	26,964	18.80	505,938
2012	126.6	209	26,482	17.40	461,861
2013	113.2	219	24,785	24.10	597,217
2014	135.2	219	29,584	23.90	706,916
2015	153.1	203	31,016	21.80	675,583

[1] Obtained by weighting State prices by production.
NASS, Crops Branch, (202) 720–2127.

Table 4-68 .—Sweet Potatoes: Area, production, and season average price per hundredweight received by farmers, by State and United States, 2013–2015

State	Area harvested			Production			Market year average price per cwt		
	2013	2014	2015	2013	2014	2015	2013	2014	2015
	1,000 acres	1,000 acres	1,000 acres	1,000 cwt	1,000 cwt	1,000 cwt	Dollars	Dollars	Dollars
Alabama	2.4	2.0	2.5	415	440	550	28.50	24.10	21.40
Arkansas	3.9	3.9	3.8	702	780	741	32.20	(D)	27.20
California	19.0	19.0	18.5	6,840	5,225	6,290	24.60	30.10	23.70
Florida	5.9	5.9	5.4	838	1,180	1,107	(D)	(D)	(D)
Louisiana	7.5	8.8	9.0	1,650	2,024	1,980	17.90	17.40	21.60
Mississippi	19.5	21.5	26.0	3,510	3,763	3,770	16.40	19.30	20.90
New Jersey	1.2	1.2	1.2	150	192	168	29.50	36.90	31.50
North Carolina	53.0	72.0	86.0	10,600	15,840	16,340	24.90	22.00	19.40
Texas	0.8	0.9	0.7	80	140	(D)	(D)	27.60	(D)
Other States	(X)	(X)	(X)	(X)	(X)	70	42.50	36.60	43.10
United States	113.2	135.2	153.1	24,785	29,584	31,016	24.10	23.90	21.80

(D) Withheld to avoid disclosing data for individual operations. (X) Not applicable.
NASS, Crops Branch, (202) 720–2127.

Table 4-69.—Squash for fresh market and processing: Area, production, and value per hundredweight, by State and United States, 2013–2015

State	Area harvested			Production			Value per unit		
	2013	2014	2015	2013	2014	2015	2013	2014	2015
	Acres	Acres	Acres	1,000 cwt	1,000 cwt	1,000 cwt	Dollars per cwt	Dollars per cwt	Dollars per cwt
California	6,800	5,400	5,400	1,224	918	864	35.40	32.90	34.50
Florida	7,800	6,800	5,900	975	800	600	72.10	50.80	45.80
Georgia	2,600	3,200	3,000	390	480	420	31.50	31.80	40.30
Michigan	6,100	6,000	5,600	1,220	1,200	1,114	14.50	15.00	17.50
New Jersey	2,600	2,700	2,800	338	354	252	39.30	32.80	37.80
New York	4,500	4,300	4,200	866	691	594	43.90	45.40	42.00
North Carolina	1,900	3,000	2,400	97	162	235	47.00	51.90	28.30
Ohio	1,400	1,600	1,800	238	304	324	40.00	34.00	40.00
Oregon	2,400	3,000	3,500	429	780	1,230	11.70	9.80	8.00
South Carolina	1,300	1,000	1,500	169	150	165	35.00	37.50	25.90
Tennessee	600	830	590	32	43	38	38.00	30.30	53.00
Texas	1,900	1,500	2,000	171	135	180	46.00	55.00	57.00
United States	39,900	39,330	38,690	6,149	6,017	6,016	37.20	31.20	29.00

NASS, Crops Branch, (202) 720–2127.

Table 4-70.—Squash for fresh market and processing, commercial crop: Area, yield, production, value and total value, United States, 2006-2015

Year	Area harvested	Yield per acre	Production	Value	
				Per cwt	Total
	Acres	Cwt	1,000 cwt	Dollars	1,000 dollars
2006	48,200	165	7,946	24.20	192,459
2007	41,600	151	6,266	27.80	173,917
2008	42,400	158	6,687	30.50	204,283
2009	42,900	165	7,089	27.90	197,614
2010	42,600	153	6,514	30.40	198,025
2011	42,900	164	7,043	34.30	241,877
2012	41,800	176	7,369	31.80	234,101
2013	39,900	154	6,149	37.20	228,985
2014	39,330	153	6,017	31.20	187,820
2015	38,690	155	6,016	29.00	174,259

NASS, Crops Branch, (202) 720–2127.

Table 4-71.—Taro: Area, production, price, and value, Hawaii, 2006–2015

Year	Area harvested	Production	Price per pound	Value of production
	Acres	1,000 pounds	Dollars	1,000 dollars
2006	380	4,500	0.570	2,565
2007	380	4,000	0.590	2,360
2008	390	4,300	0.620	2,666
2009	445	4,000	0.610	2,440
2010	475	3,900	0.645	2,516
2011	485	4,100	0.670	2,747
2012	400	3,500	0.670	2,345
2013	400	3,120	0.620	1,934
2014	360	3,240	0.600	1,944
2015	340	3,502	0.680	2,381

NASS, Crops Branch, (202) 720–2127.

Table 4-72.—Tomatoes: Foreign trade, United States, 2006–2015

Year beginning July	Imports				Domestic exports [2]			
	Fresh	Canned, whole and piece [1]	Catsup and sauces	Paste	Fresh	Canned, whole and piece	Catsup and sauces	Paste
	1,000 cwt	*1,000 cwt*	*1,000 cwt*	*1,000 cwt*	*1,000 cwt*	*1,000 cwt*	*1,000 cwt*	*1,000 cwt*
2006/2007	23,056	358	4,544	1,022	3,317	1,069	4,414	2,026
2007/2008	23,905	239	4,097	288	3,641	1,256	4,493	4,721
2008/2009	25,014	336	3,924	179	3,720	1,421	4,971	6,338
2009/2010	32,093	405	3,684	124	3,133	1,128	5,423	5,011
2010/2011	30,904	270	3,839	113	2,794	1,303	5,845	6,426
2011/2012	33,744	265	3,533	111	2,531	1,531	6,064	6,764
2012/2013	33,717	358	3,451	87	2,533	1,312	6,302	9,032
2013/2014	34,880	385	3,526	103	2,345	1,673	7,247	11,035
2014/2015	33,488	306	3,425	129	2,459	1,295	8,748	9,930
2015/2016	37,910	367	3,844	216	1,937	1,219	9,106	8,184

[1] Includes all canned tomato and tomato product imports except paste and juice, on a product-weight-basis. [2] Includes exports for military-civilian feeding abroad.
ERS, Market and Trade Economics Division, (202) 694–5326. Compiled from reports of the U.S. Department of Commerce.

Table 4-73.—Tomatoes, commercial crop: Area, yield, production, value per hundredweight and per ton, and total value, United States, 2006–2015

Year	For fresh market					For processing				
	Area harvested	Yield per acre	Production	Value [1]		Area harvested	Yield per acre	Production	Value [2]	
				Per cwt	Total				Per ton	Total
	Acres	*Cwt*	*1,000 cwt*	*Dollars*	*1,000 dollars*	*Acres*	*Tons*	*Tons*	*Dollars*	*1,000 dollars*
2006 ...	120,200	302	36,274	43.70	1,584,708	299,400	35.44	10,611,820	66.40	704,669
2007 ...	108,100	311	33,627	34.80	1,168,693	313,600	40.37	12,659,890	71.20	901,761
2008 ...	105,250	296	31,137	45.50	1,415,297	296,500	41.50	12,305,820	79.80	982,373
2009 ...	108,700	306	33,235	40.40	1,344,217	327,800	42.62	13,970,560	87.20	1,218,912
2010 ...	103,000	271	27,961	48.40	1,352,315	288,900	44.22	12,776,280	72.50	926,692
2011 ...	94,210	300	28,231	37.00	1,043,491	267,800	46.29	12,396,150	75.60	936,861
2012 ...	101,000	284	28,660	30.50	874,195	276,300	47.70	13,178,750	76.70	1,010,545
2013 ...	99,600	265	26,391	44.60	1,177,592	277,000	45.60	12,631,500	90.10	1,138,462
2014 ...	97,600	280	27,280	41.50	1,133,070	306,100	47.82	14,637,300	99.30	1,452,823
2015 ...	94,300	287	27,026	46.00	1,243,113	310,600	47.50	14,754,350	97.00	1,430,783

[1] Price and value at point of first sale. [2] Price and value at processing plant door.
NASS, Crops Branch, (202) 720–2127.

Table 4-74.—Tomatoes, commercial crop: Area, production, and value per hundredweight and per ton, by State and United States, 2013–2015 [1]

Utilization and State	Area harvested			Production			Value per unit		
	2013	2014	2015	2013	2014	2015	2013	2014	2015
Fresh market:	*Acres*	*Acres*	*Acres*	*1,000 cwt*	*1,000 cwt*	*1,000 cwt*	*Dollars per cwt*	*Dollars per cwt*	*Dollars per cwt*
Alabama	1,100	1,200	1,100	297	420	385	33.60	28.30	51.50
Arkansas	1,000	900	800	150	153	152	48.00	62.00	59.00
California	34,000	32,300	30,400	10,200	10,175	9,424	36.20	34.80	34.90
Florida	34,000	33,000	32,200	9,010	9,240	9,499	50.60	47.30	47.70
Georgia	2,200	2,300	2,700	440	575	918	35.80	30.40	39.40
Indiana	900	820	700	153	152	50	77.00	48.80	55.00
Michigan	2,400	1,900	2,600	576	466	780	55.00	35.70	53.60
New Jersey	2,800	2,900	2,900	588	624	653	52.60	61.00	81.10
New York	2,700	2,600	2,300	446	312	299	72.60	76.80	105.00
North Carolina ..	3,200	3,400	3,300	1,040	850	1,023	43.50	42.00	46.00
Ohio	3,400	3,600	3,500	595	954	886	75.00	36.60	54.40
Pennsylvania	2,200	2,500	2,200	310	415	389	81.00	65.40	85.40
South Carolina	3,000	3,200	3,200	630	896	864	25.60	32.40	50.50
Tennessee	3,400	3,900	3,400	1,122	1,326	1,003	36.00	43.00	56.00
Texas	900	780	900	90	78	81	45.00	57.30	60.00
Virginia	2,400	2,300	2,100	744	644	620	50.00	44.60	55.00
United States ...	99,600	97,600	94,300	26,391	27,280	27,026	44.60	41.50	46.00
Processing:	*Acres*	*Acres*	*Acres*	*Tons*	*Tons*	*Tons*	*Dollars per ton*	*Dollars per ton*	*Dollars per ton*
California	260,000	289,000	296,000	12,100,000	14,010,000	14,361,000	88.80	98.60	96.40
Indiana	8,700	9,300	6,700	269,700	344,100	191,620	121.00	114.00	120.00
Michigan	3,200	3,400	3,200	108,800	129,200	95,040	125.00	117.00	117.00
Ohio	5,100	4,400	4,700	153,000	154,000	106,690	116.00	111.00	115.00
United States ...	277,000	306,100	310,600	12,631,500	14,637,300	14,754,350	90.10	99.30	97.00

[1] Cherry, grape, tomatillo, and greenhouse tomatoes are excluded.
NASS, Crops Branch, (202) 720–2127.

Table 4-75.—Watermelon for fresh market: Area, production, and value per hundredweight, by State and United States, 2013–2015

State	Area harvested			Production			Value per unit		
	2013	2014	2015	2013	2014	2015	2013	2014	2015
	Acres	Acres	Acres	1,000 cwt	1,000 cwt	1,000 cwt	Dollars per cwt	Dollars per cwt	Dollars per cwt
Alabama	2,900	2,700	2,800	377	456	420	11.00	14.20	14.50
Arizona	3,600	2,900	3,600	1,800	1,334	1,584	16.20	19.40	16.80
Arkansas	1,600	1,600	1,500	336	320	338	12.00	13.00	14.80
California	10,000	11,200	10,400	5,800	6,384	5,512	13.00	14.60	14.50
Delaware	2,700	2,500	2,300	864	833	761	12.00	13.00	13.00
Florida	20,200	19,700	21,000	6,262	4,827	5,880	25.00	16.60	15.00
Georgia	18,600	19,000	19,000	5,580	5,130	5,510	10.50	15.60	14.80
Indiana	7,100	7,600	6,900	2,414	2,964	2,415	12.50	9.00	13.00
Maryland	3,300	3,300	3,200	1,056	1,089	1,040	12.00	12.00	13.00
Mississippi	2,000	2,100	1,800	400	378	315	10.50	11.70	12.00
Missouri	3,000	2,700	2,600	843	837	572	10.30	9.40	8.70
North Carolina	5,700	5,500	5,800	1,710	1,155	1,798	13.50	12.00	18.00
Oklahoma	2,200	2,600	4,000	242	364	540	13.75	11.80	13.50
South Carolina	7,100	7,300	7,200	2,734	1,862	2,736	13.90	14.80	13.50
Texas	23,000	20,000	23,000	5,520	5,200	5,520	9.70	9.60	9.40
Virginia	630	650	650	164	130	163	12.50	15.00	22.00
United States	113,630	111,350	115,750	36,102	33,263	35,104	14.20	13.50	13.80

NASS, Crops Branch, (202) 720–2127.

Table 4-76.—Watermelon for fresh market, commercial crop: Area, yield, production, value per hundredweight, and total value, United States, 2006-2015

Year	Area harvested	Yield per acre	Production	Value [1]	
				Per cwt	Total
	Acres	Cwt	1,000 cwt	Dollars	1,000 dollars
2006	131,000	304	39,865	10.40	414,111
2007	129,000	290	37,349	11.30	422,546
2008	125,550	319	40,003	12.50	499,633
2009	123,900	314	38,911	11.60	450,713
2010	134,300	311	41,736	12.00	499,800
2011	121,400	307	37,215	13.90	518,787
2012	115,800	312	36,153	13.20	476,768
2013	113,630	318	36,102	14.20	513,987
2014	111,350	299	33,263	13.50	450,324
2015	115,750	303	35,104	13.80	483,003

[1] Price and value at point of first sale.
NASS, Crops Branch, (202) 720–2127.

Table 4-77.—Vegetables and melons, fresh: Total reported domestic rail, truck, and air shipments, 2015

Commodity	Jan.	Feb.	Mar.	Apr.	May	June	July	Aug.	Sep.	Oct.	Nov.	Dec.	Total
Vegetables:	1,000 cwt	1,000 cwt	1,000 cwt	1,000 cwt	1,000 cwt	1,000 cwt	1,000 cwt	1,000 cwt	1,000 cwt	1,000 cwt	1,000 cwt.	1,000 cwt	1,000 cwt
Artichokes	39	37	73	48	52	54	72	59	40	39	40	26	579
Asparagus			75	113	138	65							391
Asparagus-organic					1								1
Beans	277	202	343	356	221	166	56	54	78	153	264	228	2,398
Beans-organic		1		4	3					2			10
Broccoli	565	626	638	528	487	606	418	395	390	475	408	448	5,984
Brussels sprouts	34	10	7	8	11	19	30	47	63	98	74	70	471
Beets	1	1	2	4	4	7	9	8	6	6	3		51
Cabbage	1,188	1,138	1,396	1,044	722	373	423	555	574	686	796	1,078	9,973
Carrots	605	544	643	623	552	591	617	543	540	563	538	606	6,965
Cauliflower	456	427	364	331	278	362	340	284	326	343	208	240	3,959
Celery	1,511	1,337	1,385	1,306	1,182	1,170	1,050	1,037	1,141	1,266	1,491	1,344	15,220
Celery-organic		1	2	1									4
Chinese cabbage	74	71	75	61	65	52	45	44	37	24	41	35	624
Corn, sweet	414	357	684	1,879	2,638	2,788	861	670	324	303	416	310	11,644
Corn, sweet-organic					2	15	4						21
Cucumbers	87	1	89	457	638	568	402	511	347	351	352	219	4,022
Cucumbers-organic											1	1	2
Cucumbers, greenhouse	28	24	34	27	39	49	38	56	40	33	15	12	395
Eggplant	60	27	46	91	108	140	41	28	55	86	89	59	830
Eggplant-organic	1			1	2					2	2		8
Endive	21	18	25	13	5	4	4	3	4	4	11	12	123
Escarole	28	24	27	17	6	5	4	4	4	6	14	18	157
Greens	451	395	384	229	229	102	90	102	110	178	268	402	2,940
Greens-organic											1	10	11
Lettuce, iceberg	2,478	2,610	3,026	2,328	2,163	2,203	2,391	2,089	1,979	2,171	2,041	2,560	28,039
Lettuce, other	378	386	426	261	200	202	220	211	213	244	333	490	3,564
Lettuce, romaine	1,957	1,985	2,292	1,632	1,456	1,376	1,515	1,522	1,522	1,630	1,703	2,081	20,671
Lettuce, romaine-organic											6	11	17
Lettuce processed	413	361	440	830	1,189	1,107	950	1,185	1,095	1,173	716	664	10,123
Onions, dry	3,966	3,120	3,506	3,303	3,359	3,098	3,526	3,748	3,902	4,132	3,699	3,947	43,306
Onions, dry-organic	1	1	1		12	4			1	1	1		24
Onions processed	599	488	562	611	824	570	481	335	367	566	529	549	6,481
Onions, green	12	7	10	18	22	32	30	29	28	29	23	14	254
Parsley	35	40	45	36	26	32	38	36	29	36	36	25	414
Peas, green				1	1	3	2	1	2	3	5		18
Peppers, bell	572	499	806	817	1,137	1,062	772	666	725	669	761	584	9,070
Peppers, bell-organic	1			4	3	7	1			1	4	1	22
Peppers, other	28	21	28	19	29	40	38	7		22	29	19	280
Potatoes, table	8,579	7,430	8,556	7,881	7,882	8,130	7,531	7,481	7,818	8,268	8,646	8,193	96,395
Potatoes, table-organic	113	46	54	32	24	8	7	47	83	100	127	101	742
Potatoes, chipper	5,301	3,947	4,160	4,088	4,051	3,953	4,723	5,518	4,490	5,017	3,940	4,216	53,404
Potatoes, seed	318	964	3,227	7,220	3,486	124			29	190	185	224	15,967
Radishes	44	39	52	47	21	13	14	10	12	21	36	21	330
Spinach	339	310	359	97	62	67	67	59	57	74	170	323	1,984
Spinach-organic				3							2	9	14
Squash	126	93	118	223	414	206	135	152	110	157	112	81	1,927
Sweet potatoes	777	706	848	819	763	622	630	622	633	819	1,289	836	9,364
Sweet potatoes-organic	7	7	11	8	7	4	1	3	1	2	9	4	64
Tomatoes	1,532	993	1,457	1,396	1,717	1,501	1,318	1,216	1,073	963	1,436	1,130	15,732
Tomatoes, greenhouse	171	109	111	162	272	218	201	242	170	186	93	91	2,026
Toms, Grape Type	157	105	192	202	211	95	77	85	90	127	220	141	1,702
Toms, Grape Type-greenhouse	17	16	19	20	31	27	28	33	25	25	12	10	263
Toms, Grape Type-organic	12	7	9	12	14	17	7		2	8	22	9	116
Toms, Cherry	61	48	72	65	52	14	14	34	24	24	46	35	489
Toms, Cherry-greenhouse				2	1	3	8	7	6	8	4	4	43
Toms, Plum Type	247	192	260	271	265	145	293	302	288	188	229	202	2,882
Toms, Plum Type-greenhouse	10	7	9	13	14	11	8	11	7	10	4	4	108
Total	34,091	29,778	36,948	39,562	37,091	32,030	29,530	30,052	28,859	31,480	31,502	31,695	392,618
Melons:													
Cantaloupe				55	1,877	2,738	2,769	2,687	1,906	1,108	465		13,605
Cantaloupe-organic						23	32	25	32	1			113
Honeydews				109	300	655	852	797	314	75			3,102
Honeydews-organic					1	5	2	2					10
Mixed & misc. melons				38	104	185	112	22					461
Watermelons, seedless			3	826	5,423	9,157	7,540	7,393	2,757	401	4		33,504
Watermelons, seeded				213	726	880	722	435	111	23			3,110
Watermelons, seeded-organic					5	2		4					11
Total			3	1,094	8,178	13,205	11,908	11,510	5,627	1,847	544		53,916

STATISTICS OF VEGETABLES AND MELONS

Table 4-78.—Vegetables (fresh), melons, potatoes, sweet potatoes: Per capita civilian utilization (farm-weight basis), United States, 2006–2015 [1]

Year	Cabbage	Cucumbers	Tomatoes[2]	Asparagus	Broccoli	Carrots	Head Lettuce	Leaf/ romaine
	Pounds	Pounds	Pounds	Pounds	Pounds	Pounds	Pounds	Pounds
2006	7.8	6.1	19.8	1.1	5.8	8.1	20.1	12.0
2007	8.0	6.4	19.2	1.2	5.6	8.1	18.4	11.5
2008	8.1	6.4	18.5	1.2	6.0	8.1	16.9	10.4
2009	7.3	6.8	19.6	1.3	6.2	7.4	16.1	10.0
2010	7.5	6.7	20.6	1.4	5.6	7.8	15.9	12.0
2011	6.6	6.4	21.0	1.4	5.9	7.5	15.8	11.7
2012	6.3	7.1	20.8	1.4	6.3	7.9	15.9	11.9
2013	6.9	7.3	20.2	1.4	6.9	8.0	14.1	11.4
2014	6.7	7.4	20.5	1.6	6.6	8.5	14.5	10.8
2015	6.3	7.5	20.6	1.5	6.6	8.3	13.5	11.0

Year	Snap beans	Garlic	Cauliflower	Celery	Sweet Corn	Onions	Spinach	Bell peppers
	Pounds	Pounds	Pounds	Pounds	Pounds	Pounds	Pounds	Pounds
2006	2.1	2.7	1.7	6.1	8.3	19.9	2.0	9.5
2007	2.2	2.7	1.7	6.3	9.2	21.6	1.6	9.4
2008	2.0	2.8	1.6	6.2	9.1	20.2	1.8	9.6
2009	1.7	2.4	1.7	6.2	9.2	19.6	2.0	9.8
2010	1.9	2.3	1.3	6.1	9.2	19.6	1.7	10.3
2011	1.7	2.3	1.2	6.0	8.2	19.1	1.8	10.6
2012	1.6	2.3	1.2	6.0	8.7	19.5	1.6	10.7
2013	1.6	2.0	1.3	5.5	8.9	18.5	1.6	10.0
2014	1.5	1.9	1.3	5.5	7.6	18.3	1.7	10.7
2015	1.6	1.9	1.4	5.4	8.6	18.6	1.7	11.1

Year	Others[3]	Total vegetables[4]	Watermelon	Cantaloupe	Honeydew melons	Potatoes	Sweet potatoes
	Pounds	Pounds	Pounds	Pounds	Pounds	Pounds	Pounds
2006	15.46	148.4	15.1	9.3	1.9	38.6	4.6
2007	15.02	148.1	14.4	9.6	1.8	38.7	5.1
2008	14.31	143.1	15.6	8.9	1.7	37.8	5.1
2009	13.98	141.3	14.9	9.0	1.6	36.7	5.3
2010	14.40	144.3	15.7	8.5	1.7	36.8	6.7
2011	14.78	141.9	13.8	8.6	1.6	34.1	7.5
2012	15.50	144.7	13.9	7.5	1.5	34.4	7.3
2013	15.00	140.6	14.5	8.3	1.6	34.8	6.7
2014	16.65	141.7	....	....	....	....	....
2015	14.03	139.7	....	....	....	....	....

[1] Fresh vegetable consumption computed for total commercial production for fresh market.   Does not include production for home use.   Consumption obtained by dividing the total apparent consumption by total July 1 population as reported by the Bureau of the Census.   All data for calendar year.   [2] After 1996, includes an ERS estimate of domestically produced hothouse tomatoes.   Hothouse imports included in all years.   [3] Includes artichokes, eggplant, radishes, brussels sprouts, squash, green limas, and escarole/endive.   Beginning in 2000, also includes collards, mustard greens, turnip greens, kale, okra, and pumpkins.   [4] Excludes melons, mushrooms, potatoes, and sweet potatoes.
ERS, Market and Trade Economics Division, (202) 694–5326.

Table 4-79.—Vegetables, canning: Per capita utilization (farm weight), United States, 2006–2015

Year	Cabbage for kraut	Asparagus	Snap beans	Carrots	Green peas
	Pounds	Pounds	Pounds	Pounds	Pounds
2006	1.2	0.2	3.9	1.0	1.2
2007	1.0	0.1	3.5	0.9	1.2
2008	0.9	0.2	3.3	1.0	1.1
2009	0.9	0.2	3.6	0.9	1.3
2010	1.0	0.1	3.7	0.7	1.1
2011	1.0	0.1	3.2	0.8	0.8
2012	1.2	0.1	2.9	0.8	0.8
2013	1.0	0.1	2.9	0.8	0.9
2014	1.0	0.1	2.8	0.7	0.7
2015	1.0	0.1	2.9	0.7	0.8

Year	Tomatoes	Sweet corn	Pickles	Other [1]	Total [2]
	Pounds	Pounds	Pounds	Pounds	Pounds
2006	64.5	8.4	3.0	8.9	92.2
2007	68.7	6.9	3.7	8.4	94.5
2008	67.1	6.7	3.5	8.8	92.8
2009	70.3	7.6	5.1	8.9	98.7
2010	71.1	6.9	3.7	9.1	97.4
2011	65.7	5.8	2.8	9.1	89.4
2012	66.5	5.9	3.0	9.8	90.9
2013	65.9	5.8	3.2	9.6	90.1
2014	67.2	5.8	3.9	9.7	91.8
2015	56.2	5.3	3.4	9.6	80.1

[1] Includes beets, chile peppers (all uses), green lima beans and spinach.   [2] Total excludes mushrooms and potatoes. May not add due to rounding.
ERS, Market and Trade Economics Division, (202) 694–5326.

Table 4-80.—Vegetables, freezing: Per capita utilization (farm weight basis), United States, 2006–2015

Year	Leafy, green, and yellow vegetables					
	Asparagus	Snap beans	Carrots	Peas	Broccoli	Cauliflower
	Pounds	Pounds	Pounds	Pounds	Pounds	Pounds
2006	0.1	1.9	2.1	1.6	2.3	0.4
2007	0.1	2.1	1.5	1.8	2.7	0.4
2008	0.1	2.1	1.5	1.8	2.7	0.4
2009	0.1	1.9	1.5	1.7	2.5	0.4
2010	0.1	2.0	1.5	1.5	2.5	0.4
2011	0.1	1.5	1.6	1.6	2.7	0.4
2012	0.1	1.9	1.2	1.9	2.6	0.3
2013	0.1	2.1	1.7	1.5	2.5	0.3
2014	0.1	1.8	1.2	1.6	2.6	0.4
2015	0.1	1.9	1.4	1.5	2.6	0.3

Year	Sweet Corn	Other [1]	Total vegetables excluding potatoes	Potato products	Grand total [2]
	Pounds	Pounds	Pounds	Pounds	Pounds
2006	9.4	4.1	21.8	53.2	75.0
2007	10.0	4.1	22.6	53.2	75.8
2008	9.2	4.0	21.9	51.5	73.4
2009	9.1	4.3	21.4	50.4	71.8
2010	8.5	4.5	20.9	50.2	71.1
2011	9.8	4.3	21.9	48.2	70.1
2012	9.8	4.5	22.3	48.4	70.7
2013	7.0	4.3	19.5	49.3	68.8
2014	7.7	4.6	19.8	....	....
2015	8.0	4.3	20.1	....	....

[1] Includes green lima beans, spinach, and miscellaneous freezing vegetables.   [2] Totals may not add due to rounding.
ERS, Market and Trade Economics Division, (202) 694–5326.

Table 4-81.—Commercially produced vegetables: Per capita utilization, United States, 2006–2015

Year	Farm weight equivalent						
	Total fresh and processed	Fresh [2]	Processed [3]				
			Total	Canning	Freezing	Potatoes	Others [1]
	Pounds	Pounds	Pounds	Pounds	Pounds	Pounds	Pounds
2006	396.4	148.4	248.0	92.2	21.8	123.7	10.3
2007	399.5	148.1	251.5	94.5	22.6	124.3	10.1
2008	386.5	143.1	243.4	92.8	21.9	118.3	10.5
2009	385.5	141.3	244.2	98.7	21.4	113.3	10.8
2010	387.7	144.3	243.3	97.4	20.9	113.8	11.2
2011	375.8	141.9	233.8	89.4	21.9	110.3	12.2
2012	384.8	144.7	240.0	90.9	22.3	114.7	12.2
2013	374.6	140.6	234.0	90.1	19.5	113.3	11.0
2014	378.1	141.7	236.3	91.8	19.8	112.1	12.7
2015	366.5	139.7	226.9	80.1	20.1	113.7	12.9

Year	Percentage of annual total					
	Fresh [2]	Processed [3]				
		Total	Canning	Freezing	Potatoes	Others [1]
	Percent	Percent	Percent	Percent	Percent	Percent
2006	37.4	62.6	23.3	5.5	31.2	2.6
2007	37.1	62.9	23.6	5.7	31.1	2.5
2008	37.0	63.0	24.0	5.7	30.6	2.7
2009	36.7	63.3	25.6	5.5	29.4	2.8
2010	37.2	62.8	25.1	5.4	29.4	2.9
2011	37.8	62.2	23.8	5.8	29.4	3.3
2012	37.6	62.4	23.6	5.8	29.8	3.2
2013	37.5	62.5	24.0	5.2	30.2	2.9
2014	37.5	62.5	24.3	5.2	29.6	3.4
2015	38.1	61.9	21.9	5.5	31.0	3.5

[1] Includes potatoes, sweet potatoes, mushrooms, and onions for dehydrating market.   [2] See table 4-78 for items included under Fresh category. Excludes melons.
ERS, Market and Trade Economics Division, (202) 694–5326.

Table 4-82.—Frozen Vegetables and potato products: Cold storage holdings, end of month, United States, 2014 and 2015

Month	Asparagus		Lima beans		Green beans, regular cut		Green beans, french style	
	2014	2015	2014	2015	2014	2015	2014	2015
	1,000 pounds	*1,000 pounds*	*1,000 pounds*	*1,000 pounds*	*1,000 pounds*	*1,000 pounds*	*1,000 pounds*	*1,000 pounds*
January	9,701	12,620	58,462	58,804	162,305	161,510	13,479	13,214
February	9,497	11,115	54,696	53,249	149,444	153,916	11,607	10,323
March	9,038	10,608	51,382	51,849	139,195	135,211	11,680	10,177
April	7,946	10,084	43,074	47,874	125,882	123,221	10,675	8,250
May	10,252	13,037	39,818	42,782	109,855	119,151	8,790	7,246
June	15,185	14,457	34,333	41,133	96,376	113,449	8,194	8,025
July	15,656	14,209	31,632	37,376	147,470	155,410	15,306	12,653
August	16,324	14,270	35,991	44,819	225,135	226,850	18,840	12,695
September	16,916	13,283	52,979	62,195	253,066	255,661	20,852	19,787
October	14,873	12,579	73,497	66,999	235,830	243,986	18,439	19,142
November	14,572	12,883	69,543	59,313	205,861	225,001	16,975	16,305
December	13,603	11,926	64,530	57,841	182,136	199,994	14,754	15,279

Month	Broccoli, spears		Broccoli, chopped & cut		Brussels sprouts		Carrots, diced	
	2014	2015	2014	2015	2014	2015	2014	2015
	1,000 pounds	*1,000 pounds*	*1,000 pounds*	*1,000 pounds*	*1,000 pounds*	*1,000 pounds*	*1,000 pounds*	*1,000 pounds*
January	19,728	25,796	34,253	34,751	15,327	15,090	153,372	155,485
February	20,956	27,826	33,570	38,107	15,343	14,120	143,648	144,853
March	23,558	29,124	31,421	37,416	14,071	13,448	130,512	134,587
April	30,509	29,731	30,447	39,532	14,283	13,170	114,796	121,669
May	34,882	29,475	32,993	39,932	13,452	13,658	109,991	115,221
June	34,093	27,821	33,394	42,211	12,480	14,691	97,279	99,334
July	34,054	26,272	38,547	41,556	10,911	13,401	81,749	87,345
August	32,216	27,564	41,970	50,159	10,142	12,241	74,944	84,143
September	30,285	28,387	41,131	50,323	9,798	12,352	71,990	83,391
October	30,697	26,873	43,775	44,935	12,149	14,338	130,075	148,160
November	29,339	27,317	37,347	41,373	13,585	16,636	173,870	172,525
December	27,108	25,023	36,143	39,398	15,960	19,843	168,289	161,755

Month	Carrots, other		Cauliflower		Corn, cut		Corn, cob	
	2014	2015	2014	2015	2014	2015	2014	2015
	1,000 pounds	*1,000 pounds*	*1,000 pounds*	*1,000 pounds*	*1,000 pounds*	*1,000 pounds*	*1,000 pounds*	*1,000 pounds*
January	146,956	155,156	21,680	20,172	492,643	464,026	225,426	206,000
February	134,187	138,003	21,139	18,289	460,752	431,593	200,562	175,910
March	121,009	120,999	18,386	17,903	397,092	395,330	174,569	156,425
April	118,973	117,545	17,768	17,761	345,504	348,564	156,311	132,387
May	114,263	102,967	18,035	17,424	296,765	313,388	121,262	110,912
June	102,942	88,645	16,497	18,203	250,726	262,028	101,087	88,817
July	97,360	83,838	15,461	16,963	273,738	287,729	108,477	112,873
August	103,813	79,673	14,427	15,146	462,352	446,215	175,199	165,657
September	130,307	106,742	15,581	14,226	596,616	621,237	258,604	239,922
October	171,014	157,115	22,014	24,622	631,669	640,668	268,087	244,687
November	195,425	173,385	21,897	24,833	576,419	582,672	240,784	229,113
December	180,493	166,317	20,567	26,448	529,586	520,804	223,053	220,295

Month	Mixed vegetables		Okra		Onion rings		Onions, other	
	2014	2015	2014	2015	2014	2015	2014	2015
	1,000 pounds	*1,000 pounds*	*1,000 pounds*	*1,000 pounds*	*1,000 pounds*	*1,000 pounds*	*1,000 pounds*	*1,000 pounds*
January	56,257	55,838	14,964	34,631	9,700	9,276	40,725	34,589
February	54,208	56,039	10,651	29,153	9,135	8,980	37,919	39,075
March	47,594	57,851	7,129	23,998	8,114	9,862	38,309	39,098
April	48,023	59,040	7,353	19,361	8,032	11,918	38,143	40,784
May	47,670	58,132	7,670	16,410	9,154	11,810	40,892	44,569
June	49,422	58,099	14,800	16,337	11,141	13,135	44,632	43,538
July	52,108	56,960	26,511	31,938	9,933	11,734	54,319	41,733
August	58,514	60,109	36,605	48,184	9,755	11,923	51,107	41,505
September	60,248	62,186	46,743	66,862	9,981	11,242	42,755	36,864
October	62,359	58,359	48,448	70,851	10,489	12,145	39,793	36,790
November	59,702	55,274	44,192	63,695	9,127	11,784	36,664	35,137
December	53,082	55,067	39,375	56,080	9,057	12,004	35,944	33,020

See footnote(s) at end of table.

STATISTICS OF VEGETABLES AND MELONS

Table 4-82.—Frozen Vegetables and potato products: Cold storage holdings, end of month, United States, 2014 and 2015—Continued

Month	Blackeye peas		Green peas		Peas & carrots mixed		Spinach	
	2014	2015	2014	2015	2014	2015	2014	2015
	1,000 pounds	*1,000 pounds*	*1,000 pounds*	*1,000 pounds*	*1,000 pounds*	*1,000 pounds*	*1,000 pounds*	*1,000 pounds*
January	1,925	1,969	192,134	205,887	6,051	7,299	41,909	37,933
February	1,620	1,846	168,608	189,997	7,608	7,326	43,140	40,421
March	2,072	1,822	136,081	154,385	7,234	7,315	51,886	48,253
April	1,684	2,364	111,179	127,621	7,106	6,964	48,413	51,884
May	2,145	2,816	95,529	127,985	6,494	7,459	54,644	57,732
June	1,655	2,869	284,293	314,155	7,274	7,580	52,579	60,413
July	1,830	2,975	427,634	414,632	7,813	7,637	49,247	53,984
August	1,484	2,121	395,401	397,020	7,840	8,292	46,541	48,267
September	1,784	2,330	354,716	372,987	7,703	8,225	38,924	40,430
October	1,870	2,159	314,577	335,536	7,483	8,135	39,770	38,540
November	1,929	2,493	285,406	308,934	7,525	7,751	39,672	40,500
December	1,565	2,434	243,251	274,778	7,671	7,052	38,024	42,629

Month	Squash		Southern greens		Other vegetables		Total frozen vegetables	
	2014	2015	2014	2015	2014	2015	2014	2015
	1,000 pounds	*1,000 pounds*	*1,000 pounds*	*1,000 pounds*	*1,000 pounds*	*1,000 pounds*	*1,000 pounds*	*1,000 pounds*
January	57,592	59,715	11,763	11,941	371,053	357,718	2,157,405	2,144,859
February	56,111	54,447	13,778	15,065	366,772	335,823	2,024,951	1,995,476
March	50,334	49,520	12,126	17,021	359,294	327,915	1,842,086	1,850,117
April	48,985	43,978	13,295	20,894	311,776	319,475	1,660,157	1,714,071
May	47,022	43,088	12,849	18,052	285,450	307,475	1,519,877	1,620,721
June	42,958	39,517	14,873	19,717	281,639	302,891	1,607,852	1,697,065
July	40,448	41,401	13,599	14,741	310,021	321,753	1,863,824	1,889,113
August	53,729	49,943	12,353	12,050	348,093	364,047	2,232,775	2,222,893
September	56,780	53,728	10,597	10,231	402,271	397,496	2,530,627	2,570,087
October	61,649	63,065	9,610	11,566	415,446	407,400	2,663,613	2,688,650
November	63,870	65,214	10,643	16,227	392,454	391,876	2,546,801	2,580,241
December	62,828	65,550	11,253	18,405	381,070	391,886	2,359,342	2,423,828

Month	Frozen potatoes					
	French fries		Other frozen potatoes		Total frozen potatoes	
	2014	2015	2014	2015	2014	2015
	1,000 pounds	*1,000 pounds*	*1,000 pounds*	*1,000 pounds*	*1,000 pounds*	*1,000 pounds*
January	913,426	899,617	191,376	192,123	1,104,802	1,091,740
February	924,730	924,506	199,400	208,415	1,124,130	1,132,921
March	840,697	925,691	203,650	220,737	1,044,347	1,146,428
April	806,282	908,125	202,949	219,040	1,009,231	1,127,165
May	780,688	919,142	202,730	216,455	983,418	1,135,597
June	807,376	916,294	205,458	235,411	1,012,834	1,151,705
July	725,507	832,304	198,740	233,141	924,247	1,065,445
August	742,033	828,168	194,555	218,845	936,588	1,047,013
September	837,281	840,315	202,414	217,728	1,039,695	1,058,043
October	890,062	874,165	209,443	218,787	1,099,505	1,092,952
November	900,485	850,663	205,070	204,894	1,105,555	1,055,557
December	842,464	809,569	187,967	197,327	1,030,431	1,006,896

NASS, Livestock Branch, (202) 720-3570.

CHAPTER V

STATISTICS OF FRUITS, TREE NUTS, AND HORTICULTURAL SPECIALTIES

For most fruits, production is estimated at two levels—total and utilized. Total production is the quantity of fruit harvested plus quantities which would have been acceptable for fresh market or processing but were not harvested or utilized because of economic and other reasons. Utilized production is the amount sold plus the quantities used on farms where grown and quantities held in storage. The difference between total and utilized production is the quantity of marketable fruit not harvested and fruit harvested but not utilized because of economic and other reasons. Production relates to the crop produced on all farms, except for apples and strawberries. In accordance with Congressional enactment, the Department's estimates of apple production since 1938 have related only to commercial production. The estimates for strawberries cover production on area grown primarily for sale. Statistics on utilization of fruit by commercial processors refer to first utilization, not necessarily final utilization. For example, frozen fruit includes fruit which may later be used for preserves.

The price shown for each crop is a marketing year average price for all methods of sales. Prices for most fresh fruit are the average prices producers received at the point of first sale, commonly referred to as the "average price as sold." Since the point of first sale is not the same for all producers, prices for the various methods of sale are weighted by the proportionate quantity sold. For example, if in a given State part of the fruit crop is sold f.o.b. packed by growers, part sold as bulk fruit at the packinghouse door, and some sold retail at roadside stands, the fresh fruit average price as sold is a weighted average of the average price for each method of sale.

The annual estimates are checked and adjusted at the end of each marketing season on the basis of shipment and processing records from transportation agencies, processors, cooperative marketing associations, and other industry organizations. The estimates are reviewed (and revised if necessary) at 5-year intervals, when the Census of Agriculture data become available. The Department's available statistics are limited to the major tree fruits and nuts and to grapes, cranberries, and strawberries, and exclude some States where census data indicate production is of only minor importance.

Table 5-1.—Fruits and planted nuts: Bearing area, United States, 2006–2015

Year	Citrus fruits [1]	Major deciduous fruits [2]	Miscellaneous Noncitrus [3]	Nuts [4]	Total
	1,000 acres	*1,000 acres*	*1,000 acres*	*1,000 acres*	*1,000 acres*
2006	886.8	1,752.5	309.1	981.2	3,929.6
2007	866.2	1,729.7	292.1	1,016.6	3,904.5
2008	850.9	1,731.5	303.4	1,101.3	3,987.1
2009	844.8	1,739.7	306.3	1,159.7	4,050.5
2010	825.2	1,750.3	310.0	1,206.0	4,091.5
2011	809.2	1,744.2	315.5	1,261.5	4,116.0
2012	801.8	1,734.6	260.7	1,316.0	4,113.1
2013	791.2	1,749.6	339.4	1,379.0	4,259.1
2014	777.7	1,733.8	361.6	1,427.0	4,300.0
2015	764.9	1,712.2	364.0	1,469.0	4,310.2

[1] Grapefruit, lemons, oranges, tangelos, and tangerines. [2] Commercial apples, apricots, cherries, grapes, nectarines, peaches, pears, plums, and prunes. [3] Avocados, bananas, berries, coffee, cranberries, dates, figs, guavas, kiwifruit, olives, papayas, and strawberries. [4] Almonds, hazelnuts, macadamia nuts, pistachios, and walnuts.
NASS, Crops Branch, (202) 720–2127.

Table 5-2.—Fruits: Total production in tons, United States, 2006–2015 [1]

Year	Apples, commercial crop [2]	Peaches	Pears	Grapes (fresh basis)	Sweet cherries	Tart cherries	Apricots
	1,000 tons	*1,000 tons*	*1,000 tons*	*1,000 tons*	*1,000 tons*	*1,000 tons*	*1,000 tons*
2006	4,912	1,010	842	6,378	294	131	45
2007	4,545	1,127	873	7,057	·311	127	89
2008	4,812	1,135	870	7,279	248	107	82
2009	4,844	1,104	957	7,268	443	180	69
2010	4,641	1,149	814	7,432	313	95	66
2011	4,720	1,071	966	7,409	334	116	67
2012	4,496	968	851	7,531	424	43	61
2013	5,216	904	877	8,632	332	147	61
2014	5,907	853	832	7,884	364	152	65
2015	5,002	847	821	7,677	338	126	42

Year	Figs (fresh basis)	Plums (CA)	Prunes (fresh basis) (CA)	Prunes & Plums (ID,MI,OR,WA)	Olives	Strawberries [3]	Avocados [3]
	1,000 tons	*1,000 tons*	*1,000 tons*	*1,000 tons*	*1,000 tons*	*1,000 tons*	*1,000 tons*
2006	43	158	634	22	24	1,202	147
2007	48	152	241	12	133	1,223	193
2008	43	160	368	16	67	1,266	116
2009	44	112	496	19	46	1,401	299
2010	41	142	390	12	206	1,426	174
2011	39	160	444	13	71	1,451	263
2012	35	115	436	13	160	1,526	(NA)
2013	33	95	255	13	166	1,524	183
2014	33	113	324	15	95	1,511	173
2015	30	106	319	10	179	1,161	224

Year	Nectarines	Oranges [4]	Tangerines and Mandarins [4]	Grapefruit [4]	Lemons [4]	Tangelos [4]	Cranberries [5]
	1,000 tons	*1,000 tons*	*1,000 tons*	*1,000 tons*	*1,000 tons*	*1,000 tons*	*1,000 tons*
2006	232	9,020	417	1,232	980	63	345
2007	283	7,625	361	1,627	798	56	328
2008	303	10,076	527	1,548	619	68	393
2009	220	9,128	443	1,304	912	52	346
2010	233	8,243	596	1,238	882	41	340
2011	225	8,905	657	1,264	920	52	386
2012	189	8,268	682	1,204	912	45	402
2013	162	6,768	732	1,047	824	40	448
2014	205	6,353	863	910	904	30	420
2015	168	5,911	935	803	890	18	312

Year	Bananas	Kiwifruit	Dates	Papayas	Berries [6]	Guavas
	1,000 tons	*1,000 tons*	*1,000 tons*	*1,000 tons*	*1,000 tons*	*1,000 tons*
2006	11	26	18	14	285	4
2007	13	25	16	17	288	2
2008	9	23	21	17	320	2
2009	9	26	24	16	359	1
2010	9	33	29	15	348	1
2011	9	38	33	14	384	1
2012	(NA)	30	31	(NA)	391	(NA)
2013	7	28	31	12	435	1
2014	6	28	33	12	480	1
2015	6	24	44	14		(NA)

(NA) Not available.   [1] For some crops in certain years, production includes some quantities unharvested for economic reasons or excess cullage fruit.   [2] Estimates of the commercial crop refer to production in orchards of 100 or more bearing-age trees.   [3] Year of bloom.   [4] Year harvest was complete.   [5] Temples included in early, midseason, and navel oranges beginning with the 2006-07 season.   [6] Excludes strawberries and cranberries.
NASS, Crops Branch, (202) 720–2127.

Table 5-3.—Apples, commercial crop: Production and season average price per pound, by State and United States, 2013–2015

State	Total production			Utilized production			Price per pound [1] for crop of—		
	2013	2014	2015	2013	2014	2015	2013	2014	2015
	Million pounds	*Million pounds*	*Million pounds*	*Million pounds*	*Million pounds*	*Million pounds*	*Dollars*	*Dollars*	*Dollars*
Arizona	16.5	7.1	(D)	16.3	7.1	(D)	0.274	0.421	(D)
California	270.0	240.0	146.0	270.0	240.0	145.0	0.233	0.238	0.278
Colorado	5.6	8.9	(D)	5.3	8.2	(D)	0.363	0.892	(D)
Connecticut	27.0	19.9	25.1	26.5	19.6	24.3	0.741	0.632	0.585
Idaho	70.6	63.3	46.1	70.0	60.0	46.0	0.331	0.146	0.326
Illinois	16.0	21.0	20.5	15.0	20.2	19.9	0.528	0.764	0.558
Indiana	30.0	17.1	22.5	27.0	16.0	20.5	0.387	0.430	0.382
Iowa	8.1	4.5	4.8	6.8	3.8	4.0	0.699	0.779	0.863
Maine	27.0	38.0	35.6	25.5	36.8	35.1	0.516	0.421	0.504
Maryland	33.0	45.4	41.0	32.6	45.2	40.9	0.187	0.245	0.202
Massachusetts	43.5	43.3	43.1	40.5	40.3	42.4	0.520	0.492	0.530
Michigan	1,260.0	1,025.0	995.0	1,250.0	1,025.0	990.0	0.204	0.216	0.224
Minnesota	26.0	25.0	26.1	22.0	23.0	21.9	0.833	0.842	0.810
Missouri	16.9	20.9	28.3	15.0	19.8	28.1	0.368	0.283	0.440
New Hampshire	25.5	16.9	20.2	24.5	16.5	20.0	0.429	0.616	0.638
New Jersey	29.0	37.0	36.7	28.5	36.0	36.0	0.451	0.847	0.905
New York	1,410.0	1,260.0	1,360.0	1,390.0	1,250.0	1,350.0	0.171	0.200	0.203
North Carolina	155.0	125.0	105.0	145.0	125.0	103.0	0.222	0.246	0.207
Ohio	54.0	44.0	50.5	51.0	43.0	49.0	0.418	0.453	0.421
Oregon	141.0	155.0	125.4	137.0	155.0	125.0	0.364	0.279	0.355
Pennsylvania	469.0	534.0	519.0	455.0	523.0	515.0	0.186	0.167	0.188
Rhode Island	2.5	1.8	2.4	2.3	1.7	2.2	0.863	0.740	0.831
Tennessee	6.9	5.5	4.6	6.3	5.0	4.5	0.444	0.438	0.295
Utah	16.5	23.0	15.0	15.8	22.4	14.9	0.481	0.219	0.329
Vermont	34.0	29.4	36.2	31.0	28.2	35.9	0.451	0.407	0.432
Virginia	195.0	205.0	195.2	190.0	205.0	195.0	0.176	0.173	0.177
Washington	5,900.0	7,650.0	5,950.0	5,900.0	7,050.0	5,910.0	0.362	0.268	0.405
West Virginia	95.0	94.0	90.2	95.0	94.0	90.0	0.137	0.150	0.150
Wisconsin	48.0	54.0	51.5	46.0	51.0	48.1	0.508	0.612	0.578
Other States	(X)	(X)	7.9	(X)	(X)	7.7	(X)	(X)	0.462
United States	10,431.6	11,814.0	10,003.9	10,339.9	11,170.8	9,924.4	0.303	0.257	0.342

(X) Not applicable. [1] Fresh fruit prices are equivalent packinghouse-door returns for California, Michigan, New York, and Washington; prices at point of first sale for other States. Processing prices are equivalent at processing plant door. NASS, Crops Branch, (202) 720–2127.

Table 5-4.—Apples: Production and value, United States, 2006–2015

Year	Apples, commercial crop			
	Total production	Utilized production	Marketing year average price [1]	Value
	Million pounds	*Million pounds*	*Cents per pound*	*1,000 dollars*
2006	9,823.4	9,730.2	22.7	2,213,155
2007	9,089.4	9,045.4	28.8	2,608,220
2008	9,623.3	9,531.7	23.2	2,210,997
2009	9,687.7	9,435.6	23.1	2,182,003
2010	9,282.1	9,205.0	25.1	2,310,982
2011	9,439.6	9,328.8	30.3	2,830,784
2012	8,992.3	8,926.5	37.1	3,314,996
2013	10,431.6	10,339.9	30.3	3,132,936
2014	11,814.0	11,170.8	25.7	2,870,745
2015	10,003.9	9,924.4	34.2	3,394,185

[1] Fresh fruit prices are equivalent packinghouse-door returns for California, Michigan, New York, and Washington; prices at point of first sale for other States. Processing prices are equivalent at processing plant door. NASS, Crops Branch, (202) 720–2127.

Table 5-5.—International Apples, fresh: Production in specified countries, 2013/2014–2015/2016

Country	Production		
	2013/2014	2014/2015	2015/2016
	metric tons	metric tons	metric tons
Brazil	1,377,400	1,266,000	1,240,000
Chile	1,310,000	1,350,000	1,350,000
China	39,680,000	40,920,000	43,000,000
European Union	11,864,900	13,618,500	12,220,000
India	2,200,000	2,200,000	2,200,000
Iran	1,693,000	1,693,000	1,693,000
Russia	1,416,500	1,408,600	1,390,000
South Africa	792,500	860,000	865,000
Turkey	2,930,000	2,289,000	2,740,000
Ukraine	1,211,000	1,211,000	1,211,000
Others	4,649,600	4,560,800	4,548,800
Total Foreign	69,124,900	71,376,900	72,457,800
United States	4,690,101	5,074,793	4,561,000
Total	73,815,001	76,451,693	77,018,800

FAS, Office of Global Analysis, (202) 720-6301. Prepared or estimated on the basis of official USDA production, supply, and distribution statistics from foreign governments.

Table 5-6.—Apples, commercial crop: Production and utilization, United States, 2006–2015

Crop of—	Total production	Utilized production	Utilization				
			Fresh	Processed (fresh basis)			
				Canned	Dried	Frozen	Juice, cider & other [1]
	Million pounds	Million pounds	Million pounds	Million pounds	Million pounds	Million pounds	Million pounds
2006	9,823.4	9,730.2	6,308.5	1,167.3	252.8	271.8	1,729.8
2007	9,089.4	9,045.4	6,077.3	1,091.2	203.7	257.7	1,415.5
2008	9,623.3	9,531.7	6,265.9	1,253.4	212.7	211.2	1,588.5
2009	9,687.7	9,435.6	6,296.4	1,158.2	161.2	236.2	1,583.6
2010	9,282.1	9,205.0	6,248.8	1,088.3	175.5	206.0	1,486.4
2011	9,439.6	9,328.8	6,312.9	1,123.7	183.5	190.5	1,517.8
2012	8,992.3	8,926.5	6,594.9	748.9	223.0	66.6	1,292.8
2013	10,431.6	10,339.9	6,895.3	1,264.4	161.0	239.3	1,779.9
2014	11,814.0	11,170.8	7,882.3	1,135.0	171.0	250.5	1,732.0
2015	10,003.9	9,924.4	6,855.7	1,085.0	179.0	232.5	1,572.2

[1] Includes vinegar, wine, and slices for pie making.
NASS, Crops Branch, (202) 720-2127.

Table 5-7.—Apples, commercial crop: Production and utilization, by State and United States, crop of 2014 and 2015

State	Total production	Utilized production	Fresh	Canned	Dried	Frozen	Juice, cider & other [1]
				Utilization			
				Processed (fresh basis)			
	Million pounds	Million pounds	Million pounds	Million pounds	Million pounds	Million pounds	Million pounds
				2014			
Arizona	7.1	7.1	(D)				
California	240.0	240.0	(D)				
Colorado	8.9	8.2	(D)				
Connecticut	19.9	19.6	19.1				
Idaho	63.3	60.0	34.0				
Illinois	21.0	20.2	(D)				
Indiana	17.1	16.0	9.0				
Iowa	4.5	3.8	(D)				
Maine	38.0	36.8	29.6				
Maryland	45.4	45.2	15.2	(D)			
Massachusetts	43.3	40.3	28.3				
Michigan	1,025.0	1,025.0	425.0	260.0			
Minnesota	25.0	23.0	(D)				
Missouri	20.9	19.8	11.6				
New Hampshire	16.9	16.5	(D)				
New Jersey	37.0	36.0	26.0				
New York	1,260.0	1,250.0	625.0	249.0			
North Carolina	125.0	125.0	65.0	(D)			
Ohio	44.0	43.0	34.0				
Oregon	155.0	155.0	132.0				
Pennsylvania	534.0	523.0	191.0	234.0			
Rhode Island	1.8	1.7	(D)				
Tennessee	5.5	5.0	(D)				
Utah	23.0	22.4	(D)				
Vermont	29.4	28.2	23.8				
Virginia	205.0	205.0	75.0				
Washington	7,650.0	7,050.0	5,900.0	(D)			
West Virginia	94.0	94.0	20.0	(D)			
Wisconsin	54.0	51.0	42.3				
Other States	(X)	(X)	176.4	392.0	171.0	250.5	1,732.0
United States	11,814.0	11,170.8	7,882.3	1,135.0	171.0	250.5	1,732.0
				2015			
Arizona	(D)	(D)	(D)				
California	146.0	145.0	(D)				
Colorado	(D)	(D)	(D)				
Connecticut	25.1	24.3	23.3				
Idaho	46.1	46.0	30.0				
Illinois	20.5	19.9	(D)				
Indiana	22.5	20.5	12.5				
Iowa	4.8	4.0	(D)				
Maine	35.6	35.1	33.1				
Maryland	41.0	40.9	(D)	(D)			
Massachusetts	43.1	42.4	29.7				
Michigan	995.0	990.0	410.0	210.0			
Minnesota	26.1	21.9	(D)				
Missouri	28.3	28.1	23.6				
New Hampshire	20.2	20.0	(D)				
New Jersey	36.7	36.0	26.0				
New York	1,360.0	1,350.0	715.0	264.0			
North Carolina	105.0	103.0	50.0	(D)			
Ohio	50.5	49.0	40.0				
Oregon	125.0	125.0	106.0				
Pennsylvania	519.0	515.0	198.0	226.0			
Rhode Island	2.4	2.2	(D)				
Tennessee	4.6	4.5	(D)				
Utah	15.0	14.9	(D)				
Vermont	36.2	35.9	30.9				
Virginia	195.2	195.0	65.0				
Washington	5,950.0	5,910.0	4,840.0	(D)			
West Virginia	90.2	90.0	17.0	(D)			
Wisconsin	51.5	48.1	39.3				
Other States	7.9	7.7	166.3	385.0	179.0	232.5	1,572.2
United States	10,003.9	9,924.4	6,855.7	1,085.0	179.0	232.5	1,572.2

[1] Includes vinegar, wine, and slices for pie making.
NASS, Crops Branch, (202) 720–2127.

Table 5-8.—Fruits, fresh: United States exports by country of destination and imports by country of origin, 2011–2013

Country	2011	2012	2013
	Metric tons	*Metric tons*	*Metric tons*
Fresh fruits, deciduous:			
Mexico	346,910	410,382	453,255
Canada	354,472	378,790	384,904
Taiwan	111,712	115,852	115,059
Hong Kong	118,203	103,192	111,600
India	79,631	91,160	63,423
United Arab Emirates	34,526	39,640	58,558
Indonesia	82,620	81,230	48,693
Vietnam	19,803	17,553	30,596
Malaysia	23,835	26,776	29,052
Philippines	22,499	24,128	28,863
Thailand	32,865	30,457	28,720
China	21,509	26,182	28,536
Colombia	29,502	28,494	24,830
Russia	18,751	15,254	24,368
Saudi Arabia	22,872	15,437	20,871
Korea, South	13,011	20,762	20,521
Singapore	16,368	14,854	19,361
Australia(*)	10,739	17,018	19,252
Japan	18,842	22,091	18,839
Dominican Republic	15,920	18,022	18,189
United Kingdom	15,540	18,076	17,649
Guatemala	15,576	15,569	15,803
Costa Rica	15,193	14,014	14,816
Honduras	12,969	12,317	11,125
Panama	9,470	8,826	10,117
El Salvador	11,497	10,841	10,028
New Zealand(*)	11,579	11,303	8,608
Israel(*)	8,090	6,959	7,744
Peru	4,280	4,079	7,471
Rest of World	82,499	70,109	59,938
World Total	1,581,281	1,669,367	1,710,789
Fresh fruit, other:			
Canada	199,822	224,529	232,049
Mexico	13,804	18,521	23,469
Japan	12,281	17,492	15,684
Korea, South	10,026	17,524	14,739
United Kingdom	6,138	4,597	6,701
Netherlands	4,386	3,822	5,031
Hong Kong	2,483	3,514	4,119
Chile	140	155	3,275
Bermuda	2,243	2,315	3,052
Taiwan	5,897	2,057	2,661
Russia	1,346	1,456	2,453
Belgium-Luxembourg(*)	1,106	1,404	2,428
Australia(*)	3,498	3,447	2,369
United Arab Emirates	4,224	1,682	1,859
France(*)	1,934	1,874	1,623
New Zealand(*)	710	666	1,349
Singapore	622	1,241	895
China	966	856	704
Saudi Arabia	409	305	618
Germany(*)	965	108	583
Bahamas, The	517	582	554
Cayman Islands	432	839	515
Brazil	604	627	474
Sweden	171	139	446
Costa Rica	137	80	338
Trinidad and Tobago	789	470	336
Colombia	159	214	336
Dominican Republic	598	528	283
Kuwait	686	273	277
Rest of World	5,677	3,576	3,006
World Total	282,768	314,892	332,225

See footnote(s) at end of table.

Table 5-8.—Fruits, fresh: United States exports by country of destination and imports by country of origin, 2011–2013—Continued

Country	2011	2012	2013
	Metric tons	*Metric tons*	*Metric tons*
Fresh melons:			
Canada	239,612	232,092	228,859
Mexico	32,787	17,526	16,141
Japan	13,546	12,905	13,122
Hong Kong	1,747	2,082	3,078
Taiwan	369	849	2,045
Korea, South	2,010	2,054	1,685
United Arab Emirates	326	51	613
Bahamas, The	817	699	554
Panama	79	84	169
Chile	10	53	100
Costa Rica	90	0	33
Ecuador	0	0	33
Netherlands Antilles(*)	16	65	28
Saudi Arabia	0	1	23
Cayman Islands	46	6	22
Turks and Caicos Islands	3	15	17
Singapore	313	65	16
Barbados	21	13	16
Leeward-Windward Islands(*)	7	33	13
Bermuda	176	114	8
China	0	19	7
Bahrain	0	27	6
Trinidad and Tobago	0	0	4
French Pacific Islands(*)	0	0	4
Bulgaria	0	0	4
Haiti	0	0	3
Dominican Republic	4	0	2
Brazil	0	1	0
Guatemala	0	30	0
Rest of World	82	56	0
World Total	292,059	268,837	299,603

(*) Denotes a country that is a summarization of its component countries. All zeroes for a data item may show that statistics exist in the other import type. Consumption or General. Users should use cautious interpretation on Quantity reports using mixed units of measure. Quantity line items will only include statistics on the units of measure that are equal to, or are able to be converted to, the assigned unit of measure of the grouped commodities.

FAS, Office of Global Analysis, (202) 720-6301. Data Source: Department of Commerce, U.S. Census Bureau, Foreign Trade Statistics.

Table 5-9.—Apples: Foreign trade, United States, 2006–2015[1]

Year beginning October	Imports, fresh and dried, in terms of fresh	Domestic exports	
		Fresh	Dried, in terms of fresh[1]
	Metric tons	*Metric tons*	*Metric tons*
2006	255,320	652,827	32,926
2007	238,333	680,618	31,679
2008	206,667	810,984	26,849
2009	225,717	748,054	22,127
2010	184,475	817,906	26,802
2011	219,305	856,819	28,492
2012	243,093	908,751	22,282
2013	243,875	848,671	23,741
2014	209,498	1,057,852	26,092
2015	233,623	747,267	27,962

[1] Dried converted to terms of fresh apples on following basis; 1 pound dried is equivalent to 8 pounds fresh. No re-exports reported.
ERS, Crops Branch, (202) 694–5255.

Table 5-10.—Apricots: Production and value, United States, 2006–2015

Year	Total production	Utilized production	Market year average price per ton	Value
	Tons	*Tons*	*Dollars*	*1,000 dollars*
2006	44,480	44,455	665.00	29,563
2007	88,460	88,460	477.00	42,227
2008	81,610	77,480	532.00	41,196
2009	68,720	68,690	654.00	44,912
2010	66,380	66,350	722.00	47,876
2011	66,650	66,620	616.00	41,056
2012	60,800	60,770	673.00	40,879
2013	61,035	61,028	737.00	44,987
2014	64,928	64,918	818.00	53,096
2015	41,657	41,507	1,010.00	41,730

NASS, Crops Branch, (202) 720–2127.

Table 5-11.—Apricots: Production and marketing year average price per ton, by State and United States, 2013–2015

State	Total production			Utilized production			Price for crop of—		
	2013	2014	2015	2013	2014	2015	2013	2014	2015
	Tons	*Tons*	*Tons*	*Tons*	*Tons*	*Tons*	*Dollars*	*Dollars*	*Dollars*
California	54,400	55,400	34,500	54,400	55,400	34,500	682.00	777.00	1,020.00
Utah	135	228	7	128	218	7	1,010.00	1,510.00	(D)
Washington	6,500	9,300	7,150	6,500	9,300	7,000	1,190.00	1,050.00	(D)
United States ..	61,035	64,928	41,657	61,028	64,918	41,507	737.00	818.00	1,010.00

NASS, Crops Branch, (202) 720–2127.

Table 5-12.—Apricots: Production and utilization, United States, 2006–2015

Crop of—	Total production	Utilized production	Utilization of quantities sold			
			Fresh	Processed[1]		
				Canned	Dried (fresh basis)	Frozen
	Tons	*Tons*	*Tons*	*Tons*	*Tons*	*Tons*
2006	44,480	44,455	13,755	14,900	5,500	(2)
2007	88,460	88,460	29,270	24,000	13,000	(2)
2008	81,610	77,480	25,760	22,000	14,000	(2)
2009	68,720	68,690	25,170	23,100	9,000	(2)
2010	66,380	66,350	23,510	20,100	10,000	(2)
2011	66,650	66,620	23,990	17,150	12,000	(2)
2012	60,800	60,770	23,130	16,500	9,500	(2)
2013	61,035	61,028	25,228	14,550	10,000	(2)
2014	64,928	64,918	25,812	17,550	9,500	(2)
2015	41,657	41,507	17,704	7,100	8,500	(2)

[1] California only. [2] Withheld to avoid disclosure of individual operations.
NASS, Crops Branch, (202) 720–2127.

Table 5-13.—Apricots: Production and utilization, by State and United States, crop of 2015

State	Total production	Utilized production	Utilization			
			Fresh	Processed[1]		
				Canned	Dried (fresh basis)	Frozen
	Tons	*Tons*	*Tons*	*Tons*	*Tons*	*Tons*
California	34,500	34,500	12,600	(2)	(2)	(2)
Utah	7	7	(2)	(2)	(2)	(2)
Washington	7,150	7,000	(2)	(2)	(2)	(2)
United States	41,657	41,507	17,704	7,100	8,500	(2)

[1] California only. [2] Withheld to avoid disclosure of individual operations.
NASS, Crops Branch, (202) 720–2127.

Table 5-14.—Apricots: Foreign trade, United States, 2006–2015

Year beginning October	Domestic exports				
	Fresh	Canned[1]	Dried[1]	Dried, in fruit salad[2]	Total, in terms of fresh[3]
	Metric tons	*Metric tons*	*Metric tons*	*Metric tons*	*Metric tons*
2006	7,061	1,201	615	364	11,083
2007	7,357	1,217	679	405	11,722
2008	5,047	1,289	602	355	9,064
2009	5,555	1,308	1,011	389	11,691
2010	5,646	1,389	708	470	10,283
2011	7,459	660	634	519	11,191
2012	7,774	646	533	547	10,978
2013	6,859	755	473	504	9,831
2014	5,401	621	370	409	7,749
2015	7,277	508	669	214	11,080

[1] Net processed weight. [2] Dried apricots are 12⅓ percent of total dried fruit for salad. [3] Dried fruit converted to unprocessed dry weight by dividing by 1.07. Unprocessed dry weight converted to terms of fresh fruit on the basis that 1 pound dried equals 5.5 pounds fresh. Canned apricots converted to terms of fresh on the basis that 1 pound canned equals 0.717 pounds fresh.
ERS, Crops Branch, (202) 694–5255.

Table 5-15.—Avocados: Foreign trade, United States, 2006–2015

Year beginning October	Imports
	Metric tons
2006	338,559
2007	307,167
2008	413,936
2009	365,146
2010	385,434
2011	475,630
2012	565,127
2013	699,388
2014	832,822
2015	874,254

ERS, Crops Branch, (202) 694–5255.

Table 5-16.—Avocados: Total production, marketing year average price per ton, and value of utilized production, by State and United States, 2006/2007 to 2015/2016

Season	California			Florida		
	Production	Price per ton	Value	Production	Price per ton	Value
	Tons	*Dollars*	*1,000 1,000 dollars*	*Tons*	*Dollars*	*1,000 1,000 dollars*
2006/07	132,000	1,890.00	249,480	14,000	912.00	12,768
2007/08	165,000	1,990.00	328,350	27,500	440.00	12,100
2008/09	88,000	2,280.00	200,640	27,450	480.00	13,176
2009/10	274,800	1,510.00	414,948	23,200	600.00	13,920
2010/11	151,500	3,040.00	460,560	22,500	800.00	18,000
2011/12	231,500	1,650.00	381,975	31,100	756.00	23,512
2012/13	(NA)	(NA)	(NA)	(NA)	(NA)	(NA)
2013/14	149,000	2,240.00	333,760	33,800	796.00	26,905
2014/15	140,000	2,170.00	303,800	32,900	656.00	21,582
2015/16	196,000	1,400.00	274,400	27,600	745.00	20,562

Season	Hawaii			United States		
	Production	Price per ton	Value	Production	Price per ton	Value
	Tons	*Dollars*	*1,000 dollars*	*Tons*	*Dollars*	*1,000 dollars*
2006/07	510	1,360.00	694	146,510	1,800.00	262,942
2007/08	580	1,360.00	789	193,080	1,770.00	341,239
2008/09	500	1,460.00	730	115,950	1,850.00	214,546
2009/10	520	1,380.00	718	298,520	1,440.00	429,586
2010/11	330	1,540.00	508	174,330	2,750.00	479,068
2011/12	350	1,600.00	560	262,950	1,540.00	406,047
2012/13	(NA)	(NA)	(NA)	(NA)	(NA)	(NA)
2013/14	320	2,000.00	640	183,120	1,970.00	361,305
2014/15	400	2,860.00	1,144	173,300	1,880.00	326,526
2015/16	410	2,320.00	835	224,010	1,320.00	295,797

(NA) Not available.
NASS, Crops Branch, (202) 720–2127.

Table 5-17.—Bananas: Bearing acreage, yield, utilized production, marketing year average price, and value, Hawaii, 2006–2015

Year	Bearing acreage	Yield per acre	Utilized Production	Price per pound	Value
	Acres	1,000 pounds	1,000 pounds	Cents	1,000 dollars
2006	1,100	20.0	22,000	49.0	10,780
2007	1,300	19.7	25,600	41.0	10,496
2008	1,100	15.8	17,400	46.0	8,004
2009	1,100	16.8	18,500	55.0	10,175
2010	1,100	16.2	17,800	60.0	10,680
2011	1,000	17.4	17,400	65.0	11,310
2012	(NA)	(NA)	(NA)	(NA)	(NA)
2013	900	16.1	14,500	90.0	13,050
2014	860	14.0	12,000	82.0	9,840
2015	830	14.5	11,900	91.3	10,866

(NA) Not available.
NASS, Crops Branch, (202) 720–2127.

Table 5-18.—Kiwifruit: Area, yield, utilized production, marketing year average price, and value, California, 2006–2015

Year	Bearing acreage	Yield [1]	Utilized Production	Price per ton	Value
	Acres	Tons	Tons	Dollars	1,000 dollars
2006	4,200	6.21	25,400	911	23,148
2007	4,200	5.83	23,700	950	22,517
2008	4,200	5.48	22,000	888	19,545
2009	4,200	6.10	24,900	847	21,084
2010	4,200	7.79	32,500	768	24,961
2011	4,000	9.43	36,700	775	28,439
2012	3,700	8.00	27,100	1,020	27,508
2013	3,900	7.08	26,900	1,110	29,812
2014	4,100	6.95	27,900	1,190	33,333
2015	4,000	5.93	23,100	1,340	30,893

[1] Yield based on total production.
NASS, Crops Branch, (202) 720-2127.

Table 5-19.—Cherries: Foreign trade, United States, 2006–2015

Year beginning October	Imports		Domestic exports	
	Fresh	Dried and preserved	Fresh	Canned
	Metric tons	Metric tons	Metric tons	Metric tons
2006	13,940	5,303	51,182	14,637
2007	22,125	5,403	45,462	14,427
2008	19,410	5,390	64,590	12,678
2009	13,601	5,036	58,737	13,758
2010	23,126	5,076	71,542	13,544
2011	17,250	6,289	101,073	16,127
2012	10,362	6,350	66,937	14,491
2013	12,033	6,614	87,626	17,761
2014	16,096	10,517	74,711	19,035
2015	11,463	8,089	75,931	17,465

ERS, Crops Branch, (202) 694–5255.

Table 5-20.—Sweet cherries: Production and value, United States, 2006–2015

Year	Total production	Utilized production	Marketing year average price per ton [1]	Value
	Tons	Tons	Dollars	1,000 dollars
2006	294,160	287,520	1,620.00	465,225
2007	310,680	306,210	1,820.00	557,056
2008	248,060	240,720	2,390.00	574,043
2009	442,870	385,625	1,330.00	513,330
2010	313,220	307,630	2,330.00	715,684
2011	334,415	330,290	2,530.00	834,585
2012	424,000	418,415	2,020.00	843,311
2013	332,090	295,950	2,610.00	771,798
2014	363,640	358,910	2,140.00	766,551
2015	338,430	336,836	2,250.00	758,915

[1] Fresh fruit prices are equivalent packinghouse-door returns for Western States, and the average price as sold for other States. Quantities processed are priced at the equivalent processing plant door level.
NASS, Crops Branch, (202) 720–2127.

Table 5-21.—Tart cherries: Production and value, United States, 2006–2015

Year	Total production	Utilized production	Marketing year average price per pound [1]	Value
	Million pounds	Million pounds	Dollars	1,000 dollars
2006	262.0	248.6	0.215	53,454
2007	253.2	248.7	0.273	67,923
2008	214.4	213.2	0.377	80,344
2009	359.2	320.8	0.192	61,628
2010	190.4	183.2	0.222	40,741
2011	231.8	230.3	0.300	69,072
2012	85.2	85.0	0.594	50,520
2013	294.2	291.1	0.359	104,395
2014	304.2	300.6	0.355	106,810
2015	252.5	251.1	0.347	87,037

[1] Fresh fruit prices are equivalent packinghouse-door returns for Western States, and the average price as sold for other States. Quantities processed are priced at the equivalent processing plant door level.
NASS, Crops Branch, (202) 720–2127.

Table 5-22.—Sweet cherries: Production and season average price, by State and United States, 2013–2015

State	Total production			Utilized production			Price [1]		
	2013	2014	2015	2013	2014	2015	2013	2014	2015
	Tons	Tons	Tons	Tons	Tons	Tons	Dollars per ton	Dollars per ton	Dollars per ton
California	82,000	33,200	60,100	78,500	29,200	59,500	3,390.00	4,840.00	3,900.00
Idaho	2,300	2,120	1,810	2,200	2,000	1,810	2,550.00	2,630.00	2,010.00
Michigan	22,900	29,860	13,430	21,800	29,460	13,330	964.00	941.00	1,020.00
Montana	2,015	2,090	740	1,785	1,930	700	2,070.00	2,070.00	2,530.00
New York	1,045	630	770	845	610	720	3,750.00	3,290.00	4,070.00
Oregon	52,000	57,900	38,700	46,000	57,900	38,650	1,980.00	1,430.00	1,750.00
Utah	830	840	230	820	810	226	2,490.00	1,500.00	854.00
Washington	169,000	237,000	222,650	144,000	237,000	221,900	2,630.00	2,120.00	1,970.00
United States	332,090	363,640	338,430	295,950	358,910	336,836	2,610.00	2,140.00	2,250.00

[1] Fresh fruit prices are equivalent packinghouse-door returns for California, Oregon, and Washington, and the average price as sold for other States. Quantities processed are priced at the equivalent processing plant door level.
NASS, Crops Branch, (202) 720–2127.

Table 5-23.—Tart cherries: Production and season average price, by State and United States, 2013–2015

State	Total production			Utilized production			Price [1]		
	2013	2014	2015	2013	2014	2015	2013	2014	2015
	Million pounds	Million pounds	Million pounds	Million pounds	Million pounds	Million pounds	Dollars per Lb.	Dollars per Lb.	Dollars per Lb.
Michigan	218.7	203.0	158.0	215.8	201.0	157.4	0.345	0.342	0.340
New York	12.0	10.3	10.5	12.0	10.3	10.4	0.358	0.302	0.253
Oregon	4.3	2.4	1.5	4.3	2.3	1.5	0.344	0.372	0.354
Pennsylvania	2.2	0.9	7.5	2.2	0.7	7.3	0.390	0.631	0.385
Utah	26.8	51.0	40.7	26.8	49.8	40.3	0.476	0.432	0.336
Washington	17.9	24.3	25.0	17.9	24.3	25.0	0.344	0.300	0.440
Wisconsin	12.3	12.3	9.3	12.1	12.2	9.2	0.357	0.394	0.331
United States	294.2	304.2	252.5	291.1	300.6	251.1	0.359	0.355	0.347

[1] Fresh fruit prices are equivalent packinghouse-door returns for Oregon and Washington, and the average price as sold for other States. Quantities processed are priced at the equivalent processing plant door level.
NASS, Crops Branch, (202) 720–2127.

Table 5-24.—Sweet cherries: Production and utilization, by State and United States, crop of 2015[1]

State	Total production	Utilized production	Utilization		
			Fresh[1]	Processed	
				Canned	Brined
	Tons	*Tons*	*Tons*	*Tons*	*Tons*
California	60,100	59,500	53,300	....	....
Idaho	1,810	1,810	(D)	....	....
Michigan	13,430	13,330	430	....	7,900
Montana	740	700	(D)	....	....
New York	770	720	(D)	....	....
Oregon	38,700	38,650	26,600	....	9,600
Utah	230	226	(D)	....	(D)
Washington	222,650	221,900	171,600	2,300	19,200
Other States[2]	(X)	(X)	2,976	1,000	6,390
United States	338,430	336,836	254,906	3,300	43,090

(D) Withheld to avoid disclosing data for individual operations.   (X) Not applicable.   [1] Includes "Home use."   [2] Includes data witheld above and/or data for States not listed in this table.
NASS, Crops Branch, (202) 720-2127.

Table 5-25.—Tart cherries: Production and utilization, by State and United States, crop of 2015[1]

State	Total production	Utilized production	Utilization		
			Fresh[1]	Processed	
				Canned and otherwise processed[2]	Frozen
	Million pounds	*Million pounds*	*Million pounds*	*Million pounds*	*Million pounds*
Michigan	158.0	157.4	0.1	26.0	114.3
New York	10.5	10.4	(D)	....	....
Oregon	1.5	1.5	(D)	....	....
Pennsylvania	7.5	7.3	(D)	....	....
Utah	40.7	40.3	....	....	....
Washington	25.0	25.0	(D)	....	....
Wisconsin	9.3	9.2	0.5	....	....
Other States[2]	....	....	0.6	10.3	62.1
United States	252.5	251.1	1.2	36.3	176.4

(D) Withheld to avoid disclosing data for individual operations.   [1] Includes "Home use."   [2] Includes data withheld above and/or data for States not listed in this table.
NASS, Crops Branch, (202) 720-2127.

Table 5-26.—Sweet cherries: Production and utilization, United States, 2006–2015

Crop of—	Total production	Utilized production	Utilization of quantities sold		
			Fresh [1]	Processed	
				Other [2]	Brined
	Tons	*Tons*	*Tons*	*Tons*	*Tons*
2006	294,160	287,520	190,770	40,520	56,230
2007	310,680	306,210	222,560	28,930	48,160
2008	248,060	240,720	175,320	20,970	39,750
2009	442,870	385,625	296,750	29,053	51,571
2010	313,220	307,630	244,340	21,905	38,234
2011	334,415	330,290	258,920	25,495	39,625
2012	424,000	418,415	332,555	35,795	47,065
2013	332,090	295,950	217,950	28,535	44,945
2014	363,640	358,910	268,010	40,360	47,041
2015	338,430	336,836	254,906	....	43,090

[1] Includes "Home use."   [2] Includes canned utilization and other processed utilizations from all States.
NASS, Crops Branch, (202) 720–2127.

Table 5-27.—Tart cherries: Production and utilization, United States, 2006–2015

Crop of—	Total production	Utilized production	Utilization of quantities sold		
			Fresh [1]	Processed	
				Other [2]	Frozen
	Million pounds	*Million pounds*	*Million pounds*	*Million pounds*	*Million pounds*
2006	262.0	248.6	1.4	90.4	156.8
2007	253.2	248.7	1.6	20.1	179.3
2008	214.4	213.2	1.0	17.4	145.0
2009	359.2	320.8	1.3	35.3	232.0
2010	190.4	183.2	0.8	21.2	125.9
2011	231.8	230.3	0.5	37.3	154.1
2012	85.2	85.0	0.4	18.3	59.8
2013	294.2	291.1	1.2	92.4	158.4
2014	304.2	300.6	1.9	63.3	198.9
2015	252.5	251.1	1.2	....	176.4

[1] Includes "Home use."   [2] Includes canned utilization and other processed utilizations from all states.
NASS, Crops Branch, (202) 720–2127.

Table 5-28.—Citrus fruit: Utilized production and value, United States, for season of 2006/2007 to 2015/2016

Season [1]	Production	Marketing year average returns per box [2]	Value	Quantities processed [3]
		Oranges [4]		
	1,000 boxes	*Dollars*	*1,000 dollars*	*1,000 boxes*
2006/2007	177,280	12.56	2,216,471	142,030
2007/2008	234,376	9.36	2,198,828	179,745
2008/2009	210,709	9.22	1,970,070	163,277
2009/2010	192,835	10.24	1,997,188	137,628
2010/2011	204,949	10.90	2,230,412	148,237
2011/2012	206,119	12.70	2,621,620	149,972
2012/2013	189,893	10.85	2,073,638	134,799
2013/2014	155,977	14.29	2,254,303	109,376
2014/2015	146,602	13.29	1,963,353	102,388
2015/2016	137,491	12.36	1,711,100	89,802
		Grapefruit		
2006/2007	39,900	7.69	311,914	20,579
2007/2008	37,900	7.15	273,076	18,677
2008/2009	32,025	6.93	224,098	14,661
2009/2010	30,400	9.65	291,424	13,231
2010/2011	30,360	9.29	283,597	14,276
2011/2012	27,650	10.08	279,033	13,121
2012/2013	28,950	8.87	256,524	14,340
2013/2014	25,200	9.60	241,686	12,389
2014/2015	21,950	9.85	216,258	9,755
2015/2016	19,400	12.93	251,036	8,935
		Lemons		
2006/2007	21,000	21.40	449,417	8,105
2007/2008	16,300	32.12	523,528	5,918
2008/2009	24,000	13.96	335,065	9,986
2009/2010	23,200	17.04	395,339	10,482
2010/2011	23,000	16.80	386,514	7,987
2011/2012	21,250	21.12	448,698	4,869
2012/2013	22,800	17.56	400,295	7,394
2013/2014	20,600	31.13	641,259	4,840
2014/2015	22,600	30.83	696,835	7,302
2015/2016	22,250	33.00	734,209	5,338
		Tangerines and Mandarins [6]		
2006/2007	8,400	18.30	156,198	2,530
2007/2008	12,600	17.91	236,193	3,411
2008/2009	10,800	18.44	207,202	2,221
2009/2010	14,700	18.30	274,519	3,204
2010/2011	15,550	20.87	330,503	3,260
2011/2012	15,290	22.34	349,167	3,029
2012/2013	16,440	25.60	426,101	2,785
2013/2014	17,750	31.05	557,357	2,869
2014/2015	21,135	22.13	468,083	5,614
2015/2016	23,115	27.52	637,412	3,457
		Tangelos (FL)		
2006/2007	1,250	11.00	13,755	822
2007/2008	1,500	5.76	8,638	1,068
2008/2009	1,150	4.81	5,528	646
2009/2010	900	7.51	6,761	485
2010/2011	1,150	8.63	9,930	707
2011/2012	1,150	12.43	14,299	716
2012/2013	1,000	12.99	12,986	526
2013/2014	880	11.18	9,839	486
2014/2015	665	13.87	9,221	319
2015/2016	390	22.23	8,672	150

[1] The crop year begins with the bloom of the first year shown and ends with completion of harvest the following year.   [2] Equivalent packing-house door returns.   [3] Includes quantities used for juice, concentrates, grapefruit segments, and other citrus products. In some seasons, includes appreciable quantities of oranges and lemons in California delivered to processing plants which were not utilized, but for which growers received payment.   [4] Includes small quantities of tangerines in Texas and Temples in Florida.   [5] Included in early, midseason, and navel orange varieties beginning with the 2006-07 season.   [6] Arizona and California tangelos and tangors included.
NASS, Crops Branch, (202) 720-2127.

Table 5-29.—Citrus fruit: Utilized production and marketing year average returns per box, by State, 2013/2014 and 2014/2015 [1]

Crop and State	Utilized production		Market year average price [2]	
	2013/2014	2014/2015	2013/2014	2014/2015
	1,000 boxes [3]	*1,000 boxes [3]*	*Dollars*	*Dollars*
Oranges				
Early, midseason, and Navel varieties: [4]				
California	38,700	39,000	19.13	16.29
Florida	53,300	47,400	11.01	11.04
Texas	1,401	1,170	13.72	11.28
Total	93,401	87,570	14.19	13.23
Valencia:				
California	10,800	9,200	18.68	15.00
Florida	51,400	49,550	13.65	13.12
Texas	376	282	11.31	11.72
Total	62,576	59,032	14.42	13.38
All oranges:				
California	49,500	48,200	19.03	16.04
Florida	104,700	96,950	12.31	12.10
Texas	1,777	1,452	13.21	11.37
United States	155,977	146,602	14.29	13.29
Grapefruit				
California	3,850	4,800	9.82	10.50
Florida, all	15,650	12,900	9.83	9.87
Colored [5]	11,500	9,650	10.18	10.47
White	4,150	3,250	8.85	8.09
Texas	5,700	4,250	8.79	9.07
United States	25,200	21,950	9.60	9.85
Lemons				
Arizona	1,800	2,000	(D)	(D)
California	18,800	20,600	(D)	(D)
United States	20,600	22,600	(D)	(D)
Tangelos				
Florida	880	665	11.18	13.87
Tangerines and Mandarins				
Arizona [6]	150	170	(D)	(D)
California [6]	14,700	18,700	(D)	(D)
Florida	2,900	2,265	19.66	21.15
United States	17,750	21,135	31.05	22.13

(D) Withheld to avoid disclosing data for individual operations. [1] The crop year begins with the bloom of the first year shown and ends with completion of harvest the following year. [2] Equivalent packinghouse-door returns. [3] Net lbs. per box: oranges—California, 80; Florida, 90; and Texas, 85; grapefruit—California, 80; Florida, 85; Texas, 80; lemons—80; tangelos, tangerines and mandarins—Arizona and California, 80; tangelos—Florida, 90. [4] Includes small quantities of tangerines in Texas and Temples in Florida. [5] Includes seedy grapefruit. [6] Includes tangelos and tangors.
NASS, Crops Branch, (202) 720–2127.

Table 5-30.—International Citrus: Area and production in specified countries, 2013/2014-2015/2016

Country	Area			Production		
	2013/2014	2014/2015	2015/2016	2013/2014	2014/2015	2015/2016
	1,000 hectares	*1,000 hectares*	*1,000 hectares*	*1,000 metric tons*	*1,000 metric tons*	*1,000 metric tons*
Oranges, fresh:						
Argentina	40,000	40,000	40,000	800	1,000	1,000
Brazil	612,600	596,100	588,600	17,870	16,320	16,728
China				7,600	6,900	7,000
Egypt	115,000	117,000	122,800	2,570	2,630	2,750
European Union	277,436	272,786	273,373	6,550	5,959	6,107
Mexico	321,683	323,000	320,000	4,533	4,158	3,534
Morocco	45,000	48,067	50,100	1,001	868	920
South Africa	38,100	37,900	37,000	1,715	1,700	1,690
Turkey	50,000	50,000	53,000	1,700	1,650	1,700
Vietnam				520	520	520
Others	25,070	24,770	24,800	1,137	1,153	1,197
Total Foreign	1,524,889	1,509,623	1,509,673	45,996	42,858	43,146
United States				6,140	5,786	4,758
Total	1,524,889	1,509,623	1,509,673	52,136	48,644	47,904
Tangerines/Mandarines:						
Argentina	30,000	30,000	30,000	370	450	450
China				17,850	19,400	20,000
European Union	148,818	153,719	152,440	3,231	3,483	2,917
Israel	6,900	7,300	7,600	139	205	240
Japan	60,600	58,400	57,600	1,124	1,070	1,115
Korea, South	19,342	19,322	19,309	672	697	640
Morocco	42,000	47,140	50,100	1,160	1,005	1,055
South Africa	6,300	6,600	6,600	195	200	205
Thailand				375	375	375
Turkey	27,450	27,450	37,795	880	960	1,040
Others				20	20	20
Total Foreign	341,410	349,931	361,444	26,016	27,865	28,057
United States				700	793	839
Total	341,410	349,931	361,444	26,716	28,658	28,896
Lemons/Limes:						
Argentina	41,000	41,000	41,000	780	1,300	1,450
European Union	66,133	65,906	66,012	1,308	1,599	1,286
Israel	1,870	1,800	1,800	64	65	60
Japan	498	500	500	10	10	10
Mexico	154,803	155,000	156,000	2,187	2,260	2,270
Morocco	3,000	2,800	2,920	43	30	35
South Africa	5,313	5,400	5,600	312	330	330
Turkey	22,900	22,900	24,000	760	725	668
Others						
Total Foreign	295,517	295,306	297,832	5,464	6,319	6,109
United States				748	816	784
Total	295,517	295,306	297,832	6,212	7,135	6,893
Grapefruit:						
China				3,717	3,900	4,300
European Union	3,288	2,416	2,494	92	109	95
Israel	3,750	3,050	3,070	236	186	185
Mexico	16,201	16,600	16,700	424	430	432
South Africa	7,943	8,000	8,100	413	400	405
Turkey	5,000	5,000	5,300	235	238	200
Others						
Total Foreign	36,182	35,066	35,664	5,117	5,263	5,617
United States				950	789	736
World Total	36,182	35,066	35,664	6,067	6,052	6,353

FAS, Office of Global Analysis, (202) 720-6301. Prepared or estimated on the basis of official USDA production, supply, and distribution statistics from foreign governments.

Table 5-31.—Fresh fruits, citrus: United States exports by country of destination, 2011–2013

Country of destination	2011	2012	2013
	Metric tons	Metric tons	Metric tons
Canada	259,245	245,110	235,313
Japan	255,641	252,301	199,258
Korea, South	166,863	213,767	188,650
Hong Kong	100,252	81,823	106,174
China	64,206	50,305	47,146
Australia(*)	33,641	24,932	36,030
Malaysia	39,699	32,627	34,039
Mexico	21,177	27,007	25,736
Singapore	22,838	18,945	20,839
France(*)	23,848	19,606	17,426
Netherlands	21,868	15,568	16,477
Taiwan	13,407	11,595	15,253
New Zealand(*)	11,070	9,473	11,700
Philippines	9,335	8,180	10,991
United Arab Emirates	14,404	3,054	8,674
Belgium-Luxembourg(*)	6,115	4,741	7,273
Chile	4,973	2,834	5,747
Vietnam	1,742	1,227	3,821
Peru	137	877	3,446
Indonesia	7,000	3,771	3,183
Thailand	1,489	1,876	2,875
Ecuador	1,022	2,404	2,710
United Kingdom	3,823	3,178	2,671
Germany(*)	2,792	2,392	1,873
Bahamas, The	1,517	1,518	1,622
Switzerland(*)	3,861	1,425	1,473
Sri Lanka	4,480	1,981	1,416
Sweden	3,230	912	1,228
Saudi Arabia	2,945	469	1,221
India	14,662	5,902	1,168
Other Partners	21,713	7,653	7,999
World Total	1,138,996	1,057,447	1,023,431

(*) Denotes a country that is a summarization of its component countries. Users should use cautious interpretation on quantity reports using mixed units of measure. Quantity line items will only include statistics on the units of measure that are equal to, or are able to be converted to, the assigned unit of measure of the grouped commodities.
FAS, Office of Global Analysis,(202) 720–6301. Data Source: Department of Commerce, U.S. Census Bureau, Foreign Trade Statistics.

Table 5-32.—Fresh citrus fruits: Foreign trade, United States, 2006–2015

Year [1]	Oranges		Grapefruit		Lemons		Limes		Tangerines	
	Imports	Domestic exports	Imports	Domestic exports	Imports	Domestic exports	Imports	Domestic exports	Imports	Domestic exports
	Metric tons	Metric tons	Metric tons	Metric tons	Metric tons	Metric tons	Metric tons	Metric tons	Metric tons	Metric tons
2006	112,108	346,936	21,531	331,538	65,656	115,862	337,356	3,413	6,148	12,700
2007	81,033	613,155	15,216	270,363	64,214	155,668	359,020	2,772	7,681	19,321
2008	90,546	495,662	11,697	246,865	37,565	91,459	358,289	2,364	8,027	17,496
2009	101,180	665,157	11,370	245,696	40,848	91,445	362,317	2,317	7,748	18,195
2010	92,159	752,613	7,103	225,836	56,278	97,275	352,956	2,550	6,101	30,419
2011	131,519	694,530	2,869	207,537	50,939	92,081	423,350	3,434	4,202	22,786
2012	130,812	679,663	11,862	182,725	38,152	104,031	438,830	4,389	3,614	23,529
2013	156,261	507,665	13,777	146,673	56,996	122,420	438,273	4,569	2,573	17,628
2014	152,211	523,392	9,291	140,724	85,544	109,663	499,221	3,954	4,565	18,713
2015	166,074	655,839	21,931	123,470	83,851	107,193	542,132	4,400	7,882	12,612

[1] Year beginning October for all commodities.
ERS, Crops Branch, (202) 694–5255.

Table 5-33.—Concentrated citrus juices: Annual packs, Florida, 2006–2015

Season beginning December	Frozen concentrated juice [1]		
	Orange [2]	Grapefruit [2]	Tangerine
	1,000 gallons	1,000 gallons	1,000 gallons
2006	79,054	15,782	446
2007	135,196	13,678	686
2008	120,800	10,731	466
2009	82,260	7,904	740
2010	82,106	9,297	1,374
2011	106,432	9,059	1,083
2012	76,132	7,437	928
2013	35,654	5,745	825
2014	28,878	4,504	1,046
2015	22,269	3,275	....

[1] Net pack.   [2] Frozen orange juice reported in 42.0° Brix; Grapefruit 40.0° Brix. Includes concentrated juice for manufacture.
ERS, Crops Branch, (202) 694–5255.

Table 5-34.—Dates: Area, yield, total production, marketing year average price per ton, and value, California, 2006–2015

Year	Bearing acreage	Yield per acre	Production	Price per ton	Value
	Acres	Tons	Tons	Dollars	1,000 dollars
2006	5,500	3.22	17,700	2,140	37,878
2007	5,300	3.08	16,300	2,290	37,327
2008	5,700	3.67	20,900	1,260	26,334
2009	6,700	3.54	23,700	1,180	27,966
2010	7,700	3.77	29,000	1,270	36,830
2011	8,400	3.96	33,300	1,320	43,956
2012	8,800	3.53	31,100	1,340	41,674
2013	8,200	3.72	30,500	1,220	37,210
2014	10,000	3.34	33,400	1,510	50,434
2015	10,000	4.36	43,600	1,560	68,016

NASS, Crops Branch, (202) 720–2127.

Table 5-35.—Dates: Foreign trade, United States, 2006–2015

Year beginning October	Imports
	Metric tons
2006	9,446
2007	5,123
2008	13,079
2009	12,094
2010	20,113
2011	22,875
2012	23,945
2013	28,974
2014	47,754
2015	48,712

ERS, Crops Branch, (202) 694–5255.

Table 5-36.—Cranberries: Area, yield, production, season average price per barrel, value and quantities processed, United States, 2006–2015 [1]

Year	Area harvested	Yield per acre [2]	Total production [3]	Utilized production	Price [4]	Value	Quantities processed [5]
	Acres	Barrels [6]	Barrels [6]	Barrels [6]	Dollars	1,000 dollars	Barrels [6]
2006	38,500	179.0	6,890,000	6,785,000	41.10	278,888	6,429,500
2007	38,100	172.0	6,554,000	6,554,000	50.70	332,092	6,194,000
2008	38,200	205.9	7,865,000	7,865,000	58.10	457,192	7,494,000
2009	38,500	179.6	6,913,000	6,913,000	44.20	305,669	6,580,000
2010	38,500	176.8	6,808,200	6,808,200	43.90	299,123	6,589,000
2011	38,500	200.4	7,713,700	7,713,700	44.80	345,561	7,473,200
2012	40,300	199.6	8,045,000	8,045,000	47.90	385,506	7,791,000
2013	42,000	211.4	8,957,400	8,880,700	32.40	287,322	8,559,300
2014	40,600	202.9	8,400,000	8,236,000	30.90	254,412	7,968,000
2015	40,900	206.7	8,563,000	8,455,000	31.60	267,527	8,016,800

[1] Estimates relate to Massachusetts, New Jersey, Oregon, Washington, and Wisconsin. [2] Derived from total production. [3] Differences between utilized and total production are quantities unharvested for economic reasons or excess cullage and/or set-aside production under provisions of the Cranberry Marketing Order. [4] Average price of utilized production. Equivalent returns at first delivery point, screened basis of utilized production. [5] Mainly for canning. [6] Barrels of 100 pounds.
NASS, Crops Branch, (202) 720–2127.

Table 5-37.—Cranberries: Area, yield, production, and season average price per barrel, by State and United States, 2013–2015

State	Area harvested			Yield per acre		
	2013	2014	2015	2013	2014	2015
	Acres	Acres	Acres	Bbl.[2]	Bbl.[2]	Bbl.[2]
Massachusetts	13,200	12,400	13,200	138.5	164.3	177.3
New Jersey	3,000	3,000	3,000	180.8	204.7	189.7
Oregon	3,000	2,900	2,900	130.0	162.8	191.4
Washington	1,700	1,600	1,600	89.4	97.5	123.8
Wisconsin	21,100	20,700	20,200	282.8	239.5	237.3
United States	42,000	40,600	40,900	211.4	202.9	206.7

State	Total production			Price per barrel [1]		
	2013	2014	2015	2013	2014	2015
	Bbl.[2]	Bbl.[2]	Bbl.[2]	Dollars	Dollars	Dollars
Massachusetts	1,852,300	2,070,000	2,352,000	31.60	37.10	32.80
New Jersey	547,500	652,000	595,000	37.50	36.90	37.30
Oregon	390,000	500,000	562,000	30.60	23.10	25.90
Washington	152,000	156,000	198,000	43.70	44.60	44.20
Wisconsin	6,015,600	5,022,000	4,856,000	32.00	27.90	30.50
United States	8,957,400	8,400,000	8,563,000	32.40	30.90	31.60

[1] Average price of utilized production. Equivalent returns at first delivery point, screened basis of utilized production. [2] Barrels of 100 pounds.
NASS, Crops Branch, (202) 720–2127.

Table 5-38.—Figs: Total production, marketing year average price per ton, and value, California, 2006–2015

Year	Dried				Total		
	Production (dry basis)			Price per ton	Production	Price per ton	Value
	Total	Standard	Substandard				
	Tons	1,000 tons	Tons	Dollars	Tons	Dollars	1,000 dollars
2006	13,000	11,100	1,900	829	42,800	426	18,253
2007	14,500	11,100	3,400	873	47,800	401	19,145
2008	13,100	11,800	1,300	1,200	43,300	599	25,954
2009	13,300	12,000	1,300	1,540	43,750	695	30,422
2010	12,320	10,900	1,420	1,220	40,910	542	22,185
2011	11,560	10,140	1,420	1,230	38,660	526	20,336
2012	10,400	9,600	800	1,350	35,200	555	19,520
2013	9,300	8,600	700	1,500	33,000	610	20,143
2014	9,470	8,840	630	1,640	33,400	719	23,998
2015	8,840	8,250	590	1,660	30,200	724	21,853

NASS, Crops Branch, (202) 720–2127.

Table 5-39.—Figs, dried: Foreign trade, United States, 2006–2015

Year beginning October	Imports for consumption	Domestic exports
	Metric tons	Metric tons
2006	5,487	3,019
2007	2,978	3,775
2008	6,042	5,019
2009	4,035	5,795
2010	4,042	5,065
2011	4,236	5,449
2012	5,816	4,895
2013	10,294	4,537
2014	7,137	4,670
2015	11,822	3,959

ERS, Crops Branch, (202) 694–5255.

Table 5-40.—Grapes: Production, price, and value, United States, 2006–2015

Year	Production (fresh basis)		Market year average price per ton [1]	Value
	Total	Utilized		
	Tons	Tons	Dollars	1,000 dollars
2006	6,377,470	6,366,170	519.00	3,304,631
2007	7,057,250	7,056,250	489.00	3,453,124
2008	7,279,360	7,265,650	497.00	3,613,420
2009	7,267,890	7,240,310	566.00	4,101,600
2010	7,432,120	7,429,490	542.00	4,023,967
2011	7,408,740	7,401,370	576.00	4,262,945
2012	7,530,883	7,524,443	752.00	5,661,096
2013	8,631,790	8,620,810	712.00	6,135,301
2014	7,883,830	7,871,580	740.00	5,821,629
2015	7,677,150	7,677,080	724.00	5,561,719

[1] Fresh fruit prices are equivalent packinghouse-door returns for California and Washington, and the average price as sold for other States. Quantities processed are priced at the equivalent processing plant door level.
NASS, Crops Branch, (202) 720–2127.

Table 5-41.—Grapes: Production and marketing year average price per ton, by State and United States, 2013–2015

State	Total production			Utilized production		
	2013	2014	2015	2013	2014	2015
	Tons	Tons	Tons	Tons	Tons	Tons
Arkansas	1,790	1,490	1,500	1,640	1,390	1,500
California	7,742,000	6,934,000	6,847,000	7,742,000	6,934,000	6,847,000
Wine	4,245,000	3,895,000	3,705,000	4,245,000	3,895,000	3,705,000
Table [1]	1,227,000	1,165,000	1,135,000	1,227,000	1,165,000	1,135,000
Raisin [1]	2,270,000	1,874,000	2,007,000	2,270,000	1,874,000	2,007,000
Georgia	4,600	4,000	5,000	3,300	3,900	4,950
Michigan	94,000	63,300	80,600	94,000	63,300	80,600
Missouri	5,970	4,030	5,650	5,970	4,030	5,650
New York	206,000	188,000	145,000	202,000	180,000	145,000
North Carolina	5,200	6,000	7,300	4,700	5,400	7,300
Ohio	6,490	3,810	3,500	6,160	3,560	3,480
Oregon	49,000	58,000	65,000	49,000	58,000	65,000
Pennsylvania	110,000	91,000	77,000	106,000	89,000	77,000
Texas	7,240	9,400	11,400	7,040	8,800	11,400
Virginia	7,500	8,800	9,200	7,000	8,200	9,200
Washington	392,000	512,000	419,000	392,000	512,000	419,000
Wine	210,000	227,000	230,000	210,000	227,000	230,000
Juice	182,000	285,000	189,000	182,000	285,000	189,000
United States	8,631,790	7,883,830	7,677,150	8,620,810	7,871,580	7,677,080

State	Price per ton [2]		
	2013	2014	2015
	Dollars	Dollars	Dollars
Arkansas	1,010.00	902.00	762.00
California	719.00	756.00	724.00
Wine	753.00	759.00	679.00
Table [1]	1,260.00	1,350.00	1,530.00
Raisin [1]	364.00	381.00	347.00
Georgia	1,110.00	1,470.00	1,510.00
Michigan	378.00	306.00	313.00
Missouri	728.00	790.00	880.00
New York	370.00	385.00	390.00
North Carolina	843.00	807.00	762.00
Ohio	593.00	476.00	410.00
Oregon	2,190.00	2,040.00	2,270.00
Pennsylvania	320.00	318.00	319.00
Texas	1,560.00	1,500.00	1,600.00
Virginia	1,800.00	1,850.00	1,950.00
Washington	708.00	590.00	708.00
Wine	1,110.00	1,110.00	1,140.00
Juice	244.00	175.00	183.00
United States	712.00	740.00	724.00

[1] Fresh equivalent of dried and not dried. [2] Fresh fruit prices are equivalent packinghouse-door returns for California and Washington, and the average price as sold for other States. Quantities processed are priced at the equivalent processing plant door level.
NASS, Crops Branch, (202) 720–2127.

Table 5-42.—Grapes: Production and utilization, United States, 2006–2015

Crop of—	Total production [1]	Utilized production	Utilization of quantities sold				
			Fresh	Processed			
				Canned	Dried (fresh basis)	Crushed for wine	Crushed for juice, etc.[2]
	Tons	Tons	Tons	Tons	Tons	Tons	Tons
2006	6,377,470	6,366,170	797,590	21,000	1,424,000	3,725,380	398,200
2007	7,057,250	7,056,250	920,330	21,000	1,621,000	3,920,520	573,400
2008	7,279,360	7,265,650	945,000	25,000	1,873,000	3,943,490	479,130
2009	7,267,890	7,240,310	898,940	20,000	1,510,000	4,373,070	438,300
2010	7,432,120	7,429,490	953,290	25,000	1,777,000	4,270,300	403,900
2011	7,408,740	7,401,370	944,740	25,000	1,807,000	4,154,530	470,100
2012	7,530,883	7,524,443	962,983	20,000	1,488,000	4,706,060	346,950
2013	8,631,790	8,620,810	1,116,940	22,000	1,909,000	5,067,610	504,160
2014	7,883,830	7,871,580	1,053,040	21,000	1,721,000	4,526,020	549,920
2015	7,677,150	7,677,080	1,043,970	22,000	1,919,000	4,255,330	435,780

[1] Total production includes utilized production plus production not harvested and harvested not sold.   [2] Mostly juice, but includes some quantities used for jam, jelly, etc.
NASS, Crops Branch, (202) 720–2127.

Table 5-43.—Grapes: Production and utilization, by State and United States, crop of 2015

State	Total production	Utilized production	Utilization				
			Fresh	Processed			
				Canned	Dried (fresh basis)	Crushed for—	
						Wine	Juice, etc.
	Tons	Tons	Tons	Tons	Tons	Tons	Tons
Arkansas	1,500	1,500					
California	6,847,000	6,847,000	1,038,000	22,000	1,919,000	3,868,000	
Wine	3,705,000	3,705,000				3,705,000	
Table	1,135,000	1,135,000	1,003,000		61,000	71,000	
Raisin	2,007,000	2,007,000	35,000	22,000	1,858,000	92,000	
Georgia	5,000	4,950					
Michigan	80,600	80,600				3,300	76,300
Missouri	5,650	5,650				5,650	
New York	145,000	145,000	2,000			35,000	108,000
North Carolina	7,300	7,300	440			6,860	
Ohio	3,500	3,480	80			1,230	2,170
Oregon [1]	65,000	65,000				65,000	
Pennsylvania	77,000	77,000	1,500			15,200	60,300
Texas	11,400	11,400	100			11,300	
Virginia	9,200	9,200				9,200	
Washington	419,000	419,000				230,000	189,000
Other States [2]			1,850			4,590	10
United States	7,677,150	7,677,080	1,043,970	21,000	1,919,000	4,255,330	435,780

[1] Wine includes small quantities for other uses.   [2] Includes data withheld above and/or data for States not listed in this table.
NASS, Crops Branch, (202) 720–2127.

Table 5-44.—Grapes and raisins: Foreign trade, United States, 2006–2015

Year beginning October	Grapes		Raisins [1]	
	Imports, fresh	Domestic exports, fresh	Imports for consumption	Domestic exports
	Metric tons	*Metric tons*	*Metric tons*	*Metric tons*
2006	625,778	294,670	29,806	122,349
2007	567,079	299,891	22,647	175,112
2008	600,908	341,205	21,467	160,451
2009	602,273	284,092	24,289	173,049
2010	557,397	334,609	18,813	154,541
2011	557,310	378,504	19,336	142,997
2012	566,038	366,088	17,102	131,980
2013	500,975	421,560	14,904	174,024
2014	554,674	366,111	20,588	132,274
2015	532,494	330,220	25,421	124,663

[1] Raisins converted to sweatbox or production basis by multiplying by 1.08.
ERS, Crops Branch, (202) 694–5255.

Table 5-45.—Guavas: Area harvested, yield, utilized production, marketing year average price, and value, Hawaii, 2006–2015

Year	Area harvested	Yield per acre	Utilized Production	Price per pound	Value
	Acres	*1,000 pounds*	*1,000 pounds*	*Cents*	*1,000 dollars*
2006	365	20.3	7,400	14.2	1,051
2007	170	25.3	4,300	15.7	675
2008	160	21.9	3,500	15.8	553
2009	135	15.6	2,100	14.0	294
2010	115	11.3	1,300	16.9	220
2011	110	17.3	1,900	17.0	323
2012	(NA)	(NA)	(NA)	(NA)	(NA)
2013	100	22.0	2,200	17.0	374
2014	100	17.0	1,700	18.0	306
2015 [1]	(NA)	(NA)	(NA)	(NA)	(NA)

(NA) Not available. [1] Estimates discontinued in 2015.
NASS, Crops Branch, (202) 720–2127.

Table 5-46.—Nectarines: Total Production, utilization, and value, United States, 2006–2015

Crop of—	Total Production	Utilization		Marketing year average price per ton	Value
		Fresh	Processed (fresh basis)		
	Tons	*Tons*	*Tons*	*Dollars*	*1,000 dollars*
2006	231,900	231,900	([1])	522.00	121,004
2007	283,000	283,000	([1])	340.00	96,305
2008	302,500	302,500	([1])	367.00	110,915
2009	219,800	219,800	([1])	631.00	138,611
2010	233,200	233,200	([1])	553.00	129,075
2011	225,200	225,200	([1])	582.00	130,973
2012	188,900	188,900	([1])	767.00	144,906
2013	162,100	148,300	13,800	772.00	125,113
2014	205,300	196,000	9,300	869.00	178,421
2015	167,700	158,000	6,600	914.00	150,413

[1] Small quantities of processed nectarines are included in fresh to avoid disclosure of individual operations.
NASS, Crops Branch, (202) 720–2127.

Table 5-47.—Olives: Total production, marketing year average price, value, and processed utilization, California, 2006–2015

Year	Total Production	Marketing year average price per ton	Value	Processed utilization			
				Crushed for oil	Canned	Limited	Undersized
	Tons	*Dollars*	*1,000 dollars*	*Tons*	*Tons*	*Tons*	*Tons*
2006	23,500	771	18,119	4,000	17,000	1,500	500
2007	132,500	654	86,694	12,000	96,000	20,000	4,000
2008	66,800	697	46,587	14,000	45,500	6,000	1,300
2009	46,300	696	32,209	20,000	24,500	1,500	300
2010	206,000	664	136,796	36,000	125,000	37,000	8,000
2011	71,200	733	52,168	42,000	26,500	2,200	500
2012	160,000	813	130,038	74,000	78,500	6,400	1,100
2013	166,000	813	134,881	75,000	78,800	10,500	1,700
2014	95,000	774	73,559	57,700	30,500	5,900	900
2015	179,000	894	160,043	101,000	60,000	14,600	3,400

NASS, Crops Branch, (202) 720–2127.

Table 5-48.—Olives and olive oil: Foreign trade, United States, 2006–2015

Year beginning October	Imports			
	Olives		Olive oil	
	In brine	Dried	Edible	Inedible
	Metric tons	*Metric tons*	*Metric tons*	*Metric tons*
2006	118,375	1,043	260,398	1,607
2007	118,085	133	262,716	1,575
2008	109,230	289	275,611	594
2009	137,533	184	268,069	114
2010	117,915	206	290,226	258
2011	118,614	145	314,937	788
2012	119,035	131	296,029	379
2013	118,729	202	311,142	351
2014	119,471	130	310,112	163
2015	127,508	48	330,319	153

ERS, Crops Branch, (202) 694–5255.

Table 5-49.—Peaches: Production and value, United States, 2006–2015

Year	Total production	Utilized production	Marketing year average price[1]	Value
	1,000 tons	1,000 tons	Dollars per ton	1,000 dollars
2006	1,010.3	987.2	520	513,093
2007	1,127.2	1,115.9	450	502,087
2008	1,135.0	1,113.2	490	545,499
2009	1,103.5	1,082.4	548	593,387
2010	1,149.4	1,129.8	547	617,459
2011	1,071.2	1,042.3	563	587,228
2012	968.1	955.9	647	618,369
2013	903.9	888.7	617	547,978
2014	852.9	838.0	751	629,524
2015	847.2	825.4	734	605,794

[1] Fresh fruit prices are equivalent packinghouse-door returns for California and Washington except equivalent returns for bulk fruit at the first delivery point for California Clingstone, and the average price as sold for other States. Quantities processed are priced at the equivalent processing plant door level.
NASS, Crops Branch, (202) 720–2127.

Table 5-50.—Peaches: Production and utilization, United States, 2006–2015

Crop of—	Total production[1]	Utilized production	Utilization of quantities sold				
			Fresh[2]	Processed (fresh basis)			
				Canned	Dried	Frozen	Other[3]
	1,000 tons	1,000 tons	1,000 tons	1,000 tons	1,000 tons	1,000 tons	1,000 tons
2006	1,010.3	987.2	481.6	374.1	13.1	96.2	22.2
2007	1,127.2	1,115.9	441.2	484.8	12.7	135.4	41.8
2008	1,135.0	1,113.2	529.8	426.3	9.5	111.4	36.5
2009	1,103.5	1,082.4	502.9	463.7	7.1	92.0	16.9
2010	1,149.4	1,129.8	566.4	428.5	12.7	104.4	17.8
2011	1,071.2	1,042.3	531.1	387.2	4.6	100.6	18.9
2012	968.1	955.9	480.8	364.6	9.8	90.2	10.5
2013	903.9	888.7	397.0	369.8	4.2	103.7	14.0
2014	852.9	838.0	393.3	330.2	4.7	95.3	14.5
2015	847.2	825.4	357.7	339.5	6.9	108.4	

[1] Includes harvested not sold and unharvested production for California Clingstone peaches. [2] Includes "Home use." [3] Used for jams, preserves, pickles, wine, brandy, baby food, etc. Includes small quantities frozen for some years.
NASS, Crops Branch, (202) 720–2127.

Table 5-51.—Peaches: Foreign trade, United States, 2006–2015

Year beginning October	Domestic exports				
	Fresh	Canned	Canned, in fruit salad[1]	Dried, in fruit salad[2][3]	Total, in terms of fresh[4]
	Metric tons	Metric tons	Metric tons	Metric tons	Metric tons
2006	105,559	18,359	5,100	619	132,458
2007	112,325	37,266	10,019	690	163,446
2008	98,911	20,235	9,549	604	132,050
2009	103,038	16,914	11,701	662	135,329
2010	103,523	25,636	15,568	801	149,175
2011	95,798	28,295	17,605	884	146,609
2012	102,140	25,526	21,811	931	154,648
2013	80,241	20,847	20,144	859	126,004
2014	84,678	18,007	20,112	696	126,665
2015	77,837	10,688	18,403	364	108,952

[1] Canned peaches are 40 percent of total canned fruit for salad. [2] Net processed weight. [3] Dried peaches are 21 percent of total dried fruit for salad. [4] Dried fruit converted to unprocessed dry weight by dividing by 1.08. Unprocessed dry weight converted to terms of fresh fruit on the basis that 1 pound dried equals 6.0 pounds fresh. Canned peaches converted to terms of fresh on basis that 1 pound canned equals about 1 pound fresh.
ERS, Crops Branch, (202) 694–5255.

Table 5-52.—Peaches: Production and season average price per pound, by State and United States, 2013–2015

State	Total production			Utilized production			Price per ton[1]		
	2013	2014	2015	2013	2014	2015	2013	2014	2015
	Tons	Tons	Tons	Tons	Tons	Tons	Dollars	Dollars	Dollars
Alabama	4,140	3,400	5,550	3,840	1,900	5,520	1,170.00	1,200.00	1,120.00
Arkansas	1,300	650	1,100	1,130	570	1,070	1,610.00	1,840.00	1,840.00
California	648,000	620,000	607,600	648,000	620,000	604,600	429.00	574.00	562.00
Freestone ...	280,000	288,000	267,000	280,000	288,000	264,000	516.00	811.00	680.00
Clingstone ..	368,000	332,000	340,600	368,000	332,000	340,600	364.00	369.00	470.00
Colorado	7,330	13,260	11,200	7,080	12,650	10,570	1,870.00	2,430.00	2,250.00
Connecticut	1,350	1,770	1,365	1,330	1,750	1,360	2,770.00	2,500.00	2,080.00
Georgia	35,250	35,500	40,600	34,810	33,000	33,700	826.00	1,090.00	1,040.00
Idaho	6,100	8,200	6,830	5,500	7,800	6,630	945.00	1,070.00	943.00
Illinois	3,780	3,700	3,340	3,570	3,550	3,020	1,340.00	1,270.00	1,470.00
Maryland	3,870	3,810	3,860	3,590	3,070	3,800	1,060.00	1,040.00	1,030.00
Massachusetts	1,400	1,249	1,485	1,300	1,187	1,455	2,770.00	2,240.00	2,330.00
Michigan	20,600	8,860	7,180	19,790	8,440	7,080	706.00	919.00	834.00
Missouri	4,160	4,090	2,480	4,040	3,940	2,250	1,850.00	1,140.00	1,370.00
New Jersey	18,120	22,450	21,170	18,000	21,050	21,100	1,510.00	1,320.00	1,310.00
New York	7,740	7,270	6,990	7,050	7,070	6,900	815.00	1,790.00	1,250.00
North Carolina	6,000	4,400	5,400	5,200	4,300	5,290	1,150.00	1,450.00	1,320.00
Ohio	5,370	230	1,200	4,960	220	1,180	1,480.00	1,680.00	1,520.00
Pennsylvania ..	19,500	14,940	18,190	18,300	14,460	18,100	1,030.00	1,190.00	1,170.00
South Carolina	69,650	65,700	68,900	64,150	60,800	59,700	1,070.00	1,120.00	1,070.00
Texas	8,250	3,800	5,100	7,500	3,300	4,700	2,570.00	2,000.00	1,800.00
Utah	5,421	6,500	3,900	5,141	6,200	3,880	1,080.00	981.00	1,080.00
Virginia	7,740	5,300	5,120	6,040	5,020	5,100	926.00	1,170.00	1,340.00
Washington	13,100	12,500	12,850	12,900	12,500	12,750	770.00	868.00	1,070.00
West Virginia ..	5,700	5,360	5,800	5,460	5,250	5,660	900.00	1,020.00	1,070.00
United States ..	903,871	852,939	847,210	888,681	838,027	825,415	617.00	751.00	734.00

[1] Fresh fruit prices are equivalent packing house-door returns for California and Washington except equivalent returns for bulk fruit at the first delivery point for California Clingstone, and the average price as sold for other States. Quantities processed are priced at the equivalent processing plant door level.
NASS, Crops Branch, (202) 720–2127.

Table 5-53.—Peaches: Production and utilization, by State and United States, crop of 2015[1]

State	Total production	Utilized production[1]	Utilization			
			Fresh[2]	Processed (fresh basis)		
				Canned	Dried	Frozen
	Tons	Tons	Tons	Tons	Tons	Tons
California, all	607,600	604,600	....	....	....	....
Freestone	267,000	264,000	151,000	....	6,900	98,000
Clingstone	340,600	340,000	....	327,000	....	....
Connecticut	1,365	1,360	1,360	....	....	....
Massachusetts	1,485	1,455	(D)	....	....	....
Missouri	2,480	2,250	(D)	....	....	....
South Carolina	68,900	59,700	59,150	....	....	....
Texas	5,100	4,700	4,700	....	....	....
Virginia	5,120	12,750	5,100	....	....	....
West Virginia	5,800	5,660	5,660	....	....	....
Other States	....	....	130,765	12,540	....	10,400
United States	847,210	825,415	357,735	339,540	6,900	108,400

(D) Withheld to avoid disclosing data for individual operations. [1] Difference between total and utilized production is harvested not sold and unharvested production. [2] Includes "Home use."
NASS, Crops Branch, (202) 720–2127.

Table 5-54.—Fruit: Exports, 2011–2013

Country	2011	2012	2013
	Metric tons	Metric tons	Metric tons
Fruit, processed:			
Canada	76,092	79,711	83,753
Netherlands	34,610	36,597	42,671
Mexico	31,129	29,488	29,103
China	27,077	37,103	14,861
Indonesia	3,543	5,118	5,783
Japan	7,395	6,910	5,550
Korea, South	3,341	4,377	5,284
Australia(*)	4,295	3,924	4,589
Taiwan	5,402	2,941	3,559
Thailand	3,531	3,515	3,493
United Kingdom	2,575	3,290	3,078
Malaysia	1,530	1,966	2,465
Singapore	2,714	3,345	2,379
Philippines	774	2,057	2,344
Poland	2,019	1,253	2,160
Israel(*)	2,337	1,028	2,015
New Zealand(*)	1,485	1,752	1,991
United Arab Emirates	1,337	1,277	1,833
Chile	614	26	1,689
Morocco	25	500	1,392
Germany(*)	2,191	1,202	1,356
Hong Kong	1,124	1,667	1,352
India	704	875	1,351
Russia	144	458	1,020
Panama	1,113	714	860
Guatemala	564	336	827
Belgium-Luxembourg(*)	1,404	896	684
Brazil	112	514	593
Vietnam	1,122	395	528
Rest of World	7,659	8,253	6,998
World Total	227,960	241,486	235,564
Fruit, prepared, misc:			
Canada	47,466	51,888	54,339
Mexico	9,271	8,411	6,893
Vietnam	2,136	3,508	4,930
China	36,199	15,151	4,821
Australia(*)	1,046	1,104	4,509
Saudi Arabia	7,074	7,919	4,065
Japan	3,833	3,710	3,411
Korea, South	9,981	8,003	2,659
Taiwan	1,290	1,166	2,235
United Kingdom	2,691	2,372	1,980
Malaysia	2,184	2,116	1,707
Hong Kong	1,803	1,802	1,683
Panama	3,604	3,157	1,636
Ecuador	126	126	1,084
United Arab Emirates	1,373	1,235	1,053
Costa Rica	896	815	1,039
Philippines	1,588	880	1,009
Thailand	577	1,162	985
Colombia	465	651	885
Bahamas, The	292	808	846
Israel(*)	749	547	719
Singapore	945	1,227	712
Kuwait	330	822	683
Guatemala	751	408	676
Leeward-Windward Islands(*)	263	275	617
Dominican Republic	649	1,099	571
Indonesia	638	517	520
Netherlands Antilles(*)	216	252	465
South Africa	35	141	446
Rest of World	9,333	7,458	6,501
World Total	147,805	128,726	113,677

(*) Denotes a country that is a summarization of its component countries. All zeroes for a data item may show that statistics exist in the other import type. Consumption or General. Users should use cautious interpretation on Quantity reports using mixed units of measure. Quantity line items will only include statistics on the units of measure that are equal to, or are able to be converted to, the assigned unit of measure of the grouped commodities.

FAS, Office of Global Analysis, (202) 720-6301. Data Source: Department of Commerce, U.S. Census Bureau, Foreign Trade Statistics.

Table 5-55.—Pears: Foreign trade, United States, 2006–2015

Year beginning October	Imports for consumption, fresh	Domestic exports				
		Fresh [1]	Canned	Dried, in fruit salad [1] [2]	Canned, in fruit salad [3]	Total, in terms of fresh fruit [4]
	Metric tons	Metric tons	Metric tons	Metric tons	Metric tons	Metric tons
2006	108,587	132,730	6,669	492	4,462	146,964
2007	85,855	165,436	8,028	548	8,767	185,688
2008	82,097	146,252	7,378	479	8,356	165,011
2009	62,808	159,677	7,112	525	10,238	180,342
2010	77,992	154,025	7,958	635	13,622	179,615
2011	64,000	197,633	8,681	702	15,404	226,146
2012	79,034	186,855	7,535	739	19,084	218,137
2013	81,586	200,928	6,283	682	17,626	229,139
2014	89,741	174,793	4,425	553	17,598	200,303
2015	79,573	148,483	5,824	289	16,103	172,235

[1] Net processed weight. [2] Dried pears are 16⅔ percent of total dried fruit for salad. [3] Canned pears are 35 percent of total canned fruit for salad. [4] Dried converted to unprocessed dry weight by dividing by 1.03. Unprocessed dry weight converted to terms of fresh on the basis that 1 pound dried equals about 6.5 pounds fresh. Canned converted to terms of fresh on basis that 1 pound of canned equals about 1 pound fresh.
ERS, Crops Branch, (202) 694-5255.

Table 5-56.—Pears: Production and value, United States 2006–2015

Year	Total production	Utilized production	Marketing year average price [1]	Value
	Tons	Tons	Dollars per ton	1,000 dollars
2006	842,035	831,120	397.00	329,928
2007	872,950	871,850	416.00	363,092
2008	869,850	868,880	456.00	396,081
2009	957,220	955,820	372.00	355,662
2010	813,600	813,400	476.00	386,955
2011	965,720	965,110	380.00	366,552
2012	851,240	851,130	509.00	432,988
2013	877,130	876,520	491.00	430,660
2014	831,610	831,250	562.00	457,194
2015	820,520	807,130	620.00	500,416

[1] Fresh fruit prices are equivalent packinghouse-door returns for California, Oregon, and Washington, and the average price as sold for other States. Quantities processed are priced at the equivalent processing plant door level.
NASS, Crops Branch, (202) 720-2127.

Table 5-57.—Pears: Production and season average price per ton, by State and United States, 2013–2015

State and Variety	Total production			Utilized production			Price per ton [1]		
	2013	2014	2015	2013	2014	2015	2013	2014	2015
	Tons	Tons	Tons	Tons	Tons	Tons	Dollars	Dollars	Dollars
California	220,000	189,000	202,000	220,000	189,000	195,000	391.00	469.00	513.00
Bartlett	177,000	154,000	172,000	177,000	154,000	165,000	348.00	416.00	503.00
Other	43,000	35,000	30,000	43,000	35,000	30,000	571.00	701.00	567.00
Michigan	5,350	2,720	2,070	5,180	2,610	2,040	348.00	441.00	605.00
New York	9,200	54,350	6,480	9,100	5,210	6,350	565.00	666.00	756.00
Oregon	207,000	216,000	228,000	207,000	216,000	226,000	537.00	590.00	675.00
Bartlett	55,000	53,000	57,500	55,000	53,000	56,000	511.00	572.00	639.00
Other	152,000	163,000	170,500	152,000	163,000	170,000	546.00	596.00	687.00
Pennsylvania	1,580	2,540	1,970	1,240	2,430	1,740	923.00	1,100.00	1,210.00
Washington	434,000	416,000	380,000	434,000	416,000	376,000	519.00	586.00	638.00
Bartlett	185,000	181,000	178,000	185,000	181,000	177,000	435.00	514.00	530.00
Other	249,000	235,000	202,000	249,000	235,000	199,000	582.00	642.00	733.00
United States	877,130	831,610	820,520	876,520	831,250	807,130	491.00	562.00	620.00

[1] Fresh fruit prices are equivalent packinghouse-door returns for California, Oregon, and Washington, and the average price as sold for other States. Quantities processed are priced at the equivalent processing plant door level.
NASS, Crops Branch, (202) 720-2127.

Table 5-58.—International Fruits: Production in specified countries, 2013/2014-2015/2016

Country	Production		
	2013/2014	2014/2015	2015/2016
	Metric tons	Metric tons	Metric tons
Apples, fresh:			
Brazil	1,377,400	1,266,000	1,240,000
Chile	1,310,000	1,350,000	1,350,000
China	39,680,000	40,920,000	43,000,000
European Union	11,864,900	13,618,500	12,220,000
India	2,200,000	2,200,000	2,200,000
Iran	1,693,000	1,693,000	1,693,000
Russia	1,416,500	1,408,600	1,390,000
South Africa	792,500	860,000	865,000
Turkey	2,930,000	2,289,000	2,740,000
Ukraine	1,211,000	1,211,000	1,211,000
Others	4,649,600	4,560,800	4,548,800
Total Foreign	69,124,900	71,376,900	72,457,800
United States	4,690,101	5,074,793	4,561,000
Total	73,815,001	76,451,693	77,018,800
Fresh Cherries, sweet and sour:			
Azerbaijan	32,300	30,000	30,000
Chile	85,900	124,184	142,660
China	180,000	220,000	250,000
European Union	723,581	745,417	745,895
Russia	278,000	278,000	278,000
Serbia	126,400	126,000	126,000
Syria	62,400	62,000	62,000
Turkey	510,000	470,000	540,000
Ukraine	282,000	280,000	280,000
Uzbekistan	140,000	140,000	140,000
Others	65,600	65,500	65,500
Total Foreign	2,486,181	2,541,101	2,660,055
United States	400,522	461,984	396,497
Total	2,886,703	3,003,085	3,056,552
Fresh Peaches & nectarines:			
Argentina	291,804	290,000	290,000
Australia	100,000	90,000	90,000
Brazil	217,706	220,000	220,000
Chile	91,000	141,000	151,000
China	11,900,000	13,000,000	13,600,000
European Union	3,730,508	4,182,344	4,026,700
Japan	124,700	136,954	137,000
Mexico	161,268	160,000	160,000
South Africa	174,230	170,000	170,000
Turkey	550,000	500,000	520,000
Others	214,770	198,577	197,900
Total Foreign	17,555,986	19,088,875	19,562,600
United States	953,253	946,491	903,055
Total	18,509,239	20,035,366	20,465,655
Pears, fresh:			
Argentina	690,000	580,000	650,000
Chile	267,000	290,000	300,000
China	17,300,000	18,000,000	19,000,000
European Union	2,523,100	2,594,300	2,450,000
India	340,000	340,000	340,000
Japan	300,000	300,000	300,000
Korea, South	282,200	302,700	259,000
Russia	132,500	160,451	160,000
South Africa	413,600	400,000	410,000
Turkey	415,000	305,000	415,000
Others	339,500	363,800	363,200
Total Foreign	23,002,900	23,636,251	24,647,200
United States	795,166	754,098	665,000
Total	23,798,066	24,390,349	25,312,200

FAS, Office of Global Analysis, (202) 720-6301. Prepared or estimated on the basis of official USDA production, supply, and distribution statistics from foreign governments.

Table 5-59.—Pears: Production and utilization, by State and United States, crop of 2015

State and variety	Total production	Utilized production	Utilization	
			Fresh[1]	Processed
	Tons	*Tons*	*Tons*	*Tons*
California	202,000	195,000	(D)	(D)
Bartlett	172,000	165,000	46,000	119,000
Other	30,000	30,000	(D)	(D)
Michigan	2,070	2,040	(D)	(D)
New York	6,480	6,350	(D)	(D)
Oregon	228,000	226,000	(D)	(D)
Bartlett	57,500	56,000	38,000	18,000
Other	170,500	170,000	(D)	(D)
Pennsylvania	1,970	1,740	(D)	(D)
Washington	380,000	376,000	(D)	(D)
Bartlett	178,000	177,000	76,000	101,000
Other	202,000	199,000	(D)	(D)
United States[2]	820,520	807,130	527,780	279,350

(D) Withheld to avoid disclosing data for individual operations.   [1] Includes "Home use."   [2] Mostly canned, but includes small quantities dried, juiced, and other uses.
NASS, Crops Branch, (202) 720-2127.

Table 5-60.—Pears: Production and utilization, United States, 2006-2015

Crop of—	Total production	Utilized production	Utilization of quantities sold—Fresh[1]
	Tons	*Tons*	*Tons*
2006	842,035	831,120	500,720
2007	872,950	871,850	551,960
2008	869,850	868,880	548,930
2009	957,220	955,820	603,800
2010	813,600	813,400	531,430
2011	965,720	965,110	643,710
2012	851,240	851,130	551,820
2013	877,130	876,520	584,730
2014	831,610	831,250	550,080
2015	820,520	807,130	527,780

[1] Includes "Home use."
NASS, Crops Branch, (202) 720-2127.

Table 5-61.—Papayas: Area, utilized production, utilization, marketing year average price, and value, Hawaii, 2006-2015

Year	Area harvested	Utilized production	Utilization		Price per pound	Value
			Fresh	Processed		
	Acres	*1,000 pounds*	*1,000 pounds*	*1,000 pounds*	*Cents*	*dollars*
2006	1,530	28,700	26,600	2,100	38.5	11,049
2007	1,310	33,400	31,200	2,200	39.2	13,094
2008	1,380	33,500	31,500	2,000	43.0	14,393
2009	1,325	31,500	30,300	1,200	45.0	14,186
2010	1,350	30,100	29,200	900	37.0	11,123
2011	1,300	28,600	27,700	900	34.0	9,722
2012	(NA)	(NA)	(NA)	(NA)	(NA)	(NA)
2013	1,300	24,200	23,000	1,200	34.4	8,316
2014	1,500	23,500	23,000	500	48.0	11,285
2015	1,500	25,200	(D)	(D)	41.9	10,570

(D) Withheld to avoid disclosing data for individual operations.   (NA) Not available.
NASS, Crops Branch, (202) 720-2127.

Table 5-62.—Plums, California: Production, value, and utilization, 2006–2015

Season	Total production	Utilized production	Marketing year average price per ton	Value
	Tons	Tons	Dollars	1,000 dollars
2006	158,000	158,000	688.00	108,648
2007	152,000	152,000	665.00	101,077
2008	160,000	160,000	356.00	56,960
2009	112,000	112,000	514.00	57,568
2010	142,000	142,000	555.00	78,810
2011	160,000	160,000	402.00	64,320
2012	115,000	115,000	695.00	79,940
2013	95,400	93,400	664.00	62,043
2014	113,000	113,000	913.00	103,167
2015	106,000	105,000	998.00	104,760

NASS, Crops Branch, (202) 720–2127.

Table 5-63.—Prunes (dried basis): Production, price and value, California, 2006–2015

Season	Total production	Utilized production	Marketing year average price per ton	Value
	Tons	Tons	Dollars	1,000 dollars
2006	198,000	189,000	1,390.00	262,710
2007	83,000	81,000	1,450.00	117,450
2008	129,000	129,000	1,500.00	193,500
2009	166,000	166,000	1,230.00	204,180
2010	130,000	130,000	1,350.00	175,500
2011	137,000	137,000	1,310.00	179,470
2012	138,000	138,000	1,330.00	183,540
2013	85,000	85,000	2,000.00	170,000
2014	108,000	108,000	2,470.00	266,760
2015	112,000	110,000	2,010.00	221,100

NASS, Crops Branch, (202) 720–2127.

Table 5-64.—Prunes and plums: Production, value, and utilization, 4-States [1], 2006–2015

Year	Total production	Utilized production	Marketing year average price per ton	Value	Utilization of quantities sold			
					Fresh	Processed (fresh basis)		
						Dried and other	Canned	Frozen
	Tons	Tons	Dollars	1,000 dollars	Tons	Tons	Tons	Tons
2006	21,500	19,200	452.00	8,678	9,550	4,300	3,250	2,100
2007	12,100	10,920	454.00	4,956	6,420	1,300	2,550	650
2008	15,500	15,480	382.00	5,918	8,700	3,540	2,130	1,110
2009	18,600	17,700	327.00	5,787	9,750	5,040	2,130	780
2010	12,100	11,200	439.00	4,915	7,700	1,650	1,395	455
2011	13,300	12,900	370.00	4,767	7,030	3,850	1,460	560
2012	12,935	11,835	531.00	6,288	7,030	3,500	755	550
2013	13,440	12,790	482.00	6,169	6,690	3,980	1,380	740
2014	14,800	14,500	487.00	7,060	7,600	(NA)	690	390
2015	9,680	9,650	553.00	5,337	5,950	(NA)	1,450	300

(NA) Not available. [1] Idaho, Michigan, Oregon,and Washington.
NASS, Crops Branch, (202) 720–2127.

Table 5-65.—Prunes and plums (fresh basis): Production and season average price per ton, by State, 2013–2015

State	Total production			Utilized production			Price per ton		
	2013	2014	2015	2013	2014	2015	2013	2014	2015
	Tons	*Tons*	*Tons*	*Tons*	*Tons*	*Tons*	*Dollars*	*Dollars*	*Dollars*
Idaho	2,400	2,600	2,000	2,350	2,400	2,000	440.00	415.00	511.00
Michigan	1,890	1,700	760	1,890	1,600	750	510.00	535.00	585.00
Oregon	6,800	7,800	4,220	6,250	7,800	4,200	488.00	442.00	494.00
Washington	2,350	2,700	2,700	2,300	2,700	2,700	489.00	652.00	667.00
Total, 4 States	13,440	14,800	9,680	12,790	14,500	9,650	482.00	487.00	553.00

NASS, Crops Branch, (202) 720–2127.

Table 5-66.—Prunes and plums: Utilization and marketing year average price per ton, by State, 2009–2015

State and season	Quantity				Price			
	Fresh	Dried and other	Canned	Frozen	Fresh	Dried and other	Canned	Frozen
	Tons	*Tons*	*Tons*	*Tons*	*Dollars*	*Dollars*	*Dollars*	*Dollars*
Michigan:								
2009	1,000	(1)	(1)	(1)	880.00	(1)	(1)	(1)
2010	1,000	(1)	(1)	(1)	870.00	(1)	(1)	(1)
2011	(1)	(1)	(1)	(1)	(1)	(1)	(1)	(1)
2012	(1)	(1)	(1)	(1)	(1)	(1)	(1)	(1)
2013	(1)	(1)	(1)	(1)	(1)	(1)	(1)	(1)
2014	(1)	(1)	(1)	(1)	(1)	(1)	(1)	(1)
2015	(1)	(1)	(1)	(1)	(1)	(1)	(1)	(1)
Oregon:								
2009	4,200	(1)	(1)	(1)	322.00	(1)	(1)	(1)
2010	2,800	(1)	(1)	(1)	498.00	(1)	(1)	(1)
2011	1,700	(1)	(1)	(1)	441.00	(1)	(1)	(1)
2012	3,220	(1)	(1)	(1)	813.00	(1)	(1)	(1)
2013	2,620	(1)	(1)	(1)	644.00	(1)	(1)	(1)
2014	3,900	(1)	(1)	(1)	554.00	(1)	(1)	(1)
2015	2,700	(1)	(1)	(1)	561.00	(1)	(1)	(1)
Total States:[2]								
2009	9,750	5,040	2,130	780	472.00	129.00	178.00	204.00
2010	7,700	1,650	1,395	455	551.00	163.00	209.00	251.00
2011	7,030	3,850	1,460	560	482.00	240.00	203.00	280.00
2012	7,030	3,500	755	550	739.00	220.00	234.00	265.00
2013	6,690	3,980	1,380	740	608.00	(NA)	380.00	400.00
2014	7,600	(NA)	690	390	656.00	(NA)	414.00	367.00
2015	5,950	(NA)	1,450	300	699.00	(NA)	388.00	410.00

(NA) Not available. [1] Not published to avoid disclosure of individual operations, but is included in total.   [2] Includes Idaho, Michigan, Oregon, and Washington.
NASS, Crops Branch, (202) 720–2127.

Table 5-67.—Prunes: Foreign trade, United States, 2006–2015

Year beginning October	Imports				Domestic exports			
	Fresh prunes and plums	Otherwise prepared or preserved	Dried prunes [1]	Total, in terms of fresh [2]	Fresh prunes and plums	Dried prunes [1]	Dried, in fruit salad [1][3]	Total, in terms of fresh [2]
	Metric tons	Metric tons	Metric tons	Metric tons	Metric tons	Metric tons	Metric tons	Metric tons
2006	36,434	866	840	39,392	46,633	66,253	1,268	221,930
2007	29,104	678	828	31,863	50,551	61,345	1,414	213,483
2008	29,409	829	3,906	40,293	46,215	54,706	1,236	191,451
2009	27,226	805	620	29,559	47,213	65,070	1,355	219,664
2010	29,768	831	497	31,805	59,653	69,110	1,639	243,328
2011	24,105	822	565	26,311	56,788	67,529	1,810	236,801
2012	30,417	680	1,159	34,036	52,426	68,642	1,906	235,579
2013	13,625	821	6,844	32,129	41,741	48,498	1,759	172,216
2014	25,259	798	11,514	55,869	36,506	48,015	1,426	164,862
2015	26,533	986	19,183	77,222	29,881	38,307	746	131,270

[1] Net processed weight. [2] Exports and imports of dried prunes converted to unprocessed dry weight by dividing by 1.04. Unprocessed dry weight converted to terms of fresh fruit by dividing that 1 pound dried equals 2.7 pounds fresh. "Otherwise prepared or preserved" converted to terms of fresh fruit on the basis that 1 pound equals 0.899 pound fresh. [3] Dried prunes in salad estimated at 43 percent of total dried fruit for salad.
ERS, Crops Branch, (202) 694–5255.

Table 5-68.—Strawberries, commercial crop: Production and value per hundredweight, by State and United States, 2013–2015

Utilization, season, and State	Production			Value per unit		
	2013	2014	2015	2013	2014	2015
	1,000 cwt	1,000 cwt	1,000 cwt	Dollars per cwt	Dollars per cwt	Dollars per cwt
Fresh market:						
California	22,398	22,100	21,821	90.40	100.00	73.00
Florida	2,332	2,071	2,442	143.00	148.00	119.00
Michigan	(D)	(D)	(D)	(D)	(D)	(D)
New York	38	32	29	203.40	235.00	254.00
North Carolina	110	154	143	148.00	152.00	164.00
Ohio	29	18	23	209.00	193.00	170.00
Oregon	22	25	20	167.00	161.00	170.00
Pennsylvania	51	33	34	223.00	160.00	213.00
Washington	(D)	(D)	(D)	(D)	(D)	(D)
Wisconsin	39	38	34	165.00	165.00	170.00
Other States	66	72	47	175.00	180.00	173.00
United States	25,085	24,543	24,593	96.50	105.00	79.00
Processing:						
California	5,175	5,492	6,088	34.00	41.50	43.10
Michigan	(D)	(D)	(D)	(D)	(D)	(D)
Oregon	145	130	135	48.00	70.00	73.00
Washington	(D)	(D)	(D)	(D)	(D)	(D)
Other States	65	72	51	73.20	75.20	80.00
United States	5,385	5,694	6,274	34.90	42.60	44.00

(D) Withheld to avoid disclosing data for individual operations.
NASS, Crops Branch, (202) 720–2127.

Table 5-69.—Strawberries, commercial crop: Area, yield, production, value per hundred weight, and total value, United States, 2006–2015

Year	Fresh market and processing				
	Area for harvest	Yield per acre	Production	Value[2]	
				Per cwt	Total
	Acres	Cwt	1,000 cwt	Dollars per cwt	1,000 dollars
2006	53,460	450	24,038	63.20	1,519,494
2007	52,180	469	24,453	71.60	1,751,108
2008	54,470	465	25,317	75.80	1,918,288
2009	58,080	482	28,013	76.00	2,129,585
2010	56,890	501	28,520	79.30	2,260,733
2011	57,250	507	29,016	82.40	2,391,406
2012	58,310	523	30,520	80.40	2,453,039
2013	60,410	504	30,470	85.60	2,609,038
2014	59,895	505	30,237	93.30	2,821,854
2015	58,010	532	30,867	71.90	2,219,144

Year	Fresh market[1]			Processing		
	Production	Value[2]		Production	Value[2]	
		Per cwt	Total		Per cwt	Total
	1,000 cwt	Dollars per cwt	1,000 dollars	1,000 cwt	Dollars per cwt	1,000 dollars
2006	19,109	72.20	1,379,658	4,929	28.40	139,836
2007	19,733	82.10	1,620,241	4,720	27.70	130,867
2008	20,911	84.10	1,759,564	4,406	36.00	158,724
2009	22,880	86.10	1,970,920	5,133	30.90	158,665
2010	23,196	90.80	2,105,492	5,324	29.20	155,241
2011	23,324	94.00	2,193,379	5,692	34.80	198,027
2012	24,552	91.80	2,253,854	5,968	33.40	199,185
2013	25,085	96.50	2,421,370	5,385	34.90	187,668
2014	24,543	105.00	2,579,425	5,694	42.60	242,429
2015	24,593	79.00	1,942,814	6,274	44.00	276,330

[1] Fresh market price and value at point of first sale. Processing price and value at processing plant door.   [2] Mostly for fresh market, but includes some quantities used for processing in States for which processing estimates are not prepared. NASS, Crops Branch, (202) 720–2127.

Table 5-70.—Strawberries, commercial crop: Area harvested, production, value per hundred weight, by State and United States, 2013–2015[1]

Season and State	Area harvested			Production			Value per unit		
	2013	2014	2015	2013	2014	2015	2013	2014	2015
	Acres	Acres	Acres	1,000 cwt	1,000 cwt	1,000 cwt	Dollars per cwt	Dollars per cwt	Dollars per cwt
California	41,500	41,500	40,500	27,573	27,592	27,909	79.80	88.40	66.50
Florida	10,600	10,900	10,900	2,332	2,071	2,442	143.00	148.00	119.00
Michigan	750	750	650	44	45	28	161.00	161.00	156.00
New York	1,400	1,000	800	38	32	29	203.00	235.00	254.00
North Carolina	1,100	1,100	1,100	110	154	143	148.00	152.00	164.00
Ohio	590	480	540	29	18	23	209.00	193.00	170.00
Oregon	1,700	1,700	1,400	167	155	155	63.70	84.70	86.00
Pennsylvania	950	735	600	51	33	34	223.00	160.00	213.00
Washington ...	1,100	1,100	910	87	99	70	106.00	112.00	112.00
Wisconsin	720	630	610	39	38	34	165.00	165.00	170.00
United States	60,410	59,895	58,010	30,470	30,237	30,867	85.60	93.30	71.90

[1] Includes quantities used for fresh market and processing. NASS, Crops Branch, (202) 720–2127.

Table 5-71.—Fruits, noncitrus: Production, utilization, and value, United States, 2006–2015 [1]

Year	Utilized production	Fresh [2]	Processed						Value of utilized production
			Canned	Dried	Juice	Frozen	Wine	Other	
	1,000 tons	*1,000 tons*	*1,000 tons*	*1,000 tons*	*1,000 tons*	*1,000 tons*	*1,000 tons*	*1,000 tons*	*1,000 dollars*
2006	16,816	6,930	1,400	2,219	1,256	710	3,726	235	10,510,417
2007	17,023	7,008	1,453	2,030	1,257	748	3,921	278	11,397,171
2008	17,603	7,248	1,406	2,413	1,228	682	3,944	290	11,270,860
2009	18,069	7,562	1,394	2,148	1,235	742	4,373	269	11,811,298
2010	17,879	7,458	1,386	2,318	1,103	706	4,270	298	12,367,607
2011	18,147	7,702	1,301	2,399	1,153	737	4,154	316	13,910,375
2012	17,635	7,443	1,100	2,091	1,213	707	4,707	374	15,611,441
2013	19,433	7,718	1,373	2,321	1,630	827	5,068	497	16,220,440
2014	19,151	8,267	1,208	2,194	1,617	857	4,526	483	16,409,428
2015	18,358	7,684	1,232	2,376	1,366	898	4,256	546	16,051,187

[1] Includes the following crops: Apples, apricots, avocados, bananas, berries, cherries, cranberries, dates, figs, grapes, guavas, kiwifruit, nectarines, olives, papayas, peaches, pears, pineapples, plums, prunes, and strawberries. [2] Includes local and roadside sales.
NASS, Crops Branch, (202) 720–2127.

Table 5-72.—Fruits, fresh: Total reported domestic rail, truck, and air shipments, 2015

Commodity	Jan.	Feb.	Mar.	Apr.	May	Jun.	Jul.	Aug.	Sep.	Oct.	Nov.	Dec.	Total
	1,000 cwt	*1,000 cwt*	*1,000 cwt*	*1,000 cwt*	*1,000 cwt*	*1,000 cwt*	*1,000 cwt*	*1,000 cwt*	*1,000 cwt*	*1,000 cwt*	*1,000 cwt*	*1,000 cwt*	*1,000 cwt*
Citrus:													
Grapefruit	1,207	1,092	803	317	83	10	10	8	23	508	948	919	5,928
Lemons	44	38	40	50	42	48	45	49	22	38	47	73	536
Oranges	859	641	727	757	559	327	159	70	97	416	783	882	6,277
Tangelos		105	17	1							26	85	234
Tangerines	174	135	121	25					36	134	131	93	849
Temples	19	26	12	1								4	62
Total	2,303	2,037	1,720	1,151	684	385	214	127	178	1,096	1,935	2,056	13,886
Noncitrus:													
Apples	6,729	5,540	5,686	5,483	6,303	3,903	3,688	5,396	4,997	6,491	5,280	5,064	64,560
Apples-organic	589	436	433	338	282	81	26	149	361	466	343	349	3,853
Apricots					31	68	46	2					147
Apricots-organic						9	5						14
Avocados	145	155	373	511	580	435	449	405	120	63	44	44	3,324
Blueberries			22	258	558	762	459	234	106	5			2,404
Blueberries-organic ..			5	27	38	63	38	18	4				193
Cherries				33	1,007	2,354	1,533	43					4,970
Cherries-organic					13	62	18	1					94
Cranberries									4	40	114	17	175
Grapes	65			8	416	458	2,587	3,157	3,196	3,478	2,251	927	16,543
Grapes-organic					8	11	4					2	25
Kiwifruit	17	6								1	17	23	64
Nectarines					266	620	673	483	96				2,138
Peaches					391	1,211	1,393	1,226	724	79			5,024
Peaches-organic							8	53	16				77
Pears	1,385	1,299	704	619	663	351	463	586	938	1,551	1,271	1,030	10,860
Pears-organic	86	40	11	4			9	42	80	264	60	58	654
Persimmons										5	25	15	45
Plums					39	309	409	326	209	2			1,294
Pomegranates								9	38	31	24	5	107
Prunes							1	24	3				28
Raspberries	45	47	57	76	169	220	243	186	178	147	79	68	1,515
Raspberries-organic ..	9	9	7	10	23	30	37	28	25	17	7	9	211
Strawberries	841	1,418	2,334	1,885	2,275	2,337	1,621	1,240	1,161	775	246	422	16,555
Strawberries-organic	17	35	78	94	136	188	156	110	106	50	10	11	991
Total	9,928	8,985	9,710	9,346	13,198	13,472	13,866	13,718	12,362	13,465	9,771	8,044	135,865
Grand total	12,231	11,022	11,430	10,497	13,882	13,857	14,080	13,845	12,540	14,561	11,706	10,100	149,751

AMS, Fruit and Vegetable Programs, Market News Division, (202) 720–9936.

Table 5-73.—Fruits, dried: Production (dry basis), California, 2006–2015

Year	Apricots	Figs[1]	Peaches[2]	Prunes	Grapes[3]
	Tons	Tons	Tons	Tons	Tons
2006	640	13,000	1,290	189,000	309,500
2007	1,970	14,500	1,365	81,000	360,000
2008	1,830	13,100	1,050	129,000	390,300
2009	1,090	13,300	850	166,000	335,500
2010	1,400	12,320	2,010	130,000	394,800
2011	1,810	11,560	470	137,000	384,300
2012	1,210	10,400	1,230	138,000	345,900
2013	1,080	9,300	420	85,000	406,100
2014	1,500	9,470	610	108,000	366,200
2015	990	8,840	880	110,000	383,700

[1] Standard and substandard.   [2] Freestone only.   [3] Raisin and table type.
NASS, Crops Branch, (202) 720-2127.

**Table 5-74.—International Raisins: Production in specified countries,
2013/2014–2015/2016**

Country	2013/2014	2014/2015	2015/2016
	Metric tons	Metric tons	Metric tons
Afghanistan	34,300	40,000	42,000
Argentina	20,500	33,000	32,000
Australia	10,000	10,000	10,000
Chile	69,200	64,000	65,000
China	165,000	180,000	190,000
European Union	10,000	10,000	10,000
Iran	160,000	130,000	150,000
South Africa	46,000	58,000	55,000
Turkey	242,635	320,000	200,000
Uzbekistan	18,000	51,000	35,000
Others	10,000	10,000	10,000
Total Foreign	785,635	906,000	799,000
United States	368,408	311,255	340,000
Total	1,154,043	1,217,255	1,139,000

FAS, Office of Global Analysis, (202) 720-6301. Prepared or estimated on the basis of official USDA production, supply, and distribution statistics from foreign governments.

Table 5-75.—Fruits: Per capita consumption, United States, 2006–2015 [1]

Year	Fruits used fresh		Canned fruits [4]	Juice [5]	Frozen fruit [6]	Dried fruits [7]
	Citrus fruit [2]	Noncitrus fruits [3]				
	Pounds	*Pounds*	*Pounds*	*Gallons*	*Pounds*	*Pounds*
2006	21.6	79.5	13.2	7.8	5.2	2.3
2007	17.9	79.4	14.1	7.6	5.5	2.2
2008	20.6	79.2	13.4	7.6	5.1	2.2
2009	20.7	77.2	13.6	7.1	5.1	2.1
2010	21.6	80.7	13.1	7.2	5.7	2.3
2011	22.8	82.0	12.1	6.2	5.3	2.4
2012	23.6	84.8	11.5	6.4	5.0	2.4
2013	24.0	87.3	12.9	6.2	5.2	2.5
2014	23.3	90.0	11.7	5.9	6.5	2.4
2015	22.6	88.6	12.2	5.9	6.6	2.6

[1] Fresh citrus fruits, canned fruit, and fruit juices are on a crop-year basis. Dried fruits are on a pack-year basis. The per capita consumption was obtained by dividing the total consumption by total population. [2] Oranges and temples, tangerines and tangelos, lemons, limes, and grapefruit. [3] Apples, apricots, avocados, bananas, cherries, cranberries, grapes, kiwifruit, mangoes, peaches and nectarines, pears, pineapples, papayas, plums and prunes, and strawberries. [4] Apples, apricots, cherries, olives, peaches, pears, pineapples, and plums and prunes. [5] Orange, grapefruit, lemon, lime, apple, grape, pineapple, prune, and cranberry. [6] Blackberries, blueberries, raspberries, strawberries, other berries, apples, apricots, cherries, and peaches. [7] Apples, apricots, dates, figs, peaches, pears, prunes, and raisins. Dried data in terms of processed weight.
ERS, Crops Branch, (202) 694–5255.

Table 5-76.—All tree nuts: Supply and utilization, United States, 2006/2007–2015/2016

Market year[1]	Beginning stocks	Marketable production [2]	Imports	Total supply	Exports	Ending stocks	Domestic consumption	
							Total	Per capita Pounds
	—Million pounds (shelled)—							
2006/2007	267.7	1,655.8	438.3	2,361.8	1,129.8	243.1	988.9	3.29
2007/2008	243.1	2,062.5	489.8	2,795.4	1,311.8	405.9	1,077.8	3.55
2008/2009	405.9	2,259.5	439.2	3,104.6	1,466.3	543.7	1,094.6	3.57
2009/2010	543.7	2,113.3	468.9	3,125.9	1,546.4	422.1	1,157.3	3.75
2010/2011	422.2	2,482.2	485.1	3,389.4	1,786.1	407.3	1,196.1	3.85
2011/2012	407.3	2,787.2	449.2	3,643.8	1,977.1	480.1	1,186.6	3.79
2012/2013	480.1	2,762.3	515.6	3,758.0	1,979.6	462.1	1,316.3	4.17
2013/2014	462.1	2,806.7	582.1	3,850.9	2,075.2	505.4	1,270.2	3.99
2014/2015	505.4	2,738.4	657.4	3,901.3	2,030.7	585.4	1,285.2	4.01
2015/2016 [3] ..	585.4	2,694.3	679.2	3,958.9	2,056.8	560.5	1,341.6	4.15

[1] Marketing season begins July 1 for hazelnuts and macadamias; August 1 for almonds and walnuts; September 1 for pistachios; and October 1 for pecans. [2] Utilized production (NASS data) minus inedibles and noncommercial useage. [3] Preliminary.
ERS, Crops Branch, (202) 694–5255.

Table 5-77.—International Nuts: Production in specified countries, 2013/2014–2015/2016

Country	Production		
	2013/2014	2014/2015	2015/2016
	Metric tons	Metric tons	Metric tons
Almonds:			
Australia	65,100	75,000	82,000
Chile	3,900	11,000	12,000
China	7,000	9,500	10,000
European Union	58,800	79,700	85,000
India	1,100	1,200	1,100
Turkey	18,000	13,000	14,000
Total Foreign	153,900	189,400	204,100
United States	911,720	848,220	816,470
Total	1,065,620	1,037,620	1,020,570
Walnuts, Inshell basis:			
Australia	2,500	2,500	2,500
Brazil	5,200	6,000	6,000
Chile	60,000	54,000	60,000
China	780,000	900,000	1,000,000
European Union	110,000	111,000	117,000
India	43,000	35,000	38,000
Moldova	23,100	32,000	35,000
Turkey	75,000	40,000	60,000
Ukraine	115,790	102,740	102,000
Total Foreign	1,214,590	1,283,240	1,420,500
United States	446,335	517,095	521,631
Total	1,660,925	1,800,335	1,942,131

FAS, Office of Global Analysis, (202) 720-6301. Prepared or estimated on the basis of official USDA production, supply, and distribution statistics from foreign governments.

Table 5-78.—Almonds (shelled basis): Bearing acreage, yield, production, price, and value, California, 2006–2015 [1]

Year	Bearing Acreage	Yield per acre [2]	Utilized production	Price per pound	Value
	Acres	Pounds	1,000 pounds	Dollars	1,000 dollars
2006	610,000	1,840	1,120,000	2.06	2,258,790
2007	640,000	2,170	1,390,000	1.75	2,401,875
2008	710,000	2,300	1,630,000	1.45	2,343,200
2009	750,000	1,880	1,410,000	1.65	2,293,500
2010	770,000	2,130	1,640,000	1.79	2,903,380
2011	800,000	2,540	2,030,000	1.99	4,007,860
2012	820,000	2,310	1,890,000	2.58	4,816,860
2013	850,000	2,360	2,010,000	3.21	7,388,000
2014	870,000	2,150	1,870,000	4.00	5,325,000
2015	890,000	2,130	1,900,000	2.84	2,525,909

[1] Price and value are based on edible portion of the crop only. Production includes inedible quantities of no value as follows (million pounds): 2006-23.5; 2007-17.5; 2008-14.0; 2009-20.0; 2010-18.0; 2011-16.0; 2012-23; 2013-21.0; 2014-23.0; 2015-25.0. [2] Yield is based on utilized production.
NASS, Crops Branch, (202) 720–2127.

Table 5-79.—Almonds (shelled basis): Foreign trade, United States, 2006–2015 [1]

Year beginning October	Imports	Domestic exports
	Metric tons	Metric tons
2006	1,515	342,046
2007	1,722	397,105
2008	796	439,759
2009	776	462,276
2010	2,391	527,228
2011	5,729	615,275
2012	15,653	579,553
2013	14,724	581,390
2014	10,026	546,081
2015	9,521	563,026

[1] Imports of unshelled nuts converted to shelled basis at ratio of 1.67 to 1. Exports of unshelled nuts converted to shelled basis at ratio of 1.67 to 1.0.
ERS, Crops Branch, (202) 694–5255.

Table 5-80.—Macadamia nuts (in-shell basis): Bearing acreage, yield, production, price, and value, Hawaii, 2006–2015

Year	Bearing Acreage	Yield per acre	Utilized production	Price per pound	Value
	Acres	Pounds	1,000 pounds	Cents	1,000 dollars
2006	15,000	3,870	58,000	67.0	38,860
2007	15,000	2,730	41,000	60.0	24,600
2008	15,000	3,330	50,000	67.0	33,500
2009	15,000	2,800	42,000	70.0	29,400
2010	15,000	2,670	40,000	75.0	30,000
2011	15,000	3,270	49,000	78.0	38,220
2012	15,000	2,930	44,000	80.0	35,200
2013	16,000	2,560	41,000	87.0	35,670
2014	16,000	2,880	46,000	87.0	40,020
2015	16,000	2,940	47,000	97.0	45,590

NASS, Crops Branch, (202) 720–2127.

Table 5-81.—Hazelnuts (in-shell basis): Bearing acreage, yield, production, price, and value, Oregon, 2006–2015

Year	Bearing Acreage	Yield per acre	Utilized production	Price per ton	Value
	Acres	Tons	Tons	Dollars	1,000 dollars
2006	28,200	1.52	43,000	1,080	46,440
2007	28,600	1.29	37,000	2,040	75,480
2008	28,300	1.13	32,000	1,620	51,840
2009	28,700	1.64	47,000	1,690	79,430
2010	29,000	0.97	28,000	2,410	67,480
2011	28,500	1.35	38,500	2,330	89,705
2012	29,000	1.22	35,500	1,830	64,965
2013	30,000	1.50	45,000	2,680	120,600
2014	30,000	1.20	36,000	3,600	129,600
2015	30,000	1.03	31,000	2,800	86,800

NASS, Crops Branch, (202) 720–2127.

Table 5-82.—Hazelnuts (shelled basis): Foreign trade, United States, 2006–2015 [1]

Year beginning October	Imports	Domestic exports
	Metric tons	Metric tons
2006	4,344	11,187
2007	4,979	10,034
2008	2,969	9,715
2009	2,533	13,917
2010	4,417	7,691
2011	3,118	10,031
2012	5,392	11,243
2013	4,747	14,862
2014	1,761	20,309
2015	2,748	8,766

[1] Imports of unshelled nuts converted to shelled basis at ratio of 2.22 to 1. Exports of unshelled nuts converted to shelled basis at ratio of 2.50 to 1.
ERS, Crops Branch, (202) 694–5255.

Table 5-83.—Pecans (in-shell basis): Production, price per pound, and value, United States, 2006–2015

Year	Improved varieties [1]			Native and seedling			All pecans		
	Utilized production	Price per pound	Value	Utilized production	Price per pound	Value	Utilized production	Price per pound	Value
	1,000 pounds	Dollars	1,000 dollars	1,000 pounds	Dollars	1,000 dollars	1,000 pounds	Dollars	1,000 dollars
2006	152,130	1.730	262,544	55,170	1.090	59,949	207,300	1.560	322,493
2007	303,462	1.230	373,131	83,843	0.722	60,513	387,305	1.120	433,644
2008	173,660	1.420	246,590	28,420	0.883	25,097	202,080	1.340	271,687
2009	249,720	1.530	381,550	52,300	0.934	48,838	302,020	1.430	430,388
2010	232,560	2.490	578,149	61,180	1.580	96,679	293,740	2.300	674,828
2011	227,030	2.590	587,064	42,670	1.610	68,825	269,700	2.430	655,889
2012	247,100	1.730	427,473	55,200	0.879	48,518	302,300	1.570	475,991
2013	220,000	1.900	417,758	46,330	0.920	42,632	266,330	1.730	460,390
2014	227,410	2.120	482,323	36,740	0.933	34,268	264,150	1.960	516,591
2015	225,180	2.310	520,529	29,110	1.360	39,687	254,290	2.200	560,216

[1] Budded, grafted or topworked varieties.
NASS, Crops Branch, (202) 720–2127.

Table 5-84.—Pecans (in-shell basis): Production and marketing year average price per pound, by State and United States, 2013–2015

Item and State	Utilized production			Price per pound		
	2013	2014	2015	2013	2014	2015
	1,000 pounds	*1,000 pounds*	*1,000 pounds*	*Dollars*	*Dollars*	*Dollars*
Improved Varieties [1]						
Alabama	2,500	1,500	1,600	1.940	2.060	2.170
Arizona	22,500	21,000	22,500	1.900	2.000	2.400
Arkansas	2,000	2,200	1,200	1.600	1.910	1.980
California	5,000	5,000	3,960	2.060	2.140	2.180
Florida	700	100	190	1.720	1.750	2.170
Georgia	83,000	74,000	90,000	1.960	2.340	2.180
Louisiana	1,500	2,500	1,000	1.400	1.520	1.710
Mississippi	3,800	700	1,000	1.230	1.570	1.540
Missouri	500	210	310	1.340	1.930	2.010
New Mexico	72,000	67,000	73,000	1.900	2.100	2.500
Oklahoma	3,000	4,000	3,000	2.050	1.630	2.090
South Carolina	1,500	200	420	2.000	2.160	1.650
Texas	22,000	49,000	27,000	1.790	1.960	2.300
United States	220,000	227,410	225,180	1.900	2.120	2.310
Native and Seedling						
Alabama	770	200	300	1.020	1.140	1.230
Arkansas	700	1,300	1,000	0.830	0.860	1.100
Florida	(D)	(D)	(D)	(D)	(D)	(D)
Georgia	6,000	2,000	3,000	1.240	1.380	1.360
Kansas	(D)	(D)	(D)	(D)	(D)	(D)
Louisiana	9,500	11,500	4,000	0.860	0.790	1.000
Mississippi	1,700	300	300	0.950	0.900	1.240
Missouri	2,240	460	1,200	1.010	1.040	1.500
Oklahoma	17,000	8,000	10,000	0.801	0.923	1.450
South Carolina	60	50	30	1.500	1.270	1.260
Texas	6,000	12,000	1,000	0.900	0.980	1.470
Other States	2,360	930	1,280	1.130	1.210	1.300
United States	46,330	36,740	29,110	0.920	0.933	1.360
All Pecans						
Alabama	3,270	1,700	1,900	1.720	1.950	2.020
Arizona	22,500	21,000	22,500	1.900	2.000	2.400
Arkansas	2,700	3,500	2,200	1.400	1.520	1.580
California	5,000	5,000	3,960	2.060	2.140	2.180
Florida	(D)	(D)	(D)	(D)	(D)	(D)
Georgia	89,000	76,000	93,000	1.910	2.310	2.150
Kansas	(D)	(D)	(D)	(D)	(D)	(D)
Louisiana	11,000	14,000	5,000	0.934	0.920	1.140
Mississippi	5,500	1,000	1,300	1.140	1.370	1.470
Missouri	2,740	670	1,510	1.070	1.320	1.600
New Mexico	72,000	67,000	73,000	1.900	2.100	2.500
Oklahoma	20,000	12,000	13,000	0.988	1.160	1.600
South Carolina	1,560	250	450	1.980	1.980	1.620
Texas	28,000	61,000	35,000	1.600	1.770	2.110
Other States	3,060	1,030	1,470	1.270	1.260	1.410
United States	266,330	264,150	254,290	1.730	1.960	2.200

(D) Withheld to avoid disclosing data for individual operations. [1] Budded, grafted or topworked varieties.
NASS, Crops Branch, (202) 720–2127.

Table 5-85.—Pecans (shelled basis): Foreign trade, United States, 2006–2015 [1]

Year beginning October	Imports	Domestic exports
	Metric tons	*Metric tons*
2006	23,923	19,145
2007	33,689	29,409
2008	26,831	22,037
2009	33,735	30,607
2010	35,906	25,531
2011	30,449	31,141
2012	33,525	37,873
2013	39,000	33,098
2014	43,736	40,727
2015	46,629	35,507

[1] Imports of unshelled nuts converted to shelled basis at ratio of 2.50 to 1. Exports of unshelled nuts converted to shelled basis at ratio of 2.50 to 1.
ERS, Crops Branch, (202) 694–5255.

Table 5-86.—Pistachios (in-shell basis): Bearing acreage, yield, production, price, and value, California, 2006–2015

Year	Bearing Acreage	Yield per acre	Utilized production	Price per pound	Value
	Acres	Pounds	1,000 pounds	Dollars	1,000 dollars
2006	112,000	2,130	238,000	1.89	449,820
2007	115,000	3,620	416,000	1.41	586,560
2008	118,000	2,360	278,000	2.05	569,900
2009	126,000	2,820	355,000	1.67	592,850
2010	137,000	3,810	522,000	2.22	1,158,840
2011	153,000	2,900	444,000	1.98	879,120
2012	182,000	3,030	551,000	2.61	1,438,110
2013	203,000	2,320	470,000	3.48	1,635,600
2014	221,000	2,330	514,000	3.57	1,834,980
2015	233,000	1,160	270,000	2.48	669,600

NASS, Crops Branch, (202) 720–2127.

Table 5-87.—Walnuts, English (in-shell basis): Bearing acreage, yield, production, price, and value, California, 2006–2015

Year	Bearing Acreage	Yield per acre [1]	Utilized production	Price per ton	Value
	Acres	Tons	Tons	Dollars	1,000 dollars
2006	216,000	1.60	346,000	1,630	563,980
2007	218,000	1.50	328,000	2,290	751,120
2008	230,000	1.90	436,000	1,280	558,080
2009	240,000	1.82	437,000	1,710	747,270
2010	255,000	1.98	504,000	2,040	1,028,160
2011	265,000	1.74	461,000	2,900	1,336,900
2012	270,000	1.84	497,000	3,030	1,505,910
2013	280,000	1.76	492,000	3,710	1,825,320
2014	290,000	1.97	571,000	3,340	1,907,140
2015	300,000	2.01	603,000	1,620	976,860

[1]Yield is based on utilized production.
NASS, Crops Branch, (202) 720–2127.

Table 5-88.—Walnuts (shelled basis): Foreign trade, United States, 2006–2015 [1]

Year beginning October	Imports	Domestic exports
	Metric tons	Metric tons
2006	974	69,062
2007	4,059	64,467
2008	837	96,313
2009	1,529	101,865
2010	186	133,868
2011	2,442	117,031
2012	3,957	133,049
2013	5,266	134,379
2014	9,780	158,163
2015	5,320	188,159

[1]Imports of unshelled nuts converted to shelled basis at ratio of 2.50 to 1. Exports of unshelled nuts converted to shelled basis at ratio of 2.50 to 1.
ERS, Crops Branch, (202) 694–5255.

Table 5-89.—Coffee, green: International trade, exports from principal producing countries, 2012/2013-2014/2015

Country of origin	2012/2013	2013/2014	2014/2015
Principle exporting countries:	1,000 bags	1,000 bags	1,000 bags
Brazil	30,660	34,146	36,570
Colombia	8,855	11,040	12,125
Ethiopia	3,500	3,285	3,500
Guatemala	3,770	3,175	3,070
Honduras	4,480	3,940	4,700
India	4,858	4,983	4,700
Indonesia	8,900	7,800	7,040
Malaysia	2,150	2,110	2,780
Uganda	3,575	3,600	3,400
Vietnam	24,643	28,289	22,072
Others	23,318	20,605	18,949
Total Foreign	118,709	122,973	118,906
United States	365	580	390
Total	119,074	123,553	119,296
Principle importing countries:			
Algeria	1,945	2,300	2,155
Australia	1,660	1,615	1,775
Canada	4,230	4,605	4,505
China	1,560	1,705	1,940
European Union	45,070	44,650	45,110
Japan	8,345	7,870	8,075
Korea,South	1,825	2,160	2,305
Philippines	3,880	3,185	3,790
Russia	4,130	4,230	4,050
Switzerland	2,310	2,300	2,420
Others	17,369	16,789	16,762
Total Foreign	92,324	91,409	92,887
United States	23,700	24,915	24,010
Total	116,024	116,324	116,897

FAS, Office of Global Analysis, (202) 720-6301. Prepared or estimated on the basis of official USDA production, supply, and distribution statistics from foreign governments. Data Source: Department of Commerce, U.S. Census Bureau, Foreign Trade Statistics.

Table 5-90.—Coffee: Area, yield, production, marketing year average price, and value, Hawaii, 2006/2007–2015/2016

Year	Area Harvested	Yield per acre	Production[1]	Price per pound	Value
Hawaii	Acres	Pounds	1,000 pounds	Dollars	1,000 dollars
2006/2007	6,300	1,170	7,400	4.30	31,820
2007/2008	6,400	1,170	7,500	4.25	31,875
2008/2009	6,900	1,260	8,700	3.40	29,580
2009/2010	7,200	1,210	8,700	3.60	31,320
2010/2011	7,500	1,170	8,800	3.80	33,440
2011/2012	7,700	990	7,600	4.15	31,540
2012/2013	7,900	890	7,000	5.90	41,300
2013/2014	8,200	1,020	8,400	6.20	52,080
2014/2015	7,800	1,030	8,100	9.12	62,622
2015/2016	6,900			8.60	54,197

[1] Parchment basis.
NASS, Crops Branch, (202) 720–2127.

Table 5-91.—Coffee and tea: U.S. imports, 2011–2013

Country	2011	2012	2013
	Metric tons	*Metric tons*	*Metric tons*
Coffee and coffee products:			
Brazil	418,257	334,905	365,420
Colombia	213,147	180,563	254,441
Vietnam	202,971	261,543	225,594
Mexico	98,323	119,203	115,375
Guatemala	94,545	107,208	101,789
Indonesia	59,413	79,706	80,627
Peru	63,046	51,755	52,170
Costa Rica	42,401	44,948	45,734
Honduras	39,920	62,369	44,838
Nicaragua	39,306	44,644	42,893
Germany(*)	35,591	33,286	42,758
Canada	37,765	37,468	39,686
El Salvador	39,420	23,842	25,645
Ethiopia(*)	16,972	12,401	16,090
Papua New Guinea	13,045	16,992	15,313
Uganda	8,530	7,910	13,870
Kenya	5,191	8,492	7,013
Italy(*)	5,925	6,797	6,807
Tanzania	4,599	4,320	6,346
India	7,157	6,324	4,639
China	5,361	5,460	4,510
Rwanda	4,596	4,291	3,893
Ecuador	4,399	2,681	3,264
Switzerland(*)	2,110	2,225	2,414
Malaysia	3,131	3,605	2,366
Dominican Republic	1,547	4,403	2,067
Panama	1,528	1,451	1,366
Korea, South	1,326	1,302	1,294
Sweden	3,892	2,070	1,281
Rest of World	21,142	17,809	11,357
World Total	1,494,552	1,489,973	1,540,860
Tea, except herbal tea:			
Canada	58,870	55,723	59,842
Argentina	50,819	50,205	51,158
China	29,147	29,527	28,080
India	14,719	13,906	15,790
Vietnam	4,638	8,164	9,549
Mexico	4,783	9,225	7,610
Germany(*)	8,891	7,160	7,286
Sri Lanka	3,984	4,141	4,542
Malawi	4,437	3,729	4,474
Indonesia	5,690	4,435	4,452
Kenya	3,725	3,502	3,501
Thailand	829	805	2,851
Japan	1,521	1,326	1,776
Chile	1,653	1,610	1,395
United Kingdom	1,794	1,090	1,332
Taiwan	1,098	1,035	1,167
Brazil	1,627	1,199	1,003
Netherlands	432	273	979
Zimbabwe	599	1,676	910
Poland	606	537	689
Egypt	677	736	687
Ecuador	544	629	669
Nigeria	495	429	637
Hong Kong	466	515	576
Korea, South	683	670	549
South Africa	325	183	473
Papua New Guinea	869	434	462
Pakistan	321	333	411
Mozambique	121	468	395
Rest of World	7,166	7,288	3,920
World Total	211,529	210,956	217,164

Data Source: Department of Commerce, U.S. Census Bureau, Foreign Trade Statistics. (*) Denotes a country that is a summarization of its component countries. All zeroes for a data item may show that statistics exist in the other import type. Consumption or General. Users should use cautious interpretation on Quantity reports using mixed units of measure. Quantity line items will only include statistics on the units of measure that are equal to, or are able to be converted to, the assigned unit of measure of the grouped commodities.
FAS, Office of Global Analysis, (202) 720-6301.

Table 5-92.—Specialty mushrooms: Number of growers, total production, volume of sales, price per pound, and value of sales, United States: 2012/2013-2014/2015

Year and variety	Growers [1]	Total Production [2]	All sales		
			Volume of sales [3]	Price per pound [4]	Value of sales
	Number	1,000 pounds	1,000 pounds	Dollars	1,000 dollars
2012/2013					
Shiitake	179	8,875	8,277	3.32	27,468
Oyster	97	7,414	6,975	3.02	21,050
Other	28	3,659	3,375	4.99	16,827
United States [5]	209	19,948	18,627	3.51	65,345
2013/2014					
Shiitake	206	9,327	8,952	3.21	28,759
Oyster	119	7,499	7,165	3.74	26,783
Other	37	1,846	1,641	6.16	10,114
United States [5]	260	18,672	17,758	3.70	65,656
2014/2015					
Shiitake	215	9,490	9,251	3.26	30,151
Oyster	124	7,996	7,724	3.19	24,610
Other	48	4,062	3,657	4.98	18,225
United States	265	21,548	20,632	3.54	72,986

[1] Growers counted only once for United States total if growing more than one specialty type mushroom. Growers growing Agaricus and specialty mushrooms are included. [2] Total production includes all fresh market and processing sales plus amount harvested but not sold (shrinkage, cullage, dumped, etc.). [3] Virtually all specialty mushroom sales are for fresh market. [4] Prices for mushrooms are the average prices producers receive at the point of first sale, commonly referred to as the average price as sold. For example, if in a given State, part of the fresh mushrooms are sold F.O.B. packed by growers, part are sold bulk to brokers or repackers, and some are sold retail at roadside stands, the mushroom average price as sold is a weighted average of the average price for each method of sale. [5] 2012-2013: Arkansas, California, Colorado, Connecticut, Florida, Georgia, Hawaii,Illinois, Indiana, Kansas, Kentucky, Maine, Maryland, Massachusetts, Michigan,Minnesota, Missouri, Montana, New York, North Carolina, Ohio, Oregon, Pennsylvania, South Carolina, Tennessee, Texas, Vermont, Virginia, Washington, West Virginia, and Wisconsin. 2013-2014: Arkansas, California, Colorado, Connecticut, Delaware, Florida, Georgia, Hawaii, Illinois, Indiana, Kansas, Kentucky, Maine, Maryland, Massachusetts, Michigan, Minnesota, Missouri, Nebraska, New Hampshire, New York, North Carolina, Ohio, Oregon, Pennsylvania, South Carolina, Tennessee, Texas, Vermont, Virginia, Washington, West Virginia, and Wisconsin.
NASS, Crops Branch, (202) 720-2127.

Table 5-93.—Agaricus mushrooms: Area, volume of sales, marketing year average price, and value of sales, United States: 2006-2015 [1]

Year	Area in production	Volume of sales	Price per pound	Value of sales		
				Total	Fresh market	Processing
	1,000 sq. ft.	1,000 pounds	Dollars	1,000 dollars	1,000 dollars	1,000 dollars
2006/2007	145,743	813,849	1.120	915,561	840,560	75,001
2007/2008	136,011	797,348	1.150	917,607	841,753	75,854
2008/2009	134,533	803,896	1.130	910,658	841,021	69,637
2009/2010	129,268	777,064	1.140	884,390	821,472	62,918
2010/2011	134,266	845,951	1.140	964,192	887,839	76,353
2011/2012	139,331	881,857	1.180	1,039,159	969,847	69,312
2012/2013	139,846	869,625	1.190	1,038,541	969,278	69,263
2013/2014	134,685	882,075	1.190	1,049,972	971,521	78,451
2014/2015	143,292	907,191	1.230	1,118,371	1,054,652	63,719

[1] Marketing year begins July 1 and ends June 30 the following year.
NASS, Crops Branch, (202) 720-2127.

Table 5-94.—Cut flowers: Sales and wholesale value for operations with $100,000+ sales, Surveyed States, 2006–2015

Year	Quantity sold	Wholesale price	Value of sales at wholesale [1]	Quantity sold	Wholesale price	Value of sales at wholesale [1]	Quantity sold	Wholesale price	Value of sales at wholesale [1]
	Alstromeria			Carnations, standard			Chrysanthemums pompon		
	1,000 stems	Cents per stem	1,000 dollars	1,000 stems	Cents per stem	1,000 dollars	1,000 bunches	Cents per bunch	1,000 dollars
2006 ...	8,595	21.9	1,885	5,428	17.6	955	10,338	1.26	12,985
2007 ...	9,879	20.8	2,057	3,328	18.8	626	18,059	0.76	13,810
2008 ...	10,774	17.9	1,927	3,343	17.0	567	10,058	1.34	13,428
2009 ...	8,800	18.5	1,629	2,837	16.9	480	7,920	1.43	11,298
2010 ...	9,868	16.7	1,650	1,893	16.7	317	8,347	1.40	11,705
2011 ...	8,764	19.1	1,676	1,441	16.2	233	7,823	1.41	11,048
2012 ...	10,964	21.0	2,307	715	23.4	167	9,167	1.43	13,115
2013 ...	10,164	20.7	2,104	624	24.5	153	9,009	1.53	13,816
2014 ...	8,890	19.1	1,702	797	30.6	244	6,677	1.73	11,518
2015 ...	7,747	19.6	1,517	896	30.5	273	7,777	1.48	11,524
	Delphinium & Larkspur			Gerbera Daisies			Gladioli		
	1,000 stems	Cents per stem	1,000 dollars	1,000 stems	Cents per stem	1,000 dollars	1,000 spikes	Cents per spike	1,000 dollars
2006 ...	26,142	23.5	6,133	112,587	30.2	33,997	95,350	23.8	22,694
2007 ...	32,158	24.4	7,842	117,403	30.6	35,939	85,471	27.0	23,081
2008 ...	31,221	24.0	7,505	120,836	29.8	35,976	76,850	25.9	19,935
2009 ...	22,373	27.1	6,071	106,805	30.9	33,027	94,951	24.1	22,880
2010 ...	27,173	26.5	7,193	107,678	30.4	32,737	89,672	25.4	22,801
2011 ...	19,639	30.5	5,981	113,655	28.5	32,385	60,167	27.1	16,299
2012 ...	19,448	37.4	7,265	107,899	30.6	33,067	60,747	25.3	15,349
2013 ...	16,929	38.6	6,542	98,535	31.0	30,552	59,435	34.1	20,244
2014 ...	17,091	35.3	6,030	105,028	31.4	32,987	61,724	32.7	20,193
2015 ...	12,671	41.0	5,199	106,680	30.4	32,423	60,778	32.3	19,608
	Iris			Lilies, all			Lisianthus		
	1,000 stems	Cents per stem	1,000 dollars	1,000 stems	Cents per stem	1,000 dollars	1,000 stems	Cents per stem	1,000 dollars
2006 ...	81,194	22.6	18,315	107,044	70.5	75,459	8,518	43.1	3,670
2007 ...	90,890	22.4	20,349	111,185	67.4	74,954	13,956	38.2	5,338
2008 ...	92,404	22.1	20,462	116,797	67.3	78,609	15,180	35.6	5,406
2009 ...	64,114	24.3	15,550	101,339	63.5	64,392	8,689	38.3	3,327
2010 ...	57,576	25.2	14,518	93,373	65.5	61,165	7,650	38.8	2,966
2011 ...	62,414	26.1	16,303	141,337	56.5	79,798	7,502	42.1	3,160
2012 ...	55,025	24.5	13,477	99,743	66.7	66,553	9,846	43.1	4,240
2013 ...	54,635	24.4	13,356	87,094	68.4	59,558	9,216	45.2	4,167
2014 ...	45,682	26.4	12,078	79,432	77.6	61,661	8,484	45.2	3,839
2015 ...	43,787	26.4	11,554	82,551	75.4	62,235	9,010	45.0	4,055
	Orchids, all			Roses, all			Snapdragons		
	1,000 blooms	Cents per bloom	1,000 dollars	1,000 stems	Cents per stem	1,000 dollars	1,000 spikes	Cents per spike	1,000 dollars
2006 ...	10,332	120.3	12,428	82,138	37.7	30,974	36,559	28.0	10,244
2007 ...	11,209	99.5	11,150	67,701	41.5	28,110	41,887	29.1	12,202
2008 ...	7,882	98.2	7,737	57,999	38.8	22,481	42,696	27.6	11,790
2009 ...	7,637	185.1	14,133	42,031	42.0	17,662	37,473	27.0	10,118
2010 ...	7,852	99.8	7,834	39,497	42.9	16,950	31,475	27.1	8,524
2011 ...	7,166	83.0	5,949	37,004	48.4	17,912	34,777	25.1	8,744
2012 ...	6,498	90.9	5,909	32,736	49.4	16,184	35,061	29.7	10,396
2013 ...	5,846	163.6	9,564	28,626	56.3	16,105	42,013	25.4	10,652
2014 ...	6,737	245.6	16,546	30,555	56.2	17,178	41,467	24.8	10,274
2015 ...	4,682	170.5	7,983	27,812	61.3	17,039	43,863	29.5	12,933

Year	Tulips			Other cut flowers
	1,000 stems	Cents per stem	1,000 dollars	1,000 dollars
2006 ...	141,893	34.1	48,391	134,198
2007 ...	157,992	35.9	56,719	133,017
2008 ...	170,854	37.6	64,285	126,990
2009 ...	150,228	38.1	57,185	103,687
2010 ...	155,667	36.6	56,900	129,466
2011 ...	169,219	35.3	59,788	99,824
2012 ...	161,802	37.2	60,172	171,159
2013 ...	161,926	36.8	59,535	116,509
2014 ...	160,990	35.5	57,198	112,006
2015 ...	167,378	34.6	57,840	130,135

[1] Equivalent wholesale value of all sales.
NASS, Crops Branch, (202) 720-2127.

Table 5-95.—Cut Cultivated Greens: Sales and wholesale value for operations with $100,000+ sales, Surveyed States, 2006–2015

Year	Leatherleaf Ferns			Other cut cultivated greens
	Quantity sold	Wholesale price	Value of sales at wholesale [1]	Value of sales at wholesale [1]
	1,000 bunches	*Dollars per bunch*	*1,000 dollars*	*1,000 dollars*
2006	44,183	1.04	45,902	51,706
2007	39,437	1.00	39,543	58,527
2008	34,761	0.98	33,924	57,824
2009	31,800	0.94	29,942	43,993
2010	28,474	1.03	29,312	47,713
2011	26,277	0.97	25,499	46,537
2012	32,063	0.99	31,706	43,811
2013	32,209	1.02	32,779	42,706
2014	32,385	0.99	32,209	40,974
2015	33,319	1.02	33,853	46,573

[1] Equivalent wholesale value of all sales.
NASS, Crops Branch, (202) 720–2127.

Table 5-96.—Foliage Plants for Indoor or Patio Use: Sales and Wholesale value for operations with $100,000+ sales, Surveyed States, 2006–2015

Year	Foliage, Hanging Baskets			Foliage, Pots
	Quantity Sold	Wholesale price	Value of sales at wholesale	Value of sales at wholesale
	1,000 baskets	*Dollars per basket*	*1,000 dollars*	*1,000 dollars*
2006	13,341	4.60	61,303	466,609
2007	14,118	4.66	65,857	589,545
2008	11,003	4.90	53,949	456,362
2009	12,072	5.29	63,881	493,148
2010	14,530	5.39	78,247	507,882
2011	14,483	5.42	78,438	534,943
2012	10,323	6.12	63,158	576,778
2013	11,493	5.92	67,996	544,562
2014	10,199	6.19	63,103	559,779
2015	11,556	5.85	67,562	646,127

NASS, Crops Branch, (202) 720–2127.

Table 5-97.— Potted Flowering Plants for Indoor or Patio Use: Sales and wholesale value for operations with $100,000+ sales, Surveyed States, 2006–2015

Year	Quantity sold		Wholesale price		Value of sales at wholesale [1]
	Less than 5 inches	5 inches or more	Less than 5 inches	5 inches or more	
			African violets		
	1,000 pots	*1,000 pots*	*Dollars per pot*	*Dollars*	*1,000 dollars*
2006	5,997	434	1.20	1.92	8,046
2007	4,357	430	1.37	1.95	6,809
2008	2,946	9	1.35	3.34	3,993
2009	2,313	38	1.49	3.48	3,580
2010	2,727	23	1.31	1.88	3,614
2011	2,582	51	1.25	3.84	3,414
2012	2,632	27	1.55	2.38	4,150
2013	2,572	33	1.47	2.53	3,870
2014	2,595	9	1.55	4.13	4,063
2015	2,641	8	1.56	3.85	4,161
			Florist azaleas		
	1,000 pots	*1,000 pots*	*Dollars per pot*	*Dollars per pot*	*1,000 dollars*
2006	2,237	6,844	2.10	4.41	34,909
2007	1,514	5,081	2.34	4.90	28,435
2008	1,095	7,188	2.24	4.65	35,897
2009	2,345	5,214	2.30	4.92	31,044
2010	1,348	3,289	2.59	5.67	22,138
2011	1,266	3,012	2.32	6.01	21,022
2012	1,129	3,544	2.68	5.28	21,734
2013	1,109	3,313	2.48	5.05	19,471
2014	22	4,063	4.25	4.65	18,977
2015	103	3,587	5.26	4.96	18,343
			Florist chrysanthemums		
	1,000 pots	*1,000 pots*	*Dollars per pot*	*Dollars*	*1,000 dollars*
2006	1,299	12,693	1.82	3.03	40,815
2007	1,810	11,363	1.65	3.15	38,777
2008	1,927	9,902	1.57	3.21	34,762
2009	1,314	6,181	1.90	3.61	24,842
2010	2,225	4,526	1.82	3.97	22,029
2011	1,351	5,210	2.43	3.54	21,720
2012	1,654	4,400	2.00	4.14	21,535
2013	613	5,343	2.82	3.99	23,045
2014	1,239	4,673	3.02	3.96	22,238
2015	457	4,119	2.02	3.83	16,699
			Easter lilies		
	1,000 pots	*1,000 pots*	*Dollars per pot*	*Dollars per pot*	*1,000 dollars*
2006	2	6,334	6.53	4.12	26,106
2007		6,546		4.05	26,512
2008	31	5,824	4.05	4.33	25,335
2009	93	6,196	3.16	4.38	27,405
2010	147	5,891	3.29	4.29	25,764
2011	99	5,094	2.73	4.35	22,436
2012	37	5,011	3.48	4.36	21,958
2013	295	5,901	2.80	4.23	25,758
2014	785	5,099	3.90	4.59	26,460
2015	155	5,370	2.54	4.45	24,317
			Potted Orchids		
	1,000 pots	*1,000 pots*	*Dollars per pot*	*Dollars per pot*	*1,000 dollars*
2006	10,140	4,615	6.90	10.96	120,521
2007	10,661	6,655	6.95	9.89	139,960
2008	10,689	5,415	7.16	9.23	126,509
2009	12,503	7,135	7.36	9.70	161,208
2010	13,395	7,933	7.10	9.71	172,196
2011	15,415	8,520	6.78	10.19	191,306
2012	15,915	12,389	7.01	9.62	230,695
2013	18,934	12,750	7.34	8.83	251,559
2014	19,085	14,474	7.65	8.80	273,481
2015	19,393	16,983	7.25	8.70	288,284

See footnote(s) at end of table.

Table 5-97.—Potted flowering for indoor or patio use: Sales and wholesale value for operations with $100,000+ sales, Surveyed States, 2006–2015—Continued

Year	Quantity sold		Wholesale price		Value of sales at wholesale [1]
	Less than 5 inches	5 inches or more	Less than 5 inches	5 inches or more	
			Poinsettias		
	1,000 pots	*1,000 pots*	*Dollars per pot*	*Dollars per pot*	*1,000 dollars*
2006	7,762	33,743	1.98	4.61	171,012
2007	7,130	31,901	2.07	4.60	161,409
2008	6,373	28,922	2.07	4.82	152,611
2009	6,679	28,649	2.13	4.69	148,579
2010	7,710	26,942	2.02	4.65	140,841
2011	7,470	27,259	1.84	4.60	139,274
2012	6,620	27,301	2.12	4.77	144,309
2013	6,610	26,797	2.06	4.82	142,805
2014	6,338	27,263	2.15	4.77	143,715
2015	6,509	25,459	2.14	4.94	139,710
			Potted florist roses		
	1,000 pots	*1,000 pots*	*Dollars per pot*	*Dollars*	*1,000 dollars*
2006	6,389	1,901	1.90	4.90	21,446
2007	6,834	2,364	1.94	5.16	25,425
2008	7,252	1,064	2.74	5.34	25,569
2009	6,924	1,031	2.57	5.14	23,115
2010	5,599	2,167	2.50	5.83	26,669
2011	5,183	2,550	2.16	4.87	23,619
2012	5,729	2,828	2.18	5.74	28,741
2013	5,976	2,750	2.40	6.14	31,214
2014	7,751	934	3.26	6.41	31,282
2015	6,341	1,784	2.42	5.25	24,696
			Potted spring flowering bulbs		
	1,000 pots	*1,000 pots*	*Dollars per pot*	*Dollars*	*1,000 dollars*
2006	13,061	8,469	1.41	3.43	47,447
2007	10,073	8,624	1.91	3.66	50,861
2008	13,317	10,958	1.69	3.56	61,532
2009	4,689	11,234	1.74	3.43	46,662
2010	9,766	12,382	1.46	3.52	57,903
2011	8,162	11,637	1.52	3.54	53,638
2012	5,095	11,434	1.74	3.63	50,376
2013	4,623	11,013	1.96	3.77	50,614
2014	5,018	10,791	1.78	3.86	50,561
2015	5,548	16,865	2.19	3.38	69,115
			Other flowering		
	1,000 pots	*1,000 pots*	*Dollars per pot*	*Dollars per pot*	*1,000 dollars*
2006	23,220	27,703	2.01	3.77	151,158
2007	27,176	36,862	2.09	4.51	223,111
2008	22,529	31,482	2.10	4.91	201,962
2009	19,531	29,173	2.13	4.62	176,570
2010	17,162	25,695	2.71	5.00	175,006
2011	24,663	21,896	1.88	5.36	163,903
2012	22,670	25,625	2.57	5.99	211,766
2013	24,868	26,835	2.67	5.87	223,931
2014	25,569	32,190	2.70	5.29	239,162
2015	23,575	29,889	1.93	5.99	224,335

[1] Equivalent wholesale value of all sales except for potted foliage which is value of sales less cost of plant material purchased from other growers for growing on. [2] Beginning in 2006, program was reduced to 15 States from 36. NASS, Crops Branch, (202) 720-2127.

Table 5-98.—Annual Bedding and Garden Hanging Baskets: Sales and wholesale value for operations with $100,000+ sales, Surveyed States, 2006–2015

Year	Quantity sold	Wholesale price	Value of sales at wholesale [1]	Quantity sold	Wholesale price	Value of sales at wholesale [1]
		Begonias			Geraniums from vegetative cuttings	
	1,000 baskets	*Dollars*	*1,000 dollars*	*1,000 baskets*	*Dollars*	*1,000 dollars*
2006	3,199	5.97	19,091	3,285	6.75	22,186
2007	1,701	5.76	9,796	3,296	6.87	22,640
2008	1,707	5.80	9,899	3,056	7.32	22,366
2009	1,531	5.84	8,944	3,564	7.44	26,523
2010	1,650	5.96	9,839	3,753	7.10	26,659
2011	1,622	5.99	9,723	3,561	7.13	25,390
2012	1,823	5.94	10,830	3,280	7.49	24,562
2013	2,016	6.33	12,770	3,126	7.39	23,113
2014	1,849	6.00	11,089	3,320	7.56	25,089
2015	1,960	6.25	12,250	3,310	7.43	24,586
		Geraniums from seeds			Impatiens, Other	
	1,000 baskets	*Dollars*	*1,000 dollars*	*1,000 baskets*	*Dollars*	*1,000 dollars*
2006	246	6.03	1,483	2,846	5.19	14,761
2007	316	5.63	1,778	2,597	5.22	13,548
2008	267	6.05	1,616	2,375	5.29	12,561
2009	455	5.95	2,708	2,098	5.79	12,146
2010	286	7.01	2,005	2,534	5.63	14,260
2011	337	6.61	2,227	2,385	5.72	13,639
2012	475	6.56	3,115	2,479	5.44	13,494
2013	415	7.28	3,020	2,065	5.33	11,015
2014	460	7.46	3,432	2,021	5.32	10,750
2015	438	7.39	3,237	2,098	5.45	11,432
		New Guinea Impatiens			Marigolds	
	1,000 baskets	*Dollars*	*1,000 dollars*	*1,000 baskets*	*Dollars*	*1,000 dollars*
2006	3,174	6.81	21,624	162	6.41	1,039
2007	2,987	6.96	20,797	184	6.08	1,118
2008	2,668	7.23	19,280	202	6.30	1,272
2009	2,556	7.09	18,122	55	6.33	348
2010	2,615	7.11	18,581	83	6.01	499
2011	2,456	7.14	17,526	77	5.77	444
2012	2,380	7.32	17,420	87	4.92	428
2013	2,254	7.37	16,613	117	4.38	513
2014	2,294	7.31	16,764	137	5.88	805
2015	2,388	7.39	17,637	139	5.99	833
		Pansies/Violas			Petunias	
	1,000 baskets	*Dollars*	*1,000 dollars*	*1,000 baskets*	*Dollars*	*1,000 dollars*
2006	510	5.15	2,625	3,673	5.74	21,081
2007	694	5.02	3,485	4,011	5.99	24,017
2008	695	5.20	3,612	3,969	5.98	23,752
2009	1,077	5.43	5,848	4,186	6.38	26,711
2010	911	5.59	5,091	5,066	6.16	31,225
2011	941	5.83	5,483	5,047	6.29	31,767
2012	1,096	5.75	6,299	5,028	6.21	31,240
2013	1,085	5.62	6,096	4,986	6.11	30,459
2014	1,182	5.92	7,002	5,557	6.28	34,906
2015	1,111	6.23	6,920	5,294	6.27	33,196
		Other flowering hanging baskets and foliar				
	1,000 baskets		*Dollars*			*1,000 dollars*
2006	14,910		6.22			92,736
2007	15,153		6.95			105,330
2008	14,718		6.92			101,856
2009	14,451		7.36			106,323
2010	13,483		7.67			103,399
2011	12,759		7.56			96,500
2012	14,865		7.33			109,030
2013	14,797		7.73			114,409
2014	13,259		8.23			109,117
2015	15,269		7.67			117,065

[1] Equivalent wholesale value of all sales.
NASS, Crops Branch, (202) 720–2127.

Table 5-99.—Annual bedding garden plants flats: Sales and wholesale value for operations with $100,000+ sales, Surveyed States, 2006–2015

Year	Quantity sold	Wholesale price	Value of sales at wholesale [1]	Quantity sold	Wholesale price	Value of sales at wholesale [1]
		Begonias			Geraniums from vegetative cuttings	
	1,000 flats	*Dollars*	*1,000 dollars*	*1,000 flats*	*Dollars*	*1,000 dollars*
2006	4,947	8.17	40,429	520	11.54	5,999
2007	4,094	8.17	33,444	417	12.32	5,138
2008	4,360	8.49	36,999	396	12.57	4,979
2009	4,055	8.75	35,490	353	15.09	5,326
2010	4,218	8.13	34,313	306	13.12	4,015
2011	3,983	7.97	31,738	264	14.08	3,716
2012	3,865	8.56	33,005	237	13.57	3,215
2013	3,758	8.15	30,640	325	11.65	3,785
2014	3,970	8.62	34,208	252	14.03	3,535
2015	3,819	8.33	31,813	248	12.83	3,182
		Geraniums from seeds			Impatiens, other	
	1,000 flats	*Dollars*	*1,000 dollars*	*1,000 flats*	*Dollars*	*1,000 dollars*
2006	398	10.99	4,373	9,884	7.77	76,771
2007	380	11.54	4,387	8,915	8.11	72,320
2008	384	11.00	4,223	8,547	8.52	72,815
2009	424	9.23	3,915	7,973	8.67	69,093
2010	465	10.29	4,784	8,641	8.24	71,160
2011	332	10.62	3,526	8,353	7.96	66,519
2012	394	9.81	3,866	6,852	9.08	62,230
2013	475	7.51	3,565	5,055	8.17	41,300
2014	372	9.26	3,445	4,775	8.38	40,034
2015	425	8.07	3,431	4,404	8.78	38,647
		New Guinea Impatiens			Marigolds	
	1,000 flats	*Dollars*	*1,000 dollars*	*1,000 flats*	*Dollars*	*1,000 dollars*
2006	305	10.14	3,093	4,032	8.13	32,788
2007	243	11.09	2,696	3,694	8.39	31,001
2008	218	11.55	2,518	3,933	8.69	34,190
2009	290	10.77	3,124	3,976	8.98	35,713
2010	276	9.69	2,675	4,100	8.89	36,444
2011	190	10.56	2,006	3,884	8.45	32,807
2012	226	9.86	2,229	3,436	8.49	29,188
2013	335	10.01	3,355	3,488	8.24	28,733
2014	625	9.84	6,148	3,671	8.36	30,706
2015	368	9.76	3,592	3,587	8.20	29,425
		Pansies/Violas			Petunias	
	1,000 flats	*Dollars*	*1,000 dollars*	*1,000 flats*	*Dollars*	*1,000 dollars*
2006	8,238	8.03	66,168	7,349	8.12	59,682
2007	8,047	8.33	67,050	7,023	8.52	59,808
2008	8,169	8.82	72,036	7,402	8.80	65,129
2009	7,430	9.15	67,957	7,006	9.32	65,317
2010	7,538	8.88	66,916	7,448	8.92	66,404
2011	7,266	8.66	62,936	7,227	8.56	61,891
2012	7,343	9.02	66,245	6,667	8.64	57,618
2013	6,369	8.94	56,926	6,581	8.36	55,034
2014	7,272	8.45	61,441	6,896	9.11	62,822
2015	6,297	8.91	56,113	6,448	8.76	56,509
		Other flowering and foliar plants			Vegetable type plants	
	1,000 flats	*Dollars*	*1,000 dollars*	*1,000 flats*	*Dollars*	*1,000 dollars*
2006	25,652	8.02	205,649	4,776	9.55	45,604
2007	21,350	8.70	185,788	4,135	9.39	38,822
2008	19,441	9.00	175,027	4,545	10.25	46,573
2009	16,331	9.45	154,342	5,025	10.53	52,911
2010	17,898	9.40	168,261	4,862	10.20	49,603
2011	17,836	9.50	169,484	5,102	10.53	53,740
2012	16,934	9.10	154,089	6,297	9.64	60,714
2013	16,275	8.71	141,740	4,998	10.18	50,858
2014	14,788	9.38	138,683	4,861	10.82	52,616
2015	14,811	8.87	131,358	5,453	10.60	57,808

[1] Equivalent wholesale value of all sales.
NASS, Crops Branch, (202) 720–2127.

Table 5-100.—Potted annual bedding and garden plants: Sales and wholesale value for operations with $100,000+ sales, Surveyed States, 2006–2015

Year	Quantity sold		Wholesale price		Value of sales at wholesale [1]
	Less than 5 inches	5 inches or more	Less than 5 inches	5 inches or more	
			Begonias		
	1,000 pots	*1,000 pots*	*Dollars*	*Dollars*	*1,000 dollars*
2006	19,939	4,387	0.85	2.28	27,004
2007	16,748	3,683	0.84	2.06	21,645
2008	16,481	4,296	0.91	2.17	24,293
2009	17,821	3,272	0.86	2.34	22,907
2010	14,709	3,163	0.92	2.22	20,617
2011	14,763	3,374	0.91	2.24	20,973
2012	15,837	3,323	0.91	2.40	22,340
2013	18,296	4,118	0.94	2.58	27,922
2014	16,901	3,619	1.02	2.48	26,248
2015	16,853	4,757	1.05	2.82	31,023
			Geraniums from cuttings		
	1,000 pots	*1,000 pots*	*Dollars*	*Dollars*	*1,000 dollars*
2006	23,991	12,246	1.58	3.29	78,244
2007	22,785	13,253	1.66	3.35	82,364
2008	23,094	13,050	1.76	3.35	84,300
2009	22,985	14,209	1.79	3.18	86,303
2010	23,339	13,431	1.81	3.54	89,770
2011	21,156	12,844	1.85	3.57	85,031
2012	21,563	13,560	1.85	3.38	85,842
2013	20,429	11,554	1.88	3.62	80,293
2014	21,093	11,119	1.93	3.70	81,970
2015	19,821	11,724	1.92	3.66	80,989
			Geraniums from seed		
	1,000 pots	*1,000 pots*	*Dollars*	*Dollars*	*1,000 dollars*
2006	31,377	456	0.81	2.20	26,545
2007	28,897	343	0.83	2.86	25,071
2008	25,114	367	0.87	3.31	23,186
2009	24,622	495	0.92	2.78	23,923
2010	18,748	1,248	1.03	4.67	25,097
2011	23,673	1,648	0.93	5.20	30,680
2012	16,449	728	0.88	3.42	16,988
2013	17,825	393	1.12	3.49	21,604
2014	17,230	340	1.11	4.06	20,552
2015	13,336	1,119	1.00	3.58	17,297
			Other impatiens		
	1,000 pots	*1,000 pots*	*Dollars*	*Dollars*	*1,000 dollars*
2006	23,555	5,804	0.73	1.87	28,130
2007	19,575	5,575	0.81	1.69	25,267
2008	20,606	4,973	0.78	1.91	25,663
2009	21,923	5,075	0.76	1.87	26,065
2010	21,955	4,033	0.77	2.43	26,702
2011	20,480	4,220	0.77	2.25	25,334
2012	13,941	4,834	0.82	2.11	21,678
2013	8,825	3,629	0.94	1.90	15,245
2014	7,115	3,776	0.95	1.88	13,880
2015	6,754	3,614	1.00	2.25	14,911
			New guinea impatiens		
	1,000 pots	*1,000 pots*	*Dollars*	*Dollars*	*1,000 dollars*
2006	12,411	5,404	1.49	2.97	34,498
2007	11,885	4,735	1.58	2.83	32,180
2008	12,630	4,283	1.57	3.18	33,459
2009	12,439	3,834	1.59	3.25	32,177
2010	10,919	3,961	1.50	3.28	29,376
2011	9,794	3,658	1.59	3.44	28,130
2012	11,640	4,236	1.47	3.27	30,956
2013	13,325	4,045	1.51	3.46	34,105
2014	12,971	4,160	1.59	3.39	34,762
2015	12,864	4,024	1.52	3.63	34,158

See footnote(s) at end of table.

Table 5-100.—Potted annual bedding and garden plants: Sales and wholesale value for operations with $100,000+ sales, Surveyed States, 2006–2015—Continued

Year	Quantity sold		Wholesale price		Value of sales at wholesale [1]
	Less than 5 inches	5 inches or more	Less than 5 inches	5 inches or more	
	Marigolds				
	1,000 pots	*1,000 pots*	*Dollars*	*Dollars*	*1,000 dollars*
2006	7,928	2,207	0.69	1.79	9,392
2007	8,372	2,150	0.68	1.59	9,125
2008	7,600	1,704	0.72	1.61	8,185
2009	8,837	1,826	0.78	1.76	10,124
2010	7,901	2,463	0.76	1.88	10,621
2011	7,180	2,272	0.81	1.82	9,947
2012	8,871	1,792	0.79	2.10	10,762
2013	8,637	1,819	0.80	2.09	10,688
2014	9,264	1,273	0.97	2.32	11,922
2015	9,100	1,745	0.97	2.14	12,596
	Pansies/violas				
	1,000 pots	*1,000 pots*	*Dollars*	*Dollars*	*1,000 dollars*
2006	27,824	7,144	0.80	1.75	34,615
2007	28,021	6,145	0.74	2.13	33,788
2008	25,980	5,989	0.75	2.09	31,966
2009	27,054	5,529	0.74	2.06	31,363
2010	23,394	6,159	0.80	2.02	31,035
2011	24,573	6,130	0.80	2.12	32,692
2012	23,416	6,918	0.81	2.19	34,078
2013	21,953	6,155	0.86	2.19	32,308
2014	22,813	6,273	0.86	2.17	33,116
2015	23,029	6,418	0.89	2.17	34,518
	Petunias				
	1,000 pots	*1,000 pots*	*Dollars*	*Dollars*	*1,000 dollars*
2006	13,630	8,106	0.91	2.04	28,894
2007	18,551	6,935	0.98	2.37	34,558
2008	16,310	6,796	1.01	2.27	31,852
2009	18,088	7,527	0.98	2.20	34,301
2010	19,152	9,142	1.07	2.11	39,732
2011	18,212	8,843	1.12	2.07	38,629
2012	20,830	8,651	1.14	2.30	43,546
2013	18,713	8,251	1.23	2.64	44,844
2014	17,627	7,763	1.27	2.70	43,273
2015	17,664	7,815	1.28	2.68	43,536
	Other flowering and foliar plants				
	1,000 pots	*1,000 pots*	*Dollars*	*Dollars*	*1,000 dollars*
2006	110,614	57,880	1.09	2.73	278,296
2007	117,723	50,605	1.13	3.08	288,890
2008	108,251	55,789	1.20	2.65	278,250
2009	93,289	48,271	1.18	3.06	257,709
2010	104,633	57,937	1.26	3.04	301,479
2011	108,661	54,133	1.25	2.96	296,030
2012	127,328	64,961	1.20	3.02	348,961
2013	111,363	50,025	1.32	3.30	309,160
2014	98,436	45,668	1.35	3.33	284,313
2015	106,082	47,534	1.38	3.39	307,799
	Vegetable type plants [2]				
	1,000 pots	*1,000 pots*	*Dollars*	*Dollars*	*1,000 dollars*
2006	18,507	4,410	0.96	2.19	27,374
2007	27,676	3,874	1.01	2.22	36,668
2008	35,998	7,497	1.06	2.49	56,697
2009	38,534	12,794	1.07	2.51	73,307
2010	38,453	14,155	1.12	2.64	80,411
2011	40,176	7,478	1.29	3.19	75,761
2012	31,100	12,211	1.22	2.98	74,478
2013	31,816	10,057	1.24	3.16	71,249
2014	44,491	10,531	1.47	3.21	99,356
2015	36,402	8,174	1.33	2.73	70,721

[1] Equivalent wholesale value of all sales. [2] Does not include vegetable transplants grown for use in commercial vegetable production.
NASS, Crops Branch, (202) 720–2127.

Table 5-101.—Potted herbaceous perennial plants: Sales and wholesale value for operations with $100,000+ sales, Surveyed States, 2006–2015

Year	Quantity sold		Wholesale price		Value of sales at wholesale [1]
	Less than 5 inches	5 inches or more	Less than 5 inches	5 inches or more	
	1,000 pots	*1,000 pots*	*Dollars*	*Dollars*	*1,000 dollars*
	Hardy/Garden Chrysanthemums				
2006	10,545	40,303	0.99	2.31	103,656
2007	9,763	35,148	1.06	2.66	103,831
2008	9,217	34,689	1.10	2.61	100,868
2009	8,049	35,833	1.25	2.81	110,747
2010	11,352	34,531	1.10	2.93	113,620
2011	11,962	36,634	1.03	2.77	113,747
2012	13,327	35,792	1.03	3.08	124,097
2013	15,558	33,176	1.08	2.89	112,552
2014	14,727	33,907	1.04	3.08	119,888
2015	9,326	37,885	1.12	2.99	123,677

Year	Quantity sold			Wholesale price			Value of sales at wholesale [1]
	Less than 1 gallon	1-2 gallons	2 gallons or more	Less than 1 gallon	1-2 gallons	2 gallons or more	
	1,000 pots	*1,000 pots*	*1,000 pots*	*Dollars*	*Dollars*	*Dollars*	*1,000 dollars*
	Hosta						
2006	1,724	6,661	399	2.10	3.25	6.66	27,924
2007	4,374	6,589	271	1.74	3.50	7.69	32,723
2008	4,451	6,793	173	1.91	3.43	9.10	33,391
2009	4,279	6,340	582	2.15	3.36	6.47	34,283
2010	1,448	6,783	337	2.11	3.26	7.54	27,704
2011	2,112	6,270	281	2.27	3.19	8.53	27,195
2012	2,365	5,535	363	2.34	3.40	8.04	27,240
2013	2,511	5,898	382	2.64	3.39	7.98	29,671
2014	5,362	3,870	398	2.73	3.75	7.78	32,242
2015	5,195	4,419	419	2.70	3.92	8.35	34,869
	1,000 pots	*1,000 pots*	*1,000 pots*	*Dollars*	*Dollars*	*Dollars*	*1,000 dollars*
	Other Herbaceous Perennials						
2006	49,602	85,660	5,075	1.55	3.14	6.97	381,511
2007	68,640	69,843	6,442	1.71	3.33	6.84	394,407
2008	65,928	74,574	7,533	1.59	3.43	6.65	410,867
2009	50,229	80,234	4,975	1.68	3.42	6.92	392,847
2010	53,613	86,180	6,257	1.63	3.38	7.11	423,266
2011	56,970	83,944	5,459	1.75	3.35	7.43	421,276
2012	56,936	65,596	10,001	1.77	3.78	7.33	422,282
2013	62,721	72,430	6,983	1.90	3.90	8.41	460,375
2014	62,508	60,111	5,823	2.19	3.90	8.47	420,547
2015	57,666	62,449	4,271	2.17	4.01	8.75	412,866

[1] Equivalent wholesale value of all sales.
NASS, Crops Branch, (202) 720–2127.

Table 5-102.—Floriculture: Number of Producers and Area Used for Production by Type of Cover, operations with $10,000+ sales, 15 Program States: 2014–2015

State	Total number of producers		Glass greenhouses		Fiberglass and other rigid greenhouses		Film plastic (single/ multi) greenhouses	
	2014	2015	2014	2015	2014	2015	2014	2015
	Number	*Number*	*1,000 square feet*	*1,000 square feet*	*1,000 square feet*	*1,000 square feet*	*1,000 square feet*	*1,000 square feet*
California	667	685	12,355	10,748	38,399	30,042	70,359	63,989
Florida	721	710	4,061	3,054	9,918	8,159	39,376	40,460
Hawaii	312	301	(D)	(D)	(D)	(D)	3,191	2,873
Illinois	255	273	3,301	2,966	1,899	2,249	10,425	9,389
Maryland	146	153	1,338	1,644	984	1,375	5,616	6,255
Michigan	510	556	3,722	3,458	3,614	3,773	40,495	39,047
New Jersey	272	273	5,458	4,675	1,057	1,031	19,328	17,530
New York	568	615	4,103	3,816	1,551	1,223	15,547	15,721
North Carolina	279	294	6,780	6,652	451	529	15,650	14,459
Ohio	397	472	7,667	8,860	1,881	1,767	17,267	16,005
Oregon	265	283	2,544	2,835	2,677	2,691	12,715	13,335
Pennsylvania	647	727	2,497	2,585	3,923	3,808	13,796	14,566
South Carolina	72	72	(D)	(D)	(D)	(D)	2,945	2,821
Texas	255	251	1,204	1,361	3,822	3,696	32,030	28,052
Washington	240	248	2,221	2,317	859	1,099	7,986	6,702
Other States ..	-	-	280	27	1,770	1,704	-	-
15 State total	5,606	5,913	57,531	54,998	72,805	63,146	306,726	291,204

State	Total greenhouse cover		Shade and temporary cover		Total covered area		Open ground [2]	
	2014	2015	2014	2015	2014	2015	2014	2015
	1,000 square feet	*1,000 square feet*	*1,000 square feet*	*1,000 square feet*	*1,000 square feet*	*1,000 square feet*	*Acres*	*Acres*
California	121,113	104,779	30,347	37,148	151,460	141,927	11,640	9,399
Florida	53,355	51,673	261,588	289,184	314,943	340,857	6,145	6,853
Hawaii	(D)	(D)	(D)	(D)	19,288	17,940	973	914
Illinois	15,625	14,604	733	664	16,358	15,268	6,223	4,272
Maryland	7,938	9,274	479	360	8,417	9,634	3,231	618
Michigan	47,831	46,278	610	711	48,441	46,989	7,197	2,063
New Jersey	25,843	23,236	364	493	26,207	23,729	4,635	4,826
New York	21,201	20,760	341	433	21,542	21,193	1,792	722
North Carolina	22,881	21,640	1,558	1,244	24,439	22,884	1,752	912
Ohio	26,815	26,632	496	956	27,311	27,588	4,039	385
Oregon	17,936	18,861	5,210	2,275	23,146	21,136	5,359	3,903
Pennsylvania	20,216	20,959	592	262	20,808	21,221	1,681	778
South Carolina	(D)	(D)	(D)	(D)	5,821	5,921	490	396
Texas	37,056	33,109	12,571	6,485	49,627	39,594	2,981	1,215
Washington	11,066	10,118	287	917	11,353	11,035	2,939	2,786
Other States [1]	8,186	7,425	16,923	16,436	-	-	-	-
15 State total	437,062	409,348	332,099	357,568	769,161	766,916	61,077	40,042

- Represents zero. (D) Withheld to avoid disclosure of individual operations. [1] Includes data withheld above. [2] Totals may not add due to rounding.
NASS, Crops Branch, (202) 720-2127.

Table 5-103.—Floriculture Crops: Wholesale value of sales by category - 15 Program States, operations with $100,000+ sales: 2014–2015

State	Annual bedding/ garden plants		Herbaceous perennial plants		Total bedding/ garden plants[1]	
	2014	2015	2014	2015	2014	2015
	1,000 dollars	1,000 dollars	1,000 dollars	1,000 dollars	1,000 dollars	1,000 dollars
California	200,642	193,209	79,225	72,361	279,867	265,570
Florida	104,133	96,402	41,042	37,023	145,175	133,425
Hawaii	2,469	2,199	1,605	1,587	4,074	3,786
Illinois	46,655	38,595	37,558	25,619	84,213	64,214
Maryland	59,937	64,704	23,407	37,527	83,344	102,231
Michigan	208,640	208,727	64,958	61,939	273,598	270,666
New Jersey	70,884	77,016	45,215	56,641	116,099	133,657
New York	77,277	86,189	27,438	24,608	104,715	110,797
North Carolina	135,076	140,596	36,047	42,873	171,123	183,469
Ohio	76,725	81,126	46,232	51,844	122,957	132,970
Oregon	49,722	52,586	29,401	30,667	79,123	83,253
Pennsylvania	72,949	67,574	17,648	22,593	90,597	90,167
South Carolina	6,620	8,036	45,833	46,786	52,453	54,822
Texas	134,703	136,112	38,232	32,118	172,935	168,230
Washington	55,552	33,511	38,836	27,226	94,388	60,737
15 State total	1,301,984	1,286,582	572,677	571,412	1,874,661	1,857,994

State	Potted flowering plants		Foliage plants for indoor or patio use		Cut flowers	
	2014	2015	2014	2015	2014	2015
	1,000 dollars	1,000 dollars	1,000 dollars	1,000 dollars	1,000 dollars	1,000 dollars
California	290,194	273,537	116,949	114,599	284,799	293,737
Florida	199,831	201,212	444,710	543,447	3,358	2,289
Hawaii	15,031	15,079	8,694	9,612	6,254	7,014
Illinois	13,015	23,875	1,373	1,121	1,424	1,103
Maryland	4,240	7,060	(D)	(D)	(D)	470
Michigan	37,397	31,147	6,411	6,305	5,359	5,312
New Jersey	36,074	29,368	(D)	(D)	12,711	14,675
New York	28,326	29,537	(D)	(D)	1,018	885
North Carolina	32,093	27,545	(D)	(D)	3,844	4,981
Ohio	59,642	68,163	7,465	4,207	(D)	(D)
Oregon	16,092	14,809	(D)	1,170	13,724	13,350
Pennsylvania	29,910	44,954	(D)	(D)	8,407	(D)
South Carolina	8,766	8,221	(D)	(D)	(D)	(D)
Texas	29,322	27,165	12,232	(D)	(D)	(D)
Washington	10,012	7,988	849	(D)	20,897	20,457
Other States[2]	-	-	24,199	33,228	1,659	10,045
15 State total	809,945	809,660	622,882	713,689	363,454	374,318

State	Cut cultivated greens		Propagative floriculture material		Total wholesale value of all plant categories[3]	
	2014	2015	2014	2015	2014	2015
	1,000 dollars	1,000 dollars	1,000 dollars	1,000 dollars	1,000 dollars	1,000 dollars
California	(D)	15,762	(D)	107,076	1,064,283	1,070,281
Florida	59,775	61,080	67,080	86,828	919,929	1,028,281
Hawaii	420	323	1,461	1,352	35,934	37,166
Illinois	(D)	-	(D)	1,359	102,812	91,672
Maryland	(D)	-	9,340	(D)	97,466	111,163
Michigan	(D)	(D)	(D)	(D)	405,644	398,113
New Jersey	(D)	(D)	(D)	18,271	186,353	197,001
New York	(D)	(D)	(D)	16,522	153,344	160,167
North Carolina	-	-	(D)	(D)	225,006	234,168
Ohio	-	-	(D)	(D)	195,522	210,102
Oregon	(D)	2,465	8,692	14,292	124,336	129,339
Pennsylvania	(D)	(D)	(D)	(D)	152,954	169,962
South Carolina	-	-	-	(D)	61,353	63,199
Texas	(D)	(D)	8,387	(D)	223,590	216,429
Washington	32	(D)	13,263	21,920	139,441	112,761
Other States[2]	12,956	796	235,619	126,097	-	-
15 State total	73,183	80,426	343,842	393,717	4,087,967	4,229,804

- Represents zero.   (D) Withheld to avoid disclosure of individual operations.   [1] Includes Annual Bedding Plants and Herbaceous Perennials.   [2] Includes data withheld above.   [3] State total wholesale value excludes plant category values denoted by (D).
NASS, Crops Branch, (202) 720-2127.

Table 5-104.—Fruit and orange juice: Cold storage holdings, end of month, United States, 2014-2015

Month	Apples		Apricots		Blackberries, IQF		Blackberries, pails & tubs	
	2014	2015	2014	2015	2014	2015	2014	2015
	1,000 pounds	1,000 pounds	1,000 pounds	1,000 pounds	1,000 pounds	1,000 pounds	1,000 pounds	1,000 pounds
January	56,312	55,349	3,532	5,246	15,649	19,656	1,549	1,480
February	60,384	53,314	3,406	4,720	15,146	17,621	1,399	1,356
March	61,220	52,079	3,380	4,162	13,441	16,119	1,247	1,140
April	60,693	50,761	2,384	4,051	12,446	14,589	936	1,079
May	58,866	49,547	1,974	3,390	13,067	14,278	1,013	1,166
June	56,046	46,985	12,488	11,760	14,845	19,095	1,220	1,319
July	55,517	42,987	9,847	11,058	28,169	27,678	1,992	1,600
August	52,050	36,586	7,852	10,059	26,634	28,470	2,107	1,422
September ...	49,676	31,854	7,533	7,931	24,528	26,472	1,972	1,587
October	53,072	35,761	6,471	6,128	22,839	24,420	1,718	1,376
November	52,828	38,917	6,191	5,640	20,939	22,634	1,632	1,618
December	55,400	37,596	5,941	4,921	20,281	21,238	1,550	1,350

Month	Blackberries, barrels		Blackberries, concentrate		Blackberries, total		Blueberries	
	2014	2015	2014	2015	2014	2015	2014	2015
	1,000 pounds	1,000 pounds	1,000 pounds	1,000 pounds	1,000 pounds	1,000 pounds	1,000 pounds	1,000 pounds
January	11,114	6,879	511	828	28,823	28,843	172,718	187,523
February	10,134	5,688	504	834	27,183	25,499	150,626	166,469
March	7,828	5,058	478	754	22,994	23,071	130,825	152,745
April	7,245	4,450	483	774	21,110	20,892	117,065	133,836
May	6,494	3,734	533	753	21,107	19,931	98,716	119,276
June	5,429	3,925	559	819	22,053	25,158	102,012	132,732
July	11,636	3,783	784	940	42,581	34,001	171,661	236,349
August	10,503	4,265	1,095	1,083	40,339	35,240	269,537	273,614
September ...	10,103	4,174	891	845	37,494	33,078	265,303	280,085
October	9,331	3,879	868	878	34,756	30,553	240,830	259,198
November	9,033	3,251	764	863	32,368	28,366	229,225	245,754
December	8,396	3,179	698	827	30,925	26,594	208,324	224,696

Month	Boysenberries		Cherries, Tart (RTP)		Cherries, Sweet		Grapes	
	2014	2015	2014	2015	2014	2015	2014	2015
	1,000 pounds	1,000 pounds	1,000 pounds	1,000 pounds	1,000 pounds	1,000 pounds	1,000 pounds	1,000 pounds
January	1,648	1,389	99,639	129,319	14,477	21,326	4,108	3,177
February	1,906	1,232	91,631	120,870	14,136	20,155	4,226	3,005
March	1,213	1,003	82,926	105,707	13,065	19,275	3,856	2,682
April	960	848	71,746	98,519	12,245	17,411	3,322	2,435
May	865	618	58,869	85,866	10,659	17,140	2,831	2,347
June	1,016	1,154	50,181	79,028	9,448	18,134	3,403	2,654
July	2,049	1,478	103,362	146,251	13,469	23,994	3,306	4,496
August	1,918	1,221	178,542	158,170	27,390	25,401	3,540	3,734
September ...	1,984	1,219	164,429	159,154	23,525	22,830	3,699	3,794
October	1,931	1,118	153,521	141,734	23,173	19,672	3,972	3,078
November	1,577	955	144,280	132,880	22,722	19,929	3,465	2,806
December	1,441	954	134,908	123,378	22,226	18,076	3,404	2,469

Month	Peaches		Raspberries, Black		Red Raspberries, IQF		Red Raspberries, pails & tubs	
	2014	2015	2014	2015	2014	2015	2014	2015
	1,000 pounds	1,000 pounds	1,000 pounds	1,000 pounds	1,000 pounds	1,000 pounds	1,000 pounds	1,000 pounds
January	53,367	44,353	710	1,089	25,341	27,494	5,801	5,473
February	47,862	38,646	677	969	22,537	24,878	4,497	4,387
March	39,621	35,803	620	944	20,300	22,869	4,202	4,250
April	32,936	30,722	592	924	18,519	20,117	3,868	4,135
May	25,714	27,566	462	867	18,294	18,658	3,712	3,055
June	19,497	26,877	1,442	2,152	18,159	21,764	3,608	4,101
July	34,001	41,488	2,027	2,555	36,855	38,655	10,779	9,152
August	54,810	63,360	1,401	1,828	36,201	37,665	9,475	8,093
September ...	63,033	78,572	1,381	1,655	34,803	38,125	8,057	8,438
October	57,543	75,817	1,353	1,625	33,491	35,218	6,709	8,131
November	52,384	75,253	1,245	1,444	30,861	33,319	6,024	7,829
December	47,465	74,308	1,145	1,356	29,517	31,991	6,027	7,033

See footnotes(s) at end of table.

Table 5-104.—Fruit and orange juice: Cold storage holdings, end of month, United States, 2014 and 2015—Continued

Month	Red Raspberries, barrels		Red Raspberries, concentrate		Red Raspberries, total		Strawberries, IQF & Poly	
	2014	2015	2014	2015	2014	2015	2014	2015
	1,000 pounds	*1,000 pounds*	*1,000 pounds*	*1,000 pounds*	*1,000 pounds*	*1,000 pounds*	*1,000 pounds*	*1,000 pounds*
January	20,019	17,660	1,569	1,506	52,730	52,133	91,367	76,820
February	16,643	15,637	1,394	2,358	45,071	47,260	81,716	75,324
March	12,497	13,399	1,419	2,258	38,418	42,776	86,100	94,274
April	10,074	10,983	1,212	2,008	33,673	37,243	108,405	96,640
May	8,133	8,769	1,214	1,929	31,353	32,411	123,274	114,524
June	7,119	12,157	1,340	2,163	30,226	40,185	154,588	139,019
July	37,278	26,538	2,580	2,292	87,492	76,637	165,509	150,237
August	33,746	22,956	2,119	1,871	81,541	70,585	148,359	133,910
September ...	29,997	21,067	2,001	2,168	74,858	69,798	131,035	131,485
October	24,924	18,025	1,971	1,969	67,095	63,343	119,620	113,082
November	23,541	18,055	1,911	1,923	62,337	61,126	101,207	102,516
December	19,996	16,692	1,758	1,646	57,298	57,362	92,172	98,091

Month	Strawberries, pails & tubs		Strawberries, barrels & drums		Strawberries, juice stock		Strawberries, total	
	2014	2015	2014	2015	2014	2015	2014	2015
	1,000 pounds	*1,000 pounds*	*1,000 pounds*	*1,000 pounds*	*1,000 pounds*	*1,000 pounds*	*1,000 pounds*	*1,000 pounds*
January	67,354	46,103	67,007	43,074	14,654	9,523	240,382	175,520
February	60,666	42,821	59,391	36,162	13,395	9,292	215,168	163,599
March	55,797	47,393	55,287	32,754	13,880	12,396	211,064	186,817
April	57,051	52,887	53,629	36,668	14,062	13,095	233,147	199,290
May	61,592	69,162	60,147	45,176	12,770	9,244	257,783	238,106
June	92,845	108,514	95,041	82,896	15,052	10,866	357,526	341,295
July	89,369	109,039	95,398	88,485	11,851	12,284	362,127	360,045
August	82,167	103,909	87,713	78,631	12,427	12,325	330,666	328,775
September ...	72,340	96,661	80,575	74,491	11,019	11,713	294,969	314,350
October	66,004	88,873	72,941	67,380	11,988	9,525	270,553	278,860
November	59,902	83,864	64,535	59,068	11,184	7,698	236,828	253,146
December	49,966	79,219	55,264	50,958	9,439	7,584	206,841	235,852

Month	Other fruit		Total frozen fruit		Orange juice	
	2014	2015	2014	2015	2014	2015
	1,000 pounds	*1,000 pounds*	*1,000 pounds*	*1,000 pounds*	*1,000 pounds*	*1,000 pounds*
January	528,790	482,436	1,261,124	1,191,799	750,467	720,747
February	475,527	434,463	1,140,370	1,084,164	799,317	728,424
March	440,532	385,477	1,051,950	1,016,361	813,007	832,345
April	389,071	348,811	981,069	949,216	877,826	857,503
May	341,283	321,532	912,114	921,622	872,936	947,948
June	305,249	297,358	972,152	1,028,135	853,656	943,782
July	286,727	309,251	1,176,038	1,294,279	815,884	868,061
August	264,490	309,356	1,317,864	1,322,557	773,000	807,795
September ...	293,231	377,612	1,286,143	1,385,630	712,015	757,074
October	542,534	659,912	1,461,405	1,580,904	721,880	695,919
November ...	576,843	665,020	1,426,813	1,535,394	676,846	641,710
December	535,775	587,967	1,315,421	1,398,611	734,784	639,347

NASS, Livestock Branch, (202) 720–3570.

Table 5-105.—Nuts: Cold storage holdings, end of month, United States, 2014-2015

Month	Pecans			
	Shelled		In-shell	
	2014	2015	2014	2015
	1,000 pounds	*1,000 pounds*	*1,000 pounds*	*1,000 pounds*
January	46,212	51,196	164,037	144,021
February	47,989	53,138	204,365	144,305
March	48,497	57,710	196,843	135,169
April	51,514	57,271	183,169	114,568
May	53,431	56,222	158,555	94,352
June	55,610	56,717	137,685	72,702
July	52,184	53,575	103,286	53,903
August	53,125	52,620	73,275	38,540
September	52,871	44,112	51,450	21,053
October	43,061	36,107	27,102	20,016
November	40,344	32,057	33,063	42,466
December	45,067	34,771	84,740	111,848

NASS, Livestock Branch, (202) 720-3570.

STATISTICS OF HAY, SEEDS, AND MINOR FIELD CROPS

Chapter VI deals with hay, pasture, seeds, and various minor field crops.

Table 6-1.—Hay, all: Area, yield, and production, by State and United States, 2013–2015

State	Area harvested			Yield per harvested acre			Production		
	2013	2014	2015	2013	2014	2015	2013	2014	2015
	1,000 acres	*1,000 acres*	*1,000 acres*	*Tons*	*Tons*	*Tons*	*1,000 tons*	*1,000 tons*	*1,000 tons*
Alabama	790	750	730	2.70	2.80	2.80	2,133	2,100	2,044
Arizona	285	300	335	7.65	8.03	7.99	2,179	2,410	2,678
Arkansas	1,335	1,225	1,125	2.10	2.01	2.00	2,810	2,458	2,254
California	1,370	1,345	1,180	5.58	5.59	5.74	7,646	7,513	6,777
Colorado	1,310	1,340	1,450	2.25	2.66	2.96	2,941	3,566	4,295
Connecticut	47	53	53	2.26	1.92	1.89	106	102	100
Delaware	18	13	14	3.28	2.62	3.14	59	34	44
Florida	300	320	290	2.20	2.60	2.80	660	832	812
Georgia	580	580	570	2.70	2.60	2.50	1,566	1,508	1,425
Idaho	1,480	1,390	1,330	3.36	3.51	3.65	4,976	4,881	4,860
Illinois	660	520	490	3.07	3.38	3.13	2,024	1,755	1,533
Indiana	640	600	560	2.80	3.25	2.96	1,792	1,950	1,656
Iowa	1,170	1,155	1,160	2.89	3.18	3.40	3,377	3,675	3,939
Kansas	2,750	2,300	2,450	2.38	2.17	2.40	6,545	5,000	5,890
Kentucky	2,400	2,265	2,370	2.29	2.10	2.40	5,500	4,761	5,689
Louisiana	400	470	430	2.20	2.70	2.50	880	1,269	1,075
Maine	135	150	135	1.46	1.55	2.02	197	233	273
Maryland	225	195	215	2.33	2.65	2.47	524	517	532
Massachusetts	84	75	92	2.12	1.72	1.73	178	129	159
Michigan	940	980	970	2.68	2.62	2.68	2,518	2,570	2,604
Minnesota	1,900	1,910	1,570	2.05	2.35	2.53	3,895	4,486	3,979
Mississippi	720	600	680	2.50	2.60	2.30	1,800	1,560	1,564
Missouri	4,030	3,480	2,960	1.97	2.04	2.16	7,921	7,100	6,398
Montana	2,800	2,730	2,500	1.95	1.97	1.87	5,460	5,381	4,680
Nebraska	2,500	2,580	2,700	1.97	2.34	2.36	4,935	6,028	6,360
Nevada	345	430	320	3.37	3.29	3.44	1,161	1,416	1,100
New Hampshire	50	54	48	2.50	1.74	2.04	125	94	98
New Jersey	97	106	102	2.42	2.46	1.76	235	261	180
New Mexico	230	305	280	4.18	3.93	3.90	962	1,198	1,091
New York	1,430	1,370	1,230	2.05	1.97	1.99	2,930	2,698	2,449
North Carolina	858	830	777	2.41	2.40	2.40	2,064	1,996	1,868
North Dakota	2,620	2,700	2,750	1.94	2.02	1.81	5,090	5,460	4,975
Ohio	1,000	960	1,080	2.50	2.82	2.34	2,495	2,710	2,532
Oklahoma	3,130	3,590	3,020	1.59	1.71	1.96	4,971	6,121	5,914
Oregon	1,020	1,030	1,060	3.14	3.08	2.90	3,204	3,172	3,072
Pennsylvania	1,260	1,400	1,290	2.32	2.28	2.33	2,918	3,185	3,010
Rhode Island	8	7	6	1.88	1.71	2.33	15	12	14
South Carolina	290	270	300	2.20	2.30	2.00	638	621	600
South Dakota	3,050	3,250	3,400	1.94	2.05	1.94	5,905	6,665	6,580
Tennessee	1,915	1,766	1,765	2.31	2.20	2.21	4,427	3,893	3,901
Texas	5,640	5,440	4,730	1.57	2.16	2.05	8,880	11,746	9,720
Utah	725	680	670	3.77	3.52	3.67	2,730	2,396	2,459
Vermont	180	185	145	1.72	1.68	1.94	310	311	281
Virginia	1,240	1,175	1,175	2.49	2.28	2.25	3,084	2,675	2,645
Washington	760	870	750	4.24	3.72	3.81	3,223	3,234	2,856
West Virginia	590	618	590	1.97	1.83	1.75	1,165	1,132	1,035
Wisconsin	1,600	1,640	1,510	2.35	2.97	2.70	3,760	4,866	4,073
Wyoming	990	1,060	1,080	2.11	2.12	2.14	2,088	2,243	2,315
United States	57,897	57,062	54,437	2.33	2.45	2.47	135,002	139,923	134,388

NASS, Crops Branch, (202) 720–2127.

Table 6-2.—Hay, all: Area, yield, production, and value, United States, 2006–2015

Year	Area harvested	Yield per acre	Production	Marketing year average price per ton received by farmers	Value of production
	1,000 acres	Tons	1,000 tons	Dollars	1,000 dollars
2006	60,632	2.32	140,783	110.00	13,633,837
2007	61,006	2.41	146,901	127.00	16,842,233
2008	60,152	2.43	146,270	152.00	18,638,748
2009	59,775	2.47	147,700	108.00	14,715,559
2010	59,574	2.43	145,000	114.00	14,606,696
2011	55,204	2.35	129,880	178.00	18,085,074
2012	54,653	2.14	117,072	191.00	18,613,183
2013	57,897	2.33	135,002	176.00	19,814,822
2014	57,062	2.45	139,923	172.00	19,099,238
2015	54,437	2.47	134,388	151.00	16,839,538

NASS, Crops Branch, (202) 720–2127.

Table 6-3.—Alfalfa and alfalfa mixtures for hay: Area, yield, and production, by State and United States, 2013–2015

State	Area harvested			Yield per harvested acre			Production		
	2013	2014	2015	2013	2014	2015	2013	2014	2015
	1,000 acres	1,000 acres	1,000 acres	Tons	Tons	Tons	1,000 tons	1,000 tons	1,000 tons
Arizona	250	260	300	8.10	8.50	8.40	2,025	2,210	2,520
Arkansas	5	5	5	3.30	3.60	2.70	17	18	14
California	830	825	790	7.00	6.90	6.90	5,810	5,693	5,451
Colorado	650	740	700	2.90	3.40	4.10	1,885	2,516	2,870
Connecticut	7	8	7	2.00	2.00	1.80	14	16	13
Delaware	6	4	4	3.50	2.70	2.70	21	11	11
Idaho	1,120	1,090	1,000	3.80	3.90	4.20	4,256	4,251	4,200
Illinois	340	270	230	3.60	4.00	3.50	1,224	1,080	805
Indiana	280	240	230	3.70	4.00	3.90	1,036	960	897
Iowa	730	810	770	3.30	3.60	3.90	2,409	2,916	3,003
Kansas	550	600	650	3.50	3.80	3.80	1,925	2,280	2,470
Kentucky	200	165	170	3.30	3.40	3.70	660	561	629
Maine	10	10	10	2.20	2.30	2.30	22	23	23
Maryland	30	35	35	3.80	3.80	4.40	114	133	154
Massachusetts	9	10	9	3.10	1.80	2.00	28	18	18
Michigan	610	640	660	3.10	2.90	3.10	1,891	1,856	2,046
Minnesota	950	1,100	1,050	2.60	2.90	2.70	2,470	3,190	2,835
Missouri	330	280	260	2.70	2.50	2.80	891	700	728
Montana	1,800	1,850	1,700	2.20	2.10	2.00	3,960	3,885	3,400
Nebraska	700	830	850	3.45	4.10	4.00	2,415	3,403	3,400
Nevada	210	280	200	4.50	4.20	4.30	945	1,176	860
New Hampshire	5	4	3	1.60	2.30	2.50	8	9	8
New Jersey	17	14	12	3.00	3.50	3.00	51	49	36
New Mexico	145	210	190	5.40	4.80	4.70	783	1,008	893
New York	350	290	280	2.20	2.60	2.30	770	754	644
North Carolina	8	10	7	3.00	2.80	2.80	24	28	20
North Dakota	1,620	1,650	1,500	2.00	2.10	1.90	3,240	3,465	2,850
Ohio	330	310	330	3.50	3.50	2.90	1,155	1,085	957
Oklahoma	230	290	220	2.70	2.90	2.70	621	841	594
Oregon	400	350	370	4.60	4.40	4.20	1,840	1,540	1,554
Pennsylvania	340	350	430	2.90	2.80	2.60	986	980	1,118
Rhode Island	1	1	1	2.00	2.50	2.00	2	3	2
South Dakota	1,800	1,900	1,900	2.10	2.30	2.20	3,780	4,370	4,180
Tennessee	15	16	15	3.80	2.70	3.40	57	43	51
Texas	140	140	130	4.50	4.40	4.00	630	616	520
Utah	550	520	510	4.20	3.90	4.10	2,310	2,028	2,091
Vermont	35	35	35	1.80	1.60	3.00	63	56	105
Virginia	90	75	75	3.60	3.40	3.00	324	255	225
Washington	410	420	390	5.30	4.70	5.20	2,173	1,974	2,028
West Virginia	20	18	20	4.10	2.90	3.30	82	52	66
Wisconsin	1,100	1,250	1,200	2.60	3.30	2.80	2,860	4,125	3,360
Wyoming	450	490	530	3.20	2.60	2.50	1,440	1,274	1,325
United States	17,673	18,395	17,778	3.24	3.34	3.32	57,217	61,451	58,974

NASS, Crops Branch, (202) 720–2127.

Table 6-4.—Hay, all other: Area, yield, and production, by State and United States, 2013–2015

State	Area harvested			Yield per harvested acre			Production		
	2013	2014	2015	2013	2014	2015	2013	2014	2015
	1,000 acres	*1,000 acres*	*1,000 acres*	*Tons*	*Tons*	*Tons*	*1,000 tons*	*1,000 tons*	*1,000 tons*
Alabama [1]	790	750	730	2.70	2.80	2.80	2,133	2,100	2,044
Arizona	35	40	35	4.40	5.00	4.50	154	200	158
Arkansas	1,330	1,220	1,120	2.10	2.00	2.00	2,793	2,440	2,240
California	540	520	390	3.40	3.50	3.40	1,836	1,820	1,326
Colorado	660	600	750	1.60	1.75	1.90	1,056	1,050	1,425
Connecticut	40	45	46	2.30	1.90	1.90	92	86	87
Delaware	12	9	10	3.20	2.50	3.30	38	23	33
Florida [1]	300	320	290	2.20	2.60	2.80	660	832	812
Georgia [1]	580	580	570	2.70	2.60	2.50	1,566	1,508	1,425
Idaho	360	300	330	2.00	2.10	2.00	720	630	660
Illinois	320	250	260	2.50	2.70	2.80	800	675	728
Indiana	360	360	330	2.10	2.75	2.30	756	990	759
Iowa	440	345	390	2.20	2.20	2.40	968	759	936
Kansas	2,200	1,700	1,800	2.10	1.60	1.90	4,620	2,720	3,420
Kentucky	2,200	2,100	2,200	2.20	2.00	2.30	4,840	4,200	5,060
Louisiana [1]	400	470	430	2.20	2.70	2.50	880	1,269	1,075
Maine	125	140	125	1.40	1.50	2.00	175	210	250
Maryland	195	160	180	2.10	2.40	2.10	410	384	378
Massachusetts	75	65	83	2.00	1.70	1.70	150	111	141
Michigan	330	340	310	1.90	2.10	1.80	627	714	558
Minnesota	950	810	520	1.50	1.60	2.20	1,425	1,296	1,144
Mississippi [1]	720	600	680	2.50	2.60	2.30	1,800	1,560	1,564
Missouri	3,700	3,200	2,700	1.90	2.00	2.10	7,030	6,400	5,670
Montana	1,000	880	800	1.50	1.70	1.60	1,500	1,496	1,280
Nebraska	1,800	1,750	1,850	1.40	1.50	1.60	2,520	2,625	2,960
Nevada	135	150	120	1.60	1.60	2.00	216	240	240
New Hampshire	45	50	45	2.60	1.70	2.00	117	85	90
New Jersey	80	92	90	2.30	2.30	1.60	184	212	144
New Mexico	85	95	90	2.10	2.00	2.20	179	190	198
New York	1,080	1,080	950	2.00	1.80	1.90	2,160	1,944	1,805
North Carolina	850	820	770	2.40	2.40	2.40	2,040	1,968	1,848
North Dakota	1,000	1,050	1,250	1.85	1.90	1.70	1,850	1,995	2,125
Ohio	670	650	750	2.00	2.50	2.10	1,340	1,625	1,575
Oklahoma	2,900	3,300	2,800	1.50	1.60	1.90	4,350	5,280	5,320
Oregon	620	680	690	2.20	2.40	2.20	1,364	1,632	1,518
Pennsylvania	920	1,050	860	2.10	2.10	2.20	1,932	2,205	1,892
Rhode Island	7	6	5	1.90	1.50	2.30	13	9	12
South Carolina [1]	290	270	300	2.20	2.30	2.00	638	621	600
South Dakota	1,250	1,350	1,500	1.70	1.70	1.60	2,125	2,295	2,400
Tennessee	1,900	1,750	1,750	2.30	2.20	2.20	4,370	3,850	3,850
Texas	5,500	5,300	4,600	1.50	2.10	2.00	8,250	11,130	9,200
Utah	175	160	160	2.40	2.30	2.30	420	368	368
Vermont	145	150	110	1.70	1.70	1.60	247	255	176
Virginia	1,150	1,100	1,100	2.40	2.20	2.20	2,760	2,420	2,420
Washington	350	450	360	3.00	2.80	2.30	1,050	1,260	828
West Virginia	570	600	570	1.90	1.80	1.70	1,083	1,080	969
Wisconsin	500	390	310	1.80	1.90	2.30	900	741	713
Wyoming	540	570	550	1.20	1.70	1.80	648	969	990
United States	40,224	38,667	36,659	1.93	2.03	2.06	77,785	78,472	75,414

[1] Alfalfa and alfalfa mixtures included in all other hay.
NASS, Crops Branch, (202) 720–2127.

Table 6-5.—Hay, all: Stocks on farms, United States, 2006–2015

Crop year	Dec. 1	May 1 [1]
	1,000 tons	*1,000 tons*
2006	96,400	14,990
2007	104,089	21,585
2008	103,658	22,065
2009	107,222	20,931
2010	101,667	22,217
2011	89,723	21,381
2012	75,175	14,156
2013	89,304	19,176
2014	92,052	24,517
2015	94,993	25,140

[1] Following year.
NASS, Crops Branch, (202) 720–2127.

Table 6-6.—Hay, all: Marketing year average price and value of production, by State and United States, 2013–2015

State	Marketing year average price per ton, baled			Value of production		
	2013	2014	2015	2013	2014	2015
	Dollars	*Dollars*	*Dollars*	*1,000 dollars*	*1,000 dollars*	*1,000 dollars*
Alabama	88.00	106.00	118.00	187,704	222,600	241,192
Arizona	201.00	218.00	161.00	437,979	524,580	434,792
Arkansas	128.00	116.00	115.00	360,904	284,200	258,020
California	199.00	232.00	176.00	1,521,832	1,710,872	1,198,053
Colorado	236.00	206.00	182.00	687,513	731,862	781,690
Connecticut	235.00	208.00	221.00	24,940	21,248	22,085
Delaware	179.00	177.00	174.00	10,533	6,028	7,634
Florida	167.00	152.00	160.00	110,220	126,464	129,920
Georgia	75.00	90.00	90.00	117,450	135,720	128,250
Idaho	191.00	198.00	174.00	943,808	961,080	836,640
Illinois	166.00	156.00	142.00	328,368	265,140	204,435
Indiana	178.00	141.00	152.00	305,508	283,800	252,816
Iowa	176.00	133.00	111.00	599,445	489,333	441,597
Kansas	164.00	152.00	109.00	893,200	653,720	574,370
Kentucky	121.00	120.00	134.00	609,840	508,395	680,457
Louisiana	93.00	100.00	121.00	81,840	126,900	130,075
Maine	172.00	113.00	124.00	33,938	26,233	33,833
Maryland	181.00	175.00	173.00	94,686	90,698	92,176
Massachusetts	200.00	171.00	195.00	35,674	22,005	31,020
Michigan	165.00	151.00	155.00	433,823	404,056	417,570
Minnesota	171.00	122.00	89.00	668,705	550,480	376,611
Mississippi	66.00	71.00	74.00	118,800	110,760	115,736
Missouri	118.00	91.50	92.00	862,088	612,500	564,487
Montana	138.00	126.00	126.00	741,360	668,427	585,400
Nebraska	148.00	96.50	86.00	698,775	569,453	539,620
Nevada	207.00	240.00	171.00	240,543	336,216	196,220
New Hampshire	172.00	153.00	167.00	21,559	14,374	16,408
New Jersey	133.00	191.00	167.00	31,225	49,807	29,988
New Mexico	242.00	248.00	203.00	231,930	295,272	219,707
New York	157.00	162.00	175.00	494,810	467,318	434,752
North Carolina	110.00	114.00	126.00	227,184	228,208	235,360
North Dakota	103.00	81.50	81.00	481,270	410,235	366,538
Ohio	171.00	139.00	150.00	426,410	375,125	365,487
Oklahoma	106.00	98.50	118.00	536,208	621,205	510,904
Oregon	200.00	219.00	198.00	636,360	685,680	604,062
Pennsylvania	184.00	208.00	226.00	541,096	669,585	686,882
Rhode Island	214.00	163.00	160.00	3,215	1,950	2,242
South Carolina	110.00	104.00	118.00	70,180	64,584	70,800
South Dakota	143.00	111.00	104.00	819,405	722,430	665,040
Tennessee	95.50	99.50	113.00	422,807	386,545	442,675
Texas	120.00	100.00	104.00	884,820	973,308	769,080
Utah	182.00	188.00	164.00	484,260	437,936	391,868
Vermont	140.00	130.00	165.00	43,450	40,463	46,320
Virginia	128.00	138.00	158.00	393,732	370,315	418,635
Washington	209.00	217.00	181.00	675,050	702,702	517,476
West Virginia	120.00	117.00	130.00	139,551	132,948	134,655
Wisconsin	192.00	143.00	91.00	722,680	697,716	377,305
Wyoming	184.00	141.00	115.00	378,144	308,762	258,655
United States	176.00	172.00	151.00	19,814,822	19,099,238	16,839,538

NASS, Crops Branch, (202) 720–2127.

Table 6-7.—Hay: Area and production, by type, United States, 2006–2015

Year	Area harvested			Production		
	Alfalfa	All other hay	All hay	Alfalfa	All other hay	All hay
	1,000 acres	*1,000 acres*	*1,000 acres*	*1,000 tons*	*1,000 tons*	*1,000 tons*
2006	21,138	39,494	60,632	70,548	70,235	140,783
2007	21,126	39,880	61,006	69,880	77,021	146,901
2008	21,060	39,092	60,152	70,180	76,090	146,270
2009	21,247	38,528	59,775	71,072	76,628	147,700
2010	19,968	39,606	59,574	67,977	77,023	145,000
2011	19,044	36,160	55,204	64,815	65,065	129,880
2012	16,795	37,858	54,653	50,600	66,472	117,072
2013	17,673	40,224	57,897	57,217	77,785	135,002
2014	18,395	38,667	57,062	61,451	78,472	139,923
2015	17,778	36,659	54,437	58,974	75,414	134,388

NASS, Crops Branch, (202) 720–2127.

Forage production is the sum of all dry hay production and haylage/greenchop production after converting the haylage/greenchop production to a dry equivalent basis (13 percent moisture) by multiplying the green weight (weight at harvest) by 0.4943. The conversion factor (0.4943) is based on the assumption that one ton of dry hay is 0.87 ton of dry matter, one ton of haylage is 0.45 ton dry matter and one ton of greenchop is 0.25 ton dry matter. The total haylage/greenchop production is assumed to be comprised of 90 percent haylage and 10 percent greenchop. Therefore, the conversion factor used to adjust haylage/greenchop production to a dry equivalent basis = ((0.45*0.9)+(0.25*0.1))/0.87 = 0.4943. The factors assumed here may vary by State and can be adjusted. Adjustments would result in a slightly different conversion factor.

Table 6-8.—All forage: Area harvested, yield, and production, by State and 18 State total, 2013–2015 [1]

State	Area harvested			Yield		
	2013	2014	2015	2013	2014	2015
	1,000 acres	*1,000 acres*	*1,000 acres*	*Tons*	*Tons*	*Tons*
California	1,565	1,530	1,375	5.98	5.89	5.87
Idaho	1,520	1,460	1,400	3.62	3.78	3.88
Illinois	680	540	510	3.12	3.54	3.29
Iowa	1,220	1,220	1,240	3.07	3.39	3.52
Kansas	2,780	2,420	2,540	2.41	2.24	2.48
Michigan	1,180	1,210	1,210	3.02	3.07	3.17
Minnesota	2,165	2,210	1,890	2.25	2.56	2.81
Missouri	4,085	3,540	3,040	2.00	2.07	2.19
Nebraska	2,530	2,610	2,720	2.00	2.37	2.38
New Mexico	244	330	305	4.16	4.00	4.00
New York	2,020	1,830	1,720	2.47	2.58	2.53
Ohio	1,050	1,030	1,180	2.70	2.96	2.56
Pennsylvania	1,540	1,720	1,620	2.79	2.65	2.71
South Dakota	3,085	3,280	3,450	1.98	2.09	1.98
Texas	5,744	5,545	4,836	1.63	2.20	2.14
Vermont	310	290	270	2.73	2.59	3.17
Washington	805	930	840	4.44	3.88	4.08
Wisconsin	2,650	2,700	2,600	2.65	3.54	3.45
18 State Total	35,173	34,395	32,746	2.53	2.78	2.79

State	Production		
	2013	2014	2015
	1,000 tons	*1,000 tons*	*1,000 tons*
California	9,362	9,008	8,070
Idaho	5,496	5,524	5,429
Illinois	2,120	1,909	1,680
Iowa	3,747	4,135	4,370
Kansas	6,710	5,413	6,293
Michigan	3,568	3,719	3,835
Minnesota	4,882	5,657	5,309
Missouri	8,158	7,319	6,651
Nebraska	5,062	6,192	6,483
New Mexico	1,014	1,320	1,220
New York	4,998	4,728	4,346
Ohio	2,838	3,044	3,018
Pennsylvania	4,294	4,562	4,393
South Dakota	6,093	6,859	6,835
Texas	9,345	12,178	10,334
Vermont	847	751	856
Washington	3,573	3,609	3,427
Wisconsin	7,022	9,570	8,967
18 State Total	89,129	95,497	91,516

[1] All forage production is the sum of the following dry equivalents: alfalfa hay harvested as dry hay, all other hay harvested as dry hay, alfalfa haylage and greenchop, all other hay haylage and greenchop; after converting alfalfa and all other haylage and greenchop to a dry equivalent basis.
NASS, Crops Branch, (202) 720–2127.

Table 6-9.—All alfalfa forage: Area harvested, yield, and production, by State and 18 State total, 2013–2015[1]

State	Area harvested			Yield		
	2013	2014	2015	2013	2014	2015
	1,000 acres	*1,000 acres*	*1,000 acres*	*Tons*	*Tons*	*Tons*
California	865	850	815	7.09	7.01	6.94
Idaho	1,135	1,120	1,030	4.10	4.20	4.45
Illinois	360	290	250	3.59	4.17	3.76
Iowa	770	860	810	3.56	3.85	4.06
Kansas	560	650	700	3.50	3.72	3.78
Michigan	840	850	890	3.43	3.48	3.62
Minnesota	1,185	1,360	1,350	2.86	3.12	3.04
Missouri	345	290	290	2.80	2.58	2.89
Nebraska	710	850	860	3.48	4.12	4.03
New Mexico	145	215	190	5.50	4.87	4.78
New York	650	500	530	3.16	3.83	3.63
Ohio	360	350	400	3.91	3.71	3.37
Pennsylvania	490	560	660	4.11	3.52	3.27
South Dakota	1,820	1,910	1,930	2.15	2.32	2.23
Texas	144	145	136	4.47	4.41	3.98
Vermont	90	70	60	3.28	3.27	3.73
Washington	425	430	425	5.25	4.80	5.16
Wisconsin	2,000	2,200	2,150	2.88	3.84	3.57
18 State Total	12,894	13,500	13,476	3.54	3.79	3.71

State	Production		
	2013	2014	2015
	1,000 tons	*1,000 tons*	*1,000 tons*
California	6,136	5,960	5,660
Idaho	4,658	4,706	4,581
Illinois	1,294	1,210	939
Iowa	2,738	3,314	3,292
Kansas	1,959	2,421	2,648
Michigan	2,879	2,961	3,220
Minnesota	3,386	4,249	4,098
Missouri	965	749	839
Nebraska	2,474	3,498	3,467
New Mexico	797	1,048	908
New York	2,055	1,914	1,925
Ohio	1,408	1,300	1,347
Pennsylvania	2,015	1,969	2,160
South Dakota	3,909	4,431	4,309
Texas	644	640	541
Vermont	295	229	224
Washington	2,232	2,064	2,191
Wisconsin	5,766	8,455	7,685
18 State Total	45,610	51,118	50,034

[1] All alfalfa forage production is the sum of alfalfa harvested as dry hay; and alfalfa haylage and greenchop production after converting it to a dry equivalent basis.
NASS, Crops Branch, (202) 720–2127.

Table 6-10.—All haylage and greenchop: Area harvested, yield, and production, by State and 18 State total, 2013–2015[1]

State	Area harvested			Yield		
	2013	2014	2015	2013	2014	2015
	1,000 acres	*1,000 acres*	*1,000 acres*	*Tons*	*Tons*	*Tons*
California	265	240	235	13.10	12.60	11.13
Idaho	90	120	110	11.68	10.83	10.45
Illinois	49	49	40	3.95	6.38	7.45
Iowa	105	155	135	7.13	6.01	6.47
Kansas	66	165	130	5.06	5.07	6.27
Michigan	295	290	295	7.20	8.02	8.44
Minnesota	331	370	380	6.03	6.40	7.08
Missouri	80	90	125	6.00	4.92	4.10
Nebraska	50	55	55	5.16	6.05	4.51
New Mexico	17	32	33	6.21	7.66	7.91
New York	760	660	650	5.51	6.22	5.91
Ohio	112	133	139	6.20	5.07	7.09
Pennsylvania	465	450	435	5.98	6.19	6.43
South Dakota	55	60	80	6.91	6.55	6.44
Texas	110	157	256	8.54	5.57	4.85
Vermont	165	140	170	6.59	6.36	6.84
Washington	83	103	130	8.53	7.36	8.88
Wisconsin	1,230	1,335	1,430	5.37	7.13	6.92
18 State Total	4,328	4,604	4,828	6.50	6.98	6.95

State	Production		
	2013	2014	2015
	1,000 tons	*1,000 tons*	*1,000 tons*
California	3,472	3,025	2,616
Idaho	1,051	1,300	1,150
Illinois	194	313	298
Iowa	749	931	873
Kansas	334	836	815
Michigan	2,123	2,326	2,491
Minnesota	1,996	2,368	2,690
Missouri	480	443	513
Nebraska	258	333	248
New Mexico	106	245	261
New York	4,184	4,106	3,839
Ohio	694	674	985
Pennsylvania	2,783	2,785	2,799
South Dakota	380	393	515
Texas	940	874	1,242
Vermont	1,087	890	1,163
Washington	708	758	1,154
Wisconsin	6,600	9,516	9,902
18 State Total	28,139	32,116	33,554

[1] Includes all types of forage harvested as haylage or greenchop (green weight). Forage harvested as dry hay and corn and sorghum silage/greenchop are not included.
NASS, Crops Branch, (202) 720–2127.

Table 6-11.—Alfalfa haylage and greenchop: Area harvested, yield, and production, by State and 18 State total, 2013–2015[1]

State	Area harvested			Yield		
	2013	2014	2015	2013	2014	2015
	1,000 acres	*1,000 acres*	*1,000 acres*	*Tons*	*Tons*	*Tons*
California	75	65	65	8.80	8.30	6.50
Idaho	65	80	70	12.50	11.50	11.00
Illinois	38	40	34	3.70	6.60	8.00
Iowa	90	130	90	7.40	6.20	6.50
Kansas	16	65	60	4.30	4.40	6.00
Michigan	270	260	270	7.40	8.60	8.80
Minnesota	285	315	350	6.50	6.80	7.30
Missouri	25	20	45	6.00	5.00	5.00
Nebraska	25	35	30	4.80	5.50	4.50
New Mexico	3	10	3	9.50	8.00	10.00
New York	400	340	360	6.50	6.90	7.20
Ohio	80	85	100	6.40	5.10	7.90
Pennsylvania	285	290	285	7.30	6.90	7.40
South Dakota	40	30	50	6.50	4.10	5.20
Texas	4	7	6	7.00	7.00	7.00
Vermont	70	50	40	6.70	7.00	6.00
Washington	23	23	45	5.20	7.90	7.30
Wisconsin	1,050	1,200	1,250	5.60	7.30	7.00
18 State Total	2,844	3,045	3,153	6.49	7.16	7.24

State	Production		
	2013	2014	2015
	1,000 tons	*1,000 tons*	*1,000 tons*
California	660	540	423
Idaho	813	920	770
Illinois	141	264	272
Iowa	666	806	585
Kansas	69	286	360
Michigan	1,998	2,236	2,376
Minnesota	1,853	2,142	2,555
Missouri	150	100	225
Nebraska	120	193	135
New Mexico	29	80	30
New York	2,600	2,346	2,592
Ohio	512	434	790
Pennsylvania	2,081	2,001	2,109
South Dakota	260	123	260
Texas	28	49	42
Vermont	469	350	240
Washington	120	182	329
Wisconsin	5,880	8,760	8,750
18 State Total	18,449	21,812	22,843

[1] Includes only alfalfa and alfalfa mixtures that were harvested as haylage or greenchop (green weight). Alfalfa harvested as dry hay is not included.
NASS, Crops Branch, (202) 720–2127.

Table 6-12.—Hay: Supply and disappearance, prices, and number of animal units fed annually, United States, 2005–2014[1]

Year beginning May	Farm carryover May 1	Production	Total supply	Disappearance	Roughage-consuming animal units	Supply per animal unit	Disappearance per animal unit	Price received per ton
	Million tons	Million tons	Million tons	Million tons	Million tons	Tons	Tons	Dollars
2005	27.8	150.5	178.2	156.9	71.6	2.49	2.19	98.20
2006	21.3	140.8	162.1	147.1	71.8	2.26	2.05	110.00
2007	15.0	146.9	161.9	140.3	71.5	2.26	1.96	127.00
2008	21.6	146.3	167.9	145.8	71.0	2.36	2.05	152.00
2009	22.1	147.7	169.8	148.8	70.5	2.41	2.11	108.00
2010	20.9	145.0	144.3	144.3	69.4	2.40	2.08	114.00
2011	22.2	129.9	132.1	132.1	68.0	2.26	1.94	178.00
2012	21.4	117.1	127.1	127.1	67.6	1.88	1.88	191.00
2013	14.2	135.0	149.2	130.0	66.8	2.23	1.95	176.00
2014	19.2	139.9	159.1	134.6	67.9	2.34	1.98	173.00

[1] Excludes trade.
ERS, Market and Trade Economics Division, (202) 694-5313.

Table 6-13.—Field seeds: Average retail price paid by farmers for seed, Mar. 15, United States, 2005–2014[1]

Kind of seed	2005	2006	2007	2008	2009	2010	2011	2012	2013	2014
				Price per 100 pounds						
	Dollars	Dollars	Dollars	Dollars	Dollars	Dollars	Dollars	Dollars	Dollars	Dollars
Alfalfa:										
Public and common	177.00	181.00	201.00	246.00	262.00	273.00	284.00	308.00	316.00	326.00
Proprietary	281.00	286.00	292.00	342.00	379.00	379.00	411.00	442.00	456.00	469.00
Clover, ladino	280.00	306.00	316.00	344.00	394.00	349.00	321.00	337.00	348.00	356.00
Clover, red	174.00	177.00	202.00	241.00	289.00	213.00	200.00	218.00	226.00	231.00
Lespedeza:										
Korean	79.30	87.00	126.00	184.00	127.00	105.00	123.00	248.00	263.00	277.00
Striate, Kobe	83.10	89.50	113.00	263.00	198.00	141.00	128.00	280.00	298.00	317.00
Sericea	220.00	181.00	(4)	(4)	(4)	(4)	(4)	(4)	(4)	(4)
Timothy	105.00	106.00	112.00	133.00	149.00	135.00	146.00	191.00	193.00	198.00
Orchardgrass	137.00	158.00	189.00	321.00	329.00	185.00	170.00	187.00	197.00	199.00
Bluegrass, Kentucky:										
Public and common	180.00	161.00	175.00	227.00	269.00	224.00	189.00	223.00	230.00	236.00
Proprietary, including Merion	235.00	224.00	232.00	251.00	326.00	284.00	259.00	260.00	266.00	270.00
Rye grass, annual	59.30	69.60	71.80	78.80	78.90	70.30	73.40	95.20	99.90	101.00
Tall fescue	100.00	124.00	146.00	158.00	143.00	99.00	102.00	146.00	147.00	150.00
Sudangrass	57.40	50.20	56.70	62.10	72.50	73.60	75.70	100.00	103.00	105.00
Potatoes	9.30	11.80	12.00	13.10	15.60	12.90	14.40	15.90	14.80	15.53
Peanuts	56.40	57.70	60.70	83.10	76.50	75.40	91.00	95.40	74.50	77.60
Sunflower	476.00	520.00	616.00	718.00	749.00	729.00	897.00	931.00	940.00	948.00
Cottonseed, all	309.00	356.00	408.00	455.00	521.00	570.00	684.00	702.00	735.00	734.00
Biotech[2]	390.00	443.00	500.00	525.00	609.00	648.00	702.00	754.00	739.00	789.00
Non-biotech	110.00	118.00	97.10	88.50	113.00	130.00	242.00	249.00	323.00	331.00
Grain sorghum, hybrid	114.00	116.00	120.00	142.00	161.00	165.00	162.00	203.00	207.00	212.00
Rice	20.80	27.30	30.20	33.50	48.60	48.90	50.60	60.30	68.10
				Price per bushel						
	Dollars	Dollars	Dollars	Dollars	Dollars	Dollars	Dollars	Dollars	Dollars	Dollars
Corn, hybrid, all[3]	111.00	118.00	133.00	165.00	217.00	229.00	237.00	258.00	274.00	283.00
Biotech[2,3]	131.00	137.00	154.00	184.00	235.00	247.00	249.00	268.00	285.00	293.00
Non-biotech[3]	93.40	95.10	100.00	115.00	139.00	152.00	163.00	182.00	197.00	197.00
Wheat, spring	7.30	7.60	8.40	20.50	11.80	10.00	14.20	15.20	15.60	16.20
Wheat, winter	9.06	9.32	10.60	14.80	16.00	13.70	15.40	17.50	18.00	18.40
Oats, spring	5.54	5.83	6.81	8.19	8.19	7.78	8.34	10.30	10.50	10.70
Barley, spring	6.72	6.58	7.18	10.10	9.78	8.52	10.30	13.10	13.90	13.90
Soybeans for seed, all	27.60	28.90	34.80	38.80	48.30	51.90	49.70	54.90	57.80	58.40
Biotech[2]	34.60	34.10	36.70	40.00	49.60	53.50	51.00	55.90	59.10	59.80
Non-biotech	19.10	21.10	20.50	26.30	33.70	33.90	33.50	38.80	40.70	40.70
Flaxseed	14.40	8.80	9.73	19.80	13.80	13.80	19.10	17.50	19.00	19.50

[1] Beginning in 2009 program changed from April 15 to March 15. [2] Biotech varieties are made to be resistant to herbicides, insects, or both. A technology fee is included within the price. [3] Price per 80,000 kernels. [4] Estimate discontinued in 2007.
NASS, Environmental, Economics, and Demographics Branch, (202) 720–6146.

Table 6-14.—Beans, dry edible (clean basis): Production, by classes, United States, 2013–2015 [1]

Class	2013	2014	2015
	1,000 cwt	*1,000 cwt*	*1,000 cwt*
Navy (pea beans)	3,402	4,273	4,559
Great northern	1,515	2,385	915
Small white	([2])	42	115
Pinto	8,486	9,889	9,551
Red kidney, light	845	1,165	1,224
Red kidney, dark	838	1,161	1,574
Pink	476	366	340
Small red	548	708	1,106
Cranberry	53	91	147
Black	2,545	3,835	5,595
Large lima (CA)	189	190	257
Baby lima (CA)	178	299	223
Blackeye	639	476	565
Small chickpeas (Garbanzo)	813	903	908
Large chickpeas (Garbanzo)	2,746	1,903	1,615
Chickpeas, all (Garbanzo)	3,559	2,806	2,523
Other	1,303	1,224	1,427
Total	24,576	28,910	30,121

[1] Excludes beans grown for garden seed. [2] Data are included in "Other" class to avoid disclosing data for individual operations.
NASS, Crops Branch, (202) 720–2127.

Table 6-15.—Beans, dry edible: Area, yield, and production, by State and United States, 2013–2015 [1]

State	Area planted			Area harvested		
	2013	2014	2015	2013	2014	2015
	1,000 acres	*1,000 acres*	*1,000 acres*	*1,000 acres*	*1,000 acres*	*1,000 acres*
Arizona	10.0	11.0	9.1	10.0	10.9	9.1
California	50.0	48.0	45.0	49.5	47.5	44.5
Colorado	39.0	46.0	50.0	36.0	44.0	46.5
Idaho	125.0	125.0	120.0	124.0	124.0	119.0
Kansas	5.0	7.5	8.0	4.8	6.9	7.8
Michigan	175.0	230.0	275.0	172.0	226.0	272.0
Minnesota	125.0	155.0	190.0	120.0	148.0	182.0
Montana	24.0	37.5	49.0	23.6	37.0	47.3
Nebraska	130.0	170.0	140.0	117.0	156.5	131.0
New Mexico	10.0	10.5	12.9	9.5	10.5	12.9
New York	9.0	8.0	8.0	8.8	7.7	7.8
North Dakota	440.0	630.0	655.0	430.0	615.0	635.0
Oregon	8.3	8.5	9.0	8.2	8.5	9.0
South Dakota	12.0	14.0	12.5	11.5	12.9	11.6
Texas	33.0	23.0	31.0	30.0	21.0	28.0
Washington	120.0	127.7	110.0	119.0	126.7	109.0
Wisconsin	5.4	7.9	7.9	5.4	7.9	7.9
Wyoming	39.0	42.0	32.0	37.0	37.6	31.0
United States	1,359.7	1,701.6	1,764.4	1,316.3	1,648.6	1,711.4

State	Yield per acre (clean basis)			Production (clean basis)		
	2013	2014	2015	2013	2014	2015
	Pounds	Pounds	Pounds	*1,000 cwt*	*1,000 cwt*	*1,000 cwt*
Arizona	1,680	1,940	2,070	168	211	188
California	2,320	2,230	2,310	1,150	1,061	1,029
Colorado	1,500	1,900	1,820	540	835	846
Idaho	1,900	1,800	1,800	2,356	2,229	2,141
Kansas	1,790	1,710	2,500	86	118	195
Michigan	1,900	1,940	2,030	3,270	4,375	5,533
Minnesota	1,950	1,950	2,140	2,340	2,886	3,896
Montana	1,920	1,630	1,340	453	603	634
Nebraska	2,350	2,500	2,380	2,750	3,916	3,117
New Mexico	2,040	1,900	2,050	194	200	264
New York	1,820	1,490	1,510	160	115	118
North Dakota	1,650	1,430	1,400	7,095	8,795	8,901
Oregon	2,260	2,250	2,300	185	191	207
South Dakota	2,000	1,880	1,770	230	243	205
Texas	1,220	1,220	1,400	366	256	392
Washington	1,820	1,480	1,450	2,165	1,881	1,582
Wisconsin	1,810	2,480	2,030	98	196	160
Wyoming	2,620	2,130	2,300	970	799	713
United States	1,867	1,754	1,760	24,576	28,910	30,121

[1] Excludes beans grown for garden seed.
NASS, Crops Branch, (202) 720–2127.

Table 6-16.—Beans, dry edible: Area, yield, production, price, and value, United States, 2006–2015 [1]

Year	Area planted	Area harvested	Yield per acre [2]	Production [2]	Marketing year average price per 100 pounds received by farmers	Value of production
	1,000 acres	*1,000 acres*	*Pounds*	*1,000 cwt.*	*Dollars*	*1,000 dollars*
2006	1,622.8	1,531.6	1,577	24,155	22.10	554,154
2007	1,527.4	1,479.2	1,730	25,586	28.80	748,680
2008	1,495.0	1,445.2	1,768	25,558	34.60	910,200
2009	1,540.0	1,464.0	1,737	25,427	30.00	790,250
2010	1,911.4	1,842.7	1,726	31,801	28.00	899,258
2011	1,217.9	1,167.9	1,703	19,890	42.10	851,131
2012	1,742.5	1,690.4	1,889	31,925	38.00	1,235,255
2013	1,359.7	1,316.3	1,867	24,576	39.10	982,385
2014	1,701.6	1,648.6	1,754	28,910	32.30	980,622
2015	1,764.4	1,711.4	1,760	30,121	28.70	883,276

[1] Excludes beans grown for garden seed.   [2] Clean basis.
NASS, Crops Branch, (202) 720–2127.

Table 6-17.—Beans, dry edible (clean basis): Marketing year average price and value of production, by State and United States, 2013–2015 [1]

State	Marketing year average price per cwt.			Value of production		
	2013	2014	2015	2013	2014	2015
	Dollars	*Dollars*	*Dollars*	*1,000 dollars*	*1,000 dollars*	*1,000 dollars*
Arizona	42.20	58.10	51.50	7,090	12,259	9,682
California	56.80	63.50	73.40	65,320	67,374	75,529
Colorado	39.70	25.90	30.20	21,438	21,627	25,549
Idaho	34.30	33.00	32.70	80,811	73,557	70,011
Kansas	36.90	(D)	20.00	3,173	(D)	3,900
Michigan	43.10	35.30	28.90	140,937	154,438	159,904
Minnesota	50.30	43.10	31.80	117,702	124,387	123,893
Montana	30.10	20.80	35.10	13,635	12,542	22,253
Nebraska	43.80	30.70	20.00	120,450	120,221	62,340
New Mexico	60.00	60.00	29.00	11,640	12,000	7,656
New York	48.70	(D)	32.40	7,792	(D)	3,823
North Dakota	35.30	28.20	23.90	250,454	248,019	212,734
Oregon	38.30	40.10	35.50	7,086	7,659	7,349
South Dakota	36.00	33.20	21.10	8,280	8,068	4,326
Texas	46.00	46.00	32.60	16,836	11,776	12,779
Washington	33.10	32.40	31.80	71,662	60,944	50,308
Wisconsin	55.00	54.00	58.00	5,390	10,584	9,280
Wyoming	33.70	33.50	30.80	32,689	26,767	21,960
Other States	(NA)	36.10	(NA)	(NA)	8,400	(NA)
United States	39.10	32.30	28.70	982,385	980,622	883,276

(D) Withheld to avoid disclosing data for individual operations. (NA) Not Available. [1] Excludes beans grown for garden seed.
NASS, Crops Branch, (202) 720–2127.

Table 6-18.—Beans, dry edible: Season average wholesale (dealer) price per 100 pounds, selected markets, 2005/2006–2014/2015

Year beginning September	F.o.b. California points			F.o.b. Northern Colorado points: Pinto	F.o.b. Western Nebraska points: Great northern	F.o.b. Southern Idaho points: Small red	F.o.b. Michigan points:		
	Baby lima	Large lima	Blackeye				Pea bean (Navy)	Black	Light red kidney
	Dollars	Dollars	Dollars	Dollars	Dollars	Dollars	Dollars	Dollars	Dollars
2005/2006	38.28	47.62	44.26	23.47	24.32	27.33	24.44	29.54	27.74
2006/2007	47.25	66.26	47.38	29.52	31.61	30.95	29.07	30.94	35.33
2007/2008	42.39	65.64	41.51	38.66	47.65	44.52	44.48	43.47	53.71
2008/2009	58.87	73.39	51.97	40.85	54.00	52.71	36.63	46.70	57.67
2009/2010	43.98	71.59	45.12	39.45	41.57	43.32	40.37	46.04	43.56
2010/2011	40.13	59.53	42.47	36.08	41.81	47.74	41.94	43.05	48.25
2011/2012	57.88	60.50	(NA)	62.72	54.75	61.25	58.61	52.09	69.30
2012/2013	(NA)	60.00	54.75	45.54	53.96	53.36	45.57	46.36	63.32
2013/2014	(NA)	(NA)	(NA)	45.65	(NA)	52.13	55.25	52.50	68.88
2014/2015	56.50	(NA)	70.00	32.59	48.60	55.20	38.71	45.45	64.85

(NA) Not available.
ERS, Specialty Crops Branch, (202) 694–5253. Compiled from the Bean Market Summary, Agricultural Marketing Service, U.S. Department of Agriculture, Greeley, Colorado.

Table 6-19.—Beans, dry edible: United States exports to specified countries, 2010–2012 [1]

Country	2010	2011	2012
	Metric tons	Metric tons	Metric tons
Mexico	134,205.1	110,246.7	214,730.2
Canada	40,737.0	48,625.7	49,751.0
United Kingdom	44,718.0	35,440.9	48,622.7
Taiwan	35,089.6	39,271.9	31,076.8
Dominican Republic	28,585.9	34,208.4	29,179.8
Italy(*)	9,464.5	19,094.4	14,031.6
Japan	16,035.5	14,083.7	12,322.0
France(*)	6,836.9	5,652.8	10,042.1
New Zealand(*)	328.2	8,104.8	9,034.0
Yemen(*)	2,130.0	205.3	6,100.3
Guatemala	8,364.1	6,902.4	5,765.7
Angola	7,975.9	7,969.5	5,382.2
Australia(*)	3,216.2	3,712.3	4,506.8
Haiti	16,919.4	4,366.4	3,846.6
Belgium-Luxembourg(*)	4,865.8	2,258.5	3,560.6
Mozambique	0.0	0.0	2,912.0
Jamaica	1,667.7	2,971.5	2,774.4
Algeria	2,322.6	1,642.6	2,739.1
Turkey	4,554.0	125.7	2,550.1
Malaysia	1,447.6	2,368.4	2,526.3
Russia	124.5	478.5	2,170.3
Netherlands	2,328.5	827.2	2,035.7
Malawi	1,248.1	948.2	1,865.9
French West Indies(*)	1,799.5	1,988.2	1,605.5
Cote d'Ivoire	0.0	0.0	1,520.0
Africa, not elsewhere specified(*)	1,867.6	1,482.3	1,414.2
Korea, South	2,315.2	1,443.3	1,399.6
Hong Kong	758.4	904.4	1,137.9
South Africa	162.2	97.5	1,067.4
Spain	1,855.6	556.2	992.7
Rest of World	30,389.1	31,549.6	12,360.8
World Total	412,312.4	387,527.2	489,024.3

(*) Denotes a country that is a summarization of its component countries. [1] Excluding seed bean exports.
FAS, Office of Global Analysis, (202) 720-6301. Compiled from U.S. Census data.

Table 6-20.—Chickpeas & lentils, dried: United States exports by class and quantity, 2010–2012 [1]

Country	2010	2011	2012
	Metric tons	*Metric tons*	*Metric tons*
Dried chickpeas:			
Spain	14,846.60	15,304.00	19,137.00
India	6,596.40	6,350.60	8,457.60
Turkey	968.00	1,282.50	7,391.70
Canada	3,695.10	7,502.20	4,484.30
Pakistan	2,213.80	3,180.70	3,664.00
Portugal	0.00	362.00	3,447.70
Italy(*)	3,237.50	6,896.10	2,895.60
Peru	1,035.40	2,085.10	2,882.60
Algeria	1,857.60	1,703.10	2,617.70
Lebanon	1,545.90	1,842.00	2,144.50
Israel(*)	433.00	792.40	2,022.10
Japan	242.90	1,126.80	1,697.10
New Zealand(*)	351.10	697.90	1,635.00
Philippines	252.20	438.50	1,610.70
United Arab Emirates	105.30	180.30	1,034.50
United Kingdom	513.30	172.90	843.50
Malaysia	217.40	221.30	808.90
France(*)	0.00	0.00	717.90
Colombia	1,178.90	3,410.90	635.80
Belgium-Luxembourg(*)	0.00	408.50	468.30
Sri Lanka	0.00	0.00	408.00
Taiwan	222.80	348.40	318.30
Trinidad and Tobago	108.70	315.30	301.90
Indonesia	0.00	88.70	261.40
Tunisia	0.00	372.80	231.90
Thailand	0.00	62.00	229.60
Russia	0.00	172.00	222.60
Vietnam	0.00	139.20	191.70
Greece	0.00	21.80	185.50
Venezuela	0.00	2.80	185.50
Rest of World	1,916.40	2,294.80	1,302.60
World Total	41,538.40	57,776.00	72,435.20
Dried lentils:			
India	40,164.00	12,011.80	31,212.80
Spain	26,173.30	20,647.90	25,411.30
Peru	10,403.50	14,725.90	10,084.60
Sudan(*)	10,051.80	22,444.20	9,539.10
Germany(*)	2,441.40	3,990.70	9,247.50
Ethiopia(*)	4,857.60	368.90	8,131.00
Canada	14,676.50	13,573.00	7,542.70
Italy(*)	3,698.80	10,038.70	7,321.00
Mexico	5,093.10	3,207.10	7,244.10
Belgium-Luxembourg(*)	1,328.80	3,904.10	3,737.30
Turkey	10,467.50	2,252.80	3,411.00
Djibouti	0	0	3,334.00
Somalia	0	0	3,255.40
Syria	0	0	2,800.00
Cameroon	0	0	2,440.00
Bangladesh	127.80	543.10	2,071.20
Israel(*)	2	51.90	1,875.30
Greece	3,117.90	1,887.30	1,874.90
Jordan	0	136.60	1,872.30
French Pacific Islands(*)	1,329.60	1,158.20	1,698.90
Netherlands	1,000.50	1,995.80	1,417.00
Sierra Leone	144.60	180.00	1,185.70
Chad	0	0	1,107.30
Pakistan	6,117.50	2,211.30	1,048.00
Burkina Faso	224.40	188.40	935.90
Egypt	3,244.90	2,193.80	752.40
Sri Lanka	8,867.60	1,391.70	731.90
Mali	0	0	609.10
Haiti	4,662.40	1,859.10	578.00
Algeria	4,198.50	1,019.00	574.80
Rest of World	27,430.70	16,616.10	5,186.60
World Total	189,824.60	138,597.20	158,231.00

(*) Denotes a country that is a summarization of its component countries. [1] Excluding seed pea exports.
FAS, Office of Global Analysis, (202) 720-6301.

**Table 6-21.—Peas, dry: United States exports to specified countries,
2010–2012[1]**

Country	2010	2011	2012
	Metric tons	*Metric tons*	*Metric tons*
India	166,003	65,038	127,120
China	56,869	12,433	26,301
Kenya	22,008	32,575	26,169
Philippines	24,488	16,324	17,079
Pakistan	43,658	10,197	14,874
Peru	9,637	10,524	12,945
Ethiopia(*)	27,131	28,774	11,741
Canada	27,962	13,432	11,431
Djibouti	8,133	13,004	8,055
South Africa	12,645	3,431	7,712
Korea, South	3,935	4,881	5,829
Mexico	6,125	3,759	4,584
Mozambique	87	749	4,289
Indonesia	8,708	7,572	3,867
Colombia	6,176	2,970	3,684
Taiwan	6,152	6,696	3,602
Belgium-Luxembourg(*)	186	70	3,508
Somalia	0	3,109	2,948
Chad	0	5,209	2,833
Mali	488	550	2,655
Sri Lanka	3,282	6,545	2,393
Spain	1,001	438	2,365
Zimbabwe	1,338	2,306	2,063
Cameroon	4,625	3,860	1,970
Congo (Kinshasa)	0	2,088	1,956
Uganda	220	3,506	1,631
Togo	0	1,248	1,610
Russia	1,943	1,363	1,487
Tanzania	4,076	1,731	1,360
Bangladesh	4,424	2,152	1,359
Rest of World	45,693	38,376	12,977
World Total	496,994	304,908	332,398

(*) Denotes a country that is a summarization of its component countries. [1] Excluding seed pea exports.
FAS, Office of Global Analysis, (202) 720-6301.

Table 6-22.—Hops: Area, yield, production, price, value, and Sept. 1 stocks, United States, 2006–2015

Year	Area harvested	Yield per acre	Production	Marketing year average price per pound received by farmers	Value of production	Stocks Sept. 1	Stocks Mar. 1
	1,000 acres	*Pounds*	*1,000 pounds*	*Dollars per pound*	*1,000 dollars*	*1,000 pounds*	*1,000 pounds*
2006	29.4	1,964	57,672	2.05	118,008	49,000	69,000
2007	30.9	1,949	60,253	2.99	179,978	47,000	70,000
2008	40.9	1,971	80,630	4.03	325,092	47,000	66,000
2009	39.7	2,383	94,678	3.57	337,874	65,000	82,000
2010	31.3	2,093	65,493	3.28	214,589	83,000	102,000
2011	29.8	2,175	64,782	3.14	203,378	87,000	121,000
2012	29.7	1,985	58,911	3.17	186,876	96,000	120,000
2013	35.3	1,962	69,246	3.35	232,308	(NA)	115,000
2014	38.0	1,868	70,996	3.67	260,627	90,000	121,000
2015	43.6	1,807	78,846	4.38	345,388	83,000	119,000

(NA) Not available.
NASS, Crops Branch, (202) 720–2127.

Table 6-23.—Hops: Area, yield, and production, by State and United States, 2013–2015

State	Area harvested			Yield per acre			Production		
	2013	2014	2015	2013	2014	2015	2013	2014	2015
	1,000 acres	*1,000 acres*	*1,000 acres*	*Pounds*	*Pounds*	*Pounds*	*1,000 pounds*	*1,000 pounds*	*1,000 pounds*
Idaho	3,356	3,743	4,863	1,740	1,847	1,794	5,837.9	6,913.8	8,724.9
Oregon	4,835	5,410	6,612	1,764	1,520	1,613	8,530.5	8,221.0	10,667.8
Washington	27,097	28,858	32,158	2,025	1,936	1,849	54,877.7	55,861.1	59,453.3
United States ...	35,288	38,011	43,633	1,962	1,868	1,807	69,246.1	70,995.9	78,846.0

NASS, Crops Branch, (202) 720–2127.

Table 6-24.—Hops: Marketing year average price and value of production, by State and United States, 2013–2015

State	Marketing year average price per pound			Value of production		
	2013	2014	2015	2013	2014	2015
	Dollars	*Dollars*	*Dollars*	*1,000 dollars*	*1,000 dollars*	*1,000 dollars*
Idaho	2.62	2.75	3.53	15,295	18,806	30,799
Oregon	3.76	4.07	3.24	32,075	33,459	34,564
Washington	3.37	3.73	4.71	184,938	208,362	280,025
United States ...	3.35	3.67	4.38	232,308	260,627	345,388

NASS, Crops Branch, (202) 720–2127.

Table 6-25.—Hops: United States exports by country of destination and imports by country of origin, 2010–2012

Country	2010	2011	2012
	Metric tons	Metric tons	Metric tons
United Kingdom	1,344	1,206	1,561
Mexico	1,192	1,156	1,258
Germany(*)	1,328	1,001	1,200
Brazil	1,147	1,031	1,047
Belgium-Luxembourg(*)	1,331	1,086	939
Colombia	803	781	804
Canada	1,613	760	742
Japan	755	663	525
China	522	426	334
Argentina	249	239	330
Venezuela	135	135	297
Hong Kong	214	272	271
Australia(*)	222	214	203
Korea, South	154	195	176
South Africa	97	86	162
Peru	222	269	143
Philippines	277	179	116
Chile	69	119	111
Ecuador	307	134	97
Netherlands	41	44	97
Dominican Republic	71	106	85
Bolivia	118	23	79
India	208	178	58
Guatemala	57	38	50
Russia	169	70	49
Nigeria	64	55	44
Vietnam	184	112	44
Tanzania	48	33	38
Ireland	2	60	38
Thailand	66	93	34
Indonesia	8	21	33
Panama	60	32	31
Costa Rica	26	12	26
Kenya	52	9	26
Uruguay	0	21	23
New Zealand(*)	8	9	22
Malaysia	74	36	18
El Salvador	26	17	16
Cambodia	8	7	15
Singapore	5	18	14
Rest of World	388	314	119
World Total	13,661	11,257	11,274

(*) Denotes a country that is a summarization of its component countries.
FAS, Office of Global Analysis, (202) 720-6301.

CHAPTER VII

STATISTICS OF CATTLE, HOGS, AND SHEEP

This chapter contains information about most kinds of farm livestock and livestock products, with the exception of dairy and poultry. The information relates to inventories, production, disposition, prices, and income for farm animals, and to livestock slaughter, meat production, and market statistics for meat animals.

Table 7-1.—All cattle and calves: Operations, inventory, and value, United States, Jan. 1, 2006–2015

Year	Inventory	Value	
		Per head	Total
	Thousands	Dollars	1,000 dollars
2006	96,342	1,009	97,230,415
2007	96,573	922	89,063,310
2008	96,035	990	95,112,820
2009	94,721	872	82,628,170
2010	94,081	833	78,335,410
2011	92,887	947	87,988,337
2012	91,160	1,111	101,252,540
2013	90,095	1,139	101,731,742
2014	88,526	1,223	108,255,950
2015	89,143	1,584	141,211,070

NASS, Livestock Branch, (202) 720–3570.istiller

Table 7-2.—All cattle and calves: Number by class, United States, Jan. 1, 2006–2015

Year	All cattle and calves [1]	Cows and heifers that have calved		500 pounds and over					Calves under 500 pounds
		Beef cows	Milk cows	Heifers			Steers	Bulls	
				Beef cow replacements	Milk cow replacements	Other			
	Thousands	Thousands	Thousands	Thousands	Thousands	Thousands	Thousands	Thousands	Thousands
2006	96,342	32,703	9,104	5,864	4,298	9,788	16,988	2,258	15,339
2007	96,573	32,644	9,145	5,835	4,325	9,914	17,185	2,214	15,311
2008	96,035	32,435	9,257	5,647	4,415	9,793	17,163	2,207	15,118
2009	94,721	31,794	9,333	5,550	4,410	9,644	16,809	2,188	14,994
2010	94,081	31,440	9,087	5,443	4,551	9,784	16,568	2,190	15,019
2011	92,887	30,913	9,156	5,135	4,577	9,938	16,394	2,165	14,610
2012	91,160	30,282	9,236	5,281	4,618	9,546	15,957	2,100	14,141
2013	90,095	29,631	9,221	5,429	4,546	9,281	15,931	2,074	13,983
2014	88,526	29,085	9,208	5,551	4,549	8,869	15,668	2,038	13,558
2015	89,143	29,302	9,307	6,086	4,710	8,464	15,630	2,109	13,534

[1] Totals may not add due to rounding.
NASS, Livestock Branch, (202) 720–3570.

Table 7-3.—All cattle and calves: Inventory and value, by State and United States, Jan. 1, 2014–2015

State	Inventory		Value			
	2014	2015	Value per head		Total value	
			2014	2015[1]	2014	2015[1]
	Thousands	Thousands	Dollars	Dollars	1,000 dollars	1,000 dollars
Alabama	1,270.0	1,200.0	1,000	1,380	1,270,000	1,656,000
Alaska	10.0	10.0	1,120	1,220	11,200	12,200
Arizona	920.0	880.0	1,150	1,470	1,058,000	1,293,600
Arkansas	1,650.0	1,630.0	1,110	1,510	1,831,500	2,461,300
California	5,250.0	5,100.0	1,140	1,520	5,985,000	7,752,000
Colorado	2,550.0	2,550.0	1,310	1,670	3,340,500	4,258,500
Connecticut	47.0	47.0	1,110	1,430	52,170	67,210
Delaware	16.0	17.0	1,200	1,330	19,200	22,610
Florida	1,670.0	1,680.0	1,090	1,530	1,820,300	2,570,400
Georgia	1,040.0	1,030.0	1,050	1,380	1,092,000	1,421,400
Hawaii	133.0	133.0	980	1,070	130,340	142,310
Idaho	2,240.0	2,280.0	1,240	1,650	2,777,600	3,762,000
Illinois	1,130.0	1,120.0	1,210	1,590	1,367,300	1,780,800
Indiana	860.0	870.0	1,110	1,430	954,600	1,244,100
Iowa	3,800.0	3,850.0	1,250	1,530	4,750,000	5,890,500
Kansas	5,800.0	5,900.0	1,250	1,490	7,250,000	8,791,000
Kentucky	2,110.0	2,050.0	1,060	1,400	2,236,600	2,870,000
Louisiana	790.0	790.0	1,170	1,530	924,300	1,208,700
Maine	85.0	85.0	1,110	1,300	94,350	110,500
Maryland	182.0	185.0	1,160	1,460	211,120	270,100
Massachusetts	39.0	38.0	1,050	1,300	40,950	49,400
Michigan	1,130.0	1,120.0	1,220	1,570	1,378,600	1,758,400
Minnesota	2,300.0	2,320.0	1,170	1,470	2,691,000	3,410,400
Mississippi	930.0	900.0	1,000	1,360	930,000	1,224,000
Missouri	3,850.0	4,000.0	1,190	1,570	4,581,500	6,280,000
Montana	2,550.0	2,500.0	1,530	2,040	3,901,500	5,100,000
Nebraska	6,250.0	6,250.0	1,350	1,600	8,437,500	10,000,000
Nevada	460.0	430.0	1,250	1,740	575,000	748,200
New Hampshire	32.0	30.0	1,150	1,390	36,800	41,700
New Jersey	27.0	28.0	1,260	1,380	34,020	38,640
New Mexico	1,310.0	1,330.0	1,260	1,690	1,650,600	2,247,700
New York	1,450.0	1,440.0	1,160	1,490	1,682,000	2,145,600
North Carolina	810.0	800.0	950	1,240	769,500	992,000
North Dakota	1,750.0	1,640.0	1,640	2,060	2,870,000	3,378,400
Ohio	1,250.0	1,240.0	1,180	1,410	1,475,000	1,748,400
Oklahoma	4,300.0	4,550.0	1,180	1,590	5,074,000	7,234,500
Oregon	1,280.0	1,300.0	1,210	1,650	1,548,800	2,145,000
Pennsylvania	1,610.0	1,530.0	1,180	1,500	1,899,800	2,295,000
Rhode Island	5.0	5.0	1,040	1,380	5,200	6,900
South Carolina	335.0	335.0	1,030	1,340	345,050	448,900
South Dakota	3,700.0	3,700.0	1,510	2,000	5,587,000	7,400,000
Tennessee	1,760.0	1,720.0	1,060	1,420	1,865,600	2,442,400
Texas	11,100.0	11,700.0	1,150	1,530	12,765,000	17,901,000
Utah	810.0	780.0	1,350	1,750	1,093,500	1,365,000
Vermont	260.0	260.0	1,210	1,510	314,600	392,600
Virginia	1,510.0	1,470.0	1,070	1,400	1,615,700	2,058,000
Washington	1,110.0	1,150.0	1,180	1,530	1,309,800	1,759,500
West Virginia	385.0	370.0	1,150	1,460	442,750	540,200
Wisconsin	3,400.0	3,500.0	1,240	1,630	4,216,000	5,705,000
Wyoming	1,270.0	1,300.0	1,530	2,130	1,943,100	2,769,000
United States	88,526.0	89,143.0	1,223	1,584	108,255,950	141,211,070

[1] Preliminary.
NASS, Livestock Branch, (202) 720–3570.

Table 7-4.—Cattle and calves, Jan. 1: Number by class, State and United States, 2014–2015

| State | Cows and heifers that have calved | | | | Heifers, 500 pounds and over | | | | | |
| | Beef cows | | Milk cows | | Beef cow replacements | | Milk cow replacements | | Other | |
	2014	2015	2014	2015	2014	2015	2014	2015	2014	2015
	1,000 head	*1,000 head*	*1,000 head*	*1,000 head*	*1,000 head*	*1,000 head*	*1,000 head*	*1,000 head*	*1,000 head*	*1,000 head*
Alabama	681.0	652.0	9.0	8.0	110.0	112.0	4.0	3.0	31.0	30.0
Alaska	4.3	4.3	0.3	0.3	0.8	0.9	0.1	0.1	0.1	0.1
Arizona	178.0	175.0	192.0	195.0	30.0	34.0	74.0	65.0	21.0	21.0
Arkansas	862.0	863.0	8.0	7.0	137.0	150.0	5.0	4.0	78.0	51.0
California	600.0	590.0	1,780.0	1,780.0	110.0	130.0	750.0	770.0	230.0	140.0
Colorado	710.0	725.0	140.0	145.0	150.0	170.0	100.0	100.0	490.0	470.0
Connecticut	4.0	5.0	19.0	19.0	1.5	2.5	9.0	8.0	0.5	0.5
Delaware	2.8	2.5	4.7	5.0	0.5	0.7	2.6	2.5	0.6	0.7
Florida	907.0	906.0	123.0	124.0	115.0	130.0	35.0	35.0	30.0	30.0
Georgia	500.0	479.0	80.0	81.0	82.0	90.0	25.0	27.0	29.0	30.0
Hawaii	71.8	68.8	2.2	2.2	9.0	11.0	1.0	1.0	4.0	5.0
Idaho	465.0	461.0	565.0	579.0	110.0	120.0	270.0	320.0	190.0	170.0
Illinois	355.0	366.0	95.0	94.0	63.0	64.0	45.0	52.0	102.0	99.0
Indiana	187.0	199.0	178.0	181.0	39.0	50.0	68.0	80.0	53.0	44.0
Iowa	895.0	900.0	205.0	210.0	160.0	180.0	120.0	130.0	640.0	630.0
Kansas	1,414.0	1,427.0	136.0	143.0	240.0	290.0	100.0	90.0	1,400.0	1,330.0
Kentucky	992.0	997.0	68.0	63.0	150.0	140.0	45.0	45.0	125.0	105.0
Louisiana	450.0	466.0	15.0	14.0	84.0	74.0	5.0	5.0	21.0	23.0
Maine	11.0	11.0	30.0	30.0	3.0	4.5	17.0	16.0	2.0	2.5
Maryland	38.0	42.0	50.0	49.0	9.5	9.0	25.0	25.0	7.5	9.0
Massachusetts	6.0	5.5	12.0	12.5	2.0	2.0	7.5	7.0	0.5	1.0
Michigan	119.0	107.0	381.0	403.0	29.0	23.0	164.0	167.0	47.0	35.0
Minnesota	340.0	340.0	460.0	460.0	80.0	90.0	280.0	280.0	170.0	160.0
Mississippi	477.0	468.0	13.0	12.0	91.0	97.0	6.0	6.0	43.0	33.0
Missouri	1,820.0	1,851.0	90.0	89.0	305.0	345.0	50.0	60.0	215.0	225.0
Montana	1,476.0	1,496.0	14.0	14.0	430.0	435.0	9.0	7.0	211.0	198.0
Nebraska	1,807.0	1,756.0	53.0	54.0	400.0	420.0	20.0	20.0	1,280.0	1,280.0
Nevada	231.0	212.0	29.0	28.0	36.0	37.0	9.0	9.0	31.0	29.0
New Hampshire	3.0	3.0	13.5	14.0	1.0	1.0	7.0	5.5	0.5	0.5
New Jersey	7.0	7.5	7.0	7.0	1.0	1.3	3.0	3.0	1.0	0.9
New Mexico	407.0	407.0	323.0	323.0	70.0	85.0	120.0	110.0	70.0	70.0
New York	105.0	105.0	615.0	615.0	45.0	41.0	355.0	350.0	40.0	44.0
North Carolina	355.0	363.0	45.0	47.0	72.0	70.0	20.0	18.0	24.0	27.0
North Dakota	923.0	894.0	17.0	16.0	170.0	169.0	10.0	6.0	210.0	185.0
Ohio	293.0	282.0	267.0	268.0	55.0	50.0	130.0	125.0	60.0	50.0
Oklahoma	1,795.0	1,880.0	45.0	40.0	325.0	420.0	20.0	25.0	340.0	365.0
Oregon	516.0	525.0	124.0	125.0	105.0	110.0	60.0	60.0	120.0	105.0
Pennsylvania	160.0	150.0	530.0	530.0	50.0	55.0	315.0	305.0	70.0	55.0
Rhode Island	1.5	1.5	0.9	0.9	0.5	0.5	0.5	0.5	0.2	0.2
South Carolina	169.0	170.0	16.0	15.0	30.0	30.0	6.0	5.0	12.0	12.0
South Dakota	1,635.0	1,611.0	95.0	99.0	340.0	405.0	50.0	65.0	490.0	490.0
Tennessee	864.0	873.0	46.0	47.0	130.0	140.0	30.0	25.0	85.0	65.0
Texas	3,910.0	4,130.0	440.0	470.0	660.0	750.0	220.0	250.0	1,420.0	1,390.0
Utah	340.0	324.0	95.0	96.0	70.0	78.0	46.0	48.0	69.0	64.0
Vermont	12.0	12.0	132.0	132.0	4.5	4.0	56.0	56.0	5.5	5.0
Virginia	637.0	637.0	93.0	93.0	115.0	110.0	40.0	43.0	70.0	42.0
Washington	214.0	198.0	266.0	277.0	50.0	54.0	125.0	136.0	130.0	135.0
West Virginia	191.0	185.0	9.0	9.0	35.0	33.0	5.0	4.0	24.0	20.0
Wisconsin	250.0	275.0	1,270.0	1,275.0	70.0	75.0	680.0	730.0	70.0	45.0
Wyoming	694.0	694.0	6.0	6.0	175.0	193.0	4.0	5.0	106.0	142.0
United States	29,085.4	29,302.1	9,207.6	9,306.9	5,551.3	6,086.4	4,548.7	4,710.4	8,869.4	8,464.4

See footnote(s) at end of table.

**Table 7-4.—Cattle and calves, Jan. 1: Number by class,
State and United States, 2014–2015—Continued**

State	Steers, 500 pounds and over		Bulls, 500 pounds and over		Calves under 500 pounds	
	2014	2015	2014	2015	2014	2015
	Thousands	*Thousands*	*Thousands*	*Thousands*	*Thousands*	*Thousands*
Alabama	60.0	50.0	50.0	45.0	325.0	300.0
Alaska	0.4	0.3	1.9	2.4	2.1	1.6
Arizona	305.0	275.0	20.0	20.0	100.0	95.0
Arkansas	145.0	130.0	55.0	55.0	360.0	370.0
California	640.0	550.0	70.0	70.0	1,070.0	1,070.0
Colorado	780.0	780.0	45.0	55.0	135.0	105.0
Connecticut	2.0	2.0	0.5	0.5	10.5	9.5
Delaware	2.0	2.2	0.3	0.4	2.5	3.0
Florida	25.0	25.0	60.0	60.0	375.0	370.0
Georgia	36.0	45.0	28.0	28.0	260.0	250.0
Hawaii	7.0	9.0	4.0	4.0	34.0	32.0
Idaho	260.0	280.0	40.0	40.0	340.0	310.0
Illinois	250.0	210.0	25.0	25.0	195.0	210.0
Indiana	108.0	116.0	17.0	17.0	210.0	183.0
Iowa	1,270.0	1,290.0	60.0	60.0	450.0	450.0
Kansas	1,770.0	1,890.0	90.0	95.0	650.0	635.0
Kentucky	200.0	200.0	70.0	65.0	460.0	435.0
Louisiana	24.0	24.0	30.0	30.0	161.0	154.0
Maine	5.0	4.5	1.5	1.5	15.5	15.0
Maryland	15.0	15.0	4.0	4.0	33.0	32.0
Massachusetts	2.0	2.0	1.0	1.0	8.0	7.0
Michigan	175.0	170.0	15.0	15.0	200.0	200.0
Minnesota	490.0	500.0	35.0	35.0	445.0	455.0
Mississippi	62.0	56.0	38.0	38.0	200.0	190.0
Missouri	380.0	400.0	100.0	110.0	890.0	920.0
Montana	220.0	175.0	100.0	100.0	90.0	75.0
Nebraska	2,300.0	2,330.0	95.0	95.0	295.0	295.0
Nevada	44.0	44.0	13.0	12.0	67.0	59.0
New Hampshire	1.0	1.0	0.5	0.5	5.5	4.5
New Jersey	2.0	2.0	1.0	1.0	5.0	4.5
New Mexico	90.0	95.0	35.0	35.0	195.0	205.0
New York	32.0	35.0	18.0	15.0	240.0	235.0
North Carolina	44.0	36.0	30.0	29.0	220.0	210.0
North Dakota	270.0	215.0	55.0	50.0	95.0	105.0
Ohio	195.0	195.0	25.0	25.0	225.0	245.0
Oklahoma	910.0	870.0	125.0	140.0	740.0	810.0
Oregon	160.0	165.0	36.0	40.0	159.0	170.0
Pennsylvania	185.0	145.0	25.0	25.0	275.0	265.0
Rhode Island	0.5	0.5	0.1	0.1	0.8	0.8
South Carolina	12.0	9.0	14.0	14.0	76.0	80.0
South Dakota	690.0	680.0	95.0	100.0	305.0	250.0
Tennessee	135.0	120.0	60.0	60.0	410.0	390.0
Texas	2,370.0	2,520.0	280.0	320.0	1,800.0	1,870.0
Utah	85.0	78.0	23.0	22.0	82.0	70.0
Vermont	4.0	4.0	3.0	3.0	43.0	44.0
Virginia	170.0	155.0	40.0	40.0	345.0	350.0
Washington	160.0	175.0	19.0	18.0	146.0	157.0
West Virginia	45.0	44.0	14.0	13.0	62.0	62.0
Wisconsin	380.0	375.0	30.0	35.0	650.0	690.0
Wyoming	150.0	135.0	40.0	40.0	95.0	85.0
United States	15,667.9	15,629.5	2,037.8	2,109.4	13,557.9	13,533.9

NASS, Livestock Branch, (202) 720–3570.

Table 7-5.—Cows and calf crop: Cows and heifers that have calved, Jan. 1, 2014–2015, and calves born, by State and United States, 2013–2014

State	Cows and heifers that have calved Jan. 1		Calves born	
	2014	2015	2013	2014
	Thousands	Thousands	Thousands	Thousands
Alabama	690.0	660.0	600.0	560.0
Alaska	4.6	4.6	3.3	3.3
Arizona	370.0	370.0	280.0	290.0
Arkansas	870.0	870.0	750.0	730.0
California	2,380.0	2,370.0	2,010.0	1,930.0
Colorado	850.0	870.0	740.0	750.0
Connecticut	23.0	24.0	18.5	19.0
Delaware	7.5	7.5	5.0	6.0
Florida	1,030.0	1,030.0	830.0	820.0
Georgia	580.0	560.0	480.0	480.0
Hawaii	74.0	71.0	60.0	55.0
Idaho	1,030.0	1,040.0	940.0	940.0
Illinois	450.0	460.0	385.0	400.0
Indiana	365.0	380.0	325.0	325.0
Iowa	1,100.0	1,110.0	1,020.0	1,040.0
Kansas	1,550.0	1,570.0	1,280.0	1,280.0
Kentucky	1,060.0	1,060.0	950.0	940.0
Louisiana	465.0	480.0	370.0	355.0
Maine	41.0	41.0	32.0	32.0
Maryland	88.0	91.0	67.0	70.0
Massachusetts	18.0	18.0	18.0	17.0
Michigan	500.0	510.0	395.0	385.0
Minnesota	800.0	800.0	780.0	770.0
Mississippi	490.0	480.0	400.0	375.0
Missouri	1,910.0	1,940.0	1,730.0	1,730.0
Montana	1,490.0	1,510.0	1,470.0	1,470.0
Nebraska	1,860.0	1,810.0	1,680.0	1,580.0
Nevada	260.0	240.0	210.0	190.0
New Hampshire	16.5	17.0	13.5	13.5
New Jersey	14.0	14.5	9.5	9.0
New Mexico	730.0	730.0	560.0	550.0
New York	720.0	720.0	540.0	500.0
North Carolina	400.0	410.0	355.0	365.0
North Dakota	940.0	910.0	890.0	840.0
Ohio	560.0	550.0	480.0	460.0
Oklahoma	1,840.0	1,920.0	1,670.0	1,760.0
Oregon	640.0	650.0	620.0	630.0
Pennsylvania	690.0	680.0	570.0	520.0
Rhode Island	2.4	2.4	2.2	2.2
South Carolina	185.0	185.0	150.0	155.0
South Dakota	1,730.0	1,710.0	1,680.0	1,610.0
Tennessee	910.0	920.0	830.0	830.0
Texas	4,350.0	4,600.0	3,800.0	3,950.0
Utah	435.0	420.0	380.0	385.0
Vermont	144.0	144.0	116.0	105.0
Virginia	730.0	730.0	640.0	650.0
Washington	480.0	475.0	425.0	405.0
West Virginia	200.0	194.0	180.0	180.0
Wisconsin	1,520.0	1,550.0	1,350.0	1,400.0
Wyoming	700.0	700.0	640.0	660.0
United States	38,293.0	38,609.0	34,730.0	33,522.0

NASS, Livestock Branch, (202) 720–3570.

Table 7-6.—Cattle and calves: All cattle on feed, United States, Jan. 1, 2006–2015 [1]

Year	Inventory
	Thousands
2006	14,391.9
2007	14,646.7
2008	14,826.7
2009	13,899.7
2010	13,727.6
2011	14,131.5
2012	14,355.3
2013	13,703.3
2014	12,949.3
2015	13,025.0

[1] Cattle and calves on feed are steers and heifers for slaughter market being fed a ration of grain or other concentrates and are expected to produce a carcass that will grade select or better.
NASS, Livestock Branch, (202) 720–3570.

Table 7-7.—Cattle and calves: Total number on feed by State and United States, Jan. 1, 2014–2015

State	2014	2015
	1,000 head	1,000 head
Arizona	274.0	252.0
California	510.0	425.0
Colorado	960.0	930.0
Idaho	235.0	245.0
Illinois	260.0	230.0
Indiana	110.0	100.0
Iowa	1,230.0	1,220.0
Kansas	2,130.0	2,180.0
Kentucky	18.0	15.0
Maryland	10.0	10.0
Michigan	150.0	160.0
Minnesota	375.0	385.0
Missouri	75.0	70.0
Montana	54.0	40.0
Nebraska	2,420.0	2,530.0
Nevada	4.0	4.0
New York	23.0	26.0
North Dakota	38.0	44.0
Ohio	160.0	170.0
Oklahoma	265.0	265.0
Oregon	75.0	85.0
Pennsylvania	110.0	90.0
South Dakota	375.0	385.0
Tennessee	3.0	....
Texas	2,450.0	2,510.0
Utah	26.0	24.0
Virginia	20.0	20.0
Washington	195.0	210.0
West Virginia	5.0	4.0
Wisconsin	260.0	260.0
Wyoming	75.0	75.0
Other States [1]	54.3	61.0
United States	12,949.3	13,025.0

[1] Individual state estimates not available for states not shown, but are included in Other States.
NASS, Livestock Branch, (202) 720–3570.

Table 7-8.—Cattle: Average price per 100 pounds, by grades, at Nebraska Direct, Worthing, SD and Louisville, KY, 2005–2015

Year	Nebraska		Worthing, SD [1]	Louisville, KY		
	Choice		Cows [2]	Cows		
	Steers [3]	Heifers [4]	Commercial	Breaking utility	Breaking utility	85-95% Lean
	Dollars	Dollars	Dollars	Dollars	Dollars	Dollars
2005 65-80%	86.54	87.35	61.89	57.82	....	....
2006 65-80%	85.55	86.58	58.72	54.19	....	....
2007 65-80%	91.87	91.86	60.91	56.85	47.02	45.22
2008	93.07	93.04	63.15	61.32	52.80	42.99
2009	82.70	82.71	53.94	54.32	46.61	37.95
2010	95.81	95.41	NA	61.49	53.98	41.70
2011	111.52	111.55	74.25	71.34	67.20	53.92
2012	123.11	124.18	83.95	78.90	76.55	62.71
2013	127.13	127.37	84.83	78.52	79.22	71.00
2014	154.03	154.67	108.79	(NA)	(NA)	83.56
2015	148.71	148.09	100.39	(NA)	(NA)	78.27

(NA) Not available.   [1] 2004-2007 Sioux Falls, SD; 2008 to present Worthing, SD.   [2] All weights; simple average of price range.   [3] 1,100 to 1,500 pound weight range; weighted average of price range.   [4] 1,000 to 1,300 pound weight range; simple average of price range.
AMS, Livestock and Grain Market News, (202) 720–7316.

Table 7-9.—Cattle and calves: Production, disposition, cash receipts, and gross income, United States, 2006–2015

| Year | Calf crop [1] | Death loss | | Marketings [2] | | Inshipments [3] | Farm slaughter |
		Cattle	Calves	Cattle	Calves		Cattle and calves
	1,000 head	1,000 head	1,000 head	1,000 head	1,000 head	1,000 head	1,000 head
2006	37,016	1,818	2,348	44,789	8,856	21,213	187
2007	36,759	1,856	2,394	45,008	8,956	21,104	188
2008	36,158	1,760	2,314	44,238	8,752	19,769	176
2009	35,939	1,741	2,323	43,369	8,530	19,550	167
2010	35,740	1,736	2,265	44,658	8,874	20,757	157
2011	35,357	1,787	2,230	44,875	8,401	20,357	148
2012	34,469	1,752	2,171	43,578	8,337	20,443	139
2013	33,730	1,730	2,140	43,259	8,007	19,965	128
2014	33,522	1,720	2,130	41,526	7,849	20,443	124
2015	34,087	1,736	2,144	40,605	7,454	20,599	115

| Year | Production (live weight) [4] | Value of production | Cash receipts [5] | Value of home consumption | Gross income [6] | Average price per 100 pounds received by farmers | |
						Cattle	Calves
	1,000 pounds	1,000 dollars	1,000 dollars	1,000 dollars	1,000 dollars	Dollars	Dollars
2006	41,824,568	35,490,732	49,110,330	447,857	49,558,187	87.20	133.00
2007	41,437,021	35,973,068	49,843,322	441,051	50,284,373	89.90	119.00
2008	41,586,134	35,597,898	48,394,131	404,745	48,798,876	89.10	110.00
2009	41,033,359	31,897,074	43,720,324	368,285	44,088,609	80.30	105.00
2010	41,379,839	36,850,474	51,246,136	405,789	51,651,925	92.20	117.00
2011	41,483,945	45,059,200	62,321,217	478,871	62,800,088	113.00	142.00
2012	40,920,479	48,059,904	66,090,126	525,630	66,615,756	121.00	172.00
2013	40,695,232	48,478,687	67,457,442	509,175	67,966,617	125.00	181.00
2014	40,171,339	59,921,694	81,478,368	610,809	82,089,177	152.00	261.00
2015	41,456,124	59,857,570	78,228,639	587,013	78,815,652	147.00	247.00

[1] Calves born during the year. [2] Includes custom slaughter for use on farms where produced and State outshipments, but excludes interfarm sales within the State. [3] Includes cattle shipped in from other States, but excludes cattle for immediate slaughter. [4] Adjustments made for changes in inventory and for inshipments. [5] Receipts from marketings and sale of farm slaughter. [6] Cash receipts from sales of cattle, calves, beef, and veal plus value of cattle and calves slaughtered for home consumption.
NASS, Livestock Branch, (202) 720–3570.

Table 7-10.—Cattle: Weighted average weight and price per 100 pounds, Texas-Oklahoma, Kansas, Colorado, Nebraska, Iowa-So. Minnesota Feedlots, 2005–2015 [1]

| Year | Steers SE/CH 65-80% | | | Steers SE/CH 35-65% | | |
	Price	Average Weight	Number of Head	Price	Average Weight	Number of Head
	Dollars	Pounds		Dollars	Pounds	
2005 ...	86.28	1,336	326,751	87.66	1,265	2,492,108
2006 ...	85.51	1,358	317,732	85.75	1,282	2,319,896
2007 ...	92.21	1,366	359,511	92.10	1,289	2,081,144
2008 ...	93.37	1,360	344,633	93.16	1,294	1,757,049
2009 ...	82.81	1,376	476,823	83.59	1,314	1,474,925
2010 ...	95.00	1,359	561,173	95.95	1,301	1,329,021
2011 ...	114.45	1,364	518,689	114.92	1,304	1,101,808
2012 ...	122.84	1,402	458,382	123.06	1,312	743,859
2013 ...	126.27	1,408	384,642	126.35	1,338	517,782
2014 ...	154.25	1,410	548,667	152.15	1,339	386,051
2015 ...	147.64	1,432	314,078	149.00	1,366	228,126

| Year | Heifers SE/CH 65-80% | | | Heifers SE/CH 35-65% | | |
	Price	Average Weight	Number of Head	Price	Average Weight	Number of Head
	Dollars	Pounds		Dollars	Pounds	
2005 ...	87.23	1,219	313,240	87.90	1,145	1,901,730
2006 ...	86.57	1,245	247,902	86.06	1,165	1,781,077
2007 ...	92.01	1,241	357,880	93.09	1,165	1,582,183
2008 ...	92.95	1,229	342,547	93.22	1,170	1,317,211
2009 ...	82.88	1,252	498,481	83.63	1,180	1,136,234
2010 ...	95.41	1,227	562,788	96.22	1,166	1,159,191
2011 ...	114.05	1,222	508,487	114.64	1,167	928,633
2012 ...	123.38	1,255	412,483	123.31	1,173	577,535
2013 ...	126.09	1,266	320,285	126.36	1,198	335,205
2014 ...	154.37	1,279	322,272	152.18	1,204	218,658
2015 ...	148.20	1,289	220,073	147.73	1,235	97,078

[1] Sales FOB the feedyard. Estimated net weights after 3-4% shrink.
AMS, Livestock and Grain Market News, (202) 720–7316.

Table 7-11.—Cattle: Receipts at selected markets, 2005–2015

Year	Oklahoma City	Fort Collins [1]	Amarillo	South St. Joseph	Sioux Falls
	Thousands	Thousands	Thousands	Thousands	Thousands
2005	491	77	87	97	198
2006	471	85	69	114	179
2007	422	82	65	52	115
2008	465	80	43	84	90
2009	501	67	56	96	43
2010	492	63	74	105	153
2011	468	30	60	94	104
2012	399	72	43	121	153
2013	409	46	35	90	156
2014	382	(NA)	37	86	140
2015	368	(NA)	21	88	139

[1] Switch to Fort Collins market 2005.
AMS, Livestock & Grain Market News, (202) 720–7316.

Table 7-12.—Cattle and calves: Number slaughtered, United States, 2006–2015

Year	Cattle slaughter					Calf slaughter				
	Commercial			Farm	Total	Commercial			Farm	Total
	Federally inspected	Other	Total [1]			Federally inspected	Other	Total [1]		
	Thousands	Thou-sands	Thousands	Thou-sands	Thousands	Thousands	Thou-sands	Thou-sands	Thou-sands	Thou-sands
2006 ...	33,145	553	33,698	150	33,849	699	13	711	37	748
2007 ...	33,721	543	34,264	150	34,414	745	13	758	37	795
2008 ...	33,805	560	34,365	140	34,505	942	15	957	36	993
2009 ...	32,765	573	33,338	132	33,470	930	14	944	35	979
2010 ...	33,702	547	34,249	124	34,373	864	14	879	33	912
2011 ...	33,555	532	34,087	119	34,205	839	14	853	30	882
2012 ...	32,426	525	32,951	112	33,062	760	12	772	27	800
2013 ...	31,947	515	32,462	103	32,565	751	11	762	26	788
2014 ...	29,682	486	30,168	98	30,266	558	8	566	26	591
2015 ...	28,296	455	28,752	92	28,843	446	7	453	24	476

[1] Totals are based on unrounded numbers.
NASS, Livestock Branch, (202) 720–3570.

Table 7-13.—Cattle and calves: Number slaughtered under Federal inspection, and average live weight, 2006–2015

Year	Cattle		Calves	
	Number slaughtered	Average live weight [1]	Number slaughtered	Average live weight [1]
	Thousands	Pounds	Thousands	Pounds
2006	33,145	1,277	699	344
2007	33,721	1,275	745	304
2008	33,805	1,284	942	255
2009	32,765	1,296	930	248
2010	33,702	1,282	864	260
2011	33,555	1,280	839	260
2012	32,426	1,305	760	257
2013	31,947	1,317	751	248
2014	29,682	1,333	558	281
2015	28,296	1,364	446	309

[1] Excludes postmortem condemnations.
NASS, Livestock Branch, (202) 720–3570.

Table 7-14.—Cattle and calves: Number slaughtered commercially, total and average live weight, by State and United States, 2015[1]

State	Cattle			Calves		
	Number slaughtered	Total live weight[2]	Average live weight[2]	Number slaughtered	Total live weight[2]	Average live weight[2]
	Thousands	1,000 pounds	Pounds	Thousands	1,000 pounds	Pounds
Alabama	5.0	5,232	1,044	(Y)	(Y)	(Y)
Alaska	0.6	597	1,085	(Y)	(Y)	(Y)
Arizona	544.7	750,861	1,382	(Y)	(Y)	(Y)
Arkansas	4.5	4,616	1,021	(D)	(D)	(D)
California	1,186.2	1,591,923	1,354	96.8	9,363	98
Colorado	2,351.0	3,264,610	1,390	(D)	(D)	(D)
Delaware-Maryland	36.0	47,948	1,332	(D)	(D)	(D)
Florida	143.9	178,638	1,249	0.4	216	498
Georgia	(D)	(D)	(D)	0.8	409	544
Hawaii	9.6	10,896	1,139	(D)	(D)	(D)
Idaho	34.8	40,489	1,233	(D)	(D)	(D)
Illinois	(D)	(D)	(D)	4.8	2,057	434
Indiana	34.8	37,455	1,075	4.6	1,286	281
Iowa	(D)	(D)	(D)	(Y)	(Y)	(Y)
Kansas	5,701.3	7,759,958	1,361	0.1	52	428
Kentucky	13.9	14,493	1,046	0.3	175	502
Louisiana	5.7	5,262	917	0.4	189	487
Michigan	489.3	677,276	1,393	1.5	430	299
Minnesota	537.4	796,178	1,489	(D)	(D)	(D)
Mississippi	2.4	2,009	846	(D)	(D)	(D)
Missouri	41.0	46,161	1,127	0.3	103	398
Montana	19.8	22,529	1,139	(D)	(D)	(D)
Nebraska	6,580.3	9,394,238	1,429	(Y)	(Y)	(Y)
Nevada	1.4	1,462	1,032	(D)	(D)	(D)
New England[3]	22.5	24,988	1,113	3.1	514	166
New Jersey	32.5	35,766	1,107	59.6	25,736	432
New Mexico	3.5	3,796	1,081	(Y)	(Y)	(Y)
New York	33.6	38,651	1,162	37.1	5,800	159
North Carolina	60.1	72,413	1,212	0.5	213	445
North Dakota	9.1	11,493	1,267	(D)	(D)	(D)
Ohio	83.6	96,417	1,164	0.2	106	456
Oklahoma	26.4	26,891	1,019	0.2	106	456
Oregon	62.7	77,708	1,253	(D)	(D)	(D)
Pennsylvania	954.1	1,155,985	1,216	85.9	37,351	435
South Carolina	169.5	205,963	1,225	(D)	(D)	(D)
South Dakota	(D)	(D)	(D)	(D)	(D)	(D)
Tennessee	54.8	50,343	943	0.2	113	483
Texas	5,064.8	6,578,030	1,302	3.9	2,079	534
Utah	543.3	743,073	1,370	(D)	(D)	(D)
Virginia	11.1	11,891	1,069	0.4	185	475
Washington	1,007.9	1,387,265	1,380	9.9	697	71
West Virginia	7.9	8,074	1,020	(D)	(D)	(D)
Wisconsin	1,149.1	1,586,034	1,389	63.9	30,218	474
Wyoming	5.7	6,291	1,106	(D)	(D)	(D)
United States	28,751.6	39,030,183	1,360	452.6	139,370	310

(Y) Less than level of precision shown.   (D) Withheld to avoid disclosing data for individual operations.   [1] Includes slaughter in federally inspected and other slaughter plants; excludes animals slaughtered on farms.   Average live weight is based on unrounded numbers.   Totals may not add due to rounding.   [2] Excludes postmortem condemnations.   [3] New England includes Connecticut, Maine, Massachusetts, New Hampshire, Rhode Island, and Vermont.
NASS, Livestock Branch, (202) 720-3570.

Table 7-15.—Cattle and calves: Production, disposition, cash receipts, and gross income, by State and United States, 2015

State	Marketings [1]		Inship-ments [2]	Farm slaugh-ter [3]	Production (live weight) [4]	Value of production	Cash receipts [5]	Value of home con-sump-tion	Gross income [6]
	Cattle	Calves							
	1,000 head	*1,000 head*	*1,000 head*	*1,000 head*	*1,000 pounds*	*1,000 dollars*	*1,000 dollars*	*1,000 dollars*	*1,000 dollars*
Alabama	416.0	71.0	7.0	2.0	483,702	603,038	558,383	3,983	562,366
Alaska	1.0	0.4	0.1	0.2	1,934	3,092	1,935	253	2,188
Arizona	507.0	100.0	375.0	1.0	438,765	644,700	891,738	2,313	894,051
Arkansas	565.0	213.0	160.0	1.0	518,619	635,468	699,631	7,387	707,018
California	1,932.0	491.0	779.0	6.0	1,930,783	2,473,512	3,395,175	11,471	3,406,646
Colorado	2,080.0	102.0	1,648.0	1.0	1,686,559	2,473,508	4,040,475	12,623	4,053,098
Connecticut	9.0	8.1	2.0	0.4	15,341	24,175	20,036	1,436	21,472
Delaware	4.6	1.9	0.7	0.2	3,761	5,825	8,336	370	8,706
Florida	236.0	550.5	63.0	1.5	429,580	786,127	859,164	4,045	863,209
Georgia	212.0	196.0	22.0	2.0	349,632	530,716	474,723	6,028	480,751
Hawaii	16.3	31.0	-	0.2	37,785	68,251	59,826	786	60,612
Idaho	1,055.0	138.0	470.0	4.0	1,170,650	1,688,405	1,951,005	14,490	1,965,495
Illinois	314.0	81.0	95.0	6.0	428,180	684,767	683,355	21,266	704,621
Indiana	258.0	112.0	100.0	4.0	250,807	363,481	428,440	18,429	446,869
Iowa	2,138.5	104.5	1,460.0	2.0	1,904,150	2,828,728	4,414,172	21,664	4,435,836
Kansas	4,398.0	5.0	3,630.0	2.0	3,854,080	4,711,927	8,839,391	18,360	8,857,751
Kentucky	446.0	373.0	86.0	3.0	596,259	980,781	927,210	15,104	942,314
Louisana	131.0	197.0	2.0	1.0	193,043	352,325	360,861	2,158	363,019
Maine	16.9	16.1	3.0	0.2	15,251	24,806	33,205	1,766	34,971
Maryland	45.9	22.1	7.0	0.2	69,725	103,094	97,146	2,085	99,231
Massachusetts	7.5	8.1	1.0	0.4	9,720	15,186	13,870	1,462	15,332
Michigan	318.0	41.0	59.0	3.0	442,890	616,092	624,664	19,331	643,995
Minnesota	952.0	109.0	525.0	4.0	1,199,982	1,678,792	2,043,960	31,068	2,075,028
Mississippi	211.0	112.0	12.0	2.0	206,572	288,222	258,371	6,394	264,765
Missouri	822.0	781.0	129.0	6.0	1,212,920	2,051,951	2,073,096	29,168	2,102,264
Montana	1,239.0	150.5	160.0	2.5	1,119,577	1,659,653	1,639,423	12,119	1,651,542
Nebraska	5,936.0	83.0	4,760.0	1.0	5,351,112	7,405,289	12,551,966	14,471	12,566,437
Nevada	186.0	87.0	95.0	0.5	184,430	286,085	374,858	2,310	377,168
New Hampshire	5.5	3.0	1.5	0.1	9,301	14,727	9,292	1,369	10,661
New Jersey	4.6	5.1	0.9	-	6,958	11,948	11,548	877	12,425
New Mexico	667.5	96.0	325.0	1.5	531,680	797,506	987,573	3,833	991,406
New York	183.1	280.7	25.0	2.2	260,313	410,986	376,928	14,653	391,581
North Carolina	207.0	130.5	6.0	1.5	283,776	404,242	409,746	4,942	414,688
North Dakota ...	753.0	98.6	135.0	1.4	694,606	1,047,838	1,102,292	8,867	1,111,159
Ohio	444.0	25.0	54.0	4.0	516,150	713,071	740,853	28,378	769,231
Oklahoma	2,121.5	320.0	1,200.0	3.5	2,077,888	3,224,841	3,884,210	27,847	3,912,057
Oregon	495.0	197.0	150.0	7.0	596,150	893,603	985,233	19,320	1,004,553
Pennsylvania ...	356.0	210.3	129.0	4.7	460,546	708,866	741,105	25,739	766,844
Rhode Island ...	1.1	1.0	0.2	-	1,241	1,850	1,784	330	2,114
South Carolina	121.5	11.0	8.0	2.0	155,102	180,041	178,599	3,981	182,580
South Dakota ...	1,469.0	420.0	620.0	1.0	1,482,523	2,390,934	2,632,480	15,574	2,648,054
Tennessee	426.0	274.0	26.0	3.0	551,914	783,440	703,710	13,861	717,571
Texas	5,760.0	252.5	2,500.0	7.5	6,244,675	9,157,228	11,363,762	32,056	11,395,818
Utah	445.5	36.0	177.0	1.5	319,895	454,749	642,075	9,066	651,141
Vermont	37.0	63.3	3.4	1.1	55,010	92,141	88,725	4,845	93,570
Virginia	354.5	233.5	11.0	3.0	473,581	682,752	676,037	9,388	685,425
Washington	534.0	7.0	165.0	5.0	629,440	858,169	1,025,112	15,103	1,040,215
West Virginia ...	101.0	66.5	19.0	1.5	136,673	220,334	210,960	7,674	218,634
Wisconsin	900.0	401.2	110.0	5.8	1,307,017	1,924,332	1,968,104	49,110	2,017,214
Wyoming	764.0	136.5	280.0	0.5	555,876	895,976	1,164,096	7,860	1,171,956
United States ...	40,604.5	7,453.9	20,596.8	115.1	41,456,124	59,857,570	78,228,639	587,013	78,815,652

- Represents zero. [1] Includes custom slaughter for use on farms where produced and State outshipments, but excludes interfarm sales within the State. [2] Includes cattle shipped in from other states, but excludes cattle for immediate slaughter. [3] Excludes custom slaughter for farmers at commercial establishments. [4] Adjustments made for changes in inventory and for inshipments. [5] Includes receipts from marketings and sales of farm-slaughter. [6] Includes cash receipts from sales of cattle, calves, beef, and veal plus value of cattle and calves slaughtered for home consumption.
NASS, Livestock Branch, (202) 720–3570.

Table 7-16.—Cattle and calves: Inventory Jan 1, 2014–2015, and number of operations, 2012, by State and United States [1]

State	January 1 Cattle inventory		Operations with cattle [2]
	2014	2015	2012
	1,000 head	*1,000 head*	*Number*
Alabama	1,270.0	1,200.0	21,149
Alaska	10.0	10.0	134
Arizona	920.0	880.0	6,029
Arkansas	1,650.0	1,630.0	25,866
California	5,250.0	5,100.0	16,764
Colorado	2,550.0	2,550.0	13,970
Connecticut	47.0	47.0	1,452
Delaware	16.0	17.0	431
Florida	1,670.0	1,680.0	21,255
Georgia	1,040.0	1,030.0	17,188
Hawaii	133.0	133.0	1,314
Idaho	2,240.0	2,280.0	10,957
Illinois	1,130.0	1,120.0	16,550
Indiana	860.0	870.0	17,370
Iowa	3,800.0	3,850.0	26,827
Kansas	5,800.0	5,900.0	27,568
Kentucky	2,110.0	2,050.0	40,141
Louisana	790.0	790.0	13,687
Maine	85.0	85.0	2,423
Maryland	182.0	185.0	3,499
Massachusetts	39.0	38.0	1,423
Michigan	1,130.0	1,120.0	13,626
Minnesota	2,300.0	2,320.0	23,702
Mississippi	930.0	900.0	15,940
Missouri	3,850.0	4,000.0	53,401
Montana	2,550.0	2,500.0	11,845
Nebraska	6,250.0	6,250.0	23,280
Nevada	460.0	430.0	1,822
New Hampshire	32.0	30.0	1,091
New Jersey	27.0	28.0	1,224
New Mexico	1,310.0	1,330.0	12,796
New York	1,450.0	1,440.0	13,559
North Carolina	810.0	800.0	19,548
North Dakota	1,750.0	1,640.0	9,868
Ohio	1,250.0	1,240.0	25,501
Oklahoma	4,300.0	4,550.0	51,043
Oregon	1,280.0	1,300.0	14,351
Pennsylvania	1,610.0	1,530.0	25,189
Rhode Island	5.0	5.0	300
South Carolina	335.0	335.0	8,121
South Dakota	3,700.0	3,700.0	15,583
Tennessee	1,760.0	1,720.0	38,826
Texas	11,100.0	11,700.0	151,362
Utah	810.0	780.0	8,625
Vermont	260.0	260.0	2,784
Virginia	1,510.0	1,470.0	23,911
Washington	1,110.0	1,150.0	11,861
West Virginia	385.0	370.0	12,067
Wisconsin	3,400.0	3,500.0	29,908
Wyoming	1,270.0	1,300.0	6,115
United States	88,526.0	89,143.0	913,246

[1] An operation is any place having one or more head of cattle on hand on December 31.   [2] State level estimates only available in conjunction with the Census of Agriculture every 5 years.
NASS, Livestock Branch, (202) 720–3570.

Table 7-17.—Cattle: Number slaughtered under Federal inspection and percentage distribution, by classes, 2006–2015 [1]

Year	Number						Percentage of total					
	Steers	Heifers	Cows			Bulls	Steers	Heifers	Cows			Bulls
			Dairy cows	Other cows	Total cows				Dairy cows	Other cows	Total cows	
	Thousands	Thousands	Thousands	Thousands	Thousands	Thousands	Percent	Percent	Percent	Percent	Percent	Percent
2006 ...	17,478	9,820	2,354	2,983	5,336	511	52.7	29.6	7.1	9.0	16.1	1.5
2007 ...	17,285	10,207	2,497	3,178	5,675	554	51.3	30.3	7.4	9.4	16.8	1.6
2008 ...	16,949	10,091	2,591	3,569	6,161	605	50.1	29.9	7.7	10.6	18.2	1.8
2009 ...	16,312	9,743	2,815	3,325	6,140	570	49.8	29.7	8.6	10.1	18.7	1.7
2010 ...	16,596	10,047	2,807	3,630	6,437	622	49.2	29.8	8.3	10.8	19.1	1.8
2011 ...	16,539	9,726	2,914	3,798	6,712	579	49.3	29.0	8.7	11.3	20.0	1.7
2012 ...	16,160	9,269	3,102	3,344	6,446	551	49.8	28.6	9.6	10.3	19.9	1.7
2013 ...	16,003	9,132	3,125	3,130	6,255	557	50.1	28.6	9.8	9.8	19.6	1.7
2014 ...	15,378	8,377	2,816	2,565	5,380	548	51.8	28.2	9.5	8.6	18.1	1.8
2015 ...	15,331	7,351	2,915	2,236	5,151	462	54.2	26.0	10.3	7.9	18.2	1.6

[1] Totals and percentages based on unrounded data and may not equal sum of classes due to rounding.
NASS, Livestock Branch, (202) 720-3570.

Table 7-18.—Cattle and calves: Average dressed weight under Federal inspection, 2006–2015 [1]

Year	Cattle					Calves
	All cattle	Steers	Heifers	Cows	Bulls	
	Pounds	Pounds	Pounds	Pounds	Pounds	Pounds
2006	781	833	767	622	914	207
2007	776	830	764	617	893	182
2008	778	838	772	609	888	150
2009	784	847	782	610	878	147
2010	773	835	768	607	875	154
2011	773	841	773	596	868	154
2012	790	859	792	608	877	152
2013	796	863	794	620	880	146
2014	808	872	800	627	898	166
2015	829	892	818	644	915	183

[1] Excludes postmortem condemnations.
NASS, Livestock Branch, (202) 720-3570.

Table 7-19.—Beef cows: Inventory Jan 1, 2014 and 2015, and number of operations, 2012, by State and United States [1]

State	January 1 beef cow inventory		Operations with beef cows [2]
	2014	2015	2012
	1,000 head	*1,000 head*	*Number*
Alabama	681	652	19,685
Alaska	4.3	4.3	98
Arizona	178	175	4,851
Arkansas	862	863	23,385
California	600	590	10,925
Colorado	710	725	11,267
Connecticut	4	5	1,003
Delaware	2.8	2.5	296
Florida	907	906	18,433
Georgia	500	479	15,175
Hawaii	71.8	68.8	1,173
Idaho	465	461	˙8,336
Illinois	355	366	12,646
Indiana	187	199	11,218
Iowa	895	900	19,677
Kansas	1,414	1,427	23,272
Kentucky	992	997	33,823
Louisiana	450	466	12,115
Maine	11	11	1,354
Maryland	38	42	2,403
Massachusetts	6	5.5	849
Michigan	119	107	8,016
Minnesota	340	340	13,547
Mississippi	477	468	14,644
Missouri	1,820	1,851	46,161
Montana	1,476	1,496	10,598
Nebraska	1,807	1,756	19,313
Nevada	231	212	1,508
New Hampshire	3	3	683
New Jersey	7	7.5	871
New Mexico	407	407	11,004
New York	105	105	6,579
North Carolina	355	363	16,059
North Dakota	923	894	8,964
Ohio	293	282	16,922
Oklahoma	1,795	1,880	44,106
Oregon	516	525	11,557
Pennsylvania	160	150	11,880
Rhode Island	1.5	1.5	212
South Carolina	169	170	7,232
South Dakota	1,635	1,611	13,327
Tennessee	864	873	33,556
Texas	3,910	4,130	133,924
Utah	340	324	6,827
Vermont	12	12	1,295
Virginia	637	637	19,596
Washington	214	198	9,285
West Virginia	191	185	10,156
Wisconsin	250	275	13,020
Wyoming	694	694	5,080
United States	29,085.4	29,302.1	727,906

[1] An operation is any place having one or more beef cows on hand on December 31.   [2] State level estimates only available in conjunction with the Census of Agriculture every 5 years.
NASS, Livestock Branch, (202) 720–3570.

Table 7-20.—Hogs and pigs: Inventory and value, Dec. 1, 2014–2015, and number of operations, 2012, by State and United States

State	Inventory		Value				Operations [1]
	2014	2015	Value per head		Total value		2012
			2014	2015	2014	2015	
	Thousands	*Thousands*	*Dollars*	*Dollars*	*1,000 dollars*	*1,000 dollars*	*Number*
Alabama	110.0	100.00	150	100	16,500	10,000	689
Alaska	1.2	1.40	205	195	246	273	37
Arizona	139.0	132.00	150	100	20,850	13,200	509
Arkansas	115.0	166.00	165	90	18,975	14,940	752
California	110.0	95.00	185	125	20,350	11,875	1,437
Colorado	700.0	700.00	140	99	98,000	69,300	1,001
Connecticut	2.6	2.50	185	125	481	313	318
Delaware	4.0	3.00	150	100	600	300	59
Florida	17.0	16.00	150	100	2,550	1,600	1,642
Georgia	155.0	160.00	120	82	18,600	13,120	866
Hawaii	9.0	9.00	170	150	1,530	1,350	231
Idaho	(D)	(D)	(D)	(D)	(D)	(D)	680
Illinois	4,700.0	5,100.00	140	99	658,000	504,900	2,045
Indiana	3,700.0	3,850.00	145	98	536,500	377,300	2,757
Iowa	21,300.0	20,900.00	150	98	3,195,000	2,048,200	6,266
Kansas	1,840.0	1,940.00	125	87	230,000	168,780	1,010
Kentucky	340.0	425.00	105	79	35,700	33,575	1,284
Louisiana	8.0	8.00	150	100	1,200	800	658
Maine	4.5	4.50	150	100	675	450	752
Maryland	21.0	21.00	150	100	3,150	2,100	333
Massachusetts	9.0	11.00	150	100	1,350	1,100	478
Michigan	1,170.0	1,120.00	155	110	181,350	123,200	2,198
Minnesota	8,100.0	8,100.00	170	110	1,377,000	891,000	3,355
Mississippi	575.0	515.00	150	100	86,250	51,500	540
Missouri	2,850.0	3,050.00	125	87	356,250	265,350	2,128
Montana	176.0	175.00	150	100	26,400	17,500	406
Nebraska	3,200.0	3,300.00	145	100	464,000	330,000	1,476
Nevada	2.0	1.00	185	125	370	125	81
New Hampshire	3.6	3.90	160	105	576	410	359
New Jersey	10.0	12.00	160	105	1,600	1,260	298
New Mexico	1.3	1.50	150	100	195	150	211
New York	70.0	76.00	130	87	9,100	6,612	1,912
North Carolina	8,800.0	8,900.00	120	77	1,056,000	685,300	2,217
North Dakota	140.0	137.00	150	100	21,000	13,700	218
Ohio	2,230.0	2,500.00	155	105	345,650	262,500	3,494
Oklahoma	2,120.0	2,110.00	130	100	275,600	211,000	1,947
Oregon	9.0	10.00	150	100	1,350	1,000	1,124
Pennsylvania	1,140.0	1,160.00	130	91	148,200	105,560	3,097
Rhode Island	1.6	1.50	150	100	240	150	77
South Carolina	255.0	235.00	125	81	31,875	19,035	838
South Dakota	1,270.0	1,360.00	155	105	196,850	142,800	681
Tennessee	210.0	220.00	130	87	27,300	19,140	1,297
Texas	810.0	860.00	125	82	101,250	70,520	4,905
Utah	610.0	680.00	150	100	91,500	68,000	669
Vermont	3.5	4.00	185	125	648	500	450
Virginia	280.0	270.00	115	80	32,200	21,600	1,265
Washington	(D)	(D)	(D)	(D)	(D)	(D)	934
West Virginia	4.0	5.00	150	100	600	500	725
Wisconsin	310.0	320.00	125	92	38,750	29,440	2,270
Wyoming	83.0	96.00	160	105	13,280	10,080	270
Idaho and Washington	57.0	52.00	150	100	8,550	5,200	....
United States	67,776.3	68,919.30	144	96	9,754,191	6,626,607	63,246

(D) Withheld to avoid disclosing data for individual operations.   [1] State level estimates only available in conjunction with the Census of Agriculture every 5 years.   An operation is any place having one or more hogs and pigs on hand December 31.   Totals may not add due to rounding.
NASS, Livestock Branch, (202) 720–3570.

Table 7-21.—Hogs and pigs: Operations, inventory and value, United States, Dec. 1, 2006–2015

Year	Inventory	Value	
		Per head	Total
	Thousands	Dollars	1,000 dollars
2006	62,516	90.00	5,598,613
2007	68,177	73.00	4,986,206
2008	67,048	89.00	5,948,761
2009	64,687	83.00	5,399,951
2010	64,725	106.00	6,876,276
2011	66,259	123.00	8,145,260
2012	66,224	116.00	7,683,152
2013	64,775	138.00	8,921,842
2014	67,776	144.00	6,566,642
2015	68,919	96.00	6,626,607

[1] An operation is any place having one or more hogs and pigs on hand December 31. [2] Operation estimates discontinued after 2012.

NASS, Livestock Branch, (202) 720–3570.

Table 7-22.—Sows farrowing and pig crop: Number, United States 2006-2015

Year	Sows farrowing		Pig crop		
	Dec.-May	June-Nov.	Dec.-May	June-Nov.	Total
	Thousands	Thousands	Thousands	Thousands	Thousands
2006	5,769	5,861	52,259	53,374	105,633
2007	5,935	6,312	54,266	58,608	112,874
2008	6,077	6,103	56,583	58,010	114,593
2009	5,982	5,874	57,111	56,977	114,088
2010	5,749	5,824	55,770	57,358	113,128
2011	5,694	5,846	56,469	58,616	115,085
2012	5,759	5,810	57,749	58,906	116,655
2013	5,595	5,670	57,020	58,115	115,135
2014	5,573	5,985	53,821	61,035	114,856
2015	5,749	5,946	59,219	62,191	121,410

NASS, Livestock Branch, (202) 720–3570.

Table 7-23.—Hogs and pigs: Number for breeding and market, United States, 2006–2015

Year	All hogs and pigs	Kept for breeding	Market hogs by weight groups				
			Under 50 pounds	50 to 119 pounds	120 to 179 pounds	180 pounds and over	Total
			June 1				
	Thousands	Thousands	Thousands	Thousands	Thousands	Thousands	Thousands
2006	61,701	6,080			11,483	9,642	55,621
2007	63,947	6,169			11,789	9,920	57,777
2008	67,100	6,131	19,782	17,637	12,790	10,761	60,969
2009	66,709	5,968	19,514	17,778	12,604	10,847	60,742
2010	64,650	5,788	19,224	16,832	12,154	10,653	58,862
2011	65,320	5,803	19,423	17,241	12,254	10,599	59,517
2012	66,469	5,862	19,746	17,924	12,243	10,694	60,607
2013	65,188	5,884	19,145	17,068	12,414	10,678	59,304
2014	61,568	5,855	18,254	15,801	11,491	10,166	55,713
2015	67,165	5,926	19,365	17,461	12,985	11,429	61,240
			Dec. 1				
	Thousands	Thousands	Thousands	Thousands	Thousands	Thousands	Thousands
2006	62,516	6,116			11,274	10,738	56,399
2007	68,177	6,233			12,658	11,569	61,944
2008	67,048	6,062	19,428	17,396	12,656	11,508	60,987
2009	64,687	5,850	18,705	16,782	12,199	11,152	58,837
2010	64,725	5,778	18,864	16,519	12,183	11,381	58,947
2011	66,259	5,803	19,447	16,618	12,473	11,918	60,456
2012	66,224	5,819	19,299	16,752	12,614	11,741	60,405
2013	64,775	5,757	18,389	16,080	12,576	11,972	59,018
2014	67,776	5,939	19,801	17,366	13,000	11,671	61,838
2015	68,919	6,002	20,008	17,262	13,370	12,276	62,917

NASS, Livestock Branch, (202) 720–3570.

Table 7-24.—Cattle and swine: Production, 2013–2015

Country	2013	2014	2015
	1,000 head	*1,000 head*	*1,000 head*
Cattle:			
Argentina	14,000	13,300	13,700
Australia	10,783	11,063	9,664
Brazil	50,185	49,600	48,220
Canada	4,516	4,606	4,400
China	48,800	47,900	49,000
European Union	29,050	29,280	29,550
India	65,000	66,000	67,000
Mexico	6,700	6,750	6,850
New Zealand	4,923	5,440	5,040
Russia	6,820	6,670	6,635
Others	14,914	14,429	10,334
Total Foreign	255,691	255,038	250,393
United States	33,730	33,900	34,300
Total	289,421	288,938	284,693
Swine:			
Belarus	5,325	4,850	5,150
Brazil	37,900	38,470	39,050
Canada	27,376	27,078	28,630
China	720,971	729,927	696,600
European Union	257,000	261,750	266,500
Japan	17,350	17,050	17,150
Korea, South	16,953	16,812	17,600
Mexico	17,800	17,600	18,000
Russia	36,000	37,000	39,760
Ukraine	9,465	9,527	9,200
Others	4,737	4,985	4,850
Total Foreign	1,150,877	1,165,049	1,142,490
United States	115,135	114,856	120,079
Total	1,266,012	1,279,905	1,262,569

FAS, Office of Global Analysis, (202) 720-6301. Prepared or estimated on the basis of official USDA production, supply, and distribution statistics from foreign governments.

Table 7-25.—Hogs: Number slaughtered, United States, 2006–2015

Year	Commercial			Farm	Total
	Federally inspected	Other	Total [1]		
	Thousands	*Thousands*	*Thousands*	*Thousands*	*Thousands*
2006	103,689	1,048	104,737	105	104,842
2007	108,138	1,033	109,172	106	109,278
2008	115,421	1,031	116,452	106	116,558
2009	112,613	1,006	113,619	114	113,732
2010	109,315	945	110,260	107	110,367
2011	109,956	904	110,860	96	110,956
2012	112,265	898	113,163	83	113,247
2013	111,248	829	112,077	84	112,161
2014	106,123	753	106,876	82	106,876
2015	114,616	810	115,425	87	115,512

[1] Totals are based on unrounded number.
NASS, Livestock Branch, (202) 720-3570.

Table 7-26.—Sows farrowing and pig crop: Number by State and United States, 2014–2015

State	Sows farrowing							
	December[1]–February		March–May		June–August		September–November	
	2014	2015	2014	2015	2014	2015	2014	2015
	Thousands	Thousands	Thousands	Thousands	Thousands	Thousands	Thousands	Thousands
Colorado	71	69	73	70	72	71	70	70
Illinois	245	245	250	230	255	260	250	235
Indiana	120	140	125	130	140	145	145	135
Iowa	480	510	470	495	510	510	485	530
Kansas	80	88	82	84	84	90	85	86
Michigan	49	52	52	50	50	52	51	51
Minnesota	275	285	290	275	280	305	280	285
Missouri	180	205	185	200	195	215	195	195
Nebraska	165	170	170	170	170	180	175	180
North Carolina ..	435	445	440	445	460	470	445	450
Ohio	90	88	80	89	85	92	87	91
Oklahoma	180	195	185	195	195	195	190	195
Pennsylvania	44	46	48	49	48	49	48	46
South Dakota ...	82	85	90	86	88	91	84	91
Texas	41	43	43	50	42	49	47	47
Utah	40	35	41	41	41	43	37	39
Other States[2] ...	186	194	186	195	191	200	197	203
United States[3]	2,763	2,895	2,810	2,854	2,906	3,017	2,871	2,929

State	Pig crop							
	December[1]–February		March–May		June–August		September–November	
	2014	2015	2014	2015	2014	2015	2014	2015
	Thousands	Thousands	Thousands	Thousands	Thousands	Thousands	Thousands	Thousands
Colorado	625	676	679	700	680	717	679	697
Illinois	2,450	2,499	2,475	2,415	2,601	2,639	2,563	2,432
Indiana	1,104	1,414	1,156	1,365	1,372	1,537	1,479	1,404
Iowa	4,752	5,457	4,982	5,247	5,457	5,508	5,190	5,830
Kansas	732	880	759	836	832	927	876	894
Michigan	451	546	463	535	505	572	520	546
Minnesota	2,888	3,135	3,074	3,039	3,038	3,401	3,066	3,164
Missouri	1,638	2,009	1,647	2,080	1,911	2,118	1,853	2,028
Nebraska	1,741	1,870	1,785	1,904	1,853	1,998	1,925	2,007
North Carolina ..	3,654	4,228	3,828	4,139	4,370	4,559	4,450	4,568
Ohio	720	920	760	943	850	966	901	956
Oklahoma	1,647	1,979	1,776	2,018	1,892	2,028	1,938	2,028
Pennsylvania	458	469	494	505	485	510	494	478
South Dakota ...	873	918	963	942	937	987	911	987
Texas	353	389	404	475	416	441	395	442
Utah	400	305	431	402	426	413	248	371
Other States[2] ...	1,840	1,933	1,819	2,048	1,909	2,022	1,885	2,016
United States[3]	26,326	29,627	27,495	29,593	29,534	31,343	29,373	30,848

[1] December preceding year.　[2] Individual State estimates not available for the 34 other States.　[3] Totals may not add due to rounding.

NASS, Livestock Branch, (202) 720–3570.

Table 7-27.—Hogs: Production, disposition, cash receipts, and gross income, United States, 2006–2015

Year	Mar-ketings [1]	Inshipments [2]	Farm slaugh-ter [3]	Production (live weight) [4]	Value of produc-tion [5]	Cash receipts [6]	Value of home consump-tion	Gross income	Average price per 100 pounds received by farmers
	1,000 head	1,000 head	1,000 head	1,000 pounds	1,000 dollars	1,000 dollars	1,000 dollars	1,000 dollars	Dollars
2006 ...	132,384	36,323	105	28,182,382	12,714,218	14,105,864	31,344	14,137,208	46.00
2007 ...	137,519	39,433	106	29,606,420	13,468,332	14,750,490	32,148	14,782,638	46.60
2008 ...	148,813	42,221	106	31,335,313	14,419,170	16,027,805	33,526	16,061,331	47.00
2009 ...	150,034	42,598	114	31,235,019	12,536,325	14,640,502	31,121	14,671,623	41.60
2010 ...	144,362	40,004	107	30,284,630	16,012,200	17,898,063	38,306	17,936,369	54.10
2011 ...	145,420	40,522	96	30,836,171	20,014,277	21,631,894	44,004	21,675,898	65.30
2012 ...	151,835	44,027	83	32,060,619	20,285,027	22,092,108	40,007	22,132,115	64.20
2013 ...	154,923	47,864	84	32,620,264	21,666,015	23,761,227	37,428	23,798,655	67.20
2014 ...	149,069	47,185	82	32,182,241	24,221,588	26,517,815	43,550	26,561,365	76.50
2015 ...	159,754	49,681	87	35,127,658	19,283,155	21,032,491	32,015	21,064,506	55.30

[1] Includes custom slaughter for use on farms where produced and State outshipments, but excludes interfarm sales within the State. [2] Includes hogs and pigs shipped in from other states but excludes animals for immediate slaughter. [3] Excludes custom slaughter for farmers at commercial establishments. [4] Adjustments made for changes in inventory and for inshipments. [5] Includes allowance for higher average price of State inshipments and outshipments of feeder pigs. [6] Receipts from marketings and sale of farm slaughter.
NASS Livestock Branch, (202) 720–3570.

Table 7-28.—Hogs: Direct receipts at interior markets, 2005–2015

Year	All receipts (live and carcass basis)			
	National	Iowa/ Southern Minnesota [1]	Western Region	Eastern Region
	Thousands	Thousands	Thousands	Thousands
2005	77,905	42,609	49,757	25,174
2006	76,527	41,773	48,089	25,405
2007	78,475	41,446	49,929	25,740
2008	80,890	42,546	51,846	24,830
2009	81,443	44,628	53,331	23,065
2010	75,104	40,143	48,413	22,896
2011	75,514	41,479	49,649	22,643
2012	75,459	42,343	50,607	22,192
2013	71,599	40,744	48,202	20,929
2014	73,182	41,687	49,180	21,341
2015	78,445	44,997	53,076	22,298

[1] Iowa / Southern Minnesota is a subset of the Western Region.
AMS, Livestock & Grain Market News, (202) 720–7316.

Table 7-29.—Hogs and corn: Hog-corn price ratio and average price received by farmers for corn, United States, 2006–2015

Year	Hog-corn price ratio [1]	Price of corn per bushel [2]
		Dollars
2006	20.3	2.28
2007	13.8	3.39
2008	10.0	4.78
2009	11.2	3.75
2010	14.4	3.83
2011	11.0	6.02
2012	9.6	6.67
2013	11.3	6.15
2014	18.9	4.11
2015	14.5	3.71

[1] Number of bushels of corn equal in value to buy 100 pounds of live hogs at local markets, based on average prices received by farmers for hogs and corn. Annual average is a simple average of monthly ratios for the calendar year. [2] Annual average is a simple average of monthly prices for the calendar year.
NASS, Environmental, Economics, and Demographics Branch, (202) 720–6146.

Table 7-30.—Hogs: Production, disposition, cash receipts, and gross income, by State and United States, 2015

State	Mar-ketings [1]	Inship-ments [2]	Farm slaugh-ter [3]	Production (live weight) [4]	Value of production [5]	Cash receipts [5][6]	Value of home consump-tion	Gross income
	1,000 head	*1,000 head*	*1,000 head*	*1,000 pounds*	*1,000 dollars*	*1,000 dollars*	*1,000 dollars*	*1,000 dollars*
Alabama	327.0	127.0	1.0	56,874	32,397	36,518	309	36,827
Alaska	2.8	2.0	0.7	810	436	422	93	515
Arizona	271.0	9.0	1.0	67,074	40,102	40,778	235	41,013
Arkansas	1,276.0	20.0	1.0	83,809	66,386	68,844	439	69,283
California	204.0	151.0	3.0	40,272	17,104	29,269	556	29,825
Colorado	2,921.0	245.0	2.0	237,274	183,875	193,107	335	193,442
Connecticut	3.6	0.5	0.3	890	483	409	125	534
Delaware	5.8	1.0	0.1	1,422	779	911	52	963
Florida	30.9	6.0	1.0	4,101	2,292	2,480	101	2,581
Georgia	683.0	200.0	2.0	96,612	54,533	60,072	654	60,726
Hawaii	18.8	8.5	3.0	5,022	2,742	2,792	521	3,313
Idaho	(D)	(D)	(D)	(D)	(D)	(D)	(D)	(D)
Illinois	11,462.0	2,356.0	3.0	2,082,566	1,216,457	1,263,882	872	1,264,754
Indiana	8,867.0	3,559.0	2.0	1,844,296	960,563	1,112,300	861	1,113,161
Iowa	46,616.0	27,500.0	6.0	12,510,642	6,584,559	7,511,940	2,768	7,514,708
Kansas	3,264.0	252.0	2.0	854,114	472,676	469,918	771	470,689
Kentucky	737.0	81.0	3.0	199,850	110,436	108,091	1,328	109,419
Louisiana	7.5	2.0	1.0	1,567	876	797	118	915
Maine	7.4	1.3	0.2	2,031	1,093	933	206	1,139
Maryland	55.0	11.0	0.2	11,679	6,510	6,875	153	7,028
Massachusetts	20.3	3.4	0.1	4,997	2,704	2,596	93	2,689
Michigan	2,424.0	250.0	2.0	616,632	339,063	352,848	619	353,467
Minnesota	19,233.0	7,600.0	5.0	4,146,644	2,189,747	2,528,304	2,046	2,530,350
Mississippi	1,164.0	4.0	4.0	168,775	102,302	105,652	674	106,326
Missouri	8,981.5	1,380.0	1.5	1,561,009	894,023	938,560	948	939,508
Montana	480.0	6.0	2.0	95,696	59,655	59,623	390	60,013
Nebraska	7,651.0	446.0	3.0	1,301,597	778,171	783,052	1,463	784,515
Nevada	6.1	3.0	0.1	1,233	594	889	46	935
New Hampshire	5.8	2.2	0.3	1,400	730	762	69	831
New Jersey	11.8	10.0	0.5	1,393	486	593	140	733
New Mexico	3.0	0.9	0.2	653	365	366	57	423
New York	155.6	18.0	1.8	35,308	19,482	19,653	931	20,584
North Carolina	16,346.0	743.0	1.0	3,960,947	2,229,386	2,284,355	515	2,284,870
North Dakota	859.6	130.0	0.4	56,791	48,891	64,082	150	64,232
Ohio	4,250.0	900.0	9.0	1,149,046	618,198	629,203	3,328	632,531
Oklahoma	8,515.0	829.0	2.0	1,559,673	863,110	878,317	460	878,777
Oregon	13.8	2.0	0.9	3,626	2,156	1,904	286	2,190
Pennsylvania	2,125.5	384.5	2.0	534,109	291,753	304,952	863	305,815
Rhode Island	2.7	-	0.1	603	336	327	24	351
South Carolina	278.5	99.0	1.5	53,654	29,767	32,941	1,095	34,036
South Dakota	4,257.0	666.0	2.0	673,536	413,253	451,705	1,472	453,177
Tennessee	453.0	81.0	2.0	115,117	63,534	63,311	805	64,116
Texas	2,570.0	1,123.0	6.0	372,791	205,525	226,344	2,403	228,747
Utah	1,255.5	3.0	0.5	254,698	161,658	156,596	145	156,741
Vermont	6.5	1.3	0.4	1,733	930	813	139	952
Virginia	323.5	285.0	1.5	77,775	34,591	47,352	582	47,934
Washington	(D)	(D)	(D)	(D)	(D)	(D)	(D)	(D)
West Virginia	4.2	0.5	0.7	1,420	787	500	256	756
Wisconsin	787.0	150.0	1.0	146,680	97,095	106,282	599	106,881
Wyoming	651.0	4.0	1.0	89,915	57,437	56,388	311	56,699
Idaho & Washington	159.0	23.8	2.0	39,302	23,127	23,883	609	24,492
United States	159,753.7	49,680.9	87.0	35,127,658	19,283,155	21,032,491	32,015	21,064,506

- Represents zero.　(D) Withheld to avoid disclosing data for individual operations.　[1] Includes custom slaughter for use on farms where produced and State outshipments,but excludes interfarm sales within the State.　[2] Includes hogs and pigs shipped in from other states but excludes animals for immediate slaughter.　[3] Excludes custom slaughter for farmers at commercial establishments.　[4] Adjustments made for changes in inventory and for inshipments.　[5] Includes allowance for higher average price of State inshipments and outshipments of feeder pigs.　[6] Receipts from marketings and sale of farm-slaughter.
NASS, Livestock Branch, (202) 720–3570.

Table 7-31.—Hogs: Number slaughtered commercially, total and average live weight, by State and United States, 2015 [1]

State	Number slaughtered	Total live weight [2]	Average live weight [2]
	Thousands	*1,000 pounds*	*Pounds*
Alabama	23.1	8,659	376
Alaska	1.5	376	258
Arizona	1.9	492	258
Arkansas	8.6	2,328	269
California	2,325.7	579,006	249
Colorado	18.0	4,354	242
Delaware & Maryland ...	19.4	4,902	253
Florida	51.1	7,312	143
Georgia	72.2	17,615	244
Hawaii	12.3	3,116	254
Idaho	128.4	33,421	261
Illinois	11,669.8	3,357,971	288
Indiana	8,539.8	2,375,100	278
Iowa	31,102.3	8,801,753	283
Kansas	(D)	(D)	(D)
Kentucky	(D)	(D)	(D)
Louisiana	12.2	2,797	230
Michigan	184.2	71,557	389
Minnesota	10,677.5	2,905,180	272
Mississippi	38.9	6,031	155
Missouri	8,832.1	2,555,129	289
Montana	17.0	3,752	220
Nebraska	7,938.2	2,216,246	279
Nevada	1.3	341	257
New England [3]	30.8	7,809	253
New Jersey	103.6	9,974	96
New Mexico	2.8	753	273
New York	45.5	10,960	241
North Carolina	(D)	(D)	(D)
North Dakota	4.4	1,258	287
Ohio	953.2	267,580	281
Oklahoma	5,640.5	1,557,104	276
Oregon	167.7	43,814	261
Pennsylvania	3,140.4	852,806	272
South Carolina	(D)	(D)	(D)
South Dakota	(D)	(D)	(D)
Tennessee	751.4	343,691	458
Texas	298.4	73,448	246
Utah	50.7	8,777	173
Virginia	(D)	(D)	(D)
Washington	(D)	(D)	(D)
West Virginia	8.6	2,364	275
Wisconsin	548.7	243,602	445
Wyoming	4.2	1,126	266
United States [4]	115,425.2	32,674,805	283

(D) Withheld to avoid disclosing data for individual operations. [1] Includes slaughter in federally inspected and other slaughter plants; excludes animals slaughtered on farms. Average live weight is based on unrounded numbers. Totals may not add due to rounding. [2] Excludes postmortem condemnations. [3] Connecticut, Maine, Massachusetts, New Hampshire, Rhode Island, and Vermont. [4] States with no data printed are still included in US total. Data are not printed to avoid disclosing individual operations.

NASS, Livestock Branch, (202) 720-3570.

Table 7-32.—Hogs: Number slaughtered, average dressed and live weights, Federally inspected, 2006–2015 [1]

Year	Barrows and gilts			Sows			Boars			Total		
	Head	Percent of total	Avg. dressed weight [2]	Head	Percent of total	Avg. dressed weight [2]	Head	Percent of total	Avg. dressed weight [2]	Head	Avg. dressed weight [2]	Avg. live weight [2]
	1,000		Pounds	1,000		Pounds	1,000		Pounds	1,000	Pounds	Pounds
2006 ..	100,113	96.6	198	3,227	3.1	309	348	0.3	227	103,689	202	269
2007 ..	104,352	96.5	198	3,309	3.1	308	477	0.4	213	108,138	202	269
2008 ..	111,461	96.6	198	3,502	3.0	308	458	0.4	208	115,421	201	268
2009 ..	108,951	96.7	200	3,243	2.9	306	419	0.4	199	112,613	203	271
2010 ..	105,983	97.0	201	2,966	2.7	305	366	0.3	200	109,315	204	273
2011 ..	106,600	96.9	203	3,027	2.8	305	329	0.3	210	109,956	206	275
2012 ..	108,912	97.0	203	3,009	2.7	306	344	0.3	208	112,265	206	275
2013 ..	107,965	97.0	205	2,932	2.6	304	351	0.3	204	111,248	207	277
2014 ..	102,998	97.1	212	2,788	2.6	305	337	0.3	204	106,123	214	285
2015 ..	111,488	97.3	210	2,852	2.5	309	276	0.2	207	114,616	213	283

[1] All weights calculated using unrounded totals. Totals and percentages based on unrounded data and may not equal sum of classes due to rounding. [2] Excludes postmortem condemnations.
NASS, Livestock Branch, (202) 720-3570.

Table 7-33.—Lard: Supply and disappearance, United States, 2006–2015

Calendar year	Supply				Disappearance		
	Stocks Jan 1	Production	Imports	Total	Domestic	Exports	Total
	Million lbs	Million lbs	Million lbs	Million lbs	Million lbs	Million lbs	Million lbs
2006	13	788	7	805	721	72	793
2007	11	821	9	844	761	73	834
2008	9	874	7	894	795	81	877
2009	18	860	17	889	775	83	859
2010	30	840	15	873	780	72	851
2011	22	852	13	891	789	77	866
2012	25	871	14	905	830	55	885
2013	20	867	14	901	816	65	881
2014	20	852	16	888	821	47	868
2015	20	915	14	949	894	47	940

ERS, Market and Trade Economics Division, Field Crops Branch, (202) 694–5300.

Table 7-34.—Lard: United States exports by country of destination, 2011–2013

Country	2011	2012	2013 [1]
	Metric tons	Metric tons	Metric tons
Mexico	32,878	23,487	28,347
Canada	1,016	598	596
Trinidad and Tobago	236	488	234
Micronesia	0	3	131
Marshall Islands	3	2	40
Barbados	4	0	34
Korea, South	0	0	19
Cayman Islands	4	0	13
Bermuda	65	49	11
Palau	0	0	9
Argentina	0	3	5
Panama	0	0	4
Mayotte	0	0	2
Colombia	0	0	1
Bahamas, The	107	0	0
Belize	2	0	0
China	0	29	0
Ireland	0	18	0
France(*)	5	20	0
Guyana	0	13	0
Haiti	70	64	0
Hong Kong	2	2	0
Honduras	3	0	0
Japan	16	0	0
Jamaica	4	4	0
Lebanon	79	0	0
Netherlands Antilles(*)	167	6	0
Nicaragua	2	0	0
Saudi Arabia	25	0	0
United Kingdom	0	10	0
Rest of World	90	30	0
World Total	34,776	24,826	29,445

(*) Denotes a country that is a summarization of its component countries.   [1] 2013 data does not reflect 13 month changes.   Users should use cautious interpretation on quantity reports using mixed units of measure.   Quantity line items will only include statistics on the units of measure that are equal to, or are able to be converted to, the assigned unit of measure of the grouped commodities.
FAS, Office of Global Analysis, (202) 720-6301.   Data Source: Department of Commerce, U.S. Census Bureau, Foreign Trade Statistics.

Table 7-35.—Sheep and lambs: Operations, inventory, and value, United States, Jan. 1, 2006–2015

Year	Operations [1]	Inventory	Value	
			Per head	Total
	Number	1,000 head	Dollars	1,000 dollars
2006	69,180	6,200	141.00	872,351
2007	83,130	6,120	134.00	818,491
2008	82,500	5,950	138.00	823,424
2009	82,000	5,747	133.00	765,194
2010	81,000	5,620	135.00	761,115
2011	80,000	5,470	170.00	929,453
2012	79,500	5,375	221.00	1,187,730
2013		5,360	177.00	951,020
2014		5,245	188.00	984,236
2015		5,280	214.00	1,129,447

[1] Operations discontinued.
NASS, Livestock Branch, (202) 720–3570.

Table 7-36.—Sheep and lambs: Number by class, United States, Jan. 1, 2006–2015

Year	All sheep and lambs	Breeding sheep			
		Total [1]	Replacement lambs	1 year and over	
				Ewes	Rams
	1,000 head	1,000 head	1,000 head	1,000 head	1,000 head
2006	6,200	4,616	786	3,661	200
2007	6,120	4,553	735	3,620	199
2008	5,950	4,432	697	3,540	195
2009	5,747	4,247	647	3,405	196
2010	5,620	4,185	655	3,335	195
2011	5,470	4,070	665	3,215	190
2012	5,375	3,995	660	3,165	170
2013	5,360	3,985	670	3,135	180
2014	5,245	3,900	635	3,090	175
2015	5,280	3,935	650	3,110	175

[1] Categories may not add to total due to rounding.
NASS, Livestock Branch, (202) 720–3570.

Table 7-37.—Lamb mutton, goat, etc. meat: U.S. imports, 2011–2013

Country	2011	2012	2013
	1,000 metric tons	1,000 metric tons	1,000 metric tons
Australia(*)	57,389	59,011	63,312
New Zealand(*)	23,569	19,227	21,192
Canada	672	805	952
Chile	0	63	354
Iceland	138	185	181
Mexico	129	99	82
Ireland	22	0	0
Finland	13	0	0

(*) Denotes a country that is a summarization of its component countries. Users should use cautious interpretation on quantity reports using mixed units of measure. Quantity line items will only include statistics on the units of measure that are equal to, or are able to be converted to, the assigned unit of measure of the grouped commodities.
FAS, Office of Global Analysis, (202) 720-6301. Data Source: Department of Commerce, U.S. Census Bureau, Foreign Trade Statistics.

Table 7-38.—Breeding sheep: Number by class, State and United States, Jan. 1, 2014–2015

State	Under one year old		One year and over			
	Replacement lambs		Ewes		Rams	
	2014	2015	2014	2015	2014	2015
	1,000 head	1,000 head	1,000 head	1,000 head	1,000 head	1,000 head
Arizona	22.0	24.0	80.0	76.0	7.0	7.0
California	45.0	45.0	270.0	275.0	10.0	10.0
Colorado	23.0	30.0	142.0	184.0	5.0	6.0
Idaho	30.0	31.0	155.0	150.0	5.0	6.0
Illinois	8.0	7.0	36.0	38.0	3.0	3.0
Indiana	6.0	6.0	34.0	34.0	3.0	3.0
Iowa	15.0	20.0	90.0	100.0	5.0	5.0
Kansas	7.0	6.0	39.0	36.0	2.0	2.0
Kentucky	7.0	6.0	31.0	30.0	2.0	2.0
Michigan	13.0	12.0	43.0	40.0	3.0	3.0
Minnesota	15.0	15.0	79.0	73.0	4.0	4.0
Missouri	12.0	11.0	57.0	58.0	3.0	4.0
Montana	37.0	34.0	155.0	153.0	6.0	6.0
Nebraska	8.0	9.0	54.0	55.0	3.0	3.0
Nevada	12.0	10.0	55.0	48.0	2.0	2.0
New England [1]	7.0	6.0	27.0	26.0	2.0	2.0
New Mexico	12.0	14.0	53.0	58.0	3.0	4.0
New York	14.0	16.0	47.0	51.0	3.0	3.0
North Carolina	4.0	4.0	15.0	18.0	2.0	2.0
North Dakota	7.0	6.0	41.0	42.0	2.0	2.0
Ohio	15.0	17.0	72.0	73.0	6.0	6.0
Oklahoma	9.0	8.0	36.0	31.0	3.0	3.0
Oregon	22.0	24.0	97.0	110.0	6.0	6.0
Pennsylvania	11.0	10.0	63.0	56.0	5.0	5.0
South Dakota	31.0	32.0	172.0	162.0	7.0	6.0
Tennessee	6.0	6.0	25.0	28.0	3.0	3.0
Texas	105.0	105.0	440.0	435.0	35.0	30.0
Utah	37.0	40.0	215.0	220.0	8.0	10.0
Virginia	10.0	9.0	53.0	50.0	3.0	3.0
Washington	8.0	8.0	34.0	31.0	3.0	3.0
West Virginia	4.0	4.0	22.0	22.0	1.0	1.0
Wisconsin	14.0	12.0	52.0	50.0	3.0	3.0
Wyoming	38.0	43.0	220.0	210.0	7.0	7.0
Other States [2]	21.0	20.0	86.0	87.0	10.0	10.0
United States	635.0	650.0	3,090.0	3,110.0	175.0	175.0

[1] New England includes Connecticut, Maine, Massachusetts, New Hampshire, Rhode Island, and Vermont. [2] Individual state estimates not available for states not shown, but are included in Other States.
NASS, Livestock Branch, (202) 720–3570.

Table 7-39.—Sheep and lambs: Average price per 100 pounds at San Angelo, 2005–2015[1]

Year	Sheep			Slaughter lambs choice & prime			
	Good	Utility	Cull	Wooled		Shorn	
				90-110 lbs	110-130 1bs	90-110 lbs	110-130 1bs
2005	54.21	56.59	41.39	98.26	97.69	98.24	97.50
2006	42.16	42.33	26.41	80.41	78.99	78.17	78.16
2007	41.06	41.31	25.16	85.36	85.36	85.13	85.18
2008	36.91	37.62	18.01	86.02	85.62	86.06	85.66
2009	40.27	40.53	21.20	90.49	90.24	90.55	90.32
2010	59.51	59.50	41.12	114.66	115.12	115.18	115.68
2011	66.22	66.62	41.43	161.09	161.08	161.53	161.24
2012	64.30	65.80	42.84	113.64	114.01	114.56	114.68
2013	47.68	49.68	28.21	107.34	107.50	107.44	107.69
2014	65.66	67.88	43.76	157.08	157.14	157.19	157.30
2015	79.11	80.63	54.70	143.97	143.97	144.02	144.12

[1] Simple average of monthly bulk-of-sales prices from data of the livestock reporting service. AMS, Livestock & Grain Market News, (202) 720–7316.

Table 7-40.—Sheep and lambs: Number of breeding and market sheep, by State and United States, Jan. 1, 2014–2015

State	Breeding sheep and lambs		Market sheep and lambs	
	2014	2015	2014	2015
	1,000 head	*1,000 head*	*1,000 head*	*1,000 head*
Arizona	109.0	107.0	41.0	43.0
California	325.0	330.0	265.0	270.0
Colorado	170.0	220.0	195.0	200.0
Idaho	190.0	187.0	60.0	73.0
Illinois	47.0	48.0	9.0	9.0
Indiana	43.0	43.0	7.0	7.0
Iowa	110.0	125.0	45.0	50.0
Kansas	48.0	44.0	27.0	22.0
Kentucky	40.0	38.0	9.0	10.0
Michigan	59.0	55.0	22.0	21.0
Minnesota	98.0	92.0	37.0	38.0
Missouri	72.0	73.0	11.0	12.0
Montana	198.0	193.0	22.0	22.0
Nebraska	65.0	67.0	11.0	14.0
Nevada	69.0	60.0	11.0	9.0
New England[1]	36.0	34.0	8.0	9.0
New Mexico	68.0	76.0	13.0	14.0
New York	64.0	70.0	11.0	10.0
North Carolina	21.0	24.0	6.0	6.0
North Dakota	50.0	50.0	16.0	14.0
Ohio	93.0	96.0	24.0	25.0
Oklahoma	48.0	42.0	11.0	11.0
Oregon	125.0	140.0	70.0	55.0
Pennsylvania	79.0	71.0	15.0	15.0
South Dakota	210.0	200.0	60.0	55.0
Tennessee	34.0	37.0	5.0	7.0
Texas	580.0	570.0	150.0	150.0
Utah	260.0	270.0	20.0	20.0
Virginia	66.0	62.0	17.0	13.0
Washington	45.0	42.0	10.0	10.0
West Virginia	27.0	27.0	5.0	6.0
Wisconsin	69.0	65.0	14.0	12.0
Wyoming	265.0	260.0	90.0	85.0
Other States[2]	117.0	117.0	28.0	28.0
United States	3,900.0	3,935.0	1,345.0	1,345.0

[1] New England includes Connecticut, Maine, Massachusetts, New Hampshire, Rhode Island, and Vermont. [2] Individual state estimates not available for states not shown, but are included in Other States.
NASS, Livestock Branch, (202) 720–3570.

Table 7-41.—Lamb crop: Per 100 ewes 1+, number and percent of previous year, by State and United States, 2014–2015

State	Breeding ewes 1 year & older, Jan. 1		Lambs per 100 ewes, Jan. 1		Lamb crop[1]		
	2014	2015	2014	2015	2014	2015	2015 as % of 2014
	1,000 head	1,000 head	Number	Number	1,000 head	1,000 head	Percent
Arizona	80.0	76.0	61	67	49.0	51.0	104
California	270.0	275.0	89	85	240.0	235.0	98
Colorado	142.0	184.0	141	109	200.0	200.0	100
Idaho	155.0	150.0	119	120	185.0	180.0	97
Illinois	36.0	38.0	139	118	50.0	45.0	90
Indiana	34.0	34.0	126	121	43.0	41.0	95
Iowa	90.0	100.0	167	124	150.0	124.0	83
Kansas	39.0	36.0	113	119	44.0	43.0	98
Kentucky	31.0	30.0	129	117	40.0	35.0	88
Michigan	43.0	40.0	140	123	60.0	49.0	82
Minnesota	79.0	73.0	158	123	125.0	90.0	72
Missouri	57.0	58.0	125	124	71.0	72.0	101
Montana	155.0	153.0	129	124	200.0	190.0	95
Nebraska	54.0	55.0	135	118	73.0	65.0	89
Nevada	55.0	48.0	85	102	47.0	49.0	104
New England[2]	27.0	26.0	107	108	29.0	28.0	97
New Mexico	53.0	58.0	85	78	45.0	45.0	100
New York	47.0	51.0	117	118	55.0	60.0	109
North Carolina	15.0	18.0	127	106	19.0	19.0	100
North Dakota	41.0	42.0	134	119	55.0	50.0	91
Ohio	72.0	73.0	136	121	98.0	88.0	90
Oklahoma	36.0	31.0	100	100	36.0	31.0	86
Oregon	97.0	110.0	132	108	128.0	119.0	93
Pennsylvania	63.0	56.0	97	116	61.0	65.0	107
South Dakota	172.0	162.0	125	120	215.0	195.0	91
Tennessee	25.0	28.0	136	121	34.0	34.0	100
Texas	440.0	435.0	77	80	340.0	350.0	103
Utah	215.0	220.0	109	105	235.0	230.0	98
Virginia	53.0	50.0	104	116	55.0	58.0	105
Washington	34.0	31.0	121	110	41.0	34.0	83
West Virginia	22.0	22.0	132	114	29.0	25.0	86
Wisconsin	52.0	50.0	131	120	68.0	60.0	88
Wyoming	220.0	210.0	109	112	240.0	235.0	98
Other States[3]	86.0	87.0	93	92	80.0	80.0	100
United States	3,090.0	3,110.0	111	105	3,440.0	3,275.0	95

[1] Lamb crop is defined as lambs born in the Eastern States and lambs docked or branded in the Western States. [2] New England includes Connecticut, Maine, Massachusetts, New Hampshire, Rhode Island, and Vermont. [3] Unpublished states. NASS, Livestock Branch, (202) 720–3570.

Table 7-42.—Sheep and lambs: Receipts at selected markets, 2006–2015

Year	Sioux Falls, SD	Billings, MT	San Angelo, TX	New Holland,PA
	Thousands	Thousands	Thousands	Thousands
2006	48	58	217	98
2007	40	58	186	108
2008	42	50	192	107
2009	28	54	144	102
2010	41	48	143	92
2011	38	42	156	94
2012	35	53	89	101
2013	48	52	110	109
2014	52	49	107	80
2015	82	63	1,098	84

AMS, Livestock & Grain Market News, (202) 720–7316.

Table 7-43.—Sheep and lambs: Number slaughtered, United States, 2006–2015

Year	Commercial			Farm	Total
	Federally inspected	Other	Total [1]		
	Thousands	Thousands	Thousands	Thousands	Thousands
2006	2,547	151	2,699	80	2,766
2007	2,529	165	2,694	85	2,779
2008	2,394	162	2,556	92	2,647
2009	2,323	193	2,516	95	2,611
2010	2,261	196	2,458	95	2,553
2011	2,000	164	2,164	93	2,258
2012	2,012	171	2,183	93	2,275
2013	2,120	199	2,319	94	2,412
2014	2,105	205	2,310	95	2,404
2015	1,998	225	2,224	95	2,319

[1] Totals are based on unrounded numbers.
NASS, Livestock Branch, (202) 720–3570.

Table 7-44.—Sheep and lambs: Number slaughtered commercially, total and average live weight, by State and United States, 2015 [1]

State	Number slaughtered	Total live weight	Average live weight [2]
	Thousands	1,000 pounds	Pounds
Alabama	0.2	23	95
Alaska	0.1	14	124
Arizona	3.5	460	130
Arkansas	0.8	78	97
California	291.6	41,557	143
Colorado	864.5	138,772	161
Delaware-Maryland	42.7	4,155	97
Florida	11.9	763	64
Georgia	10.3	688	67
Hawaii	1.5	205	140
Idaho	2.2	290	132
Illinois	(D)	(D)	(D)
Indiana	47.2	5,295	112
Iowa	2.8	438	154
Kansas	5.7	497	88
Kentucky	16.8	2,004	119
Louisiana	5.0	404	80
Michigan	203.0	27,857	137
Minnesota	4.5	493	109
Mississippi	3.9	234	60
Missouri	9.7	871	89
Montana	5.2	645	123
Nebraska	1.0	134	136
Nevada	1.3	166	125
New England [3]	35.6	3,485	98
New Jersey	126.6	9,853	78
New Mexico	8.8	1,309	148
New York	60.5	6,134	102
North Carolina	14.5	1,056	73
North Dakota	0.5	67	128
Ohio	17.0	1,885	111
Oklahoma	3.3	344	104
Oregon	34.2	4,896	143
Pennsylvania	61.1	6,343	104
South Carolina	(D)	(D)	(D)
South Dakota	3.4	548	159
Tennessee	12.6	897	71
Texas	109.3	10,754	98
Utah	24.0	3,341	139
Virginia	4.3	408	95
Washington	13.1	1,753	133
West Virginia	(D)	(D)	(D)
Wisconsin	14.5	2,005	138
Wyoming	1.4	208	144
United States	2,223.5	301,254	136

(D) Withheld to avoid disclosing data for individual operations.   [1] Includes slaughter in federally inspected and in other slaughter plants; exludes animals slaughtered on farms.   Excludes postmortem condemnations.   [2] Averages are based on unrounded data.   [3] New England includes Connecticut, Maine, Massachusetts, New Hampshire, Rhode Island, and Vermont.
NASS, Livestock Branch, (202) 720–3570.

Table 7-45.—Sheep and lambs: Number slaughtered, average dressed and live weights, percentage distribution, by class, Federally inspected, 2006–2015 [1]

Year	Federally inspected								
	Lambs and yearlings			Mature sheep			Total		
	Head	Pct. of total	Avg. dressed weight [2]	Head	Pct. of total	Avg. dressed weight [2]	Head	Avg. dressed weight [2]	Avg. live weight [2]
	1,000		Pounds	1,000		Pounds	1,000	Pounds	Pounds
2006	2,429	95.4	70	118	4.6	67	2,547	70	138
2007	2,413	95.4	69	116	4.6	67	2,529	69	138
2008	2,271	94.9	69	122	5.1	67	2,394	69	138
2009	2,165	93.2	70	158	6.8	64	2,323	70	139
2010	2,105	93.1	69	156	6.9	65	2,261	68	137
2011	1,860	93.0	71	141	7.0	66	2,000	70	141
2012	1,869	92.9	74	143	7.1	64	2,012	74	147
2013	1,988	93.7	69	133	6.3	64	2,120	69	137
2014	1,968	93.5	69	136	6.5	66	2,105	69	137
2015	1,885	94.3	70	113	5.7	67	1,998	70	139

[1] All percents and weights calculated using unrounded totals. [2] Excludes postmortem condemnations.
NASS, Livestock Branch, (202) 720–3570.

Table 7-46.—Sheep and lambs: Inventory Jan 1, 2014–2015, and number of operations, 2012, by State and United States [1]

State	January 1 Sheep inventory		Operations with sheep [2]
	2014	2015	2012
	1,000 head	1,000 head	Number
Alabama			712
Alaska			50
Arizona	150.0	150.0	7,447
Arkansas			778
California	590.0	600.0	4,224
Colorado	365.0	420.0	1,509
Delaware			69
Florida			1,161
Georgia			688
Hawaii			353
Idaho	250.0	260.0	1,241
Illinois	56.0	57.0	1,751
Indiana	50.0	50.0	2,109
Iowa	155.0	175.0	2,904
Kansas	75.0	66.0	1,160
Kentucky	49.0	48.0	1,743
Louisiana			643
Maryland			663
Michigan	81.0	76.0	2,312
Minnesota	135.0	130.0	2,171
Mississippi			499
Missouri	83.0	85.0	2,454
Montana	220.0	215.0	1,338
Nebraska	76.0	81.0	1,464
Nevada	80.0	69.0	508
New England [3]	44.0	43.0	3566
New Jersey			819
New Mexico	81.0	90.0	3,385
New York	75.0	80.0	2,017
North Carolina	27.0	30.0	1,311
North Dakota	66.0	64.0	661
Ohio	117.0	121.0	3,568
Oklahoma	59.0	53.0	1,779
Oregon	195.0	195.0	2,753
Pennsylvania	94.0	86.0	3,590
South Carolina			549
South Dakota	270.0	255.0	1,798
Tennessee	39.0	44.0	1,476
Texas	730.0	720.0	10,674
Utah	280.0	290.0	1,755
Virginia	83.0	75.0	2,315
Washington	55.0	52.0	1,967
West Virginia	32.0	33.0	1,043
Wisconsin	83.0	77.0	2,590
Wyoming	355.0	345.0	771
Other States [4]	145.0	145.0	
United States	5,245.0	5,280.0	88,338

[1] An operation is any place having one or more head of sheep on hand December 31. [2] State level estimates only available in conjunction with the Census of Agriculture every 5 years. [3] New England includes Connecticut, Maine, Massachusetts, New Hampshire, Rhode Island, and Vermont. [4] Individual state estimates not available for states not shown, but are included in Other States.
NASS, Livestock Branch, (202) 720–3570.

Table 7-47.—Wool: Number of sheep shorn, weight per fleece, production, average price per pound received by farmers, value of production, exports, imports, total new supply of apparel wool, and imports of carpet wool, United States, 2006–2015

Year	Sheep and lambs shorn [1]	Weight per fleece	Shorn wool production	Price per pound [2]	Value of production [3]
	Thousands	Pounds	1,000 pounds	Cents	1,000 dollars
2006	4,847	7.4	35,899	0.68	24,300
2007	4,657	7.5	34,723	0.87	30,242
2008	4,434	7.4	32,963	0.99	32,486
2009	4,195	7.4	30,860	0.79	24,337
2010	4,180	7.3	30,360	1.15	35,013
2011	4,030	7.3	29,280	1.67	48,920
2012	3,790	7.3	27,630	1.52	41,972
2013	3,700	7.3	26,990	1.45	39,209
2014	3,680	7.3	26,680	1.46	38,909
2015	3,675	7.4	27,015	1.45	39,205

Year	Shorn wool production	Raw wool supply (clean)				Total new supply [8]
		Domestic production [4]	Exports [5]	Imports for consumption		
				48's and Finer [6]	Not Finer than 46's [7]	
	1,000 pounds	1,000 pounds	1,000 pounds	1,000 pounds	1,000 pounds	
2006	35,899	18,955	17,998	7,324	9,929	18,210
2007	34,723	18,334	17,077	5,245	9,025	15,527
2008	32,963	17,404	10,307	4,551	8,631	20,279
2009	30,860	16,294	10,207	3,306	6,046	15,439
2010	30,360	16,035	9,973	3,108	4,928	14,098
2011	29,280	15,465	9,586	3,791	4,694	14,364
2012	27,630	14,467	7,741	4,564	4,551	15,841
2013	26,990	14,251	9,998	3,858	3,746	11,857
2014	26,680	14,087	7,916	3,917	3,363	13,247
2015	27,015	14,264	7,847	3,980	3,321	13,718

[1] Includes sheep shorn at commercial feeding yards.   [2] Price computed by weighting State average prices for all wool sold during the year by sales of shorn wool.   [3] Production by States multiplied by annual average price.   [4] Conversion factor from grease basis to clean basis are as follows: Shorn wool production—52.8 percent (Stat. Bull. 616) from 1987-1997.   [5] Includes carpet wool exports.   [6] Prior to 1989, known as dutiable imports.   [7] Prior to 1989, known as duty-free imports. In 1994 includes 24,645,306 pounds of imported raw wool not finer than 46's and 2,182,576 pounds of miscellaneous imported raw wool.   [8] Production minus exports plus imports; stocks not taken into consideration.
NASS, Livestock Branch, (202) 720–3570 and ERS, Field Crops Branch, (202) 694–5300.   Imports and exports from reports of the U.S. Department of Commerce.

Table 7-48.—Wool: Price-support operations, United States, 2006–2015 [1]

Year	Program price levels per pound		Put under loan		Acquired by CCC under loan program [2]	Owned by CCC at end of marketing year
	Graded wool loan	Nongraded loan	Quantity	Percentage of production		
	Dollars	Dollars	1,000 pounds	Percent	1,000 pounds	1,000 pounds
2006	1.00	0.40	3.2	0.01	0.0	0.0
2007	1.00	0.40	3.0	0.01	0.0	0.0
2008	1.00	0.40	8.6	0.03	0.0	0.0
2009	1.00	0.40	28.9	0.09	0.0	0.0
2010	1.15	0.40	27.8	0.09	0.0	0.0
2011	1.15	0.40	0.0	0.00	0.0	0.0
2012	1.15	0.40	0.0	0.00	0.0	0.0
2013	1.15	0.40	14.0	0.05	0.0	0.0
2014	1.15	0.40	11.0	0.04	0.0	0.0
2015	1.15	0.40	0.0	0.0	0.0	0.0

[1] Nonrecourse Marketing Loan Program authorized by the Farm Security and Rural Investment Act of 2002, as amended, for marketing years 2002-2007, the Food, Conservation, and Energy Act of 2008, as amended, for marketing years 2008-2013, and the Agricultural Act of 2014, as amended, for marketing years 2014-2018.   [2] Acquisitions as of July 31, 2016.
FSA, Fibers, Peanuts, Tobacco, (202) 401–9255.

Table 7-49.—Wool: United States imports (for consumption), clean content, by grades, 2007–2016 [1][2]

Grade	2007	2008	2009	2010	2011	2012	2013	2014	2015	2016
	Million pounds	Million pounds	Million pounds	Million pounds	Million pounds	Million pounds	Million pounds	Million pounds	Million pounds	Million pounds
48's and finer:										
Finer than 58's [3]	4.7	4.0	3.0	2.7	3.2	3.7	2.9	3.0	2.9	2.7
48's–58's [4]	0.5	0.5	0.3	0.4	0.6	0.9	1.0	0.9	1.1	1.1
Total	5.2	4.6	3.3	3.1	3.8	4.6	3.9	3.9	4.0	3.8
Not Finer than 46's:										
Wool for special use [5]	0.7	0.6	0.2	0.2	0.3	0.4	0.2	0.3	0.2	0.2
Not finer than 40's [6]	2.6	1.8	1.5	1.1	1.2	1.2	0.9	1.3	0.8	0.4
Finer than 40's– 44's [7]	4.4	5.0	3.6	3.2	2.3	2.5	1.4	1.3	0.9	0.5
46's [8]	1.4	1.3	0.7	0.5	0.9	0.5	1.2	0.5	1.41	1.1
Total	9.0	8.6	6.0	4.9	4.7	4.6	3.7	3.4	3.3	2.2
Miscellaneous [9]	0	0	0	0	0	0	0	0	0	0
Grand total	14.3	13.2	9.4	8.0	8.5	9.1	7.6	7.3	7.3	6.0

[1] Natural fiber grown by sheep or lambs. [2] Beginning 1989 the following Harmonized Tariff Schedule numbers are in the above 7 wool import groups: 5101.19.606060, 5101.19.6060, 5101.21.4000, 5101.21.4000, 5101.29.4060, 0.5(5101.30.4000). [4] 5101.11.6030, 5101.19.6030, 5101.21.4030, 5101.29.4030, 0.5(5101.30.4000). [5] 5101.11.1000, 5101.19.1000, 5101.21.1000, 5101.29.1000. [6] 5101.11.2000, 5101.19.2000, 5101.21.1500, 5101.29.1500, 5101.30.1000. [7] 5101.11.4000, 5101.19.4000, 5101.21.3000, 5101.29.3000, 5101.30.1500. [8] 5101.11.5000, 5101.19.5000, 5101.21.3500, 5101.29.3500, 5101.30.3000. [9] 5101.21.6000, 5101.29.6000, 5101.30.6000. They include wool not carded or combed but processed beyond the scoured or carbonized condition, e.g. dyed. This wool is not identified by use or grade.
ERS, Field Crops Branch, (202) 694–5300. Compiled from reports of the U.S. Department of Commerce.

Table 7-50.—Wool: United States imports (for consumption), clean content, by country of origin, 2007–2016 [1]

Country of origin	2007	2008	2009	2010	2011	2012	2013	2014	2015	2016
	Million pounds	Million pounds	Million pounds	Million pounds	Million pounds	Million pounds	Million pounds	Million pounds	Million pounds	Million pounds
48's and finer:										
Argentina	0.1	0.0	0.0	0.0	0.3	0.0	0.0	0.0	0.0	0.0
Australia	3.7	2.9	2.0	2.2	2.0	2.5	2.3	1.6	2.0	1.4
Canada	0.5	0.5	0.9	0.3	0.4	0.9	0.6	0.9	0.9	1.0
Chile	0.0	0.0	0.0	0.0	0.0	0.0	0.0	0.0	0.0	0.0
New Zealand	0.2	0.2	0.0	0.2	0.1	0.3	0.2	0.4	0.3	0.4
South Africa	0.3	0.4	0.1	0.0	0.3	0.2	0.0	0.0	0.0	0.0
United Kingdom	0.0	0.1	0.0	0.1	0.0	0.1	0.0	0.0	0.0	0.0
Uruguay	0.3	0.2	0.1	0.1	0.3	0.4	0.3	0.4	0.2	0.3
Other	0.1	0.3	0.2	0.2	0.4	0.2	0.5	0.6	0.6	0.7
Total	5.2	4.6	3.3	3.1	3.8	4.6	3.9	3.9	4.0	3.8
Not finer than 46's:										
Argentina	0.2	0.1	0.4	0.4	0.2	0.1	0.0	0.0	0.0	0.0
Australia	0.0	0.2	0.1	0.4	1.1	0.3	0.2	0.1	0.3	0.3
Canada	0.0	0.0	0.0	0.0	0.0	0.0	0.0	0.0	0.0	0.0
New Zealand	7.2	6.5	4.6	3.3	2.3	3.0	2.8	2.6	1.8	1.1
Uruguay	0.0	0.0	0.0	0.0	0.0	0.0	0.0	0.0	0.0	0.0
South Africa	0.1	0.2	0.2	0.3	0.2	0.2	0.1	0.1	0.2	0.0
United Kingdom	1.4	1.4	0.5	0.3	0.7	1.0	0.5	0.4	0.3	0.2
Other	0.1	0.2	0.2	0.2	0.2	0.0	0.1	0.2	0.7	0.6
Total	9.0	8.6	6.0	4.9	4.7	4.6	3.7	3.4	3.3	2.2
Grand total	14.3	13.2	9.4	8.0	8.5	9.1	7.6	7.3	7.3	6.0

[1] Wool not advanced in any manner or by any process of manufacture beyond washed, scoured, or carbonized condition.
ERS, Field Crops Branch, (202) 694–5300. Compiled from reports of the U.S. Department of Commerce.

Table 7-51.—Wool: Average price per pound, clean basis, delivered to United States mills, 2007–2016 [1]

Year	Territory [2]		Australian 64's good topmaking (in bond, American yield)
	64's (20.60–22.04 microns)	Avg. 58's–56's (24.95–27.84 microns)	
	Cents	Cents	Cents
2007	265	157	373
2008	309	204	347
2009	227	155	302
2010	327	203	410
2011	425	374	630
2012	281	346	590
2013	423	259	534
2014	417	274	489
2015	358	284	450
2016	394	282	489

[1] Beginning January 1976 the unit designation terminology for wool prices changed to microns. For example 64's (20.60–22.04 microns) formerly was fine good French combing and staple. Two designations 56's (26.40–27.84 microns) and 58's (24.95–26.39 microns) have been averaged in the price data shown here and together were formerly the category fleece ⅜ blood good French combing and staple. [2] Wool grown in the range areas of California, Oregon, Washington, Texas, the intermountain States (including Arizona and New Mexico), and parts of the Dakotas, Kansas, Nebraska, and Oklahoma. These wools vary considerably in shrinkage and color.
ERS, Field Crops Branch, (202) 694–5300 and AMS.

Table 7-52.—Wool: Number of sheep shorn, weight per fleece, and production, by State and United States, 2014–2015

State	Sheep		Weight per fleece		Shorn wool production	
	2014	2015	2014	2015	2014	2015
	1,000 head	1,000 head	Pounds	Pounds	1,000 pounds	1,000 pounds
Arizona	115.0	110.0	5.9	5.6	680	620
California	460.0	430.0	6.3	6.6	2,900	2,850
Colorado	340.0	320.0	7.1	7.5	2,400	2,410
Idaho	200.0	200.0	8.5	8.3	1,700	1,650
Illinois	43.0	44.0	7.1	7.0	305	310
Indiana	35.0	34.0	6.0	6.2	210	210
Iowa	165.0	165.0	5.5	5.5	900	900
Kansas	39.0	42.0	6.2	6.5	240	275
Kentucky	11.0	12.0	6.5	6.7	71	80
Michigan	63.0	66.0	6.0	6.0	380	395
Minnesota	125.0	120.0	6.4	6.4	800	770
Missouri	44.0	44.0	6.0	6.0	265	265
Montana	185.0	205.0	9.0	9.0	1,660	1,840
Nebraska	62.0	63.0	7.2	7.3	445	460
Nevada	54.0	47.0	9.8	9.4	530	440
New England [1]	30.0	31.0	6.8	6.5	205	200
New Mexico	81.0	82.0	7.8	7.9	630	645
New York	40.0	53.0	6.5	6.1	260	325
North Carolina	7.0	8.0	5.0	5.0	35	40
North Dakota	58.0	64.0	7.9	7.6	460	485
Ohio	83.0	81.0	5.9	6.0	490	490
Oklahoma	21.0	17.0	5.5	5.6	115	95
Oregon	158.0	152.0	6.3	6.6	1,000	1,000
Pennsylvania	48.0	48.0	6.7	6.6	320	315
South Dakota	230.0	239.0	7.6	7.6	1,750	1,820
Tennessee	18.0	19.0	5.6	5.5	100	105
Texas	290.0	270.0	7.2	7.2	2,100	1,950
Utah	245.0	255.0	9.2	9.4	2,260	2,390
Virginia	27.0	27.0	5.6	5.4	150	145
Washington	36.0	33.0	7.5	7.9	270	260
West Virginia	20.0	22.0	5.7	5.5	114	120
Wisconsin	50.0	51.0	7.0	7.2	350	365
Wyoming	250.0	270.0	9.2	9.2	2,300	2,480
Other States [2]	47.0	51.0	6.1	6.1	285	300
United States	3,680.0	3,675.0	7.3	7.4	26,680	27,015

[1] New England includes Connecticut, Maine, Massachusetts, New Hampshire, Rhode Island, and Vermont. [2] Unpublished states.
NASS, Livestock Branch, (202) 720–3570.

Table 7-53.—Wool: Price and value, by State and United States, 2014–2015

State	Price per pound		Value [1]	
	2014	2015	2014	2015
	Dollars	Dollars	1,000 Dollars	1,000 dollars
Arizona	1.10	1.15	748	713
California	1.35	1.55	3,915	4,418
Colorado	1.85	1.80	4,440	4,338
Idaho	1.54	1.53	2,618	2,525
Illinois	0.57	0.75	174	233
Indiana	0.46	0.75	97	158
Iowa	0.58	0.79	522	711
Kansas	0.95	0.87	228	239
Kentucky	0.75	0.80	53	64
Michigan	0.86	0.72	327	284
Minnesota	0.82	0.79	656	608
Missouri	0.56	0.73	148	193
Montana	2.05	1.90	3,403	3,496
Nebraska	1.06	0.90	472	414
Nevada	1.95	2.10	1,034	924
New England [2]	0.85	0.80	174	160
New Mexico	1.65	1.40	1,040	903
New York	0.70	0.60	182	195
North Carolina	1.30	1.20	46	48
North Dakota	1.45	1.28	667	621
Ohio	0.59	0.74	289	363
Oklahoma	0.70	0.70	81	67
Oregon	1.36	1.73	1,360	1,730
Pennsylvania	0.50	0.60	160	189
South Dakota	1.52	1.26	2,660	2,293
Tennessee	0.96	0.84	96	88
Texas	1.57	1.64	3,297	3,198
Utah	1.80	1.70	4,068	4,063
Virginia	0.80	1.05	120	152
Washington	1.80	1.90	486	494
West Virginia	0.86	1.00	98	120
Wisconsin	0.75	0.75	263	274
Wyoming	1.97	1.80	4,531	4,464
Other States [3]	1.60	1.50	456	465
United States	1.46	1.45	38,909	39,205

[1] Production multiplied by marketing year average price. [2] New England includes Connecticut, Maine, Massachusetts, New Hampshire, Rhode Island, and Vermont. [3] Unpublished states.
NASS, Livestock Branch, (202) 720–3570.

Table 7-54.—Mohair: Price-support operations, United States, 2006–2015 [1]

Year	Program price levels per pound	Put under loan	
	Loan	Quantity	Percentage of production
	Dollars	1,000 pounds	Percent
2006	4.20	25.0	1.85
2007	4.20	20.4	1.79
2008	4.20	8.7	0.73
2009	4.20	14.5	1.37
2010	4.20	4.2	0.38
2011	4.20	0.0	0.0
2012	4.20	0.0	0.0
2013	4.20	0.0	0.0
2014	4.20	0.0	0.0
2015	4.20	0.0	0.0

[1] Nonrecourse Marketing Loan Program authorized by the Farm Security and Rural Investment Act of 2002, as amended, for marketing years 2002-2007, the Food , Conservation, and Energy Act of 2008, as amended, for marketing years 2008-2013, and the Agricultural Act of 2014, as amended, for marketing years 2014-2018. [2] Acquisitions as of July 31, 2016.
FSA, Fibers, Peanuts, Tobacco, (202) 401–9255.

Table 7-55.—Mohair: Goats clipped, production, price, and value, by State and United States, 2014–2015

State	Goats clipped		Average clip per goat		Production		Price per pound		Value [1]	
	2014	2015	2014	2015	2014	2015	2014	2015	2014	2015
	Head	Head	Pounds	Pounds	1,000 pounds	1,000 pounds	Dollars	Dollars	1,000 dollars	1,000 dollars
Arizona	33,000	30,000	3.6	4.0	120	115	1.15	1.60	138	184
California	2,000	2,000	5.0	5.0	10	10	6.00	6.00	60	60
New Mexico ...	9,000	8,000	3.9	3.8	35	30	1.20	1.40	42	42
Texas	92,000	80,000	6.3	6.2	580	480	6.30	7.10	3,654	3,408
Other States [2]	23,000	21,000	5.9	5.9	135	130	2.80	2.75	378	358
United States	159,000	141,000	5.5	5.5	880	765	4.85	5.30	4,272	4,052

[1] Production multiplied by marketing year average price.   U.S. value is summation of State values.   [2] Includes data for States not published in this table.
NASS, Livestock Branch, (202) 720–3570.

Table 7-56—Angora goats: Inventory Jan 1, 2014–2015, and number of operations, 2012, by State and United States [1]

State	January 1 angora goats inventory		Operations with angora goats [2]
	2014	2015	2012
	Head	Head	Number
Alabama			152
Alaska			5
Arizona	33,000	36,000	3,218
Arkansas			25
California	3,600	3,400	314
Colorado			182
Delaware			3
Florida			165
Georgia			140
Hawaii			23
Idaho			70
Illinois			84
Indiana			85
Iowa			67
Kansas			45
Kentucky			143
Louisiana			49
Maryland			59
Michigan			177
Minnesota	1,000	(NA)	125
Mississippi			77
Missouri	1,500	(NA)	152
Montana			59
Nebraska			47
Nevada			27
New England [3]			664
New Jersey			75
New Mexico	10,000	11,000	904
New York			176
North Carolina			195
North Dakota			24
Ohio			189
Oklahoma			61
Oregon	2,500	(NA)	166
Pennsylvania			268
South Carolina			121
South Dakota			22
Tennessee			28
Texas	76,000	85,000	629
Utah			275
Virginia			169
Washington			151
Wisconsin	900	(NA)	75
West Virginia			158
Wyoming			16
Other States [4]	19,500	24,600	
United States	148,000	160,000	9,859

(NA) Not Available. [1] An operation is any place having one or more head of angora goats on hand December 31. [2] State level estimates only available in conjunction with the Census of Agriculture every 5 years. [3] New England includes Connecticut, Maine, Massachusetts, New Hampshire, Rhode Island, and Vermont. [4] Unpublished states. NASS, Livestock Branch, (202) 720–3570.

Table 7-57.—Milk goats: Inventory Jan 1, 2014–2015, and number of operations, 2012, by State and United States [1]

State	January 1 milk goats inventory		Operations with milk goats [2]
	2014	2015	2012
	Head	Head	Number
Alabama	3,500	3,800	359
Alaska	(NA)	(NA)	34
Arizona	3,000	(NA)	642
Arkansas	4,200	4,400	567
California	38,000	40,000	1,416
Colorado	9,000	10,000	756
Delaware	(NA)	(NA)	24
Florida	6,200	6,400	852
Georgia	4,500	4,000	396
Hawaii	(NA)	(NA)	79
Idaho	4,000	3,900	435
Illinois	3,400	5,700	593
Indiana	11,800	11,500	1,123
Iowa	30,000	31,000	639
Kansas	4,800	4,200	518
Kentucky	4,500	5,700	822
Louisiana	1,200	(NA)	234
Maryland	2,200	(NA)	234
Michigan	11,400	11,300	1,194
Minnesota	13,000	13,500	575
Mississippi	2,000	(NA)	234
Missouri	9,600	9,900	1,083
Montana	3,000	(NA)	311
Nebraska	3,800	3,700	361
Nevada	(NA)	(NA)	143
New England [3]	14,600	14,500	1,460
New Jersey	2,100	(NA)	182
New Mexico	2,500	(NA)	377
New York	13,100	(NA)	1,091
North Carolina	7,100	13,400	803
North Dakota	(NA)	8,000	154
Ohio	9,500	10,000	1,341
Oklahoma	5,900	6,900	929
Oregon	11,200	9,500	842
Pennsylvania	12,000	13,000	1,288
South Carolina	3,100	3,400	453
South Dakota	3,300	(NA)	276
Tennessee	7,200	6,300	587
Texas	20,000	23,000	2,591
Utah	2,500	(NA)	298
Virginia	3,300	4,500	708
Washington	8,400	7,200	934
West Virginia	2,000	(NA)	380
Wisconsin	46,000	44,000	994
Wyoming	2,000	(NA)	258
Other States [4]	9,100	32,300	
United States	358,000	365,000	29,570

(NA) Not Available.   [1] An operation is any place having one or more head of milk goats on hand December 31.   [2] State level estimates only available in conjunction with the Census of Agriculture every 5 years.   [3] New England includes Connecticut, Maine, Massachusetts, New Hampshire, Rhode Island, and Vermont.   [4] Unpublished states.
NASS, Livestock Branch, (202) 720–3570.

Table 7-58.—Meat and other goats: Inventory Jan 1, 2014–2015, and number of operations, 2012, by State and United States [1]

State	January 1 meat and other goats inventory		Operations with Meat goats [2]
	2014	2015	2012
	Head	Head	Number
Alabama	50,000	47,000	2,469
Alaska	(NA)	(NA)	35
Arizona	25,000	28,000	2,387
Arkansas	38,000	38,000	2,083
California	85,000	85,000	3,330
Colorado	22,000	25,000	1,591
Delware	(NA)	(NA)	136
Florida	37,000	41,000	3,022
Georgia	58,000	67,000	2,951
Hawaii	10,500	(NA)	336
Idaho	14,000	(NA)	859
Illinois	18,000	16,700	1,698
Indiana	31,400	31,000	2,026
Iowa	25,000	25,500	1,383
Kansas	37,000	40,000	1,653
Kentucky	58,000	60,000	3,092
Louisiana	19,000	20,000	1,225
Maryland	11,600	(NA)	687
Michigan	16,500	(NA)	1,669
Minnesota	23,000	20,500	1,121
Mississippi	19,000	19,000	1,329
Missouri	85,200	78,600	3,161
Montana	8,000	(NA)	435
Nebraska	19,500	(NA)	1,247
Nevada	7,300	(NA)	406
New England [3]	11,000	(NA)	1,734
New Jersey	6,000	(NA)	581
New Mexico	20,000	(NA)	1,629
New York	22,000	18,700	1,535
North Carolina	45,000	43,000	3,659
North Dakota	2,800	(NA)	227
Ohio	47,000	46,000	3,401
Oklahoma	75,000	95,000	4,020
Oregon	23,000	24,000	1,672
Pennsylvania	39,000	37,000	2,989
South Carolina	36,000	34,000	2,447
South Dakota	11,500	(NA)	514
Tennessee	110,000	118,000	4,514
Texas	810,000	800,000	23,107
Utah	11,000	(NA)	812
Virginia	45,000	46,000	2,742
Washington	24,000	21,000	1,792
West Virginia	19,000	21,000	1,276
Wisconsin	22,000	(NA)	1,518
Wyoming	4,700	(NA)	410
Other States [4]	3,000	179,000	....
United States	2,105,000	2,125,000	100,910

(NA) Not Available. [1] An operation is any place having one or more head of meat goats on hand December 31. [2] State level estimates only available in conjunction with the Census of Agriculture every 5 years. [3] New England includes Connecticut, Maine, Massachusetts, New Hampshire, Rhode Island, and Vermont. [4] Unpublished states.
NASS, Livestock Branch, (202) 720–3570.

Table 7-59.—All goats: Number of operations, 2012, by State and United States [1]

State	2012 [2]
	Number
Alabama	2,832
Alaska	56
Arizona	5,525
Arkansas	2,450
California	4,474
Colorado	2,168
Delware	154
Florida	3,746
Georgia	3,329
Hawaii	420
Idaho	1,208
Illinios	2,193
Indiana	2,883
Iowa	1,921
Kansas	1,995
Kentucky	3,797
Louisiana	1,412
Maryland	903
Michigan	2,623
Minnesota	1,580
Mississippi	1,533
Missouri	3,955
Montana	698
Nebraska	1,509
Nevada	523
New England [3]	3,096
New Jersey	766
New Mexico	2,638
New York	2,449
North Carolina	4,387
North Dakota	350
Ohio	4,485
Oklahoma	4,629
Oregon	2,350
Pennsylvania	4,088
South Carolina	2,861
South Dakota	748
Tennessee	4,929
Texas	25,063
Utah	1,249
Virginia	3,376
Washington	2,506
West Virginia	1,586
Wisconsin	2,419
Wyoming	594
United States	128,456

[1] An operation is any place having one or more head of goats on hand December 31.   [2] State level estimates only available in conjunction with the Census of Agriculture every 5 years.   [3] New England includes Connecticut, Maine, Massachusetts, New Hampshire, Rhode Island, and Vermont.
NASS, Livestock Branch, (202) 720–3570.

Table 7-60.—Red meat: Production, by class of slaughter, United States, 2006–2015

Year	Commercial			Farm	Total
	Federally inspected	Other [1]	Total		
			Beef		
	Million pounds	Million pounds	Million pounds	Million pounds	Million pounds
2006	25,792	360	26,152	104	26,256
2007	26,070	351	26,421	103	26,524
2008	26,200	361	26,561	96	26,657
2009	25,598		25,966	90	26,056
2010	25,954		· 26,305	84	26,389
2011	25,854		26,195	75	26,270
2012	25,573		25,913	76	25,989
2013	25,385		25,720	70	25,790
2014	23,932		24,249	66	24,315
2015	23,395		23,698	62	23,760
			Pork, excluding lard		
	Million pounds	Million pounds	Million pounds	Million pounds	Million pounds
2006	20,877	177	21,054	20	21,073
2007	21,768	175	21,943	20	21,962
2008	23,170	177	23,347	20	23,367
2009	22,827		22,999	21	23,020
2010	22,275		22,437	19	22,456
2011	22,599		22,758	17	22,775
2012	23,090		23,253	15	23,268
2013	23,040		23,187	16	23,203
2014	22,710		22,843	15	22,858
2015	24.355		24,501	16	24,517
			Veal		
	Million pounds	Million pounds	Million pounds	Million pounds	Million pounds
2006	144	3	147	9	155
2007	134	3	137	9	146
2008	140	3	143	9	152
2009	135		138	8	147
2010	131		134	8	143
2011	126		130	7	136
2012	115		118	7	125
2013	109		111	6	117
2014	92		94	6	100
2015	81		83	5	88
			Lamb and Mutton		
	Million pounds	Million pounds	Million pounds	Million pounds	Million pounds
2006	177	8	185	5	190
2007	175	8	183	6	189
2008	166	8	174	6	180
2009	162		171	5	175
2010	154		164	5	168
2011	141		149	5	153
2012	148		156	5	161
2013	146		156	5	161
2014	145		156	5	161
2015	139		151	5	156
			All meat, excluding lard		
	Million pounds	Million pounds	Million pounds	Million pounds	Million pounds
2006	46,990	547	47,537	137	47,675
2007	48,147	537	48,684	137	48,820
2008	49,675	550	50,225	131	50,355
2009	48,721		49,274	124	49,398
2010	48,514		49,039	116	49,155
2011	48,721		49,232	103	49,335
2012	48,925		49,440	103	49,543
2013	48,679		49,174	97	49,271
2014	46,879		47,342	91	47,433
2015	47,970		48,432	88	48,520

[1] Other class no longer reported.
NASS, Livestock Branch, (202) 720–3570.

Table 7-61.—Meat: United States exports and imports into the United States, carcass weight equivalent, 2006–2015 [1]

Year	Exports				Imports			
	Beef and veal	Lamb and mutton	Pork [2]	All meat	Beef & veal	Lamb and mutton	Pork [2]	All meat
	Million pounds	*Million pounds*	*Million pounds*	*Million pounds*	*Million pounds*	*Million pounds*	*Million pounds*	*Million pounds*
2006	1,145	18	2,995	4,158	3,085	190	990	4,265
2007	1,434	9	3,141	4,585	3,052	203	968	4,223
2008	1,887	12	4,667	6,566	2,538	183	832	3,553
2009	1,935	16	4,095	6,046	2,626	171	834	3,631
2010	2,300	16	4,223	6,539	2,298	166	859	3,323
2011	2,785	19	5,196	8,000	2,057	162	803	3,022
2012	2,452	11	5,379	7,842	2,220	154	802	3,175
2013	2,588	7	4,986	7,582	2,250	173	880	3,302
2014	2,572	7	5,092	7,670	2,947	195	1,011	4,153
2015	2,265	4	5,009	7,279	3,371	214	1,116	4,700

[1] Carcass weight equivalent of all meat, including the meat content of minor meats and of mixed products. Shipments to U.S. Territories are included in domestic consumption.   [2] The pork series has been revised to a dressed weight equivalent rather than "Pork, excluding lard."

ERS, Market and Trade Economics Division, Animal Products and Cost of Production Branch, (202) 694–5409. Data on imports and commercial exports are computed from records of the U.S. Department of Commerce, those on exports by the U.S. Department of Agriculture are separately estimated from deliveries and stocks.

Table 7-62.—International Meat: Production, 2013–2015

Country	2013	2014	2015
	1,000 metric tons	*1,000 metric tons*	*1,000 metric tons*
Beef and veal:			
Argentina	2,850	2,700	2,740
Australia	2,359	2,595	2,547
Brazil	9,675	9,723	9,425
Canada	1,049	1,099	1,050
China	6,730	6,890	6,700
European Union	7,388	7,443	7,670
India	3,800	4,100	4,100
Mexico	1,807	1,827	1,850
Pakistan	1,630	1,675	1,725
Russia	1,380	1,370	1,355
Others	9,063	9,232	8,402
Total Foreign	47,731	48,654	47,564
United States	11,751	11,076	10,861
Total	59,482	59,730	58,425
Swine:			
Brazil	3,335	3,400	3,519
Canada	1,822	1,805	1,890
China	54,930	56,710	54,870
European Union	22,359	22,540	23,350
Japan	1,309	1,264	1,254
Korea, South	1,252	1,200	1,217
Mexico	1,284	1,290	1,323
Philippines	1,340	1,353	1,370
Russia	2,400	2,510	2,615
Vietnam	2,349	2,425	2,450
Others	5,918	5,692	5,337
Total Foreign	98,298	100,189	99,195
United States	10,525	10,370	11,121
Total	108,823	110,559	110,316

FAS, Office of Global Analysis, (202) 720-6301.   Prepared or estimated on the basis of official USDA production, supply, and distribution statistics from foreign governments.

Table 7-63.—Meat: U.S. exports, 2011–2013

Country	2011	2012	2013
	Metric tons	Metric tons	Metric tons
Beef & veal, fr/ch/fz:			
Japan	138,945	137,769	207,586
Canada	145,535	132,954	130,652
Mexico	156,083	111,614	128,437
Hong Kong	44,020	63,765	120,869
Korea, South	135,553	111,991	94,404
Taiwan	35,354	19,365	32,046
Egypt	33,814	32,169	17,341
Netherlands	14,589	11,131	11,710
Chile	3,988	8,906	10,235
Philippines	6,780	5,962	7,049
Italy(*)	6,397	5,308	5,719
United Arab Emirates	7,999	5,113	4,930
Vietnam	43,306	39,644	3,891
Dominican Republic	4,142	3,579	3,772
Germany(*)	3,812	2,975	3,230
Bahamas, The	3,470	3,197	2,887
Kuwait	3,905	2,084	2,560
Indonesia	5,241	1,030	2,045
Guatemala	2,519	2,022	1,870
Netherlands Antilles(*)	1,932	1,841	1,851
Jamaica	2,135	1,835	1,533
Singapore	2,894	1,216	1,371
Peru	896	1,744	1,353
Cayman Islands	1,377	1,332	1,332
Qatar	2,539	868	1,198
Colombia	1,087	331	1,148
Angola	2,471	317	1,117
Trinidad and Tobago	1,095	1,067	1,103
Honduras	627	646	961
Rest of World	68,943	61,525	11,456
World Total	881,448	773,297	815,656
Beef & veal, prep/pres:			
Canada	30,715	32,898	34,306
Hong Kong	2,811	458	4,010
Mexico	521	866	893
Germany(*)	533	285	840
Australia(*)	923	793	710
Panama	330	396	561
Philippines	30	548	252
Brazil	352	0	184
Guatemala	122	135	136
New Zealand(*)	54	78	115
United Arab Emirates	130	128	111
Suriname	123	162	88
Turks and Caicos Islands	188	180	71
Netherlands Antilles(*)	76	76	70
United Kingdom	103	25	67
Angola	0	8	65
Bahamas, The	242	200	63
Colombia	7	5	61
Costa Rica	6	31	60
Italy(*)	21	0	55
Lebanon	37	38	51
Leeward-Windward Islands(*)	10	47	46
Trinidad and Tobago	25	27	42
Palau	28	32	35
Kuwait	63	18	33
Jamaica	25	40	33
Peru	73	0	31
Dominican Republic	31	38	29
Netherlands	95	32	28
Rest of World	2,385	522	345
World Total	40,058	38,065	43,390

See footnote(s) at end of table.

Table 7-63.—Meat: U.S. exports, 2011–2013—Continued

Country	2011	2012	2013
	Metric tons	*Metric tons*	*Metric tons*
Pork, fr/ch/fz:			
Mexico	286,080	409,868	427,968
Japan	465,636	421,235	401,950
China	219,502	198,995	153,405
Canada	120,932	130,672	111,999
Korea, South	148,885	132,603	87,239
Australia(*)	56,080	59,788	48,706
Hong Kong	36,672	38,295	45,753
Colombia	9,444	16,198	29,969
Philippines	29,561	26,000	28,169
Honduras	17,827	17,850	20,824
Dominican Republic	12,054	13,892	17,028
Chile	8,192	12,223	16,815
Ukraine	3,791	2,236	14,456
Taiwan	14,583	9,690	11,927
Guatemala	5,885	7,776	8,465
New Zealand(*)	4,587	7,643	7,411
Panama	2,756	3,014	7,006
Singapore	8,375	7,549	6,800
Russia	63,679	89,324	5,315
Albania	1,137	5,128	3,498
Bahamas, The	2,899	2,831	3,374
Costa Rica	850	1,342	3,276
El Salvador	1,541	2,469	2,943
Trinidad and Tobago	2,353	2,518	2,442
Ecuador	1,733	1,737	2,265
Cuba	3,007	2,692	2,212
Leeward-Windward Islands(*)	1,306	1,236	1,346
Uruguay	617	747	1,242
Vietnam	2,855	1,766	1,226
Rest of World	15,528	17,639	13,375
World Total	1,548,345	1,644,953	1,488,403
Pork, hams/shldrs,cured:			
Canada	25,984	35,135	36,134
China	5,045	5,545	6,796
Mexico	70,518	2,041	2,456
Panama	534	1,432	2,102
Japan	1,543	1,294	2,025
Trinidad and Tobago	750	858	798
Belize	338	473	589
Colombia	277	642	576
Bahamas, The	1,131	411	384
Netherlands Antilles(*)	314	413	330
Philippines	278	74	198
Australia(*)	74	0	148
Honduras	71	72	123
Leeward-Windward Islands(*)	90	162	119
Haiti	147	149	109
Barbados	138	198	88
Korea, South	78	94	52
Bolivia	33	0	48
Italy(*)	8	0	35
United Kingdom	54	15	33
Bermuda	0	13	33
Netherlands	160	35	24
Chile	0	34	18
Singapore	87	30	17
Guyana	0	3	15
Costa Rica	63	7	9
Suriname	4	2	6
Dominican Republic	3	1	4
El Salvador	0	0	3
Rest of World	2,662	3,437	11
World Total	110,382	52,571	53,284

See footnote(s) at end of table.

Table 7-63.—Meat: U.S. exports, 2011-2013—Continued

Country	2011	2012	2013
	Metric tons	Metric tons	Metric tons
Pork, bacon, cured:			
Mexico	15,973	13,291	12,596
Canada	3,343	5,794	3,513
Japan	4,610	3,415	1,951
Korea, South	426	977	1,552
Dominican Republic	518	494	547
Bahamas, The	367	404	370
Colombia	120	226	363
Guatemala	397	373	339
Costa Rica	275	360	304
Australia(*)	130	225	273
Ecuador	114	110	272
Netherlands	109	80	258
Panama	125	159	216
Chile	99	79	214
Singapore	392	358	209
Hong Kong	164	316	184
Netherlands Antilles(*)	149	85	149
Honduras	175	151	140
Philippines	60	64	119
Italy(*)	110	12	118
China	6	20	114
Leeward-Windward Islands(*)	113	104	112
New Zealand(*)	238	141	108
El Salvador	37	74	101
United Arab Emirates	66	35	80
Bermuda	74	79	73
Trinidad and Tobago	98	190	67
Peru	25	1	60
Nicaragua	84	92	59
Rest of World	596	689	370
World Total	28,992	28,395	24,827
Pork, prep/pres, nt/cn:			
Canada	32,653	42,651	56,414
Mexico	6,109	11,604	17,998
Chile	25	850	880
Colombia	210	550	858
Dominican Republic	301	215	349
Panama	112	137	326
Netherlands Antilles(*)	131	151	301
Philippines	726	492	279
Ecuador	198	211	276
Bahamas, The	106	258	271
Singapore	261	72	211
Korea, South	265	321	208
Honduras	410	228	201
Japan	565	723	187
Micronesia	146	140	174
Hong Kong	234	195	173
Trinidad and Tobago	46	94	138
Costa Rica	224	196	120
Guatemala	318	402	111
Peru	14	93	91
Ukraine	0	168	82
Palau	64	61	80
Marshall Islands	82	97	78
Belize	0	1	66
Leeward-Windward Islands(*)	35	39	61
El Salvador	34	35	50
Netherlands	4	4	32
Italy(*)	0	0	29
Nicaragua	12	15	27
Rest of World	1,004	219	171
World Total	44,287	60,223	80,241

See footnote(s) at end of table.

Table 7-63.—Meat: U.S. exports, 2011–2013—Continued

Country	2011	2012	2013
	Metric tons	*Metric tons*	*Metric tons*
Pork, prep/pres, canned:			
Philippines	4,310	4,463	6,572
Japan	6,467	5,927	4,134
Australia(*)	2,524	2,983	3,011
Canada	2,198	2,561	2,168
Hong Kong	1,116	1,823	1,987
Mexico	521	1,174	1,181
Korea, South	1,065	1,059	1,052
China	205	258	1,035
Panama	709	705	808
New Zealand(*)	459	411	560
Guatemala	166	368	470
Singapore	260	310	416
Colombia	241	132	351
Costa Rica	81	73	279
Honduras	70	217	264
United Arab Emirates	93	98	249
Netherlands Antilles(*)	65	169	213
Thailand	59	134	179
Dominican Republic	72	147	175
Bahamas, The	202	187	143
Trinidad and Tobago	49	17	106
French Pacific Islands(*)	67	72	89
El Salvador	38	18	81
Taiwan	58	30	66
Vietnam	35	63	61
Jamaica	14	19	46
Haiti	69	16	44
Palau	31	37	41
Bermuda	31	15	39
Rest of World	1,352	886	235
World Total	22,626	24,372	26,055
Lamb & mutton; fr/ch/fz:			
Mexico	3,027	3,140	1,909
Canada	560	369	383
Leeward-Windward Islands(*)	419	172	130
Bahamas, The	564	248	97
Dominican Republic	128	148	81
Bermuda	149	18	63
Honduras	12	7	59
Trinidad and Tobago	166	108	48
Panama	99	42	30
Netherlands Antilles(*)	520	124	29
Germany(*)	28	6	28
Costa Rica	124	36	23
Spain	29	33	20
Singapore	89	8	17
Philippines	0	17	16
Hong Kong	130	14	14
Australia(*)	7	6	13
Barbados	92	8	12
Turks and Caicos Islands	22	15	10
Guatemala	115	18	10
United Arab Emirates	84	27	9
Malaysia	0	0	7
Netherlands	943	12	5
Mayotte	0	0	5
Equatorial Guinea	27	5	4
Jamaica	99	15	4
Saudi Arabia	43	0	4
Italy(*)	296	9	4
Uruguay	5	0	4
Rest of World	333	79	23
World Total	8,108	4,680	3,060

See footnote(s) at end of table.

Table 7-63.—Meat: U.S. exports, 2011–2013—Continued

Country	2011	2012	2013
	Metric tons	Metric tons	Metric tons
Sausages & bologna:			
Canada	29,908	33,435	37,369
Japan	29,103	24,631	26,751
China	5,150	15,341	18,543
Mexico	6,914	8,266	8,787
Hong Kong	23,018	11,456	3,128
Korea, South	3,011	2,329	2,332
Philippines	1,579	1,553	2,186
Belize	631	770	854
Bahamas, The	941	1,001	768
Panama	332	495	683
Guatemala	432	403	538
United Kingdom	151	322	420
Uruguay	311	447	403
Taiwan	811	247	337
Singapore	223	216	311
El Salvador	265	283	302
Peru	102	262	291
Netherlands Antilles(*)	314	356	278
Dominican Republic	317	253	254
Marshall Islands	170	202	252
Australia(*)	111	97	193
Trinidad and Tobago	100	94	163
Chile	104	104	148
Colombia	55	67	130
Bermuda	159	157	126
Netherlands	100	12	79
Thailand	58	67	76
Costa Rica	60	44	72
Ecuador	87	91	68
Rest of World	887	790	604
World Total	105,402	103,790	106,443
Other meat products, fr/ch/fz:			
Mexico	61	0	244
Hong Kong	37	143	35
Leeward-Windward Islands(*)	34	40	32
Dominican Republic	0	5	20
Singapore	21	21	16
Brazil	0	0	6
Cayman Islands	21	2	6
Netherlands Antilles(*)	27	14	4
Barbados	1	0	2
Costa Rica	9	1	1
Bahamas, The	0	19	1
Bermuda	0	0	1
Trinidad and Tobago	1	0	1
Netherlands	0	0	1
Turks and Caicos Islands	0	2	1
Panama	2	10	0
Bahrain	3	1	0
China	0	25	0
Ireland	0	2	0
Germany(*)	0	1	0
Guatemala	0	5	0
Israel(*)	0	1	0
Jamaica	0	1	0
Jordan	0	3	0
Malaysia	52	30	0
Saudi Arabia	50	0	0
Timor-Leste	1	0	0
Taiwan	24	0	0
United Kingdom	0	1	0
Rest of World	143	252	0
World Total	488	578	369

See footnote(s) at end of table.

Table 7-63.—Meat: U.S. exports, 2011–2013—Continued

Country	2011	2012	2013
	Metric tons	*Metric tons*	*Metric tons*
Variety meats, beef:			
Egypt	113,381	105,299	122,973
Mexico	100,144	80,324	87,542
Japan	19,738	14,921	27,071
Peru	9,931	11,155	16,005
Korea, South	18,469	13,622	10,966
Canada	14,766	13,711	9,656
Angola	8,903	10,220	7,188
Hong Kong	4,931	3,516	5,018
Philippines	5,938	5,127	4,077
Gabon	2,688	3,155	3,634
Jamaica	5,347	3,703	3,129
Indonesia	12,555	617	2,986
Cote d'Ivoire	4,522	5,935	2,690
Chile	260	2,481	2,624
Moldova	1,834	2,954	2,527
Ukraine	998	779	2,025
Colombia	331	581	1,526
Ecuador	1,210	1,248	1,239
Trinidad and Tobago	528	523	608
Guatemala	864	1,621	571
Vietnam	272	497	567
Costa Rica	353	540	493
Bahrain	543	654	420
Mozambique	515	516	418
Panama	72	58	311
Zambia	1,039	1,333	308
Honduras	389	385	278
Libya	0	0	268
Nicaragua	194	179	225
Rest of World	32,589	34,258	2,308
World Total	363,303	319,909	319,653
Variety meats, pork:			
China	136,784	134,633	156,841
Mexico	144,853	148,940	148,751
Hong Kong	69,353	34,846	35,481
Canada	18,421	17,794	16,268
Philippines	3,970	6,478	14,327
Japan	15,594	22,713	14,108
Singapore	963	259	9,375
Korea, South	37,869	14,212	6,603
Taiwan	11,764	7,744	5,782
Chile	4,338	3,333	5,741
Panama	1,499	2,968	2,860
Haiti	1,270	1,141	2,494
Colombia	1,124	1,868	1,926
Cuba	468	497	1,789
Guatemala	1,337	726	1,120
Peru	158	186	1,043
Albania	231	1,039	988
Honduras	908	1,626	957
Trinidad and Tobago	1,015	643	668
Benin	0	0	544
Dominican Republic	871	942	544
Australia(*)	4,676	3,975	477
El Salvador	301	350	367
Germany(*)	7	0	328
Vietnam	733	821	306
Netherlands Antilles(*)	401	286	289
Jamaica	53	25	288
Gabon	0	104	226
Uruguay	0	144	167
Rest of World	7,936	4,784	959
World Total	466,895	413,076	431,617

See footnote(s) at end of table.

Table 7-63.—Meat: U.S. exports, 2011–2013—Continued

Country	2011	2012	2013
	Metric tons	Metric tons	Metric tons
Variety meats, other:			
Mexico	11,405	11,836	12,129
Netherlands	568	2,780	3,340
Canada	2,974	2,302	3,169
Hong Kong	3,159	2,526	2,045
Korea, South	572	1,149	1,092
Venezuela	0	38	699
Taiwan	9	0	330
Vietnam	0	59	282
Peru	13	0	264
China	219	360	221
Saudi Arabia	1	8	209
Japan	4	34	171
Trinidad and Tobago	12	0	163
Dominican Republic	283	31	140
Colombia	67	25	89
Philippines	13	40	74
Netherlands Antilles(*)	280	111	68
Germany(*)	180	248	68
Nigeria	24	51	49
Brazil	1	0	41
Barbados	15	21	26
Leeward-Windward Islands(*)	66	26	25
Singapore	14	17	22
Bahamas, The	43	21	21
Denmark(*)	127	15	21
Uruguay	28	12	20
Bermuda	39	37	19
United Arab Emirates	75	48	19
Australia(*)	20	3	16
Rest of World	834	158	93
World Total	21,042	21,954	24,922

(*) Denotes a country that is a summarization of its component countries. Users should use cautious interpretation on quantity reports using mixed units of measure. Quantity line items will only include statistics on the units of measure that are equal to, or are able to be converted to, the assigned unit of measure of the grouped commodities.
FAS, Office of Global Analysis, (202) 720-6301. Data Source: Department of Commerce, U.S. Census Bureau, Foreign Trade Statistics.

Table 7-64.—Meat, beef, veal, and swine: International trade, imports and exports, 2013–2015

Country	2013	2014	2015
	1,000 metric tons	*1,000 metric tons*	*1,000 metric tons*
Principle exporters, beef and veal:			
Australia	1,593	1,851	1,854
Belarus	220	184	213
Brazil	1,849	1,909	1,705
Canada	332	378	391
European Union	244	301	303
India	1,765	2,082	1,806
Mexico	166	194	228
New Zealand	529	579	639
Paraguay	326	389	380
Uruguay	340	350	372
Others	588	605	627
Total Foreign	7,952	8,822	8,518
United States	1,174	1,167	1,035
Total	9,126	9,989	9,553
Principle importers, beef and veal:			
Canada	296	284	280
Chile	245	241	245
China	412	417	663
Egypt	195	270	360
European Union	376	372	365
Hong Kong	473	646	339
Japan	760	739	708
Korea, South	375	392	414
Malaysia	194	205	235
Russia	1,023	929	625
Others	2,096	2,066	1,814
Total Foreign	6,445	6,561	6,048
United States	1,020	1,337	1,559
Total	7,465	7,898	7,607
Principle exporters, swine:			
Australia	36	37	36
Belarus	585	556	627
Brazil	1,246	1,218	1,234
Canada	164	163	180
Chile	244	276	231
China	2,227	2,164	2,400
European Union	111	117	128
Mexico	4	25	19
Norway	7	13	36
Vietnam	40	40	40
Others	100	62	46
Total Foreign	4,764	4,671	4,977
United States	2,262	2,203	2,242
Total	7,026	6,874	7,219
Principle importers, swine:			
Australia	183	191	220
Canada	220	214	216
China	770	761	1,029
Hong Kong	399	347	397
Japan	1,223	1,332	1,269
Korea, South	388	480	599
Mexico	783	818	981
Philippines	172	199	175
Russia	868	515	407
Ukraine	98	117	116
Others	1,085	891	756
Total Foreign	6,189	5,865	6,165
United States	399	457	504
Total	6,588	6,322	6,669

FAS, Office of Global Analysis, (202) 720-6301. Prepared or estimated on the basis of official USDA production, supply, and distribution statistics from foreign governments.

Table 7-65.—Meats and lard: Production and consumption, United States, 2006–2015 [1]

Year	Beef			Veal			Lamb and mutton		
	Production	Consumption		Produc-tion	Consumption		Produc-tion	Consumption	
		Total	Per capita		Total	Per capita		Total	Per capita
	Million pounds	Million pounds	Pounds	Million pounds	Million pounds	Pounds	Million pounds	Million pounds	Pounds
2006	26,256	28,137	94.2	156	155	0.5	190	356	1.2
2007	26,523	28,141	93.3	146	145	0.5	189	385	1.3
2008	26,664	27,303	89.6	152	150	0.5	180	343	1.1
2009	26,068	26,836	87.3	147	147	0.5	177	338	1.1
2010	26,412	26,390	85.2	145	150	0.5	168	318	1.0
2011	26,281	25,538	81.8	136	137	0.4	153	295	0.8
2012	25,995	25,755	81.9	125	123	0.4	161	299	0.8
2013	25,790	25,476	80.5	117	119	0.4	161	324	0.9
2014	24,316	24,683	77.4	100	98	0.3	161	340	0.9
2015	23,760	24,773	77.1	88	88	0.3	155	357	1.0

Year	Pork			All meats		
	Production	Consumption		Production	Consumption	
		Total	Per capita		Total	Per capita
	Million pounds	Million pounds	Pounds	Million pounds	Million pounds	Pounds
2006	21,074	19,055	63.8	47,675	47,703	160
2007	21,962	19,763	65.5	48,817	48,434	160
2008	23,367	19,415	63.8	50,362	47,211	155
2009	23,020	19,870	64.7	49,412	47,191	154
2010	22,456	19,076	61.6	49,181	45,934	148
2011	22,775	18,382	58.9	49,232	44,351	142
2012	23,268	18,607	59.2	49,439	44,784	143
2013	23,204	19,195	60.4	49,174	45,024	142
2014	22,858	18,836	59.1	47,344	43,957	139
2015	24,517	20,593	64.1	48,432	45,811	142

[1] Carcass weight equivalent or dressed weight. Beginning 1977, pork production was no longer reported as "pork, excluding lard." This series has been revised to reflect pork production in prior years on a dressed weight basis that is comparable with the method used to report beef, veal, and lamb and mutton. Edible offals are excluded. Shipments to the U.S. territories are included in domestic consumption.

ERS, Animal Products and Cost of Production Branch, (202) 694–5409.

Table 7-66.—Hides, packer: Average price per piece, Central U.S., 2004–2013

Year	Steers				Heifers	
	Heavy native	Heavy Texas	Butt branded	Colorado branded	Heavy native	Branded
	Dollars	Dollars	Dollars	Dollars	Dollars	Dollars
2004	67.09	64.91	64.39	61.48	57.07	54.02
2005	65.64	63.50	63.53	60.90	57.89	54.20
2006	68.87	67.76	67.79	65.99	60.30	57.52
2007	72.01	70.51	70.72	67.79	65.70	61.85
2008	63.94	63.22	62.62	59.35	58.35	57.18
2009	45.29	44.10	42.70	31.84	29.00	31.86
2010	71.93	72.30	70.77	67.31	63.44	67.57
2011	82.17	81.70	80.77	76.42	74.39	71.17
2012	84.53	84.23	84.28	79.83	71.77	70.19
2013	98.73	95.74	98.56	94.38	84.68	82.90

AMS, Livestock & Grain Market News, (202) 720–7316.

Table 7-67.—Hides and skins: United States imports by country of origin, 2011–2013

Country of origin	2011	2012	2013
	Pieces	Pieces	Pieces
Hides and skins, mixed:			
Canada	931,947	1,209,375	1,007,569
Mexico	196,290	240,792	432,885
China	16,484	36,891	80,638
Turkey	48,515	17,273	22,030
Brazil	4,676	36,896	15,216
Italy(*)	18,929	3,939	12,227
Belgium-Luxembourg(*)	10,002	9,951	11,367
India	4	5,658	7,134
Thailand	6,868	6,833	5,324
Argentina	333	1,432	5,115
New Zealand(*)	16,621	7,538	4,377
Germany(*)	782	17	3,088
Australia(*)	321	7,165	2,647
Pakistan	24,660	2,555	2,514
Singapore	1,659	410	2,391
Taiwan	0	0	1,800
Lebanon	900	800	1,000
Colombia	6,204	10,336	889
France(*)	931	241	786
United Kingdom	145	1,916	625
Malaysia	40	1,661	575
Uruguay	524	31	547
Guinea	1,205	200	540
Korea, South	1,539	26	421
Dominican Republic	0	0	360
Spain	3,365	1,071	343
South Africa	4,980	258	203
Cote d'Ivoire	0	1,760	200
Denmark(*)	0	6	48
Rest of World	7,734	4,634	133
World Total	1,305,658	1,609,665	1,622,992
	Number	Number	Number
Furskins:			
Canada	1,874,531	2,543,217	2,288,602
Netherlands	387,273	483,686	390,630
Belgium-Luxembourg(*)	263,860	75,336	190,823
Poland	16,335	35,765	190,022
Sweden	133,255	129,624	124,520
Ukraine	0	97,000	50,148
Spain	28,418	35,162	44,123
Germany(*)	34,632	167	43,623
Finland	35,487	28,613	24,441
Denmark(*)	9,308	5,373	17,145
Greece	81,799	4,539	13,955
Russia	5,969	7,666	12,375
Mexico	27,696	3,477	11,025
Italy(*)	172	2	3,358
Ireland	0	0	3,019
Bosnia and Herzegovina	1,728	1,739	1,937
Switzerland(*)	4,350	8	574
China	1,026	8,540	497
Latvia	0	4	233
United Kingdom	0	0	59
Lithuania	0	138	35
South Africa	21	38	30
Uruguay	0	0	22
United Arab Emirates	0	0	13
Namibia	6	0	8
France(*)	1,446	864	6
Hong Kong	0	746	5
Austria	0	0	4
New Zealand(*)	7	0	4
Rest of World	93,759	96,776	0
World Total	3,001,078	3,558,480	3,411,236

(*) Denotes a country that is a summarization of its component countries.
FAS, Office of Global Analysis, (202) 720–6301.

Table 7-68.—Hides: U.S. trade exports, 2011–2013

Country	2011	2012	2013
	Pieces	*Pieces*	*Pieces*
Cattle hides, whole:			
China	10,539,676	12,009,559	17,052,725
Korea, South	4,383,022	4,137,583	4,356,831
Taiwan	1,496,401	1,098,726	1,870,784
Mexico	1,299,132	1,506,203	1,740,158
Italy(*)	2,347,259	1,154,127	1,129,983
Hong Kong	1,522,844	495,545	844,279
Japan	605,802	711,187	625,791
Vietnam	1,138,348	650,162	478,776
Thailand	692,901	500,532	309,296
Belgium-Luxembourg(*)	42,240	78,298	220,809
Turkey	292,970	270,603	213,326
Israel(*)	95,806	123,957	91,471
Netherlands	88,688	35,248	89,113
Indonesia	5,700	4,183	80,795
Germany(*)	72,251	40,842	77,561
Canada	184,116	99,987	72,476
India	14,128	11,506	41,713
Pakistan	18,790	0	30,321
Croatia	19,179	8,560	25,890
Spain	54,403	7,318	22,101
Portugal	4,800	1,680	18,605
Haiti	4,000	3,800	12,358
France(*)	42,528	5,480	7,932
Macau	5,879	0	6,469
Slovenia	0	0	4,059
Poland	2,217	0	3,525
Philippines	0	0	3,318
Austria	0	0	2,559
Chile	2,550	0	2,420
Rest of World	122,044	67,615	6,806
World Total	25,097,674	23,022,701	29,442,250
Sheep & lambskins:			
China	1,173,215	1,010,317	980,503
Mexico	43,255	222,824	151,017
Turkey	63,307	129,514	144,851
Russia	4,860	66,075	85,690
Vietnam	0	0	34,125
Pakistan	0	8,036	31,321
India	0	7,400	10,200
France(*)	0	349	9,450
Morocco	0	0	9,000
Canada	13,498	6,451	8,790
Germany(*)	1,400	3,000	7,900
Guatemala	0	0	7,474
Ethiopia(*)	0	0	5,098
United Kingdom	207	3,466	3,600
Argentina	0	0	700
Netherlands Antilles(*)	15	508	150
United Arab Emirates	100	0	0
Barbados	0	10	0
Belgium-Luxembourg(*)	111	0	0
Guyana	30	0	0
Italy(*)	0	10,311	0
Japan	782	0	0
Lebanon	0	53	0
Poland	0	7,425	0
Spain	0	3,296	0
Switzerland(*)	1,192	391	0
Leeward-Windward Islands(*)	0	630	0
World Total	1,301,972	1,480,056	1,489,869
Pig and hog skins, pieces:			
Mexico	2,204,647	2,359,767	2,523,738
Taiwan	1,367,474	1,079,508	1,470,387
Vietnam	550,345	442,702	643,800
Thailand	114,426	560,787	425,200
United Kingdom	56,317	43,745	85,905
Japan	20,939	12,665	72,125
Slovenia	14,631	15,245	61,997
Hong Kong	390,447	143,878	24,276
Italy(*)	9,270	1,360	6,989
Greece	6,931	0	6,344
China	117,555	2,600	3,721
Ecuador	0	0	2,688
El Salvador	0	0	1,995
Canada	5,393	0	0
Dominican Republic	1,067	0	0
France(*)	0	2,400	0
Germany(*)	1,650	0	0
Korea, South	107,561	66,730	0
World Total	4,968,653	4,731,387	5,329,165

See footnote(s) at end of table.

Table 7-68.—Hides: U.S. trade exports, 2011–2013—Continued

Country	2011	2012	2013
	Number	Number	Number
Mink furskins, undressed:			
China	3,678,292	3,626,923	4,942,895
Hong Kong	1,107,203	2,743,323	3,664,569
Canada	1,833,404	2,316,857	3,285,175
Korea, South	1,290,403	2,504,559	1,816,284
Greece	394,578	625,060	610,844
Germany(*)	196,763	45,971	269,869
Italy(*)	51,089	61,612	76,373
Denmark(*)	0	7,580	65,201
United Kingdom	20,781	12,070	54,696
France(*)	5,327	25,901	53,542
United Arab Emirates	10,258	48,988	42,316
Poland	9,224	50,441	17,732
Malaysia	22,840	18,236	16,363
Russia	7,172	6,016	4,804
Japan	3,767	4,084	4,147
Haiti	0	0	4,006
Finland	459	0	3,450
Switzerland(*)	1,272	0	1,538
Ukraine	6,828	9,805	1,458
Turkey	1,491	0	1,409
Mexico	316	2,423	1,400
Brazil	0	0	1,292
Australia(*)	0	374	0
Belize	7,878	0	0
Cyprus	23,526	0	0
Czech Republic	25,177	0	0
Lithuania	2,282	0	0
Macau	4,912	0	0
New Zealand(*)	0	2,832	0
World Total	8,705,242	12,113,055	14,939,363
Other furskins, whole:			
Canada	1,639,155	1,945,633	2,035,197
China	837,207	1,052,347	1,103,397
Germany(*)	100,935	162,938	125,626
Italy(*)	38,613	56,435	119,973
Greece	16,594	3,801	83,534
Hong Kong	43,831	182,017	79,071
Poland	78,049	102,855	70,587
Czech Republic	59,226	73,521	31,010
Estonia	12,473	19,007	23,512
Portugal	5,986	2,471	11,954
Korea, South	690	942	6,897
Turkey	12,740	7,264	5,117
Venezuela	0	128	3,766
Belgium-Luxembourg(*)	1,915	10,442	3,468
Taiwan	0	0	2,723
Pakistan	1,277	1,571	1,509
Austria	500	0	1,116
France(*)	0	0	973
United Kingdom	2,000	5,898	785
Brazil	0	0	698
Guatemala	3,400	5,100	611
Suriname	0	0	576
Panama	0	0	547
Lithuania	2,000	0	500
Malaysia	0	0	334
Netherlands	0	2,092	256
Singapore	65	0	51
United Arab Emirates	1,866	2,718	0
Australia(*)	2,204	746	0
Rest of World	16,689	27,057	0
World Total	2,877,415	3,664,983	3,713,788

See footnote(s) at end of table.

Table 7-68.—Hides: U.S. trade exports, 2011–2013—Continued

Country	2011	2012	2013
	Pieces	Pieces	Pieces
Other hides & skins, mixed:			
China	5,378,381	5,606,765	6,606,123
Korea, South	2,298,612	1,941,925	2,473,377
Taiwan	904,180	549,675	469,238
Thailand	351,699	236,534	266,663
Vietnam	376,378	232,105	242,790
France(*)	133,990	136,165	214,068
Italy(*)	266,746	142,152	190,938
Germany(*)	156,706	82,726	152,641
Hong Kong	350,766	159,310	141,698
Singapore	41,706	24,363	114,521
Haiti	76,179	85,870	66,126
Turkey	186,624	90,016	47,671
Pakistan	31,061	30,934	43,524
Mexico	116,508	25,528	39,364
Canada	39,600	16,903	28,556
Japan	11,320	18,365	24,600
Malaysia	0	0	12,062
India	3,216	0	6,101
Israel(*)	0	0	6,056
Portugal	1,600	855	2,632
Indonesia	0	0	1,680
Belgium-Luxembourg(*)	4,080	5,425	1,360
Kuwait	19	0	1,080
Spain	9	8,568	477
Estonia	0	0	333
United Kingdom	17	40	298
Brazil	0	2,115	249
Colombia	398	100	157
Sri Lanka	0	0	150
Rest of World	24,725	24,018	138
World Total	10,754,520	9,420,457	11,154,671

(*) Denotes a country that is a summarization of its component countries.
FAS, Office of Global Analysis, (202) 720-6301.

Table 7-69.—Mink: Farms, pelts produced, average price, and value, United States, 2006–2015

Year	Mink farms	Pelts produced	Average marketing price	Value of mink pelts
	Number	Number	Dollars	Dollars
2006	279	2,858,800	48.40	138,365,920
2007	283	2,828,200	65.70	185,812,740
2008	274	2,820,700	41.60	117,341,120
2009	278	2,866,700	65.10	186,622,170
2010	265	2,840,200	81.90	232,612,380
2011	268	3,091,470	94.30	291,525,621
2012 [1]	(NA)	(NA)	(NA)	(NA)
2013	(NA)	3,544,610	56.30	199,561,543
2014	(NA)	3,741,150	57.70	215,864,355
2015	(NA)	3,749,450	31.10	116,607,895

(NA) Not available.   [1] Due to sequestration the Mink report was suspended.
NASS, Livestock Branch, (202) 720–3570.

Table 7-70.—Mink pelts: Pelts produced by color class, selected States, and United States, 2015

State	Black	Demi/Wild	Pastel	Sapphire	Blue Iris	Mahogany
	Number	Number	Number	Number	Number	Number
Idaho	129,000	10,000	8,000	14,500	(D)	193,000
Illinois	68,000	-	-	(D)	-	(D)
Iowa	50,000	(D)	(D)	(D)	(D)	37,000
Michigan	44,000	-	(D)	-	(D)	(D)
Minnesota	100,000	27,000	(D)	(D)	(D)	78,000
Montana	3,700	(D)	(D)	(D)	(D)	(D)
Ohio	57,000	(D)	-	(D)	(D)	(D)
Oregon	220,000	-	(D)	(D)	77,000	(D)
Pennsylvania	35,000	(D)	(D)	9,000	17,000	(D)
Utah	310,000	34,000	(D)	41,000	2,200	395,000
Washington	58,000	-	(D)	(D)	(D)	(D)
Wisconsin	780,000	(D)	34,000	33,000	129,000	135,000
Other States [1]	9,000	8,280	68,900	17,980	53,220	58,840
United States	1,863,700	79,280	110,900	115,480	278,420	896,840

State	Pearl	Lavender	Violet	White	Other	Total [2]
	Number	Number	Number	Number	Number	Number
Idaho	-	(D)	(D)	(D)	(D)	372,700
Illinois	(D)	-	-	(D)	-	98,700
Iowa	-	-	-	(D)	-	96,300
Michigan	-	(D)	(D)	8,000	-	60,250
Minnesota	-	-	-	11,500	-	265,520
Montana	-	-	-	(D)	(D)	14,900
Ohio	-	-	-	(D)	-	77,800
Oregon	-	-	(D)	-	-	332,900
Pennsylvania	(D)	(D)	(D)	(D)	(D)	86,770
Utah	(D)	(D)	(D)	(D)	(D)	933,880
Washington	-	-	-	-	-	71,800
Wisconsin	(D)	14,500	26,000	158,000	6,000	1,316,610
Other States [1]	95,700	7,960	33,250	28,410	15,510	21,320
United States	95,700	22,460	59,250	205,910	21,510	3,749,450

- Represents zero.   (D) Withheld to avoid disclosing data for individual operations.   [1] Other States include New Hampshire, New Jersey, New York, North Carolina, South Dakota, Virginia, and some pelts from the above listed States that were not published to avoid disclosing individual operations.   [2] Published color classes may not add to the State total to avoid disclosing individual operations.
NASS, Livestock Branch, (202) 720–3570.

Table 7-71.—Livestock: Number of animals slaughtered under Federal inspection and number of whole carcasses condemned, 2006–2015 [1]

Year	Cattle		Calves		Sheep and lambs	
	Total head	Condemned	Total head	Condemned	Total head	Condemned
	1,000	*1,000*	*1,000*	*1,000*	*1,000*	*1,000*
2006	32,861	143.1	682	11.1	2,534	4.7
2007	33,473	141.5	769	13.6	2,497	4.1
2008	34,220	146.8	866	24.0	2,447	5.2
2009	32,714	143.1	951	23.0	2,297	3.3
2010	33,295	149.4	908	22.7	2,285	3.5
2011	29,530	143.2	740	33.0	1.684	3.1
2012	32,062	142.3	742	26.7	1,978	4.2
2013	31,631	154.0	761	36.7	2,129	5.0
2014	30,721	157.9	645	36.7	2,171	4.1
2015	27,738	161.2	395	21.9	1,994	2.9

Year	Goats		Hogs		Horses [2]	
	Total head	Condemned	Total head	Condemned	Total head	Condemned
	1,000	*1,000*	*1,000*	*1,000*	*1,000*	*1,000*
2006	561	0.9	103,600	417.0	102	0.9
2007	613	0.7	105,611	404.8	58	0.4
2008	654	0.9	115,600	393.0	0	0
2009	660	0.8	113,395	333.6	0	0
2010	617	0.7	109,346	316.6	0	0
2011	509	0.5	91,541	266.2	0	0
2012	576	0.5	111,440	256.7	0	0
2013	540	0.5	111,935	296.3	0	0
2014	499	0.7	108,650	320.5	0	0
2015	448	0.5	113,259	307.2	0	0

[1] Data are reported by the Food Safety and Inspection Service, USDA for the fiscal year ending September 30. Condemnations include ante-mortem and post-mortem inspection. [2] Equine slaughter was discontinued during the week of September 22, 2007.
NASS, Livestock Branch, (202) 720–3570.

Table 7-72.—Livestock: Inventory and value, United States, Jan. 1, 2013–2015

Class of livestock and poultry	Inventory			Value					
	2013	2014	2015	Per head [1]			Total		
				2013	2014	2015	2013	2014	2015
	1,000	*1,000*	*1,000*	*Dollars*	*Dollars*	*Dollars*	*1,000 Dollars*	*1,000 Dollars*	*1,000 Dollars*
Angora goats [2]	136	148	160	87.70	105	124	12,694	12,807	16,850
Cattle	90,095	88,526	89,143	1,139.00	1,223.00	1,584.00	102,671,904	108,255,950	141,211,070
Chickens [3]	474,902	483,530	470,915	4.15	4.11	4.36	1,971,590	1,989,495	2,055,503
Hogs [3]	64,775	67,776	68,299	138	144	96	8,921,822	9,754,191	6,566,642
Sheep and lambs	5,360	5,245	5,280	221.00	188	214	946,194	984,236	1,129,447

[1] Based on reporters' estimates of average price per head in their localities. [2] Four state total for angora goats (Arizona, California, New Mexico, Texas). [3] Dec. 1 of preceding year.
NASS, Livestock Branch, (202) 720–3570.

Table 7-73.—Livestock: Market year for specified commodities

Cattle market year average price received: United States marketing years 2006–2015 [1]

Year	Cows [2]	Steers and heifers	Calves	Beef cattle [3]	Milks cows [4]
	Dollars	Dollars	Dollars	Dollars	Dollars
2006	46.60	92.30	133.00	87.20	1,730.00
2007	47.90	95.40	119.00	89.90	1,830.00
2008	50.60	94.50	110.00	89.10	1,950.00
2009	44.80	85.40	105.00	80.30	1,390.00
2010	54.80	97.70	117.00	92.20	1,330.00
2011	71.20	117.00	142.00	113.00	1,420.00
2012	81.70	123.00	172.00	121.00	1,430.00
2013	82.30	126.00	181.00	125.00	1,380.00
2014	107.00	153.00	261.00	152.00	1,830.00
2015	103.00	148.00	247.00	147.00	1,990.00

Hog market year average price received: United States marketing years 2005–2014 [5]

Year	Barrows and gilts	Sows	All Hogs
	Dollars	Dollars	Dollars
2005	50.70	41.30	50.20
2006	46.60	33.20	46.00
2007	47.40	32.00	46.60
2008	47.90	29.20	47.00
2009	42.00	36.10	41.60
2010	54.40	48.70	54.10
2011	65.70	57.50	65.30
2012	64.60	51.30	64.20
2013	67.30	62.90	67.20
2014	76.50	77.70	76.50

[1] Market year for cattle January 1-December 31 for United States and all States. [2] Cows includes beef cows and cull dairy cows sold for slaughter. [3] All beef includes steer and heifers combined and cows. [4] Milk cow prices are calendar year average. [5] Market year for hogs December 1-November 30 for United States and all States.
NASS, Livestock Branch, (202) 720–3570.

Table 7-74.—Frozen meat: Cold storage holdings, end of month, United States, 2014–2015

Month	Boneless beef		Beef cuts		Total beef	
	2014	2015	2014	2015	2014	2015
	1,000 pounds	*1,000 pounds*	*1,000 pounds*	*1,000 pounds*	*1,000 pounds*	*1,000 pounds*
January	389,031	446,120	40,266	45,877	429,297	491,997
February	365,804	450,530	43,703	41,346	409,507	491,876
March	364,771	440,402	41,016	41,071	405,787	481,473
April	360,079	444,422	42,202	39,872	402,281	484,294
May	336,099	435,436	41,544	39,171	377,643	474,607
June	320,350	436,811	37,819	37,463	358,169	474,274
July	328,572	421,179	39,370	38,914	367,942	460,093
August	308,350	429,885	38,206	40,434	346,556	470,319
September ...	337,357	459,080	40,813	39,266	378,170	498,346
October	342,898	470,891	38,030	38,243	380,928	509,134
November	365,351	472,272	35,474	38,364	400,825	510,636
December	406,484	473,967	37,892	38,557	444,376	512,524

Month	Picnics		Bellies		Butts	
	2014	2015	2014	2015	2014	2015
	1,000 pounds	*1,000 pounds*	*1,000 pounds*	*1,000 pounds*	*1,000 pounds*	*1,000 pounds*
January	9,920	12,811	87,171	53,507	21,179	25,104
February	11,716	19,339	87,675	67,794	26,094	35,641
March	9,080	19,582	79,721	68,297	20,163	33,656
April	13,204	16,239	83,579	70,412	19,771	25,234
May	9,413	10,867	85,888	64,805	21,732	20,177
June	7,638	8,757	83,936	44,432	19,999	22,018
July	6,909	9,680	64,644	23,634	21,116	22,876
August	8,505	11,767	45,562	13,738	23,778	20,053
September ...	10,513	10,490	34,311	10,872	22,035	18,793
October	10,893	8,816	29,006	17,853	23,458	19,648
November	10,860	10,972	35,894	41,160	22,549	18,897
December	8,827	10,634	47,455	53,392	19,567	15,029

Month	Hams					
	Bone-in		Boneless		Total	
	2014	2015	2014	2015	2014	2015
	1,000 pounds	*1,000 pounds*	*1,000 pounds*	*1,000 pounds*	*1,000 pounds*	*1,000 pounds*
January	42,570	43,432	63,613	66,972	106,183	110,404
February	48,511	46,630	68,953	81,082	117,464	127,712
March	37,661	41,669	51,791	56,362	89,452	98,031
April	34,580	59,102	47,205	77,120	81,785	136,222
May	44,989	63,780	65,010	95,102	109,999	158,882
June	56,394	78,663	70,099	101,810	126,493	180,473
July	69,161	98,547	79,744	107,002	148,905	205,549
August	86,106	123,713	93,377	112,818	179,483	236,531
September ...	96,916	132,723	97,161	114,422	194,077	247,145
October	83,095	104,303	78,940	92,157	162,035	196,460
November	49,067	54,527	47,330	54,515	96,397	109,042
December	26,898	26,780	39,451	41,033	66,349	67,813

Month	Loins					
	Bone-in		Boneless		Total	
	2014	2015	2014	2015	2014	2015
	1,000 pounds	*1,000 pounds*	*1,000 pounds*	*1,000 pounds*	*1,000 pounds*	*1,000 pounds*
January	16,424	16,612	26,372	23,163	42,796	39,775
February	16,290	18,456	29,083	27,428	45,373	45,884
March	15,218	14,676	26,252	27,822	41,470	42,498
April	16,239	17,013	29,711	27,621	45,950	44,634
May	12,788	14,488	26,674	24,872	39,462	39,360
June	9,327	13,668	18,264	23,740	27,591	37,408
July	7,908	11,734	16,367	22,234	24,275	33,968
August	6,859	11,125	14,054	20,274	20,913	31,399
September ...	7,508	12,059	12,759	17,953	20,267	30,012
October	10,006	10,726	13,519	18,179	23,525	28,905
November	15,310	18,477	18,059	25,556	33,369	44,033
December	16,543	17,419	21,256	28,802	37,799	46,221

See end of table.

**Table 7-74.—Frozen meat: Cold storage holdings, end of month,
United States, 2014–2015—Continued**

Month	Ribs		Trimmings		Other frozen pork	
	2014	2015	2014	2015	2014	2015
	1,000 pounds	*1,000 pounds*	*1,000 pounds*	*1,000 pounds*	*1,000 pounds*	*1,000 pounds*
January	112,246	94,863	41,577	59,328	110,199	102,072
February	122,859	106,781	37,448	67,454	110,102	105,917
March	120,606	115,372	36,871	69,262	94,722	102,269
April	115,957	113,832	42,023	64,853	97,321	107,884
May	89,596	89,290	47,940	50,740	91,231	98,086
June	59,943	73,022	43,581	47,263	86,547	101,562
July	55,140	75,533	42,413	44,202	87,745	98,703
August	49,621	72,626	40,440	42,321	90,271	109,803
September ...	53,957	85,335	35,143	39,951	90,627	106,304
October	69,374	98,871	37,337	36,666	97,433	100,310
November	76,897	117,148	41,110	36,932	98,139	91,917
December	87,020	134,777	50,943	42,655	100,625	89,434

Month	Variety meats		Unclassified pork		Total pork	
	2014	2015	2014	2015	2014	2015
	1,000 pounds	*1,000 pounds*	*1,000 pounds*	*1,000 pounds*	*1,000 pounds*	*1,000 pounds*
January	32,442	41,834	55,033	55,975	618,746	595,673
February	36,807	44,124	59,174	65,417	654,712	686,063
March	35,427	45,365	48,027	78,099	575,539	672,431
April	38,733	47,153	45,568	74,620	583,891	701,083
May	35,819	45,409	44,738	77,685	575,818	655,301
June	32,170	44,251	49,549	75,339	537,447	634,525
July	31,906	48,157	50,206	70,912	533,259	633,214
August	33,587	49,939	51,506	65,583	543,666	653,760
September ...	35,068	40,126	54,624	66,902	550,622	655,930
October	28,769	43,045	51,246	52,880	533,076	603,454
November	32,769	41,313	44,768	49,501	492,752	560,915
December	39,177	43,170	46,030	42,571	503,792	545,696

Month	Veal		Lamb & mutton		Total red meat	
	2014	2015	2014	2015	2014	2015
	1,000 pounds	*1,000 pounds*	*1,000 pounds*	*1,000 pounds*	*1,000 pounds*	*1,000 pounds*
January	3,285	8,289	25,658	35,206	1,076,986	1,131,165
February	3,190	9,614	26,191	36,771	1,093,600	1,224,324
March	2,921	8,159	28,076	34,250	1,012,323	1,196,313
April	3,212	6,576	26,536	37,004	1,015,920	1,228,957
May	2,752	4,676	25,208	38,360	981,421	1,172,944
June	3,324	3,291	31,119	35,470	930,059	1,147,560
July	3,952	3,850	33,968	39,064	939,121	1,136,221
August	3,581	4,072	40,157	41,883	933,960	1,170,034
September ...	2,326	5,924	39,693	41,921	970,811	1,202,121
October	3,797	6,172	38,686	40,742	956,487	1,159,502
November	5,315	6,576	31,366	44,693	930,258	1,122,820
December	6,435	6,235	33,942	41,452	988,545	1,105,907

NASS, Livestock Branch, (202) 720–3570.

CHAPTER VIII
DAIRY AND POULTRY STATISTICS

Dairy statistics in this chapter include series relating to many phases of production, movement, prices, stocks, and consumption of milk and its products. Two series of number of milk cows on farms are included in this publication. One series is an inventory number of a specific classification estimated as one of the major groups making up the total cattle population on January 1. The other series identified as "milk cows" is an annual average number of milk cows during the year (excluding any not yet fresh) and is used in estimating milk production.

In comparing the several series of milk prices, it is important to note that prices received by farmers for all whole milk sold are for milk or milkfat content as actually sold, while certain prices paid by dealers for milk for fluid purposes or for specified manufacturing purposes may be quoted on a 3.5 percent butterfat basis, or for some types of manufacturing milk on the test of the milk used for that particular purpose.

Poultry and poultry products statistics include inventory numbers of chickens by classes; the production, disposition, cash receipts, and gross income from chickens and eggs; poultry and egg receipts at principal markets; commercial broiler production; turkey production, disposition, and gross income; poultry and eggs under Federal inspection; and the National Poultry Improvement Plan. Estimates relating to inventories, production, and income exclude poultry and eggs produced on places not classified as farms.

Table 8-1.—Milk cows and heifers: Number that have calved and heifers 500 pounds and over kept for milk cow replacements, United States, Jan. 1, 2006–2015

Year	Milk cows and heifers that have calved	Heifers 500 pounds and over kept for milk cow replacements
	Thousands	*Thousands*
2006	9,104	4,298
2007	9,145	4,325
2008	9,257	4,415
2009	9,333	4,410
2010	9,087	4,551
2011	9,156	4,577
2012	9,236	4,618
2013	9,221	4,546
2014	9,208	4,549
2015	9,307	4,710

NASS, Livestock Branch, (202) 720–3570.

Table 8-2.—Milk cows and heifers: Number that have calved and heifers 500 pounds and over kept for milk cow replacements, by State and United States, Jan. 1, 2014 and 2015

State	Milk cows and heifers that have calved		Heifers 500 pounds and over kept for milk cow replacements	
	2014	2015	2014	2015
	Thousands	*Thousands*	*Thousands*	*Thousands*
Alabama	9.0	8.0	4.0	3.0
Alaska	0.3	0.3	0.1	0.1
Arizona	192.0	195.0	74.0	65.0
Arkansas	8.0	7.0	5.0	4.0
California	1,780.0	1,780.0	750.0	770.0
Colorado	140.0	145.0	100.0	100.0
Connecticut	19.0	19.0	9.0	8.0
Delaware	4.7	5.0	2.6	2.5
Florida	123.0	124.0	35.0	35.0
Georgia	80.0	81.0	25.0	27.0
Hawaii	2.2	2.2	1.0	1.0
Idaho	565.0	579.0	270.0	320.0
Illinois	95.0	94.0	45.0	52.0
Indiana	178.0	181.0	68.0	80.0
Iowa	205.0	210.0	120.0	130.0
Kansas	136.0	143.0	100.0	90.0
Kentucky	68.0	63.0	45.0	45.0
Louisiana	15.0	14.0	5.0	5.0
Maine	30.0	30.0	17.0	16.0
Maryland	50.0	49.0	25.0	25.0
Massachusetts	12.0	12.5	7.5	7.0
Michigan	381.0	403.0	164.0	167.0
Minnesota	460.0	460.0	280.0	280.0
Mississippi	13.0	12.0	6.0	6.0
Missouri	90.0	89.0	50.0	60.0
Montana	14.0	14.0	9.0	7.0
Nebraska	53.0	54.0	20.0	20.0
Nevada	29.0	28.0	9.0	9.0
New Hampshire	13.5	14.0	7.0	5.5
New Jersey	7.0	7.0	3.0	3.8
New Mexico	323.0	323.0	120.0	110.0
New York	615.0	615.0	355.0	350.0
North Carolina	45.0	47.0	20.0	18.0
North Dakota	17.0	16.0	10.0	6.0
Ohio	267.0	268.0	130.0	125.0
Oklahoma	45.0	40.0	20.0	25.0
Oregon	124.0	125.0	60.0	60.0
Pennsylvania	530.0	530.0	315.0	305.0
Rhode Island	0.9	0.9	0.5	0.5
South Carolina	16.0	15.0	6.0	5.0
South Dakota	95.0	99.0	50.0	65.0
Tennessee	46.0	47.0	30.0	25.0
Texas	440.0	470.0	220.0	250.0
Utah	95.0	96.0	46.0	48.0
Vermont	132.0	132.0	56.0	56.0
Virginia	93.0	93.0	40.0	43.0
Washington	266.0	277.0	125.0	136.0
West Virginia	9.0	9.0	5.0	4.0
Wisconsin	1,270.0	1,275.0	680.0	730.0
Wyoming	6.0	6.0	4.0	5.0
United States	9,207.6	9,306.9	4,548.7	4,710.4

NASS, Livestock Branch, (202) 720–3570.

Table 8-3.—Milk and milkfat production: Number of milk cows, production per cow, and total quantity produced, by State and United States, 2014 [1]

State	Number of milk cows [2]	Production of milk and milkfat [3]				
		Per milk cow		All milk percent of fat	Total	
		Milk	Milkfat		Milk	Milkfat
	Thousands	Pounds	Pounds	Percent	Million pounds	Million pounds
Alabama	8.0	13,625	511	3.75	109.0	4.1
Alaska	0.3	11,667	462	3.96	3.5	0.1
Arizona	193.0	24,368	845	3.47	4,703.0	163.1
Arkansas	7.0	13,714	495	3.61	96.0	3.5
California	1,780.0	23,786	880	3.70	42,339.0	1,566.5
Colorado	144.0	24,951	886	3.55	3,593.0	127.6
Connecticut	19.0	20,158	782	3.88	383.0	14.9
Delaware	4.8	20,104	743	3.69	96.5	3.6
Florida	123.0	20,390	734	3.60	2,508.0	90.3
Georgia	81.0	20,877	759	3.65	1,691.0	61.5
Hawaii	2.2	13,591	472	3.47	29.9	1.0
Idaho	575.0	24,127	905	3.75	13,873.0	520.2
Illinois	94.0	19,681	744	3.78	1,850.0	69.9
Indiana	178.0	21,865	809	3.70	3,892.0	144.0
Iowa	207.0	22,449	842	3.75	4,647.0	174.2
Kansas	141.0	22,085	812	3.68	3,114.0	114.5
Kentucky	63.0	15,905	574	3.61	1,002.0	36.2
Louisiana	15.0	13,600	496	3.65	204.0	7.4
Maine	30.0	19,967	754	3.77	599.0	22.6
Maryland	50.0	19,740	740	3.75	987.0	37.0
Massachusetts	13.0	17,923	686	3.83	233.0	8.9
Michigan	390.0	24,638	912	3.70	9,609.0	355.5
Minnesota	460.0	19,841	772	3.89	9,127.0	355.0
Mississippi	13.0	14,462	538	3.72	188.0	7.0
Missouri	89.0	15,539	580	3.73	1,383.0	51.6
Montana	14.0	21,500	808	3.76	301.0	11.3
Nebraska	54.0	22,130	841	3.80	1,195.0	45.4
Nevada	29.0	23,793	849	3.57	690.0	24.6
New Hampshire	14.0	20,143	776	3.85	282.0	10.9
New Jersey	7.0	18,143	679	3.74	127.0	4.7
New Mexico	323.0	25,093	891	3.55	8,105.0	287.7
New York	615.0	22,325	842	3.77	13,730.0	517.7
North Carolina	46.0	20,891	754	3.61	961.0	34.7
North Dakota	16.0	20,250	765	3.78	324.0	12.2
Ohio	267.0	20,318	770	3.79	5,425.0	205.6
Oklahoma	40.0	18,150	629	3.61	726.0	25.2
Oregon	124.0	20,565	799	3.88	2,550.0	99.1
Pennsylvania	530.0	20,121	754	3.74	10,664.0	399.5
Rhode Island	0.9	19,000	766	4.03	17.1	0.7
South Carolina	16.0	16,438	626	3.82	263.0	10.0
South Dakota	97.0	21,753	857	3.94	2,110.0	83.1
Tennessee	46.0	16,196	589	3.61	745.0	27.1
Texas	463.0	22,268	853	3.83	10,310.0	394.9
Utah	95.0	22,968	854	3.72	2,184.0	81.2
Vermont	132.0	20,197	767	3.80	2,666.0	101.3
Virginia	93.0	19,129	701	3.66	1,779.0	65.1
Washington	273.0	24,088	936	3.88	6,576.0	255.5
West Virginia	9.0	15,556	576	3.70	140.0	5.2
Wisconsin	1,271.0	21,869	827	3.78	27,795.0	1,050.7
Wyoming	6.0	21,583	822	3.81	129.5	4.9
United States	9,257.0	22,259	832	3.74	206,054.0	7,698.5

[1] May not add due to rounding. [2] Average number during year, excluding heifers not yet fresh. [3] Excludes milk sucked by calves.

NASS, Livestock Branch, (202) 720–3570.

Table 8-4.—Milk and milkfat production: Number of milk cows, production per cow, and total quantity produced, by State and United States, 2015 [1]

State	Number of milk cows [2]	Production of milk and milkfat [3]				
		Per milk cow		All milk percent of fat	Total	
		Milk	Milkfat		Milk	Milkfat
	Thousands	*Pounds*	*Pounds*	*Percent*	*Million pounds*	*Million pounds*
Alabama	8.0	12,625	477	3.78	101.0	3.8
Alaska	0.3	11,667	460	3.94	3.5	0.1
Arizona	195.0	24,477	849	3.47	4,773.0	165.6
Arkansas	7.0	13,000	471	3.62	91.0	3.3
California	1,778.0	23,002	853	3.71	40,898.0	1,517.3
Colorado	146.0	25,685	925	3.60	3,750.0	135.0
Connecticut	19.0	20,842	823	3.95	396.0	15.6
Delaware	5.0	19,700	743	3.77	98.5	3.7
Florida	125.0	20,656	744	3.60	2,582.0	93.0
Georgia	83.0	21,651	782	3.61	1,797.0	64.9
Hawaii	2.2	15,909	544	3.42	35.0	1.2
Idaho	585.0	24,126	919	3.81	14,114.0	537.7
Illinois	94.0	20,128	759	3.77	1,892.0	71.3
Indiana	182.0	22,143	813	3.67	4,030.0	147.9
Iowa	211.0	22,943	870	3.79	4,841.0	183.5
Kansas	143.0	22,231	823	3.70	3,179.0	117.6
Kentucky	61.0	17,607	637	3.62	1,074.0	38.9
Louisiana	14.0	13,429	498	3.71	188.0	7.0
Maine	30.0	19,800	748	3.78	594.0	22.5
Maryland	49.0	20,061	750	3.74	983.0	36.8
Massachusetts ..	12.0	18,083	702	3.88	217.0	8.4
Michigan	408.0	25,130	920	3.66	10,253.0	375.3
Minnesota	460.0	20,578	794	3.86	9,466.0	365.4
Mississippi	11.0	15,000	563	3.75	165.0	6.2
Missouri	88.0	15,511	597	3.85	1,365.0	52.6
Montana	14.0	21,357	794	3.72	299.0	11.1
Nebraska	57.0	22,930	869	3.79	1,307.0	49.5
Nevada	29.0	23,069	824	3.57	669.0	23.9
New Hampshire	14.0	20,143	767	3.81	282.0	10.7
New Jersey	7.0	18,143	680	3.75	127.0	4.8
New Mexico	323.0	24,245	866	3.57	7,831.0	279.6
New York	618.0	22,816	862	3.78	14,100.0	533.0
North Carolina ...	47.0	20,979	759	3.62	986.0	35.7
North Dakota	16.0	20,750	780	3.76	332.0	12.5
Ohio	267.0	20,573	774	3.76	5,493.0	206.5
Oklahoma	39.0	18,462	672	3.64	720.0	26.2
Oregon	125.0	20,408	794	3.89	2,551.0	99.2
Pennsylvania	530.0	20,387	760	3.73	10,805.0	403.0
Rhode Island	0.9	17,667	728	4.12	15.9	0.7
South Carolina ..	15.0	17,400	679	3.90	261.0	10.2
South Dakota	106.0	22,255	892	4.01	2,359.0	94.6
Tennessee	45.0	16,489	597	3.62	742.0	26.9
Texas	463.0	22,235	863	3.88	10,295.0	399.4
Utah	96.0	23,146	863	3.73	2,222.0	82.9
Vermont	132.0	20,197	774	3.83	2,666.0	102.1
Virginia	91.0	19,462	716	3.68	1,771.0	65.2
Washington	277.0	23,848	928	3.89	6,606.0	257.0
West Virginia	9.0	15,667	578	3.69	141.0	5.2
Wisconsin	1,279.0	22,697	853	3.76	29,030.0	1,091.5
Wyoming	6.0	22,567	779	3.45	135.4	4.7
United States	9,317.0	22,393	837	3.74	208,633.0	7,810.7

[1] May not add due to rounding.   [2] Average number during year, excluding heifers not yet fresh.   [3] Excludes milk sucked by calves.

NASS, Livestock Branch, (202) 720–3570.

Table 8-5.—Milk: Quantities used and marketed by producers, by State and United States, 2014[1]

State	Milk used where produced			Milk marketed by producers	
	Fed to calves[2]	Used for milk cream & butter	Total	Total quantity[3]	Fluid grade[4]
	Million pounds	Million pounds	Million pounds	Million pounds	Percent
Alabama	0.7	0.3	1.0	108.0	100
Alaska	0.1	0.1	0.2	3.3	100
Arizona	12.0	1.0	13.0	4,690.0	100
Arkansas	1.7	0.3	2.0	94.0	100
California	29.0	4.0	33.0	42,306.0	99
Colorado	21.0	1.0	22.0	3,571.0	100
Connecticut	2.5	0.5	3.0	380.0	100
Delaware	0.9	0.1	1.0	95.5	100
Florida	5.0	1.0	6.0	2,502.0	100
Georgia	7.0	1.0	8.0	1,683.0	100
Hawaii	0.1	0.1	0.2	29.7	100
Idaho	30.0	1.0	31.0	13,842.0	100
Illinois	10.0	2.0	12.0	1,838.0	98
Indiana	23.0	4.0	27.0	3,865.0	100
Iowa	13.0	1.0	14.0	4,633.0	99
Kansas	12.0	1.0	13.0	3,101.0	100
Kentucky	8.0	1.0	9.0	993.0	100
Louisiana	3.0	1.0	4.0	200.0	100
Maine	3.0	1.0	4.0	595.0	100
Maryland	6.0	1.0	7.0	980.0	100
Massachusetts	1.5	0.5	2.0	231.0	100
Michigan	24.0	2.0	26.0	9,583.0	100
Minnesota	100.0	5.0	105.0	9,022.0	99
Mississippi	1.0	1.0	2.0	186.0	100
Missouri	17.0	4.0	21.0	1,362.0	97
Montana	3.0	2.0	5.0	296.0	100
Nebraska	6.0	1.0	7.0	1,188.0	100
Nevada	5.0	1.0	6.0	684.0	100
New Hampshire	1.5	0.5	2.0	280.0	100
New Jersey	1.5	0.5	2.0	125.0	100
New Mexico	39.0	3.0	42.0	8,063.0	100
New York	40.0	2.0	42.0	13,688.0	100
North Carolina	5.0	1.0	6.0	955.0	100
North Dakota	5.5	0.5	6.0	318.0	97
Ohio	18.0	4.0	22.0	5,403.0	97
Oklahoma	6.0	1.0	7.0	719.0	100
Oregon	20.0	1.0	21.0	2,529.0	100
Pennsylvania	54.0	12.0	66.0	10,598.0	100
Rhode Island	0.2	-	0.2	16.9	100
South Carolina	2.0	1.0	3.0	260.0	100
South Dakota	5.0	1.0	6.0	2,104.0	99
Tennessee	4.0	1.0	5.0	740.0	100
Texas	27.0	1.0	28.0	10,282.0	100
Utah	12.0	1.0	13.0	2,171.0	100
Vermont	13.0	2.0	15.0	2,651.0	100
Virginia	6.0	2.0	8.0	1,771.0	100
Washington	16.0	1.0	17.0	6,559.0	100
West Virginia	1.0	1.0	2.0	138.0	100
Wisconsin	248.0	16.0	264.0	27,531.0	99
Wyoming	1.3	0.2	1.5	128.0	100
United States	872.0	92.0	964.0	205,091.0	99

- Represents zero. [1] May not add due to rounding. [2] Excludes milk sucked by calves. [3] Milk sold to plants and dealers as whole milk and equivalent amounts of milk for cream. Includes milk produced by dealers' own herds and milk sold directly to consumers. Also includes milk produced by institutional herds. [4] Percentage of milk sold that is eligible for fluid use (Grade A in most States). Includes fluid grade milk used in manufacturing dairy products.
NASS, Livestock Branch, (202) 720–3570.

Table 8-6.—Milk: Quantities used and marketed by producers, by State and United States, 2015[1]

State	Milk used where produced			Milk marketed by producers	
	Fed to calves[2]	Used for milk cream & butter	Total	Total quantity[3]	Fluid grade[4]
	Million pounds	Million pounds	Million pounds	Million pounds	Percent
Alabama	0.7	0.3	1.0	100.0	100
Alaska	0.2	0.2	0.4	3.1	100
Arizona	12.0	1.0	13.0	4,760.0	100
Arkansas	1.7	0.3	2.0	89.0	100
California	29.0	4.0	33.0	40,865.0	99
Colorado	21.0	1.0	22.0	3,728.0	100
Connecticut	2.5	0.5	3.0	393.0	100
Delaware	0.9	0.1	1.0	97.5	100
Florida	5.0	1.0	6.0	2,576.0	100
Georgia	7.0	1.0	8.0	1,789.0	100
Hawaii	0.1	0.1	0.2	34.8	100
Idaho	32.0	1.0	33.0	14,081.0	100
Illinois	10.0	2.0	12.0	1,880.0	98
Indiana	23.0	4.0	27.0	4,003.0	100
Iowa	14.0	1.0	15.0	4,826.0	99
Kansas	11.0	1.0	12.0	3,167.0	100
Kentucky	6.0	1.0	7.0	1,067.0	100
Louisiana	3.0	1.0	4.0	184.0	100
Maine	3.0	1.0	4.0	590.0	100
Maryland	6.0	1.0	7.0	976.0	100
Massachusetts	1.5	0.5	2.0	215.0	100
Michigan	26.0	2.0	28.0	10,225.0	100
Minnesota	100.0	5.0	105.0	9,361.0	99
Mississippi	1.0	1.0	2.0	163.0	100
Missouri	17.0	4.0	21.0	1,344.0	97
Montana	3.0	2.0	5.0	294.0	100
Nebraska	7.5	0.5	8.0	1,299.0	100
Nevada	5.0	1.0	6.0	663.0	100
New Hampshire	1.5	0.5	2.0	280.0	100
New Jersey	1.5	0.5	2.0	125.0	100
New Mexico	39.0	3.0	42.0	7,789.0	100
New York	45.0	2.0	47.0	14,053.0	100
North Carolina	5.0	1.0	6.0	980.0	100
North Dakota	4.5	0.5	5.0	327.0	97
Ohio	20.0	5.0	25.0	5,468.0	97
Oklahoma	5.0	1.0	6.0	714.0	100
Oregon	20.0	1.0	21.0	2,530.0	100
Pennsylvania	57.0	12.0	69.0	10,736.0	100
Rhode Island	0.2	-	0.2	15.7	100
South Carolina	2.0	1.0	3.0	258.0	100
South Dakota	6.0	1.0	7.0	2,352.0	99
Tennessee	4.0	1.0	5.0	737.0	100
Texas	24.0	1.0	25.0	10,270.0	100
Utah	12.0	1.0	13.0	2,209.0	100
Vermont	13.0	2.0	15.0	2,651.0	100
Virginia	6.0	2.0	8.0	1,763.0	100
Washington	17.0	1.0	18.0	6,588.0	100
West Virginia	1.0	1.0	2.0	139.0	100
Wisconsin	244.0	15.0	259.0	28,771.0	99
Wyoming	1.3	0.2	1.5	133.9	100
United States	878.0	91.0	969.0	207,663.0	99

- Represents zero.   [1] May not add due to rounding.   [2] Excludes milk sucked by calves.   [3] Milk sold to plants and dealers as whole milk and equivalent amounts of milk for cream.   Includes milk produced by dealers' own herds and milk sold directly to consumers.   Also includes milk produced by institutional herds.   [4] Percentage of milk sold that is eligible for fluid use (Grade A in most States).   Includes fluid grade milk used in manufacturing dairy products.
NASS, Livestock Branch, (202) 720–3570.

Table 8-7.—Milk and cream: Marketings and income, by State and United States, 2014 [1]

State	Milk utilized	All milk Average returns per cwt [2]	Returns per lb milkfat	Cash receipts from marketings
	Million pounds	Dollars	Dollars	1,000 dollars
Alabama	108.0	27.80	7.41	30,024
Alaska	3.3	21.90	5.53	723
Arizona	4,690.0	23.20	6.69	1,088,080
Arkansas	94.0	25.50	7.06	23,970
California	42,306.0	22.12	5.98	9,358,087
Colorado	3,571.0	24.00	6.76	857,040
Connecticut	380.0	26.10	6.73	99,180
Delaware	95.5	24.10	6.53	23,016
Florida	2,502.0	28.20	7.83	705,564
Georgia	1,683.0	26.40	7.23	444,312
Hawaii	29.7	35.40	10.20	10,514
Idaho	13,842.0	23.10	6.16	3,197,502
Illinois	1,838.0	24.70	6.53	453,986
Indiana	3,865.0	24.30	6.57	939,195
Iowa	4,633.0	24.60	6.56	1,139,718
Kansas	3,101.0	24.00	6.52	744,240
Kentucky	993.0	25.80	7.15	256,194
Louisiana	200.0	26.20	7.18	52,400
Maine	595.0	26.80	7.11	159,460
Maryland	980.0	24.80	6.67	245,000
Massachusetts ..	231.0	26.10	6.81	60,291
Michigan	9,583.0	24.10	6.51	2,309,503
Minnesota	9,022.0	24.80	6.38	2,237,456
Mississippi	186.0	26.80	7.20	49,848
Missouri	1,362.0	24.60	6.60	335,052
Montana	296.0	22.00	5.85	65,120
Nebraska	1,188.0	25.00	6.58	297,000
Nevada	684.0	24.80	6.95	169,632
New Hampshire	280.0	25.90	6.73	72,520
New Jersey	125.0	24.90	6.66	31,125
New Mexico	8,063.0	22.30	6.28	1,798,049
New York	13,688.0	25.40	6.74	3,476,752
North Carolina ..	955.0	25.70	7.12	245,435
North Dakota	318.0	23.60	6.24	75,048
Ohio	5,403.0	24.60	6.49	1,329,138
Oklahoma	719.0	25.90	7.17	186,221
Oregon	2,529.0	25.70	6.62	649,953
Pennsylvania	10,598.0	25.70	6.87	2,723,686
Rhode Island	16.9	26.40	6.55	4,462
South Carolina ..	260.0	27.30	7.15	70,980
South Dakota	2,104.0	24.70	6.27	519,688
Tennessee	740.0	25.80	7.15	190,920
Texas	10,282.0	24.60	6.42	2,529,372
Utah	2,171.0	23.70	6.37	514,527
Vermont	2,651.0	25.50	6.71	676,005
Virginia	1,771.0	27.00	7.38	478,170
Washington	6,559.0	24.70	6.37	1,620,073
West Virginia	138.0	24.40	6.59	33,672
Wisconsin	27,531.0	24.50	6.48	6,745,095
Wyoming	128.0	23.40	6.14	29,952
United States	205,091.0	24.07	6.44	49,352,950

[1] May not add due to rounding. [2] Cash receipts divided by milk or milkfat in combined marketings.
NASS, Livestock Branch, (202) 720–3570.

Table 8-8.—Milk and cream: Marketings and income, by State and United States, 2015 [1]

State	Milk utilized	All milk Average returns per cwt [2]	Returns per lb milkfat	Cash receipts from marketings
	Million pounds	Dollars	Dollars	1,000 dollars
Alabama	100.0	19.50	5.16	19,500
Alaska	3.1	22.00	5.58	682
Arizona	4,760.0	16.00	4.61	761,600
Arkansas	89.0	18.10	5.00	16,109
California	40,865.0	15.40	4.15	6,293,210
Colorado	3,728.0	17.80	4.94	663,584
Connecticut	393.0	18.60	4.71	73,098
Delaware	97.5	17.40	4.62	16,965
Florida	2,576.0	21.30	5.92	548,688
Georgia	1,789.0	20.00	5.54	357,800
Hawaii	34.8	28.50	8.33	9,918
Idaho	14,081.0	16.70	4.38	2,351,527
Illinois	1,880.0	17.70	4.69	332,760
Indiana	4,003.0	17.20	4.69	688,516
Iowa	4,826.0	17.30	4.56	834,898
Kansas	3,167.0	16.90	4.57	535,223
Kentucky	1,067.0	18.80	5.19	200,596
Louisiana	184.0	19.00	5.12	34,960
Maine	590.0	20.00	5.29	118,000
Maryland	976.0	17.70	4.73	172,752
Massachusetts ..	215.0	18.60	4.79	39,990
Michigan	10,225.0	16.60	4.54	1,697,350
Minnesota	9,361.0	17.60	4.56	1,647,536
Mississippi	163.0	19.40	5.17	31,622
Missouri	1,344.0	18.50	4.81	248,640
Montana	294.0	14.90	4.01	43,806
Nebraska	1,299.0	17.80	4.70	231,222
Nevada	663.0	17.70	4.96	117,351
New Hampshire .	280.0	18.50	4.86	51,800
New Jersey	125.0	17.60	4.69	22,000
New Mexico	7,789.0	16.10	4.51	1,254,029
New York	14,053.0	18.20	4.81	2,557,646
North Carolina ..	980.0	18.60	5.14	182,280
North Dakota	327.0	17.10	4.55	55,917
Ohio	5,468.0	17.90	4.76	978,772
Oklahoma	714.0	18.60	5.11	132,804
Oregon	2,530.0	18.60	4.78	470,580
Pennsylvania	10,736.0	18.50	4.96	1,986,160
Rhode Island	15.7	18.90	4.59	2,967
South Carolina ..	258.0	19.80	5.08	51,084
South Dakota	2,352.0	18.50	4.61	435,120
Tennessee	737.0	18.80	5.19	138,556
Texas	10,270.0	17.70	4.56	1,817,790
Utah	2,209.0	17.00	4.56	375,530
Vermont	2,651.0	18.60	4.86	493,086
Virginia	1,763.0	19.40	5.27	342,022
Washington	6,588.0	17.20	4.42	1,133,136
West Virginia	139.0	17.50	4.74	24,325
Wisconsin	28,771.0	17.80	4.73	5,121,238
Wyoming	133.9	18.30	5.30	24,504
United States	207,663.0	17.21	4.60	35,739,249

[1] May not add due to rounding.   [2] Cash receipts divided by milk or milkfat in combined marketings.
NASS, Livestock Branch, (202) 720–3570.

Table 8-9.—Milk production: Value, by State and United States, 2014 [1]

State	Used for milk, cream, & butter where produced		Gross producer income [3]	Value of milk produced [2] [4]
	Milk utilized	Value [2]		
	Million pounds	1,000 Dollars	1,000 Dollars	1,000 Dollars
Alabama	0.3	83	30,107	30,302
Alaska	0.1	22	745	767
Arizona	1.0	232	1,088,312	1,091,096
Arkansas	0.3	77	24,047	24,480
California	4.0	885	9,358,972	9,365,387
Colorado	1.0	240	857,280	862,320
Connecticut	0.5	131	99,311	99,963
Delaware	0.1	24	23,040	23,257
Florida	1.0	282	705,846	707,256
Georgia	1.0	264	444,576	446,424
Hawaii	0.1	35	10,549	10,585
Idaho	1.0	231	3,197,733	3,204,663
Illinois	2.0	494	454,480	456,950
Indiana	4.0	972	940,167	945,756
Iowa	1.0	246	1,139,964	1,143,162
Kansas	1.0	240	744,480	747,360
Kentucky	1.0	258	256,452	258,516
Louisiana	1.0	262	52,662	53,448
Maine	1.0	268	159,728	160,532
Maryland	1.0	250	245,250	246,750
Massachusetts	0.5	131	60,422	60,813
Michigan	2.0	482	2,309,985	2,315,769
Minnesota	5.0	1,240	2,238,696	2,263,496
Mississippi	1.0	268	50,116	50,384
Missouri	4.0	984	336,036	340,218
Montana	2.0	440	65,560	66,220
Nebraska	1.0	250	297,250	298,750
Nevada	1.0	248	169,880	171,120
New Hampshire	0.5	130	72,650	73,038
New Jersey	0.5	125	31,250	31,623
New Mexico	3.0	669	1,798,718	1,807,415
New York	2.0	508	3,477,260	3,487,420
North Carolina	1.0	257	245,692	246,977
North Dakota	0.5	118	75,166	76,464
Ohio	4.0	984	1,330,122	1,334,550
Oklahoma	1.0	259	186,480	188,034
Oregon	1.0	257	650,210	655,350
Pennsylvania	12.0	3,084	2,726,770	2,740,648
Rhode Island	-	-	4,462	4,514
South Carolina	1.0	273	71,253	71,799
South Dakota	1.0	247	519,935	521,170
Tennessee	1.0	258	191,178	192,210
Texas	1.0	246	2,529,618	2,536,260
Utah	1.0	237	514,764	517,608
Vermont	2.0	510	676,515	679,830
Virginia	2.0	540	478,710	480,330
Washington	1.0	247	1,620,320	1,624,272
West Virginia	1.0	244	33,916	34,160
Wisconsin	16.0	3,920	6,749,015	6,809,775
Wyoming	0.2	47	29,999	30,303
United States	92.0	22,699	49,375,649	49,589,494

- Represents zero. [1] May not add due to rounding. [2] Value at average returns per 100 pounds of milk in combined marketings of milk and cream. [3] Cash receipts from marketings of milk and cream plus value of milk used for home consumption. [4] Includes value of milk fed to calves.
NASS, Livestock Branch, (202) 720–3570.

Table 8-10.—Milk production: Value, by State and United States, 2015[1]

State	Used for milk, cream, & butter by producers		Gross producer income[3]	Value of milk produced[2][4]
	Milk utilized	Value[2]		
	Million pounds	1,000 Dollars	1,000 Dollars	1,000 Dollars
Alabama	0.3	59	19,559	19,695
Alaska	0.2	44	726	770
Arizona	1.0	160	761,760	763,680
Arkansas	0.3	54	16,163	16,471
California	4.0	616	6,293,826	6,298,292
Colorado	1.0	178	663,762	667,500
Connecticut	0.5	93	73,191	73,656
Delaware	0.1	17	16,982	17,139
Florida	1.0	213	548,901	549,966
Georgia	1.0	200	358,000	359,400
Hawaii	0.1	29	9,947	9,975
Idaho	1.0	167	2,351,694	2,357,038
Illinois	2.0	354	333,114	334,884
Indiana	4.0	688	689,204	693,160
Iowa	1.0	173	835,071	837,493
Kansas	1.0	169	535,392	537,251
Kentucky	1.0	188	200,784	201,912
Louisiana	1.0	190	35,150	35,720
Maine	1.0	200	118,200	118,800
Maryland	1.0	177	172,929	173,991
Massachusetts	0.5	93	40,083	40,362
Michigan	2.0	332	1,697,682	1,701,998
Minnesota	5.0	880	1,648,416	1,666,016
Mississippi	1.0	194	31,816	32,010
Missouri	4.0	740	249,380	252,525
Montana	2.0	298	44,104	44,551
Nebraska	0.5	89	231,311	232,646
Nevada	1.0	177	117,528	118,413
New Hampshire	0.5	93	51,893	52,170
New Jersey	0.5	88	22,088	22,352
New Mexico	3.0	483	1,254,512	1,260,791
New York	2.0	364	2,558,010	2,566,200
North Carolina	1.0	186	182,466	183,396
North Dakota	0.5	86	56,003	56,772
Ohio	5.0	895	979,667	983,247
Oklahoma	1.0	186	132,990	133,920
Oregon	1.0	186	470,766	474,486
Pennsylvania	12.0	2,220	1,988,380	1,998,925
Rhode Island	-	-	2,967	3,005
South Carolina	1.0	198	51,282	51,678
South Dakota	1.0	185	435,305	436,415
Tennessee	1.0	188	138,744	139,496
Texas	1.0	177	1,817,967	1,822,215
Utah	1.0	170	375,700	377,740
Vermont	2.0	372	493,458	495,876
Virginia	2.0	388	342,410	343,574
Washington	1.0	172	1,133,308	1,136,232
West Virginia	1.0	175	24,500	24,675
Wisconsin	15.0	2,670	5,123,908	5,167,340
Wyoming	0.2	37	24,541	24,778
United States	91.0	16,291	35,755,540	35,910,597

- Represents zero.   [1] May not add due to rounding.   [2] Value at average returns per 100 pounds of milk in combined marketings of milk and cream.   [3] Cash receipts from marketings of milk and cream plus value of milk used for home consumption.   [4] Includes value of milk fed to calves.
NASS, Livestock Branch, (202) 720–3570.

Table 8-11.—Milk cows, milk, and fat in cream: Average prices received by farmers, United States, 2006–2015

Year	Milk cows, per head [1]	Milk per 100 pounds [2]					
		Eligible for fluid market [3]		Of manufacturing grade		All milk wholesale	
		Price per 100 lb.	Fat test	Price per 100 lb.	Fat test	Price per 100 lb.	Fat test
	Dollars	*Dollars*	*Percent*	*Dollars*	*Percent*	*Dollars*	*Percent*
2006	1,730.00	12.96	3.68	12.19	3.93	12.96	3.69
2007	1,830.00	19.22	3.68	18.31	3.99	19.21	3.68
2008	1,950.00	18.45	3.68	17.91	4.01	18.45	3.68
2009	1,390.00	12.94	3.67	12.03	4.00	12.93	3.67
2010	1,330.00	16.37	3.65	14.56	4.00	16.35	3.66
2011	1,420.00	(NA)	(NA)	(NA)	(NA)	20.24	3.71
2012	1,430.00	(NA)	(NA)	(NA)	(NA)	18.56	3.72
2013	1,380.00	(NA)	(NA)	(NA)	(NA)	20.11	3.76
2014	1,830.00	(NA)	(NA)	(NA)	(NA)	24.07	3.74
2015	1,990.00					17.21	3.74

(NA) Not available. [1] Simple average of quarterly prices, by States, weighted by the number of milk cows on farms Jan. 1 of the current year. [2] Average price at average fat test for all milk sold at wholesale to plants and dealers, based on reports from milk-market administrators, cooperative milk-market associations, whole-milk distributors, and milk-products manufacturing plants, f.o.b. plant or receiving station (whichever is the customary place for determining prices) before hauling costs are deducted and including all premiums. [3] Includes fluid milk surplus diverted to manufacturing.
NASS, Livestock Branch, (202) 720–3570.

Table 8-12.—Milk-feed price ratios: All milk price; Dairy feed price, 16% protein; Milk-feed price ratios, United States, annual 2006–2015

Year	All milk price	Dairy feed price 16% protein [1]	Milk-feed price ratio [2]
	Dollars/cwt	*Dollars/ton*	*Pounds*
2006	12.96	210	2.57
2007	19.21	249	2.80
2008	18.45	313	2.01
2009	12.93	293	1.78
2010	16.35	274	2.26
2011	20.24	369	1.90
2012	18.56	442	1.52
2013	20.11	481	1.75
2014	24.07	502	2.54
2015	17.21		

[1] Commercially prepared 16%protein dairy ration price collected from March annual Prices Paid Survey. [2] Annual ratios based on average of monthly ratios. Pounds of 16% protein mixed dairy feed equal in value to one pound of whole milk. Feed price component (16% protein) calculated using U.S. price received for corn (51 lbs), soybeans (8 1b), and alfalfa hay (41 lbs).
NASS, Environmental, Economics, and Demographics Branch, (202) 720–6146.

Table 8-13.—Fluid beverage milk: Total and per capita consumption, United States, 2006–2015

Year	Consumption	
	Total	Per capita
	Billion pounds	*Pounds*
2006	54.8	184
2007	54.6	181
2008	54.5	179
2009	54.7	178
2010	54.9	177
2011	54.2	174
2012	53.3	170
2013	52.2	165
2014	50.7	159
2015	49.9	155

ERS, Animal Products and Cost of Production Branch, (202) 694–5171.

Table 8-14.—Federal milk order markets: Measures of growth, 2006–2014 [1]

Year	Number of markets[2]	Population of Federal milk marketing areas	Number of handlers[2]	Number of producers[3]	Receipts of producer milk	Producer milk used in Class I	Percentage of producer milk used in Class I
	Number	Thousands	Number	Number	Million pounds	Million pounds	Percent
2006	10	239,142	314	52,725	120,618	45,304	37.6
2007	10	241,000	312	49,782	114,407	45,226	39.5
2008	10	242,988	333	47,859	115,867	44,989	38.8
2009	10	245,445	251	46,677	123,430	45,262	36.7
2010	10	247,031	251	45,918	126,909	44,970	35.4
2011	10	247,675	241	43,654	126,879	44,383	35.0
2012	10	250,724	237	40,750	122,388	43,492	35.5
2013	10	251,201	225	40,048	132,100	42,742	32.4
2014	10	255,184	223	38,391	129,420	41,420	32.0

Year	Prices at 3.5 percent butterfat content per hundredweight[4]		Receipts as percentage of milk sold to plants and dealers		Daily deliveries of milk per producer	Gross value of receipts of producer milk[5]	
	Class I	Blend	Fluid grade	All milk		Per producer	All producer
	Dollars	Dollars	Percent	Percent	Pounds	Dollars	1,000 dollars
2006	14.59	12.86	68	67	6,264	303,429	15,998,288
2007	20.81	19.19	63	62	6,297	452,097	22,507,219
2008	20.78	18.24	61	62	6,613	453,886	21,772,538
2009	14.40	12.44	66	66	7,242	339,698	15,856,077
2010	18.25	16.07	67	66	7,572	444,038	20,389,201
2011	21.97	19.87	66	65	7,963	577,538	25,211,996
2012	20.39	18.05	62	61	8,229	542,121	22,091,337
2013	21.80	19.44	67	66	9,047	641,295	25,682,588
2014	26.14	23.54	63	63	9,236	793,728	30,472,016

[1] Over this period, handler selected periodically not to pool substantial volumes of milk that normally would have been pooled under Federal orders. This decision resulted from disadvantageous blend/class price relationships and qualification circumstances. This fact should be kept in mind if year-to-year comparisons are made using the various "producer deliveries" measures of growth. [2] End of year. [3] Average for year. [4] Prices are weighted averages. [5] Based on blend (uniform) price adjusted for butterfat content, and in later years, other milk components of producer milk.
AMS, Dairy Programs, (202) 720-7461.

Table 8-15.—Milk production: Marketings, income and value, United States, 2006–2015

Year	Combined marketings of milk and cream				Used for milk, cream, and butter on farms where produced		Gross farm income from dairy products[3]	Farm value of all milk produced[2][4]
	Milk utilized	Average returns[1]		Cash receipts from marketings	Milk utilized	Value[2]		
		Per 100 pounds milk	Per pound milkfat					
	Million pounds	Dollars	Dollars	1,000 dollars	Million pounds	1,000 dollars	1,000 dollars	1,000 dollars
2006	180,700	12.96	3.51	23,412,552	138	18,591	23,431,143	23,556,102
2007	184,565	19.21	5.22	35,453,399	137	27,073	35,480,472	35,665,894
2008	188,911	18.45	5.01	34,846,236	124	23,743	34,869,979	35,048,242
2009	188,189	12.93	3.52	24,321,032	112	15,224	24,336,256	24,455,900
2010	191,897	16.35	4.47	31,372,346	107	18,269	31,390,615	31,535,575
2011	195,290	20.24	5.46	39,531,306	98	20,561	39,551,867	39,730,346
2012	199,687	18.56	4.99	37,064,731	98	18,779	37,083,510	37,247,542
2013	200,254	20.11	5.35	40,276,790	100	20,629	40,297,419	40,476,608
2014	205,091	24.07	6.44	49,352,950	92	22,699	49,375,649	49,589,494
2015	207,663	17.21	4.60	35,739,249	91	16,291	35,755,540	35,910,597

[1] Cash receipts divided by milk or milkfat represented in combined marketings [2] Valued at average returns per 100 pounds of milk in combined marketings of milk and cream. [3] Cash receipts from marketings of milk and cream plus value of milk used for home consumption. [4] Includes value of milk fed to calves.
NASS, Livestock Branch, (202) 720-3570.

Table 8-16.—Official Dairy Herd Information test plans: Numbers of herds and cows and milk, fat, and protein production, United States, 2005–2014

Year	Herds	Cows	Cows per herd	Average production			Cows with protein information	Average protein production	Average protein production
				Milk	Fat	Fat			
	Number	*Number*	*Number*	*Pounds*	*Percent*	*Pounds*	*Percent*	*Percent*	*Pounds*
2005 ...	18,349	3,537,857	192.8	22,077	3.67	812	95	3.08	680
2006 ...	17,606	3,602,719	204.6	22,282	3.69	825	95	3.09	688
2007 ...	17,174	3,749,257	218.3	22,371	3.68	826	95	3.09	693
2008 ...	16,602	3,804,216	229.1	22,437	3.69	830	96	3.10	696
2009 ...	15,331	3,665,911	239.1	22,501	3.68	831	98	3.09	698
2010 ...	15,067	3,746,177	248.6	22,765	3.66	836	98	3.09	706
2011 ...	14,490	3,800,410	262.3	22,961	3.71	855	98	3.11	714
2012 ...	13,894	3,762,925	270.8	23,328	3.72	871	98	3.12	728
2013 ...	13,329	3,712,080	278.5	23,695	3.77	894	98	3.13	742
2014 ...	13,043	3,829,273	293.6	24,101	3.76	907	98	3.14	757

Council on Dairy Cattle Breeding, (301) 525–2006, https://www.cdcb.us.

Table 8-17.—Milk and milkfat production: Number of producing cows, production per cow, and total quantity produced, United States, 2005–2014

Year	Number of milk cows [1]	Production of milk and milkfat [2]				
		Per milk cow		Percentage of fat in all milk produced	Total	
		Milk	Milkfat		Milk	Milkfat
	Thousands	*Pounds*	*Pounds*	*Percent*	*Million pounds*	*Million pounds*
2006	9,137	19,895	734	3.69	181,782	6,700
2007	9,189	20,204	744	3.68	185,654	6,832
2008	9,314	20,397	751	3.68	189,978	6,998
2009	9,202	20,561	755	3.67	189,202	6,944
2010	9,123	21,142	774	3.66	192,877	7,054
2011	9,199	21,334	791	3.71	196,255	7,284
2012	9,237	21,722	808	3.72	200,642	7,463
2013	9,224	21,816	820	3.76	201,231	7,557
2014	9,257	22,259	832	3.74	206,054	7,699
2015	9,317	22,393	837	3.74	208,633	7,812

[1] Average number during year, excluding heifers not yet fresh. [2] Excludes milk sucked by calves.
NASS, Livestock Branch, (202) 720–3570.

Table 8-18.—Milk: Quantities used and marketed by farmers, United States, 2006–2015

Year	Milk used on farms where produced			Milk marketed by producers	
	Fed to calves [1]	Consumed as fluid milk or cream	Total	Total [2]	Fluid grade [3]
	Million pounds	*Million pounds*	*Million pounds*	*Million pounds*	*Percent*
2006	943	138	1,081	180,700	99
2007	952	137	1,089	184,565	99
2008	944	124	1,068	188,911	99
2009	901	112	1,013	188,189	98
2010	873	107	980	191,897	98
2011	867	98	965	195,290	99
2012	858	98	956	199,687	99
2013	877	100	977	200,254	99
2014	872	92	964	205,091	99
2015	878	91	969	207,663	99

[1] Excludes milk sucked by calves. [2] Milk sold to plants and dealers as whole milk and equivalent amounts of milk for cream. Includes milk produced by dealers' own herds and small amounts sold directly to consumers. Also includes milk produced by institutional herds. [3] Percentage of milk sold that is eligible for fluid use (Grade A in most States). Includes fluid-grade milk used in manufacturing dairy products.
NASS, Livestock Branch, (202) 720–3570.

Table 8-19.—Milk markets under Federal order program: Whole milk and fat-reduced milk products sold for fluid consumption within defined marketing areas, 2011–2013 [1]

Federal milk order marketing area	Whole milk products [2]		Fat-reduced milk products [3]		Total fluid milk products	
	Quantity	Butterfat content	Quantity	Butterfat content	Quantity	Butterfat content
	Million pounds	Percent	Million pounds	Percent	Million pounds	Percent
2011						
Northeast	2,999	3.29	6,110	1.08	9,109	1.81
Appalachian	1,041	3.30	2,538	1.29	3,580	1.87
Southeast	1,607	3.34	3,291	1.33	4,898	1.99
Florida	1,021	3.33	1,879	1.18	2,900	1.94
Mideast	1,263	3.32	4,725	1.31	5,988	1.24
Upper Midwest	578	3.31	3,643	1.10	4,221	1.40
Central	1,005	3.32	3,543	1.24	4,548	1.70
Southwest	1,522	3.32	2,953	1.36	4,476	2.03
Arizona	308	3.26	872	1.20	1,179	1.74
Pacific Northwest	457	3.40	1,757	1.30	2,214	1.73
Combined areas	11,802	3.32	31,311	1.23	43,113	1.80
2012						
Northeast	2,976	3.30	6,055	1.08	9,030	1.81
Appalachian	1,022	3.27	2,477	1.29	3,498	1.87
Southeast	1,538	3.34	3,232	1.31	4,770	1.96
Florida	1,016	3.35	1,839	1.18	2,855	1.95
Mideast	1,254	3.32	4,620	1.29	5,874	1.73
Upper Midwest	568	3.33	3,551	1.10	4,120	1.41
Central	1,002	3.33	3,525	1.23	4,527	1.70
Southwest	1,482	3.34	2,960	1.35	4,442	2.01
Arizona	306	3.23	848	1.16	1,154	1.71
Pacific Northwest	462	3.41	1,708	1.31	2,170	1.76
Combined areas	11,625	3.32	30,814	1.22	42,440	1.79
2013						
Northeast	2,979	3.29	5,892	1.10	8,871	1.83
Appalachian	1,019	3.25	2,357	1.31	3,376	1.89
Southeast	1,512	3.34	3,133	1.32	4,646	1.97
Florida	1,013	3.34	1,785	1.20	2,797	1.97
Mideast	1,244	3.32	4,415	1.30	5,659	1.74
Upper Midwest	577	3.32	3,411	1.12	3,988	1.44
Central	1,000	3.34	3,418	1.25	4,417	1.72
Southwest	1,495	3.34	2,982	1.36	4,477	2.02
Arizona	311	3.24	816	1.17	1,128	1.74
Pacific Northwest	470	3.42	1,618	1.32	2,089	1.79
Combined areas	11,620	3.32	29,828	1.23	41,448	1.82

[1] In-area sales include total sales in each of the areas by handlers regulated under the respective order, by handlers regulated under other orders, by partially regulated handlers, by exempt handlers, and by producer-handlers. Sales routes of handlers may extend outside defined marketing areas; therefore, some handlers' in-area sales are partially estimated. [2] Plain, organic, flavored, and miscellaneous whole milk products, and eggnog. [3] Plain, fortified, organic, and flavored reduced fat milk (2%), low fat milk (1%), and fat-free milk (skim), and miscellaneous fat-reduced milk products, and buttermilk.

AMS, Dairy Programs, (202) 720–7461.

Table 8-20.—Milk markets under Federal order program: Uniform and Class I milk prices at 3.5 percent fat test, number of producers, producer milk receipts, producer milk used in Class I, Class I percentage, daily milk deliveries per producer, average fat test of producer milk receipts, by markets, 2013–2014

Federal milk order marketing area	Class I $ / cwt[1]	Uniform $ / cwt[1][2]	Average number of producers	Receipts of producer milk	Producer milk used in Class I	Class I utilization	Daily milk deliveries per producer	Average fat test of producer milk
	Dollars	Dollars	Number	Million pounds	Million pounds	Percent	Pounds	Percent
				2013				
Northeast[3]	22.09	20.23	12,353	25,420	9,508	37.40	5,638	3.76
Appalachian[4]	22.24	21.34	2,307	5,729	3,845	67.12	6,816	3.71
Florida[5]	24.24	23.53	262	2,833	2,424	85.54	30,731	3.64
Southeast[6]	22.64	21.74	2,320	6,129	4,163	67.92	7,336	3.72
Upper Midwest[7][8]	20.64	18.29	12,467	34,315	3,686	10.74	7,545	3.81
Central[8][9]	20.85	18.82	3,040	15,199	4,867	32.02	13,707	3.75
Mideast[8][10]	20.85	19.17	6,155	16,719	6,448	38.57	7,447	3.76
Pacific Northwest[8][11]	20.74	18.83	566	8,239	2,120	25.74	40,104	3.88
Southwest[8][12]	21.85	19.59	487	12,901	4,324	33.51	72,981	3.69
Arizona[13]	21.18	19.41	92	4,615	1,357	29.40	137,341	3.52
All markets combined	21.80	19.44	40,048	132,100	42,742	32.36	9,047	3.76
				2014				
Northeast[3]	26.53	24.28	12,178	25,793	9,123	35.37	5,803	3.78
Appalachian[4]	26.68	25.62	2,244	5,593	3,783	67.63	6,834	3.65
Florida[5]	28.67	27.82	214	2,771	2,343	84.56	37,006	3.60
Southeast[6]	27.07	26.20	2,109	5,289	3,905	73.83	6,870	3.66
Upper Midwest[7][8]	25.08	22.54	12,219	32,785	3,587	10.94	7,552	3.80
Central[8][9]	25.28	22.90	3,093	15,063	4,816	31.97	13,364	3.73
Mideast[8][10]	25.28	23.13	5,958	17,279	6,245	36.10	7,954	3.75
Pacific Northwest[11]	25.18	22.71	537	7,892	2,021	25.61	39,849	3.88
Southwest[8][12]	26.29	23.66	504	12,137	4,310	35.51	65,929	3.67
Arizona[13]	25.62	23.25	93	4,800	1,287	26.81	142,220	3.47
All markets combined	26.14	23.54	38,391	129,420	41,420	32.00	9,239	3.74

[1] Prices are for milk of 3.5 percent butterfat content and for the principal pricing point of the market. See footnotes 3-13. [2] For those orders that use the component pricing system for paying producers (orders 1, 30, 32, 33, 124,and 126), the figures are the statistical uniform price (the sum of the producer price differential and the Class III price). For those orders that use the skim milk/butterfat pricing system for paying producers (orders 5, 6, 7, and 131), the figures are the uniform price (the sum of the uniform butterfat price times 3.5 and the uniform skim milk price times 0.965). [3] Suffolk Co. (Boston), MA. [4] Mecklenburg Co. (Charlotte), NC. [5] Hillsborough Co. (Tampa), FL. [6] Fulton Co. (Atlanta), GA. [7] Cook Co. (Chicago), IL. [8] Due to the disadvantageous intraorder class and uniform price relationships in some months in these markets, handlers elected not to pool milk that normally would have been pooled under these orders. [9] Jackson Co. (Kansas City), MO. [10] Cuyahoga Co. (Cleveland), OH. [11] King Co. (Seattle), WA. [12] Dallas Co. (Dallas), TX. [13] Maricopa Co. (Phoenix), AZ.

AMS, Dairy Programs, (202) 720–7461.

Table 8-21.—Dairy products: Quantities manufactured, United States, 2011–2015

Product	2011	2012	2013	2014	2015
	1,000 pounds	*1,000 pounds*	*1,000 pounds*	*1,000 pounds*	*1,000 pounds*
Butter	1,809,751	1,859,537	1,862,516	1,855,315	1,857,998
Cheese:					
American types	4,226,670	4,355,252	4,419,848	4,588,023	4,694,886
Cheddar	3,096,350	3,143,443	3,189,859	3,317,038	3,393,464
Other American	1,130,320	1,211,809	1,230,009	1,270,985	1,301,422
Blue and Gorgonzola	85,867	87,932	92,401	92,845	94,477
Brick	11,420	12,482	9,312	2,942	3,449
Cream and Neufchatel	714,594	807,681	842,264	851,713	876,285
Feta	107,323	109,310	103,609	105,622	112,534
Gouda	16,158	37,227	47,799	55,911	58,317
Hispanic	224,396	223,893	241,420	249,579	254,344
Italian types	4,585,327	4,632,825	4,735,476	4,950,235	5,088,755
Mozzarella	3,574,428	3,614,848	3,699,989	3,924,925	3,995,810
Parmesan	287,705	296,963	319,741	301,848	338,995
Provolone	351,460	356,079	360,069	361,088	379,316
Ricotta	260,229	247,853	241,117	245,284	243,285
Romano	44,705	48,481	45,052	45,021	55,034
Other Italian types	66,800	68,601	69,608	72,069	76,315
Muenster cheese	146,585	152,516	163,220	163,650	177,151
Swiss cheese	329,145	320,599	294,495	297,802	312,040
All other types	147,521	146,524	151,894	153,783	166,187
Total cheese [1]	10,595,006	10,886,241	11,101,669	11,512,105	11,838,425
Cottage cheese:					
Curd [2]	423,683	423,900	389,396	381,146	398,693
Creamed [3]	322,099	323,220	307,391	303,113	317,528
Lowfat [4]	381,499	386,105	370,323	364,614	363,005
Sour cream	1,255,008	1,281,432	1,281,366	1,302,900	1,316,454
Yogurt plain & flavored	4,271,395	4,416,986	4,717,076	4,756,562	4,742,087
Bulk condensed milk:					
Skim, sweetened	42,503	39,635	38,190	43,593	52,935
Skim, unsweetened	1,575,211	1,619,260	1,623,159	1,661,446	1,668,369
Whole, sweetened	82,650	87,322	89,681	88,607	99,867
Whole, unsweetened	64,440	95,642	75,645	65,596	173,701
Condensed or evaporated buttermilk	76,502	81,278	94,499	109,044	105,676
Canned milk:					
Evaporated and condensed whole [5]	494,853	470,615	495,844	315,290	452,780
Evaporated skim	18,946	20,903	20,387	27,856	43,940
Dry milk products:					
Dry buttermilk, total	100,138	109,132	119,273	111,097	98,364
Dry skim milk animal	8,535	9,792	10,350	11,303	10,117
Dry whole milk	65,787	58,132	72,053	103,122	108,522
Milk protein concentrate, total	94,489	102,318	101,747	126,017	157,305
Nonfat dry milk, human	1,499,477	1,764,449	1,477,864	1,764,632	1,822,183
Skim milk powder, total [6]	446,017	380,672	630,689	543,504	446,324
Dry whey, total	1,010,117	998,898	961,020	869,701	974,905
	1,000 gallons	*1,000 gallons*	*1,000 gallons*	*1,000 gallons*	*1,000 gallons*
Ice cream, regular, total	888,378	894,033	897,363	865,753	897,676
Ice cream, lowfat, total [7]	415,464	457,273	400,925	411,701	431,701
Ice cream, nonfat, total	17,159	17,166	14,225	18,303	15,863
Sherbet, total	45,387	43,563	46,096	48,353	46,059
Frozen yogurt, total	62,715	57,555	74,483	66,456	68,715

[1] Excluding cottage cheese. [2] Mostly used for processing into creamed or lowfat cottage cheese. [3] Fat content 4 percent or more. [4] Fat content less than 4 percent. [5] Combined to avoid disclosing individual plant operations. [6] Includes protein standardized and blends. [7] Includes freezer-made milkshake.
NASS, Livestock Branch, (202) 720–3570.

Table 8-22.—Dairy Products: Factory production of specified items, by State and United States, 2014–2015

State	Butter		Total American cheese [1]		Total cheese [2]	
	2014	2015	2014	2015	2014	2015
	1,000 pounds	*1,000 pounds*	*1,000 pounds*	*1,000 pounds*	*1,000 pounds*	*1,000 pounds*
California	612,658	580,549	647,081	652,773	2,444,260	2,435,632
Idaho			587,502	611,114	895,189	941,683
Illinois					75,157	77,773
Iowa			180,525	141,959	263,627	244,487
Minnesota					666,537	679,467
New Jersey					54,239	58,778
New Mexico					757,990	768,028
New York	21,865	34,962	130,582	134,185	784,663	801,436
Ohio					202,350	211,382
Oregon			200,838	196,983		
Pennsylvania	93,844	93,221	701	1,014	415,246	408,624
South Dakota					263,618	278,343
Vermont					123,534	131,053
Wisconsin			848,879	909,940	2,912,144	3,070,202
Other States [3]	1,126,948	1,149,266	1,991,915	2,046,918	1,653,551	1,731,537
United States	1,855,315	1,857,998	4,588,023	4,694,886	11,512,105	11,838,425

State	Ice cream, regular (hard)		Nonfat dry milk for human food	
	2014	2015	2014	2015
	1,000 Gallons	*1,000 Gallons*	*1,000 Pounds*	*1,000 Pounds*
California	129,002	129,050	719,044	702,164
Connecticut	13,078	14,334		
Missouri	26,451	25,620		
Ohio	21,027	24,740		
Oregon	12,393	12,616		
Pennsylvania	36,236	38,866	162,489	179,917
Wisconsin	18,803	18,254		
Other States [3]	510,092	514,084	883,099	940,102
United States	767,082	777,564	1,764,632	1,822,183

[1] Includes Cheddar, Colby, washed curd, stirred curd, Monterey, and Jack. [2] Excluding cottage cheese. [3] States not shown when fewer than 3 plants reported or individual plant operations could be disclosed.
NASS, Livestock Branch, (202) 720–3570.

Table 8-23.—Dairy products: Average price per pound for specified products, 2010–2014

Item and market	2010	2011	2012	2013	2014
	Dollars	*Dollars*	*Dollars*	*Dollars*	*Dollars*
Butter, Chicago Mercantile Exchange: Grade AA:					
High [1]	2.2350	2.1800	1.9525	1.7875	3.0600
Low [1]	1.3100	1.5950	1.3000	1.3600	1.5325
Butter, Agricultural Marketing Service, Grade AA: [2]	1.7020	1.9498	1.5943	1.5451	2.1341
Cheese, Cheddar, Chicago Mercantile Exchange, Barrels:					
High [1]	1.7350	2.1350	2.0800	1.9700	2.4900
Low	1.2500	1.3400	1.4075	1.5300	1.4200
Cheese, Cheddar, Chicago Mercantile Exchange, 40-lb blocks:					
High [1]	1.7700	2.1550	2.1200	2.0000	2.4500
Low [1]	1.2675	1.3425	1.4600	1.5500	1.5200
Cheese, Cheddar, Agricultural Marketing Service, Barrels: [2]	1.5033	1.8146	1.6888	1.7456	2.1283
Cheese, Cheddar, Agricultural Marketing Service, 40-lb blocks: [2]	1.5138	1.8084	1.6972	1.7620	2.1341
Nonfat dry milk, Agricultural Marketing Service: Low/medium heat [2]	1.1687	1.5058	1.3279	1.7066	1.7592
Dry whey, Agricultural Marketing Service: Edible nonhygroscopic [2]	0.3716	0.5325	0.5935	0.5902	0.6526

[1] Figures are the high and low prices for any trading day during the year. [2] Prices used in Federal milk order price formulas. Prior to April 2012, the product prices were published by the National Agricultural Statistics Service.
AMS, Dairy Programs, (202) 720–9351.

Table 8-24.—Dairy products: Manufacturers' stocks, end of month, United States, 2014–2015 [1]

Month	Evaporated and condensed whole milk [2]		Dry whole milk		Nonfat dry milk, human	
	2014	2015	2014	2015	2014	2015
	1,000 pounds	1,000 pounds	1,000 pounds	1,000 pounds	1,000 pounds	1,000 pounds
January	52,954	36,002	6,925	23,237	148,795	240,210
February	60,456	42,714	9,196	18,282	181,382	237,571
March	62,094	42,917	7,381	17,651	216,478	250,569
April	78,889	52,594	5,757	25,258	239,529	248,076
May	86,239	96,400	10,497	27,788	220,868	261,416
June	85,172	102,608	14,405	25,050	228,179	262,322
July	96,381	102,950	14,817	23,211	252,137	269,897
August	99,668	107,206	19,860	19,633	239,623	231,140
September	90,157	99,109	24,968	9,383	171,163	211,576
October	67,282	72,913	21,657	6,039	189,545	180,038
November	50,320	36,624	21,957	7,404	219,755	198,812
December	32,402	43,053	23,244	11,370	239,624	204,289

[1] Stocks held by manufacturers at all points and in transit.   [2] Combined to avoid disclosing individual plant operations.
NASS, Livestock Branch, (202) 720–3570.

Table 8-25.—Dairy products: Total disappearance, and total and per capita consumption, United States, 2006–2015 [1]

Year	Butter			Cheese [2]			Condensed and evaporated milk [3]		
	Total dis-appear-ance	Consumption		Total dis-appear-ance	Consumption		Total dis-appear-ance	Consumption	
		Total	Per capita		Total	Per capita		Total	Per capita
	Million pounds	Million pounds	Pounds	Million pounds	Million pounds	Pounds	Million pounds	Million pounds	Pounds
2006	1,418	1,395	4.7	9,841	9,617	32.2	732	678	2.3
2007	1,503	1,414	4.7	10,150	9,868	32.7	716	628	2.1
2008	1,693	1,516	5.0	10,162	9,799	32.1	787	712	2.3
2009	1,574	1,522	5.0	10,248	9,917	32.3	771	703	2.3
2010	1,623	1,521	4.9	10,605	10,136	32.7	718	607	2.0
2011	1,795	1,676	5.4	10,893	10,301	33.0	711	598	1.9
2012	1,829	1,730	5.5	11,123	10,461	33.3	715	627	2.0
2013	1,915	1,732	5.5	11,364	10,580	33.4	723	594	1.9
2014	1,887	1,753	5.5	11,784	10,887	34.1	537	446	1.4
2015	1,840	1,800	5.6	12,056	11,275	35.1	782	697	2.2

Year	Reg hard ice cream (product weight)			Dry whole milk			Nonfat dry milk (human food)		
	Total dis-appear-ance	Consumption		Total dis-appear-ance	Consumption		Total dis-appear-ance	Consumption	
		Total	Per capita		Total	Per capita		Total	Per capita
	Million pounds	Million pounds	Pounds	Million pounds	Million pounds	Pounds	Million pounds	Million pounds	Pounds
2006	4,122	4,063	13.6	60	33	0.11	1,592	951	3.2
2007	4,016	3,959	13.1	73	50	0.16	1,442	862	2.9
2008	3,949	3,893	12.8	86	29	0.09	1,812	930	3.0
2009	3,860	3,804	12.4	97	78	0.25	1,789	1,227	4.0
2010	3,689	3,618	11.7	86	64	0.21	1,864	998	3.2
2011	3,603	3,522	11.3	84	67	0.21	1,926	942	3.0
2012	3,608	3,504	11.1	79	53	0.17	2,132	1,137	3.6
2013	3,565	3,430	10.8	88	54	0.17	2,159	918	2.9
2014	3,455	3,313	10.4	101	63	0.20	2,209	997	3.1
2015	3,494	3,362	10.5	141	105	0.33	2,306	1,057	3.3

[1] Total disappearance is based on production, imports, and change in stocks during the year.   Production statistics for these commodities appear in other tables in this chapter.   The total apparent consumption was obtained by subtracting ending stocks, shipments, and exports, from the total supply.   The per capita consumption for each year was obtained by dividing the total apparent consumption by the number of persons.   [2] Includes all kinds of cheese except cottage cheese.   [3] The evaporated milk is unskimmed, unsweetened, case goods.   The condensed milk is unsweetened, unskimmed, bulk goods; and sweetened condensed milk, unskimmed, case and bulk goods.
ERS, Animal Products and Cost of Production Branch, (202) 694–5171.

Table 8-26.—Dairy products: Dec. 31 stocks, United States, 2006–2015

Year	Butter [1][2]	Cheese [1][3]	Canned milk	Dry whole milk	Nonfat dry milk for human consumption [1]
	Million pounds	*Million pounds*	*Million pounds*	*Million pounds*	*Million pounds*
2006	109	817	31	2	107
2007	155	798	37	4	166
2008	119	852	42	5	247
2009	133	967	45	7	192
2010	82	1,048	52	8	145
2011	107	992	37	7	165
2012	153	1,023	38	7	183
2013	112	1,009	37	8	135
2014	105	1,018	32	23	240
2015	155	1,146	43	11	204

[1] Includes Government holdings. [2] Includes anhydrous milkfat and butter oil. [3] Excludes cottage cheese.
ERS, Animal Products and Cost of Production Branch, (202) 694–5171.

Table 8-27.—International dairy: Butter production, 2013–2015

Country	2013	2014	2015
	1,000 metric tons	*1,000 metric tons*	*1,000 metric tons*
Australia	117	125	122
Belarus	99	105	120
Brazil	83	85	87
Canada	95	88	91
European Union	2,100	2,250	2,310
India	4,745	4,887	5,035
Mexico	190	192	195
New Zealand	535	580	570
Russia	219	252	265
Ukraine	93	115	105
Others	128	116	118
Total Foreign	8,404	8,795	9,018
United States	845	842	830
Total	9,249	9,637	9,848

FAS, Office of Global Analysis, (202) 720-6301. Prepared or estimated on the basis of official USDA production, supply, and distribution statistics from foreign governments.

Table 8-28.—International dairy: Cheese production, 2013–2015

Country	2013	2014	2015
	1,000 metric tons	*1,000 metric tons*	*1,000 metric tons*
Argentina	556	564	570
Australia	320	320	330
Belarus	140	170	190
Brazil	722	736	751
Canada	388	396	400
European Union	9,368	9,560	9,610
Mexico	270	275	282
New Zealand	311	325	347
Russia	713	760	850
Ukraine	140	104	100
Others	73	72	68
Total Foreign	13,001	13,282	13,498
United States	5,036	5,194	5,299
Total	18,037	18,476	18,797

FAS, Office of Global Analysis, (202) 720-6301. Prepared or estimated on the basis of official USDA production, supply, and distribution statistics from foreign governments.

Table 8-29.—Dairy products: United States imports by country of origin, 2011–2013

Commodity and country of origin	2011	2012	2013
	Metric tons	*Metric tons*	*Metric tons*
Licensed cheese items 1:			
Netherlands	4,740	5,221	5,502
United Kingdom	2,472	2,566	3,124
Denmark(*)	1,844	1,962	2,182
Ireland	879	1,442	1,484
Canada	1,869	958	1,023
Germany(*)	664	670	707
New Zealand(*)	711	5,249	683
Italy(*)	340	436	457
Australia(*)	331	560	386
France(*)	245	233	260
Israel(*)	0	3	146
Austria	34	72	137
Dominican Republic	109	211	118
Jamaica	116	12	108
Bahrain	29	141	80
Spain	34	81	61
India	37	58	42
Lithuania	30	46	40
Finland	70	65	38
Mexico	0	0	38
Poland	24	56	27
Argentina	26	21	18
Philippines	5	12	15
Portugal	3	3	3
Egypt	8	0	2
Rest of World	75	4	1
World Total	14,694	20,080	16,682
Licensed cheese items 2:			
France(*)	18,712	18,088	19,344
Italy(*)	15,634	16,157	16,209
Switzerland(*)	5,374	6,090	8,069
Norway(*)	6,286	6,375	6,421
Ireland	5,111	6,101	5,537
Denmark(*)	5,125	5,166	5,536
Netherlands	6,668	6,025	5,238
Canada	3,021	3,309	4,304
Finland	7,164	6,134	3,699
Spain	1,081	1,922	3,562
Nicaragua	3,207	3,647	3,554
Germany(*)	1,901	1,695	2,973
Mexico	4,301	3,565	2,935
United Kingdom	1,880	1,901	2,146
Australia(*)	209	1,146	1,872
Poland	1,769	1,651	1,456
Uruguay	1,309	666	927
Austria	872	941	894
Argentina	2,989	2,510	853
Belgium-Luxembourg(*)	968	689	821
Egypt	496	573	642
Sweden	592	687	620
Lithuania	200	365	387
Portugal	445	334	351
Israel(*)	600	388	263
Rest of World	3,738	9,437	2,121
World Total	99,650	105,560	100,732
Licensed dairy, misc :			
Mexico	19,026	21,510	25,679
Canada	7,130	10,084	14,249
Chile	2,146	1,874	2,035
New Zealand(*)	3,409	3,333	1,675
Germany(*)	267	2,411	1,520
Netherlands	835	599	1,212
France(*)	211	346	889
Belgium-Luxembourg(*)	989	994	883
Australia(*)	882	2,204	658
Korea, South	452	449	589
Honduras	0	200	403
Ireland	515	1	394
Spain	60	26	325
Peru	120	205	264
Colombia	117	279	250
Argentina	158	361	234
United Kingdom	177	140	178
Denmark(*)	170	167	163
Italy(*)	160	54	150
Pakistan	0	0	136
Brazil	132	187	118
Ecuador	13	28	95
Japan	22	23	52
Israel(*)	200	1,753	49
Costa Rica	0	12	45
Rest of World	471	1,003	164
World Total	37,660	48,241	52,406

See footnote(s) at end of table.

Table 8-29.—Dairy products: United States imports by country of origin, 2011–2013—Continued

Commodity and country of origin	2011	2012	2013
	Metric tons	*Metric tons*	*Metric tons*
Non-licensed dairy, misc:			
Germany(*)	1,356	8,877	15,050
Mexico	15,409	18,513	15,016
Canada	4,526	5,722	6,471
Peru	3,123	3,935	5,936
Chile	4,842	4,468	5,221
New Zealand(*)	2,240	5,967	3,527
Denmark(*)	1,699	1,679	1,309
Greece	328	228	674
Netherlands	1,031	1,451	587
Israel(*)	44	425	343
Belgium-Luxembourg(*)	340	340	308
Ukraine	50	78	170
Brazil	287	134	131
United Kingdom	130	586	111
Norway(*)	46	181	100
Australia(*)	295	107	90
Russia	207	96	78
Nicaragua	76	40	70
Ireland	47	13	66
Poland	55	63	62
Argentina	273	20	39
Colombia	0	2	32
Spain	381	401	19
Italy(*)	0	8	19
Yemen(*)	0	0	18
Rest of World	63	275	76
World Total	36,849	53,609	55,520
Non-licensed cheese:			
Italy(*)	12,978	14,015	12,984
Spain	3,966	3,587	4,244
France(*)	3,165	2,757	3,399
Greece	2,519	2,827	2,986
Bulgaria	2,672	2,756	2,685
Turkey	315	585	995
Lithuania	0	40	769
Israel(*)	474	565	571
United Kingdom	486	431	483
Romania	169	157	319
Cyprus	298	244	277
Norway(*)	156	140	172
Netherlands	142	115	171
Macedonia	26	17	66
Egypt	18	29	44
Portugal	31	15	19
El Salvador	0	30	13
Jordan	31	12	9
Switzerland(*)	0	0	4
Croatia	5	4	4
Germany(*)	25	2	3
Poland	3	5	2
Bosnia and Herzegovina	0	0	1
Austria	1	4	1
Guatemala	0	0	1
Rest of World	1,194	34	0
World Total	28,672	28,368	30,222
Casein:			
New Zealand(*)	30,262	31,136	30,222
Netherlands	8,001	10,191	13,797
Poland	6,856	5,689	8,007
India	4,730	5,534	7,480
Ireland	13,641	10,664	7,143
Argentina	10,552	4,142	4,237
Germany(*)	5,868	4,472	4,135
France(*)	3,294	5,468	2,688
Uruguay	146	2,959	1,914
Denmark(*)	2,353	2,156	1,664
Australia(*)	1,439	565	996
China	74	631	237
Ukraine	183	253	155
Malaysia	0	0	71
Russia	0	0	20
Portugal	0	0	1
Belgium-Luxembourg(*)	0	19	0
Belarus	2,150	195	0
Canada	0	0	0
Japan	0	-	0
Mexico	5	0	0
United Kingdom	85	249	0
World Total	89,640	84,322	82,767

See footnote(s) at end of table.

Table 8-29.—Dairy products: United States imports by country of origin, 2011–2013—Continued

Commodity and country of origin	2011	2012	2013
	Metric tons	*Metric tons*	*Metric tons*
Lactose:			
Canada	3,193	2,435	2,798
Germany(*)	2,878	2,618	1,737
Netherlands	1,377	252	873
New Zealand(*)	127	102	137
France(*)	1	7	45
China	208	124	22
Belarus	66	44	22
Russia	9	0	7
Bosnia and Herzegovina	0	5	4
Thailand	0	0	3
Ireland	0	0	3
South Africa	0	2	3
Hong Kong	0	0	2
United Kingdom	7	1	2
Argentina	0	0	2
Japan	17	0	2
India	1	1	0
Belgium-Luxembourg(*)	2	0	0
Guatemala	15	0	0
Honduras	0	51	0
Israel(*)	-	0	0
Italy(*)	0	0	0
Singapore	0	0	0
Spain	-	-	0
Ukraine	0	6	0
World Total	7,900	5,646	5,661
Butter:			
New Zealand(*)	8,823	10,328	5,725
Ireland	1,856	2,391	3,391
Mexico	6	1,125	1,544
Australia(*)	627	1,860	1,123
France(*)	313	1,121	901
Germany(*)	239	182	352
Canada	228	279	285
Denmark(*)	242	230	219
India	277	171	141
Malaysia	0	14	135
Poland	122	114	133
United Kingdom	112	102	116
Italy(*)	61	55	61
Belgium-Luxembourg(*)	14	159	38
Argentina	10	139	37
Israel(*)	0	0	26
Turkey	10	13	17
Lithuania	13	15	17
Netherlands	5	5	13
Iceland	48	56	12
Portugal	12	11	10
Other Pacific Islands, NEC(*)	10	2	6
Bangladesh	0	0	6
Pakistan	0	0	5
Kenya	0	1	4
Rest of World	99	227	5
World Total	13,125	18,600	14,322

- Represents zero.   (*) Denotes a country that is a summarization of its component countries.   Note: All zeroes for a data item may show that statistics exist in the other importtype.   Consumption or General.   Users should use cautious interpretation on quantity reports using mixed units of measure.   Quantity line items will only include statistics on the units of measure that are equal to, or are able to be converted to the assigned unit of measure of the grouped commodities.

FAS, Office of Global Analysis, (202) 720-6301.   Data Source: Department of Commerce, U.S. Census Bureau, Foreign Trade Statistics

Table 8-30.—Dairy products: United States exports by country of destination, 2011–2013

Commodity and country of destination	2011	2012	2013
	Metric tons	*Metric tons*	*Metric tons*
Condensed & evap milk:			
Mexico	15,002	12,585	13,643
Vietnam	413	554	11,056
China	347	1,016	3,572
Philippines	55	882	3,497
Indonesia	0	194	2,035
Algeria	0	0	1,573
Canada	1,054	1,690	1,269
Bahamas, The	1,978	692	761
Hong Kong	553	458	759
Panama	117	266	583
Malaysia	48	633	571
Korea, South	270	435	562
Venezuela	38	32	518
Dominican Republic	367	341	424
Taiwan	540	388	382
Trinidad and Tobago	127	189	270
Bermuda	114	189	211
Australia(*)	452	539	199
El Salvador	18	6	185
Singapore	50	154	152
Honduras	1,057	166	135
Sweden	0	14	122
Belize	10	124	106
Thailand	0	141	100
Japan	8	39	94
Barbados	101	44	91
Leeward-Windward Islands(*)	74	60	86
Colombia	30	146	81
Netherlands Antilles(*)	110	115	73
Peru	9	3	69
Rest of World	1,249	2,151	432
World Total	24,188	24,243	43,610
Non-fat dry milk:			
Mexico	173,107	195,718	182,779
China	14,843	15,425	60,686
Philippines	51,363	45,689	57,708
Indonesia	44,790	31,543	56,145
Vietnam	41,868	24,335	37,403
Malaysia	23,303	16,494	30,712
Egypt	8,793	12,420	19,358
Algeria	0	8,335	17,554
Thailand	11,081	6,935	10,593
Peru	6,213	11,128	9,374
Pakistan	9,332	11,409	7,752
Dominican Republic	4,795	8,396	6,125
Korea, South	374	1,480	5,524
Japan	7,974	6,784	5,330
Chile	4,901	9,568	4,683
Singapore	5,629	3,418	4,263
Sri Lanka	416	2,621	3,997
Morocco	1,158	4,941	3,844
Canada	2,416	2,271	2,733
Libya	0	552	2,329
Colombia	422	2,330	2,197
Venezuela	1,527	2,530	2,141
Panama	1,293	1,821	1,827
Saudi Arabia	760	573	1,538
Russia	256	411	1,401
Taiwan	878	898	1,353
Hong Kong	130	561	1,197
Jamaica	115	740	1,116
Nigeria	1,349	824	1,033
Ukraine	471	19	993
Rest of World	15,543	14,181	11,433
World Total	435,098	444,349	555,122

See footnote(s) at end of table.

Table 8-30.—Dairy products: United States exports by country of destination, 2011–2013—Continued

Commodity and country of destination	2011	2012	2013
	Metric tons	*Metric tons*	*Metric tons*
Dry whole milk & cream:			
Algeria	0	0	12,828
China	63	1,591	6,179
Mexico	6,858	5,326	4,177
Vietnam	694	1,836	3,487
Philippines	65	1,005	1,803
Canada	1,036	650	1,271
Israel(*)	2,005	909	1,035
Japan	1,019	743	911
Thailand	256	1	759
Korea, South	114	44	572
Panama	232	290	507
Egypt	1,854	313	500
Djibouti	0	0	480
Malaysia	7	329	387
Colombia	325	552	364
Australia(*)	0	26	323
Hong Kong	491	837	323
Singapore	119	51	320
Indonesia	764	394	318
Guatemala	216	404	226
Tanzania	0	0	200
Pakistan	916	1,219	172
Trinidad and Tobago	3	123	145
Nicaragua	200	877	138
United Arab Emirates	22	32	130
Peru	370	283	128
South Africa	57	591	121
Italy(*)	0	4	115
Belize	41	102	114
Saudi Arabia	170	318	112
Rest of World	3,658	2,081	998
World Total	21,555	20,932	39,141
Butter and milkfat:			
Saudi Arabia	13,065	16,643	19,182
Iran	2,871	6,561	11,791
Morocco	3,747	3,878	8,081
Ukraine	0	0	7,817
Egypt	3,661	3,063	7,066
Turkey	478	19	5,905
Korea, South	3,631	507	3,693
Canada	5,625	1,539	3,307
Mexico	6,785	3,226	3,201
United Arab Emirates	445	1,566	2,178
Bahrain	557	927	1,514
Lithuania	0	4	1,512
France(*)	894	459	1,448
Vietnam	5	1	1,373
Denmark(*)	2,966	334	1,267
Belgium-Luxembourg(*)	3,295	148	991
China	503	240	946
Dominican Republic	1,074	1,106	795
Israel(*)	630	537	709
Lebanon	213	44	636
Algeria	0	94	631
Nigeria	132	5	505
Uruguay	257	105	482
Indonesia	499	0	458
Bahamas, The	169	1,603	440
Taiwan	489	195	412
Moldova	0	0	396
Netherlands	546	67	385
Panama	773	285	358
Georgia	19	0	323
Rest of World	10,341	5,575	3,850
World Total	63,669	48,730	91,649

See footnote(s) at end of table.

Table 8-30.—Dairy products: United States exports by country of destination, 2011–2013—Continued

Commodity and country of destination	2011	2012	2013
	Metric tons	*Metric tons*	*Metric tons*
Ice cream:			
Mexico	18,857	21,265	22,798
Saudi Arabia	62	1,446	6,471
United Arab Emirates	451	2,750	4,948
Canada	3,643	3,599	3,876
Australia(*)	965	1,955	3,370
China	283	782	1,641
Malaysia	164	673	1,442
Trinidad and Tobago	1,007	1,035	1,262
Korea, South	833	903	1,220
Bahamas, The	1,218	1,160	1,213
Kuwait	85	384	937
Chile	540	1,150	837
Indonesia	14	230	781
Jamaica	828	1,185	769
Netherlands Antilles(*)	687	691	741
Singapore	789	849	719
Qatar	66	256	601
Philippines	423	519	576
Dominican Republic	451	409	527
Netherlands	685	299	456
Russia	15	13	424
Leeward-Windward Islands(*)	482	445	415
Bahrain	45	134	381
Bermuda	583	394	372
Nigeria	148	170	356
Taiwan	290	327	332
Japan	159	310	321
United Kingdom	81	168	238
Panama	93	114	230
Cayman Islands	252	289	227
Rest of World	2,706	2,927	2,637
World Total	36,904	46,829	61,116
Cheese and curd:			
Mexico	48,407	65,483	82,234
Korea, South	35,431	39,411	49,215
Japan	22,882	25,913	31,189
Canada	11,177	13,946	14,314
Australia(*)	9,973	11,059	12,105
Saudi Arabia	10,989	11,333	11,881
China	6,678	8,798	11,343
Egypt	7,443	4,003	8,509
Panama	5,182	6,542	7,651
Taiwan	5,071	6,355	7,028
Philippines	5,545	4,913	6,469
Chile	2,764	4,594	6,122
Dominican Republic	4,264	3,542	4,993
Morocco	3,751	3,483	4,961
Bahrain	1,424	3,588	4,546
Indonesia	4,598	5,377	4,338
Guatemala	3,084	3,383	3,133
Honduras	2,420	2,331	2,635
Hong Kong	1,906	2,462	2,298
Peru	2,099	2,246	2,176
Colombia	581	1,213	2,128
Kuwait	1,323	1,440	2,093
United Arab Emirates	1,628	1,653	2,055
United Kingdom	336	491	1,953
Netherlands	1,486	372	1,952
Algeria	0	180	1,879
Libya	486	1,284	1,874
Bahamas, The	1,946	1,783	1,634
Singapore	1,769	1,396	1,599
Costa Rica	1,127	1,625	1,566
Rest of World	19,295	19,765	20,315
World Total	225,064	259,962	316,185

See footnote(s) at end of table.

Table 8-30.—Dairy products: United States exports by country of destination, 2011–2013—Continued

Commodity and country of destination	2011	2012	2013
	Liters	*Liters*	*Liters*
Fluid milk and cream:			
Canada	30,188,599	27,065,590	32,347,609
Mexico	26,687,323	24,992,377	32,177,942
Peru	32,164	2,269,686	5,059,489
Taiwan	255,755	165,260	3,209,351
Hong Kong	853,282	1,823,650	2,856,649
China	430,502	1,408,906	2,393,649
Singapore	340,419	602,266	2,311,769
Dominican Republic	67,330	2,917	1,387,323
Bahamas, The	1,447,509	1,448,556	1,040,446
Philippines	26,718	358,912	780,511
Vietnam	1,683,492	747,290	485,296
Malaysia	618,444	339,584	413,033
Turks and Caicos Islands	226,227	191,817	276,883
Netherlands Antilles(*)	419,894	152,575	272,314
Leeward-Windward Islands(*)	170,578	196,311	271,720
Cayman Islands	630,190	458,872	269,753
Korea, South	10,735,241	197,858	226,629
Pakistan	232,200	303,180	200,047
Saudi Arabia	156,000	321,753	156,000
Palau	200,433	133,431	150,200
Panama	18,368	5,604	103,948
Colombia	4,526	6,839	81,929
Guatemala	20,859	76,935	73,073
Equatorial Guinea	42,942	17,208	69,974
United Arab Emirates	6,259	0	57,342
Australia(*)	0	2,781	52,946
Uruguay	0	16,227	47,257
Thailand	14,934	36,337	43,604
French Pacific Islands(*)	66,522	435	41,026
Bangladesh	8,724	42,885	36,883
Rest of World	535,507	274,665	171,326
World Total	76,120,941	63,660,707	87,065,921
	Metric tons	*Metric tons*	*Metric tons*
Whey:			
China	152,214	154,558	183,079
Mexico	63,340	65,823	59,126
Canada	41,726	38,666	39,691
Japan	29,155	26,181	26,154
Indonesia	22,218	21,504	25,585
New Zealand(*)	3,227	11,517	22,627
Philippines	14,509	22,396	21,866
Korea, South	16,341	21,173	20,654
Vietnam	19,816	27,043	18,976
Malaysia	20,945	14,612	15,087
Morocco	6,578	8,590	14,287
Thailand	14,141	8,944	10,808
Taiwan	5,753	6,725	8,732
Australia(*)	7,027	3,878	8,573
Brazil	6,424	8,576	4,490
Singapore	5,083	4,902	4,110
Egypt	2,930	2,050	4,083
Peru	1,024	3,408	3,540
Dominican Republic	2,591	3,113	2,481
Guatemala	2,101	2,638	2,375
Venezuela	2,059	1,672	2,102
Chile	2,358	3,076	1,714
El Salvador	2,355	2,108	1,621
Switzerland(*)	272	1,079	1,535
Uruguay	526	358	1,380
Belgium-Luxembourg(*)	1,599	2,803	1,242
United Kingdom	379	1,463	1,158
Ghana	0	60	1,120
Panama	759	749	1,112
Colombia	422	1,073	1,008
Rest of World	14,905	17,247	11,080
World Total	462,772	487,984	521,396
Grand Total	1,269,249	1,333,029	1,628,219

(*) Denotes a country that is a summarization of its component countries.　Note: Users should use cautious interpretation on quantity reports using mixed units of measure.　Quantity line items will only include statistics on the units of measure that are equal to, or are able to be converted to, the assigned unit of measure of the grouped commodities.

FAS, Office of Global Analysis, (202) 720-6301.　Data Source: Department of Commerce, U.S. Census Bureau, Foreign Trade Statistics.

Table 8-31.—Chickens: Layers, pullets, and other chickens, by State and United States, December 1, 2014 and 2015

State	Total layers		Total pullets		Other Chickens	
	2014	2015	2014	2015	2014	2015
	Thousands	Thousands	Thousands	Thousands	Thousands	Thousands
Alabama	9,279	9,267	5,820	5,763	961	1,004
Arkansas	12,529	14,141	7,284	8,230	1,097	1,129
California	14,890	12,091	2,704	4,010	19	15
Colorado	4,791	4,521	1,052	1,235	50	64
Connecticut [1]	2,418		558		9	
Florida	8,998	9,167	2,053	2,508	26	25
Georgia	18,797	19,317	8,442	7,739	909	1,060
Illinois	5,026	5,079	212	211	24	24
Indiana	26,913	30,353	7,522	9,849	115	98
Iowa	59,889	37,821	14,038	13,996	69	93
Kentucky	4,583	5,059	2,306	2,276	161	209
Louisiana	2,086	2,077	651	684	100	135
Maine [1]	3,535		24		-	
Maryland	2,844	2,805	676	949	7	8
Massachusetts	144	146	8	14	-	-
Michigan	13,192	13,249	2,819	3,704	-	-
Minnesota	11,277	9,377	3,379	3,590	37	37
Mississippi	5,571	5,600	3,446	3,663	492	561
Missouri	9,138	11,211	3,139	4,352	167	179
Montana	458	439	162	150	-	-
Nebraska	9,571	7,019	1,983	2,423	-	-
New York	5,155	5,754	1,346	1,622	8	9
North Carolina	14,441	14,869	6,523	6,987	850	1,089
Ohio	31,542	34,437	7,584	8,231	40	41
Oklahoma	3,072	2,884	1,186	1,069	179	184
Oregon	2,253	2,222	554	693	-	-
Pennsylvania	25,900	26,596	4,666	5,187	143	176
South Carolina	4,218	4,397	1,443	1,529	110	147
South Dakota	2,715	1,575	716	476	-	-
Tennessee	1,495	1,668	1,076	644	132	183
Texas	19,116	20,421	5,564	5,959	380	403
Utah	4,473	4,532	773	1,328	-	-
Vermont	141	122	11	4	1	1
Virginia	2,939	2,979	979	1,055	135	153
Washington	7,094	7,358	1,320	2,038	-	1
West Virginia	1,207	1,231	715	780	102	101
Wisconsin	5,626	4,894	1,408	1,437	36	35
Other States [2]	8,729	16,978	2,460	3,729	44	47
United States	366,045	351,656	106,602	118,114	6,403	7,211

- Represents zero. [1] 2015 included in Other States. [2] Alaska, Arizona, Connecticut, Delaware, Hawaii, Idaho, Kansas, Maine, New Hampshire, New Jersey, New Mexico, Nevada, North Dakota, and Rhode Island combined to avoid disclosing individual operations.
NASS, Livestock Branch, (202) 720–3570.

Table 8-32.—Chickens: Inventory number and value, United States, Dec. 1, 2006–2015 [1]

Year	Total layers	Total pullets	Other chickens	All chickens	Value per head	Total value
	Thousands	Thousands	Thousands	Thousands	Dollars	1,000 dollars
2006	352,316	97,459	8,038	457,813	2.60	1,189,978
2007	346,613	103,816	8,164	458,593	2.95	1,351,549
2008	340,811	101,016	7,589	449,416	3.38	1,518,547
2009	342,058	104,069	8,487	454,614	3.32	1,510,038
2010	342,782	107,025	7,390	457,197	3.58	1,637,016
2011	343,283	104,159	6,922	454,364	3.79	1,721,671
2012	354,977	105,140	6,827	466,944	4.04	1,885,904
2013	361,403	106,646	6,853	474,902	4.15	1,971,590
2014	366,045	106,602	6,403	479,050	4.10	1,962,126
2015	351,656	118,114	7,211	476,981	4.37	2,085,151

[1] Does not include commercial broilers.
NASS Livestock Branch, (202) 720-3570.

Table 8-33.—Chicken inventory: Number, value per head, and total value, by State and United States, December 1, 2014 and 2015 [1]

State	Number		Value per bird		Total value	
	2014	2015	2014	2015	2014	2015
	1,000 Head	1,000 Head	Dollars	Dollars	1,000 Dollars	1,000 Dollars
Alabama	16,060	16,034	7.60	8.30	122,056	133,082
Arkansas	20,910	23,500	7.70	7.70	161,007	180,950
California	17,613	16,116	2.80	3.40	49,316	54,794
Colorado	5,893	5,820	2.30	2.50	13,554	14,550
Connecticut [2]	2,985		2.90		8,657	
Florida	11,077	11,700	2.30	2.40	25,477	28,080
Georgia	28,148	28,116	5.50	6.80	154,814	191,189
Illinois	5,262	5,314	3.90	3.80	20,522	20,193
Indiana	34,550	40,300	2.90	2.80	100,195	112,840
Iowa	73,996	51,910	2.60	2.70	192,390	140,157
Kentucky	7,050	7,544	5.00	4.90	35,250	36,966
Louisiana	2,837	2,896	5.10	5.30	14,469	15,349
Maine [2]	3,559		2.70		9,609	
Maryland	3,527	3,762	3.30	3.20	11,639	12,038
Massachusetts	152	160	6.80	6.60	1,034	1,056
Michigan	16,011	16,953	2.20	4.00	35,224	67,812
Minnesota	14,693	13,004	3.90	3.10	57,303	40,312
Mississippi	9,509	9,824	7.20	8.00	68,465	78,592
Missouri	12,444	15,742	5.30	4.90	65,953	77,136
Montana	620	589	5.00	8.50	3,100	5,007
Nebraska	11,554	9,442	3.10	3.90	35,817	36,824
New York	6,509	7,385	2.90	2.10	18,876	15,509
North Carolina	21,814	22,945	8.10	7.90	176,693	181,266
Ohio	39,166	42,709	3.80	4.00	148,831	170,836
Oklahoma	4,437	4,137	7.00	7.00	31,059	28,959
Oregon	2,807	2,915	2.90	2.90	8,140	8,454
Pennsylvania	30,709	31,959	3.40	2.70	104,411	86,289
South Carolina	5,771	6,073	4.40	4.20	25,392	25,507
South Dakota	3,431	2,051	2.30	3.20	7,891	6,563
Tennessee	2,703	2,495	7.70	8.20	20,813	20,459
Texas	25,060	26,783	3.80	4.50	95,228	120,524
Utah	5,246	5,860	2.50	3.10	13,115	18,166
Vermont	153	127	5.40	3.70	826	470
Virginia	4,053	4,187	6.30	6.60	25,534	27,634
Washington	8,414	9,397	3.30	2.80	27,766	26,312
West Virginia	2,024	2,112	8.00	7.80	16,192	16,474
Wisconsin	7,070	6,366	2.60	2.70	18,382	17,188
Other States [3]	11,233	20,754	3.31	3.26	37,126	67,614
United States	479,050	476,981	4.10	4.37	1,962,126	2,085,151

Totals may not add due to rounding.   [1] Excludes commercial broilers.   [2] 2015 included in Other States.   [3] Alaska, Arizona, Connecticut, Delaware, Hawaii, Idaho, Kansas, Maine, New Hampshire, New Jersey, New Mexico, Nevada, North Dakota, Rhode Island and Wyoming combined to avoid disclosing data for individual operations.
NASS, Livestock Branch, (202) 720-3570.

Table 8-34.—Chickens: Lost, sold for slaughter, and value of sales, by State and United States, 2014 and 2015 [1]

State	Number lost [2]		Number sold		Pounds sold		Value of sales	
	2014	2015	2014	2015	2014	2015	2014	2015
	1,000 head	1,000 head	1,000 head	1,000 head	1,000 pounds	1,000 pounds	1,000 dollars	1,000 dollars
Alabama	2,739	2,696	10,736	11,466	82,667	82,667	13,475	14,568
Arkansas	3,916	3,372	13,389	14,387	100,418	100,418	15,966	16,699
California	6,461	6,446	4,196	5,033	13,847	13,847	208	48
Colorado	484	673	3,536	3,025	14,144	14,144	834	826
Connecticut ...	....	....			....	....	....	....
Florida	3,378	5,345	1,964	352	7,463	7,463	343	417
Georgia	4,641	6,037	15,191	16,296	106,337	106,337	16,057	16,159
Illinois	722	978	2,410	2,514	8,194	8,194	197	244
Indiana	7,803	8,676	11,796	12,442	38,927	38,927	506	550
Iowa	25,766	43,125	15,484	11,293	49,549	49,549	149	149
Kentucky	1,591	807	4,278	4,089	25,668	25,668	3,363	3,136
Louisiana	340	342	2,014	2,120	12,688	12,688	1,738	1,843
Maine	....	....	1,225	....	....	....	....	....
Maryland	234	226	892	1,301	3,300	3,300	129	150
Massachusetts	7	8	97	95	310	310	(Z)	(Z)
Michigan	2,232	1,104	6,206	9,305	19,859	19,859	20	31
Minnesota	1,952	4,980	7,457	5,923	25,354	25,354	482	518
Mississippi	1,410	1,419	6,062	5,583	44,253	44,253	6,859	6,480
Missouri	1,428	1,322	8,243	8,023	39,566	39,566	3,759	3,174
Montana	323	323	55	54	176	176	(Z)	(Z)
Nebraska	1,054	4,222	5,624	4,094	17,997	17,997	18	13
New York	708	631	3,801	3,282	12,163	12,163	36	21
North Carolina	2,900	3,225	12,524	13,578	83,911	83,911	12,251	14,241
Ohio	11,624	11,874	8,733	10,345	29,692	29,692	475	457
Oklahoma	709	711	2,176	2,653	16,538	16,538	2,663	3,371
Oregon	1,045	748	207	819	642	642	1	3
Pennsylvania ..	2,794	2,394	10,438	10,452	36,533	36,533	950	1,470
South Carolina	1,337	1,549	2,626	2,743	16,281	16,281	2,214	2,242
South Dakota ..	760	1,517	1,142	336	3,654	3,654	4	1
Tennessee	424	349	1,987	1,904	15,300	15,300	2,479	2,495
Texas	4,138	2,983	12,091	12,783	59,246	59,246	5,747	6,138
Utah	1,208	863	1,593	2,484	5,098	5,098	5	8
Vermont	18	17	92	108	405	405	31	39
Virginia	890	611	2,268	2,440	16,103	16,103	2,448	2,576
Washington ...	5,505	3,484	146	300	496	496	9	1
West Virginia ..	284	263	1,599	1,568	12,632	12,632	2,072	2,093
Wisconsin	378	1,651	3,495	4,205	12,932	12,932	504	529
Other States [3]	4,415	4,505	6,035	6,125	19,793	19,793	159	209
United States	105,618	129,476	190,583	193,520	952,136	952,136	96,151	100,899

(Z) Less than half of the unit shown. [1] Annual estimates cover the period December 1 previous year through November 30. Exclude broilers. Totals may not add due to rounding. [2] Includes rendered, died, destroyed, composted, or disappeared for any reason except sold during the 12-month period. [3] Alaska, Arizona, Delaware, Hawaii, Idaho, Kansas, Nevada, New Hampshire, New Jersey, New Mexico, North Dakota, Rhode Island, and Wyoming combined to avoid disclosing data for individual operations.
NASS, Livestock Branch, (202) 720–3570.

Table 8-35.—Mature chickens: Lost, sold for slaughter, price, and value, United States, 2006–2015 [1]

Year	Number		Pounds (live weight) sold [3]	Price per pound live weight [3]	Value of sales [3]
	Lost [2]	Sold [3]			
	1,000 head	1,000 head	1,000 pounds	Dollars	1,000 dollars
2006	101,611	173,883	924,993	0.059	54,141
2007	101,152	168,283	912,875	0.056	51,498
2008	102,071	176,023	940,221	....	62,456
2009	99,596	176,266	910,333	....	65,381
2010	108,356	173,404	903,750	....	73,099
2011	106,361	182,153	943,142	....	81,110
2012	103,758	179,162	901,411	....	79,208
2013	101,166	185,837	922,546	....	87,939
2014	105,618	190,583	952,136	....	96,151
2015	129,476	193,520	978,612	....	100,899

[1] Annual estimates cover the period December 1 previous year through November 30. Excludes broilers. Totals may not add due to rounding. [2] Includes rendered, died, destroyed, composted, or disappeared for any reason (excluding sold for slaughter) during the 12-month period. [3] Sold for slaughter.
NASS, Livestock Branch, (202) 720–3570.

Table 8-36.—Poultry, meat, and broiler: International trade, 2013–2015

Country	2013	2014	2015
	1,000 metric tons	*1,000 metric tons*	*1,000 metric tons*
Principle exporting:			
Argentina	334	278	187
Belarus	105	113	135
Brazil	3,482	3,558	3,841
Canada	150	137	133
Chile	88	87	99
China	420	430	401
European Union	1,083	1,133	1,177
Thailand	504	546	622
Turkey	337	379	321
Ukraine	141	167	159
Others	298	340	331
Total Foreign	6,942	7,168	7,406
United States	3,332	3,312	2,866
Total	10,274	10,480	10,272
Principle importing:			
China	244	260	268
European Union	671	712	728
Hong Kong	272	299	312
Iraq	673	730	630
Japan	854	888	936
Mexico	682	722	790
Russia	535	465	249
Saudi Arabia	838	775	930
South Africa	355	369	436
United Arab Emirates	217	192	262
Others	3,288	3,474	3,059
Total Foreign	8,629	8,886	8,600
United States	55	53	59
Total	8,684	8,939	8,659

FAS, Office of Global Analysis, (202) 720-6301.   Prepared or estimated on the basis of official USDA production, supply, and distribution statistics from foreign governments.

Table 8-37.—Broilers: Production and value, United States, 2006–2015 [1] [2] [3]

Year	Number produced	Pounds produced	Value of production
	1,000 head	1,000 pounds	1,000 dollars
2006	8,867,800	48,829,900	17,739,234
2007	8,906,700	49,330,700	21,513,536
2008	9,009,300	50,441,600	23,203,136
2009	8,550,200	47,752,300	21,822,804
2010	8,623,600	49,152,600	23,691,553
2011	8,607,600	50,082,400	22,987,822
2012	8,463,000	49,655,600	24,827,800
2013	8,533,800	50,678,200	30,761,669
2014	8,545,000	51,378,700	32,728,234
2015	8,686,600	53,364,000	28,709,834

[1] December 1, previous year through November 30, current year.　[2] Broiler production including other domestic meat-type strains.　[3] Excludes States producing less than 500,000 broilers.
NASS, Livestock Branch, (202) 720–3570.

Table 8-38.—Chickens: Supply, distribution, and per capita consumption, ready-to-cook basis, United States, 2006–2015

Year	Production			Commercial storage at beginning of year	Exports	Commercial storage at end of year	Consumption	
	Commercial broilers	Other chickens	Total [1]				Total [1] [2]	Per capita
	Million pounds	Million pounds	Million pounds	Million pounds	Million pounds	Million pounds	Million pounds	Pounds
2006	35,120	504	35,624	913	5,365	738	30,484	102
2007	35,772	498	36,270	737	6,072	721	30,280	100
2008	36,511	559	37,070	721	7,110	748	30,036	99
2009	35,131	500	35,631	748	7,226	618	28,948	94
2010	36,515	503	37,018	618	6,841	777	30,130	97
2011	36,804	521	37,325	777	7,061	592	30,558	98
2012	36,643	517	37,160	592	7,364	652	29,850	95
2013	37,425	523	37,948	652	7,453	675	30,596	97
2014	38,153	521	38,674	675	7,407	683	7,407	98
2015	39,620	522	40,142	683	6,466	840	6,466	105

[1] Totals may not add due to rounding.　[2] Shipments to territories now included in total consumption.
ERS Markets and Trade Economics Division, Animal Products and Cost of Production Branch, (202) 694–5308.

Table 8-39.—Poultry: Feed-price ratios, United States, 2006–2014

Year	Ratios [1]		
	Egg-feed	Broiler-feed	Turkey-feed
	Pounds	Pounds	Pounds
2006	7.5	5.7	7.8
2007	10.2	5.0	6.0
2008	8.6	3.7	4.6
2009	7.2	4.1	5.0
2010	7.5	4.5	6.2
2011	6.0	3.1	4.9
2012	5.4	3.1	4.7
2013	6.5	3.7	4.6
2014	10.2	5.0	6.3

[1] Number of pounds of poultry feed equivalent in value at local market prices to 1 dozen market eggs, or 1 pound of broiler or 1 pound of turkey live weight.　Simple average of monthly feed-price ratios.　Egg feed= corn (75 lbs) and soybeans (25 lbs); broiler feed= corn (58 lbs); soybeans (42 lbs); turkey feed= corn (51 lbs), soybeans (28 lbs), and wheat (21 lbs).　Monthly equivalent prices of commercial prepared feeds are based on current U.S. prices received for corn, soybeans, and wheat.
NASS, Environmental, Economics, and Demographics Branch, (202) 720–6146.

Table 8-40.—Broilers: Production and value, States, 19 States, and United States, 2014 and 2015 [1]

State	2014			2015		
	Number produced	Pounds produced	Value of production	Number produced	Pounds produced	Value of production
	1,000 head	*1,000 pounds*	*1,000 dollars*	*1,000 head*	*1,000 pounds*	*1,000 dollars*
Alabama	1,061,500	6,050,600	3,854,232	1,082,900	6,172,500	3,320,805
Arkansas	970,100	6,014,600	3,831,300	962,000	6,156,800	3,312,358
Delaware	244,100	1,733,100	1,103,985	244,300	1,759,000	946,342
Florida	66,700	386,900	246,455	65,100	377,600	203,149
Georgia	1,324,200	7,547,900	4,808,012	1,339,600	7,903,600	4,252,137
Kentucky	308,000	1,724,800	1,098,698	307,700	1,784,700	960,169
Maryland	287,900	1,554,700	990,344	303,500	1,730,000	930,740
Minnesota	46,800	280,800	178,870	46,400	278,400	149,779
Mississippi	727,200	4,508,600	2,871,978	722,500	4,551,800	2,448,868
Missouri	288,500	1,384,800	882,118	294,600	1,414,100	760,786
North Carolina ..	795,200	6,043,500	3,849,710	822,700	6,417,100	3,452,400
Ohio	75,600	430,900	274,483	80,400	450,200	242,208
Oklahoma	205,800	1,337,700	852,115	217,000	1,432,200	770,524
Pennsylvania	181,300	997,200	635,216	190,400	1,066,200	573,616
South Carolina ..	232,500	1,650,800	1,051,560	242,600	1,746,700	939,725
Tennessee	180,600	939,100	598,207	185,200	981,600	528,101
Texas	591,800	3,550,800	2,261,860	608,700	3,773,900	2,030,358
Virginia	262,000	1,441,000	917,917	262,800	1,471,700	791,775
West Virginia	95,300	371,700	236,773	93,700	356,100	191,582
Wisconsin	53,400	224,300	142,879	53,900	226,400	121,803
Other States [2] ...	546,500	3,204,900	2,041,522	560,600	3,313,400	1,782,609
United States	8,545,000	51,378,700	32,728,234	8,686,600	53,364,000	28,709,834

[1] Annual estimates cover the period December 1 previous year through November 30. Broiler production including other domestic meat-type strains. Excludes States producing less than 500,000 broilers. [2] California, Illinois, Indiana, Iowa, Louisiana, Michigan, Nebraska, New York, Oregon, and Washington combined to avoid disclosing individual operations. NASS, Livestock Branch, (202) 720–3570.

Table 8-41.—Chicks hatched by commercial hatcheries: Number, average price, and value, United States, 2006–2015

Year	Chicks hatched			Average price of baby chicks per 100		Value of chick production
	Broiler-type	Egg-type	All	Broiler-type	Egg-type	
	Thousands	*Thousands*	*Thousands*	*Dollars*	*Dollars*	*1,000 dollars*
2006	9,414,070	427,373	9,841,443	22.90	66.50	2,297,924
2007	9,590,018	446,562	10,036,580	25.60	69.40	2,610,002
2008	9,468,133	467,763	9,935,896	26.30	75.60	2,666,933
2009	9,116,802	467,981	9,584,783	26.60	81.20	2,615,070
2010	9,276,240	484,393	9,760,633	27.00	83.00	2,705,608
2011	9,057,461	478,606	9,536,067	28.20	86.70	2,761,680
2012	8,965,425	483,443	9,448,868	29.50	89.70	2,861,625
2013	9,078,764	514,479	9,593,243	29.90	93.40	2,954,812
2014	9,160,978	518,346	9,679,324	31.00	74.80	3,033,765

NASS, Environmental, Economics, and Demographics Branch, (202) 720–6146 and Livestock Branch, (202) 720–3570.

Table 8-42.—Poultry: Slaughtered under Federal inspection, by class, United States, 2013–2015

Class	Number slaughtered			Total live weight		
	2013	2014	2015	2013	2014	2015
	Thousands	*Thousands*	*Thousands*	*Thousands*	*Thousands*	*Thousands*
Young chickens	8,503,750	8,525,393	8,688,462	50,357,463	51,225,964	53,169,030
Mature chickens	145,006	144,235	134,230	812,167	822,537	807,037
Total chickens	8,648,756	8,669,628	8,822,692	51,169,630	52,048,501	53,976,067
Young turkeys	237,964	235,189	230,812	7,220,540	7,150,782	6,972,127
Old turkeys	1,440	1,428	1,577	39,347	39,442	44,542
Total turkeys	239,404	236,617	232,389	7,259,887	7,190,224	7,016,669
Ducks	24,575	26,368	27,749	168,223	181,068	190,062
Other poultry [1]	(NA)	(NA)	(NA)	3,518	3,143	3,678
Total poultry	(NA)	(NA)	(NA)	58,601,258	59,422,936	61,186,476

Class	Pounds condemned					
	Ante-mortem (live weight)			Post-mortem (Carcass and parts)		
	2013	2014	2015	2013	2014	2015
	Thousands	*Thousands*	*Thousands*	*Thousands*	*Thousands*	*Thousands*
Young chickens	99,945	108,109	121,020	316,386	341,079	364,274
Mature chickens	7,844	8,853	9,206	24,825	27,984	28,779
Total chickens	107,789	116,962	130,226	341,211	369,063	393,053
Young turkeys	15,728	16,491	16,873	79,430	77,912	79,937
Old turkeys	434	536	568	1,640	1,417	1,628
Total turkeys	16,162	17,027	17,441	81,070	79,329	81,565
Ducks	531	469	481	3,391	3,342	3,738
Other poultry	7	4	3	22	23	26
Total poultry	124,489	134,462	148,151	425,694	451,757	478,382

Class	Pounds certified (ready-to-cook)		
	2013	2014	2015
	Thousands	*Thousands*	*Thousands*
Young chickens	37,830,041	38,565,192	40,048,276
Mature chickens	523,332	521,176	522,433
Total chickens	38,535,373	39,086,368	40,570,709
Young turkeys	5,775,814	5,725,668	5,593,008
Old turkeys	29,725	29,963	33,640
Total turkeys	5,805,539	5,755,631	5,626,648
Ducks	121,398	130,950	137,558
Other poultry [2]	2,346	2,098	2,413
Total poultry	44,282,656	44,975,047	46,337,328

(NA) Not available.　[1] Includes geese, guineas, ostriches, emus, rheas, and squab.　[2] Includes geese, guineas, and squab.

NASS, Livestock Branch, (202) 720–3570.

Table 8-43.—Turkeys: Supply, distribution, and per capita consumption, ready-to-cook basis, United States, 2006–2015

Year	Production	Commercial storage at beginning of year	Exports	Commercial storage at end of year	Consumption Total[1][2]	Consumption Per capita
	Million pounds	*Million pounds*	*Million pounds*	*Million pounds*	*Million pounds*	*Pounds*
2006	5,607	206	547	218	5,064	16.9
2007	5,873	218	547	261	5,300	17.5
2008	6,165	261	676	396	5,367	17.6
2009	5,589	396	534	262	5,210	17.0
2010	5,570	262	581	192	5,084	16.4
2011	5,715	192	703	211	5,013	16.1
2012	5,889	211	797	296	5,028	16.0
2013	5,729	296	741	237	5,068	16.0
2014	5,756	237	775	193	5,052	15.8
2015	5,627	193	529	201	5,136	16.0

[1] Totals may not add due to rounding.   [2] Shipments to territories now included in consumption.
ERS Markets and Trade Economics Division, Animal Products and Cost of Production Branch, (202) 694–5409.

Table 8-44.—Poultry, meat, and turkey: International trade, 2013–2014

Country	2013	2014
	1,000 metric tons	*1,000 metric tons*
Principle exporting:		
Others	329	281
Total Foreign	329	281
United States	344	366
Total	673	647
Principle importing:		
Others	339	276
Total Foreign	339	276
United States	10	13
Total	349	289

FAS, Office of Global Analysis, (202) 720-6301. Prepared or estimated on the basis of official USDA production, supply, and distribution statistics from foreign governments.

Table 8-45.—Turkeys: Production and value, United States, 2006–2015

Year	Number raised	Pounds (live weight) produced	Price per pound live weight [1]	Value of production
	1,000 head	1,000 pounds	Cents	1,000 dollars
2006	256,334	7,223,675	48.0	3,467,534
2007	266,828	7,566,315	52.3	3,954,472
2008	273,088	7,911,797		4,471,046
2009	247,359	7,149,314		3,573,324
2010	244,188	7,108,222		4,372,438
2011	248,500	7,313,205		4,987,608
2012	253,500	7,561,905		5,452,135
2013	240,000	7,277,536		4,839,562
2014	237,500	7,217,006		5,304,501
2015	233,100	7,038,065		5,707,871

[1] Price per pound discontinued in 2008.
NASS, Livestock Branch, (202) 720–3570.

Table 8-46.—Turkeys: Production and value, by State and United States, 2014–2015 [1]

State	2013			2014		
	Number raised	Pounds produced	Value of production	Number raised	Pounds produced	Value of production
	1,000 head	1,000 pounds	1,000 dollars	1,000 head	1,000 pounds	1,000 dollars
Arkansas	30,000	612,000	449,820	27,500	561,000	454,971
California	11,000	310,200	227,997	11,500	338,100	274,199
Indiana	19,000	754,300	554,411	19,300	752,700	610,440
Iowa	10,500	435,750	320,276	9,100	353,990	287,086
Michigan	5,100	205,530	151,065	5,200	204,880	166,158
Minnesota	45,500	1,178,450	866,161	41,000	988,100	801,349
Missouri	17,000	545,700	401,090	19,000	600,400	486,924
North Carolina	28,500	997,500	733,163	31,000	1,091,200	884,963
Ohio	5,100	209,100	153,689	5,200	209,040	169,531
Pennsylvania	6,900	173,190	127,295	6,500	163,150	132,315
South Dakota	4,500	187,650	137,923	4,300	181,460	147,164
Utah	4,000	96,800	71,148	3,600	91,440	74,158
Virginia	16,800	443,520	325,987	17,000	460,700	373,628
West Virginia	3,100	81,840	60,152	3,000	90,600	73,477
Other States [2]	30,500	985,476	724,324	29,900	951,305	771,508
United States	237,500	7,217,006	5,304,501	233,100	7,038,065	5,707,871

(D) Withheld to avoid disclosure of individual operations. [1] Based on turkeys placed September 1 through August 31. Excludes young turkeys lost. [2] Includes State estimates not shown and States withheld to avoid disclosing data for individual operations.
NASS, Livestock Branch, (202) 720–3570.

Table 8-47.—Turkeys: Net poults placements, United States, Monthly, 2013–2014 [1]

Month	Total all breeds		Percent of Previous Year
	2014	2015	
	Thousands	Thousands	Percent
January	22,107	22,283	101
February	20,902	20,110	96
March	21,562	21,880	101
April	22,630	21,852	97
May	23,320	21,331	91
June	23,462	23,030	98
July	24,241	22,788	94
August	23,224	21,863	94
September	21,180	19,041	90
October	21,695	20,213	93
November	20,379	20,946	103
December	22,516	22,562	100
Total	267,218	257,899	97

[1] Includes imports and excludes exports.
NASS, Livestock Branch, (202) 720–3570.

Table 8-48.—Turkeys: Poults hatched by commercial hatcheries, United States by Month, 2014–2015

Month	2014	2015	Percent of Previous Year
	Thousands	Thousands	Percent
January	23,441	23,921	102
February	21,499	21,712	101
March	22,391	24,178	108
April	22,818	23,263	102
May	24,208	22,164	92
June	23,667	23,148	98
July	24,419	22,976	94
August	23,500	22,006	94
September	22,855	19,528	85
October	23,744	20,621	87
November	21,936	20,996	96
December	24,660	22,642	92
Total	279,138	267,155	96

NASS, Livestock Branch, (202) 720–3570.

Table 8-49.—Eggs: Supply, distribution, and per capita consumption, United States, 2006–2015 [1]

Year	Total egg production	Storage at beginning of the year [1]	Imports [2]	Exports [2]	Eggs used for hatching	Consumption		
						Storage at end of the year [2]	Total [3]	Per capita
	Million dozen	Million dozen	Million dozen	Million dozen	Million dozen	Million dozen	Million dozen	Number
2006	7,650	16	9	202	992	13	6,472	260
2007	7,588	13	14	250	1,016	11	6,351	252
2008	7,519	11	14	206	996	17	6,338	250
2009	7,569	17	11	242	955	18	6,390	250
2010	7,656	18	12	259	982	19	6,436	249
2011	7,715	19	21	277	950	28	6,501	250
2012	7,883	28	19	302	941	21	6,666	254
2013	8,149	21	17	388	964	23	6,812	258
2014	8,433	23	35	379	981	23	7,108	268
2015	8,053	23	124	314	996	31	6,859	256

[1] Calendar years. [2] Shell eggs and the approximate shell-egg equivalent of egg product. [3] Shipments to territories now included in total consumption.
ERS Markets and Trade Economics Division, Animal Products and Cost of Production Branch, (202) 694–5409.

Table 8-50.—Eggs, shell: Average price per dozen on consumer Grade A cartoned white eggs to volume buyers, store-door delivery, New York, 2005–2015

Year	Large
	Cents
2005	65.51
2006	71.76
2007	114.36
2008	128.32
2009	102.97
2010	106.29
2011	115.31
2012	117.46
2013	124.69
2014	142.33
2015	181.74

AMS, Livestock, Poultry and Seed Programs, Livestock, Poultry and Grain Market News Division, (515) 284-4460.

Table 8-51.—All layers and egg production: Annual average number of layers, eggs per layer, and total production, by State and United States, 2014–2015 [1]

State	Average number of layers		Eggs per layer [2]		Total egg production	
	2014	2015	2014	2015	2014	2015
	Thousands	Thousands	Number	Number	Millions	Millions
Alabama	9,280	9,403	232	232	2,148	2,185.0
Arkansas	12,131	13,373	244	245	2,962	3,275.4
California	16,173	11,870	281	278	4,551	3,302.8
Colorado	4,845	4,544	299	291	1,450	1,322.8
Connecticut [3]	2,367		283		669	
Florida	8,610	9,028	278	273	2,390	2,463.1
Georgia	18,270	18,794	259	257	4,723	4,837.5
Illinois	4,832	4,816	292	282	1,409	1,356.4
Indiana	27,400	28,437	283	281	7,747	8,004.7
Iowa	59,154	45,491	278	274	16,449	12,470.6
Kentucky	4,717	4,943	258	259	1,219	1,279.4
Louisiana	2,006	2,072	270	271	541	560.8
Maine [3]	3,524		281		989	
Maryland	2,782	2,656	282	291	785	773.3
Massachusetts	140	144	315	310	44	44.7
Michigan	13,044	12,951	297	297	3,867	3,844.7
Minnesota	10,926	10,155	281	283	3,071	2,874.9
Mississippi	5,626	5,719	240	246	1,351	1,405.5
Missouri	8,666	10,956	278	279	2,407	3,056.4
Montana	455	456	314	312	143	142.1
Nebraska	9,466	7,696	302	298	2,860	2,290.7
New York	5,048	5,117	296	298	1,493	1,523.2
North Carolina	13,458	14,459	251	253	3,381	3,659.3
Ohio	30,624	33,980	285	286	8,731	9,708.4
Oklahoma	3,054	2,993	233	234	712	699.0
Oregon	2,323	2,363	313	309	727	731.1
Pennsylvania	25,529	25,934	297	300	7,570	7,778.2
South Carolina	4,260	4,360	262	267	1,117	1,163.5
South Dakota	2,632	2,076	286	291	752	604.5
Tennessee	1,530	1,566	223	223	341	349.1
Texas	18,932	19,426	270	266	5,109	5,161.1
Utah	4,144	4,409	285	284	1,180	1,252.2
Vermont	143	139	252	276	36	38.4
Virginia	2,977	2,939	257	256	765	752.0
Washington	6,928	7,179	282	292	1,950	2,094.9
West Virginia	1,188	1,221	227	229	270	279.2
Wisconsin	5,105	4,918	284	286	1,449	1,406.3
Other States [4]	8,585	15,828	281	285	2,410	4,517.0
United States	360,873	352,411	277	276	99,768	97,208.2

[1] Annual estimates cover the period December 1 previous year through November 30. Totals may not add due to rounding. [2] Total egg production divided by average number of layers on hand. [3] 2015 included in Other States. [4] Alaska, Arizona, Delaware, Hawaii, Idaho, Kansas, New Hampshire, New Jersey, New Mexico, Nevada, North Dakota, Rhode Island, and Wyoming combined to avoid disclosing data for individual operations.
NASS, Livestock Branch, (202) 720–3570.

Table 8-52.—Eggs: Broken under Federal inspection, United States, 2013–2014

Item	Quantity	
	2013	2014
	1,000 dozen	1,000 dozen
Shell eggs broken	2,134,068	2,262,138
	1,000 pounds	1,000 pounds
Edible product from shell eggs broken		
Whole	1,663,149	1,696,633
White	730,059	821,741
Yolk	366,399	405,616
Total	2,759,607	2,923,990
Inedible product from shell eggs broken	220,874	233,471

NASS, Livestock Branch, (202) 720–3570.

Table 8-53.—Eggs: Number, rate of lay, production, and value, United States, 2006–2015 [1]

Year	Layers average number	Eggs per layer [2]	Total egg production	Price per dozen [3][4]	Value of production
	Thousands	Number	Millions	Dollars	1,000 dollars
2006	349,700	263	91,788	0.583	4,460,211
2007	346,498	263	91,101	0.885	6,718,853
2008	339,933	266	90,239	....	8,239,313
2009	338,846	268	90,737	....	6,191,040
2010	341,505	269	91,811	....	6,553,087
2011	340,598	271	92,450	....	7,355,873
2012	394,844	274	94,364	....	7,929,104
2013	354,844	275	97,555	....	8,678,859
2014	360,873	277	99,768	....	10,257,972
2015	352,411	276	97,208	....	13,499,904

[1] Annual estimates cover the period December 1 previous year through November 30. [2] Total egg production divided by average number of layers on hand. [3] Average mid-month price of all eggs sold by producers including hatching eggs. [4] Price per dozen discontinued in 2008.
NASS, Livestock Branch, (202) 720–3570.

Table 8-54.—All Eggs: Production and value by State and United States, 2014–2015 [1]

State	Eggs produced		Value of production	
	2014	2015	2014	2015
	Millions	Millions	1,000 dollars	1,000 dollars
Alabama	2,126	2,185	396,045	404,148
Arkansas	3,057	3,277	490,121	572,613
California	4,589	3,306	422,607	528,471
Colorado	1,449	1,325	130,350	177,461
Florida	2,390	2,463	219,087	315,615
Georgia	4,729	4,832	666,920	763,068
Illinois	1,409	1,359	124,361	181,327
Indiana	7,762	8,005	676,030	1,066,202
Iowa	16,449	12,463	1,404,761	1,533,821
Kentucky	1,219	1,280	154,883	191,743
Louisiana	541	561	72,790	89,256
Maryland	785	772	70,726	98,876
Massachusetts	44	47	3,842	5,921
Michigan	3,866	3,846	325,518	495,170
Minnesota	3,071	2,876	266,139	366,883
Mississippi	1,354	1,407	235,306	248,452
Missouri	3,019	3,058	302,887	421,590
Montana	143	142	12,965	22,477
Nebraska	2,860	2,291	240,640	281,171
New York	1,493	1,527	133,207	194,339
North Carolina	3,381	3,661	501,063	588,679
Ohio	9,059	9,316	772,430	1,211,830
Oklahoma	713	702	102,284	115,393
Oregon	727	732	65,778	116,161
Pennsylvania	7,595	7,751	717,181	1,004,626
South Carolina	1,117	1,166	130,092	167,836
South Dakota	752	603	63,348	73,589
Tennessee	341	347	67,998	66,755
Texas	5,109	5,118	525,954	720,036
Utah	1,187	1,252	107,255	199,439
Vermont	36	36	4,274	5,258
Virginia	766	752	114,471	122,456
Washington	1,953	2,095	177,074	331,824
West Virginia	270	274	55,887	53,872
Wisconsin	1,449	1,405	129,992	187,368
Other States [2]	4,069	4,205	373,706	576,178
United States	100,879	96,437	10,257,972	13,499,904

Totals may not add due to rounding.   [1] Annual estimates cover the period December 1 previous year through November 30.   Includes hatching and market (table) eggs.   [2] Alaska, Arizona, Connecticut, Delaware, Hawaii, Idaho, Kansas, Maine, Nevada, New Hampshire, New Jersey, New Mexico, North Dakota, Rhode Island, and Wyoming combined to avoid disclosing individual operations.
NASS, Livestock Branch, (202) 720–3570.

Table 8-55.—Poultry and poultry products: Cold storage holdings, end of month, United States, 2014– 2015

Month	Frozen eggs							
	Whites		Yolks		Whole & mixed		Unclassified	
	2014	2015	2014	2015	2014	2015	2014	2015
	1,000 pounds	1,000 pounds	1,000 pounds	1,000 pounds	1,000 pounds	1,000 pounds	1,000 pounds	1,000 pounds
January	4,099	4,933	2,176	1,489	12,367	11,168	16,045	17,080
February	4,478	4,239	1,727	1,417	12,162	12,843	16,264	17,149
March	4,117	5,001	1,906	949	10,437	10,586	12,584	15,442
April	3,427	4,753	1,627	960	9,279	10,632	13,097	14,915
May	3,462	3,099	1,302	785	10,172	11,274	13,364	13,058
June	3,590	3,094	1,478	767	10,649	10,472	14,425	12,557
July	3,912	2,526	1,760	927	10,576	10,891	15,205	13,064
August	3,615	2,453	1,756	1,011	10,585	13,257	13,864	13,851
September ...	4,146	2,660	1,791	1,035	10,385	15,302	14,804	13,327
October	4,447	4,766	1,540	1,172	10,171	15,611	15,772	15,937
November	4,036	3,634	1,458	1,096	10,685	16,419	14,170	14,864
December	3,574	3,817	1,471	1,169	10,148	19,023	15,525	16,887

Month	Frozen eggs, total		Frozen chicken					
	2014	2015	Broilers (Whole)		Hens		Breast and breast meat	
			2014	2015	2014	2015	2014	2015
	1,000 pounds	1,000 pounds	1,000 pounds	1,000 pounds	1,000 pounds	1,000 pounds	1,000 pounds	1,000 pounds
January	34,687	34,670	16,750	10,827	7,084	3,888	113,677	155,526
February	34,631	35,648	13,053	14,022	7,492	2,668	122,830	158,716
March	29,044	31,978	11,117	14,836	5,099	1,888	113,032	151,960
April	27,430	31,260	8,448	17,204	5,472	3,776	108,812	150,514
May	28,300	28,216	7,138	17,116	4,850	4,954	117,089	146,329
June	30,142	26,890	7,932	19,943	4,541	5,496	113,318	146,728
July	31,453	27,408	13,578	20,867	5,536	7,320	108,948	139,178
August	29,820	30,572	12,603	20,097	6,946	10,100	110,520	133,678
September ...	31,126	32,324	11,092	12,939	6,934	10,579	114,033	139,304
October	31,960	37,486	11,723	10,522	4,972	15,029	126,303	162,707
November	30,349	36,013	10,771	10,955	2,257	10,382	146,185	171,403
December	30,718	40,896	11,425	15,031	3,030	7,657	159,121	171,896

Month	Frozen chicken							
	Drumsticks		Leg quarters		Legs		Thigh and thigh quarters	
	2014	2015	2014	2015	2014	2015	2014	2015
	1,000 pounds	1,000 pounds	1,000 pounds	1,000 pounds	1,000 pounds	1,000 pounds	1,000 pounds	1,000 pounds
January	27,248	28,315	158,922	148,456	8,846	15,926	9,590	10,868
February	23,320	22,963	152,883	162,890	8,193	15,199	11,281	12,532
March	23,383	26,303	106,551	189,412	9,514	13,528	8,647	12,187
April	25,127	32,571	100,723	180,833	13,320	15,889	8,403	10,764
May	28,396	29,550	118,697	166,213	13,957	14,604	8,128	10,667
June	22,836	28,315	99,556	138,577	13,613	13,262	9,104	10,704
July	17,934	25,461	124,248	145,187	13,816	17,030	8,810	12,500
August	18,030	18,733	127,116	146,540	11,327	16,117	8,115	13,723
September ...	17,445	20,167	118,616	164,192	11,859	16,591	8,434	14,097
October	21,308	28,031	129,762	163,454	12,270	19,454	8,877	18,102
November	26,001	31,377	128,192	156,951	10,597	23,599	11,211	17,407
December	26,331	37,153	128,638	140,098	15,748	19,406	13,088	18,121

Month	Frozen chicken							
	Thigh meat		Wings		Paws and feet		Other	
	2014	2015	2014	2015	2014	2015	2014	2015
	1,000 pounds	1,000 pounds	1,000 pounds	1,000 pounds	1,000 pounds	1,000 pounds	1,000 pounds	1,000 pounds
January	22,169	25,398	62,433	45,811	23,416	38,837	239,006	246,876
February	22,265	23,251	58,935	44,923	23,491	32,917	237,200	242,009
March	17,618	20,085	55,699	42,596	22,102	30,311	218,807	248,743
April	17,759	21,722	61,373	44,842	21,380	28,518	213,276	258,998
May	16,130	19,358	66,719	44,175	22,727	26,641	208,234	259,782
June	17,659	22,406	64,938	48,995	22,930	27,894	214,923	262,155
July	19,593	27,293	64,462	58,649	26,232	31,732	255,592	281,025
August	17,690	25,974	63,781	68,117	25,228	34,546	224,565	291,475
September ...	16,970	27,038	62,014	71,861	25,356	26,617	232,493	296,820
October	21,880	28,025	64,322	78,458	27,671	21,392	229,532	313,424
November	23,657	27,846	70,126	84,073	30,252	20,209	237,866	327,693
December	24,547	24,267	58,720	78,254	29,346	21,650	242,512	328,106

See end of table.

Table 8-55.—Poultry and poultry products: Cold storage holdings, end of month, United States, 2014–2015—Continued

Month	Total chicken		Frozen turkey					
	2014	2015	Toms		Hens		Total whole	
			2014	2015	2014	2015	2014	2015
	1,000 pounds	1,000 pounds	1,000 pounds	1,000 pounds	1,000 pounds	1,000 pounds	1,000 pounds	1,000 pounds
January	689,141	730,728	53,681	58,677	53,873	58,029	107,554	116,706
February	680,943	732,090	72,426	80,220	67,950	79,668	140,376	159,888
March	591,569	751,849	77,979	82,683	85,038	96,487	163,017	179,170
April	584,093	765,631	88,096	97,828	104,883	118,656	192,979	216,484
May	612,065	739,389	107,122	113,518	129,404	139,178	236,526	252,696
June	591,350	724,475	118,638	115,019	150,261	158,570	268,899	273,589
July	628,749	766,242	131,617	125,577	157,002	171,682	288,619	297,259
August	625,921	779,100	136,686	129,926	164,098	160,607	300,784	290,533
September ...	625,246	800,205	141,964	136,648	155,259	129,159	297,223	265,807
October	658,440	858,598	112,734	103,743	112,045	87,912	224,779	191,655
November	697,115	881,895	39,855	24,341	28,047	34,334	67,902	58,675
December	712,506	861,639	30,715	22,375	28,841	31,098	59,556	53,473

Month	Frozen turkey							
	Breasts		Legs		Mechanically deboned meat		Other	
	2014	2015	2014	2015	2014	2015	2014	2015
	1,000 pounds	1,000 pounds	1,000 pounds	1,000 pounds	1,000 pounds	1,000 pounds	1,000 pounds	1,000 pounds
January	56,457	38,368	10,001	11,950	7,809	7,481	28,399	32,081
February	56,448	42,939	8,018	9,053	5,779	6,598	24,782	27,088
March	53,024	47,509	8,127	10,414	5,919	5,712	22,880	25,710
April	51,436	52,368	7,471	13,510	5,079	6,250	25,924	25,338
May	53,904	54,981	7,979	15,262	6,180	7,733	26,994	26,732
June	54,594	54,542	8,326	14,286	9,022	6,505	29,771	24,550
July	54,820	59,672	9,852	14,171	9,566	5,619	31,157	25,260
August	50,394	58,102	9,350	12,886	6,261	6,055	27,545	25,023
September ...	48,599	56,302	7,309	11,603	6,107	4,633	25,134	21,607
October	36,441	41,813	7,827	10,357	7,750	5,291	27,474	21,566
November	27,884	30,415	9,835	11,200	8,122	7,189	26,355	22,658
December	29,860	35,805	9,964	14,061	6,582	8,098	28,465	26,713

Month	Frozen turkey				Ducks		Total frozen poultry	
	Unclassified		Total turkey		2014	2015	2014	2015
	2014	2015	2014	2015				
	1,000 pounds	1,000 pounds	1,000 pounds	1,000 pounds	1,000 pounds	1,000 pounds	1,000 pounds	1,000 pounds
January	65,567	73,814	275,787	280,400	2,019	2,579	966,947	1,013,707
February	75,353	75,771	310,756	321,337	2,021	2,810	993,720	1,056,237
March	83,416	77,063	336,383	345,578	1,782	3,280	929,734	1,100,707
April	92,124	80,465	375,013	394,415	1,793	5,178	960,899	1,165,224
May	90,919	83,984	422,502	441,388	1,426	6,743	1,035,993	1,187,520
June	91,974	88,548	462,586	462,020	1,824	10,818	1,055,760	1,197,313
July	95,744	92,056	489,758	494,037	1,853	7,884	1,120,360	1,268,163
August	100,452	84,980	494,786	477,579	2,588	8,103	1,123,295	1,264,782
September ...	100,175	89,670	484,547	449,622	2,887	7,491	1,112,680	1,257,318
October	86,383	81,146	390,654	351,828	2,317	5,806	1,051,411	1,216,232
November	47,465	60,131	187,563	190,268	2,296	6,488	886,974	1,078,651
December	59,002	62,861	193,429	201,011	2,231	6,579	908,166	1,069,229

NASS, Livestock Branch, (202) 720–3570.

Table 8-56.—Dairy products: Cold storage holdings, end of month, United States, 2014–2015

Month	Butter		American cheese	
	2014	2015	2014	2015
	1,000 pounds	*1,000 pounds*	*1,000 pounds*	*1,000 pounds*
January	143,890	148,885	630,820	636,019
February	171,773	179,003	628,679	645,670
March	191,755	184,373	639,067	634,270
April	186,914	232,372	648,900	644,113
May	209,430	265,198	656,446	669,464
June	199,248	256,000	655,239	685,745
July	180,834	254,347	660,438	698,029
August	172,789	212,189	648,784	709,029
September	152,361	187,528	631,279	698,875
October	147,956	178,834	623,336	696,781
November	107,566	132,740	635,776	699,794
December	104,728	155,082	627,769	701,073

Month	Swiss cheese		Other	
	2014	2015	2014	2015
	1,000 pounds	*1,000 pounds*	*1,000 pounds*	*1,000 pounds*
January	25,421	22,411	358,812	389,813
February	26,942	23,587	354,511	397,803
March	27,425	24,573	351,798	409,802
April	28,316	23,986	360,370	417,810
May	30,538	21,424	378,556	420,966
June	27,972	20,841	372,231	435,655
July	24,593	21,591	369,861	442,176
August	27,085	22,203	365,539	436,161
September	25,684	22,037	356,819	431,534
October	25,600	21,404	346,730	428,006
November	24,419	22,665	356,997	425,601
December	21,282	24,587	368,885	420,426

Month	Total natural cheese	
	2014	2015
	1,000 pounds	*1,000 pounds*
January	1,015,053	1,048,243
February	1,010,132	1,067,060
March	1,018,290	1,068,645
April	1,037,586	1,085,909
May	1,065,540	1,111,854
June	1,055,442	1,142,241
July	1,054,892	1,161,796
August	1,041,408	1,167,393
September	1,013,782	1,152,446
October	995,666	1,146,191
November	1,017,192	1,148,060
December	1,017,936	1,146,086

NASS, Livestock Branch, (202) 720–3570.

CHAPTER IX

FARM RESOURCES, INCOME, AND EXPENSES

The statistics in this chapter deal with farms, farm resources, farm income, and expenses. Many of the series are estimates developed in connection with economic research activities of the Department.

Table 9-1.—Economic trends: Data relating to agriculture, United States, 2004–2013

Year	Prices paid by farmers [1]			Farm income [2]		
	Total including interest, taxes, and wage rates (PPITW)	Production items	Prices received by farmers [1]	Gross farm income [3]	Production expenses	Net farm income
	Index numbers 2011=100	Index numbers 2011=100	Index numbers 2011=100	Billion dollars	Billion dollars	Billion dollars
2004	66	62	73	294.9	207.5	87.4
2005	70	65	71	298.5	219.7	78.8
2006	74	69	71	290.2	232.7	57.4
2007	79	75	84	339.6	269.5	70.0
2008	90	88	92	377.7	292.6	85.1
2009	87	85	81	343.3	280.3	63.0
2010	90	88	82	365.6	285.2	80.4
2011	100	100	100	428.5	310.6	117.9
2012	104	105	105	....	....	....
2013	106	107	107	....	....	....

Year	National income [5]	Personal income [5]	Industrial production [6]	Consumer prices all items [7]	Producer prices consumer foods [7]
	Billion dollars	Billion dollars	Index numbers 2007=100	Index numbers 1982–84=100	Index numbers 1982=100
2004	10,541.9	10,049.2	92.5	188.9	152.7
2005	11,240.8	10,610.3	95.5	195.3	155.7
2006	12,005.6	11,389.8	97.6	201.6	156.7
2007	12,322.3	11,995.7	100.0	207.3	167.0
2008	12,430.8	12,430.6	96.6	215.3	178.3
2009	12,124.5	12,082.1	85.7	214.5	175.5
2010	12,739.5	12,435.2	90.6	218.1	182.4
2011	13,395.7	13,191.3	93.6	224.9	....
2012	13,971.6	13,743.8	97.1	229.6	....
2013	14,542.4	14,134.7	99.9	233.0	....

[1] U.S. Department of Agriculture - NASS. [2] U.S. Department of Agriculture - ERS. [3] Includes cash receipts from farm marketings, government payments, nonmoney income (gross rental value of dwelling and value of home consumption), other income (machine hire custom work and recreational income), and value of change in farm inventories. [4] Forecast. [5] Department of Commerce, Bureau of Economic Analysis. [6] Federal Reserve Board. [7] U.S. Department of Labor, Bureau of Labor Statistics.
ERS, Farm and Rural Business Branch, (202) 694–5446. E-mail contact is Timothy Park at tapark@ers.usda.gov. For National Income, Personal Income, Industrial Production and Consumer Price Indexes, Contact David Torgerson at (202) 694-5334. E-mail contact is dtorg@ers.usda.gov.

Table 9-2.—Farms: Number, land in farms, and average size of farm, United States, 2006–2015

Year	Farms [1]	Land in farms	Average size farm
	Number	1,000 acres	Acres
2006	2,088,790	925,790	443
2007	2,204,600	921,460	418
2008	2,184,500	918,600	421
2009	2,169,660	917,590	423
2010	2,149,520	915,660	426
2011	2,131,240	914,420	429
2012	2,109,810	914,600	433
2013	2,102,010	914,030	435
2014	2,085,000	913,000	438
2015	2,068,000	912,000	441

[1] A farm is any establishment from which $1,000 or more of agricultural products were sold or would normally be sold during the year.
NASS, Environmental, Economics, and Demographics Branch, (202) 720–6146.

Table 9-3.—Farms: Percent of farms, land in farms, and average size, by economic sales class, United States, 2014–2015

Economic sales class	Percent of total				Average size farm	
	Farms		Land		2014	2015
	2014	2015	2014	2015		
	Percent	Percent	Percent	Percent	Acres	Acres
$ 1,000 - $ 9,999	50.6	50.1	9.9	9.6	86	85
$ 10,000 - $ 99,999	29.9	30.1	21.2	21.0	311	308
$100,000 - $249,999	7.0	7.0	14.1	14.3	889	900
$250,000 - $499,999	4.7	4.7	13.7	13.8	1,290	1,288
$500,000 - $999,999	4.0	4.0	17.1	17.1	1,884	1,884
$1,000,000 or more	3.9	4.0	23.9	24.1	2,652	2,662
Total	100.0	100.0	100.0	100.0	438	441

NASS, Environmental, Economics, and Demographics Branch, (202) 720–6146.

Table 9-4.—Land in farms: Classification by tenure of operator, United States, 1935–2012

Year	Land in farms	Tenure of operator			
		Full owners	Part owners	Managers	All tenants
	Acres	Percent	Percent	Percent	Percent
1935 [1]	1,054,515,111	37.1	25.2	5.8	31.9
1940	1,065,113,774	35.9	28.2	6.5	29.4
1945 [1]	1,141,615,364	36.1	32.5	9.3	22.0
1950	1,161,419,720	36.1	36.4	9.2	18.3
1954 [1]	1,158,191,511	34.2	40.7	8.6	16.5
1959	1,123,507,574	31.0	44.0	9.8	14.8
1964	1,110,187,000	28.7	48.0	10.2	13.1
1969	1,062,892,501	35.3	51.8	....	13.0
1974	1,017,030,357	35.3	52.6	....	12.0
1978	1,014,777,234	32.7	55.3	....	12.0
1982	986,796,579	34.7	53.8	....	11.5
1987	964,470,625	32.9	53.9	....	13.2
1992	945,531,506	31.3	55.7	....	13.0
1997	931,795,255	33.9	54.5	....	11.6
2002 [2]	938,279,056	38.0	52.8	....	9.2
2007 [2]	922,095,843	37.3	53.8	....	8.9
2012 [2]	914,527,657	36.8	53.7	....	9.5

[1] Excludes Alaska and Hawaii.   [2] The 2002 Census of Agriculture introduced new methodology to account for all farms in the United States.   All 2002 published census items were reweighted for undercoverage.   Strictly speaking, data for 2002 and subsequent years are not fully comparable with data from earlier years.
ERS, Resource and Rural Economics Division, (202) 694–5572.   Data from the Census of Agriculture, National Agricultural Statistics Service.

Table 9-5.—Number of farms and land in farms by economic sales class by State and United States, 2014–2015

State	Economic Sales Class: $1,000-$9,999			
	Number of Farms		Land in Farms	
	2014	2015	2014	2015
	Number	Number	1,000 Acres	1,000 Acres
Alabama	25,600	25,000	2,750	2,650
Alaska	380	380	270	270
Arizona	14,900	14,700	800	600
Arkansas	22,000	21,700	1,650	1,700
California	30,300	30,400	1,400	1,300
Colorado	19,100	18,500	2,400	2,200
Connecticut	4,000	4,000	150	150
Delaware	870	870	20	20
Florida	27,400	27,100	1,260	1,240
Georgia	23,000	23,000	2,150	2,400
Hawaii	3,850	3,850	100	100
Idaho	12,200	12,200	800	800
Illinois	30,000	29,300	1,400	1,200
Indiana	27,900	27,500	1,000	930
Iowa	27,100	26,600	1,200	900
Kansas	22,800	22,000	2,500	2,400
Kentucky	42,700	42,000	3,000	2,800
Louisiana	16,600	16,200	1,300	1,180
Maine	5,000	5,000	480	480
Maryland	6,600	6,500	300	300
Massachusetts	5,000	5,000	210	210
Michigan	25,800	25,600	1,100	1,100
Minnesota	26,600	26,500	2,100	2,100
Mississippi	21,700	21,100	2,800	2,600
Missouri	45,500	44,200	3,900	3,900
Montana	11,900	11,600	2,900	2,700
Nebraska	13,000	12,500	900	900
Nevada	2,300	2,300	150	140
New Hampshire	3,100	3,100	240	240
New Jersey	5,600	5,600	160	160
New Mexico	16,200	16,200	4,200	4,100
New York	16,700	16,700	1,400	1,600
North Carolina	28,000	27,500	1,500	1,300
North Dakota	9,300	9,000	1,600	1,600
Ohio	33,200	33,300	1,650	1,600
Oklahoma	42,500	41,100	5,150	5,000
Oregon	21,000	21,000	1,500	1,440
Pennsylvania	26,700	26,100	1,800	1,800
Rhode Island	710	710	20	20
South Carolina	16,700	16,800	1,700	1,800
South Dakota	8,300	8,500	1,200	1,300
Tennessee	43,400	42,400	3,050	2,900
Texas	156,000	151,000	18,200	16,500
Utah	10,600	10,600	650	700
Vermont	4,050	4,050	340	340
Virginia	26,300	25,800	1,850	1,800
Washington	22,200	21,500	1,000	1,020
West Virginia	15,100	14,600	1,600	1,600
Wisconsin	30,500	30,500	2,000	1,950
Wyoming	4,900	5,000	1,000	900
United States	1,055,160	1,036,660	90,800	86,940

See footnote(s) at end of table.

Table 9-5.—Number of farms and land in farms by economic sales class by State and United States, 2014–2015—Continued

State	Economic Sales Class: $10,000-$99,999			
	Number of Farms		Land in Farms	
	2014	2015	2014	2015
	Number	*Number*	*1,000 Acres*	*1,000 Acres*
Alabama	12,700	13,300	3,300	3,400
Alaska	290	290	450	450
Arizona	3,200	3,350	8,200	8,200
Arkansas	14,000	14,100	3,600	3,600
California	26,500	26,400	4,300	4,200
Colorado	10,200	10,000	7,200	7,000
Connecticut	1,550	1,550	110	110
Delaware	(D)	(D)	(D)	(D)
Florida	14,700	14,600	1,900	1,900
Georgia	10,500	10,500	2,200	2,200
Hawaii	2,600	2,600	260	260
Idaho	7,000	7,000	2,050	2,050
Illinois	18,700	18,300	2,700	2,700
Indiana	15,300	15,500	1,800	1,830
Iowa	23,600	23,600	3,100	3,100
Kansas	21,500	21,500	8,000	8,000
Kentucky	26,900	27,400	4,700	4,700
Louisiana	7,300	7,300	1,800	1,800
Maine	2,400	2,400	400	400
Maryland	3,500	3,500	400	400
Massachusetts	2,000	2,000	140	140
Michigan	14,800	14,700	1,700	1,650
Minnesota	21,900	21,900	3,700	3,700
Mississippi	10,600	10,800	2,950	2,900
Missouri	38,200	38,300	9,000	8,700
Montana	8,000	7,800	9,100	8,800
Nebraska	13,900	13,900	5,000	4,800
Nevada	900	850	330	280
New Hampshire	1,050	1,050	150	150
New Jersey	2,350	2,350	160	160
New Mexico	6,200	6,200	11,500	11,300
New York	11,600	11,600	1,780	1,800
North Carolina	12,800	12,800	1,500	1,500
North Dakota	6,600	6,200	3,000	2,600
Ohio	25,300	25,300	2,650	2,650
Oklahoma	28,300	28,000	11,500	11,300
Oregon	9,100	9,100	2,860	2,860
Pennsylvania	19,900	19,400	2,300	2,300
Rhode Island	(D)	(D)	(D)	(D)
South Carolina	5,600	5,500	1,350	1,300
South Dakota	9,700	9,400	5,100	4,900
Tennessee	19,500	20,200	3,700	3,700
Texas	70,000	71,000	41,700	41,700
Utah	5,500	5,500	1,900	1,900
Vermont	2,100	2,100	320	320
Virginia	14,700	14,200	2,900	2,800
Washington	8,400	8,400	1,900	1,900
West Virginia	5,300	5,400	1,400	1,400
Wisconsin	21,000	21,000	2,900	2,900
Wyoming	3,900	3,700	4,800	4,800
Other States[1]	1,010	1,010	60	60
United States	622,650	622,850	193,820	191,570

See footnote(s) at end of table.

Table 9-5.—Number of farms and land in farms by economic sales class by State and United States, 2014–2015—Continued

State	Economic Sales Class: $100,000-$249,999			
	Number of Farms		Land in Farms	
	2014	2015	2014	2015
	Number	*Number*	*1,000 Acres*	*1,000 Acres*
Alabama	1,050	1,100	700	750
Alaska	50	50	70	70
Arizona	610	590	11,200	11,200
Arkansas	1,700	1,600	1,050	1,000
California	6,300	6,100	3,800	3,700
Colorado	2,500	2,600	5,100	5,400
Connecticut	(D)	(D)	(D)	(D)
Delaware	(D)	(D)	(D)	(D)
Florida	2,200	2,300	940	940
Georgia	1,400	1,400	630	630
Hawaii	280	280	160	160
Idaho	1,900	1,900	1,650	1,650
Illinois	8,700	8,500	3,000	2,900
Indiana	5,400	5,300	1,560	1,580
Iowa	10,400	10,400	3,000	3,000
Kansas	6,600	6,500	7,300	7,200
Kentucky	3,100	3,300	1,200	1,200
Louisiana	710	850	450	520
Maine	340	340	100	100
Maryland	600	600	230	230
Massachusetts	380	380	50	50
Michigan	4,200	4,100	1,050	1,050
Minnesota	8,600	8,500	2,500	2,500
Mississippi	900	1,000	550	600
Missouri	6,700	7,400	4,100	4,400
Montana	3,100	3,100	9,800	9,700
Nebraska	7,400	7,300	5,900	5,900
Nevada	270	270	610	600
New Hampshire	(D)	(D)	(D)	(D)
New Jersey	450	450	80	80
New Mexico	1,100	1,050	7,400	7,600
New York	3,500	3,300	900	850
North Carolina	2,000	1,800	600	600
North Dakota	3,500	3,600	3,900	3,800
Ohio	7,200	7,100	1,900	1,950
Oklahoma	4,400	4,450	5,700	5,750
Oregon	1,700	1,700	2,900	2,900
Pennsylvania	5,500	5,600	1,050	1,050
Rhode Island	(D)	(D)	(D)	(D)
South Carolina	600	600	270	270
South Dakota	4,400	4,200	5,500	5,300
Tennessee	1,750	2,000	750	800
Texas	8,400	8,700	20,200	21,600
Utah	900	900	1,400	1,400
Vermont	490	490	140	140
Virginia	1,850	1,950	950	1,000
Washington	2,200	2,150	1,600	1,550
West Virginia	400	400	240	240
Wisconsin	7,400	7,300	1,950	1,900
Wyoming	1,300	1,300	4,700	4,900
Other States[1]	530	530	100	100
United States	144,960	145,330	128,930	130,810

See footnote(s) at end of table.

Table 9-5.—Number of farms and land in farms by economic sales class by State and United States, 2014–2015—Continued

State	Economic Sales Class: $250,000-$499,999			
	Number of Farms		Land in Farms	
	2014	2015	2014	2015
	Number	*Number*	*1,000 Acres*	*1,000 Acres*
Alabama	950	900	500	450
Alaska	(D)	(D)	(D)	(D)
Arizona	240	230	2,500	2,600
Arkansas	1,100	1,200	600	650
California	4,500	4,600	2,500	2,500
Colorado	1,400	1,300	5,500	5,800
Connecticut	(D)	(D)	(D)	(D)
Delaware	(D)	(D)	(D)	(D)
Florida	1,250	1,250	950	970
Georgia	900	900	470	470
Hawaii	120	120	60	60
Idaho	1,300	1,300	1,400	1,400
Illinois	7,100	7,300	4,400	4,500
Indiana	4,000	3,900	2,100	2,100
Iowa	9,600	9,700	5,000	5,100
Kansas	4,300	4,200	7,600	7,400
Kentucky	1,500	1,500	1,000	1,000
Louisiana	540	550	450	450
Maine	220	220	110	110
Maryland	650	650	250	250
Massachusetts	(D)	(D)	(D)	(D)
Michigan	2,700	2,900	1,200	1,250
Minnesota	6,400	6,300	3,300	3,300
Mississippi	800	800	500	500
Missouri	3,200	3,000	3,500	3,400
Montana	2,400	2,500	12,200	12,200
Nebraska	5,100	5,300	7,300	7,500
Nevada	420	440	960	970
New Hampshire	(D)	(D)	(D)	(D)
New Jersey	270	270	90	90
New Mexico	550	550	6,300	6,300
New York	1,700	1,800	800	800
North Carolina	1,500	1,500	700	700
North Dakota	2,900	3,000	5,000	5,100
Ohio	3,850	3,800	1,900	1,900
Oklahoma	2,050	2,050	4,550	4,550
Oregon	1,000	1,000	2,290	2,290
Pennsylvania	3,550	3,600	950	950
Rhode Island	(D)	(D)	(D)	(D)
South Carolina	300	300	230	230
South Dakota	3,500	3,300	7,200	7,200
Tennessee	950	950	700	700
Texas	4,500	4,600	16,700	16,700
Utah	600	600	1,550	1,550
Vermont	350	350	150	150
Virginia	1,200	1,200	700	700
Washington	1,250	1,250	2,250	2,250
West Virginia	150	150	110	110
Wisconsin	4,900	4,900	2,050	2,050
Wyoming	850	850	6,600	6,400
Other States[1]	560	560	150	150
United States	97,170	97,640	125,320	125,800

See footnote(s) at end of table.

Table 9-5.—Number of farms and land in farms by economic sales class by State and United States, 2014–2015—Continued

State	Economic Sales Class: $500,000-$999,999			
	Number of Farms		Land in Farms	
	2014	2015	2014	2015
	Number	*Number*	*1,000 Acres*	*1,000 Acres*
Alabama	1,650	1,500	500	500
Alaska	(D)	(D)	(D)	(D)
Arizona	300	300	400	400
Arkansas	1,500	1,500	1,200	1,300
California	3,300	3,200	2,600	2,600
Colorado	850	900	5,600	5,400
Connecticut	(D)	(D)	(D)	(D)
Delaware	370	370	100	100
Florida	820	820	650	650
Georgia	2,900	2,900	1,630	1,680
Hawaii	70	70	240	240
Idaho	800	800	1,900	1,900
Illinois	5,500	5,500	6,300	6,300
Indiana	3,000	3,000	2,840	2,880
Iowa	10,000	9,900	8,700	8,700
Kansas	3,000	3,000	9,600	9,600
Kentucky	1,100	1,100	1,240	1,340
Louisiana	950	900	1,000	1,000
Maine	110	110	90	90
Maryland	400	400	270	250
Massachusetts	(D)	(D)	(D)	(D)
Michigan	2,250	2,350	1,800	1,800
Minnesota	5,300	5,100	4,800	4,800
Mississippi	1,200	1,100	1,000	950
Missouri	2,200	2,300	3,700	4,000
Montana	1,450	1,500	13,500	13,600
Nebraska	4,900	4,900	9,800	9,800
Nevada	170	180	1,000	1,050
New Hampshire	(D)	(D)	(D)	(D)
New Jersey	210	210	80	80
New Mexico	320	370	8,100	8,100
New York	1,000	1,000	650	650
North Carolina	1,600	1,600	950	950
North Dakota	4,600	4,700	10,600	10,800
Ohio	2,850	2,850	2,400	2,400
Oklahoma	1,350	1,350	4,000	4,000
Oregon	850	850	2,650	2,650
Pennsylvania	1,900	1,950	770	750
Rhode Island	(D)	(D)	(D)	(D)
South Carolina	400	400	450	450
South Dakota	3,300	3,500	11,200	11,300
Tennessee	800	850	850	850
Texas	3,250	3,400	15,500	15,100
Utah	220	240	4,500	4,500
Vermont	150	160	90	90
Virginia	1,050	1,050	830	830
Washington	1,050	1,050	2,950	2,930
West Virginia	170	170	110	110
Wisconsin	2,900	2,850	2,050	2,050
Wyoming	500	500	6,800	6,800
Other States[1]	260	260	80	80
United States	82,820	83,010	156,070	156,400

See footnote(s) at end of table.

Table 9-5.—Number of farms and land in farms by economic sales class by State and United States, 2014–2015—Continued

State	Economic Sales Class: $1,000,000 or more			
	Number of Farms		Land in Farms	
	2014	2015	2014	2015
	Number	*Number*	*1,000 Acres*	*1,000 Acres*
Alabama	1,450	1,500	1,150	500
Alaska	(D)	(D)	(D)	(D)
Arizona	350	300	2,900	400
Arkansas	3,700	1,500	5,700	1,300
California	6,500	3,200	10,900	2,600
Colorado	950	900	6,000	5,400
Connecticut	90	(D)	100	(D)
Delaware	380	370	270	100
Florida	1,230	820	3,800	650
Georgia	2,400	2,900	2,320	1,680
Hawaii	80	70	300	240
Idaho	1,200	800	4,000	1,900
Illinois	4,500	5,500	9,100	6,300
Indiana	2,600	3,000	5,400	2,880
Iowa	7,300	9,900	9,500	8,700
Kansas	2,800	3,000	11,000	9,600
Kentucky	1,100	1,100	1,860	1,340
Louisiana	1,100	900	2,800	1,000
Maine	130	110	270	90
Maryland	550	400	580	250
Massachusetts	80	(D)	50	(D)
Michigan	1,850	2,350	3,100	1,800
Minnesota	5,200	5,100	9,500	4,800
Mississippi	1,900	1,100	3,100	950
Missouri	1,900	2,300	4,100	4,000
Montana	950	1,500	12,200	13,600
Nebraska	4,800	4,900	16,300	9,800
Nevada	140	180	2,900	1,050
New Hampshire	(D)	(D)	(D)	(D)
New Jersey	220	210	150	80
New Mexico	330	370	5,700	8,100
New York	1,000	1,000	1,650	650
North Carolina	3,600	1,600	3,150	950
North Dakota	3,400	4,700	15,200	10,800
Ohio	2,100	2,850	3,500	2,400
Oklahoma	1,000	1,350	3,400	4,000
Oregon	950	850	4,200	2,650
Pennsylvania	1,250	1,950	850	750
Rhode Island	(D)	(D)	(D)	(D)
South Carolina	800	400	1,000	450
South Dakota	2,500	3,500	13,100	11,300
Tennessee	900	850	1,850	850
Texas	3,350	3,400	17,700	15,100
Utah	280	240	1,000	4,500
Vermont	160	160	210	90
Virginia	800	1,050	970	830
Washington	1,600	1,050	5,000	2,930
West Virginia	180	170	140	110
Wisconsin	2,300	2,850	3,550	2,050
Wyoming	250	500	6,500	6,800
Other States[1]	40	260	40	80
United States	82,240	83,010	218,060	156,400

(D) Withheld to avoid disclosing data for individual operations.   [1] Includes data withheld above.
NASS, Environmental, Economics, and Demographics Branch, (202) 720–6146.

Table 9-6.—Number of farms and land in farms by State and United States, 2014–2015

State	Number of Farms [1]		Land in farms		Average farm size	
	2014	2015	2014	2015	2014	2015
	Number	*Number*	*1,000 acres*	*1,000 acres*	*Acres*	*Acres*
Alabama	43,400	43,200	8,900	8,900	205	206
Alaska	760	760	830	830	1,092	1,092
Arizona	19,600	19,500	26,000	26,000	1,327	1,333
Arkansas	44,000	43,500	13,800	13,800	314	317
California	77,400	77,400	25,500	25,400	334	328
Colorado	35,000	34,200	31,800	31,700	909	927
Connecticut	6,000	6,000	440	440	73	73
Delaware	2,500	2,500	500	500	200	200
Florida	47,600	47,300	9,500	9,450	200	200
Georgia	41,100	41,000	9,400	9,400	229	229
Hawaii	7,000	7,000	1,120	1,120	160	160
Idaho	24,400	24,400	11,800	11,800	484	484
Illinois	74,500	73,400	26,900	26,800	361	365
Indiana	58,200	57,700	14,700	14,700	253	255
Iowa	88,000	87,500	30,500	30,500	347	349
Kansas	61,000	60,400	46,000	46,000	754	762
Kentucky	76,400	76,400	13,000	13,000	170	170
Louisiana	27,200	26,900	7,800	7,750	287	288
Maine	8,200	8,200	1,450	1,450	177	177
Maryland	12,300	12,200	2,030	2,030	165	166
Massachusetts	7,800	7,800	520	520	67	67
Michigan	51,600	51,500	9,950	9,950	193	193
Minnesota	74,000	73,600	25,900	25,900	350	352
Mississippi	37,100	36,700	10,900	10,800	294	294
Missouri	97,700	97,100	28,300	28,300	290	291
Montana	27,800	27,500	59,700	59,700	2,147	2,171
Nebraska	49,100	48,700	45,200	45,200	921	928
Nevada	4,200	4,200	5,950	5,950	1,417	1,417
New Hampshire	4,400	4,400	470	470	107	107
New Jersey	9,100	9,100	720	720	79	79
New Mexico	24,700	24,700	43,200	43,200	1,749	1,749
New York	35,500	35,500	7,180	7,200	202	203
North Carolina	49,500	48,800	8,400	8,300	170	170
North Dakota	30,300	30,000	39,300	39,200	1,297	1,307
Ohio	74,500	74,400	14,000	14,000	188	188
Oklahoma	79,600	78,000	34,300	34,200	431	438
Oregon	34,600	34,600	16,400	16,400	474	474
Pennsylvania	58,800	57,900	7,720	7,700	131	133
Rhode Island	1,240	1,240	70	70	56	56
South Carolina	24,400	24,400	5,000	5,000	205	205
South Dakota	31,700	31,300	43,300	43,300	1,366	1,383
Tennessee	67,300	67,300	10,900	10,900	162	162
Texas	245,500	242,000	130,000	130,000	530	537
Utah	18,100	18,100	11,000	11,000	608	608
Vermont	7,300	7,300	1,250	1,250	171	171
Virginia	45,900	45,000	8,200	8,100	179	180
Washington	36,700	36,000	14,700	14,700	401	408
West Virginia	21,300	20,900	3,600	3,600	169	172
Wisconsin	69,000	68,900	14,500	14,400	210	209
Wyoming	11,700	11,600	30,400	30,400	2,598	2,621
United States	2,085,000	2,068,000	913,000	912,000	438	441
Puerto Rico	10,400	....	450	....	43	....

[1] A farm is any establishment from which $1,000 or more of agricultural products were sold or would normally be sold during the year.
NASS, Environmental, Economics, and Demographics Branch, (202) 720–6146.

Table 9-7.—Farms: Classification by tenure of operator, United States, 1935–2012

Year	Farms	Tenure of operator			
		Full owners	Part owners	Managers	All tenants
	Number	Percent	Percent	Percent	Percent
1935 [1]	6,812,350	47.1	10.1	0.7	42.1
1940	6,102,417	50.6	10.1	0.6	38.8
1945 [1]	5,859,169	56.4	11.3	0.7	31.7
1950	5,388,437	57.4	15.3	0.4	26.9
1954 [1]	4,783,021	57.4	18.2	0.4	24.0
1959	3,710,503	57.1	21.9	0.6	20.5
1964	3,157,857	57.6	24.8	0.6	17.1
1969	2,730,250	62.5	24.6		12.9
1974	2,314,013	61.5	27.2		11.3
1978	2,257,775	57.5	30.2		12.3
1982	2,240,976	59.2	29.3		11.6
1987	2,087,759	59.3	29.2		11.5
1992	1,925,300	57.7	31.0		11.3
1997	1,911,859	60.0	30.0		10.0
2002 [2]	2,128,982	67.1	25.9		7.0
2007 [2]	2,204,792	69.0	24.6		6.4
2012 [2]	2,109,303	67.7	25.3		7.0

[1] Excludes Alaska and Hawaii.   [2] The 2002 Census of Agriculture introduced new methodology to account for all farms in the United States.   All 2002 published census items were reweighted for undercoverage.   Strictly speaking, data for 2002 and subsequent years are not fully comparable with data from earlier years.
ERS, Resource and Rural Economics Division, (202) 694–5572.   Data from the Census of Agriculture, National Agricultural Statistics Service.

Table 9-8.—Farmland Rented: Classification by tenants and part owners, United States, 1910–2012

Year	Land in farms	Land rented by tenure of operator [1]			Percentage of land rented
		Tenants	Part owners	Total	
	Million acres	Million acres	Million acres	Million acres	Percent
1910	878.8	225.5	[2] 51.3	277.8	31.6
1920	958.7	[3] 265.0	[4] 54.7	319.7	33.3
1925	924.3	264.9	96.3	361.2	39.1
1930	990.1	307.3	125.2	432.5	43.7
1935	1,054.5	336.8	134.3	471.1	44.7
1940	1,065.1	313.2	155.9	469.1	44.0
1945	1,141.6	251.6	178.9	430.5	37.7
1950	1,161.4	212.2	196.2	408.4	35.2
1954	1,158.2	192.6	212.3	404.9	35.0
1959	1,123.0	166.8	234.1	400.9	35.7
1964	1,110.2	144.9	248.1	[5] 393.0	35.4
1969	1,063.3	137.6	241.8	379.4	35.7
1974	1,017.0	122.3	258.4	380.7	37.4
1978	1,029.7	124.1	282.2	406.2	39.4
1982	986.2	113.6	269.9	383.5	38.9
1987	964.5	126.9	275.4	402.3	41.7
1992	945.5	122.7	282.2	404.9	42.8
1997	931.8	108.1	270.0	378.1	40.6
2002 [6]	938.3	86.5	266.8	353.3	37.7
2007 [6]	922.1	81.8	269.0	350.8	38.0
2012 [6]	914.5	82.9	268.1	351.0	38.4

[1] Columns 3, 4, and 5 refer only to land rented from others and operated, so subleased land is not included.   Acres of land rented are comparable in the same year, but definitions change over time.   Basic sources are 1969 Census of Agriculture, table 5, p.14; 1974 Census of Agriculture, table 3, pp.1-6; 1978 Census of Agriculture, vol. 1, part 51, table 5, pp. 124-127; 1982 Census of Agriculture, vol. 1, part 51, table 48, p. 49; 1987 Census of Agriculture vol. 1 part 51, table 48, p. 49; 1992 Census of Agriculture vol. 1, part 51, table 46, p. 53; 1997 Census of Agriculture, vol. 1, part 51, chapter 1, table 46, p. 57; 2002 Census of Agriculture, vol. 1, part 51, chapter 1, table 61, p. 214; 2007 Census of Agriculture, vol. 1, part 51, chapter 1, table 65, p. 262; 2012 Census of Agriculture, vol. 1, part 51, table 70, p. 231; and earlier census volumes as noted.   [2] Assumes land leased by part-owners is the difference between the average size of full-owner and part-owner farms.   Acreage leased by part-owners is this difference times the number of part-owners.   1910 Census of Agriculture, chapter 11, table 1 and 3, pp.97-99.   [3] 1920 Census of Agriculture, vol. VI, part 1, table 5, p. 19.   [4] Assumes same proportion of owner and part-owner as in 1910.   [5] 1964 Census of Agriculture, vol. II, chapter 8, p.757.   [6] The 2002 Census of Agriculture introduced new methodology to account for all farms in the United States.   All 2002 published census items were reweighted for undercoverage.   Strictly speaking, data for 2002 and subsequent years are not fully comparable with data from earlier years.
ERS, Resource and Rural Economics Division, (202) 694–5572.   Data from the Census of Agriculture, National Agricultural Statistics Service.

Table 9-9.—Land: Utilization, by State and United States, 2007

State	Cropland			Grassland pasture	Forest land
	Used for crops	Idle	Used only for pasture		
	1,000 acres	*1,000 acres*	*1,000 acres*	*1,000 acres*	*1,000 acres*
Alabama	2,070	406	627	2,642	22,587
Alaska	31	48	7	738	93,801
Arizona	818	96	0	40,648	16,780
Arkansas	7,409	94	736	3,293	18,596
California	8,084	657	809	27,524	26,983
Colorado	8,110	2,076	1,242	28,871	18,236
Connecticut	111	13	13	33	1,413
Delaware	404	10	9	23	383
Florida	2,098	90	571	5,558	15,649
Georgia	3,665	364	590	1,292	24,267
Hawaii	72	49	23	738	1,552
Idaho	4,680	789	511	18,082	17,455
Illinois	22,778	970	309	1,940	4,363
Indiana	12,201	262	284	1,642	4,533
Iowa	24,277	1,608	845	2,460	2,864
Kansas	24,649	2,608	1,292	16,438	2,104
Kentucky	5,395	676	1,550	3,516	11,686
Louisiana	3,340	460	635	1,860	14,142
Maine	341	85	37	105	17,355
Maryland	1,208	71	0	464	2,386
Massachusetts	139	14	16	54	2,240
Michigan	7,145	553	316	1,697	19,019
Minnesota	19,857	1,745	740	3,020	15,572
Mississippi	4,411	376	769	2,055	19,579
Missouri	13,301	1,372	1,887	8,423	14,838
Montana	12,631	3,558	1,678	46,051	19,875
Nebraska	19,495	1,221	896	23,191	1,234
Nevada	516	30	185	46,850	10,436
New Hampshire ...	76	10	17	90	4,422
New Jersey	415	24	40	51	1,472
New Mexico	1,174	545	648	52,122	14,977
New York	3,594	266	280	2,414	16,168
North Carolina	4,258	247	339	1,231	18,037
North Dakota	23,290	3,569	817	11,935	699
Ohio	10,141	431	352	1,924	7,666
Oklahoma	9,169	890	2,781	18,707	7,620
Oregon	3,560	696	677	22,726	27,813
Pennsylvania	4,325	513	427	1,123	16,119
Rhode Island	18	2	2	8	313
South Carolina	1,536	201	264	795	12,646
South Dakota	17,026	1,516	1,311	23,263	1,640
Tennessee	4,406	410	1,203	2,093	13,913
Texas	21,515	4,663	7,938	101,735	17,159
Utah	1,137	198	403	26,120	16,058
Vermont	413	27	47	261	4,504
Virginia	2,570	194	487	2,463	15,350
Washington	5,733	1,521	372	6,789	19,225
West Virginia	678	51	192	1,249	11,833
Wisconsin	9,069	738	395	2,768	16,168
Wyoming	1,657	141	420	44,653	7,661
United States [1]	334,996	37,154	35,989	613,733	671,390

[1] Distributions may not add to totals due to rounding.
ERS, Resource and Rural Economics Division, (202) 694–5626. See notes to Table 9-13 for definitions and data sources.
Estimates developed for years coinciding with a Census of Agriculture.

Table 9-10.—Land in farms: Irrigated land, by State and United States, 1969–2007 [1]

State	1969	1974	1978 [2]	1982	1987	1992	1997	2002	2007
	1,000 acres	1,000 acres	1,000 acres	1,000 acres	1,000 acres	1,000 acres	1,000 acres	1,000 acres	1,000 acres
Alabama	11	14	59	66	84	82	80	109	113
Alaska	1	1	1	1	2	2	3	3	4
Arizona	1,178	1,153	1,196	1,098	914	956	1,075	932	876
Arkansas ...	1,010	949	1,683	2,022	2,406	2,702	3,785	4,150	4,461
California ...	7,240	7,749	8,506	8,461	7,596	7,571	8,887	8,709	8,016
Colorado ...	2,895	2,874	3,431	3,201	3,014	3,170	3,374	2,591	2,868
Connecticut	9	7	7	7	7	6	8	10	10
Delaware ...	20	20	34	44	61	62	75	97	105
Florida	1,365	1,559	1,980	1,585	1,623	1,783	1,874	1,815	1,552
Georgia	79	112	463	575	640	725	773	871	1,018
Hawaii	146	142	159	146	149	134	77	69	59
Idaho	2,761	2,859	3,475	3,450	3,219	3,260	3,544	3,289	3,300
Illinois	51	54	130	166	208	328	352	391	474
Indiana	34	33	75	132	170	241	256	313	397
Iowa	21	39	101	91	92	116	133	142	190
Kansas	1,522	2,010	2,686	2,675	2,463	2,680	2,696	2,678	2,763
Kentucky ...	20	11	14	23	38	28	60	37	59
Louisiana ...	702	702	681	694	647	898	961	939	954
Maine	6	6	7	6	6	10	22	20	21
Maryland ...	22	23	28	39	51	57	69	81	93
Massachu- setts	19	19	17	17	20	20	27	24	23
Michigan	77	97	226	286	315	366	407	456	500
Minnesota	36	78	272	315	354	370	403	455	506
Mississippi	150	162	309	431	637	883	1,110	1,176	1,369
Missouri	156	150	320	403	535	709	921	1,033	1,200
Montana	1,841	1,759	2,070	2,023	1,997	1,978	2,102	1,976	2,013
Nebraska ...	2,857	3,967	5,683	6,039	5,682	6,312	7,066	7,625	8,559
Nevada	753	778	881	830	779	556	764	747	691
New Hamp- shire	2	2	2	1	3	2	3	2	3
New Jersey	72	89	77	83	91	80	94	97	95
New Mex- ico	823	867	891	807	718	738	852	845	830
New York ..	55	55	56	52	51	47	74	75	68
North Caro- lina	59	51	90	81	138	113	156	264	232
North Da- kota	63	71	141	163	168	187	183	203	236
Ohio	22	22	25	28	32	29	35	41	38
Oklahoma ..	524	515	602	492	478	512	509	518	535
Oregon	1,519	1,561	1,881	1,808	1,648	1,622	1,963	1,908	1,845
Pennsyl- vania	19	18	15	18	30	23	40	43	38
Rhode Is- land	2	2	3	2	4	3	3	4	4
South Caro- lina	15	10	32	81	81	76	89	96	132
South Da- kota	148	152	335	376	362	371	367	401	374
Tennessee	12	10	13	18	38	37	47	61	81
Texas	6,888	6,594	6,947	5,576	4,271	4,912	5,764	5,075	5,010
Utah	1,025	970	1,169	1,082	1,161	1,143	1,218	1,091	1,134
Vermont	(3)	1	1	1	2	2	3	2	2
Virginia	37	28	42	43	79	62	86	99	82
Washington	1,224	1,309	1,639	1,638	1,519	1,641	1,787	1,823	1,736
West Vir- ginia	3	2	1	1	3	3	4	2	2
Wisconsin ..	106	128	235	259	285	331	358	386	377
Wyoming ...	1,523	1,460	1,662	1,565	1,518	1,465	1,750	1,542	1,551
United States	39,122	41,243	50,350	49,003	46,386	49,404	56,289	55,316	56,599
Puerto Rico	91	70	54	42	36	46	35	43	40
Virgin Is- lands	(3)	(3)	(4)	(4)	(4)	(4)	(4)	(3)	(3)
Total	39,213	41,313	50,350	49,002	46,386	49,404	55,058	55,360	56,639

[1] Data may not add because of rounding.   [2] Data for 1978 not directly comparable with earlier censuses as it includes estimates from the direct enumeration sample for farms not represented on the mail list.   [3] Less than 500 acres.   [4] Not available.   Note: Data from the Census of Agriculture, U.S. Department of Commerce. Beginning in 1997 Census of Agriculture, U.S. Department of Agriculture. Estimates developed for years coinciding with a Census of Agriculture. ERS, Resource and Rural Economics Division, (202) 694–5626.

Table 9-11.—Land utilization, United States, selected years, 1959–2007

Major land uses	1959	1969	1978	1987	1992	1997	2002	2007
	Million acres	Million acres	Million acres	Million acres	Million acres	Million acres	Million acres	Million acres
Cropland used for crops [1]	359	333	369	331	338	349	340	335
Idle cropland	34	51	26	68	56	39	40	37
Cropland used for pasture [2]	66	88	76	65	67	68	62	36
Grassland pasture [3]	633	604	587	591	591	580	587	614
Forest land [4]	745	723	703	648	648	641	651	671
Special uses [5]	115	143	158	279	281	286	297	313
Urban areas [6]	27	31	45	57	59	66	60	61
Other land [7]	293	291	301	227	224	236	228	197
Total land area [8]	2,271	2,264	2,264	2,265	2,263	2,263	2,264	2,264

[1] Cropland harvested, crop failure, and cultivated summer fallow. [2] The 2007 estimate declined due to a change in the methodology for determining cropland used for pasture for non-respondents. [3] Grassland and other nonforest pasture and range. [4] Excludes reserved and other forest land duplicated in parks and special uses of land. Includes forested grazing land. [5] Includes rural transportation areas, Federal and State areas used primarily for recreation and wildlife purposes, military areas, farmsteads, and farm roads and lanes. [6] The 2002 urban acreage estimate is not directly comparable to estimates in prior years due to a change in the definition of urban areas in the 2000 Census of Population and Housing. The apparent change in "urban" acreage between 1997 and 2002 reflects a definitional change, rather than a decline in acreage. [7] Miscellaneous areas such as marshes, open swamps, bare rock areas, deserts, and other uses not inventoried. [8] Remeasurement and increases in reservoirs account for changes in total land areas except for the major increase in 1949 when data for Alaska and Hawaii were added.

ERS, Rural and Resource Economics Division, (202) 694–5626. Estimates based on reports and records of the U.S. Agriculture and Commerce Departments, and public land administering and conservation agencies. Estimates developed for years coinciding with a Census of Agriculture. See http://www.ers.usda.gov/data/majorlanduses for data and more information.

Table 9-12.—Farm real estate: Value of farmland and buildings, Region, State, and United States, 2011–2015 [1]

Region and State	Total value of land and buildings				
	2011	2012	2013	2014	2015
	Million dollars	Million dollars	Million dollars	Million dollars	Million dollars
Northeast:					
Connecticut	4,872	4,928	4,884	4,928	4,972
Delaware	4,151	4,157	4,085	4,090	4,090
Maine	2,912	3,002	3,150	3,016	3,031
Maryland	14,015	13,723	14,207	14,007	14,140
Massachusetts	5,777	5,408	5,408	5,408	5,408
New Hampshire	2,098	2,087	2,026	2,012	2,012
New Jersey	9,216	8,856	9,216	9,216	9,360
New York	17,395	19,027	18,720	19,386	21,600
Pennsylvania	38,380	40,810	41,811	43,232	42,350
Rhode Island	1,008	966	966	959	966
Vermont	3,695	3,825	4,000	4,075	4,125
Lake:					
Michigan	36,000	38,706	42,785	46,765	48,755
Minnesota	82,476	97,240	111,370	123,025	121,730
Wisconsin	56,648	60,006	59,860	63,800	67,680
Corn Belt:					
Illinois	145,530	167,049	190,990	202,288	201,750
Indiana	74,529	85,848	94,080	102,165	105,105
Iowa	165,546	199,818	235,620	259,250	244,000
Missouri	68,970	76,693	80,940	87,730	94,805
Ohio	57,408	64,960	71,400	77,700	80,500
Northern Plains:					
Kansas	57,164	69,611	80,675	94,300	93,380
Nebraska	83,536	109,626	126,840	141,024	137,860
North Dakota	36,735	45,588	60,915	71,526	75,264
South Dakota	47,197	57,589	73,177	89,631	100,456
Appalachian:					
Kentucky	36,300	36,920	39,260	40,950	42,250
North Carolina	36,022	36,415	36,456	37,884	37,350
Tennessee	37,908	38,368	38,913	39,240	39,785
Virginia	35,670	35,358	35,773	35,424	34,992
West Virginia	9,288	9,169	9,180	9,216	9,360
Southeast:					
Alabama	20,826	21,271	22,500	23,140	23,144
Florida	48,479	49,278	49,660	49,875	51,030
Georgia	35,017	31,296	31,350	31,020	30,411
South Carolina	14,900	14,960	14,811	15,050	15,050
Delta States:					
Arkansas	33,428	36,156	37,260	39,330	42,090
Louisiana	17,270	18,960	20,018	20,826	21,700
Mississippi	23,980	24,525	24,743	25,506	26,136
Southern Plains:					
Oklahoma	43,344	47,128	49,880	54,194	58,140
Texas	219,605	220,038	218,568	240,500	252,200
Mountain:					
Arizona [2]	13,796	17,960	18,506	23,922	24,125
Colorado	34,760	37,323	40,576	42,930	45,648
Idaho	23,896	25,252	26,196	27,848	29,146
Montana	43,097	45,448	47,163	51,342	53,133
Nevada [2]	4,838	5,071	5,113	5,566	5,817
New Mexico [2]	18,814	20,641	19,929	20,663	20,310
Utah [2]	12,773	12,773	13,439	15,528	15,669
Wyoming	16,800	17,936	18,120	19,152	20,064
Pacific:					
California	170,752	176,128	175,950	186,150	196,350
Oregon	30,780	31,948	32,505	33,620	34,768
Washington	30,076	33,222	34,040	36,750	39,690
United States [3]	2,027,677	2,223,067	2,397,034	2,595,159	2,651,697

[1] Total value of land and buildings is derived by multiplying average value per acre of farm real estate by the land in farms.   [2] Value of all land and buildings adjusted to include American Indian reservation land value.   [3] Excludes Alaska and Hawaii.
NASS, Environmental, Economics, and Demographics Branch, (202) 720–6146.

Table 9-13.—Farm real estate: Average value per acre - Region, State, and United States, 2012–2015

Region and State	2012	2013	2014	2015
	Dollars	*Dollars*	*Dollars*	*Dollars*
Northeast:	4,790	4,850	4,930	5,020
Connecticut	11,200	11,100	11,200	11,300
Delaware	8,150	8,170	8,180	8,180
Maine	2,070	2,100	2,080	2,090
Maryland	6,760	6,930	6,900	7,000
Massachusetts	10,400	10,400	10,400	10,400
New Hampshire	4,440	4,310	4,280	4,280
New Jersey	12,300	12,800	12,800	13,000
New York	2,650	2,600	2,700	3,000
Pennsylvania	5,300	5,430	5,600	5,500
Rhode Island	13,800	13,800	13,700	13,800
Vermont	3,060	3,200	3,260	3,300
Lake:	3,880	4,240	4,640	4,740
Michigan	3,890	4,300	4,700	4,900
Minnesota ¹	3,740	4,300	4,750	4,700
Wisconsin	4,110	4,100	4,400	4,700
Corn Belt:	5,190	5,880	6,370	6,350
Illinois	6,210	7,100	7,520	7,500
Indiana	5,840	6,400	6,950	7,150
Iowa	6,530	7,700	8,500	8,000
Missouri	2,710	2,850	3,100	3,350
Ohio	4,640	5,100	5,550	5,750
Northern Plains:	1,620	1,960	2,280	2,340
Kansas	1,510	1,750	2,050	2,030
Nebraska	2,420	2,800	3,120	3,050
North Dakota	1,160	1,550	1,820	1,920
South Dakota	1,330	1,690	2,070	2,320
Appalachian:	3,530	3,610	3,690	3,730
Kentucky	2,840	3,020	3,150	3,250
North Carolina	4,330	4,340	4,510	4,500
Tennessee	3,520	3,570	3,600	3,650
Virginia	4,260	4,310	4,320	4,320
West Virginia	2,540	2,550	2,560	2,600
Southeast:	3,530	3,590	3,630	3,670
Alabama	2,390	2,500	2,600	2,630
Florida	5,160	5,200	5,250	5,400
Georgia	3,260	3,300	3,300	3,270
South Carolina	3,010	2,980	3,010	3,010
Delta:	2,440	2,520	2,640	2,780
Arkansas	2,620	2,700	2,850	3,050
Louisiana	2,400	2,550	2,670	2,800
Mississippi	2,250	2,270	2,340	2,420
Southern Plains:	1,620	1,630	1,790	1,890
Oklahoma	1,370	1,450	1,580	1,700
Texas	1,690	1,680	1,850	1,940
Mountain:	953	1,010	1,070	1,100
Arizona ¹	3,370	3,500	3,740	3,780
Colorado	1,170	1,280	1,350	1,440
Idaho	2,140	2,220	2,360	2,470
Montana	760	790	860	890
Nevada ¹	1,040	1,040	1,080	1,130
New Mexico ¹	520	500	540	510
Utah ¹	1,800	1,900	2,030	2,050
Wyoming	590	600	630	660
Pacific:	4,270	4,290	4,520	4,780
California	6,880	6,900	7,300	7,700
Oregon	1,960	1,970	2,050	2,120
Washington	2,260	2,300	2,500	2,700
United States ²	2,520	2,730	2,950	3,020

¹ Excludes American Indian Reservation Land.   ² Excludes Alaska and Hawaii.
NASS, Environmental, Economics, and Demographics Branch, (202) 720–6146.

Table 9-14.—Land values, cropland and pasture: Region, State, and United States, 2012–2015

Region and State	Cropland			
	2012	2013	2014	2015
	Dollars	*Dollars*	*Dollars*	*Dollars*
Northeast:	5,280	5,260	5,260	5,330
Delaware	7,850	7,870	7,880	7,950
Maryland	6,570	6,470	6,470	6,470
New Jersey	12,400	12,900	13,000	13,500
New York	2,600	2,550	2,530	2,600
Pennsylvania	5,760	5,840	5,840	5,900
Other States [1]	7,120	7,060	7,020	7,090
Lake:	3,790	4,240	4,670	4,730
Michigan	3,660	4,120	4,500	4,550
Minnesota	3,740	4,390	4,870	4,800
Wisconsin	4,000	4,010	4,350	4,700
Corn Belt:	5,600	6,470	7,000	6,840
Illinois	6,300	7,190	7,700	7,650
Indiana	5,840	6,590	7,050	7,000
Iowa	6,810	8,000	8,750	8,200
Missouri	3,120	3,500	3,810	3,810
Ohio	4,640	5,190	5,650	5,850
Northern Plains:	2,210	2,720	3,090	3,130
Kansas	1,650	1,930	2,260	2,210
Nebraska	4,190	4,860	5,180	5,070
North Dakota	1,260	1,750	2,050	2,140
South Dakota	2,200	2,840	3,430	3,730
Appalachian:	3,550	3,690	3,780	3,830
Kentucky	3,210	3,430	3,550	3,720
North Carolina	3,850	4,050	4,200	4,100
Tennessee	3,260	3,340	3,400	3,470
Virginia	4,500	4,450	4,460	4,440
West Virginia	3,250	3,200	3,210	3,250
Southeast:	3,710	3,690	3,730	3,770
Alabama	2,750	2,750	2,850	2,850
Florida	6,420	6,450	6,500	6,560
Georgia	3,130	3,080	3,080	3,160
South Carolina	2,620	2,440	2,460	2,460
Delta:	2,160	2,380	2,510	2,600
Arkansas	2,180	2,380	2,540	2,630
Louisiana	2,120	2,260	2,380	2,500
Mississippi	2,180	2,470	2,570	2,620
Southern Plains:	1,500	1,480	1,630	1,770
Oklahoma	1,280	1,390	1,500	1,590
Texas	1,590	1,520	1,680	1,840
Mountain:	1,600	1,780	1,690	1,740
Arizona [2]	7,960	8,290	8,320	8,320
Colorado	1,450	1,770	1,840	1,910
Idaho	2,580	2,850	3,040	3,200
Montana	852	890	987	997
Nevada [2]	(D)	(D)	2,670	2,670
New Mexico [2]	1,550	1,450	1,450	1,440
Utah [2]	2,690	3,230	3,260	3,300
Wyoming	1,360	1,360	1,370	1,370
Pacific:	5,310	5,690	5,860	6,160
California	9,400	9,860	10,140	10,690
Oregon	2,340	2,400	2,500	2,600
Washington	2,190	2,420	2,560	2,630
United States [4]	3,350	3,810	4,100	4,130

See footnote(s) at end of table.

Table 9-14.—Land values, cropland and pasture: Region, State, and United States, 2012–2015—Continued

Region, State, and land type	Irrigated and Non-Irrigated Cropland			
	2012	2013	2014	2015
	Dollars	Dollars	Dollars	Dollars
Corn Belt:				
Missouri all cropland	3,120	3,500	3,810	3,810
Irrigated	3,640	4,140	4,750	5,130
Non-irrigated	3,080	3,450	3,730	3,700
Northern Plains:				
Kansas all cropland	1,650	1,930	2,260	2,210
Irrigated	2,250	2,760	3,280	3,270
Non-irrigated	1,590	1,840	2,150	2,090
Nebraska all cropland	4,190	4,860	5,180	5,070
Irrigated	5,610	6,700	7,100	6,870
Non-irrigated	3,270	3,730	4,000	3,970
South Dakota all cropland	2,200	2,840	3,430	3,730
Irrigated	(D)	(D)	(D)	(D)
Non-irrigated	2,190	2,820	3,400	3,700
Southeast:				
Florida all cropland	6,420	6,450	6,500	6,560
Irrigated	7,180	7,280	7,430	7,570
Non-irrigated	5,720	5,660	5,630	5,610
Georgia all cropland	3,130	3,080	3,080	3,160
Irrigated	2,980	3,120	3,430	3,600
Non-irrigated	3,170	3,070	2,950	3,000
Delta:				
Arkansas all cropland	2,180	2,380	2,540	2,630
Irrigated	2,530	2,790	3,000	3,100
Non-irrigated	1,790	1,760	1,840	1,900
Louisiana all cropland	2,120	2,260	2,380	2,500
Irrigated	2,000	2,150	2,270	2,400
Non-irrigated	2,150	2,300	2,420	2,530
Mississippi all cropland	2,180	2,470	2,570	2,620
Irrigated	2,440	2,760	2,930	3,030
Non-irrigated	2,090	2,330	2,390	2,420
Southern Plains:				
Oklahoma all cropland	1,280	1,390	1,500	1,590
Irrigated	(D)	(D)	(D)	(D)
Non-irrigated	1,270	1,370	1,480	1,580
Texas all cropland	1,590	1,520	1,680	1,840
Irrigated	1,660	1,700	1,880	2,050
Non-irrigated	1,580	1,490	1,650	1,800
Mountain:				
Arizona all cropland [2]	7,610	7,960	8,290	8,320
Irrigated	7,610	7,960	8,290	8,320
Colorado all cropland	1,340	1,450	1,770	1,910
Irrigated	3,160	3,400	4,100	4,650
Non-irrigated	880	960	1,200	1,230
Idaho all cropland	2,470	2,580	2,850	3,200
Irrigated	3,820	4,040	4,240	4,830
Non-irrigated	1,210	1,210	1,310	1,400
Montana all cropland	806	852	890	997
Irrigated	2,690	2,690	2,780	2,960
Non-irrigated	630	680	710	810
Nevada all cropland [2]	(D)	(D)	(D)	2,670
Irrigated	(D)	(D)	(D)	2,670
New Mexico all cropland [2]	1,740	1,550	1,450	1,440
Irrigated	5,190	4,450	3,910	3,920
Non-irrigated	410	430	400	390
Utah all cropland [2]	2,690	2,690	3,230	3,300
Irrigated	5,000	5,000	5,200	5,300
Non-irrigated	1,030	1,030	1,100	1,140
Wyoming all cropland	1,320	1,360	1,360	1,370
Irrigated	2,140	2,170	2,140	2,190
Non-irrigated	(D)	(D)	(D)	770
Pacific:				
California all cropland	9,130	9,400	9,860	10,690
Irrigated	11,100	11,500	11,800	12,700
Non-irrigated	3,480	3,390	3,400	4,000
Oregon all cropland	2,180	2,340	2,400	2,600
Irrigated	3,470	3,920	3,940	4,360
Non-irrigated	1,710	1,770	1,830	1,950
Washington all cropland	1,930	2,190	2,420	2,630
Irrigated	5,130	6,190	7,240	7,850
Non-irrigated	1,080	1,130	1,170	1,280

See footnote(s) at end of table.

Table 9-14.—Land values, cropland and pasture: Region, State, and United States, 2012–2015-Continued

Region and State	Pasture			
	2012	2013	2014	2015
	Dollars	*Dollars*	*Dollars*	*Dollars*
Northeast:	3,240	3,370	3,460	3,480
Maryland	(D)	(D)	6,000	6,000
New Jersey	13,600	13,500	13,500	13,500
New York	1,250	1,240	1,330	1,430
Pennsylvania	2,650	2,770	2,900	2,850
Other States 5	5,760	5,840	5,790	5,840
Lake:	1,740	1,870	1,950	2,060
Michigan	2,290	2,420	2,500	2,680
Minnesota	1,390	1,580	1,600	1,700
Wisconsin	2,010	2,000	2,150	2,250
Corn Belt:	2,130	2,290	2,360	2,440
Illinois	2,870	3,370	3,400	3,550
Indiana	2,450	2,500	2,550	2,600
Iowa	2,800	3,220	3,400	3,400
Missouri	1,700	1,790	1,850	1,950
Ohio	2,970	3,010	3,100	3,140
Northern Plains:	648	754	954	1,020
Kansas	938	1,150	1,300	1,390
Nebraska	617	643	900	870
North Dakota	458	578	750	850
South Dakota	561	667	860	980
Appalachian:	3,110	3,210	3,280	3,340
Kentucky	2,330	2,560	2,700	2,750
North Carolina	4,330	4,530	4,760	4,700
Tennessee	3,410	3,380	3,400	3,500
Virginia	3,920	3,930	3,930	4,000
West Virginia	1,980	1,990	2,000	2,020
Southeast:	3,700	3,770	3,790	3,790
Alabama	1,850	2,000	2,100	2,140
Florida	4,820	4,850	4,910	4,900
Georgia	3,910	3,850	3,650	3,580
South Carolina	2,960	2,820	2,900	2,940
Delta:	2,130	2,190	2,270	2,320
Arkansas	2,110	2,160	2,240	2,290
Louisiana	2,300	2,400	2,500	2,590
Mississippi	2,030	2,070	2,110	2,140
Southern Plains:	1,390	1,410	1,540	1,570
Oklahoma	1,060	1,210	1,360	1,430
Texas	1,460	1,450	1,580	1,600
Mountain:	550	594	611	614
Arizona 2	(D)	(D)	940	(D)
Colorado	640	680	760	760
Idaho	1,220	1,220	1,220	1,250
Montana	570	580	640	650
Nevada 2	(D)	(D)	(D)	(D)
New Mexico 2	330	320	360	340
Utah 2	920	950	1,050	1,050
Wyoming	480	470	490	510
Pacific:	1,590	1,590	1,610	1,630
California	2,680	2,650	2,700	2,700
Oregon	600	620	630	660
Washington	800	800	810	820
United States 4	1,110	1,170	1,300	1,330

(D) Withheld to avoid disclosing data for individual operations.
1 Includes: Connecticut, Maine, Massachusetts, New Hampshire, Rhode Island, and Vermont.   2 Excludes American Indian Reservation land.   3 Not published due to insufficient reports.   4 Excludes Alaska and Hawaii.   5 Other pasture States include Connecticut, Maine, Massachusetts, New Hampshire, Rhode Island, and Vermont.
NASS, Environmental, Economics, and Demographics Branch, (202) 720–6146.

Table 9-15.—Cash rents, cropland and pasture: By State, 2014–2015

State	2014				2015			
	Cropland	Irrigated cropland	Non-irrigated cropland	Pasture	Cropland	Irrigated cropland	Non-irrigated cropland	Pasture
	Dollars	Dollars	Dollars	Dollars	Dollars	Dollars	Dollars	Dollars
Alabama	54.50	93.00	53.00	21.00	60.50	98.00	59.00	23.00
Arizona	220.00	220.00	....	1.80	215.00	215.00	....	....
Arkansas	102.00	125.00	50.00	18.00	106.00	128.00	54.00	18.00
California	313.00	405.00	32.00	13.50	329.00	425.00	....	14.00
Colorado	74.50	140.00	28.00	4.80	74.50	140.00	28.00	5.00
Connecticut	(S)	(S)	62.00	(S)	....	....	....	....
Delaware	98.50	133.00	82.50	(S)	97.50	....	....	....
Florida	103.00	192.00	54.00	13.00	110.00	200.00	60.00	13.00
Georgia	111.00	195.00	62.00	27.00	110.00	191.00	63.00	28.00
Hawaii	254.00	375.00	140.00	15.00	....	....	....	....
Idaho	151.00	197.00	61.00	12.00	158.00	205.00	65.00	12.00
Illinois	234.00	262.00	233.00	38.00	228.00	240.00	228.00	35.00
Indiana	195.00	260.00	193.00	40.00	197.00	262.00	195.00	....
Iowa	260.00	255.00	260.00	50.00	250.00	....	250.00	50.00
Kansas	62.00	126.00	54.00	17.50	65.50	124.00	58.00	20.00
Kentucky	148.00	215.00	147.00	27.00	148.00	....	147.00	27.00
Louisiana	84.50	104.00	70.00	16.00	90.50	115.00	72.00	17.00
Maine	(S)	(S)	48.00	(S)	....	....	....	....
Maryland	94.50	143.00	92.00	43.50	....	....	95.00	....
Massachusetts	73.00	190.00	62.00	(S)	....	....	....	....
Michigan	123.00	225.00	115.00	30.00	118.00	220.00	110.00	....
Minnesota	186.00	210.00	185.00	26.00	181.00	210.00	180.00	28.00
Mississippi	113.00	135.00	83.00	17.00	116.00	137.00	87.00	19.00
Missouri	132.00	173.00	127.00	29.00	132.00	173.00	127.00	34.00
Montana	32.50	81.00	25.50	5.60	33.00	83.00	26.00	5.80
Nebraska	205.00	262.00	149.00	20.50	206.00	254.00	160.00	28.50
Nevada	150.00	150.00	....	13.00	....	....	....	....
New Hampshire	(S)	(S)	34.00	(S)	....	....	....	....
New Jersey	66.00	100.00	57.00	(S)	67.00	....	....	....
New Mexico	93.00	150.00	17.00	3.00	93.00	....	....	....
New York	55.50	160.00	54.00	20.00	59.50	....	58.00	....
North Carolina	84.50	100.00	84.00	26.00	88.50	102.00	88.00	27.00
North Dakota	68.00	166.00	67.00	16.00	69.00	161.00	68.00	18.00
Ohio	144.00	165.00	144.00	25.00	150.00	....	150.00	....
Oklahoma	33.50	64.00	32.00	12.00	33.50	67.00	32.00	12.00
Oregon	149.00	200.00	90.00	15.00	146.00	200.00	85.00	15.00
Pennsylvania	78.50	130.00	78.00	27.00	82.50	....	82.00	....
Rhode Island	(S)	(S)	(S)	(S)	....	....	....	....
South Carolina	43.00	85.00	40.00	18.00	44.00	91.00	41.00	....
South Dakota	116.00	193.00	114.00	22.00	126.00	186.00	125.00	27.00
Tennessee	101.00	165.00	98.00	20.00	101.00	165.00	98.00	20.00
Texas	38.50	87.00	27.00	6.50	39.00	82.00	29.00	7.50
Utah	68.00	91.00	25.00	5.00	68.50	92.00	....	....
Vermont	(S)	(S)	48.00	21.00	....	....	....	....
Virginia	56.00	100.00	54.00	21.00	57.50	95.00	56.00	21.00
Washington	205.00	340.00	74.00	9.00	208.00	345.00	75.00	....
West Virginia	40.00	....	40.00	13.00	40.00	....	40.00	....
Wisconsin	135.00	240.00	130.00	34.00	134.00	225.00	130.00	34.00
Wyoming	61.00	91.00	14.00	5.00	....	....	....	....

(S) Not published due to insufficient reports.
NASS, Environmental, Economics, and Demographics Branch, (202) 720–6146.

Table 9-16.—Farm assets and claims: Comparative balance sheet of the farming sector, excluding operator households, United States, Dec. 31, 2006–2015

Item	2006	2007	2008	2009	2010
Assets	Billion dollars	Billion dollars	Billion dollars	Billion dollars	Billion dollars
Physical assets:					
Real estate	1,536.6	1,549.0	1,568.6	1,560.6	1,660.1
Non-real estate:					
Livestock and poultry [1]	86.7	89.4	86.2	80.3	90.7
Machinery and motor vehicles [2]	155.3	194.8	189.1	191.6	207.6
Crops [3]	41.1	54.2	58.8	53.4	57.2
Purchased inputs	11.7	16.0	19.5	18.7	19.8
Financial	70.1	55.6	78.8	67.0	129.6
Total	1,906.7	1,961.9	2,005.9	1,976.9	2,170.8
Claims					
Liabilities:					
Real estate	113.4	131.7	147.9	146.0	154.1
Non-real estate	102.3	109.0	113.2	122.4	124.9
Total liabilities [4]	215.7	240.7	261.1	268.3	278.9
Equity	1,691.0	1,721.2	1,744.8	1,708.6	1,891.9
Ratio:	Percent	Percent	Percent	Percent	Percent
Debt/equity	12.8	14.0	15.0	15.7	14.7
Debt/assets	11.3	12.3	13.0	13.6	12.8

Item	2011	2012	2013	2014	2015
Assets	Billion dollars	Billion dollars	Billion dollars	Billion dollars	Billion dollars
Physical assets:					
Real estate	1,802.7	2,073.7	2,251.0	2,383.2	2,395.7
Non-real estate:					
Livestock and poultry [1]	101.4	95.7	110.8	127.9	118.2
Machinery and motor vehicles [2]	223.8	244.0	247.3	245.7	239.5
Crops [3]	62.5	64.8	60.5	61.7	52.5
Purchased inputs	21.0	21.5	20.6	20.1	16.4
Financial	101.5	131.9	78.2	102.8	80.2
Total	2,318.7	2,638.2	2,776.1	2,949.2	2,910.0
Claims					
Liabilities:					
Real estate	167.2	173.4	185.2	196.8	208.8
Non-real estate debt to	127.3	124.2	130.2	148.4	148.0
Total liabilities [4]	294.5	297.5	315.3	345.2	356.7
Equity	2,024.2	2,340.7	2,460.8	2,6,04.0	2,553.2
Ratio:	Percent	Percent	Percent	Percent	Percent
Debt/equity	14.5	12.7	12.8	13.3	14.0
Debt/assets	12.7	11.3	11.4	11.7	12.3

Note: Data as of February 7, 2017. [1] The U.S. total exceeds the sum of the states because NASS does not release state data for some minor producing states due to disclosure issues.   Horses and mules are excluded.   [2] Includes only farm share value for trucks and autos.   [3] All non-CCC crops held on farms plus the value above loan rate for crops held under CCC.   [4] Excludes debt for nonfarm purposes.
ERS, Farm Income Team. http://www.ers.usda.gov/data-products/farm-income-and-wealth-statistics.aspx.

Table 9-17.—Farm labor: Number of workers on farms and average wage rates, United States, 2007–2015

Year	Hired workers [1]	Wage rate [2]
	Thousands	Dollars Per Hour
2007		
Jan	([4])	([4])
Apr	736	10.20
July	843	9.99
Oct	817	10.38
Annual average	746.5	10.23
2008		
Jan	594	10.81
Apr	700	10.57
July	847	10.37
Oct	813	10.72
Annual average	738.5	10.60
2009		
Jan	603	10.93
Apr	694	10.86
July	892	10.66
Oct	829	10.93
Annual average	754.5	10.83
2010		
Jan	612	11.08
Apr	746	10.82
July	885	10.79
Oct	827	11.13
Annual average	767.5	10.95
2011		
Jan	603	11.30
Apr	([4])	([4])
July	834	10.93
Oct	828	11.15
Annual average	748.8	11.07
2012		
Jan	575	11.52
Apr	748	11.41
July	906	11.36
Oct	876	11.75
Annual average	776.3	11.51
2013		
Jan	596	12.02
Apr	736	11.89
July	906	11.68
Oct	871	11.96
Annual average	777.3	11.87
2014		
Jan	540	12.23
Apr	690	12.01
July	838	11.96
Oct	782	12.12
Annual average	712.5	12.07
2015		
Jan	549	12.53
Apr	687	12.28
July	872	12.47
Oct	841	12.82
Annual average		12.54

[1] Includes all workers, other than agricultural service workers, who were paid for at least one hour of agricultural work on a farm or ranch during the survey week. [2] Applies to hired workers only (excludes pay for agricultural service workers). [3] Annual average not computed. [4] The Farm Labor Survey was not conducted for this quarter.
NASS, Economic, Environmental and Demographics Branch, (202) 720–6146.

Table 9-18.—Farm labor: Number of hired workers on farms and average wage rates, by regions and United States, 2015[1]

Region[2]	Workers on farms	Farm wage rates			
	Hired	Type of worker			
		Field	Livestock	Field and livestock	All hired workers[3]
	Thousands	Dollars per hour	Dollars per hour	Dollars per hour	Dollars per hour
January 11–17, 2015					
Northeast I	23	11.87	11.15	11.45	12.67
Northeast II	22	12.64	11.23	11.85	12.97
Appalachian I	23	11.13	11.56	11.30	12.28
Appalachian II	20	10.82	10.48	10.65	11.80
Southeast	21	10.80	9.89	10.45	11.18
Florida	44	10.65	11.65	10.74	11.59
Lake	36	12.50	11.86	12.00	12.87
Cornbelt I	23	12.33	11.06	11.55	12.79
Cornbelt II	21	11.47	12.04	11.80	12.68
Delta	17	11.29	10.55	11.00	11.59
Northern Plains	28	14.58	13.41	13.80	14.54
Southern Plains	45	11.60	11.04	11.25	11.54
Mountain I	18	12.28	11.54	11.70	12.41
Mountain II	11	11.52	12.73	12.30	13.21
Mountain III	19	10.16	12.29	10.95	11.87
Pacific	40	11.48	12.45	11.65	12.55
California	132	11.30	12.30	11.48	12.90
Hawaii	6	12.30	13.80	12.47	14.27
United States	549	11.40	11.69	11.52	12.53
April 12–18, 2015					
Northeast I	31	11.40	11.06	11.25	12.39
Northeast II	35	11.56	11.39	11.50	12.34
Appalachian I	33	10.88	11.11	10.95	11.72
Appalachian II	26	10.82	9.84	10.30	11.35
Southeast	27	10.43	10.07	10.30	10.92
Florida	39	10.40	11.75	10.51	11.27
Lake	47	11.85	12.03	11.95	12.67
Cornbelt I	36	11.98	11.33	11.75	12.86
Cornbelt II	25	10.92	11.79	11.30	12.08
Delta	25	10.52	10.46	10.50	10.86
Northern Plains	35	14.13	13.09	13.50	14.08
Southern Plains	53	11.27	11.15	11.20	11.48
Mountain I	27	11.65	12.12	11.90	12.37
Mountain II	15	11.14	11.36	11.25	12.10
Mountain III	18	10.08	12.42	11.05	12.03
Pacific	50	11.46	12.01	11.55	12.27
California	159	11.60	11.95	11.65	12.81
Hawaii	6	12.30	14.15	12.50	14.18
United States	687	11.36	11.58	11.43	12.28

See footnote(s) at end of table.

Table 9-18.—Farm labor: Number of hired workers on farms and average wage rates, by regions and United States, 2015[1]—Continued

Region[2]	Workers on farms	Type of worker			
	Hired	Field	Livestock	Field and livestock	All hired workers[3]
	Thousands	Dollars per hour	Dollars per hour	Dollars per hour	Dollars per hour
July 12–18, 2015					
Northeast I	50	11.81	11.38	11.65	12.41
Northeast II	50	11.54	11.39	11.50	12.13
Appalachian I	48	10.23	10.95	10.35	10.85
Appalachian II	30	10.97	10.72	10.90	11.57
Southeast	30	10.53	11.15	10.65	11.05
Florida	32	10.65	11.55	10.76	12.21
Lake	80	11.76	11.61	11.70	12.56
Cornbelt I	43	12.08	11.95	12.05	13.07
Cornbelt II	25	12.27	12.68	12.45	12.89
Delta	34	10.67	11.01	10.75	11.11
Northern Plains	37	13.66	13.31	13.50	13.96
Southern Plains	55	10.77	11.32	11.05	11.64
Mountain I	27	11.95	11.41	11.70	11.93
Mountain II	21	11.12	10.56	10.90	11.38
Mountain III	19	11.26	11.72	11.45	12.19
Pacific	104	12.62	12.30	12.60	12.97
California	180	12.05	13.10	12.20	13.25
Hawaii	7	12.50	14.95	12.69	14.51
United States	872	11.73	11.80	11.75	12.47
October 11–17, 2015					
Northeast I	49	12.25	12.24	12.25	12.97
Northeast II	39	12.15	11.48	11.90	12.67
Appalachian I	41	10.65	11.13	10.75	11.35
Appalachian II	31	11.34	10.96	11.25	11.91
Southeast	33	10.72	11.34	10.85	11.22
Florida	37	10.75	11.50	10.83	12.10
Lake	73	12.53	12.20	12.40	13.28
Cornbelt I	54	12.63	11.68	12.45	13.12
Cornbelt II	34	12.82	12.48	12.70	13.03
Delta	35	10.57	10.89	10.65	10.95
Northern Plains	40	14.59	13.86	14.30	14.72
Southern Plains	45	10.91	11.25	11.10	11.87
Mountain I	28	11.79	11.59	11.70	11.98
Mountain II	18	11.49	10.64	11.15	11.70
Mountain III	20	11.05	11.86	11.35	11.95
Pacific	80	14.12	12.41	14.00	14.39
California	177	11.85	13.15	12.05	13.16
Hawaii	7	12.70	14.70	12.85	14.68
United States	841	12.11	12.02	12.09	12.82

[1] Includes all workers, other than agricultural service workers, who were paid for at least one hour of agricultural work on a farm or ranch during the survey week. [2] Multi-state regions consist of the following: Northeast I: Connecticut, Maine, Massachusetts, New Hampshire, New York, Rhode Island, Vermont; Northeast II: Delaware, Maryland, New Jersey, Pennsylvania; Appalachian I: North Carolina, Virginia; Appalachian II: Kentucky, Tennessee, West Virginia; Southeast: Alabama, Georgia, South Carolina; Lake: Michigan, Minnesota, Wisconsin; Cornbelt I: Illinois, Indiana, Ohio; Cornbelt II: Iowa, Missouri; Delta: Arkansas, Louisiana, Mississippi; Northern Plains: Kansas, Nebraska, North Dakota, South Dakota; Southern Plains: Oklahoma, Texas; Mountain I: Idaho, Montana, Wyoming; Mountain II: Colorado, Nevada, Utah; Mountain III: Arizona, New Mexico; Pacific: Oregon, Washington. [3] Includes field workers, livestock workers, supervisors, and others, excluding agricultural service workers.
NASS, Economic, Environmental and Demographics Branch, (202) 720–6146.

Table 9-19.—Farm production and output: Index numbers of total output, and production of livestock, crops, and secondary output, by groups, United States, 2000–2009

[2005=100]

Year	Total farm output	Livestock and products			
		All livestock and products [1]	Meat animals [2]	Dairy products [3]	Poultry and eggs [4]
2000	0.9662	0.9764	1.0143	0.9446	0.9040
2001	0.9681	0.9743	1.0169	0.9334	0.9187
2002	0.9509	0.9862	1.0022	0.9608	0.9555
2003	0.9730	1.0014	1.0022	0.9624	1.0088
2004	1.0128	0.9803	0.9957	0.9652	0.9680
2005	1.0000	1.0000	1.0000	1.0000	1.0000
2006	0.9882	1.0244	1.0147	1.0277	1.0299
2007	1.0200	1.0252	1.0186	1.0496	1.0388
2008	1.0326	1.0323	1.0108	1.0744	1.0552
2009	1.0590	1.0227	1.0128	1.0709	1.0068

Year	Crops				
	All crops	Food Grains	Feed crops	Oil crops [5]	Vegetables and melons
2000	0.9525	1.0127	0.9485	0.8783	1.0110
2001	0.9503	0.9330	0.9303	0.9345	0.9761
2002	0.9114	0.8004	0.8779	0.8768	1.0053
2003	0.9486	1.0662	0.9633	0.7998	0.9828
2004	1.0399	1.0291	1.0616	0.9957	1.0299
2005	1.0000	1.0000	1.0000	1.0000	1.0000
2006	0.9482	0.8626	0.9388	1.0066	0.9806
2007	1.0231	0.9613	1.0997	0.8578	0.9834
2008	1.0450	1.1408	1.0492	0.9440	0.9735
2009	1.1012	1.0504	1.1018	1.0469	0.9584

Year	Crops		Farm-related output [5]
	Fruits and nuts	Other crops	
2000	1.0808	0.9098	1.0285
2001	1.0544	0.9190	1.0999
2002	1.0603	0.9425	1.0616
2003	1.0696	0.9627	0.9921
2004	1.1021	1.0150	1.0394
2005	1.0000	1.0000	1.0000
2006	1.0910	0.8320	1.0606
2007	1.1719	1.0426	0.9241
2008	1.2539	1.1742	0.8838
2009	1.2241	1.4187	0.8416

[1] Includes wool, mohair, horses, mules, honey, beeswax, bees, goats, rabbits, aquaculture, and fur animals. These items are not included in the separate groups of livestock and products shown. [2] Cattle and calves, sheep and lambs, and hogs. [3] Butter, butterfat, wholesale milk, retail milk, and milk consumed on farms. [4] Chicken eggs, commercial broilers, chickens, and turkeys. [5] These activities are defined as activities closely linked to agriculture for which information on production and input use cannot be separately observed.
ERS, Agricultural Structure and Productivity Branch, (202) 694–5460, (202) 694–5601.

Table 9-20.—Hired farmworkers: Number of Workers and Median Weekly Earnings, 2013–2015 [1]

Characteristics	Workers			Median Weekly Earnings [2]		
	2013	2014	2015	2013	2014	2015
	Thousands	Thousands	Thousands	Dollars	Dollars	Dollars
All workers	759	828	877	400	400	440
15–19 years old	78	72	65	200	184	192
20–24 years old	112	101	88	362	400	368
25–34 years old	195	203	226	468	450	462
35–44 years old	145	161	192	450	438	441
45–54 years old	122	142	157	461	462	464
55 years old and older	107	149	148	478	470	450
Male	615	670	701	440	447	461
Female	144	158	176	330	360	360
White [3]	359	404	410	460	461	480
Black and other races [3]	45	44	46	*	*	*
Hispanic	354	380	421	400	400	420
Schooling completed						
Less than 5th grade	71	69	68	380	360	420
5th-8th grade	127	155	162	368	400	400
9th-12th grade (no diploma) ..	140	156	171	350	380	400
High school diploma	242	238	256	450	440	440
Beyond high school	179	211	220	500	550	600
Full-time (35 or more hours per week) [4]	619	681	734	462	455	480
Part-time (less than 35 hours per week) [4]	138	146	140	162	201	208

* Insufficient number of reports to publish data.　[1] Represents annual average number of persons 15 years old and over in the civilian noninstitutional population who were employed as hired farm managers, supervisors, or laborers.　Employment estimates based on 12 monthly Current Population Survey microdata files.　Earnings estimates based on 12 monthly Current Population Survey Outgoing Rotation Group microdata files.　[2] "Median weekly earnings" is the earnings value that divides farmworkers into two equal-sized groups, one group having earnings above the median and the other group having earnings below the median.　"Earnings" refers to the weekly earnings the farmworker usually earns at a farmwork job, before deductions, and includes any overtime pay or commissions.　[3] Excludes persons of Hispanic origin.　[4] The sum of full-time and part-time workers will not equal the total because usual hours worked varies for some individuals.
ERS, Rural Economy Branch, (202) 694–5416.

Table 9-21.—Crops: Area, United States, 2006–2015

Year	Principal crops			Area planted total [3]	Commercial vegetables, harvested area	Fruits and nuts, bearing area [4]
	Area harvested					
	Feed grains [1]	Food grains [2]	Total [3]			
	1,000 acres	1,000 acres	1,000 acres	1,000 acres	1,000 acres	1,000 acres
2006	80,090	49,895	294,453	315,645	3,083.2	3,929.6
2007	98,318	53,999	304,376	320,369	3,033.5	3,904.5
2008	91,071	59,298	309,225	325,632	2,935.3	3,987.1
2009	89,446	53,195	301.172	319,022	2,946.6	4,050.5
2010	89,984	50,768	303,619	315,414	2,827.2	4,091.5
2011	90,940	48,543	292,652	314,343	2,682.0	4,132.6
2012	96,639	51,687	306,855	324,304	2,762.1	4,113.1
2013	98,085	48,079	303,717	324,869	2,681.2	4,256.8
2014	93,009	49,558	308,243	326,436	2,672.9	4,301.4
2015	....	....	304,706	318,975	....	....

[1] Corn for grain, oats, barley, and sorghum for grain.　[2] Wheat, rye, and rice.　[3] Crops included in area planted and area harvested are corn, sorghum, oats, barley, winter wheat, rye, durum wheat, other spring wheat, rice, soybeans, peanuts, sunflower, cotton, dry edible beans, potatoes, canola, proso millet, and sugarbeets. Harvested acreage for all hay, tobacco, and sugarcane are used in computing total area planted.　[4] Includes the following fruits and nuts: Citrus fruits—oranges, tangerines, Temples, grapefruit, lemons and tangelos; area is for the year of harvest; deciduous fruits—commercial apples, peaches, pears, grapes, cherries, plums, prunes, apricots, bananas, nectarines, figs, kiwifruit, olives, avocados, papayas, dates, berries, guavas, cranberries, pineapples (discontinued as of 2007 crop) and strawberries; nuts—almonds, hazelnuts, macadamias, pistachios, and walnuts.
NASS, Crops Branch, (202) 720–2127.

Table 9-22.—Crops: Area harvested and yield, United States, 2014–2015

Crop	Area harvested		Yield per acre		
	2014	2015	Unit	2014	2015
	1,000 acres	*1,000 acres*			
Grains and hay:					
Barley	2,497	3,158	Bushels	72.7	69.1
Corn for grain	83,136	80,753	Bushels	171.0	168.4
Corn for silage	6,371	6,237	Tons	20.1	20.4
Hay, all	57,062	54,447	Tons	2.45	2.47
Alfalfa	18,395	17,778	Tons	3.34	3.32
All other	38,667	36,669	Tons	2.03	2.06
Oats	1,035	1,276	Bushels	67.9	70.2
Proso millet	430	418	Bushels	31.4	33.9
Rice	2,933	2,585	Pounds	7,576	7,472
Rye	258	365	Bushels	27.9	31.8
Sorghum for grain	6,401	7,851	Bushels	67.6	76.0
Sorghum for silage	315	306	Tons	13.1	14.6
Wheat, all	46,385	47,318	Bushels	43.7	43.6
Winter	32,299	32,346	Bushels	42.6	42.5
Durum	1,346	1,911	Bushels	40.2	44.0
Other spring	12,740	13,061	Bushels	46.7	46.2
Oilseeds:					
Canola	1,556.7	1,713.5	Pounds	1,614	1,680
Cottonseed	(X)	(X)	Tons	(X)	(X)
Flaxseed	302	456	Bushels	21.1	22.1
Mustard Seed	31.2	40.1	Pounds	930	671
Peanuts	1,322.5	1,560.9	Pounds	3,923	3,845
Rapeseed	2.1	1.1	Pounds	1,233	1,382
Safflower	170.2	161.1	Pounds	1,226	1,356
Soybeans for beans	82,591	81,732	Bushels	47.5	48.0
Sunflower	1,510.1	1,799.4	Pounds	1,469	1,625
Cotton, tobacco and sugar crops:					
Cotton, all	9,346.8	8,074.9	Pounds	838	766
Upland	9,157.0	7,920.0	Pounds	826	755
American Pima	189.8	154.9	Pounds	1,432	1,342
Sugarbeets	1,146.3	1,145.4	Tons	27.3	30.9
Sugarcane	868.5	874.7	Tons	35.0	36.7
Tobacco	378.4	328.7	Pounds	2,316	2,188
Dry beans, peas and lentils:					
Austrian winter peas	16.8	21.0	Cwt	1,381	1,238
Dry edible beans	1,648.6	1,708.4	Cwt	1,754	1,759
Dry edible peas	899.5	1,083.5	Cwt	1,907	1,687
Lentils	259.0	475.0	Cwt	1,331	1,108
Wrinkled seed peas	(NA)	(NA)	Cwt	(NA)	(NA)
Potatoes and miscellaneous:					
Hops	38.0	43.6	Pounds	1,868	1,807
Peppermint oil	63.5	65.3	Pounds	90	90
Potatoes, all	1,051.1	1,054.4	Cwt	421	418
Spring	71.1	72.5	Cwt	318	286
Summer	48.9	47.1	Cwt	324	334
Fall	931.1	934.8	Cwt	434	433
Spearmint oil	24.4	25.8	Pounds	114	114
Sweet potatoes	135.2	153.1	Cwt	219	203
Taro (Hawaii)	0.4	0.3	Pounds	9,000	10,300

(NA) Not available.　　(X) Not applicable.
NASS, Crops Branch, (202) 720–2127.

Table 9-23.—Crops: Production and value, United States, 2014–2015

Crop	Unit	Production 2014	Production 2015	Value of production 2014	Value of production 2015
		Thousands	*Thousands*	*1,000 dollars*	*1,000 dollars*
Grains and hay:					
Barley	Bushels	181,542	218,187	914,955	1,149,831
Corn for grain	Bushels	14,215,532	13,601,964	52,951,760	49,339,261
Corn for silage	Bushels	128,048	127,311	....	....
Hay, all	Tons	139,923	134,502	19,099,238	16,548,834
Alfalfa	Tons	61,451	58,974	10,569,103	8,471,797
All other	Tons	78,472	75,528	8,530,135	8,077,037
Oats	Bushels	70,232	89,535	241,472	214,575
Proso millet	Bushels	13,483	14,159	45,104	41,547
Rice	Cwt	222,215	193,148	3,075,618	2,421,955
Rye	Bushels	7,189	11,616	55,639	75,497
Sorghum for grain	Bushels	432,575	596,751	1,721,330	2,064,638
Sorghum for silage	Bushels	4,123	4,475	....	....
Wheat, all	Bushels	2,026,310	2,061,939	11,914,954	10,018,323
Winter	Bushels	1,377,216	1,374,690	8,036,108	6,555,216
Durum	Bushels	54,056	84,009	482,417	615,160
Other Spring	Bushels	595,038	603,240	3,396,429	2,847,947
Oilseeds:					
Canola	Pounds	2,512,645	2,878,470	423,382	448,552
Cottonseed	Tons	5,125.0	4,043.0	1,015,607	932,894
Flaxseed	Bushels	6,368	10,095	75,077	90,561
Mustard Seed	Pounds	29,004	26,927	10,089	8,567
Peanuts	Pounds	5,188,665	6,001,357	1,158,251	1,160,560
Rapeseed	Pounds	2,590	1,520	903	656
Safflower	Pounds	208,643	218,451	52,207	53,515
Soybeans for beans	Bushels	3,927,090	3,926,339	39,474,861	35,192,058
Sunflower	Pounds	2,219,050	2,923,730	497,775	574,156
Cotton, tobacco and sugar crops:					
Cotton, all	Bales	16,319.4	12,888.0	5,147,241	3,988,978
Upland	Bales	15,753.0	12,455.0	4,727,124	3,743,765
American Pima	Bales	566.4	433.0	420,117	245,213
Sugarbeets	Tons	31,285	35,371	1,440,068	1,667,874
Sugarcane	Tons	30,424	32,122	1,054,657	995,122
Tobacco [1]	Pounds	876,415	719,171	1,835,308	1,440,645
Dry Beans, peas and lentils:					
Austrian winter peas	Cwt	232	260	5,138	6,092
Dry edible beans	Cwt	28,910	30,057	980,622	866,222
Dry edible peas	Cwt	17,155	18,283	206,194	458,780
Lentils	Cwt	3,446	5,263	86,404	324,868
Wrinkled seed peas	Cwt	618	384	17,208	11,347
Potatoes and miscellaneous:					
Hops	Pounds	70,995.9	78,846.0	260,627	345,388
Peppermint oil	Pounds	5,719	5,888	131,364	128,610
Potatoes, all	Cwt	442,170	441,205	3,928,211	3,865,538
Spring	Cwt	22,608	20,770	....	....
Summer	Cwt	15,859	15,734	....	....
Fall	Cwt	403,703	404,701	....	....
Spearmint oil	Pounds	2,784	2,930	54,977	53,664
Sweet potatoes	Cwt	29,584	31,016	706,916	675,583
Taro (Hawaii)	Pounds	3,240	3,502	1,944	2,381

[1] Excludes estimated 2012 value of production for Connecticut Massachusetts.
NASS, Crops Branch, (202) 720–2127.

Table 9-24.—Fruits and nuts: Bearing acreage and yield, United States, 2014–2015

Crop	Bearing acreage		Yield per acre		
	2014	2015	Unit	2014	2015
	Acres	Acres			
Noncitrus fruits:					
Apples	318,180	318,180	Tons	18.60	15.80
Apricots	10,820	10,820	Tons	6.00	4.33
Avocados	59,000	59,000	Tons	2.94	3.78
Bananas (Hawaii) [1]	860	860	Tons	6.98	7.25
Blackberries (Oregon) [1] [2]	6,100	6,100	Tons	3.69	3.68
Blueberries					
Cultivated [1]	84,750	84,750	Tons	3.34	3.08
Wild (Maine) [3]	22,800	22,800	Tons	2.29	2.25
Boysenberries (Oregon) [1]	500	500	Tons	2.55	3.50
All (California)	89,700	89,700	Tons	4.05	3.77
Cherries, Sweet	37,850	37,850	Tons	4.02	3.32
Cherries, Tart	7,800	7,800	Tons	2.36	2.65
Cranberries	40,600	40,600	Tons	10.30	10.50
Dates (California)	10,000	10,000	Tons	3.34	4.36
Figs (California)	7,000	7,000	Tons	4.77	4.44
Grapes	1,034,400	1,034,400	Tons	7.62	7.51
Guava (Hawaii) [1]	100	100	Tons	8.50	(NA)
Kiwifruit (California)	4,100	4,100	Tons	6.95	5.93
Nectarines	22,500	22,500	Tons	9.12	8.30
Olives (California)	37,000	37,000	Tons	2.57	4.97
Papayas (Hawaii) [1]	1,500	1,500	Tons	7.83	9.10
Peaches	102,540	102,540	Tons	8.32	8.49
Pears	49,300	49,300	Tons	16.90	16.80
Plums (California)	18,000	18,000	Tons	6.28	5.96
Prunes, dried (California)	48,000	48,000	Tons	6.75	6.79
Prunes and plums [4]	2,500	2,500	Tons	5.92	4.00
Raspberries [1]					
Black (Oregon)	1,950	1,950	Tons	2.78	2.69
Red	17,600	17,600	Tons	6.17	6.77
Strawberries [1]	59,895	59,895	Tons	25.20	26.60
Tree nuts:					
Almonds (California) [5]	870,000	860,000	Tons	1.78	1.83
Hazelnuts (Oregon) [5]	30,000	30,000	Tons	1.20	1.03
Macadamia (Hawaii) [5]	16,000	16,000	Tons	1.44	1.47
Pecans [2]	(NA)	(NA)	Tons	(NA)	(NA)
Pistachios (California) [5]	221,000	233,000	Tons	1.16	0.58
Walnuts (California) [5]	290,000	300,000	Tons	1.97	2.01
Citrus fruits:					
Oranges [6]	575,900	595,700	Boxes	255	249
Grapefruit [6]	67,300	64,100	Boxes	326	303
Lemons [6]	55,300	54,500	Boxes	409	408
Tangelos (Florida) [6]	3,000	2,500	Boxes	222	156
Tangerines and Mandarins [6]	63,400	65,300	Boxes	333	354

(NA) Not available.   [1] Harvested acreage.   Yield based on utilized production.   [2] Cultivated.   [3] Bearing acreage and yield not estimated.   [4] Idaho, Michigan, Oregon, and Washington.   [5] Yield based on in-shell basis.   [6] Crop year begins with bloom in one year and ends with completion of harvest the following year.   Citrus production is for the the year of harvest.

NASS, Crops Branch, (202) 720–2127.

Table 9-25.—Fruits and nuts: Production and value, United States, 2014–2015

Crop	Total production		Value of utilized production	
	2014	2015	2014	2015
	1,000 tons [1]	*1,000 tons* [1]	*1,000 dollars*	*1,000 dollars*
Noncitrus fruits:				
Apples	5,907.0	5,002.0	2,870,745	3,394,185
Apricots	64.9	41.7	53,096	41,730
Avocados	173.3	224.0	326,526	295,797
Bananas (Hawaii) [2]	6.0	6.0	9,840	10,866
Blackberries (Oregon)	22.7	25.6	50,133	38,036
Blueberries				
Cultivated	289.4	280.0	825,759	811,992
Wild (Maine)	52.2	50.6	63,480	47,180
Boysenberries (Oregon)	1.3	1.2	3,541	2,743
Cherries, Sweet	363.6	338.4	766,551	758,915
Cherries, Tart	152.1	126.3	106,810	87,037
Cranberries	18.4	18.3	254,412	267,527
Dates (California)	420.0	428.2	50,434	68,016
Figs (California)	33.4	43.6	23,998	21,853
Grapes	33.4	30.2	5,821,629	5,561,719
Guava (Hawaii) [2]	7,883.8	7,677.2	306	
Kiwifruit (California)	0.9	(NA)	33,333	30,893
Nectarines	28.5	23.7	178,421	150,413
Olives (California)	205.3	167.7	73,559	160,043
Papayas (Hawaii) [2]	95.0	179.0	11,285	10,570
Peaches	11.8	13.7	629,524	605,794
Pears	852.9	847.2	467,194	500,416
Plums (California)	831.6	820.5	103,167	104,760
Prunes, dried (California)	113.0	106.0	266,760	221,100
Prunes and plums [3]	324.0	319.0	7,060	5,337
Raspberries	14.8	9.7	527,389	580,924
Strawberries [2]	(NA)	(NA)	2,821,854	2,239,301
Nuts:				
Almonds, shelled (California)	(NA)	(NA)	7,388,000	5,325,000
Hazelnuts, in-shell (Oregon)	(NA)	(NA)	129,600	86,800
Macadamia, in-shell (Hawaii)	(NA)	(NA)	40,020	45,590
Pecans, in-shell	(NA)	(NA)	516,591	560,216
Pistachios, in-shell (California)	(NA)	(NA)	1,834,980	669,600
Walnuts, in-shell (California)	(NA)	(NA)	1,907,140	976,860
Citrus fruits [2] [4]:				
Grapefruit	910	803	216,258	251,036
Lemons	904	890	696,835	734,209
Oranges	6,353	5,911	1,963,353	1,711,100
Tangelos (Florida)	30	18	9,221	8,672
Tangerines and mandarins	863	935	468,083	637,412

(NA) Not available.　[1] Tons refers to the 2,000 lb. short Tons.　[2] Only utilized production estimated.　[3] Production is shelled basis.　[4] Value of production is packinghouse-door equivalent.
NASS, Crops Branch, (202) 720–2127.

Table 9-26.—Vegetables, commercial: Area harvested and yield, United States, 2014–2015

Crop	Area harvested			Yield per harvested acre	
	2014	2015	Unit	2014	2015
	Acres	*Acres*			
Fresh market:					
Artichokes [1]	7,300	6,500	Cwt	130	135
Asparagus [1]	23,800	23,500	Cwt	31	29
Beans, snap	68,600	71,170	Cwt	54	56
Broccoli [1]	126,400	123,400	Cwt	163	176
Cabbage	59,650	55,920	Cwt	354	361
Cantaloups	58,200	51,600	Cwt	234	260
Carrots	74,200	71,550	Cwt	342	345
Cauliflower [1]	35,070	38,110	Cwt	181	176
Celery [1]	28,900	30,100	Cwt	636	618
Corn, sweet	211,050	229,090	Cwt	117	122
Cucumbers	37,570	37,980	Cwt	184	177
Garlic [1]	23,800	23,600	Cwt	163	162
Honeydew melons	14,450	13,600	Cwt	259	262
Lettuce					
Head	125,500	119,000	Cwt	366	362
Leaf	54,100	47,800	Cwt	240	258
Romaine	85,400	84,400	Cwt	289	303
Onions [1]	139,850	132,900	Cwt	499	503
Peppers, bell [1]	44,000	43,800	Cwt	353	371
Peppers, chile [1]	19,100	18,100	Cwt	242	223
Pumpkins [1]	49,100	40,900	Cwt	265	184
Spinach	37,540	41,190	Cwt	158	148
Squash [1]	39,330	38,690	Cwt	153	155
Tomatoes	97,600	94,300	Cwt	280	286
Watermelons	111,350	115,750	Cwt	299	305
Processing:					
Beans, lima	27,760	29,200	Tons	1.82	1.72
Beans, snap	159,070	158,920	Tons	4.28	4.81
Carrots	11,430	11,020	Tons	28.16	26.68
Corn, sweet	312,280	307,500	Tons	8.22	8.09
Cucumbers	84,430	85,110	Tons	6.36	6.27
Peas, green	187,600	166,200	Tons	2.17	2.30
Spinach	9,100	7,200	Tons	9.92	10.44
Tomatoes	306,100	310,600	Tons	47.82	47.50

[1] Includes processing total for dual usage crops.
NASS, Crops Branch, (202) 720–2127.

Table 9-27.—Vegetables, commercial: Production and value, United States, 2014–2015

Crop	Unit	Production		Value of utilized production	
		2014	2015	2014	2015
		Thousands	*Thousands*	*1,000 dollars*	*1,000 dollars*
Fresh market:					
Artichokes [1]	Cwt	949	918	54,662	80,600
Asparagus [1]	Cwt	743	629	73,782	67,864
Beans, snap	Cwt	3,720	3,952	224,052	234,934
Broccoli [1]	Cwt	20,600	21,888	824,961	1,058,770
Cabbage	Cwt	21,141	20,178	416,303	389,340
Cantaloups	Cwt	13,612	13,552	300,634	264,131
Carrots	Cwt	25,379	25,740	688,573	784,265
Cauliflower [1]	Cwt	6,361	6,559	316,937	401,747
Celery [1]	Cwt	18,393	17,110	314,134	424,664
Corn, sweet	Cwt	24,649	27,952	708,332	787,581
Cucumbers	Cwt	6,895	6,725	168,038	175,762
Garlic [1]	Cwt	3,868	4,089	266,265	321,006
Honeydew melons	Cwt	3,739	3,769	88,649	85,343
Lettuce					
Head	Cwt	45,918	43,108	1,119,736	1,253,150
Leaf	Cwt	12,984	14,233	491,602	850,238
Romaine	Cwt	24,679	26,335	763,887	1,046,571
Onions [1]	Cwt	69,815	67,184	892,325	1,039,325
Peppers, bell [1]	Cwt	15,536	15,155	626,089	732,699
Peppers, chile [1]	Cwt	4,625	4,034	216,107	135,743
Pumpkins [1]	Cwt	13,019	7,538	143,158	90,214
Spinach	Cwt	5,919	6,076	259,700	343,575
Squash [1]	Cwt	6,017	6,016	187,820	174,259
Tomatoes	Cwt	27,280	26,375	1,133,070	1,221,270
Watermelons	Cwt	33,263	35,475	450,324	488,383
Processing:					
Beans, lima	Tons	50,400	50,255	29,236	29,236
Beans, snap	Tons	681,240	764,900	170,066	170,066
Carrots	Tons	321,820	294,000	37,443	37,443
Corn, sweet	Tons	2,567,820	2,488,110	289,573	289,573
Cucumbers	Tons	536,910	533,460	202,705	202,705
Peas, green	Tons	362,860	411,320	133,261	133,261
Spinach	Tons	90,310	75,200	11,312	11,312
Tomatoes	Tons	14,637,300	14,754,350	1,452,823	1,452,823

[1] Includes processing total for dual usage crops.
NASS, Crops Branch, (202) 720–2127.

Table 9-28.—Livestock and livestock products: Production and value, United States, 2013–2015

Product	Production [1]			Value of production		
	2013	2014	2015	2013	2014	2015
	Millions	*Millions*	*Millions*	*1,000 dollars*	*1,000 dollars*	*1,000 dollars*
Eggs	97,555	100,879	94,437	8,678,859	10,257,972	13,499,904
	1,000 pounds	*1,000 pounds*	*1,000 pounds*	*1,000 dollars*	*1,000 dollars*	*1,000 dollars*
Cattle and calves	40,695,232	40,171,339	41,456,124	48,478,687	59,921,694	59,857,570
Hogs	32,620,264	32,011,688	35,127,658	21,666,015	24,152,893	19,283,155
Broilers [2]	50,678,200	51,378,700	53,364,000	30,761,669	32,728,234	28,709,834
Mature chickens	922,546	952,136	978,612	87,939	96,151	100,899
Turkeys [3]	7,277,536	7,217,006	7,038,065	4,839,562	5,304,501	5,707,871
Milk	201,231,000	206,054,00	208,633	40,476,608	49,589,494	35,910,597
Catfish [3][4]	351,314	320,741	327,789	356,734	351,940	361,458
Trout [3][4]	58,212	62,720	60,004	97,414	103,216	104,393
Honey	149,499	178,270	156,544	320,077	387,381	327,177
Wool (shorn)	26,990	26,680	27,015	39,209	38,909	39,205
Mohair	780	880	765	3,300	4,272	4,052

[1] For cattle, sheep, and hogs, the quantity of net production is the live weight actually produced during the year, adjustments having been made for animals shipped in and changes in inventory.   Estimates for broilers and eggs cover the 12-month period Dec. 1, previous year through Nov. 30.   [2] Young chickens of meat–type strains raised for meat production.   [3] Live weight.   [4] Value of fish sold, excludes eggs.
NASS, Livestock Branch, (202) 720–3570.

Table 9-29.—Total farm input: Index numbers of farm input, by major subgroups, United States, 2000–2009

[2005=100]

Year	Total farm input	Capital					Labor		
		All	Durable equipment	Service buildings	Inven-tories	Land	All	Hired labor	Self-employed
2000 ...	1.0231	1.0051	0.9159	1.0838	1.0355	0.9599	1.0720	1.0874	1.0629
2001 ...	1.0168	1.0001	0.9142	1.0708	1.0280	0.9683	1.0708	1.1009	1.0527
2002 ...	1.0073	1.0000	0.9250	1.0501	1.0202	0.9798	1.0813	1.1123	1.0627
2003 ...	1.0048	0.9934	0.9358	1.0331	1.0131	0.9482	1.0446	1.0915	1.0163
2004 ...	0.9890	0.9934	0.9596	1.0153	1.0065	0.9242	1.0066	1.0037	1.0083
2005 ...	1.0000	1.0000	1.0000	1.0000	1.0000	1.0000	1.0000	1.0000	1.0000
2006 ...	0.9823	1.0009	1.0232	0.9826	0.9934	0.9977	0.9399	0.9566	0.9298
2007 ...	1.0332	0.9941	1.0224	0.9636	0.9865	0.9732	0.9612	1.0397	0.9140
2008 ...	0.9937	1.0085	1.0362	1.1512	0.9794	0.9809	0.9351	0.9906	0.9018
2009 ...	0.9966	1.0118	1.0745	1.1378	0.9724	1.0133	0.8927	0.9772	0.8417

Year	Materials				
	All	Farm origin	Energy	Chemicals	Purchased services
2000 ...	1.0121	1.0110	1.1329	0.9276	1.0304
2001 ...	1.0033	0.9791	1.1048	0.9288	1.0539
2002 ...	0.9832	0.9736	1.2022	0.8539	0.9930
2003 ...	0.9941	1.0064	1.0045	0.9856	0.9593
2004 ...	0.9796	0.9901	1.0826	0.9713	0.9523
2005 ...	1.0000	1.0000	1.0000	1.0000	1.0000
2006 ...	0.9924	1.0114	0.9530	0.8769	1.0203
2007 ...	1.0760	1.0452	1.0757	1.0338	1.1189
2008 ...	1.0074	0.9736	0.9804	1.0362	1.0476
2009 ...	1.0254	0.9791	1.1946	1.0616	1.0084

ERS, Agricultural Structure and Productivity Branch, (202) 694–5460, (202) 694–5601.

Table 9-30.—Agricultural productivity: Index numbers (2005=100) of farm output per unit of input, United States, 2000–2009

Year	Productivity [1]
2000	0.9444
2001	0.9521
2002	0.9440
2003	0.9683
2004	1.0240
2005	1.0000
2006	1.0060
2007	0.9872
2008	1.0392
2009	1.0626

[1] Productivity is the output-input ratio.
ERS, Agricultural Structure and Productivity Branch (202) 694–5601, (202) 694–5460.

Table 9-31.—Farm product prices: Marketing year average prices received by farmers; Parity prices for January, United States, 2012–2013

Commodity	Unit	Marketing year average price [1]		Parity price [2]	
		2012	2013	2012	2013
		Dollars	*Dollars*	*Dollars*	*Dollars*
Basic commodities:					
Cotton:					
Upland	pound	.725	.779	1.99	2.12
Extra long staple(American Pima)	pound	1.24	1.72	3.89	3.91
Wheat	bushel	7.77	6.87	17.50	18.70
Rice	cwt	15.10	16.30	40.30	48.50
Corn	bushel	6.89	4.46	11.40	12.80
Peanuts	pound	0.301	0.249	0.718	0.794
Tobacco:					
Flue-cured, types (class 1)	pound	1.983	2.115	5.41	5.44
Types 21	pound	2.578	2.661	7.01	7.47
Burley, type 31	pound	1.968	2.061	5.58	5.53
Maryland, type 32 [3]	pound	1.750	1.900	4.80	4.91
Dark air-cured, types 35–36	pound	(NA)	(NA)	7.78	8.29
Sun-cured, type 37	pound	(NA)	(NA)	6.49	6.91
Pa., seedleaf, type 41	pound	1.95	2.100	5.02	5.17
Cigar binder type 51	pound	6.538	(NA)	17.70	17.90
Puerto Rican filler, type 46	pound	(NA)	(NA)	4.28	4.56
Cigar wrapper, type 61	pound	(NA)	(NA)	79.50	71.10
Designated nonbasic commodities:					
All milk, sold to plants	cwt	18.56	20.11	49.70	49.60
Honey, all	pound	199.2	214.1	3.84	4.15
Wool and mohair:					
Wool [4]	pound	1.520	1.450	2.87	3.15
Mohair [5]	pound	3.87	4.23	9.19	9.88
Other nonbasic commodities:					
Field crops and miscellaneous:					
Austrian winter peas	cwt	20.70	23.50	(NA)	(NA)
Barley	bushel	6.43	6.06	12.00	13.10
Beans, dry edible	cwt	38.00	39.10	83.10	85.30
Cottonseed	ton	252.00	246.00	480.00	504.00
Crude pine gum	barrel	(NA)	(NA)	384.00	407.00
Flaxseed	bushel	13.80	13.80	30.40	31.50
Hay, all, baled	ton	191.00	176.00	362.00	388.00
Hops	pound	3.17	3.35	8.45	8.91
Lentils	cwt	20.70	19.80	(NA)	(NA)
Oats	bushel	3.89	3.75	7.26	7.79
Peas, dry edible	cwt	15.70	14.60	(NA)	(NA)
Peppermint oil	pounds	24.10	23.90	49.40	52.60
Popcorn, shelled basis	cwt	(NA)	(NA)	50.00	53.20
Potatoes	cwt	8.63	9.75	23.60	24.20
Rye	bushel	7.69	7.95	14.50	15.70
Sorghum grain	cwt	11.30	7.64	20.10	22.30
Soybeans	bushel	14.40	13.00	27.60	30.30
Spearmint oil	pound	19.50	18.90	41.40	44.10
Sweet potatoes	cwt	17.40	24.10	60.70	61.20
Fruits:					
Citrus (equiv. on-tree): [6]					
Grapefruit	box	7.65	6.51	19.00	20.50
Lemons	box	15.13	12.05	37.80	38.80
Oranges	box	10.14	8.27	19.30	20.90
Tangelos	box	9.65	10.21	(NA)	(NA)
Tangerines	box	19.16	22.45	41.10	43.80
Temples, Florida	box	(NA)	(NA)	10.30	10.90
Deciduous and other:					
Apples:					
For fresh consumption [7]	pound	0.453	0.303	0.911	0.941
For processing [8]	ton	281.00	197.00	455.00	488.00
Apricots:					
For fresh consumption [9]	ton	1,140.00	1,190.00	3,390.00	3,260.00
Dried, California (dried basis) [8]	ton	2,470.00	2,980.00	7,150.00	7,410.00
For processing (excl dried) [8]	ton	387.00	415.00	1,020.00	1,050.00
Avocados [9]	ton	(NA)	1,970.00	5,630.00	5,360.00

See footnote(s) at end of table.

Table 9-31.—Farm product prices: Marketing year average prices received by farmers; Parity prices for January, United States, 2012–2013—Continued

Commodity	Unit	Marketing year average price [1]		Parity price [2]	
		2012	2013	2012	2013
		Dollars	Dollars	Dollars	Dollars
Deciduous and other—Continued					
Berries for processing:					
Blackberries (Oregon)	pound	0.720	0.740	17.10	18.20
Boysenberries (California & Oregon)	pound	1.03	1.07	2.43	2.59
Gooseberries	pound	(NA)	(NA)	1.02	1.09
Loganberries (Oregon)	pound	(NA)	(NA)	1.66	1.76
Raspberries, black (Oregon)	pound	2.65	(NA)	2.93	3.12
Raspberries, red (Oregon & Washington)	pound	0.745	0.945	2.32	2.47
Cherries:					
Sweet	ton	2,020.00	(NA)	5,930.00	6,030.00
Tart	pound	0.594	(NA)	0.938	0.970
Cranberries [10]	barrel	47.90	(NA)	140.00	138.00
Dates, California [9]	ton	1,340.00	(NA)	4,890.00	4,730.00
Figs, California	ton	555.00	(NA)	(NA)	(NA)
Grapes:					
For all sales	ton	752.00	712.00	(NA)	(NA)
Raisin varieties dried, California (dried basis) [8]	ton	1,880.00	1,630.00	3,450.00	3,910.00
Other dried grapes	ton	435.00	347.00	1,550.00	1,620.00
Kiwi	ton	1,020.00	1,110.00	2,600.00	2,600.00
Nectarines (California):					
For fresh consumption [17]	ton	777.00	(NA)	1,490.00	1,610.00
For processing [18]	ton	(NA)	(NA)	116.00	58.50
Olives (California): [11]					
For all sales	ton	813.00	813.00	(NA)	(NA)
Crushed for oil	ton	550.00	582.00	1,550.00	1,640.00
For all sales (excl crushed)	ton	(NA)	(NA)	2,180.00	2,320.00
For canning	ton	1,110.00	1,110.00	2,470.00	2,580.00
Papayas	pound	(NA)	(NA)	1.21	1.12
Peaches:					
For all sales	ton	647.00	(NA)	(NA)	(NA)
For fresh consumption [7]	ton	967.00	(NA)	2,260.00	2,360.00
Dried, California (dried basis) [8]	ton	327.00	(NA)	1,480.00	1,430.00
For processing California (excl dried):					
Clingstone [11]	ton	348.00	364.00	919.00	929.00
Freestone [8]	ton	242.00	272.00	729.00	738.00
Pears:					
For all sales	ton	509.00	491.00	(NA)	(NA)
For fresh consumption [7]	ton	656.00	599.00	1,530.00	1,660.00
Dried, California (dried basis) [8]	ton	(NA)	(NA)	4,720.00	4,410.00
For processing (excl dried) [8]	ton	238.00	276.00	1,080.00	1,350.00
Plums (California):					
For all sales [9]	ton	695.00	664.00	(NA)	(NA)
For fresh consumption [17]	ton	(NA)	(NA)	1,620.00	1,710.00
For processing [18]	ton	(NA)	(NA)	192.00	253.00
Prunes, dried (California) [8]	ton	421.00	(NA)	4,080.00	4,250.00
Prunes and plums (excl California):					
For fresh consumption [12]	ton	739.00	608.00	1,720.00	1,820.00
For processing (excl dried) [8]	ton	228.00	345.00	696.00	700.00
Strawberries:					
For fresh consumption [13]	cwt	91.80	96.50	254.00	258.00
For processing [8]	cwt	33.40	34.90	97.20	96.70
Sugar crops:					
Sugarbeets	ton	66.40	46.90	145.00	154.00
Sugarcane for sugar	ton	41.90	31.40	95.20	98.50
Tree nuts: [14]					
Almonds	pound	2.58	3.21	5.88	6.20
Hazelnuts	ton	1,830.00	2,180.00	5,410.00	5,620.00
Macadamia	pound	0.800	0.870	(NA)	(NA)
Pecans, all	pound	1.57	1.73	4.91	5.06
Pistachios	pound	2.61	3.48	5.41	5.70
Walnuts	ton	3,030.00	3,710.00	4,940.00	5,440.00

See footnote(s) at end of table.

Table 9-31.—Farm product prices: Marketing year average prices received by farmers; Parity prices for January, United States, 2012–2013—Continued

Commodity	Unit	Marketing year average price [1]		Parity price [2]	
		2012	2013	2012	2013
		Dollars	Dollars	Dollars	Dollars
Vegetables for fresh market: [13]					
Artichokes, California	cwt	54.30	61.00	126.00	134.00
Asparagus	cwt	110.00	121.00	326.00	326.00
Broccoli	cwt	33.80	43.20	111.00	111.00
Cabbage	cwt	18.50	20.10	38.10	40.60
Cantaloups	cwt	19.10	17.60	58.80	62.60
Carrots [15]	cwt	26.60	28.60	74.20	75.60
Cauliflower [15]	cwt	35.90	44.50	119.00	119.00
Celery [15]	cwt	18.20	25.40	55.50	57.00
Cucumbers	cwt	25.00	26.80	62.90	67.00
Eggplant	cwt	(NA)	(NA)	66.50	70.90
Escarole/Endive	cwt	(NA)	(NA)	85.60	91.10
Garlic	cwt	52.60	60.00	80.90	86.10
Green peppers [15]	cwt	32.80	41.60	99.40	106.00
Honeydew melons	cwt	20.70	20.90	57.70	57.90
Lettuce	cwt	17.70	26.70	63.20	61.40
Onions [15]	cwt	14.20	15.00	41.10	41.50
Snap beans	cwt	62.70	68.90	124.00	133.00
Spinach	cwt	42.20	43.20	109.00	116.00
Sweet corn	cwt	26.30	28.20	74.80	76.10
Tomatoes	cwt	30.50	44.60	129.00	126.00
Watermelons	cwt	13.20	14.20	23.30	24.90
Vegetables for processing: [8]					
Asparagus	ton	1,570.00	1,570.00	4,000.00	4,090.00
Beets	ton	66.40	46.90	215.00	229.00
Cabbage	ton	(NA)	(NA)	166.00	177.00
Cucumbers	ton	342.00	313.00	(NA)	67.00
Green peas	ton	434.00	427.00	930.00	976.00
Lima beans	ton	524.00	582.00	1,850.00	1,970.00
Snap beans	ton	271.00	320.00	569.00	597.00
Spinach	ton	137.00	129.00	400.00	426.00
Sweet corn	ton	127.00	140.00	273.00	290.00
Tomatoes	ton	76.70	90.10	221.00	225.00
Livestock and livestock products:					
All beef cattle	cwt	121.00	125.00	292.00	294.00
Cows	cwt	81.70	82.30	(NA)	(NA)
Steers and heifers	cwt	123.00	126.00	(NA)	(NA)
Calves	cwt	172.00	81.00	384.00	403.00
Beeswax	pound	(NA)	(NA)	8.03	8.56
Chickens:					
Broilers, live [18]	pound	0.511	0.604	1.35	(NA)
All Eggs	dozen	0.999	1.09	2.63	2.62
Hogs	cwt	64.20	67.20	160.00	161.00
Lambs	cwt	([19])	([19])	323.00	344.00
Milk cows [16]	head	1,430	1,380.00	(NA)	(NA)
Sheep	cwt	([19])	([19])	119.00	129.00
Turkeys, live	pound	0.719	0.665	1.650	1.68

(NA) Not available. [1] Unless otherwise noted, these prices are for marketing year average or calendar year average computed by weighing State prices by quantities sold, or by production for those commodities for which the production is sold. [2] Parity prices are for January of the year shown as published in the January issue of Agricultural Prices. [3] Previous year. [4] Average local market price for wool sold excluding incentive payment. [5] Average local market price for mohair sold excluding incentive payment. [6] Crop year begins with bloom in one year and ends with completion of harvest the following year. Prices refer to the year harvest begins. [7] Equivalent packinghouse-door returns for California, Oregon (pears only), Washington, and New York (apples only), and prices as sold for other States. [8] Equivalent returns at processing plant-door. [9] Equivalent returns at packinghouse-door. [10] Weighted average of co-op and independent sales. Co-op prices represent pool proceeds excluding returns from non-cranberry products and before deductions for capital stock and other retains. [11] Equivalent per unit returns for bulk fruit at first delivery point. [12] Average price as sold. [13] FOB shipping point when available. Weighted average of prices at points of first sale. [14] Prices are in-shell basis except almonds which are shelled basis. [15] Includes some processing. [16] Simple average of States weighted by estimated Jan. 1 head for U.S. average. [17] Prices for fresh and processing breakdown no longer published to avoid disclosure of individual operations. [18] Live weight equivalent price. [19] Estimates discontinued in 2011.

NASS, Environmental, Economics, and Demographics Branch (202) 720–6146.

Table 9-32.—Producer prices: Index numbers, by groups of commodities, United States, 2006–2015

[1982=100]

Year	Total finished goods	Consumer foods	Total consumer goods	Total intermediate materials	Total crude materials
2006	160.4	156.7	166.0	164.0	184.8
2007	166.6	167.0	173.5	170.7	207.1
2008	177.1	178.3	186.3	188.3	251.8
2009	172.5	175.5	179.1	172.5	175.2
2010	179.8	182.4	189.1	183.4	212.2
2011	190.5	193.9	203.3	199.8	249.4
2012	194.3	199.1	207.5	200.7	241.4
2013	196.7	203.5	210.3	200.8	246.7
2014	200.4	212.5	214.7	201.9	249.3
2015	193.9	209.8	204.4	188.1	189.1

ERS, Food Marketing Branch, (202) 694–5349. Compiled from reports of the U.S. Department of Labor.

Table 9-33.—Prices received by farmers: Index numbers by groups of commodities and parity ratio, United States, 2005–2014 [1]

[1910–14=100]

Year	Food grain	Feed grain	Cotton	Tobacco	Oilseed	Fruit & tree nut [2]	Vegetable & melon	Other crop
2005	351	338	361	1,417	579	894	932	558
2006	425	388	402	1,377	550	1,074	974	572
2007	590	541	423	1,392	748	1,103	1,128	582
2008	820	734	515	1,409	1,107	1,038	1,076	604
2009	590	579	416	1,569	971	977	1,153	609
2010	560	587	601	1,557	944	1,031	1,161	622
2011	758	902	745	1,497	1,225	1,162	1,224	645
2012	779	1,010	673	1,576	1,356	1,339	1,040	649
2013	759	943	651	1,760	1,343	1,311	1,330	645
2014	(NA)	(NA)	(NA)	(NA)	(NA)	(NA)	(NA)	(NA)

Year	Potato & dry edible bean	Crop Production	Meat animal	Dairy	Poultry & egg	Livestock Production	Agricultural Production	Parity ratio [3]
2005	554	546	1,201	931	347	910	726	38
2006	634	593	1,180	793	312	850	730	37
2007	637	706	1,204	1,177	393	994	862	40
2008	797	836	1,195	1,128	424	1,000	947	39
2009	761	747	1,075	790	390	857	832	35
2010	708	690	1,256	1,000	427	1,002	850	37
2011	867	872	1,540	1,239	422	1,161	1,032	38
2012	846	933	1,633	1,140	459	1,190	1,083	38
2013	901	922	1,669	1,230	515	1,260	1,104	38
2014	(NA)	802	(NA)	(NA)	(NA)	1,495	1,112	37

(NA) Not available.   [1] These indexes are computed using the price estimates of averages for all classes and grades for individual commodities being sold in local farm markets.   In computing the group indexes, prices of individual commodities have been compared with 2011 weighted average prices.   The resulting ratios are seasonally weighted by average quantities sold for the most recent 5-year period.   For example, 2013 indexes use quantities sold for the period 2007–11.   Then, the 2011 indexes are used to adjust the 1910–14 base period.   [2] Fresh market for noncitrus, and fresh market and processing for citrus.   [3] Ratio of Index of Prices Received to the Index of Prices Paid by Farmers for Commodities and Services, Interest, Taxes, and Farm Wage Rates.
NASS, Environmental, Economics, and Demographics Branch, (202) 720–6146.

Table 9-34.—Prices received by farmers: Index numbers by groups of commodities and ratio, United States, 2005–2014 [1]

(2011=100)

Year	Food grain	Feed grain	Cotton [2]	Tobacco [2]	Oilseed	Fruit & tree nut [3]	Vegetable & melon	Other crop
2005	46.7	38.0	48.5	90.1	49.8	76.8	83.1	86.1
2006	56.4	43.6	54.0	87.6	47.3	92.2	86.8	88.3
2007	78.4	60.8	56.8	88.5	64.4	94.7	100.5	89.9
2008	109.1	82.6	69.1	89.6	95.2	89.1	95.9	93.3
2009	78.5	65.1	55.9	99.7	83.5	83.9	102.8	94.1
2010	74.4	66.0	80.6	98.9	81.2	88.5	103.5	95.9
2011	100.0	100.0	(NA)	(NA)	100.0	100.0	100.0	100.0
2012	103.8	110.8	(NA)	(NA)	111.9	112.4	91.9	102.7
2013	101.5	102.4	(NA)	(NA)	111.6	119.1	103.8	100.5
2014	90.4	69.1	(NA)	(NA)	98.8	133.5	103.3	94.7

Year	Potato [4] & dry edible bean [5]	Crop Production	Meat animal	Dairy product	Poultry & egg	Livestock Production	Agricultural Production	Parity ratio [6]
2005	64.3	62.63	78.1	75.2	82.6	78.4	71.0	91.2
2006	73.5	68.0	76.7	64.0	74.3	73.2	71.0	86.6
2007	73.9	81.0	78.3	95.0	93.6	85.7	84.0	96.8
2008	92.5	95.9	77.6	91.0	101.0	86.1	92.0	93.3
2009	88.3	85.7	69.9	63.7	93.0	73.8	81.0	84.4
2010	82.1	87.0	81.6	80.7	101.6	85.7	82.0	88.2
2011	(NA)	100.0	100.0	100.0	100.0	100.0	100.0	100.0
2012	(NA)	107.0	103.9	92.1	109.0	102.5	105.0	101.7
2013	(NA)	105.7	106.8	99.8	123.3	108.5	107.0	100.0
2014	(NA)	92.0	131.1	119.2	133.2	128.8	107.7	96.8

[1] These indexes are computed using the price estimates of averages for all classes and grades for individual commodities being sold in local farm markets. In computing the group indexes, prices of individual commodities have been compared with 2011 weighted average prices. The resulting ratios are seasonally weighted by average quantities sold for the most recent previous 5-year period. For example, 2013 indexes use quantities sold for the period 2007–11. [2] Included in Other crop index group beginning in 2010. [3] Fresh market for noncitrus, and fresh market and processing for citrus. [4] Included in Vegetable and melon index group beginning 2010. [5] Included in Food grain index group beginning 2010. [6] Ratio of Index of Prices Received (2011=100) to Index of Prices Paid by Farmers for Commodities & Services, Interest, Taxes, and Wage Rates (2007–11=100).
NASS, Environmental, Economics, and Demographics Branch, (202) 720–6146.

Table 9-35.—Prices paid by farmers: Index numbers, by groups of commodities, United States, 2005–2014

(2011=100)

Year	Production all commodities	Feed	Livestock & Poultry	Seeds	Fertilizer	Agricultural chemicals	Fuels	Supplies & Repairs	Autos & trucks
						Production indexes [1]			
2005	65.4	51.8	89.1	50.5	49.9	84.9	59.7	84.4	98.6
2006	68.9	55.1	86.4	54.8	53.5	88.2	66.0	87.2	97.3
2007	74.6	65.8	84.7	61.6	65.6	89.2	72.9	89.9	96.0
2008	88.3	85.8	80.4	78.0	119.2	95.7	94.9	92.5	93.8
2009	84.8	82.5	74.6	90.1	83.8	102.5	63.1	94.3	94.9
2010	87.7	79.7	86.0	93.4	76.8	99.5	78.3	96.1	97.3
2011	100.0	100.0	100.0	100.0	100.0	100.0	100.00	100.0	100.0
2012	105.4	115.3	106.6	105.5	101.3	105.3	99.3	102.9	101.9
2013	107.4	117.5	108.3	110.3	96.6	108.6	98.3	103.8	103.4
2014	114.1	114.8	153.6	113.5	94.7	109.7	97.8	105.7	104.4

Year	Farm machinery	Building materials	Farm services	Rent	Interest	Taxes	Wage rates	Production, interest, taxes, & wage rates	Family living	Commodities, interest, taxes, & wage rates [2]
	Production indexes [1] - continued									
2005	70.9	83.0	81.3	62.7	76.0	69.9	86.0	67.4	86.8	69.9
2006	74.6	88.5	84.8	68.4	91.3	79.7	89.2	71.6	89.6	73.9
2007	78.5	90.5	89.1	71.8	97.6	90.2	92.4	77.2	92.2	79.0
2008	85.6	96.5	89.0	80.2	101.9	94.3	96.6	89.5	95.7	90.0
2009	91.2	95.4	95.3	89.6	94.2	91.9	97.9	86.4	95.4	87.3
2010	94.3	96.6	98.1	92.6	91.6	95.9	98.9	89.0	96.9	90.0
2011	100.0	100.0	100.0	100.0	100.0	100 0	100.0	100.0	100.0	100.0
2012	105.3	102.7	101.8	101.7	99.3	101.0	102.9	104.8	102.1	104.4
2013	108.1	104.8	105.5	105.5	93.3	103.4	106.3	106.7	103.6	106.3
2014	112.0	106.8	109.1	114.2	92.8	116.3	108.0	113.1	105.3	112.0

[1] Annual price index computed using annual weights. Base period (2011=100) are now being published to one decimal digit. [2] Family Living component included.
NASS, Environmental, Economics, and Demographics Branch, (202) 720–6146.

Table 9-36.—Prices paid by farmers: Index numbers, by groups of commodities, United States, 2005–2014 [1]

[1910–14=100]

Year	Family living	Production indexes							
		Production all commodities	Feed	Livestock & poultry	Seed	Fertilizer	Agricultural chemicals	Fuels	Supplies & repairs
2005	1,855	1,361	571	1,759	1,661	601	762	1,668	995
2006	1,915	1,434	607	1,706	1,802	644	792	1,845	1,029
2007	1,969	1,552	725	1,671	2,024	790	801	2,038	1,060
2008	2,045	1,839	945	1,587	2,563	1,436	859	2,653	1,091
2009	2,038	1,766	909	1,472	2,960	1,009	921	1,765	1,113
2010	2,071	1,825	879	1,698	3,070	925	893	2,191	1,134
2011	2,136	2,082	1,102	1,974	3,286	1,204	898	2,797	1,180
2012	2,181	2,193	1,270	2,104	3,468	1,221	946	2,779	1,214
2013	2,213	2,236	1,295	2,137	3,622	1,164	974	2,752	1,225
2014	2,249	2,376	1,266	3,032	3,729	1,138	983	2,739	1,247

Year	Production indexes—Continued				Interest	Taxes	Wage rates	Production, interest, taxes, & wage rates	Commodities, interest, taxes, & wage rates [2]
	Autos & trucks	Farm machinery	Building materials	Farm services & rent					
2005	3,031	4,329	1,930	1,569	2,772	4,150	6,158	1,900	1,891
2006	2,991	4,556	2,059	1,664	3,328	4,729	6,390	2,019	1,999
2007	2,949	4,794	2,104	1,746	3,560	5,356	6,618	2,178	2,138
2008	2,882	5,231	2,245	1,818	3,727	5,598	6,917	2,522	2,434
2009	2,917	5,575	2,218	1,976	3,436	5,453	7,007	2,435	2,364
2010	2,992	5,762	2,248	2,036	3,339	5,694	7,079	2,509	2,434
2011	3,074	6,111	2,326	2,123	3,646	5,935	7,161	2,820	2,706
2012	3,131	6,432	2,387	2,162	3,620	5,993	7,366	2,954	2,824
2013	3,179	6,609	2,434	2,240	3,402	6,137	7,607	3,010	2,876
2014	3,209	6,845	2,486	2,360	3,383	6,904	7,730	3,189	3,031

[1] Based on Consumer Price Index-Urban of Bureau of Labor Statistics.   [2] The index known as the Parity Index is the Index of Prices Paid by Farmers for Commodities and Services, Interest, Taxes, and Wage Rates expressed on the 1910–14=100 base.
  NASS, Environmental, Economics, and Demographics Branch, (202) 720–6146.

Table 9-37.—Prices paid by farmers: April prices, by commodities, United States, 2012–2014 [1]

Commodity	Unit	2012	2013	2014
		Dollars	Dollars	Dollars
Fuels and energy:				
Diesel fuel [2] [3]	Gal	3.740	3.570	3.537
Gasoline, service station, unleaded [4]	Gal	3.837	3.684	3.595
Gasoline, service station, bulk delivery [4]	Gal	3.862	3.696	3.569
L. P. gas, bulk delivery [2]	Gal	2.233	2.009	2.151
Feeds:				
Alfalfa Meal	Cwt	26.60	31.20	32.10
Alfalfa Pellets	Cwt	27.10	30.90	31.90
Bran	Cwt	26.40	29.70	30.70
Beef Cattle Concentrate.				
32-36% Protein	Ton	525.00	602.00	617.00
Corn Meal	Cwt	20.40	22.30	23.00
Cottonseed Meal, 41%	Cwt	27.80	31.40	32.90
Dairy Feed				
14% Protein	Ton	437.00	474.00	487.00
16% Protein	Ton	442.00	481.00	502.00
18% Protein	Ton	431.00	447.00	470.00
20% Protein	Ton	407.00	414.00	434.00
32% Protein Conc.	Ton	550.00	630.00	654.00
Hog Feed				
14-18% Protein	Ton	475.00	490.00	507.00
38-42% Protein Conc.	Ton	592.00	646.00	666.00
Molasses, Liquid	Cwt	23.20	24.70	25.70
Poultry Feed				
Broiler Grower	Ton	591.00	639.00	654.00
Chick Starter	Ton	597.00	628.00	644.00
Laying Feed	Ton	527.00	552.00	581.00
Turkey Grower	Ton	589.00	604.00	619.00
Soybean meal				
44%	Cwt	26.50	32.50	34.00
over 44%	Cwt	24.10	29.40	31.00
Stock Salt	50 Lb	6.33	6.43	6.60
Trace mineral blocks	50 Lb	7.59	7.80	8.07

See footnote(s) at end of table.

**Table 9-37.—Prices paid by farmers: April prices, by commodities,
United States, 2012–2014 [1]—Continued**

Commodity	Unit	2012	2013	2014
		Dollars	*Dollars*	*Dollars*
Fertilizer: [5]				
0-15-40	Ton	(NA)	(NA)	(NA)
0-18-36	Ton	621.00	595.00	595.00
0-20-20	Ton	(NA)	(NA)	(NA)
3-10-30	Ton	527.00	504.00	480.00
5-10-10	Ton	(NA)	(NA)	(NA)
5-10-15	Ton	(NA)	(NA)	(NA)
5-10-30	Ton	547.00	596.00	600.00
5-20-20	Ton	601.00	590.00	580.00
6- 6- 6	Ton	(NA)	(NA)	(NA)
6- 6-18	Ton	555.00	571.00	575.00
6-12-12	Ton	(NA)	(NA)	(NA)
6-24-24	Ton	687.00	703.00	690.00
8- 8- 8	Ton	(NA)	(NA)	(NA)
8-20- 5	Ton	(NA)	(NA)	(NA)
8-32-16	Ton	(NA)	(NA)	(NA)
9-23-30	Ton	654.00	626.00	623.00
10- 3- 3	Ton	(NA)	(NA)	(NA)
10- 6- 4	Ton	(NA)	(NA)	(NA)
10-10-10	Ton	514.00	535.00	518.00
10-20-10	Ton	(NA)	(NA)	(NA)
10-20-20	Ton	638.00	613.00	572.00
10-34- 0	Ton	755.00	629.00	611.00
11-52- 0	Ton	719.00	691.00	620.00
13-13-13	Ton	601.00	606.00	620.00
15-15-15	Ton	(NA)	(NA)	(NA)
16- 0-13	Ton	(NA)	(NA)	(NA)
16- 4- 8	Ton	585.00	593.00	616.00
16- 6-12	Ton	537.00	529.00	522.00
16-16-16	Ton	(NA)	(NA)	(NA)
16-20- 0	Ton	626.00	646.00	619.00
17-17-17	Ton	641.00	640.00	598.00
18-46- 0 (DAP)	Ton	675.00	640.00	611.00
19-19-19	Ton	662.00	637.00	590.00
24- 8- 0	Ton	443.00	427.00	436.00
Ammonium Nitrate	Ton	574.00	544.00	560.00
Anhydrous Ammonia	Ton	785.00	847.00	851.00
Aqua Ammonia	Ton	266.00	262.00	253.00
Limestone, Spread on field	Ton	57.50	54.90	56.60
Muriate of Potash, 60–62% K2O	Ton	665.00	595.00	601.00
Nitrate of Soda	Ton	(NA)	(NA)	(NA)
Nitrogen Solutions.				
28% N	Ton	388.00	397.00	372.00
30% N	Ton	403.00	410.00	359.00
32% N	Ton	425.00	441.00	416.00
Sulfate of Ammonia	Ton	503.00	522.00	533.00
Superphosphate, 44-46% P2O5	Ton	729.00	701.00	621.00
Urea, 44-46% Nitrogen	Ton	644.00	592.00	571.00
Farm Machinery:				
Baler, Pick-Up, Automatic Tie, P.T.O.				
Square Conventional, Under 200 Lb Bales	Each	23,000	24,100	24,900
Round, 1200-1500 Lb Bale	Each	28,000	28,500	29,700
Round, 1900-2200 Lb Bale	Each	38,900	40,400	42,400
Chisel Plow, Maxiumum 1 Foot Depth				
Tillage, Chisel or Sweep Type, Drawn.				
Mounted, 16-20 Foot	Each	27,800	29,100	29,600
Mounted, 21-25 Foot	Each	39,300	41,600	42,100
Combine, Self Propelled with Grain head				
Extra-large capacity	Each	363,000	371,000	388,000
Large capacity	Each	295,000	305,000	315,000
Corn Head for combine				
6 Row	Each	42,700	45,300	47,400
8 Row	Each	57,800	60,200	62,100
Cotton Picker, Self Propelled, with spindle,				
6-Row	Each	585,000	602,000	610,000
Cultivator, Row Crop				
6-Row	Each	(NA)	(NA)	(NA)
8-Row	Each	18,000	18,200	18,600
12-Row, Flexible	Each	32,700	33,000	33,800
Disk Harrow, Tandem, Drawn [6]				
15-17 Foot	Each	21,800	22,100	22,300
18-20 foot	Each	31,400	31,800	32,800
21-25 foot	Each	38,800	39,600	41,700

See footnote(s) at end of table.

Table 9-37.—Prices paid by farmers: April prices, by commodities, United States, 2012–2014 [1]—Continued

Commodity	Unit	2012	2013	2014
		Dollars	*Dollars*	*Dollars*
Farm Machinery (continued):				
Elevator, Portable, Without Power Unit,				
Auger Type, 8 Inch Diameter, 60 Foot	Each	9,030	9,350	9,740
Feed Grinder-Mixer, Trailer Mtd., P.T.O.	Each	34,700	37,200	37,800
Field Cultivator, Mounted or Drawn				
17-19 Foot	Each	21,800	23,100	23,800
20-25 Foot, Flexible	Each	30,600	31,700	32,500
Forage Harvester, P.T.O., Shear Bar,				
With Pick-Up Attachment	Each	47,600	49,500	51,600
With Row Crop Unit, 2-Row	Each	51,200	53,600	57,300
Forage Harvester, Self-propelled, Shear Bar				
With 4–6 row	Each	414,000	422,000	454,000
Front-End Loader, Hydraulic, Tractor Mounted				
1800-2500 Lb. Capacity, 60 Inch Bucket	Each	7,140	7,160	7,360
Grain Drill, Most Common Spacing				
Plain, 15-17 Openers	Each	21,000	22,100	23,600
Press, 23-25 Openers	Each	40,500	41,400	44,300
With Fertilizer Attachment, 20-24 Openers	Each	32,700	34,000	34,700
Min/No-Till W/Fert. Attach., 15 Foot	Each	42,300	44,300	45,100
Hayrake, Side-Delivery, or Wheel Rake,				
Traction Drive, 8-12 Foot Working Width	Each	7,430	7,880	7,960
Hay Tedder, 15-18 Foot	Each	7,090	7,570	7,610
Manure Spreader, Conveyor Type, P.T.O.,				
2-Wheel, with Tires.				
141-190 Bushel Capacity	Each	10,300	11,300	12,200
225-300 Bushel Capacity	Each	15,700	16,200	17,100
Mower-Conditioner, P.T.O., Pull Type, with				
8-10 Foot, Sickle (Cutter) Bar or Disc	Each	21,700	22,400	23,400
14-16 Foot, Sickle (Cutter) Bar or Disc	Each	33,500	34,200	36,200
Mower, Mounted or Drawn,				
7-8 ft Sickle (Cutter) Bar	Each	7,930	8,020	8,440
13-14 Foot, Sickle (Cutter) Bar or Disc	Each	17,600	17,900	19,000
Planter, Row Crop				
With Fertilizer Attachment, 4-Row	Each	24,300	25,700	26,500
With Fertilizer Attachment, 8-Row	Each	47,800	49,600	50,400
With Fertilizer Attachment, 24-Row	Each	164,000	173,000	190,000
12-Row Conservation (No-Till Cond), w/Fert	Each	86,100	87,400	88,900
Rotary Hoe, 20-25 Foot	Each	13,900	14,300	15,900
Rotary Cutter, 7-8 Foot	Each	4,200	4,330	4,450
Sprayer, Field Crop, Power, Boom Type				
(Excl. Self-Propelled and Orchard).				
Tractor Mounted, w/ 300 Gal. Spray Tank	Each	7,530	7,960	8,310
Trailer Type, w/ 500-700 Gal. Spray Tank	Each	17,000	17,000	17,300
Tractor, 2-Wheel Drive				
30-39 P.T.O. horsepower	Each	19,000	19,900	20,600
50-59 P.T.O. horsepower	Each	25,800	26,700	27,500
70-89 P.T.O. horsepower	Each	43,000	44,000	45,100
110 - 129 P.T.O. horsepower	Each	81,400	84,000	85,800
140 - 159 P.T.O. horsepower	Each	128,000	133,000	135,000
190 - 220 P.T.O. horsepower	Each	185,000	190,000	197,000
Tractor, 4-Wheel Drive				
200 - 280 P.T.O. horsepower	Each	217,000	226,000	233,000
281 - 350 Engine horsepower	Each	272,000	277,000	294,000
51-500 Engine horsepower	Each	301,000	309,000	321,000
Wagon, Gravity Unload, W/Box and Running				
Gear, and Tires,				
200-400 Bushel Capacity				
Without Side Extensions	Each	7,850	8,350	8,820
Wagon, Running Gear, W/O Box				
8-10 Ton Capacity	Each	2,450	2,480	2,600
Windrower, Self-Propelled,				
14-16 Foot	Each	115,000	119,000	129,000

See footnote(s) at end of table.

Table 9-37.—Prices paid by farmers: April prices, by commodities,
United States, 2012–2014 [1]—Continued

Commodity	Unit	2012	2013	2014
		Dollars	*Dollars*	*Dollars*
Agricultural Chemicals: [7]				
Fungicides:				
Calcium Polysulfide (Lime Sulfur) Liq.Conc	Gal	(NA)	(NA)	(NA)
Captan 50% WP	Lb	7.84	7.92	8.14
Captan 80% WP	Lb	8.47	8.61	8.75
Chlorothalonil (Bravo), 6 pounds/gallon EC	Gal	43.80	45.30	47.20
Copper Hydroxide (Kocide 200), 35% WP	Lb	6.40	6.64	6.88
Copper Hydroxide (Kocide 101), 77% WP	Lb	5.29	5.40	5.51
Dodine (Cyprex), 65% WP	Lb	(NA)	(NA)	(NA)
Fenarimol (Rubigan), 1 pounds/gallon EC	Gal	375.00	386.00	392.00
Ferbam (Carbamate), 76% WP	Lb	4.81	5.03	5.17
Fosethyl-AL (Aliette), 80% WP	Lb	15.10	15.30	15.80
Mancozeb (Dithane 80% WP,Manzate 75% DF)	Lb	5.11	5.23	5.40
Maneb (Manex), 4 pounds/gallon	Lb	40.10	41.80	42.40
Maneb, 80% WP, 75% DF	Lb	(NA)	(NA)	(NA)
Myclobutanil (Systhane, Nova, Rally), 40% WP	Lb	68.90	71.50	73.20
Oxytetraycline (Mycoshield), 17% WP	Lb	22.50	22.90	23.80
Sulfur, 80% - Microthiol Disperssm Kumulus DF	Lb	0.89	0.93	0.99
Triadimefon (Bayleton), 50% WP	Lb	107.00	109.00	113.00
Ziram, 76% WP	Lb	4.44	4.52	4.67
Herbicides:				
2,4–D, 4 pounds/gallon EC	Gal	20.10	20.40	20.90
Acetochlor (Surpass), 6.4–7 pounds/gallon EC	Gal	70.80	74.50	78.40
Alachlor (Lasso), 4 pounds/gallon EC	Gal	29.60	30.40	31.40
Atrazine(AAtrex), 4 pounds/gallon L	Gal	17.60	17.80	18.40
Bentazon (Basagran), 4 pounds/gallon EC	Gal	110.00	111.00	114.00
Butylate (Sutan), 6.7 pounds/gallon EC	Gal	34.40	35.40	37.30
Chlorimuron–ethyl (Classic), 25% DF	oz	16.00	16.60	17.10
Chlorsulfuron (Glean), 75%	oz	21.30	22.30	23.10
DCPA (Dacthal), 75% WP	Lb	21.50	22.00	23.10
Dicamba (Banvel), 4 pounds/gallon EC	Gal	(NA)	(NA)	(NA)
Diuron (Karmex, Diurex), 80% WP	Lb	6.38	6.52	6.67
EPTC (Eptan), 7E–(Eradicane),6.7 pounds/gallon EC ...	Gal	46.00	47.60	48.40
Glyphosate (Roundup), 4.5 pounds/gallon EC	Gal	17.90	18.20	18.70
Glyphosate (Roundup), 5 pounds/gallon EC	Gal	26.00	26.10	26.80
Linuron (Lorox, Linex), 50% DF	Lb	21.60	22.60	23.00
MCPA, 4 pounds/gallon EC	Gal	21.90	22.10	22.90
Metribuzin (Lexone or Sencor), 75% DF	Lb	17.40	17.80	18.00
Napropamide (Devrinol), 50% WP	Lb	12.10	12.40	12.60
Pendimethalin (Prowl), 3.3–3.8 pounds/gallon EC	Gal	40.40	41.60	42.60
Sethoxydim (Poast), 1.5 pounds/gallon EC	Gal	87.00	89.10	91.50
Simazine (Princep), 4 pounds/gallon EC	Gal	25.40	26.40	26.80
Terbacil (Sinbar), 80% WP	Lb	43.60	45.20	46.20
Trifluralin (Treflan), 4 pounds/gallon EC	Gal	26.20	26.80	27.60

See footnote(s) at end of table.

Table 9-37.—Prices paid by farmers: April prices, by commodities, United States, 2012–2014 [1]—Continued

Commodity	Unit	2012	2013	2014
		Dollars	*Dollars*	*Dollars*
Insecticides:				
Acephate (Orthene), 75% SP	Lb	11.50	12.20	12.60
Acephate (Orthene), 90% SP	Lb	9.13	9.41	9.62
Aldicarb (Temik), 15% G	Lb	4.03	4.27	4.32
Azinphos–methyl (Guthion), 50% WP	Lb	14.50	15.10	15.70
Bt (Dipel 2X), WP	Lb	12.50	13.50	14.00
Carbaryl, (Sevin), 80% S, SP or WP	Lb	7.29	7.45	7.61
Carbaryl, (Sevin), 4 pounds/gallon 4F or XLR Plus WP	Lb	46.90	48.80	50.40
Carbofuran (Furadan), 4F	Gal	83.50	87.30	91.70
Chlorpyrifos (Lorsban), 4 pounds/gallon EC	Gal	37.60	38.80	40.10
Cyfluthrin (Baythroid) 2 pounds/gallon EC	Gal	302.00	303.00	308.00
Dicofol, 4 pounds/gallon 4E WP	Lb	44.90	47.90	49.00
Dicrotophos (Bidrin), 8 pounds/gallon EC	Gal	134.00	138.00	143.00
Dimethoate (Cygon), 2.67 pounds/gallon EC	Gal	49.00	49.60	51.00
Disulfoton (Di-Syston), 8 pounds/gallon EC	Gal	(NA)	(NA)	(NA)
Endosulfon (Thiodan, Phaser), 3 pounds/gallon EC	Gal	32.40	33.40	34.10
Esfenvalerate (Asana XL),0.66 pounds/gallon EC	Gal	101.00	102.00	105.00
Malathion, 5 pounds/gallon EC	Gal	44.00	46.80	48.30
Malathion,(Fyfanon ULV AG), 9.9 pounds/gallon EC ..	Gal	58.10	61.50	63.70
Methidathion (Supracide), 25% WP	Lb	10.50	10.60	10.80
Methomyl (Lannate LV), 2.4 pounds/gallon EC	Gal	70.50	71.10	73.10
Methyl Parathion (Penncap–M), 2 pounds/gallon EC ..	Gal	40.20	44.80	46.40
Oil(Oil, Superior Oil, Supreme, Volck)	Gal	10.40	10.70	11.20
Oxamyl (Vydate–L), 2 pounds/gallon L	Gal	101.00	104.00	107.00
Oxydemeton–methyl (MSR Spray),. 2 pounds/gallon EC	Gal	138.00	146.00	150.00
Phorate (Thimet), 20% G	Lb	2.95	3.06	3.14
Phosmet (Imidan), 50% WSP	Lb	10.40	10.10	9.86
Phosmet (Imidan), 70% WSP	Lb	10.60	11.40	12.30
Propargite (Comite, Omite), 32% WP	Lb	7.70	7.91	8.01
Synthetic Pyrethroids,. (Pounce, Ambush) 2–3.2 pounds/gallon	Gal	86.30	93.30	95.45
Terbufos (Counter), 15% G	Lb	3.07	3.38	3.48
Zeta–Cyermethrin (Zeta –cype), 0.8 pounds/gallon EC	Gal	136.00	134.00	133.00
Zeta–Cyermethrin (Fury), 1.5 pounds/gallon EC	Gal	178.00	187.00	191.00
Other:				
Gibberellic Acid (Pro–Gibb)4.0% L	Gal	118.00	114.00	113.00
NAD Napthaleneacetamide (Amid–Thin W), 8.4% WP	Lb	69.00	84.40	63.80

(NA) Not available. [1] Prices paid by famers are collected, for the most part, from retail establishments located in smaller cities and towns in rural areas. Prior to 1995, recorded prices reflected a modified annual average based on frequency item was surveyed during the year. Recorded item values, 1995-99, are the U.S. April average price. [2] Includes Federal, State, and local per gallon taxes where applicable. [3] Excludes Federal excise tax. [4] Includes Federal, State, and local per gallon taxes. [5] Excludes cost of application, except for limestone. [6] With hydraulic lift, transport wheels, and tires. [7] Active Ingredient, (Common Names),and Formulation abbreviations: EC-Emulsifiable Concentrate, DF-Dry Flowable, DG-Dry Granular, G-Granular, L-Liquid, S-Solution, P-Soluble Powder, and WP-Wettable Powder. [8] Insufficient data.

NASS, Environmental, Economics, and Demographics Branch, (202) 720-6146.

Table 9-38.—Agricultural commodities: Support prices per unit, United States, 2011–2015 [1]

Commodity	Unit	2011	2012	2013	2014	2015 [2]
		Dollars	*Dollars*	*Dollars*	*Dollars*	*Dollars*
Basic commodities:						
Corn:						
Target/reference price	Bushel	2.63	2.63	2.63	3.70	3.70
Loan rate	do	1.95	1.95	1.95	1.95	1.95
Cotton:						
American upland:						
Target price	Cwt	71.25	71.25	71.25	(NA)	(NA)
Loan rate	do	52.00	52.00	52.00	52.00	52.00
Extra-long staple:						
Target price	do	(NA)	(NA)	(NA)	(NA)	(NA)
Loan rate	do	79.77	79.77	79.77	79.77	79.77
Peanuts:						
Target/reference price	Short tons	495.00	495.00	495.00	535.00	535.00
Loan rate	do	355.00	355.00	355.00	355.00	355.00
Rice:						
Target/reference price [3]	Cwt	10.50	10.50	10.50	14.00	14.00
Loan rate	do	6.50	6.50	6.50	6.50	6.50
Wheat:						
Target/reference price	Bushel	4.17	4.17	4.17	5.50	5.50
Loan rate	do	2.94	2.94	2.94	2.94	2.94
Barley:						
Target/reference price	Bushel	2.63	2.63	2.63	4.95	4.95
Loan rate	do	1.95	1.95	1.95	1.95	1.95
Sorghum grain:						
Target/reference price	Cwt	4.70	4.70	4.70	7.05	7.05
Loan rate	do	3.48	1.95	1.95	3.48	3.48
Oats:						
Target/reference price	Bushel	1.79	1.79	1.79	2.40	2.40
Loan rate	do	1.39	1.39	1.39	1.39	1.39
Other oilseeds: [4]						
Target/reference price	Cwt	12.68	12.68	12.68	20.15	20.15
Loan rate	do	10.09	10.09	10.09	10.09	10.09
Soybeans:						
Target/reference price	Bushel	6.00	6.00	6.00	8.40	8.40
Loan rate	do	5.00	5.00	5.00	5.00	5.00
Dry Peas:						
Target/reference price	Cwt	8.32	8.32	8.32	11.00	11.00
Loan rate	do	5.40	5.40	5.40	5.40	5.40
Small chick peas:						
Target/reference price	Cwt	10.36	10.36	10.36	19.04	19.04
Loan rate	do	7.43	7.43	7.43	7.43	7.43
Large chick peas:						
Target/reference price	Cwt	12.81	12.81	12.81	21.54	21.54
Loan rate	do	11.28	11.28	11.28	11.28	11.28
Lentils:						
Target/reference price	do	12.81	12.81	12.81	19.97	19.97
Loan rate	do	11.28	11.28	11.28	11.28	11.28
Sugar, raw cane:						
Loan rate	Pound	0.187	0.187	0.187	0.188	0.188
Sugar, refined beet:						
Loan rate	do	0.241	0.241	0.241	0.241	0.241
Honey, extracted:						
Loan rate	Pound	0.69	0.69	0.69	0.69	0.69
Mohair:						
Loan rate	do	4.20	4.20	4.20	4.20	4.20
Wool, graded:						
Loan rate	Pound	1.15	1.15	1.15	1.15	1.15
Wool, nongraded:						
Loan rate	Pound	0.40	0.40	0.40	0.40	0.40
Milk for manufacturing: [5]						
Support price	Cwt	9.35	9.35	9.35	(NA)	(NA)

(NA) Not available.　[1] National averages during the marketing years for the individual crops.　[2] Target price is applicable for crop years 2003-2013 and reference price is applicable beginning with crop year 2014.　[3] Target and reference price shown is for long grain and medium (including short) grain rice. The reference price for temperate japonica rice is $16.10 per Cwt beginning with crop year 2014. The loan rate for temperate japonica is the same as for other types of rice, at 6.50 per cwt.　[4] Other oilseeds are flaxseed, sunflower seed (oil and other types), safflower, rapeseed, canola, mustard seed, crambe, and sesame seed.　[5] Effective support price calculated from product prices specified in 2008 Farm Bill, effective January 1, 2008. the Dairy Product Price Support Program expired December 31, 2013.

FSA, Feed Grains and Oilseeds, (202) 720–2711.

Table 9-39.—Farm income: Cash receipts by commodity groups and selected commodities, United States, 2011–2015 [1]

Commodity	2011	2012	2013	2014	2015
All commodities	365,843,138	401,433,056	403,552,908	424,155,438	375,419,786
Livestock and products	164,799,514	169,818,778	182,704,665	212,792,575	189,765,870
Cattle and calves	62,321,217	66,090,126	67,457,442	81,478,368	78,228,639
Hogs	21,631,894	22,092,108	23,761,227	26,517,815	21,032,491
Sheep and lambs	(X)	(X)	(X)	(X)	(X)
Dairy products	39,531,306	37,064,731	40,276,790	49,352,950	35,739,249
Broilers	22,987,822	24,827,800	30,761,669	32,728,234	28,709,834
Farm chickens	81,110	79,208	87,939	96,151	100,899
Chicken eggs	7,355,873	7,929,104	8,678,859	10,257,972	13,499,904
Turkeys	4,987,608	5,452,135	4,839,562	5,304,501	5,707,871
Miscellaneous livestock	5,902,684	6,283,566	6,841,177	7,056,584	6,746,983
Crops	201,043,624	231,614,279	220,848,243	211,362,863	185,653,916
Food grains	16,527,336	19,292,117	17,230,685	16,050,095	12,418,473
Feed crops	71,733,544	82,136,536	70,835,571	65,873,669	57,109,253
Cotton	7,303,972	8,230,448	6,515,834	7,111,320	4,913,549
Tobacco	1,139,738	1,347,847	1,546,501	1,715,575	1,605,372
Oil crops	35,324,388	46,925,602	47,274,387	42,620,534	35,584,395
Vegetables	17,616,247	17,412,740	19,417,988	18,871,455	19,752,897
Fruits/nuts	24,166,286	28,107,818	29,904,334	31,930,203	27,064,020
All other crops	27,232,113	28,161,171	28,122,943	27,190,012	27,205,957

Data as of February 7, 2017. (X) Data are not available/applicable.   [1] USDA estimates and publishes individual cash receipt values only for major commodities and major producing States.   The U.S. receipts for individual commodities, computed as the sum of the reported States, may understate the value of sales for some commodities, with the balance included in the appropriate category labeled "other" or "miscellaneous".   The degree of underestimation in some of the minor commodities can be substantial.

ERS, Farm Income Team, (202) 694–5344. FarmIncomeTeam@ers.usda.gov. http://www.ers.usda.gov/data-products/farm-income-and-wealth-statistics.aspx.

Table 9-40.—Farm income: United States, 2006–2015

Item	2006	2007	2008	2009	2010
	Thousand dollars	Thousand dollars	Thousand dollars	Thousand dollars	Thousand dollars
Total gross farm income	290.2	339.6	364.6	336.5	356.5
Value of Production	274.4	327.7	352.3	324.4	344.1
Crops	118.7	151.1	173.8	164.7	168.1
Livestock and products	119.3	138.4	139.4	119.5	140.2
Services and forestry	36.4	38.1	39.1	40.2	35.8
Direct government payments	15.8	11.9	12.2	12.2	12.4
Total production expenses [1]	232.7	269.5	286.5	274.4	279.4
Net farm income	57.4	70.0	78.1	62.2	77.1
Gross cash income	273.2	318.0	349.5	327.3	353.6
Cash expenses	204.8	240.6	264.5	253.0	257.3
Net cash income	68.4	77.4	85.0	74.3	96.3

Item	2011	2012	2013	2014	2015
Total gross farm income	420.4	449.8	483.8	483.1	439.7
Value of production	410.0	439.2	472.8	473.3	428.9
Crops	199.3	212.9	233.6	206.0	182.8
Livestock and product	163.7	169.1	181.0	214.4	194.6
Services and forestry	46.9	57.1	58.2	52.9	51.5
Direct government payments	10.4	10.6	11.0	9.8	10.8
Total production expenses [1]	306.9	353.3	360.1	390.5	358.8
Net farm income	113.5	96.5	123.7	92.6	80.9
Gross cash income	407.0	451.3	455.5	470.5	420.6
Cash expenses	283.9	316.1	320.0	339.0	315.9
Net cash income	123.2	135.3	135.6	131.5	104.7

Data as of February 7, 2017. [1] Includes expenses associated with operator dwellings.
ERS, Farm Income Team. FarmIncomeTeam@ers.usda.gov. http://www.ers.usda.gov/data-products/farm-income-and-wealth-statistics.aspx.

Table 9-41.—Expenses: Farm production expenses, United States, 2008–2015

Item	2008	2009	2010	2011
	Thousand dollars	Thousand dollars	Thousand dollars	Thousand dollars
Total production expenses	286.5	274.4	279.4	306.9
Feed purchased	46.9	45.0	45.4	54.6
Livestock and poultry purchased	18.9	17.9	20.4	23.7
Seed purchased	15.1	15.5	16.3	17.8
Fertilizer and lime	22.5	20.1	21.0	25.1
Pesticides	11.7	11.5	10.7	11.8
Fuel and oil	16.9	13.2	13.8	16.2
Electricity	4.6	4.6	4.6	4.9
Other [1]	61.8	59.1	56.8	61.7
Interest [2]	16.8	17.6	16.9	16.0
Contract and hired labor expenses	29.9	29.0	27.5	27.0
Net rent to landlords	13.0	13.6	16.9	17.4
Capital consumption [2]	17.0	16.2	17.6	18.5
Property taxes and fees [2]	11.3	11.1	11.5	12.0

Item	2012	2013	2014	2015
Total production expenses	353.3	360.1	390.5	358.8
Feed purchased	60.5	62.4	63.7	58.5
Livestock and poultry purchased	24.8	25.5	31.2	30.4
Seed purchased	20.9	21.9	22.1	21.3
Fertilizer and lime	28.9	28.3	28.1	25.5
Pesticides	14.0	14.6	15.8	14.6
Fuel and oil	16.5	17.3	17.7	13.2
Electricity	5.4	5.5	5.9	5.7
Other [1]	66.3	64.2	70.9	66.5
Interest [2]	16.9	14.9	15.7	16.4
Contract and hired labor expenses	32.1	32.3	34.5	32.1
Net rent to landlords	20.5	22.9	20.7	20.1
Capital consumption [2]	34.2	37.8	49.7	40.9
Property taxes and fees [2]	12.2	12.4	14.3	13.5

Data as of February 7, 2017. [1] Includes repair and maintenance, machine hire and custom work, marketing, storage and transportation, insurance premiums, and miscellaneous other expenses. [2] Includes expenses associated with operator dwellings.
ERS, Farm Income Team. FarmIncomeTeam@ers.usda.gov. http://www.ers.usda.gov/data-products/farm-income-and-wealth-statistics.aspx.

Table 9-42.—Principal Farm Operator Households: Farm and Off-farm Income, United States, 2012–2015

Item	2012	2013	2014	2015
Number of family farms	2,043,483	2,045,352	2,053,008	2,032,300
	Dollars per farm operator household			
Median household income from farming	(1,480)	(644)	(118)	(765)
Median off-farm income	58,410	62,500	70,000	67,500
Earned income	38,675	40,737	42,257	38,270
Unearned income	17,500	20,742	27,439	25,013
Median household income of farm operators	68,680	73,219	81,637	76,735
Average household income of farm operators	111,524	121,120	134,164	119,880
	Dollars per farm operator household			
U.S. Median household income	50,017	53,585	53,657	56,516
U.S. Average household income	71,274	75,195	75,738	79,263
	Percent			
Median farm household incomeas a percent of median U.S. household income	134.6	136.6	152.1	135.8
Average farm household incomeas a percent of average U.S. household income	156.5	161.1	177.1	151.2

Source: USDA, Economic Research Service and National Agricultural Statistics Service, 2010-2015 Agricultural Resource Management Survey. Data as of February 7, 2017.
Estimates of U.S. Household Income are produced by the U.S. Census Bureau, Current Population Reports, p60-252, *Income and Poverty in the United States: 2015*.
ERS, Farm Economy Branch, farmincometeam@ERS.usda.gov.

Table 9-43.—Grazing fees: Rates for cattle by selected States and regions, 2013–2014

State	Monthly lease rates for private non-irrigated grazing land [1]					
	Animal unit [2]		Cow-calf		Per head	
	2013	2014	2013	2014	2013	2014
	Dollars per month	Dollars per month	Dollars per month	Dollars per month	Dollars per month	Dollars per month
Arizona	9.00	9.00	(S)	(S)	12.00	(S)
California	19.50	20.00	23.50	25.10	21.00	22.50
Colorado	17.50	17.00	20.00	20.00	19.00	19.00
Idaho	15.50	16.50	19.00	19.00	15.50	18.50
Kansas	17.00	18.00	21.00	22.00	18.50	21.50
Montana	21.00	23.00	23.60	26.00	20.80	23.00
Nebraska	33.50	38.00	40.50	44.50	34.00	35.00
Nevada	(S)	11.50	(S)	(S)	(S)	(S)
New Mexico ...	13.00	13.50	(S)	(S)	14.00	15.50
North Dakota ..	18.00	19.20	21.50	22.00	18.00	18.00
Oklahoma	(S)	(S)	(S)	(S)	(S)	14.00
Oregon	15.00	17.00	17.80	19.50	19.00	19.00
South Dakota .	27.90	30.00	33.30	35.00	28.50	32.50
Texas	(S)	(S)	(S)	(S)	(S)	14.00
Utah	14.50	15.00	18.50	19.00	16.00	16.50
Washington	13.50	13.50	15.00	16.50	14.50	15.00
Wyoming	18.70	20.00	21.00	22.00	19.40	21.00
17-State [3]	18.30	19.90	21.40	23.20	18.90	20.70
16-State [4]	20.10	21.60	23.80	25.40	21.00	22.60
11-State [5]	17.50	18.20	20.30	21.60	18.50	19.70
9-State [6]	18.50	20.30	21.50	23.50	18.90	20.90

(S) Insufficient number of reports to establish an estimate.
[1] The average rates are estimates (rates over $10.00 are rounded to the nearest dime) based on survey indications of monthly lease rates for private, non-irrigated grazing land from the January Cattle Survey. [2] Includes animal unit plus cow-calf rates. Cow-calf rate converted to animal unit (AUM) using (1 aum=cow-calf *0.833). [3] Seventeen Western States: All States listed. [4] Sixteen Western States: All States, except Texas. [5] Eleven Western States: Arizona, California, Colorado, Idaho, Montana, Nevada , New Mexico, Oregon, Utah, Washington and Wyoming. [6] Nine Great Plains States: Colorado, Kansas, Nebraska, New Mexico, North Dakota, Oklahoma, South Dakota, Texas and Wyoming.
NASS, Environmental, Economics, and Demographics Branch, (202) 720–6146.

CHAPTER X

INSURANCE, CREDIT, AND COOPERATIVES

The statistics in this chapter deal with insurance, agricultural credit and farm cooperatives. Some of the series were developed in connection with research activities of the Department, while others, such as data from agricultural credit agencies, are primarily records of operations.

Table 10-1.—Crop losses: Average percentage of indemnities attributed to specific hazards, by crops, 1948–2016

Crop	Year	Drought heat (excess)	Hail	Precip. (excess poor drainage)	Frost freeze (other cold damage)	Flood	Cyclone, tornado, wind, hot wind	In-sects	Dis-ease	All oth-ers
					Percent					
Adj. gross revenue	2001-2014	13	12	12	27	0	2	0	0	34
Adj. gross revenue-lite	2003-2014	32	1	1	9	1	34	11	0	11
Alfalfa seed	2002-2016	27	14	13	31	0	8	6	0	1
All other citrus trees	2001-2014	0	0	2	12	0	2	0	0	84
Almonds	1981-2016	9	4	53	30	0	4	0	0	0
Annual Forage	2014-2016	0	0	0	0	0	0	0	0	0
Apiculture	2009-2016	0	0	0	0	0	0	0	0	0
Apples	1963-2016	8	29	4	55	0	2	0	0	1
Avocado trees	1996-2013	0	0	4	9	0	87	0	0	0
Avocados	1998-2017	39	0	1	33	0	22	0	0	4
Banana	2012-2013	54	0	0	0	0	0	0	0	46
Barley	1956-2016	41	15	31	5	0	2	2	2	1
Blueberries	1995-2016	5	10	13	56	0	2	0	2	11
Buckwheat	2010-2016	40	24	22	10	0	3	0	0	0
Burley tobacco	1997-2016	42	3	38	2	6	2	0	5	2
Cabbage	1999-2016	15	1	38	21	1	12	6	3	4
Camelina	2012-2015	100	0	0	0	0	0	0	0	0
Canola	1995-2016	18	9	58	10	0	3	1	1	1
Cherries	1963-2016	19	4	35	30	0	3	0	0	9
Cherries	2000-2016	1	24	12	15	6	21	8	4	10
Cigar binder tobacco	1997-2016	10	14	24	2	0	2	0	45	1
Cigar filler tobacco	1998-2015	85	0	6	0	0	0	1	7	1
Cigar wrapper tobacco	1997-2015	0	0	51	3	0	0	0	47	0
Citrus	1952-1997	18	5	1	74	0	2	0	0	0
Citrus I	1998-2012	0	0	0	3	0	97	0	0	0
Citrus II	2000-2013	0	0	0	27	0	73	0	0	0
Citrus III	2001-2013	0	0	0	13	0	87	0	0	0
Citrus IV	1998-2011	0	6	0	13	0	81	0	0	0
Citrus V	1998-2013	0	1	0	13	0	85	0	0	0
Citrus VI	2005-2006	0	0	0	0	0	100	0	0	0
Citrus VII	1998-2013	0	4	0	15	0	81	0	0	0
Citrus VIII	2009-2012	0	13	0	87	0	0	0	0	0
Citrus trees	1984-1997	0	0	0	100	0	0	0	0	0
Citrus trees I	2008-2010	0	0	70	0	0	30	0	0	0
Citrus trees II	2008-2010	0	0	100	0	0	0	0	0	0
Citrus trees III	2010	0	0	99	0	0	0	0	0	1
Citrus trees IV	2004-2010	0	0	100	0	0	0	0	0	0
Citrus trees V	2010	0	0	100	0	0	0	0	0	0
Clams	2001-2016	0	0	0	19	0	23	0	3	55
Coffee	2009-2016	73	0	24	0	0	1	3	0	0
Corn	1948-2016	51	5	25	3	2	2	0	0	11
Cotton	1948-2016	56	12	13	3	1	11	1	0	2
Cotton ex long staple	1984-2016	82	2	5	8	0	2	1	0	0
Cranberries	1984-2016	16	17	16	41	2	2	4	1	1
Cucumbers	2014-2016	7	4	85	1	0	1	0	1	0
Cultivated wild rice	1999-2016	40	9	6	9	1	16	1	1	17
Dark air tobacco	1997-2016	15	4	62	1	4	5	0	7	1
Dry beans	1948-2016	16	18	47	13	0	2	0	2	1
Dry peas	1963-2016	32	29	29	5	0	1	1	1	2
Early, midseason oranges	1998-2017	0	7	18	40	0	34	0	0	0
Figs	1988-2016	19	0	44	28	0	2	0	0	6
Fire cured tobacco	1997-2016	17	8	49	3	6	6	0	4	7
Flax	1948-2016	38	8	44	4	0	3	1	1	1
Flue cured tobacco	1997-2016	33	5	29	2	0	16	0	15	0
Forage production	1979-2016	68	3	7	20	0	0	1	0	1
Forage seeding	1978-2016	30	1	21	45	1	2	0	0	0
Fresh apricots	1997-2016	10	18	22	46	0	4	0	0	1
Fresh freestone peaches	1997-2016	14	22	17	44	0	3	0	0	0
Fresh market beans	2000-2016	2	0	58	13	0	17	2	8	0
Fresh market sweet corn	1985-2016	10	2	33	33	0	18	1	2	1
Fresh market tomatoes	1984-2016	7	5	45	29	0	8	1	5	0
Fresh nectarines	1997-2016	11	34	19	34	0	2	0	0	0
Grain sorghum	1959-2016	76	2	10	4	0	4	1	0	2

See note(s) at end of table.

Table 10-1.—Crop losses: Average percentage of indemnities attributed to specific hazards, by crops, 1948–2016—Continued

Crop	Year	Drought heat (excess)	Hail	Precip. (excess poor drainage)	Frost freeze, (other cold damage)	Flood	Cyclone, tornado, wind, hot wind	In- sects	Dis- ease	All others
Grapefruit	1997-2017	28	0	1	57	0	13	0	0	0
Grapefruit trees	2001-2014	0	0	1	5	0	0	0	0	93
Grapes	1967-2016	20	5	14	57	0	2	0	0	2
Grass Seed	2012-2016	1	3	91	5	0	1	0	0	0
Green peas	1962-2016	46	5	39	7	0	2	0	1	0
Hybrid corn seed	1983-2016	58	5	21	2	2	11	0	1	0
Hybrid sorghum seed	1988-2016	25	20	8	17	0	28	0	0	1
Late oranges	1998-2016	0	12	24	25	0	38	0	0	0
Lemons	1997-2016	13	1	2	80	0	3	0	0	0
Lime trees	1998-2005	0	0	0	1	0	0	0	0	99
Macadamia nuts	1996-2016	72	0	7	0	0	4	6	0	10
Macadamia trees	2000-2012	0	0	56	0	0	36	0	0	8
Mandarins	1997-2014	16	3	1	79	0	2	0	0	0
Mandarins/tangerines	2015-2017	60	1	1	37	0	1	0	0	0
Mango trees	1997-2010	0	0	0	59	0	41	0	0	0
Maryland tobacco	1997-2016	58	8	7	4	0	6	0	17	0
Millet	1996-2016	70	15	12	2	0	1	0	0	0
Minneola tangelos	1998-2014	11	1	1	86	0	1	0	0	0
Mint	2000-2016	36	3	24	31	1	3	1	2	0
Mustard	1999-2016	47	21	22	4	0	4	1	0	0
Navel oranges	1998-2014	42	2	7	45	0	1	0	0	2
Nursery (fg&c)	2001-2017	2	13	26	19	4	29	4	1	1
Oats	1956-2016	55	10	25	5	0	2	1	1	0
Olives	2012-2016	45	4	11	29	0	11	0	0	0
Onions	1988-2016	21	8	43	8	1	2	0	16	1
Orange trees	1996-2016	0	0	30	59	0	12	0	0	0
Oranges	1958-2017	34	3	0	59	0	2	0	0	1
Orlando tangelos	1998-2012	0	0	0	100	0	0	0	0	0
Papaya	2007-2015	0	0	4	0	0	22	6	68	0
Papaya tree	2014-2015	0	0	0	0	0	84	0	5	10
Pastures/Rangeland Forage	2007-2016	0	0	0	0	0	0	0	0	0
Peaches	1957-2016	3	15	3	75	0	0	0	0	3
Peanuts	1962-2016	69	0	19	2	0	3	0	5	1
Pears	1990-2016	17	22	8	48	0	2	0	1	1
Pecans	1998-2016	22	8	21	28	0	13	0	3	4
Peppers	1984-2016	1	5	56	22	0	14	0	2	0
Pistachios	2012-2016	89	0	6	2	0	1	0	0	1
Plums	1998-2016	16	26	17	37	0	3	0	0	0
Popcorn	1984-2016	43	21	22	6	2	2	0	0	4
Potatoes	1962-2016	23	6	42	14	1	2	1	11	1
Processing apricots	1997-2016	9	7	35	44	0	5	0	0	0
Processing beans	1988-2016	47	7	35	5	0	3	0	2	1
Processing cling peaches	1997-2016	22	16	26	34	0	2	0	0	0
Processing cucumbers	2000-2005	45	1	47	2	0	1	1	4	0
Processing freestone	1998-2016	18	10	18	52	0	1	0	0	0
Prunes	1986-2016	37	2	30	19	0	11	0	0	0
Pumpkins	2009-2016	13	0	72	14	0	0	0	0	2
Raisins	1961-2015	0	0	100	0	0	0	0	0	0
Raspberry and blackberry	2002-2006	40	0	22	27	0	12	0	0	0
Rice	1960-2016	32	0	48	3	4	6	0	2	6
Rio red & star ruby	1998-2017	0	5	28	34	0	32	0	0	0
Ruby red grapefruit	1998-2017	0	12	37	19	0	33	0	0	0
Rye	1980-2016	36	15	30	15	0	1	0	0	1
Safflower	1964-2016	54	5	17	12	0	7	2	2	1
Sesame	2011-2016	54	3	34	2	0	5	0	0	3
Silage sorghum	2005-2016	82	7	2	1	0	9	1	0	0
Soybeans	1955-2016	45	6	32	4	3	1	0	0	9
Special citrus	1992-1994	6	12	0	82	0	0	0	0	0
Stonefruit	1988-1996	1	28	44	19	0	2	0	0	6
Strawberries	2000-2015	27	1	30	23	0	0	2	17	0
Sugar beets	1965-2016	11	5	47	20	1	6	1	9	1
Sugarcane	1967-2016	28	0	15	34	1	9	3	8	2
Sunflowers	1976-2016	25	7	49	7	0	3	2	3	4
Sweet corn	1978-2016	54	2	25	11	0	4	0	1	2
Sweet oranges	1998-2014	11	0	10	75	0	4	0	0	1
Sweet potatoes	1998-2016	1	12	0	35	0	0	0	3	49
Table grapes	1984-2016	34	8	29	26	0	2	0	0	1
Tangelos	1997-2017	32	1	0	56	0	11	0	0	0
Tangors	2014-2016	0	0	0	25	0	75	0	0	0
Tobacco	1948-1996	17	20	20	1	2	18	0	20	2
Tomatoes	1963-2016	33	2	41	9	1	2	3	9	1
Valencia oranges	1998-2014	34	2	3	57	0	3	0	0	0
Walnuts	1984-2016	28	6	28	32	0	5	0	0	0
Watermelons	1999	8	7	38	1	0	14	0	29	2
Wheat	1948-2016	43	11	22	13	0	3	0	2	4
Whole Farm Revenue Protection	2015-2016	35	21	9	7	0	0	0	1	28
Winter squash	1995-2005	10	13	75	0	2	0	0	0	1

GRP crops do not have any specific cause of loss.
RMA, Requirements, Analysis and Validation Branch, (816) 926–7910.

Table 10-2.—Crop insurance programs: Coverage, amount of premiums and indemnities, by crops, United States, 2013-2015

Commodity and year	Coverage				Amount of premium	Indemnities		
	County programs	Insured units [1]	Area insured [2]	Maximum insured production		Number	Area indemnified [2]	Amount
	Number	*Number*	*1,000 acres*	*1,000 dollars*	*1,000 dollars*		*Acres*	*1,000 dollars*
Adjusted gross revenue:								
2013	230	366	....	365,057	12,901	40	....	5,549
2014	230	358	....	396,739	14,351	100	....	41,220
Adjusted gross revenue-lite:								
2013	1,706	390	....	105,312	4,409	68	....	4,619
2014	1,706	435	....	125,450	5,168	150	....	14,023
Alfalfa seed:								
2013	12	321	28	27,942	2,081	98	5,235	1,856
2014	13	373	32	37,403	2,810	123	7,985	3,297
2015	13	347	32	36,489	2,776	88	6,860	3,221
All other citrus trees:								
2013	28	549	....	51,551	932	2	....	105
2014	28	498	....	47,933	850	1	....	13
2015	28	466	....	46,683	830	....	....	....
All other grapefruit:								
2013	3	1	....	1	....	....	....	....
2014	3	1	....	....	....	....	....	....
2015	3	....	....	....	....	....	....	....
Almonds:								
2013	16	5,769	696	1,516,401	55,328	201	11,648	4,573
2014	16	6,024	721	2,185,671	82,572	648	54,672	40,050
2015	16	6,113	737	2,908,986	102,620	886	60,497	62,500
Annual Forage:								
2014	648	1,356	178	22,721	5,203	844	....	6,928
2015	648	2,267	322	36,925	8,730	1,175	....	6,490
Apiculture:								
2013	2,059	2,079	....	33,797	7,154	1,351	....	9,984
2014	2,059	5,150	....	77,381	17,015	2,764	....	15,670
2015	2,059	6,851	....	101,171	21,840	3,661	....	24,785
Apples:								
2013	360	6,544	240	934,488	85,609	1,156	26,414	59,961
2014	334	6,644	249	1,089,448	101,798	1,028	19,777	48,510
2015	334	7,108	239	1,167,385	103,483	1,624	31,162	88,214
Avocado trees:								
2013	1	168	....	22,590	782	10	....	199
2014	1	161	....	21,363	764	....	....	....
2015	1	166	....	23,995	868	....	....	....
Avocados:								
2013	7	1,176	38	76,476	8,850	90	5	861
2014	7	1,143	38	84,506	9,376	299	6,346	6,458
2015	7	1,118	37	84,499	8,031	261	7,274	5,145
Banana tree:								
2013	4	1	0	674	41	....	....	....
2014	4	1	0	462	26	....	....	....
2015	4	1	0	685	19	....	....	....
Bananas:								
2013	4	6	....	1,393	35	1	1	1
2014	4	4	....	1,487	36	....	....	....
2015	4	4	....	719	8	....	....	....
Barley:								
2013	3,030	22,459	2,589	609,591	81,372	5,949	774,232	69,367
2014	3,021	19,058	2,189	442,532	53,410	5,364	819,962	61,401
2015	3,696	20,840	2,676	534,655	69,085	4,621	706,580	37,329
Blueberries:								
2013	74	1,144	63	155,633	12,443	215	4,816	7,145
2014	76	1,181	66	176,751	13,504	100	2,358	3,032
2015	77	1,219	69	194,131	11,381	235	6,916	8,676
Buckwheat:								
2013	22	133	10	1,826	362	46	3,437	248
2014	22	130	11	1,852	382	45	4,585	277
2015	22	150	12	2,114	368	62	4,425	349
Burley tobacco:								
2013	253	6,471	64	166,547	20,562	2,163	23,965	33,561
2014	253	7,905	86	259,103	35,278	2,727	39,888	78,746
2015	253	6,135	69	180,196	25,126	3,349	43,717	72,881

See footnote(s) at end of table.

Table 10-2.—Crop insurance programs: Coverage, amount of premiums and indemnities, by crops, United States, 2013–2015—Continued

Commodity and year	Coverage					Indemnities		
	County programs	Insured units [1]	Area insured [2]	Maximum insured production	Amount of premium	Number	Area indemnified [2]	Amount
	Number	*Number*	*1,000 acres*	*1,000 dollars*	*1,000 dollars*		*Acres*	*1,000 dollars*
Cabbage:								
2013	29	265	13	21,376	1,973	23	558	516
2014	29	305	14	25,852	2,221	55	1,694	1,554
2015	29	310	14	28,094	2,047	79	2,465	1,270
Camelina:								
2013	52	4		22	4			
2014	52							
2015	52	1		10	1	1	140	6
Canola:								
2013	489	13,183	1,685	419,483	73,801	6,225	807,096	118,179
2014	504	13,580	1,705	350,829	59,725	5,713	719,123	70,290
2015	552	12,722	1,719	321,688	56,780	4,054	629,680	27,333
Carambola trees:								
2013	1	6		511	16			
2014	2	11		1,716	59			
2015	2	10		1,640	56			
Cherries:								
2013	51	3,068	70	418,721	37,951	969	19,553	41,753
2014	63	3,503	90	468,225	42,430	971	24,645	58,819
2015	64	3,549	89	471,917	42,861	1,119	26,792	53,623
Chile Peppers:								
2013	3	24	3	1,351	80			
2014	3	21	2	939	54			
2015	3	20	2	904	39	1	36	14
Cigar binder tobacco:								
2013	16	372	3	17,889	4,951	145	1,143	5,330
2014	16	337	3	15,849	4,310	98	732	3,584
2015	16	331	3	17,585	4,527	47	456	2,308
Cigar filler tobacco:								
2013	3	13		273	5			
2014	3	8		231	4			
2015	3	8		235	4	1	1	
Cigar wrapper tobacco:								
2013	5	20	1	8,799	616			
2014	5	14		5,600	448			
2015	5	11		3,589	199	7	142	400
Citrus I:								
2013	29	2,366	163	158,785	3,960			
Citrus II:								
2013	29	2,314	206	233,852	8,113	8		54
Citrus III:								
2013	29	103	2	1,764	47	2		4
Citrus IV:								
2013	29	437	8	10,645	463			
Citrus V:								
2013	29	339	7	12,584	679	2		24
Citrus VI:								
2013	5							
Citrus VII:								
2013	29	882	59	88,543	4,922	3		19
Citrus VIII:								
2013	29	332	5	7,714	285			
Clams:								
2013	13	117		22,689	726	19		530
2014	13	116		20,041	657	14		1,004
2015	9	88		17,272	430	21		1,808
Coffee:								
2013	4	71	4	4,969	186	31	197	265
2014	4	79	4	8,542	269	14	130	168
2015	4	92	4	12,375	378	30	209	391
Coffee tree:								
2013	4	27		14,800	47			
2014	4	29		14,023	49			
2015	4	45		17,644	58			
Corn:								
2013	9,567	772,912	84,929	56,546,612	4,692,708	315,523	47,592,069	5,844,262
2014	9,549	735,034	79,060	43,986,245	3,649,571	248,383	38,355,086	3,842,778
2015	15,951	712,167	78,681	40,319,695	3,685,913	124,700	15,742,435	1,677,587
Cotton:								
2013	2,352	117,446	9,915	3,780,863	82,457	50,312	13,975,603	1,006,347
2014	2,352	119,489	10,372	4,048,548	23,417	41,774	11,001,930	723,241
2015	5,922	131,968	8,593	2,990,043	57,614	36,200	6,843,739	389,259
Cotton exlong staple:								
2013	31	859	213	147,806	12,445	332	61,217	26,892
2014	29	1,235	277	251,872	33,351	540	157,632	89,973
2015	29	1,233	257	213,150	33,189	635	180,789	94,042

See footnote(s) at end of table.

Table 10-2.—Crop insurance programs: Coverage, amount of premiums and indemnities, by crops, United States, 2013–2015—Continued

Commodity and year	Coverage				Amount of premium	Indemnities		
	County programs	Insured units [1]	Area insured [2]	Maximum insured production		Number	Area indemnified [2]	Amount
	Number	Number	1,000 acres	1,000 dollars	1,000 dollars		Acres	1,000 dollars
Cranberries:								
2013	30	701	32	127,433	4,122	55	1,227	1,547
2014	30	684	32	99,913	3,041	79	2,617	2,949
2015	30	682	33	86,549	2,501	86	3,287	3,061
Cucumbers:								
2014	40	379	28	19,651	1,506	66	4,772	1,595
2014	40	424	31	20,369	1,628	98	6,297	1,895
Cultivated wild rice:								
2013	11	95	25	14,853	826	12	2,268	393
2014	11	76	23	15,652	980	17	3,984	875
2015	11	91	26	21,290	1,381	15	6,743	529
Dark air tobacco:								
2013	30	684	4	13,934	487	107	622	946
2014	30	801	4	18,057	654	46	242	423
2015	30	809	5	18,513	677	156	974	1,583
Dry beans:								
2013	422	15,172	1,119	464,071	70,185	4,079	333,664	63,920
2014	484	17,793	1,359	577,726	85,099	4,618	426,141	70,219
2015	484	15,858	1,382	537,269	75,450	4,337	511,572	57,776
Dry Peas:								
2013	345	12,187	1,369	264,587	38,863	2,623	354,306	29,420
2014	372	11,496	1,353	208,394	30,404	2,936	422,921	25,688
2015	373	13,020	1,718	312,919	47,190	4,432	753,885	49,106
Early and Midseason oranges:								
2013	3	223	3	2,178	135	19		63
2014	3	210	4	2,776	160	25	207	86
2015	3	285	4	4,011	235	26	227	75
Figs:								
2013	4	41	4	4,134	178			
2014	4	64	4	5,821	295	6	555	81
2015	4	64	4	6,518	317	4	642	245
Fired cured tobacco:								
2013	41	1,158	11	50,891	2,144	195	1,752	3,578
2014	41	1,370	14	70,208	3,032	123	1,338	2,100
2015	41	1,312	14	67,901	2,866	174	1,916	4,553
Flax:								
2013	150	2,606	219	35,760	5,698	1,152	96,480	9,487
2014	142	3,362	297	47,596	7,024	794	70,779	4,794
2015	142	4,541	426	52,847	7,672	719	71,035	3,002
Flue cured tobacco:								
2013	146	11,241	221	651,841	47,536	3,659	71,905	79,056
2014	146	11,986	242	870,559	64,072	2,250	47,869	71,157
2015	146	11,222	216	663,106	48,251	3,501	72,294	106,347
Forage production:								
2013	772	30,693	3,298	606,014	62,563	3,336	301,863	26,207
2014	772	25,022	3,075	649,475	74,018	2,019	227,838	22,289
2015	780	20,381	2,716	562,738	52,558	3,268	370,559	35,856
Forage seeding:								
2013	589	4,322	190	32,267	4,211	993	38,499	3,850
2014	589	3,832	159	26,529	3,527	1,213	46,888	5,312
2015	590	4,798	215	37,269	4,923	1,260	50,378	5,139
Fresh apricots:								
2013	29	150	3	10,051	1,364	18	497	745
2014	29	146	3	9,643	1,277	14	402	668
2015	29	148	3	11,779	1,543	23	589	1,239
Fresh freestone peaches:								
2013	25	660	18	26,302	900	34	237	195
2014	25	681	19	33,067	1,214	41	311	201
2015	25	727	18	53,357	1,459	60	603	614
Fresh market beans:								
2013	15	26	4	1,812	222	1	37	4
2014	15	28	4	1,968	255			
2015	15	26	5	2,653	301	10	2,268	376
Fresh market sweet corn:								
2013	230	804	53	49,565	4,652	105	4,242	1,918
2014	230	738	51	48,527	4,533	93	4,280	2,529
2015	230	665	43	43,574	3,225	86	3,519	1,727
Fresh market tomatoes:								
2013	59	671	45	208,654	26,958	171	6,934	16,942
2014	59	586	44	177,051	20,758	177	5,609	14,033
2015	59	520	39	145,861	14,426	109	3,576	8,626

See footnote(s) at end of table.

Table 10-2.—Crop insurance programs: Coverage, amount of premiums and indemnities, by crops, United States, 2013–2015—Continued

Commodity and year	Coverage				Amount of premium	Indemnities		
	County programs	Insured units [1]	Area insured [2]	Maximum insured production		Number	Area indemnified [2]	Amount
	Number	Number	1,000 acres	1,000 dollars	1,000 dollars		Acres	1,000 dollars
Fresh nectarines:								
2013	24	714	16	27,778	2,036	35	581	313
2014	24	700	17	34,460	2,462	72	867	581
2015	24	721	15	49,916	3,005	91	1,353	808
Grain sorghum:								
2013	3,879	82,928	5,806	1,307,616	277,750	36,801	4,863,048	365,212
2014	3,852	75,944	5,304	948,986	210,250	23,889	2,813,876	127,032
2015	5,196	91,901	6,962	1,103,103	255,604	17,138	1,970,394	83,961
Grapefruit:								
2013	11	254	8	11,845	432	25		656
2014	41	888	47	79,527	4,067	55	1,554	1,017
2015	41	833	46	80,233	4,032	27	816	532
Grapefruit trees:								
2013	31	1,280		188,091	4,865	8		55
2014	31	1,229		178,677	4,773	1		5
2015	31	1,193		185,634	5,036			
Grapes:								
2013	105	16,273	594	1,263,826	46,714	688	12,743	8,128
2014	110	18,098	606	1,490,850	58,731	1,804	41,890	33,727
2015	110	17,408	585	1,455,011	54,573	3,477	61,811	62,295
Grass seed:								
2013	6	74	18	2,408	352			
2014	6	101	29	4,061	637	1	177	35
2015	6	98	31	7,074	727	12	3,705	363
Green peas:								
2013	170	2,146	149	71,271	8,128	802	58,281	11,174
2014	170	2,039	141	63,220	6,970	830	62,341	10,424
2015	170	2,034	139	62,522	6,549	507	41,585	8,722
Hybrid corn seed:								
2013	229	9,399	618	702,336	47,643	645	48,553	14,107
2014	229	5,280	356	303,713	19,381	416	32,106	9,282
2015	229	4,179	272	207,456	13,056	444	35,729	7,072
Hybrid sorghum seed:								
2013	23	875	71	51,309	5,720	37	2,850	1,110
2014	23	625	43	23,558	2,623	73	6,351	1,793
2015	26	488	36	16,230	1,748	30	3,156	885
Late oranges:								
2013	3	89	1	691	121	7		26
2014	3	88	1	889	145	5	37	17
2015	3	97	1	1,414	241	19	258	39
Lemons:								
2013	15	781	43	98,717	5,036	27		1,256
2014	19	813	43	118,210	6,442	66	2,450	2,007
2015	19	843	43	134,524	7,612	63	1,817	2,589
Macadamia nuts:								
2013	3	103	12	20,044	618	17		590
2014	3	101	12	19,955	617	36	3,900	1,523
2015	3	102	12	20,678	637	1	67	33
Macadamia trees:								
2013	3	101	12	57,554	587			
2014	3	107	13	57,698	588			
2015	3	132	13	52,399	537			
Mandarins:								
2013	11	463	27	116,648	11,717	45		2,850
2014	12	556	29	138,656	13,877	146	4,907	16,341
Mandarins/Tangerines:								
2014	29	328	5	7,022	297			
2014	41	973	40	192,912	19,020	70	2,163	3,624
Mango trees:								
2013	1	13		393	15			
2014	1	13		401	16			
2015	1	13		436	17			
Maryland tobacco:								
2013	6	24		132	7	3	10	6
2014	6	25		207	10	5	34	12
2015	6	2		2		2	3	2
Millet:								
2013	68	6,139	544	69,805	16,752	2,000	268,750	14,753
2014	68	5,195	412	43,112	10,253	729	89,434	3,088
2015	68	4,702	364	29,024	6,452	682	74,789	2,148

See footnote(s) at end of table.

Table 10-2.—Crop insurance programs: Coverage, amount of premiums and indemnities, by crops, United States, 2013–2015—Continued

Commodity and year	Coverage				Amount of premium	Indemnities		
	County programs	Insured units [1]	Area insured [2]	Maximum insured production		Number	Area indemnified [2]	Amount
	Number	Number	1,000 acres	1,000 dollars	1,000 dollars		Acres	1,000 dollars
Minneola Tangelos:								
2013	10	246	7	14,561	1,307	13	....	250
2014	10	244	7	15,974	1,407	46	1,336	1,458
Mint:								
2013	31	220	23	21,219	1,060	46	2,979	1,001
2014	31	256	25	22,203	989	56	3,531	810
2015	31	248	20	17,991	696	63	3,467	504
Mustard:								
2013	45	214	31	5,337	1,296	43	7,894	808
2014	45	119	18	2,528	544	33	6,881	447
2015	50	165	23	2,784	629	51	8,378	376
Navel Oranges:								
2013	16	2,950	120	226,192	15,490	212	....	3,137
2014	17	2,867	119	237,713	16,267	662	29,289	27,623
Nursery (FG & C):								
2013	2,802	3,766	....	1,804,900	38,402	38	....	1,748
2014	2,799	3,056	....	1,563,248	33,906	50	....	3,407
2015	2,802	2,792	....	1,460,237	31,038	36	....	678
Oats:								
2013	1,509	10,304	464	54,974	9,931	2,100	96,720	5,703
2014	1,505	10,551	493	49,967	9,039	1,899	79,636	4,135
2015	1,505	11,518	539	54,599	9,381	1,434	65,233	3,372
Olives:								
2012	13	487	26	27,375	2,641	73	2,093	1,284
2013	13	468	26	28,935	2,913	256	11,008	7,331
2014	13	456	26	37,296	3,541	80	2,010	1,630
Onions:								
2013	93	1,547	92	164,042	33,043	642	20,743	26,469
2014	101	1,843	112	213,749	39,998	679	23,830	29,786
2015	100	1,702	103	200,672	35,998	866	28,653	41,229
Orange trees:								
2013	31	3,561	....	1,325,478	22,500	16	....	329
2014	31	3,332	....	1,311,749	20,685	2	....	33
2015	31	3,050	....	1,273,440	19,903	1	....	48
Oranges:								
2014	29	4,375	369	451,339	13,735	4	216	127
2015	48	7,656	497	711,371	33,115	458	12,307	8,808
Orlando Tangelos:								
2014	5	2	....	23	2	....	....	....
2015	5	2	....	20	1	....	....	....
Papaya:								
2013	4	6	....	355	6	....	....	....
2014	4	6	....	242	4	2	17	55
2015	4	56	....	1,795	46	9	47	200
Papaya tree:								
2013	4	3	....	105	1	....	....	....
2014	4	3	....	133	4	1	....	4
2015	4	10	....	345	6	5	....	38
Pasture/Range-land/Forage:								
2013	2,059	124,772	54,287	980,553	196,672	56,657	26,881,556	177,728
2014	2,059	112,378	52,764	974,372	201,457	55,095	....	180,119
2015	2,061	120,811	54,653	1,045,626	216,375	37,982	....	126,518
Peaches:								
2013	232	1,493	35	78,948	17,554	478	8,846	14,095
2014	222	1,477	34	80,398	18,408	625	13,605	24,096
2015	222	1,456	33	81,158	19,876	491	9,620	16,873
Peanuts:								
2013	360	19,979	975	495,394	46,212	2,661	113,357	25,519
2014	355	24,813	1,261	699,282	65,255	5,737	269,340	74,004
2015	1,065	25,373	1,502	737,902	72,653	6,917	404,645	96,246
Pears:								
2013	29	1,810	34	87,648	1,900	74	1,015	839
2014	29	1,866	34	97,744	2,174	88	2,178	1,659
2015	29	1,810	33	134,626	2,840	123	1,603	2,022
Pecans:								
2013	143	1,586	169	190,116	13,461	588	44,538	15,249
2014	147	1,784	158	237,766	14,792	481	35,194	13,673
2015	147	1,736	157	234,366	14,679	332	24,923	9,322
Peppers:								
2013	13	118	7	26,334	4,382	33	819	2,433
2014	13	110	6	23,689	3,788	14	382	1,211
2015	13	114	7	21,327	2,271	34	1,014	2,757

See footnote(s) at end of table.

Table 10-2.—Crop insurance programs: Coverage, amount of premiums and indemnities, by crops, United States, 2013–2015—Continued

Commodity and year	Coverage				Amount of premium	Indemnities		
	County programs	Insured units [1]	Area insured [2]	Maximum insured production		Number	Area indemnified [2]	Amount
	Number	Number	1,000 acres	1,000 dollars	1,000 dollars		Acres	1,000 dollars
Pistachios:								
2013	24	523	89	201,453	7,477	118	9,003	4,738
2014	24	567	92	295,970	11,495	99	12,244	19,961
2015	24	613	106	431,536	15,676	332	81,285	193,093
Plums:								
2013	23	850	15	27,427	3,157	118	1,435	972
2014	23	800	14	22,972	2,745	93	1,087	731
2015	23	754	14	30,890	3,203	212	2,645	2,494
Popcorn:								
2013	660	2,252	209	131,735	10,691	252	29,183	4,363
2014	624	2,858	256	154,229	12,516	478	59,570	9,649
2015	2,634	2,225	218	132,375	10,458	542	69,588	12,467
Potatoes:								
2013	316	6,062	860	1,127,787	94,573	1,047	74,847	45,701
2014	315	6,161	865	1,179,811	98,447	788	51,484	37,263
2015	314	5,766	844	1,132,524	83,015	932	60,312	43,661
Processing apricots:								
2013	13	64	3	3,599	528	5	330	183
2014	13	67	3	4,685	721	5	335	343
2015	13	66	3	5,488	804	32	1,864	1,437
Processing beans:								
2013	148	1,354	86	42,814	4,009	232	12,810	3,157
2014	148	1,540	98	46,990	4,562	277	17,489	3,980
2015	146	1,497	94	49,684	4,545	327	16,608	4,543
Processing cling peaches:								
2013	10	933	16	42,829	1,763	48	531	417
2014	10	877	16	46,791	2,024	53	590	417
2015	10	818	15	54,825	2,115	33	478	642
Processing freestone:								
2013	8	96	3	5,988	216	2	53	95
2014	8	99	3	6,116	263	7	83	20
2015	8	89	2	7,220	231	6	466	84
Prunes:								
2013	14	838	48	55,160	10,860	446	22,641	12,657
2014	14	807	46	78,579	15,465	188	10,823	9,533
2015	14	773	44	113,650	23,347	146	8,507	9,707
Pumpkins:								
2013	11	137	6	5,386	319	55	3,676	530
2014	11	140	6	4,136	246	12	628	64
2015	11	180	8	5,150	296	142	10,546	1,916
Raisins:								
2013	7	2,361	....	252,736	13,900	22	....	162
2014	7	2,412	....	191,246	8,294	34	....	100
2015	7	2,509	....	243,625	11,125	343	....	1,667
Rice:								
2013	405	15,633	2,308	1,279,279	62,585	3,964	619,814	123,859
2014	402	18,632	2,668	1,783,211	93,808	3,395	506,728	142,153
2015	771	23,425	3,232	1,505,736	69,151	5,324	837,268	188,292
Rio Red & Star Ruby:								
2013	3	361	10	11,534	2,475	42	....	479
2014	3	340	12	14,398	3,064	50	940	397
2015	3	330	12	15,196	3,226	72	1,256	750
Ruby red grapefruit:								
2013	3	57	1	476	81	5	....	30
2014	3	51	1	475	79	8	91	15
2015	3	48	1	561	94	17	379	75
Rye:								
2013	48	329	32	3,160	616	102	7,474	485
2014	48	363	32	3,305	602	196	21,399	1,015
2015	48	568	45	5,420	976	157	10,416	613

See footnote(s) at end of table.

Table 10-2.—Crop insurance programs: Coverage, amount of premiums and indemnities, by crops, United States, 2013–2015—Continued

| Commodity and year | Coverage | | | | Amount of premium | Indemnities | | |
	County programs	Insured units [1]	Area insured [2]	Maximum insured production		Number	Area indemnified [2]	Amount
	Number	Number	1,000 acres	1,000 dollars	1,000 dollars		Acres	1,000 dollars
Safflower:								
2013	83	718	121	13,502	2,777	125	21,026	741
2014	83	740	125	13,678	3,028	189	26,582	2,063
2015	83	639	122	10,467	1,884	113	22,806	868
Sesame:								
2013	35	379	27	4,629	784	225	22,375	2,454
2014	35	717	43	7,807	1,318	215	21,220	878
2015	53	1,148	68	10,502	2,131	421	41,143	2,518
Silage Sorghum:								
2013	97	908	81	23,263	3,574	187	20,833	2,236
2014	97	948	82	19,971	2,898	123	13,179	1,421
2015	97	708	63	15,391	2,117	88	8,905	1,013
Soybeans:								
2013	8,010	694,129	67,549	27,810,209	2,494,824	161,218	15,072,204	1,205,727
2014	8,010	722,912	73,903	27,484,770	2,261,062	171,940	20,152,834	1,224,139
2015	13,263	709,924	74,839	24,281,897	2,106,819	147,763	15,737,357	1,155,244
Strawberries:								
2013	6	3		2,520	75			
2014	6	1		325	22			
2015	6	16	1	22,034	719	1	41	111
Sugar beets:								
2013	120	14,987	1,105	1,107,671	62,066	2,174	213,903	64,697
2014	120	14,201	1,062	736,568	40,307	1,896	180,310	22,865
2015	120	14,011	1,040	817,279	41,518	919	79,126	10,115
Sugarcane:								
2013	52	5,072	759	292,739	7,531	150	6,610	1,697
2014	31	4,791	584	257,586	7,183	210	14,421	3,442
2015	31	4,832	760	226,401	3,697	193	9,836	975
Sunflowers:								
2013	915	11,151	1,691	417,687	76,040	5,885	1,024,992	129,101
2014	915	10,184	1,593	334,517	56,441	4,123	757,076	73,996
2015	915	10,510	1,697	306,477	52,318	2,686	428,374	31,458
Sweet corn:								
2013	167	2,751	225	110,466	5,951	354	29,127	8,414
2014	166	2,710	226	102,611	5,248	364	25,817	4,483
2015	166	2,541	209	87,790	4,384	266	20,990	5,208
Sweet oranges:								
2013	6	69		854	67	4		12
2014	6	75		923	70	12	86	84
Sweet potatoes:								
2013	9	23	7	5,441	819			
2014	9	25	7	5,579	718			
2015	9	18	7	4,105	320	1	133	12
Table grapes:								
2013	9	1,157	85	251,429	9,695	58	1,267	2,440
2014	10	1,208	81	286,090	10,792	106	2,788	4,374
2015	10	1,144	78	294,140	8,779	135	3,617	9,612
Tangelos:								
2014	29	247	2	1,537	68			
2015	39	477	9	18,573	1,548	25	484	395
Tangerine trees:								
2013	3	15	17	390	14			
2014	3	15	17	394	15			
2015	3	13	9	319	14			
Tangors:								
2014	29	282	6	9,998	541	1	12	14
2015	29	248	5	10,115	5535			
Tomatoes:								
2013	78	2,739	269	468,186	8,873	175	18,108	8,618
2014	93	3,142	296	555,727	11,587	130	11,115	4,103
2015	93	3,191	296	677,754	14,179	163	11,814	8,254
Valencia oranges:								
2012	13	1,120	31	48,458	4,415	82		1,897
2013	13	1,152	30	47,396	4,268	214	4,384	3,930
Walnuts:								
2013	26	1,623	140	257,627	7,722	83	2,700	1,123
2014	26	1,721	149	349,066	10,396	36	1,506	1,402
2015	26	1,750	154	436,225	9,910	50	1,997	1,208
Wheat:								
2013	8,394	478,484	48,675	11,750,706	1,984,549	200,321	28,442,765	2,280,632
2014	8,376	455,446	47,968	9,270,772	1,453,541	194,289	26,787,104	1,643,091
2015	10,530	461,526	49,448	8,415,303	1,284,514	182,301	26,082,988	1,218,538
Whole Farm Revenue Protection:								
2015	2,544	1,122		1,148,025	53,284	322		66,851

[1] Number of farms on which the insured crop was planted including duplication where both the landlord and tenant are insured. Insured farms on which no insured crop was planted are not included. [2] The insured's share of the planted area on the farm.

RMA, Requirements, Analysis and Validation Branch, (816) 926–7910.

Table 10-3.—Farm real estate debt: Amount outstanding by lender, United States, Dec. 31, 2006–2015 [1]

Year	Farm Credit System	Farm Service Agency	Farmer Mac	Commercial banks [1]
	Billion dollars	Billion dollars	Billion dollars	Billion dollars
2006	49.7	2.4	1.5	37.5
2007	57.8	2.6	2.8	41.1
2008	62.8	2.5	2.4	46.4
2009	69.1	2.6	2.1	44.2
2010	72.5	3.5	3.6	51.2
2011	75.3	3.3	3.8	52.9
2012	80.3	3.7	3.8	64.6
2013	85.3	3.7	4.5	68.9
2014	88.8	4.3	4.7	73.3
2015	96.7	4.9	4.8	79.2

Year	Individuals and others [1]	Life insurance companies	Storage facility loans	Total farm mortgage debt
	Billion dollars	Billion dollars	Billion dollars	Billion dollars
2006	10.1	12.0	0.3	113.4
2007	16.3	10.8	0.3	131.7
2008	19.6	13.7	0.4	147.9
2009	14.2	13.3	0.5	146.0
2010	10.2	12.4	0.6	154.1
2011	18.1	13.0	0.7	167.2
2012	8.7	11.5	0.7	173.4
2013	10.1	12.0	0.7	185.2
2014	12.5	12.4	0.8	196.8
2015	10.0	12.5	0.8	208.8

Data as of February 7, 2017.
[1] Beginning with 2012 estimates, farm sector debt held by savings associations is reported with the commercial bank lender group instead of the individuals and others grouping.
ERS, Farm Farm Income Team, (202) 694–5586. FarmIncomeTeam@ers.usda.gov. http://www.ers.usda.gov/data-products/farm-income-and-wealth-statistics.aspx.

Table 10-4.—Nonreal estate farm debt: Amount outstanding, by lender, United States, Dec. 31, 2006–2015

Year	Commericial banks [1]	Farm Credit System	Farm Service Agency	Individuals and others [1]	Total nonreal estate debt
	Billion dollars	Billion dollars	Billion dollars	Billion dollars	Billion dollars
2006	48.4	27.0	2.7	24.2	102.3
2007	54.0	31.4	3.1	20.5	109.0
2008	56.3	36.0	3.0	17.9	113.2
2009	57.2	39.0	3.3	22.9	122.4
2010	56.2	39.2	3.6	25.8	124.9
2011	59.2	41.1	3.5	23.5	127.3
2012	59.9	42.7	3.4	18.2	124.2
2013	63.7	44.0	2.8	19.6	130.2
2014	70.7	47.9	3.6	26.2	148.4
2015	73.2	48.3	3.7	22.8	148.0

Data as of February 7, 2017.
[1] Beginning with 2012 estimates, farm sector debt held by savings associations is reported with the commercial bank lender group instead of the individuals and others grouping.
ERS, Farm Farm Income Team, (202) 694–5586. FarmIncomeTeam@ers.usda.gov. http://www.ers.usda.gov/data-products/farm-income-and-wealth-statistics.aspx.

Table 10-5.—Farm Service Agency: Loans made to individuals and associations for farming purposes, and amount outstanding, United States and Territories, 2005–2014 [1]

Year	Loans to individuals						
	Farm ownership			Soil and water			Recreation
	New borrowers	Loans made	Outstanding Jan. 1	New borrowers	Loans made	Outstanding Jan. 1	Outstanding Jan. 1
	Number	1,000 dollars	1,000 dollars	Number	1,000 dollars	1,000 dollars	1,000 dollars
2005	4,199	1,298,943	8,190,313	0	0	27,341	875
2006	3,878	1,223,725	8,343,554	0	0	21,451	714
2007	3,865	1,268,809	8,518,399	0	0	18,477	674
2008	4,335	1,552,303	8,876,232	0	0	13,954	514
2009	5,048	1,832,709	9,800,441	0	0	12,469	267
2010	6,281	2,308,813	10,875,039	0	0	10,364	4,632
2011	5,945	2,486,929	11,872,786	0	0	8,053	208
2012	4,976	2,029,778	12,498,221	0	0	6,585	191
2013	4,224	1,946,912	12,842,052	0	0	5,483	133
2014	7,418	3,012,450	13,768,000	0	0	4,617	102

Year	Loans to individuals					
	Operating			Emergency		
	New borrowers	Loans made	Outstanding Jan. 1	New borrowers	Loans made	Outstanding Jan. 1
	Number	1,000 dollars	1,000 dollars	Number	1,000 dollars	1,000 dollars
2005	8,891	1,723,953	6,404,277	235	23,569	1,150,557
2006	9,623	1,849,894	6,131,132	494	51,525	975,594
2007	8,673	1,789,590	5,732,012	691	74,898	920,453
2008	8,207	1,710,441	5,731,149	385	44,994	792,120
2009	11,778	2,611,248	6,500,532	177	30,401	726,370
2010	12,056	2,934,097	7,047,105	187	35,598	661,950
2011	10,064	2,285,131	6,953,778	176	32,610	578,846
2012	10,325	2,103,112	6,594,239	155	31,436	537,640
2013	10,685	1,960,150	6,254,820	224	23,340	479,869
2014	12,039	2,201,420	6,214,490	118	18,106	428,304

Year	Loans to associations					Economic opportunity individual loans	Economic emergency loans
	Indian tribe land acquisition			Grazing association	Irrigation, drainage, and soil conservation		
	New borrowers	Loans made	Outstanding Jan. 1	Outstanding Jan. 1	Outstanding Jan. 1	Outstanding Jan. 1	Outstanding Jan. 1
	Number	1,000 dollars	1,000 dolllars	1,000 dollars	1,000 dollars	1,000 dollars	1,000 dollars
2005	0	0	55,205	4,883	1,471	8	249,039
2006	0	360	52,134	3,613	1,263	8	198,266
2007	0	0	47,914	3,317	1,184	7	173,095
2008	0	0	43,764	2,945	1,045	6	135,303
2009	0	0	38,510	2,590	860	6	117,942
2010	0	0	34,306	2,232	707	6	100,515
2011	0	0	29,539	1,848	605	6	85,915
2012	0	0	25,112	1,618	511	6	76,843
2013	0	0	24,246	1,014	468	6	65,040
2014	0	0	19,825	798	284	6	57,053

[1] Includes loans made directly by FmHA and those guaranteed by the Agency. Amounts of loans made represent obligations and include loans to new borrowers and subsequent loans to borrowers who received an initial loan in a prior year. Amounts outstanding are loan advances less principal repayments for loans made directly by the Agency.

FSA, Loan Making Division, (202) 690–4006.

Table 10-6.—Farmers' marketing, farm supply, and related service cooperatives: Number, memberships, and business volume, United States, 2006–2015

Year[1]	Cooperatives[2]				Estimated memberships[4]			
	Marketing	Farm supply	Related service[3]	Total	Marketing	Farm supply	Related service[3]	Total
	Number	Number	Number	Number	1,000 members	1,000 members	1,000 members	1,000 members
2006	1,454	1,146	135	2,735	939	1,609	48	2,596
2007	1,385	1,094	116	2,595	814	1,605	40	2,459
2008	1,354	1,011	110	2,475	809	1,509	36	2,354
2009	1,277	992	121	2,390	754	1,448	35	2,237
2010	1,215	974	125	2,314	737	1,463	35	2,234
2011	1,222	935	128	2,285	846	1,398	36	2,279
2012	1,200	916	120	2,236	652	1,426	37	2,115
2013	1,195	871	120	2,186	655	1,284	37	1,977
2014	1,114	876	116	2,106	627	1,333	35	1,996
2015	1,079	874	94	2,047	591	1,296	34	1,921

Year[1]	Marketing volume		Farm supply volume		Service	Total marketing and farm supply volume and service receipts	
	Gross[5]	Net[6]	Gross[5]	Net[6]	Receipts[8]	Gross[5]	Net[6]
	Billion dollars	Billion dollars	Billion dollars	Billion dollars	Billion dollars	Billion dollars	Billion dollars
2006	77.613	71.484	44.916	34.871	4.225	126.754	110.580
2007	94.103	86.129	49.784	38.569	4.132	148.019	128.830
2008	118.197	111.699	70.525	51.172	4.744	193.465	167.615
2009	101.386	94.558	62.999	47.362	4.940	169.325	146.860
2010	103.031	95.756	63.842	47.118	4.930	171.803	147.805
2011	128.041	121.784	80.898	57.322	4.453	213.391	187.100
2012	140.900	133.200	92.200	64.700	4.700	237.800	202.600
2013	144.615	135.810	95.933	67.175	5.572	246.120	208.557
2014	147.731	138.340	92.624	65.638	6.315	246.670	210.293
2015	124.892	115.409	81.709	59.023	5.458	212.059	179.890

[1] Reports of cooperatives are included for the calendar year.   [2] Includes independent local cooperatives, centralized cooperatives, federations of cooperatives, and cooperatives with mixed organizational structures.   Cooperatives are classified according to their major activity.   If, for example, more than 50 percent of a cooperative's business is derived from marketing activities, it is included as a marketing cooperative.   [3] Includes cooperatives whose major activity is providing services related to marketing and farm supply activities.   [4] Includes members (those entitled to vote for directors) but does not include nonvoting patrons. (Some duplication exists because some farmers belong to more than one cooperative.)   [5] Estimated gross business includes all business reported between cooperatives, such as the wholesale business of farm supply cooperatives with other cooperatives or terminal market sales for local cooperatives.   [6] Estimated net business represents the value at the first level at which cooperatives transact business for farmers, adjusted for duplication resulting from intercooperative business.   [7] Receipts for services related to marketing or purchasing activities but not included in the volumes reported for those activities, plus other operting and non-operating income and losses and extraordinary items..
Rural Development, Cooperative Programs, (202) 720-7395.

Table 10-7.—Farmers' cooperatives: Business volume of marketing, farm supply, and related service cooperatives, United States, 2014–2015

Item	Gross business		Net business [1]	
	2014	2015	2014	2015
	Billion dollars	Billion dollars	Billion dollars	Billion dollars
Products marketed:				
Beans and peas (dry edible)	0.238	0.210	0.234	0.207
Cotton and cotton products	2.730	2.768	2.627	2.653
Dairy products	52.394	41.007	49.636	38.328
Fish	0.215	0.224	0.215	0.224
Fruits and vegetables	8.362	8.301	5.848	5.889
Grain and oilseeds [2]	58.837	49.320	57.898	48.347
Livestock and livestock products	4.948	4.793	4.947	4.792
Nuts	1.569	1.725	1.566	1.723
Poultry products	1.353	0.788	1.353	0.788
Rice	0.935	0.875	0.936	0.875
Sugar products	7.758	7.569	5.146	4.727
Tobacco	0.339	0.339	0.339	0.339
Wool and mohair	0.005	0.005	0.005	0.005
Other products [3]	8.048	6.969	7.590	6.514
Total farm products	147.731	124.893	138.340	115.409
Supplies purchased:				
Crop protectants	11.530	10.935	7.504	7.315
Feed	13.674	12.261	10.800	9.932
Fertilizer	16.251	15.051	13.116	12.326
Petroleum	39.211	32.277	25.558	21.390
Seed	5.791	5.397	3.444	3.188
Other supplies [4]	6.167	5.787	5.216	4.873
Total farm supplies	92.624	81.708	65.638	59.023
Receipts for services: [5]				
Trucking, cotton ginning, storage, grinding, locker plants, miscellaneous	6.315	5.458	6.315	5.458
Total business	246.670	212.059	210.293	179.890

[1] Represents value at the first level at which cooperatives transact business for farmers; adjusted for inter-cooperative business. [2] Excludes oilseed meal and oil. Oilseed meal is included in feed sales while oil sales are included in other products sales. [3] Includes coffee, forest products, hay, hops, seed marketed for growers, nursery stock, other farm products not separately classified, and sales of farm products not received directly from member-patrons. Also includes manufactured food products and resale items marketed by cooperatives. [4] Includes automotive supplies, building material, chicks, containers, farm machinery and equipment, hardware, meats and groceries, and other supplies not separately classified. [5] Trucking, cotton ginning, storage, grinding, locker plants, and/or miscellaneous services related to marketing or purchasing but not included in the volume reported for those activities, plus other operating and non-operating income and losses and extraordinary items.
Rural Development, Cooperative Programs, (202) 720–7395.

Table 10-8.—Farmers' cooperatives: Number of cooperatives, memberships, and business volume of marketing, farm supply, and related service cooperatives, by State, United States and Foreign, 2014–2015

State	Cooperatives headquartered in State		Memberships in State [1]		Net business [1]	
	2014	2015	2014	2015	2014	2015
	Number	*Number*	*Thousand*	*Thousand*	*Billion dollars*	*Billion dollars*
Alabama	47	46	21.89	20.83	0.86	0.81
Arkansas	32	30	34.47	32.92	2.27	2.13
California	109	107	33.80	33.91	10.36	8.92
Colorado	26	26	15.39	13.39	1.17	0.84
Florida	29	27	16.15	16.10	0.97	0.99
Georgia	12	11	1.30	1.29	0.37	0.40
Hawaii	13	12	0.58	0.59	0.01	0.01
Idaho	23	23	7.76	7.49	0.90	0.90
Illinois	104	97	113.68	107.80	17.49	15.27
Indiana	33	30	32.41	30.38	3.39	2.80
Iowa	91	88	98.96	95.21	16.23	14.20
Kansas	89	89	96.34	91.93	5.98	5.39
Kentucky	26	25	108.37	107.82	0.95	0.73
Louisiana	32	30	7.32	7.28	1.09	1.05
Maine	14	21	5.29	5.26	0.12	0.12
Maryland	14	11	22.74	22.55	0.26	0.24
Massachusetts	8	7	5.84	5.73	1.95	1.77
Michigan	40	39	13.55	13.67	3.09	2.81
Minnesota	186	184	182.88	172.81	55.42	45.46
Mississippi	47	44	88.21	54.99	2.11	2.18
Missouri	57	59	100.49	107.75	21.38	16.97
Montana	42	39	14.28	13.82	0.74	0.74
Nebraska	48	45	62.05	58.67	13.23	11.70
New Jersey	14	10	1.30	1.20	0.11	0.11
New York	51	49	5.79	4.41	3.24	1.74
North Carolina	11	12	3.05	2.68	0.51	0.46
North Dakota	142	138	57.01	57.31	5.94	5.19
Ohio	40	39	46.52	45.09	5.32	4.69
Oklahoma	50	49	34.46	34.02	0.96	0.86
Oregon	32	29	30.60	28.57	2.79	2.62
Pennsylvania	32	35	4.15	4.17	0.64	0.58
South Dakota	70	67	50.73	46.67	4.43	3.87
Tennessee	60	60	67.75	69.63	1.66	1.59
Texas	177	172	75.20	74.13	4.21	3.81
Utah	12	11	15.72	14.32	1.12	1.26
Virginia	52	52	298.03	298.65	3.66	3.16
Washington	60	60	25.25	23.87	5.15	4.81
West Virginia	13	12	46.34	46.34	0.04	0.03
Wisconsin	113	113	135.16	132.64	7.58	7.00
Wyoming	10	9	3.84	4.17	0.17	0.17
Other States [2]	73	40	11.10	11.01	2.44	1.48
United States	2,106	2,047	1,996	1,921	210.29	179.89
Foreign [3]	....	....	....	....	0.66	0.55
Total	2,106	2,047	1,996	1,921	210.29	179.89

[1] Represents value at the first level at which cooperatives transact business for farmers. Totals may not add due to rounding. [2] Dollar volume or membership is not shown to avoid disclosing operations of individual cooperatives. [3] Sales outside the United States and sales of certain products not received directly from member-patrons.

Rural Development, Cooperative Programs, (202) 720–7395.

Table 10-9.—Rural Utilities Service: Long-term electric financing approved by purpose, by States and United States as of December 31, 2011

State	Borrowers	Total financing approved						Loan estimates	
		Non-RUS financing		Financing approved by purpose					
		RUS loans [1]	With RUS guar- antee [2]	Without RUS guar- antee [3]	Distribu- tion	Genera- tion and trans- mission [3]	Consumer facilities	Miles of line	Con- sumers
	Number	1,000 dollars	1,000 dollars	1,000 dollars	1,000 dollars	1,000 dollars	1,000 dollars	Number	Number
AL	27	902,681	1,808,680	280,177	1,296,291	1,693,889	1,359	69,762	658,661
AK	18	864,376	670,188	198,024	678,309	1,053,190	1,089	11,157	197,575
AZ	15	423,646	957,481	220,448	705,799	895,285	490	23,372	243,293
AR	20	1,086,615	2,708,733	813,414	1,892,049	2,712,241	4,471	78,753	614,860
CA	10	95,586	56,513	7,216	117,757	41,501	56	6,828	80,037
CO	25	1,377,508	4,034,083	794,956	1,804,302	4,402,171	74	78,247	514,191
CT	-	-	-	-	-	-	-	-	-
DE	1	85,430	57,300	31,266	173,132	861	3	7,646	99,435
FL	18	1,467,692	2,565,326	1,153,744	2,448,286	2,735,958	2,519	77,937	1,048,309
GA	52	2,540,840	10,562,822	1,852,266	5,342,708	9,610,261	2,959	186,166	2,126,681
HI	1	215,000	142,928	8,240	282,314	83,854	-	872	31,099
ID	10	196,308	92,542	36,612	285,187	39,078	1,197	13,079	72,575
IL	29	628,023	1,352,350	638,935	709,872	1,909,224	212	55,715	256,959
IN	46	526,088	2,894,224	707,487	831,699	3,295,383	717	60,054	498,583
IA	45	857,085	1,000,326	173,640	911,243	1,119,411	397	67,365	219,974
KS	31	782,940	1,053,657	206,156	1,061,806	980,531	415	78,735	293,224
KY	26	1,766,221	5,430,379	1,581,746	2,726,967	6,050,220	1,159	94,386	969,693
LA	20	772,408	3,134,623	425,178	1,267,910	3,064,122	177	56,828	565,227
ME	5	41,737	22,350	20,396	51,425	33,014	44	2,183	20,708
MD	2	368,228	144,253	140,356	536,899	115,937	-	15,312	1 189,187
MA	-	-	-	-	-	-	-	-	-
MI	10	665,087	1,090,491	92,692	893,022	954,775	473	40,471	334,575
MN	46	1,687,889	2,749,021	556,717	2,492,135	2,497,134	4,357	124,632	717,487
MS	29	969,263	1,764,166	361,096	1,365,474	1,728,357	694	90,200	753,265
MO	48	1,863,673	3,437,021	600,270	2,346,124	3,554,019	821	127,495	790,878
MT	25	442,360	160,143	66,866	568,825	100,315	229	47,711	146,603
NE	35	483,130	38,616	42,711	449,732	114,163	562	76,004	177,923
NV	8	75,008	1,241	10,441	65,315	21,127	248	6,265	24,889
NH	1	81,213	143,839	8,696	87,335	146,380	32	4,616	64,601
NJ	2	18,173	-	5,377	22,250	1,295	4	1,000	13,334
NM	18	655,808	345,257	78,251	941,363	134,944	3,010	47,420	250,430
NY	6	46,782	30,205	20,114	87,731	9,286	85	5,298	29,533
NC	33	2,040,634	2,845,230	560,907	2,946,394	2,495,742	4,635	106,205	1,142,335
ND	23	1,179,286	4,845,541	953,122	1,060,313	5,915,651	1,986	71,164	588,129
OH	27	652,958	1,582,082	472,666	1,005,120	1,702,367	218	49,071	380,290
OK	29	1,197,869	1,621,525	328,543	1,577,044	1,568,217	2,676	101,801	538,851
OR	18	309,173	142,611	93,022	393,129	151,440	237	25,001	146,978
PA	13	462,705	689,683	132,794	658,902	626,041	239	28,276	234,175
RI	1	-	3,940	-	334	3,606	-	4	160
SC	28	1,884,073	1,684,010	379,965	3,110,289	835,527	2,231	79,910	860,908
SD	30	814,091	471,391	155,806	1,123,535	316,872	881	68,396	191,151
TN	33	782,429	491,626	190,919	1,429,703	35,048	223	94,631	1,170,176
TX	99	2,424,164	5,081,218	1,681,403	3,360,478	5,823,783	2,524	274,809	1,535,995
UT	6	72,455	1,031,811	216,171	62,920	1,257,393	124	5,860	25,033
VT	3	75,437	52,370	7,310	72,238	61,975	903	3,011	25,878
VA	19	939,353	5,107,913	251,378	5,889,353	408,803	488	57,555	585,053
WA	23	250,753	47,692	40,141	311,427	26,947	252	21,122	112,830
WV	1	26,236	3,000	1,059	29,443	847	5	970	8,663
WI	26	569,792	1,255,217	464,963	588,244	1,699,931	1,798	47,007	249,640
WY	13	384,512	235,722	27,216	524,882	122,467	101	33,270	105,968
AS	1	-	3,000	-	-	3,000	-	-	-
MH	1	-	11,857	-	-	11,857	-	161	3,426
PW	-	-	-	-	-	-	-	-	-
PR	1	300,981	-	31,424	292,851	39,554	-	16,633	624,343
VI	1	430	-	-	234	197	-	85	912
US [4]	1,058	36,354130	75,656,195	17,122,336	56,880,093	72,205,194	47,374	2,640,451	20,534,683

[1] Includes $628,992,093 discounted principal from 226 prepaid borrowers.   [2] Includes RUS Section 313A loan guarantees.   [3] Includes loans obtained by RUS borrowers' affiliates specifically organized to facillitate non-RUS finanacing.   [4] Includes figures not shown elsewhere in this table for two borrowers whose loans have been foreclosed.   The amount of these loans was $37,237.   Note: Territories are American Samoa, Marshall Island, Palau, Puerto Rico and Virgin Islands.

Rural Development, Rural Utilities Service, Electric Program, (202) 692-0163.

Table 10-10.—Rural Utilities Service: Composite revenues and patronage capital, average number of consumers and megawatt-hour sales reported by RUS electric borrowers operating distribution systems—calendar years 2009–2011

Item	2009 Amount	2009 Percent of total	2010 Amount	2010 Percent of total	2011 Amount	2011 Percent of total
Number of borrowers reporting	580	....	571	....	566	....
Average number of consumers served:						
Residential service (farm & non-farm) ...	10,952,374	88.4	10,826,192	88.3	10,885,204	88.3
Commercial & industrial, small	1,248,412	10.1	1,241,357	10.1	1,248,623	10.1
Commercial & industrial, large	9,660	0.1	8,879	0.1	9,296	0.1
Irrigation	106,753	0.9	110,902	0.9	112,587	0.9
Other electric service	79,019	0.6	73,367	0.6	73,330	0.6
To others for resale	188	*	184	*	261	*
Total	12,396,406	100.0	12,260,881	100.0	12,329,301	100.0
Megawatt-hour sales:						
Residential service (farm & non-farm) ...	152,105,577	57.3	161,245,731	57.5	156,164,256	56.1
Commercial & industrial, small	49,551,825	18.7	52,539,103	18.7	52,524,707	18.9
Commercial & industrial, large	55,247,184	20.8	58,063,409	20.7	59,679,110	21.5
Irrigation	4,350,266	1.6	4,200,967	1.5	5,297,215	1.9
Other electric service	2,707,597	1.0	2,681,929	1.0	2,527,060	0.9
To others for resale	1,330,927	0.5	1,651,962	0.6	1,962,944	0.7
Total	265,293,361	100.0	280,383,115	100.0	278,155,291	100.0
	1,000 dollars		*1,000 dollars*		*1,000 dollars*	
Revenue and patronage capital:						
Residential service (farm & non-farm) ...	16,141,422	62.7	17,085,401	62.6	17,083,438	61.3
Commercial & industrial, small	4,854,339	18.8	5,161,990	18.9	5,340,980	19.2
Commercial & industrial, large	3,495,396	13.6	3,710,526	13.6	3,948,978	14.2
Irrigation	405,528	1.6	429,483	1.6	515,794	1.9
Other electric service	289,136	1.1	297,010	1.1	288,713	1.0
To others for resale	80,712	0.3	96,606	0.4	113,155	0.4
Total from sales of electric energy	25,266,534	98.1	26,781,016	98.2	27,291,058	98.0
Other operating revenue	494,890	1.9	495,333	1.8	558,911	2.0
Total operating revenue	25,761,424	100.0	27,276,349	100.0	27,849,969	100.0

* Less than 0.05 percent.
Rural Development, Rural Utilities Service, Electric Program, (202) 692-0163.

Table 10-11.—Rural Utilities Service: Annual revenues and expenses reported by electric borrowers, United States, 2002–2011

Year	Operating revenue	Operating expense	Interest expense	Depreciation and amortization expense	Net margins	Total utility plant
	1,000 dollars	1,000 dollars	1,000 dollars	1,000 dollars	1,000 dollars	1,000 dollars
2002	27,458,144	22,568,763	1,867,431	1,992,415	1,382,964	72,481,696
2003	31,821,409	26,393,809	2,153,155	2,314,811	1,303,510	84,991,605
2004	30,649,839	25,646,721	1,919,835	2,181,541	1,340,317	79,508,979
2005	34,330,831	29,164,368	2,075,557	2,271,565	1,441,751	83,405,976
2006	36,765,064	31,213,044	2,247,071	2,375,325	1,747,997	88,112,547
2007	38,423,386	32,659,447	2,311,524	2,369,896	1,989,271	90,936,276
2008	42,087,440	36,048,847	2,372,255	2,462,420	2,068,523	97,191,799
2009	42,189,052	35,680,844	2,456,980	2,656,212	2,751,860	104,286,469
2010	45,261,487	38,315,046	2,474,526	2,821,453	2,486,432	109,572,949
2011	46,146,088	39,129,741	2,515,389	2,969,391	2,461,388	115,047,960

Rural Development, Rural Utilities Service, Electric Program, (202) 692-0163.

Table 10-12.—Loans to farmers' cooperative organizations: Outstanding amounts held by the agricultural credit bank classified by type of loan, United States, Jan. 1, 2006-2015

Year	Operating capital loans	Facility loans
	1,000 dollars	1,000 dollars
2006	12,293,156	14,004,128
2007	10,956,633	22,119,209
2008	13,263,702	27,226,859
2009	11,769,457	32,780,415
2010	10,743,647	33,430,718
2011	10,871,732	35,413,410
2012	32,789,008	39,191,450
2013	32,449,687	41,153,688
2014	32,171,519	48,210,977
2015	35,316,639	53,723,941

FCA, Office of Management Services, (703) 883–4073.

CHAPTER XI

STABILIZATION AND PRICE-SUPPORT PROGRAMS

The statistics in this chapter relate to activities of the Commodity Credit Corporation (CCC), loan and inventory acquisition and disposition programs, the CCC and Farm Service Agency payment programs, and marketing agreements and order programs for fruits and vegetables. Statistics for Federal Milk Marketing Order programs are contained in chapter VIII.

Table 11-1.—Commodity Credit Corporation: Price-supported commodities owned as of September 2015 [1] (Inventory quantity)

Year	Barley	Butter and butter oil	Cheese and products	Corn	Cotton extra long staple	Cotton upland	Sorghum and products
	Million bushels	Million pounds	Million pounds	Million bushels	1,000 bales	1,000 bales	Million bushels
2006	[2]	[2]	0	1	0	5	[2]
2007	0	0	0	1	0	0	[3]
2008	0	0	0	0	0	0	[2]
2009	0	0	0	0	0	0	0
2010	0	[3]	0	0	0	0	[3]
2011	0	0	0	0	0	0	[3]
2012	0	0	0	0	0	0	0
2013	0	0	0	0	0	0	[3]
2014	0	0	0	0	0	0	0
2015	0	0	0	0	0	0	0

Year	Milk and products	Oils and oilseeds	Oats and products	Rice and products [4]	Peanut and products	Soybeans	Beans, dry edible
	Million pounds	Million cwt	Million bushels	Million cwt	Million pounds	Million bushels	Million bushels
2006	49	[2]	[2]	[3]	51	1	[2]
2007	14	[2]	[2]	[3]	0	1	[3]
2008	0	0	0	[2]	0	0	[2]
2009	224	0	0	[2]	[2]	0	0
2010	7	0	0	[3]	0	0	[5]
2011	0	0	0	[2]	0	0	[2]
2012	0	0	0	0	0	0	0
2013	0	0	0	[3]	0	0	0
2014	0	0	0	[3]	4	0	0
2015	0	0	0	0	168	0	0

Year	Wheat	Blended Foods	Poultry	Meat	Fish	Vegetable Oil Products	Value of all commodities owned
	Million bushels	Million pounds	Million pounds	Million pounds	Million pounds	Million pounds	Million dollars
2006	43	3	0	0	0	4	226
2007	39	0	0	0	0	6	185
2008	0	8	0	0	0	2	11
2009	0	13	0	0	0	6	205
2010	[2]	16	0	0	0	5	48
2011	0	16	0	0	0	33	53
2012	0	2	0	0	0	12	14
2013	0	25	0	0	0	23	71
2014	0	18	0	0	0	32	40
2015	2	13	0	0	0	23	55

[1] Commodities which were owned by CCC in some years but not shown in this table are as follows: tobacco, honey, sugar and products, dry whole peas, potatoes, and wool and mohair.   [2] Less than 50,000 units.   [3] Less than 500,000 units.
[4] Total value of all commodities owned by CCC, including price-supported commodities not shown and commodities acquired under programs other than price-support programs, less, reserve for losses on inventory.   [5] Less than 500 units.
FSA, Office and Budget and Finance, Budget Division, (202) 720–0174.

Table 11-2.—Commodity Credit Corporation: Loans pledge made, by quantity and face amount, United States and Territories, by crop year 2014–2015 [1]

Commodity	Unit	2014		2015 [3]	
		Quantity pledged	Face amount	Quantity pledged	Face amount
		1,000	*1,000 dollars*	*1,000*	*1,000 dollars*
Barley	1,000 bushels	995	1,900	2,556	4,871
Corn	1,000 bushels	7,822	16,432	44,185	110,902
Cotton, ELS & Upland [2]	1,000 bales	270	71,822	188	51,479
Seed cotton, ELS & Upland	1,000 pounds	0	0	0	0
Sugar Cane and Beet	1,000 pounds	0	0	0	0
Flaxseed	1,000 cwt	0	0	11	108
Honey	1,000 pounds	2,082	1,437	1,666	1,149
Oats	1,000 bushels	273	362	443	596
Peanuts	1,000 pounds	89,190	15,805	330,202	68,156
Rice	1,000 cwt	1,602	10,474	4,384	31,467
Wool	1,000 pounds	0	0	0	0
Grain sorghum	1,000 bushels	98	194	31	62
Soybeans	1,000 bushels	83	422	3,990	29,951
Wheat	1,000 bushels	15,696	45,513	38,173	116,493
Sunflower Seed	1,000 cwt	0	0	52	537
Canola Seed	1,000 cwt	80	802	185	1,873
Safflower Seed	1,000 cwt	0	0	0	0
Mustard Seed	1,000 cwt	0	0	0	0
Sunflower Seed (non-oil)	1,000 cwt	0	0	32	642
Crambe Oilseed	1,000 cwt	0	0	0	0
Mohair	1,000 pounds	0	0	0	0
Chickpeas	1,000 cwt	0	0	7	81
Dry Whole Peas	1,000 cwt	106	568	372	3,586
Lentil Dry	1,000 cwt	37	413	68	2,049

[1] Includes loans pledge directly by Commodity Credit Corporation. [2] Includes extra long staple cotton and upland cotton. [3] Loan pledges are made through fiscal year 2015.

FSA, Office and Budget and Finance, Budget Division, (202) 720–0174.

Table 11-3.—Commodity Credit Corporation: Loan transactions for fiscal year 2015, by commodities [1]

Commodity	Unit	Loans outstanding Oct. 1, 2014 [2]	New loans made	Loans repayments	Collateral acquired in settlement	Loans written off and transferred to accounts receivable [3]	Loans outstanding Sept. 30, 2015 Value	Loans outstanding Sept. 30, 2015 Quantity collateral remaining pledged
		1,000 dollars	*1,000 dollars*	*1,000 dollars*	*1,000 dollars*	*1,000 dollars*	*1,000 dollars*	*1,000 units*
Basic commodities:								
Corn	Bushel	73,334	1,109,450	(1,069,848)	0	(2,021)	110,915	44,195
Cotton	Bale	37,660	2,112,484	(1,873,489)	0	(23)	51,479	188
Seed cotton	Pound	0	3,194	(3,194)	(6)	0	0	0
Peanuts	Pound	21,269	757,937	(669,856)	(54,833)	0	38,430	162,071
Rice	Cwt	16,242	274,964	(257,283)	0	(2,455)	31,467	4,384
Tobacco	Pound	0	0	0	0	0	0	0
Wheat	Bushel	47,637	186,439	(117,009)	0	(511)	116,557	38,194
Total [4]		196,142	4,444,468	(3,990,679)	(54,839)	(5,010)	348,848	249,032
Designated nonbasic commodities								
Barley	Bushel	2,214	9,715	(7,059)	0	0	4,870	2,556
Sorghum	Bushel	217	1,205	(1,360)	0	0	62	31
Honey	Pound	1,456	3,442	(3,749)	0	0	1,149	1,666
Oats	Bushel	405	812	(607)	0	(2)	608	452
Raw sugar, cane	Pound	0	290,112	(290,112)	0	0	0	0
Refined sugar, cane	Pound	0	0	0	0	0	0	0
Raw sugar beet	Pound	0	166,465	(166,465)	0	0	0	0
Refined sugar beet	Pound	0	382,370	(382,370)	0	0	0	0
FlaxSeed	Cwt	0	292	(184)	0	0	107	10
Sunflower seed (oil)	Cwt	145	4,223	(3,830)	0	0	538	52
Canola seed	Cwt	802	6,765	(5,693)	0	0	1,874	185
Safflower seed	Cwt	0	0	0	0	0	0	0
Rapeseed	Cwt	0	0	0	0	0	0	0
Mustard seed	Cwt	0	0	0	0	0	0	0
Crambe oilseed	Cwt	0	0	0	0	0	0	0
Sunflower seed, non oil	Cwt	0	2,488	(1,847)	0	0	641	32
Total [4]		5,239	867,889	(863,276)	0	(2)	9,849	4,984
Other nonbasic commodities:								
Soybeans	Bushel	6,611	405,841	(381,488)	0	(981)	29,982	3,996
Mohair	Pound	0	0	0	0	0	0	0
Chickpeas	Pound	43	314	(275)	0	0	82	7
Lentils	Pound	413	460	(740)	0	0	133	12
Peas, dry whole	Pound	703	2,295	(1,576)	0	0	1,422	260
Wool	Pound	0	4	(4)	0	0	0	0
Total [4]		7,770	408,914	(384,083)	0	(981)	31,619	4,275
Grand total [5]		209,151	5,721,271	(5,238,038)	(54,839)	(5,993)	390,316	258,291

[1] Loans made directly by Commodity Credit Corporation. [2] Book value of outstanding loans; includes face amounts and any charges paid. [3] Includes transfers to accounts receivable. [4] Totals do not include allowance for losses. [5] Table may not add due to rounding.
FSA, Office and Budget and Finance, Budget Division, (202) 720–0174.

Table 11-4.—Commodity Credit Corporation: Selected inventory transactions, programs and commodity, as of September 30, 2015

Program and commodity	Unit	Quantity					
		Inventory Oct. 1, 2014	Purchases	Collateral acquired from loans	Other addition deduction	Sales and other dis- positions [1]	Inventory Sept. 30, 2015
		Thousands	*Thousands*	*Thousands*	*Thousands*	*Thousands*	*Thousands*
Feed grains:							
Barley	Bushel	0	0	0	0	0	0
Corn	Bushel	0	603	0	0	603	0
Corn products	Pound	0	52,953	0	0	52,953	0
Grain sorghum	Bushel	0	18,328	0	0	18,328	0
Sorghum grits	Pound	0	0	0	0	0	0
Oats	Bushel	0	0	0	0	0	0
Food grains, cotton and to-bacco:							
Wheat	Bushel	0	11,777	0	0	11,777	0
Wheat flour	Pound	0	26,571	0	0	26,571	0
Bulgur	Pound	0	15,834	0	0	13,941	1,894
Rice, milled	Cwt	127	828	0	0	1,543	0
Rice, rough	Cwt	0	0	0	0	0	0
Cotton, extra long staple	Bale ...	0	0	0	0	0	0
Upland cotton	Bale ...	0	0	0	1	0	0
Tobacco Products	Pound	0	0	0	0	0	0
Dairy products:							
Butter	Pound	0	0	0	0	955	0
Cheese	Pound	0	0	0	0	0	0
Milk, dried	Pound	0	0	0	0	0	0
Milk, UHT	Pound	0	0	0	0	0	0
Dry whole milk	Pound	0	0	0	0	0	0
Non fat dry milk	Pound	0	0	0	0	0	0
Oils and oilseeds:							
Crambe oilseed	Cwt	0	0	0	0	0	0
Canola seed	Cwt	0	0	0	0	0	0
Sunflower seed	Cwt	0	0	0	0	0	0
Sunflower seed, non-oil	Cwt	0	0	0	0	0	0
Peanuts	Pound	4,482	0	309,387	(25,746)	119,992	168,132
Peanut butter	Pound	0	0	0	0	0	0
Soybeans	Bushel	0	0	0	0	0	0
Soybean products	Pound	0	21,492	0	0	21,492	0
Dry edible beans	Cwt	0	62	0	0	62	0
Flaxseed	Cwt	0	0	0	0	0	0
Blended foods	Pound	18,464	124,708	0	0	129,748	13,424
Dry whole peas and lentils	Cwt	213	2,316	0	0	2,251	213
Sugar cane and beet	Pound	0	0	0	0	170,750	0
Vegetable oil products	Pound	32,247	182,743	0	0	166,946	32,347
Wool	Pound	0	0	0	0	0	0
Mohair	Pound	0	0	0	0	0	0
Other [2]	....	1,658	21,398	0	0	10,341	1,658

See footnote(s) at end of table.

Table 11-4.—Commodity Credit Corporation: Selected inventory transactions, programs and commodity, as of September 30, 2014—Continued

Program and commodity	Unit	Value					
		Inventory Oct. 1, 2014	Purchases	Collateral acquired from loans	Other addition deduction	Sales and other disposi- tions [1]	Inventory Sept. 30, 2015
		1,000 dollars	*1,000 dollars*	*1,000 dollars*	*1,000 dollars*	*1,000 dollars*	*1,000 dollars*
Feed grains:							
Barley	Bushel	0	0	0	0	0	0
Corn	Bushel	0	2,988	0	0	2,988	0
Corn products	Pound	0	10,828	0	0	10,828	0
Grain sorghum	Bushel	0	108,240	0	0	108,240	0
Sorghum grits	Pound	0	0	0	0	0	0
Oats	Bushel	0	0	0	0	0	0
Food grains, cotton and tobacco:							
Wheat	Bushel	0	80,933	0	0	80,933	0
Wheat flour	Pound	0	5,599	0	0	5,599	0
Bulgur	Pound	0	2,894	0	0	2,653	241
Rice, milled	Cwt	3,056	19,748	0	0	22,803	0
Rice, rough	Cwt	0	0	0	0	0	0
Cotton, extra long staple	Bale ..	0	0	0	0	0	0
Upland cotton	Bale ..	0	0	0	7	(7)	0
Tobacco Products	Pound	0	0	0	0	0	0
Dairy products:							
Butter	Pound	0	0	0	0	0	0
Cheese	Pound	0	0	0	0	0	0
Milk, dried	Pound	0	0	0	0	0	0
Milk, UHT	Pound	0	0	0	0	0	0
Dry whole milk	Pound	0	0	0	0	0	0
Non fat dry milk	Pound	0	0	0	0	0	0
Oils and oilseeds:							
Crambe oilseed	Cwt	0	0	0	0	0	0
Canola seed	Cwt	0	0	0	0	0	0
Sunflower seed	Cwt	0	0	0	0	0	0
Sunflower seed, non-oil	Cwt	0	0	0	0	0	0
Peanuts	Pound	722	0	54,833	(4,538)	21,291	29,726
Peanut butter	Pound	0	0	0	0	0	0
Soybeans	Bushel	0	0	0	0	0	0
Soybean products	Pound	0	3,906	0	0	3, 906	0
Dry edible beans	Cwt	0	1,823	0	0	1,823	0
Flaxseed	Cwt	0	0	0	0	0	0
Blended foods	Pound	6,765	38,122	0	0	42,218	2,669
Dry whole peas and lentils	Cwt	4,010	58,443	0	0	58,372	4,081
Sugar, cane and beet	Pound	0	0	0	0	0	0
Vegetable oil products	Pound	20,074	91,312	0	0	99,261	12,125
Wool	Pound	0	0	0	0	0	0
Mohair	Pound	0	0	0	0	0	0
Other [2]		3,584	38,196	0	0	37,632	5,746
Total inventory operations		38,211	463,030	54,839	(4,545)	498,546	54,588

[1] Includes sales, commodity donations, transfers to other government agencies and inventory adjustment. [2] Includes to-mato, vegetable, and cartons of soup.
FSA, Office and Budget and Finance, Budget Division, (202) 720–0174.

Table 11-5.—Commodity Credit Corporation: Cost value of export and domestic commodity dispositions, by type of disposition, fiscal year 2015 [1]

(In Thousands)

Commodity	Domestic				
	Dollar sales (Costs)	Transfers to other Government agencies	Donations [1]	Inventory adjustments and other recoveries (domestic)	Total domestic
Feed grains:					
Barley	0	0	0	0	0
Corn	0	0	0	0	0
Corn products	0	0	0	0	0
Grain sorghum	0	0	0	0	0
Sorghum grits	0	0	0	0	0
Oats	0	0	0	0	0
Food grains, cotton and tobacco:					
Tobacco products	0	0	0	0	0
Bulgur	0	0	0	0	0
Wheat	0	0	0	0	0
Wheat flour	0	0	0	0	0
Wheat product, Other	0	0	0	0	0
Rice, milled	0	0	0	0	0
Rice, rough	0	0	0	0	0
Rice, brown and Textured soy	0	0	0	0	0
Cotton, extra long staple & upland	0	0	0	0	0
Dairy products:					
Butter oil	0	0	0	0	0
Butter	0	0	0	0	0
Cheese Products	0	0	0	0	0
Nonfat dry milk	0	0	0	0	0
Milk, dried. UT high temp	0	0	0	0	0
Oils and oilseeds:					
Peanuts	0	0	1,433	0	1,433
Peanut butter	0	0	4,351	0	4,351
Peanuts farmer's stock & products	0	0	0	0	0
Soya flour	0	0	0	0	0
Flaxseed	0	0	0	0	0
Sunflower Seed (oil & non-oil)	0	0	0	0	0
Soybeans & Soybean products	0	0	0	0	0
Fruit fresh apples	0	0	0	0	0
Blended foods	0	0	0	0	0
Potatoes	0	0	0	0	0
Grains and seeds:					
Feed for Government facilities	0	0	0	0	0
Foundation seeds	0	0	0	0	0
Lentils dry	0	0	0	0	0
Vegetable Seeds	0	0	0	0	0
Canola seed	0	0	0	0	0
Crambe oil seed	0	0	0	0	0
Peas, dried whole	0	0	0	0	0
Dry edible beans	0	0	0	0	0
Honey	0	0	0	0	0
Sugar	0	0	0	0	0
Vegetable oil products	0	0	0	0	0
Mohair	0	0	0	0	0
Meat (and products)	0	0	0	0	0
Veg. canned tomato sauce	0	0	0	0	0
Wool	0	0	0	0	0
Other					
(rice products, fish, canned salmon)	0	0	0	0	0
Total	0	0	5,784	0	5,784

See footnote(s) at end of table.

Table 11-5.—Commodity Credit Corporation: Cost value of export and domestic commodity dispositions, by type of disposition, fiscal year 2015[1]—Continued

(In Thousands)

Commodity	Export				Total export and domestic
	Dollar sales (Costs)	Public law 480 (Costs)	Donations[1]	Total export	
Feed grains:					
Barley	0	0	0	0	0
Corn	0	1,059	1,929	2,988	2,988
Corn products	0	10,828	0	10,828	10,828
Grain sorghum	0	108,240	0	108,240	108,240
Sorghum grits	0	0	0	0	0
Oats	0	0	0	0	0
Food grains, cotton and tobacco:					
Tobacco Products	0	0	0	0	0
Bulgur	0	2,653	0	2,653	2,653
Wheat	0	62,182	18,750	80,932	80,932
Wheat flour	0	5,599	0	5,599	5,599
Wheat product, Other	0	0	0	0	0
Rice, milled	0	20,845	1,958	22,803	22,803
Rice, rough	0	0	0	0	0
Rice, brown and textured soy	0	0	0	0	0
Cotton, extra long staple & upland	0	0	0	0	0
Dairy products:					
Butter oil	0	0	0	0	0
Butter	0	0	0	0	0
Cheese Products	0	0	0	0	0
Nonfat dry milk	0	0	0	0	0
Milk, dried UT high temp	0	0	0	0	0
Oils and oilseeds:					
Peanuts	0	0	0	0	1,433
Peanut butter	0	0	0	0	4,351
Peanuts farmer's stock & products	0	0	0	0	0
Soya flour	0	0	0	0	0
Flaxseed	0	0	0	0	0
Sunflower Seed (oil & non-oil)	0	0	0	0	0
Soybeans & Soybean products	0	0	3,906	3,906	3,906
Fruit fresh apples	0	0	0	0	0
Blended foods	0	42,218	0	42,218	42,218
Potatoes	0	9,521	0	9,521	9,521
Grains and seeds:					
Feed for Government facilities	0	0	0	0	0
Foundation seeds	0	0	0	0	0
Lentils dry	0	18,950	0	18,950	18,950
Vegetable Seeds	0	0	0	0	0
Canola seed	0	0	0	0	0
Crambe oil seed	0	0	0	0	0
Peas, dried whole	0	39,422	0	39,422	39,422
Dry edible beans	0	1,823	0	1,823	1,823
Honey	0	0	0	0	0
Sugar	0	0	0	0	0
Vegetable oil products	0	71,383	27,878	99,261	99,261
Mohair	0	0	0	0	0
Meat (and products)	0	0	0	0	0
Veg. canned tomato sauce	0	0	0	0	0
Wool	0	0	0	0	0
Other (rice products, fish, canned salmon)	0	27,702	409	28,111	28,111
Total	0	422,425	54,830	477,255	483,039

[1] Includes donations under section 202,407,416, Section 210, P.L. 85-540, miscellaneous donations under various other authorizations.

FSA, Office and Budget and Finance, Budget Division, (202) 720–0174.

Table 11-6.—Commodity Credit Corporation: Investment in price-support operations, March 31 and June 30, 2006–2015 [1]

Year Month	Inventory investment	Loan investment	Total investment
	Million dollars	*Million dollars*	*Million dollars*
2006:			
March	84	5,503	5,587
June	93	3,016	3,109
2007:			
March	72	7,031	7,103
June	48	2,902	2,950
2008:			
March	168	5,926	6,094
June	11	3,335	3,346
2009:			
March	174	4,647	4,821
June	209	2,215	2,424
2010:			
March	119	4,054	4,173
June	70	1,996	2,066
2011:			
March	41	3,259	3,300
June	66	1,588	1,654
2012:			
March	63	2,740	2,803
June	27	1,260	1,287
2013:			
March	28	2,981	3,009
June	43	1,664	1,707
2014:			
March	188	2,094	2,282
June	94	890	984
2015:			
March	32	2,734	2,766
June	41	1,572	1,613

[1] Reflects total CCC loans and inventories investment.
FSA, Office and Budget and Finance, Budget Division, (202) 720–0174.

Table 11-7.—Commodity Credit Corporation: Loans made in fiscal year 2015 as of September 30, by State and Territories [1]

State or Territory	Barley	Corn	Cotton	Flaxseed	Honey	Oats
	1,000 dollars	*1,000 dollars*	*1,000 dollars*	*1,000 dollars*	*1,000 dollars*	*1,000 dollars*
Alabama	0	1,230	35,229	0	0	0
Alaska	13	0	0	0	0	0
Arizona	0	1,824	0	0	0	0
Arkansas	0	8,667	105,652	0	0	0
California	32	0	371,442	0	109	0
Colorado	16	7,409	0	0	0	0
Connecticut	0	0	0	0	0	0
Delaware	0	1,190	0	0	0	0
Florida	0	381	0	0	56	0
Georgia	11	6,771	10,002	0	0	3
Hawaii	0	0	0	0	0	0
Idaho	1,479	0	0	0	225	3
Illinois	0	100,730	0	0	0	5
Indiana	0	124,316	0	0	0	0
Iowa	0	208,601	0	0	451	12
Kansas City Commod	0	0	0	0	0	0
Kansas	0	8,987	0	0	0	0
Kentucky	60	22,472	0	0	0	0
Louisiana	0	2,365	5,341	0	17	0
Maine	157	0	0	0	0	162
Maryland	0	5,944	0	0	0	0
Massachusetts	0	40	0	0	0	0
Michigan	4	53,070	0	0	15	6
Minnesota	703	208,582	0	0	193	125
Mississippi	0	3,356	688,372	0	0	0
Missouri	8	47,696	3,388	0	0	0
Montana	3,467	80	0	29	421	2
Nebraska	0	53,760	0	0	22	9
Nevada	0	0	0	0	0	0
New Hampshire	0	0	0	0	0	0
New Jersey	0	1,018	0	0	0	0
New Mexico	0	884	63	0	0	0
New York	4	21,855	0	0	0	11
North Carolina	67	4,231	90,108	0	0	10
North Dakota	3,494	22,913	0	220	715	44
Ohio	0	56,913	0	0	0	6
Oklahoma	0	521	0	0	0	0
Oregon	108	0	0	0	124	0
Pennsylvania	0	9,911	0	0	0	30
Rhode Island	0	0	0	0	0	0
South Carolina	0	1,481	976	0	0	23
South Dakota	0	61,527	0	44	753	338
Tennessee	0	6,323	211,623	0	8	0
Texas	0	5,360	589,884	0	145	0
Utah	20	413	0	0	33	0
Vermont	0	32	0	0	0	0
Virginia	47	5,469	407	0	0	0
Washington	16	0	0	0	68	0
West Virginia	0	1,202	0	0	0	0
Wisconsin	0	41,182	0	0	86	23
Wyoming	11	748	0	0	0	0
Adjustments	....	....	....	....	....	....
Peanut Associations	....	....	....	....	....	....
Total [2]	9,717	1,109,454	2,112,487	293	3,441	812

See footnote(s) at end of table.

Table 11-7.—Commodity Credit Corporation: Loans made in fiscal year 2015 as of September 30, by State and Territories [1]—Continued

State or Territory	Oilseeds	Peanuts	Rice	Seed cottton	Sorghum	Soybeans
	1,000 dollars	*1,000 dollars*	*1,000 dollars*	*1,000 dollars*	*1,000 dollars*	*1,000 dollars*
Alabama	0	7,151	0	0	0	318
Alaska	0	0	0	0	0	0
Arizona	0	0	0	0	48	0
Arkansas	0	0	122,484	494	22	3,242
California	0	0	98,054	0	0	0
Colorado	24	0	0	0	0	0
Connecticut	0	0	0	0	0	0
Delaware	0	0	0	0	0	614
Florida	0	21,175	0	0	0	75
Georgia	0	549,689	0	0	0	1,005
Hawaii	0	0	0	0	0	0
Idaho	85	0	0	0	0	0
Illinois	0	0	0	0	31	34,022
Indiana	0	0	0	0	0	57,680
Iowa	0	0	0	0	0	77,184
Kansas City Commod	0	0	0	0	0	0
Kansas	0	0	0	0	322	4,237
Kentucky	21	0	0	0	45	7,404
Louisiana	0	144	14,568	0	0	540
Maine	0	0	0	0	0	0
Maryland	0	0	0	0	0	2,210
Massachusetts	0	0	0	0	0	0
Michigan	0	0	0	0	0	17,424
Minnesota	158	0	0	0	0	50,063
Mississippi	0	3,207	7,951	538	0	2,376
Missouri	0	0	16,894	1,237	228	28,233
Montana	517	0	0	0	0	0
Nebraska	55	0	0	0	62	13,166
Nevada	0	0	0	0	0	0
New Hampshire	0	0	0	0	0	0
New Jersey	0	0	0	0	0	324
New Mexico	0	1,570	0	0	0	0
New York	4	0	0	0	0	6,008
North Carolina	0	43,366	0	0	0	3,779
North Dakota	8,294	0	0	0	0	5,603
Ohio	0	0	0	0	0	46,210
Oklahoma	0	1,300	0	0	118	0
Oregon	0	0	0	0	0	0
Pennsylvania	0	0	0	0	0	4,754
Rhode Island	0	0	0	0	0	0
South Carolina	0	7,128	0	0	0	2,055
South Dakota	4,318	0	0	0	233	19,721
Tennessee	0	0	0	0	0	3,073
Texas	0	59,427	15,013	926	96	0
Utah	0	0	0	0	0	0
Vermont	0	0	0	0	0	0
Virginia	0	63,778	0	0	0	4,312
Washington	0	0	0	0	0	0
West Virginia	0	0	0	0	0	717
Wisconsin	0	0	0	0	0	9,490
Wyoming	0	0	0	0	0	0
Adjustments	....	....	....	....	....	....
Peanut Associations	....	....	....	....	....	....
Total [2]	13,476	757,935	274,964	3,195	1,205	405,839

See footnote(s) at end of table.

Table 11-7.—Commodity Credit Corporation: Loans made in fiscal year 2015 as of September 30, by State and Territories [1]—Continued

State or Territory	Sugar	Wheat	Mohair	Dry whole peas	Wool
	1,000 dollars	1,000 dollars	1,000 dollars	1,000 dollars	1,000 dollars
Alabama	0	0	0	0	0
Alaska	0	0	0	0	0
Arizona	0	0	0	0	0
Arkansas	0	12	0	0	0
California	0	0	0	0	0
Colorado	110,374	4,821	0	0	0
Connecticut	0	0	0	0	0
Delaware	0	0	0	0	0
Florida	93,000	0	0	0	0
Georgia	0	0	0	0	0
Hawaii	0	0	0	0	0
Idaho	242,982	8,543	0	172	0
Illinois	0	1,996	0	0	0
Indiana	0	30	0	0	0
Iowa	0	0	0	0	0
Kansas City Commod	0	0	0	0	0
Kansas	0	2,381	0	0	0
Kentucky	0	3,085	0	0	0
Louisiana	197,112	0	0	0	0
Maine	0	0	0	0	0
Maryland	0	363	0	0	0
Massachusetts	0	0	0	0	0
Michigan	131,158	943	0	0	0
Minnesota	38,193	25,998	0	0	0
Mississippi	0	48	0	0	0
Missouri	0	1,065	0	0	0
Montana	0	46,716	0	1,930	4
Nebraska	0	1,461	0	0	0
Nevada	0	0	0	0	0
New Hampshire	0	0	0	0	0
New Jersey	0	22	0	0	0
New Mexico	0	0	0	0	0
New York	0	1,826	0	0	0
North Carolina	0	376	0	0	0
North Dakota	16,604	48,805	0	1,055	0
Ohio	0	366	0	0	0
Oklahoma	0	2,162	0	0	0
Oregon	0	1,874	0	35	0
Pennsylvania	0	475	0	0	0
Rhode Island	0	0	0	0	0
South Carolina	0	181	0	0	0
South Dakota	0	15,243	0	0	0
Tennessee	0	767	0	0	0
Texas	0	530	0	0	0
Utah	0	1,268	0	0	0
Vermont	0	0	0	0	0
Virginia	0	1,374	0	0	0
Washington	0	13,082	0	83	0
West Virginia	0	0	0	0	0
Wisconsin	0	471	0	0	0
Wyoming	9,524	155	0	0	0
Adjustments					
Peanut Associations					
Total [2]	838,947	186,439	0	3,068	4

[1] Loans made directly by Commodity Credit Corporation.   As much as possible, loans have been distributed according to the location of producers receiving the loans.   Direct loans to cooperative associations for the benefit of members have been distributed according to the location of the association.   [2] Totals may not add due to rounding.
FSA, Office and Budget and Finance, Budget Division, (202) 720–0174.

Table 11-8.—Farm Service Agency programs: Payments to producers, by program and commodity, United States, calendar year 2014–2015

Program and commodity	2014	2015
	1,000 dollars	*1,000 dollars*
Acreage grazing payments	0	0
Agricultural management assistance	0	0
Additional risk coverage	0	4,378,513
American indian - livestock feed	0	0
Aquaculture block grant	0	0
Auto conservation reserve program (crp)-cost shares	80,871	72,597
Avg crop revenue election program	266,139	18,822
Bioenergy program	0	0
Biomass crop assistance	5,795	7,919
Cotton transition program	460,320	29,056
Cottonseed payment program	0	0
Crop assistance program	7	12
Crop disaster - North Carolina	0	0
Crop disaster - Virginia	0	0
Crop disaster program	0	0
Crop disaster program - 2005	0	0
Crop hurricane damage program	0	0
Crp annual rental	1,580,605	1,580,750
Crp incentives	35,785	69,869
Crp transition incentive	2,857	4,144
Dairy economic loss assistance	15	0
Dairy indemnity	379	462
Dairy market loss assistance	0	0
Direct and counter cyclical program	75,340	11,187
Durum wheat quality program	0	0
Extra long staple special provision program	0	0
Emergency Assistance program	26,831	21,905
Emergency conservation program	24,986	21,066
Emergency forest restoration	1,964	3,769
Environment quality incentives	0	0
Feed indemnity program	0	0
Finality rule	0	0
Fl hurricane citrus disaster	0	0
Forestry conservation reserve	5,178	4,511
Forgiven interest	0	5
Fl nursery disaster	0	0
Fl vegetable disaster	0	0
Florida sugarcane program	0	0
Grasslands reserve program	10,671	9,131
Geographic disadvantaged program	1,837	2,009
Hard white winter wheat	0	0
Hawaii sugar disaster	0	0
Hurricane indemnity program	0	0
Interest payments	0	0
Lamb meat adjustment assistance	0	0
Livestock assistance grant	0	0
Livestock assistance program	0	0
Livestock compensation program	0	0
Livestock emergency assistance	0	0
Livestock forage program	4,396,383	1,296,327
Livestock indemnity program	77,188	37,708
Loan deficiency	62,378	155,734
Margin protection-Dairy	0	688
Market gains	34,308	53,234
Marketing loss assistance-asparagus	0	0
Marketing loss assistance	0	0
Milk income loss contract transitional	0	0
Milk income loss contract	0	0
Milk income loss II	1,079	1,440
Noninsured assistance program	184,410	142,304
Peanut quota buyout program	0	0
Pima cotton trust fund program	16,000	14,832
Price loss coverage	0	755,177
Speciality crop - nursery	0	0
Speciality crop - tropical fruit	0	0
Specialty crop - citrus	0	0
Specialty crop - fruit/vegetable	0	0
Storage forgiven	1,992	9,908
Sugar beet disaster program	0	0
Supplemental assistance program	31,528	4,231
Texas sugarcane storage & transportation	0	0
Tobacco quota holder-interest	23	10
Trade adjustment assistance	6,354	13
Tree assistance program	3,413	13,892
TTPP tobacco producer	286,835	307
Upland cotton assistance	47,553	48,377
Wetlands reserve	0	0
Wool manufacturers trust fund	0	29,158
01 - 02 crop disaster assistance	0	0
05 - 07 crop disaster assistance	2	0
05 - 07 dairy disaster program	0	0
05 - 07 livestock compensation	0	0
05 - 07 livestock indemnity program	0	0
Grand Total	7,729,025	8,799,065

FSA, Office and Budget and Finance, Budget Division, (202) 720–0174.

Table 11-9.—Farm Service Agency programs: Payments received, by States, calendar year 2014–2015

State	Payments	
	2014	2015
	1,000 dollars	*1,000 dollars*
Alabama	97,286	75,619
Alaska	1,146	1,055
Arizona	51,349	31,122
Arkansas	221,862	267,292
California	171,285	143,305
Colorado	235,429	164,683
Connecticut	1,620	1,371
Delaware	1,033	778
District of Columbia	0	44
Florida	60,284	52,313
Georgia	188,340	245,235
Hawaii	19,610	9,054
Idaho	47,797	62,793
Illinois	161,442	380,279
Indiana	71,066	190,489
Iowa	375,624	1,174,773
Kansas	596,887	461,665
Kentucky	109,312	107,462
Louisiana	78,673	122,169
Maine	3,909	1,940
Maryland	11,992	13,967
Massachusetts	2,234	1,172
Michigan	35,184	166,723
Minnesota	174,420	789,788
Mississippi	121,105	117,334
Missouri	443,227	157,173
Montana	107,273	79,158
Nebraska	623,533	695,213
Nevada	35,371	26,342
New Hampshire	380	730
New Jersey	1,697	3,449
New Mexico	192,026	54,022
New York	7,647	50,363
North Carolina	127,764	65,017
North Dakota	137,663	268,806
Ohio	51,670	283,950
Oklahoma	957,512	477,250
Oregon	70,596	110,226
Pennsylvania	21,471	34,664
Rhode Island	219	37
South Carolina	49,104	44,975
South Dakota	397,736	336,432
Tennessee	60,018	31,521
Texas	1,074,627	856,156
Utah	58,420	30,705
Vermont	2,383	3,426
Virginia	25,082	35,398
Washington	91,624	178,262
West Virginia	2,290	3,211
Wisconsin	50,189	252,650
Wyoming	135,753	40,910
KCCO	162,228	92,366
Puerto Rico	2,550	4,033
Virgin Islands	19	20
Guam	47	135
MI	12	35
AS	2	1
Total [1]	7,729,025	8,799,065

[1] Total may not add due to rounding.
FSA, Office and Budget and Finance, Budget Division, (202) 720–0174.

Table 11-10.—Fruit, vegetable, and tree nut and specialty crop marketing orders and peanut program, 2014-15 season

Program	Estimated number of producers	Farm value
	Number	*1,000 dollars*
Citrus fruits		
Florida oranges, grapefruit, tangerines, and tangelos ...	8,000	216,034
Texas oranges and grapefruit	170	55,066
Deciduous fruits		
California olives	1,000	72,904
California desert grapes	41	79,170
California kiwifruit	175	32,676
Florida avocados	300	21,582
Washington apricots	94	9,721
Washington sweet cherries	1,500	480,570
Washington and Oregon pears [1]	1,500	325,498
Tart cherries (7 States) [2]	600	106,745
Cranberries (10 States) [3]	1,200	254,412
Dried fruits		
California dates	70	34,510
California plums	800	232,960
California raisins	3,000	709,502
Vegetables		
Florida tomatoes	100	386,113
Idaho and Eastern Oregon onions	250	100,951
South Texas onions	85	56,160
Georgia onions (Vidalia)	80	79,458
Onions (Walla Walla, Washington, and Oregon)	21	5,135
Potatoes [4]		
Colorado	180	120,031
Idaho and Eastern Oregon	450	272,662
Washington	267	76,464
Tree Nuts		
California almonds	6,400	5,891,930
California, Arizona, and New Mexico pistachios [5]	1,040	1,593,400
California walnuts	4,100	1,841,100
Oregon and Washington hazelnuts	650	129,600
Spearmint oil [5]	130	43,241
Peanuts [7]	6,182	1,122,276
Total 28 programs [8]	....	14,349,871

[1] Includes fresh and processed pears.
[2] The tart cherry order covers the States of Michigan, New York, Pennsylvania, Oregon, Utah, Washington, and Wisconsin.
[3] The cranberry order covers Massachusetts, Rhode Island, Connecticut, New Jersey, Wisconsin, Michigan, Minnesota, Oregon, Washington, and Long Island in New York.
[4] The farm values of fresh potatoes regulated under the three potato marketing orders are based on the season average grower prices for all potatoes (fresh and processed utilization combined) in Colorado, Idaho, and Washington, respectively. NASS grower prices for fresh potato utilization were not available for all of the states regulated under the marketing order.
[5] Pistachio farm value is for California only.
[6] The marketing order regulates the handling of spearmint oil produced in the States of Washington, Idaho, Oregon and designated parts of Nevada and Utah. The farm value is the sum of crop values for Washington, Idaho and Oregon, and excludes the minor level of production in Nevada and Utah, which is not published by NASS.
[7] The Farm Security and Rural Investment Act of 2002 terminated the Peanut Administrative Committee (which locally administered Marketing Agreement No. 146). As a result, the marketing agreement was terminated and and new quality standards were established for domestic and imported peanuts, administered by the Peanut Standards Board.
[8] The total number of producers cannot be determined from totals for individual commodities; some producers may produce more than one commodity.
AMS, Specialty Crops Programs, Promotion and Economics Division, (202) 720–9915.

CHAPTER XII
AGRICULTURAL CONSERVATION AND FORESTRY STATISTICS

Statistics in this chapter concern conservation of various natural resources, particularly soil, water, timber, wetlands, wildlife, and improvement of water quality. Forestry statistics include area of private and public-owned forest land, timber production, imports and exports, pulpwood consumption and paper and board production, area burned over by forest fires, livestock grazing, and recreational use of national forest lands.

Conservation Practices on Active CRP Contracts

Practice code	Practice	Acres
CP1	Introduced grasses and legumes	3,300,553
CP2	Native grasses	6,280,904
CP3	Tree planting	959,416
CP4	Wildlife habitat with woody vegetation	1,916,100
CP5	Field windbreaks	92,326
CP6	Diversions	18
CP7	Erosion control structures	19
CP8	Grass waterways	126,346
CP9	Shallow water areas for wildlife	29,490
CP10	Existing grasses and legumes[1]	4,018,927
CP11	Existing trees	376,490
CP12	Wildlife food plots	48,158
CP15	Contour grass strips	63,628
CP16	Shelterbelts	34,445
CP17	Living snow fences	6,566
CP18	Salinity reducing vegetation	220,412
CP21	Filter strips (grass)	838,057
CP22	Riparian buffers (trees)	811,407
CP23	Wetland restoration	1,053,132
CP24	Cross wind trap strips	98
CP25	Rare and declining habitat	1,556,229
CP26	Sediment retention	63
CP27	Farmable wetland pilot (wetland)	92,062
CP28	Farmable wetland pilot (upland)	206,844
CP29	Wildlife habitat buffer (marginal pasture)	120,375
CP30	Wetland buffer (marginal pasture)	43,521
CP31	Bottomland hardwood	130,456
CP32	Hardwood trees	9,075
CP33	Upland bird habitat buffers	257,160
CP34	Flood control structure	69
CP36	Longleaf pine	139,254
CP37	Duck nesting habitat	300,415
CP38	State acres for wildlife enhancement	1,016,323
CP39	FWP--Constucted wetlands	381
CP40	FWP--Aquaculture wetlands	17,166
CP41	FWP--Flooded praire wetlands	43,750
CP42	Pollinator Habitat[2]	92,953
	Total	24,202,586

[1] Includes both introduced grasses and legumes and native grasses.   [2] Does not include about 40,000 acres from Signup 39 before CP42 was implemented.
FSA, Conservation and Environmental Programs Division, (530) 792-5594.

CRP enrollment: By sign up and initial contract year[1], as of August 2015

Sign up	Before 2007	2007	2007	2009	2010	2011
1-27	2,196,642					
28	187,008					
29	963,239	56,905				
30	375,111					
31	188,812	140,70				
32	155		2,194,765	821,337	523,332	238,008
33	0	802,781	78			
35	0	150,239	347,781			
36	22		202,282	171,831		
37	22			223,946	225,072	
38		2			216,903	382,440
39						3,703,621
40						171,335
41						
42						
43						
44						
45						
46						
47						
All	3,911,011	1,150,626	2,744,907	1,217,113	965,307	4,495,403

Sign up	2012	2013	2014	2015	Total
1-27					2,196,642
28					187,008
29					1,020,144
30					375,111
31					329,512
32					3,777,597
33					802,859
35					498,020
36					374,134
37		90			449,130
38					599,344
39					3,703,621
40	309,290				480,625
41	2,509,411				2,509,411
42	117,411	493,447			610,859
43		3,491,432			3,491,432
44	34	23,096	509,916		533,046
45			1,560,573		1,560,573
46			3,211	527,575	530,786
47				167,778	167,778
All	2,936,146	4,008,066	2,070,489	695,353	24,197,633

[1] For CRP, contract year is the same as fiscal year, which begins October 1.
General Signup Numbers: 1-13, 15, 16, 18, 20, 26, 29, 33, 39, 41, 43, 45. Continuous Sign-up Numbers: 14, 17, 19, 21-25, 27, 28, 30, 31, 35-38, 40, 42, 44, 46, 47.
FSA, Conservation and Environmental Programs Division, (530) 792–5594.

Table 12-1.—Conservation Reserve Program (CRP): Enrollment by practice, under contract, August 2015
(CP1 and CP2)

State	CP1 Establishment of permanent introduced grasses and legumes			CP2 Establishment of permanent native grasses		
	Total acres treated	Total cost share	Avg cost share per acre treated [1]	Total acres treated	Total cost share	Avg cost share per acre treated [1]
Alabama	13,722	794,388	71.09	3,674	180,872	77.89
Alaska	2,033	0	....	....	0	....
Arkansas	4,076	260,290	84.90	4,441	710,116	191.15
California	21,025	821,890	136.16	253	32,374	314.95
Colorado	67,235	3,386,169	65.68	853,863	49,405,544	70.58
Delaware	31	11,105	358.23	23	16,310	700.00
Florida	85	0	....	108	3,718	143.00
Georgia	754	53,778	97.64	177	17,165	204.83
Hawaii	....	0	....	....	0	....
Idaho	208,868	13,795,782	77.86	44,745	3,127,392	78.61
Illinois	176,613	7,775,348	66.14	53,214	8,106,413	220.06
Indiana	23,435	1,243,506	83.58	17,010	1,960,103	182.02
Iowa	191,602	5,789,865	46.05	130,723	18,802,814	219.59
Kansas	12,348	491,080	51.96	738,126	31,123,377	48.26
Kentucky	38,060	3,075,361	126.88	26,167	2,361,810	138.53
Louisiana	862	31,838	67.02	1,474	194,232	204.20
Maine	2,921	458,339	194.71	....	0	....
Maryland	8,919	672,850	161.18	2,462	282,935	208.61
Massachusetts	....	0	....	....	0	....
Michigan	26,279	2,748,049	132.95	20,597	3,042,920	198.78
Minnesota	72,772	2,087,505	44.10	59,313	2,610,754	89.55
Mississippi	20,586	252,257	113.74	2,788	71,630	133.41
Missouri	468,562	21,928,510	86.49	166,654	14,314,954	161.63
Montana	439,290	5,232,955	21.95	352,380	5,289,743	33.33
Nebraska	29,846	2,819,701	104.52	304,118	20,241,253	80.35
Nevada	*	*	*	*	*	*
New Hampshire	....	0	....	....	0	....
New Jersey	303	5,620	200.00	124	21,417,745	114.47
New Mexico	43,654	3,051,940	88.72	288,317	321,899	561.98
New York	6,792	1,569,291	273.98	757	109,285	418.56
North Carolina	1,069	58,728	86.58	465	1,830,804	79.34
North Dakota	174,137	6,237,189	48.65	29,771	4,679,930	116.83
Ohio	15,160	936,978	127.12	52,794	18,302,339	71.23
Oklahoma	112,877	5,473,133	61.08	349,891	8,437,711	91.85
Oregon	211,531	12,628,865	70.91	104,529	3,766,962	178.88
Pennsylvania	84,977	10,240,570	151.43	29,371	0	....
Puerto Rico	108	17,550	162.50	....	5,198,675	86.60
South Carolina	899	46,742	215.30	182	3,090,587	195.00
South Dakota	53,925	1,532,526	34.39	87,279	126,423,395	86.01
Tennessee	31,373	1,417,184	117.23	25,449	1,060,598	77.39
Texas	264,549	15,534,927	82.98	1,840,234	0	....
Utah	74,154	2,513,606	42.77	14,652	312,804	505.09
Vermont	9	1,080	800.00	....	44,068,435	92.18
Virginia	2,932	520,590	340.50	1,387	0	....
Washington	253,870	13,309,041	61.85	641,963	2,836,859	159.33
West Virginia	208	17,792	134.89	....	49,436	86.38
Wisconsin	38,014	2,616,877	94.88	29,666		
Wyoming	99,938	1,870,235	31.40	1,739		
United States	3,300,553	153,343,838	67.96	6,280,904	403,803,889	84.74

* Data withheld to avoid disclosure of individual operators.
[1] Not including acres which receive no cost share.
Total acres treated may not add due to rounding.
FSA, Conservation and Environmental Programs Division, (530) 792-5594.

Table 12-2.—Conservation Reserve Program (CRP): Enrollment by practice, under contract, August 2015

(CP3 and CP4)

State	CP3 Tree planting			CP4 Permanent wildlife habitat		
	Total acres treated	Total cost share	Avg cost share per acre treated [1]	Total acres treated	Total cost share	Avg cost share per acre treated [1]
Alabama	136,850	12,485,255	118.07	7,648	903,451	289.53
Alaska	0	0	....	0	0	....
Arkansas	47,985	4,309,802	111.43	2,400	136,917	76.55
California	55	0	....	9,851	1,275	250.00
Colorado	37	11,793	589.65	430,685	25,377,387	82.78
Delaware	3,392	1,154,143	361.06	991	279,826	368.04
Florida	20,192	1,232,448	139.76	1,180	0	....
Georgia	110,966	13,596,763	143.04	2,782	270,915	225.73
Hawaii	13	55,854	4,168.21	0	0	....
Idaho	6,525	540,363	102.53	83,870	2,577,173	50.99
Illinois	48,027	9,635,654	267.55	109,977	25,098,302	347.11
Indiana	18,001	3,333,550	281.96	8,182	1,042,839	244.70
Iowa	16,030	4,440,793	369.37	131,481	3,484,473	138.44
Kansas	450	45,385	191.34	318,572	16,607,861	57.32
Kentucky	4,590	878,145	222.47	331	67,372	335.52
Louisiana	81,841	7,482,194	113.54	32,713	3,170,826	125.75
Maine	129	84,326	938.00	38	0	....
Maryland	989	244,539	487.04	1,583	167,008	239.16
Massachusetts	0	0	....	0	0	....
Michigan	6,513	869,681	217.86	12,659	484,436	121.89
Minnesota	27,392	2,741,526	143.16	167,524	2,030,096	93.32
Mississippi	260,741	7,187,684	111.73	4,507	312,287	360.03
Missouri	14,702	1,104,478	220.32	3,295	228,099	147.11
Montana	45	0	....	12,543	588,147	217.70
Nebraska	664	70,385	250.83	45,482	2,973,116	75.49
Nevada	*	*	*	*	*	*
New Hampshire	0	0	....	0	0	....
New Jersey	99	12,750	250.00	0	0	....
New Mexico	0	0	....	140	16,800	120.00
New York	895	391,141	564.43	529	276,127	646.52
North Carolina	31,212	6,700,097	283.54	986	493,292	888.49
North Dakota	456	36,482	183.51	299,946	14,455,178	75.07
Ohio	8,554	2,470,915	494.86	13,257	620,992	111.14
Oklahoma	286	21,405	111.89	623	31,681	98.22
Oregon	2,196	161,252	303.28	10,678	1,050,042	151.50
Pennsylvania	1,144	948,973	1,047.43	4,342	1,446,558	405.71
Puerto Rico	16	1,510	151.00	20	0	....
South Carolina	34,675	1,690,934	87.02	1,547	62,780	182.02
South Dakota	48	946	220.00	55,905	3,234,065	73.97
Tennessee	25,378	1,785,814	161.30	3,635	128,080	122.61
Texas	1,465	49,738	95.48	20,730	799,532	76.11
Utah	0	0	....	13	0	....
Vermont	0	0	....	0	0	....
Virginia	8,947	1,120,034	176.87	488	104,512	435.46
Washington	2,185	178,703	90.83	110,211	7,972,909	126.18
West Virginia	22	4,602	464.85	0	0	....
Wisconsin	35,706	6,635,429	287.06	4,163	430,317	280.40
Wyoming	1	0	....	591	0	....
United States	959,416	93,715,489	163.34	1,916,100	116,924,670	98.88

* Data withheld to avoid disclosure of individual operators.
[1] Not including acres which receive no cost share.
Total acres treated may not add due to rounding.
FSA, Conservation and Environmental Programs Division, (530) 792-5594.

Table 12-3.—Conservation Reserve Program (CRP): Enrollment by practice, under contract, August 2015

(CP5, CP6, and CP7)

State	CP5 Establishment of field windbreaks			CP6 Diversions			CP7 Erosion control structures		
	Total acres treated	Total cost share	Avg cost share per acre treated [1]	Total acres treated	Total cost share	Avg cost share per acre treated [1]	Total acres treated	Total cost share	Avg cost share per acre treated [1]
Alabama	0.0	0	....	....	0	....	....	0	....
Alaska	0.0	0	....	....	0	....	....	0	....
Arkansas	0.0	0	....	....	0	....	....	0	....
California	0.0	0	....	....	0	....	....	0	....
Colorado	1,547	1,189,875	1,011.77	....	0	....	....	0	....
Delaware	0	0	....	....	0	....	....	0	....
Florida	0	0	....	....	0	....	....	0	....
Georgia	0	0	....	....	0	....	....	0	....
Hawaii	0	0	....	....	0	....	....	0	....
Idaho	263	621,952	2,733.85	....	0	....	....	0	....
Illinois	2,401	460,333	253.04	15	0	....	4	0	....
Indiana	1,970	374,331	215.29	....	0	....	0	0	....
Iowa	6,460	1,607,833	305.76	....	0	....	10	2,484	709.57
Kansas	1,805	740,333	523.61	....	0	....	....	0	....
Kentucky	0	0	....	....	0	....	....	0	....
Louisiana	0	0	....	....	0	....	....	0	....
Maine	0	0	....	....	0	....	....	0	....
Maryland	0	0	....	....	0	....	....	0	....
Massachusetts	0	0	....	....	0	....	....	0	....
Michigan	2,467	682,310	310.37	1	0	....	....	0	....
Minnesota	8,930	2,675,459	350.58	....	0	....	0	0	....
Mississippi	0	0	....	....	0	....	....	0	....
Missouri	82	23,330	481.03	3	0	....	3	0	....
Montana	186	74,634	477.72	....	0	....	....	0	....
Nebraska	32,942	17,855,924	707.83	....	0	....	....	0	....
Nevada	*	*	*	*	*	*	*	*	*
New Hampshire	0	0	....	....	0	....	....	0	....
New Jersey	7	37,144	5,306.29	....	0	....	....	0	....
New Mexico	0	0	....	....	0	....	....	0	....
New York	7	2,836	436.31	....	0	....	....	0	....
North Carolina	23	2,558	127.90	....	0	....	....	0	....
North Dakota	4,539	2,620,728	658.44	....	0	....	....	0	....
Ohio	3,493	1,175,264	384.18	....	0	....	....	0	....
Oklahoma	30	6,348	869.59	....	0	....	....	0	....
Oregon	4	525	145.83	....	0	....	....	0	....
Pennsylvania	0	0	....	....	0	....	....	0	....
Puerto Rico	0	0	....	....	0	....	....	0	....
South Carolina	36	2,709	90.60	....	0	....	....	0	....
South Dakota	24,652	17,267,723	833.80	....	0	....	....	0	....
Tennessee	0	0	....	....	0	....	....	0	....
Texas	41	29,371	861.32	....	0	....	....	0	....
Utah	1	2,437	1,740.71	....	0	....	....	0	....
Vermont	0	0	....	....	0	....	....	0	....
Virginia	3	500	1,250.00	....	0	....	....	0	....
Washington	12	3,000	476.19	....	0	....	....	0	....
West Virginia	0	0	....	....	0	....	....	0	....
Wisconsin	149	49,907	426.27	....	0	....	1	0	....
Wyoming	277	265,656	1,276.58	....	0	....	....	0	....
United States	92,326	47,773,019	636.25	18	....	....	19	2,484	709.57

* Data withheld to avoid disclosure of individual operators.
[1] Not including acres which receive no cost share.
Total acres treated may not add due to rounding.
FSA, Conservation and Environmental Programs Division, (530) 792-5594.

Table 12-4.—Conservation Reserve Program (CRP): Enrollment by practice, under contract, August 2015

(CP8 and CP9)

State	CP8 Grass waterways			CP9 Shallow water areas for wildlife		
	Total acres treated	Total cost share	Avg cost share per acre treated [1]	Total acres treated	Total cost share	Avg cost share per acre treated [1]
Alabama	1	0	....	102	39,771	872.94
Alaska	0	0	....	....	0	....
Arkansas	5	66	12.74	557	58,925	267.72
California	0	0	....	40	17,389	434.40
Colorado	519	6,739	157.01	18	0	....
Delaware	6	0	....	221	251,097	1,671.75
Florida	0	0	....	....	0	....
Georgia	35	17,047	906.76	23	14,424	1,109.54
Hawaii	0	0	....	....	0	....
Idaho	3	0	....	23	224	70.00
Illinois	32,217	29,313,301	1,724.97	4,727	1,291,099	646.00
Indiana	18,892	57,069,750	5,509.75	979	303,084	1,063.04
Iowa	36,632	30,532,911	1,368.87	10,285	957,365	210.44
Kansas	8,401	2,846,628	515.68	1,051	145,804	434.13
Kentucky	4,401	9,212,974	2,957.55	1,897	455,754	806.22
Louisiana	5	0	....	442	70,256	380.79
Maine	59	329,918	6,858.99	....	0	....
Maryland	252	847,999	6,307.64	1,153	1,679,852	2,437.04
Massachusetts	0	0	....	....	0	....
Michigan	598	1,261,122	4,855.14	1,856	1,031,131	744.54
Minnesota	3,547	1,765,336	1,116.83	243	14,498	233.09
Mississippi	58	300	93.75	539	31,731	480.04
Missouri	2,005	1,310,080	1,304.73	1,854	95,421	324.02
Montana	83	1,990	50.90	11	0	....
Nebraska	1,294	315,729	415.21	187	50,060	533.91
Nevada	*	*	*	*	*	*
New Hampshire	0	0	....	....	0	....
New Jersey	151	1,799,022	12,676.31	....	0	....
New Mexico	0	0	....	....	0	....
New York	42	88,343	3,424.14	3	3,462	1,018.24
North Carolina	268	620,112	2,779.40	350	241,127	887.15
North Dakota	95	22,004	449.07	1	0	....
Ohio	11,111	35,412,109	5,564.57	560	248,033	1,262.25
Oklahoma	166	8,784	358.53	79	16,109	449.97
Oregon	27	10,680	726.53	....	0	....
Pennsylvania	341	741,243	3,188.69	33	17,602	1,248.37
Puerto Rico	0	0	....	....	0	....
South Carolina	36	34,448	1,299.93	44	42,836	1,457.01
South Dakota	1,328	472,032	555.84	41	0	....
Tennessee	162	199,206	2,388.85	73	8,000	800.00
Texas	1,603	1,123,264	1,018.09	92	42,514	920.22
Utah	14	347	43.38	....	0	....
Vermont	16	12,615	1,025.61	....	0	....
Virginia	88	181,450	2,382.48	69	142,381	2,256.43
Washington	376	4,390	55.36	51	3,151	201.47
West Virginia	0	0	....	....	0	....
Wisconsin	1,513	2,021,724	2,611.68	1,884	873,267	919.49
Wyoming	0	0	....	....	0	....
United States	126,346	177,583,663	2,457.01	29,490	8,146,366	645.65

* Data withheld to avoid disclosure of individual operators.
[1] Not including acres which receive no cost share.
Total acres treated may not add due to rounding.
FSA, Conservation and Environmental Programs Division, (530) 792-5594.

Table 12-5.—Conservation Reserve Program (CRP): Enrollment by practice, under contract, August 2015

(CP10, CP11, and CP12)

State	CP10 Vegetative-cover-grass-already established			CP11 Vegetative-cover-trees-already established			CP12 Wildlife food plots		
	Total acres treated	Total cost share	Avg cost share per acre treated [1]	Total acres treated	Total cost share	Avg cost share per acre treated [1]	Total acres treated	Total cost share	Avg cost share per acre treated [1]
Alabama	22,589	667,627	73.26	40,259	2,501,994	128.51	1,077	0	....
Alaska	14,919	0	....	....	0	....	....	0	....
Arkansas	4,726	64,373	158.36	12,392	88,861	227.50	397	0	....
California	45,989	0	....	303	0	....	6	0	....
Colorado	487,854	4,464,034	28.67	22	0	....	257	0	....
Delaware	25	0	....	....	0	....	8	0	....
Florida	308	0	....	12,456	12,500	690.61	58	0	....
Georgia	1,172	24,000	568.72	24,331	480,714	133.80	777	0	....
Hawaii	....	0	....	....	0	....	....	0	....
Idaho	119,986	17,652	9.57	1,117	0	....	227	0	....
Illinois	77,033	56,675	58.34	10,947	12,219	62.92	5,508	0	....
Indiana	18,962	6,298	17.77	5,535	8,834	42.17	656	30	50.00
Iowa	188,516	2,872,708	109.00	6,485	585,354	714.89	4,750	0	....
Kansas	206,903	217,136	10.06	223	12,000	2,926.83	3,369	12	12.00
Kentucky	9,263	20,149	22.58	983	1,012	34.66	783	0	....
Louisiana	1,455	638	11.71	13,725	4,135	7.88	1,044	0	....
Maine	2,633	0	....	274	0	....	....	0	....
Maryland	1,736	8,518	116.53	310	634	36.86	37	0	....
Massachusetts	....	0	....	....	0	....	....	0	....
Michigan	15,891	19,331	53.14	2,558	8,243	32.16	1,154	0	....
Minnesota	62,961	33,137	52.21	9,063	162,192	263.56	3,964	1,013	63.28
Mississippi	32,015	1,764	32.30	161,788	1,096,965	196.29	2,705	0	....
Missouri	139,965	125,242	64.27	6,341	15,871	160.48	3,580	0	....
Montana	444,412	242,713	4.54	442	0	....	1,726	24	24.00
Nebraska	98,973	1,001,308	44.98	858	38,165	481.28	811	50	50.00
Nevada	*	*	*	*	*	*	*	*	*
New Hampshire	....	0	....	....	0	....	....	0	....
New Jersey	22	0	....	....	0	....	2	0	....
New Mexico	91,508	1,955	6.50	....	0	....	1	0	....
New York	14,974	454,994	112.22	738	45,310	216.48	77	0	....
North Carolina	1,187	0	....	10,520	1,493,353	358.14	28	0	....
North Dakota	320,912	2,134,639	23.24	342	103	42.96	2,482	0	....
Ohio	25,457	398,426	684.00	2,615	2,200	35.15	468	0	....
Oklahoma	233,579	614,035	11.63	173	0	....	1,008	0	....
Oregon	123,558	703,713	25.71	518	152,111	987.73	120	300	600.00
Pennsylvania	14,053	372,030	114.89	296	2,079	44.05	1,155	0	....
Puerto Rico	224	0	....	117	0	....	....	0	....
South Carolina	1,755	1,185	28.07	21,393	129,911	34.92	263	0	....
South Dakota	88,251	34,384	13.85	210	138	43.23	4,019	0	....
Tennessee	19,457	6,004	44.24	8,024	52	43.23	384	0	....
Texas	756,897	11,866,765	112.80	3,028	0	....	3,215	1,880	100.00
Utah	82,055	5,454	3.00	....	0	....	19	0	....
Vermont	45	0	....	....	0	....	....	0	....
Virginia	3,419	67,468	258.99	5,233	100,579	139.37	28	0	....
Washington	117,821	137,346	23.01	781	4,168	54.14	597	80	80.00
West Virginia	164	0	....	....	0	....	....	0	....
Wisconsin	50,872	462,733	76.97	12,079	96,941	195.29	1,379	0	....
Wyoming	74,425	0	....	12	0	....	20	0	....
United States	4,018,927	27,104,432	45.31	376,490	7,056,640	169.75	48,158	3,389	84.92

* Data withheld to avoid disclosure of individual operators.
[1] Not including acres which receive no cost share.
Total acres treated may not add due to rounding.
FSA, Conservation and Environmental Programs Division, (530) 792-5594.

Table 12-6.—Conservation Reserve Program (CRP): Enrollment by practice, under contract, August 2015

(CP15, CP16, and CP17)

State	CP15 Contour grass strips			CP16 Shelter belts			CP17 Living snow fences		
	Total acres treated	Total cost share	Avg cost share per acre treated [1]	Total acres treated	Total cost share	Avg cost share per acre treated [1]	Total acres treated	Total cost share	Avg cost share per acre treated [1]
Alabama	11	0	....	0.0	0	....	0.0	0	....
Alaska	0	0	....	0.0	0	....	0.0	0	....
Arkansas	0	0	....	0.0	0	....	0.0	0	....
California	0	0	....	0.0	0	....	0.0	0	....
Colorado	12	0	....	4,419	4,010,964	1,037.40	35	18,880	1,026.09
Delaware	4	1,290	300.00	0	0	....	0	0	....
Florida	0	0	....	0	0	....	0	0	....
Georgia	0	0	....	0	0	....	0	0	....
Hawaii	0	0	....	0	0	....	0	0	....
Idaho	261	6,964	37.14	101	175,876	1,843.56	62	41,357	668.13
Illinois	935	22,615	62.03	144	33,204	302.16	56	11,532	241.76
Indiana	26	674	89.87	18	3,393	273.85	1	0	....
Iowa	12,527	361,753	101.66	2,250	1,946,475	1,082.23	656	151,289	268.70
Kansas	3,435	78,569	46.73	808	336,693	541.99	67	32,182	569.59
Kentucky	42	1,901	162.48	0	0	....	0	0	....
Louisiana	0	0	....	0	0	....	0	0	....
Maine	0	0	....	0	0	....	0	0	....
Maryland	0	0	....	0	0	....	0	0	....
Massachusetts	0	0	....	0	0	....	0	0	....
Michigan	3	1,069	509.05	54	12,013	322.06	14	4,232	313.48
Minnesota	748	20,279	57.52	3,951	1,281,989	389.21	4,209	809,161	229.40
Mississippi	27	0	....	0	0	....	0	0	....
Missouri	886	14,760	84.93	70	26,696	523.35	0	0	....
Montana	0	0	....	232	136,326	650.47	41	26,264	634.40
Nebraska	326	10,505	63.62	2,121	1,126,126	601.93	114	34,255	420.30
Nevada	*	*	*	*	*	*	*	*	*
New Hampshire	0	0	....	0	0	....	0	0	....
New Jersey	0	0	....	0	0	....	0	0	....
New Mexico	0	0	....	0	0	....	0	0	....
New York	3	0	....	0	0	....	0	0	....
North Carolina	0	0	....	5	0	....	0	0	....
North Dakota	0	0	....	5,109	3,304,152	717.55	667	360,547	644.22
Ohio	14	0	....	99	25,050	288.59	0	0	....
Oklahoma	9	0	....	21	4,783	226.68	0	0	....
Oregon	0	0	....	0	0	....	0	0	....
Pennsylvania	60	3,202	119.92	0	0	....	0	0	....
Puerto Rico	60	0	....	0	0	....	0	0	....
South Carolina	0	0	....	0	0	....	0	0	....
South Dakota	33	154	33.06	14,948	11,729,874	899.97	601	361,683	767.77
Tennessee	17	161	107.33	0	0	....	0	0	....
Texas	104	1,412	129.54	5	4,833	1,074.00	0	0	....
Utah	40	0	....	0	0	....	0	0	....
Vermont	0	0	....	0	0	....	0	0	....
Virginia	4	286	130.00	0	0	....	0	0	....
Washington	43,451	531,573	70.69	8	24,588	7,025.14	0	0	....
West Virginia	0	0	....	0	0	....	0	0	....
Wisconsin	593	9,917	109.23	27	8,125	303.17	39	8,824	239.12
Wyoming	0	0	....	55	74,819	1,370.31	3	729	214.41
United States	63,628	1,067,085	75.33	34,445	24,265,979	814.24	6,566	1,860,934	339.44

* Data withheld to avoid disclosure of individual operators.
[1] Not including acres which receive no cost share.
Total acres treated may not add due to rounding.
FSA, Conservation and Environmental Programs Division, (530) 792-5594.

Table 12-7.—Conservation Reserve Program (CRP): Enrollment by practice, under contract, August 2015

(CP18 and CP21)

State	CP18 Salt tolerant grasses			CP21 Alternative perennials		
	Total acres treated	Total cost share	Avg cost share per acre treated [1]	Total acres treated	Total cost share	Avg cost share per acre treated [1]
Alabama	0	0		589	41,376	135.00
Alaska	0	0			0	
Arkansas	0	0		5,594	375,358	95.10
California	0	0			0	
Colorado	7	0		115	10,287	229.01
Delaware	0	0		1,031	337,194	352.03
Florida	0	0			0	
Georgia	0	0		314	4,025	838.54
Hawaii	0	0			0	
Idaho	0	0		797	60,559	117.84
Illinois	4	261	104.40	114,699	5,945,276	86.27
Indiana	0	0		50,471	4,436,133	151.86
Iowa	0	0		195,758	11,542,548	99.58
Kansas	875	0		26,143	1,149,329	63.34
Kentucky	0	0		23,025	1,823,014	144.34
Louisiana	0	0		484	14,321	47.03
Maine	0	0		61	850	404.76
Maryland	0	0		29,431	3,505,166	159.84
Massachusetts	0	0		10	7,074	714.55
Michigan	0	0		42,203	5,677,880	156.87
Minnesota	4,528	181,678	57.22	123,856	6,914,299	84.54
Mississippi	0	0		6,917	441,745	99.12
Missouri	0	0		32,611	1,379,873	87.84
Montana	95,493	166,287	8.91	225	4,877	25.13
Nebraska	503	9,685	37.47	16,258	850,072	66.20
Nevada	*	*	*	*	*	*
New Hampshire	0	0			0	
New Jersey	0	0		364	75,890	234.00
New Mexico	0	0			0	
New York	0	0		431	164,210	455.35
North Carolina	0	0		1,928	498,950	315.54
North Dakota	101,809	1,366,060	37.53	8,291	292,818	50.96
Ohio	0	0		70,903	4,216,488	88.29
Oklahoma	1,862	11,560	72.39	601	15,690	71.31
Oregon	0	0		1,852	141,546	149.59
Pennsylvania	0	0		1,334	183,314	205.24
Puerto Rico	0	0			0	
South Carolina	0	0		2,030	40,944	78.91
South Dakota	14,869	549,343	46.90	9,058	380,582	54.63
Tennessee	0	0		7,378	703,975	145.16
Texas	436	20	0.13	1,233	192,893	259.33
Utah	0	0		38	1,413	94.83
Vermont	0	0		215	50,584	243.08
Virginia	0	0		4,281	477,231	125.08
Washington	24	0		35,662	1,646,773	90.63
West Virginia	0	0		562	152,062	696.25
Wisconsin	0	0		21,304	2,303,668	139.66
Wyoming	0	0			0	
United States	220,412	2,284,894	32.40	838,057	56,060,288	105.01

* Data withheld to avoid disclosure of individual operators.
[1] Not including acres which receive no cost share.
Total acres treated may not add due to rounding.
FSA, Conservation and Environmental Programs Division, (530) 792-5594.

Table 12-8.—Conservation Reserve Program (CRP): Enrollment by practice, under contract, August 2015

(CP22 and CP23)

State	CP22 Riparian buffer			CP23 Wetland restoration		
	Total acres treated	Total cost share	Avg cost share per acre treated [1]	Total acres treated	Total cost share	Avg cost share per acre treated [1]
Alabama	32,784	4,316,343	153.58	5	0	....
Alaska	145	21,880	230.80	0	0	....
Arkansas	61,577	5,128,995	115.64	46,832	15,262,043	505.12
California	1,534	615,932	463.77	569	0	....
Colorado	676	728,736	1,196.04	779	24,355	95.75
Delaware	102	38,188	381.88	327	269,836	911.30
Florida	64	0	....	0	0	....
Georgia	1,040	576,492	791.94	403	0	....
Hawaii	987	3,616,588	3,664.59	0	0	....
Idaho	5,609	2,478,207	525.68	507	12,915	39.87
Illinois	107,513	18,961,456	231.01	42,431	6,539,224	195.86
Indiana	5,568	1,046,767	278.13	8,303	1,369,490	282.63
Iowa	54,436	16,037,871	378.04	112,185	17,116,171	198.04
Kansas	2,974	202,195	156.85	8,681	327,996	56.23
Kentucky	23,163	7,949,768	363.21	134	21,706	214.49
Louisiana	5,408	566,553	128.48	102,183	8,654,895	176.78
Maine	49	126,632	5,147.64	0	0	....
Maryland	15,705	5,821,553	515.58	2,669	2,197,012	1,050.90
Massachusetts	....	0	....	0	0	....
Michigan	3,155	1,104,633	403.89	22,589	5,554,994	268.69
Minnesota	40,821	6,786,915	212.63	267,786	20,913,203	104.47
Mississippi	170,943	8,930,374	73.85	12,751	673,263	155.88
Missouri	29,080	4,999,444	268.94	14,082	1,057,890	150.89
Montana	3,082	357,050	184.87	2,140	33,183	37.41
Nebraska	2,544	773,451	366.28	5,933	374,423	108.40
Nevada	*	*	*	*	*	*
New Hampshire	13	11,593	1,332.53	0	0	....
New Jersey	249	199,730	802.19	0	0	....
New Mexico	5,280	2,217,812	485.76	0	0	....
New York	12,580	10,953,125	1,003.61	95	9,634	296.43
North Carolina	22,554	2,043,615	107.11	2,535	354,815	164.67
North Dakota	399	207,202	660.98	202,878	1,969,518	44.09
Ohio	7,069	1,664,602	336.17	11,225	5,655,333	739.16
Oklahoma	1,856	549,615	427.71	2,050	95,490	59.71
Oregon	36,356	20,004,706	724.56	243	52,256	307.03
Pennsylvania	24,658	30,198,640	1,362.19	947	1,132,916	1,278.69
Puerto Rico	297	12,180	60.00	0	0	....
South Carolina	14,796	806,587	83.77	172	0	....
South Dakota	5,730	3,095,324	696.46	167,774	3,597,749	34.86
Tennessee	6,051	1,155,638	241.00	628	28,818	101.79
Texas	29,844	5,425,161	239.70	7,153	54,125	101.02
Utah	102	36,843	443.36	0	0	....
Vermont	2,588	3,261,312	1,336.99	0	0	....
Virginia	24,394	30,420,291	1,418.25	178	61,655	375.26
Washington	23,723	23,205,233	1,203.21	266	29,616	120.63
West Virginia	5,577	5,006,566	979.22	0	0	....
Wisconsin	13,663	4,616,745	397.29	5,699	781,749	268.81
Wyoming	4,639	1,398,796	308.49	0	0	....
United States	811,407	237,677,338	381.03	1,053,132	94,226,273	153.52

* Data withheld to avoid disclosure of individual operators.
[1] Not including acres which receive no cost share.
Total acres treated may not add due to rounding.
FSA, Conservation and Environmental Programs Division, (530) 792-5594.

Table 12-9.—Conservation Reserve Program (CRP): Enrollment by practice, under contract, August 2015

(CP25, CP26, and CP27)

State	CP25 Rare and declining habitat			CP26 Sediment retention			CP27 Farmable wetland pilot (wetland)		
	Total acres treated	Total cost share	Avg cost share per acre treated [1]	Total acres treated	Total cost share	Avg cost share per acre treated [1]	Total acres treated	Total cost share	Avg cost share per acre treated [1]
Alabama	422	19,234	125.71		0			0	
Alaska		0			0			0	
Arkansas		0			0			0	
California		0			0			0	
Colorado	1,802	60,639	95.77		0		2	0	
Delaware		0			0			0	
Florida		0			0			0	
Georgia		0			0			0	
Hawaii		0			0			0	
Idaho	79	4,120	100.00		0		4	606	173.14
Illinois	2,461	368,086	158.02		0		317	62,200	253.10
Indiana	2,318	428,688	185.47		0		333	517,305	1,966.19
Iowa	163,061	38,415,230	259.80		0		23,133	4,852,676	252.41
Kansas	730,864	45,772,661	70.15		0		724	16,216	124.26
Kentucky	38,695	8,669,377	231.10		0			0	
Louisiana		0			0			0	
Maine		0			0			0	
Maryland	416	117,676	282.60		0		1	1,062	885.00
Massachusetts		0			0			0	
Michigan	199	22,643	143.55	49	149,562	3,844.79	25	3,047	507.83
Minnesota	162,468	16,741,771	113.46		0		16,182	1,758,331	139.84
Mississippi		0			0		40	5,109	129.34
Missouri	56,323	7,341,520	153.79		0		6	2,258	525.12
Montana	109,770	5,946,584	64.98		0		203	344	23.24
Nebraska	176,590	17,462,619	106.67		0		1,570	69,056	144.07
Nevada	*	*	*	*	*	*	*	*	*
New Hampshire		0			0			0	
New Jersey		0			0			0	
New Mexico		0			0			0	
New York		0			0			0	
North Carolina		0			0			0	
North Dakota	11,547	992,767	89.78		0		21,772	937,226	66.89
Ohio	6,595	2,185,601	421.88		0		82	395,261	5,028.77
Oklahoma	24,294	2,063,392	91.21		0		9	7,316	2,090.29
Oregon	34,216	1,081,639	98.09		0			0	
Pennsylvania		0			0			0	
Puerto Rico		0			0			0	
South Carolina		0			0			0	
South Dakota	18,553	1,384,824	99.26	14	0		27,628	823,715	62.76
Tennessee		0			0			0	
Texas		0			0			0	
Utah		0			0			0	
Vermont		0			0			0	
Virginia		0			0			0	
Washington	1,166	88,535	156.98		0		2	0	
West Virginia		0			0			0	
Wisconsin	14,391	4,043,886	314.09		0		29	1,072	126.12
Wyoming		0			0			0	
United States	1,556,229	153,211,492	111.69	63	149,562	3,844.79	92,062	9,452,801	156.99

* Data withheld to avoid disclosure of individual operators.
[1] Not including acres which receive no cost share.
Total acres treated may not add due to rounding.
FSA, Conservation and Environmental Programs Division, (530) 792-5594.

Table 12-10.—Conservation Reserve Program (CRP): Enrollment by practice, under contract, August 2015

(CP28, CP29, and CP30)

State	CP28 Farmable wetland pilot (buffer)			CP29 Wildlife habitat buffer (marginal pastureland)			CP30 Wetland buffer (marginal pastureland)		
	Total acres treated	Total cost share	Avg cost share per acre treated [1]	Total acres treated	Total cost share	Avg cost share per acre treated [1]	Total acres treated	Total cost share	Avg cost share per acre treated [1]
Alabama	0	0	....	214	3,001	130.48	....	0	....
Alaska	....	0	....	....	0	....	422	128,957	305.51
Arkansas	0	0	....	509	236,058	497.59	3,389	6,502	6,502.00
California	....	0	....	115	34,205	297.95	....	0	....
Colorado	4	0	....	94	630	420.00	38	8,221	437.29
Delaware	....	0	....	....	0	....	....	0	....
Florida	....	0	....	....	0	....	....	0	....
Georgia	....	0	....	28	9,463	4,301.36	....	0	....
Hawaii	....	0	....	....	0	....	....	0	....
Idaho	2	62	31.00	229	62,987	593.10	162	14,845	99.76
Illinois	545	59,787	137.32	266	44,038	199.90	24	2,368	101.63
Indiana	614	74,736	136.23	67	81,387	1,285.73	51	9,398	185.00
Iowa	55,162	4,394,773	94.39	9,714	2,789,879	378.11	2,188	285,774	192.76
Kansas	1,390	51,675	49.33	19	4,056	209.07	....	0	....
Kentucky	....	0	....	71,764	11,160,245	162.42	5	1,508	295.69
Louisiana	0	0	....	....	0	....	151	0	....
Maine	....	0	....	5	18,088	36,176.00	12	37,630	3,084.43
Maryland	4	0	....	1,009	233,380	489.90	149	7,992	583.36
Massachusetts	....	0	....	....	0	....	....	0	....
Michigan	52	3,444	186.16	5	0	....	296	285,145	1,092.30
Minnesota	35,622	2,794,778	96.20	837	35,841	84.43	7,469	378,670	117.66
Mississippi	157	20,462	130.17	24	3,564	2,741.54	24	2,466	104.49
Missouri	12	2,416	525.22	1,327	459,543	451.10	2,009	1,203,996	780.65
Montana	184	1,819	33.14	98	5,876	62.38	....	0	....
Nebraska	2,493	146,892	69.28	1,399	348,643	281.52	238	34,166	193.14
Nevada	*	*	*	*	*	*	*	*	*
New Hampshire	....	0	....	....	0	....	....	0	....
New Jersey	....	0	....	....	0	....	....	0	....
New Mexico	....	0	....	....	0	....	....	0	....
New York	....	0	....	2,572	1,910,552	838.63	1,124	777,606	772.81
North Carolina	0	0	....	55	72,972	1,600.26	....	0	....
North Dakota	52,125	2,286,669	71.40	35	0	....	....	0	....
Ohio	179	17,383	99.27	2,864	352,848	155.76	131	206,106	2,605.97
Oklahoma	21	465	62.84	6	4,324	697.42	9	850	100.00
Oregon	....	0	....	13,565	2,892,996	358.58	659	905,601	1,373.18
Pennsylvania	....	0	....	1,229	513,191	573.53	467	123,820	395.21
Puerto Rico	....	0	....	545	12,180	60.00	....	0	....
South Carolina	....	0	....	47	128,607	2,730.51	77	156,170	2,030.82
South Dakota	58,228	2,990,413	63.04	5,104	348,429	103.01	24,184	539,660	48.61
Tennessee	....	0	....	36	16,884	1,110.77	....	0	....
Texas	....	0	....	2,303	400,071	183.81	2	2,973	1,351.36
Utah	....	0	....	35	6,563	321.72	....	0	....
Vermont	....	0	....	....	0	....	3	9,394	2,846.67
Virginia	....	0	....	691	961,333	1,507.19	20	0	....
Washington	8	0	....	956	285,230	729.12	190	227,527	1,224.12
West Virginia	....	0	....	....	0	....	....	0	....
Wisconsin	41	2,077	112.27	1,175	274,428	271.66	28	16,619	595.66
Wyoming	....	0	....	1,435	308,037	400.72	....	0	....
United States	206,844	12,847,851	80.47	120,375	24,019,529	234.15	43,521	5,373,965	257.54

* Data withheld to avoid disclosure of individual operators.
[1] Not including acres which receive no cost share.
Total acres treated may not add due to rounding.
FSA, Conservation and Environmental Programs Division, (530) 792-5594.

Table 12-11.—Conservation Reserve Program (CRP): Enrollment by practice, under contract, August 2015

(CP31, CP32, and CP33)

State	CP31 Bottomland hardwood			CP32 Hardwood trees			CP33 Upland bird habitat buffers		
	Total acres treated	Total cost share	Avg cost share per acre treated [1]	Total acres treated	Total cost share	Avg cost share per acre treated [1]	Total acres treated	Total cost share	Avg cost share per acre treated [1]
Alabama	1,208	89,829	117.25		0		1,179	73,231	101.78
Alaska		0			0			0	
Arkansas	15,442	1,131,161	120.75	980	6,000	594.06	5,511	500,367	132.72
California		0		160	0			0	
Colorado		0			0		344	16,357	94.57
Delaware		0			0			0	
Florida		0			0			0	
Georgia	25	2,000	103.63		0		2,221	92,280	81.25
Hawaii		0			0			0	
Idaho		0			0			0	
Illinois	4,481	870,881	212.35	636	0		66,065	7,192,256	122.51
Indiana	5,764	1,179,388	253.53	518	7,527	46.26	14,145	1,908,283	148.62
Iowa	3,208	925,669	342.39	1,546	53,465	283.48	26,546	4,056,452	178.56
Kansas	215	37,640	235.84		0		40,407	1,437,284	44.20
Kentucky	276	98,849	372.31	234	9	0.50	7,727	1,068,489	150.17
Louisiana	48,604	3,240,857	106.28	913	0		382	23,012	100.49
Maine		0			0			0	
Maryland		0			0		741	108,107	153.76
Massachusetts		0			0			0	
Michigan	11	7,300	675.93	6	610	100.00	859	117,470	191.22
Minnesota	596	16,340	246.83	1,684	12,859	82.32	503	44,279	93.59
Mississippi	45,162	1,224,550	94.56	775	0		2,113	154,959	99.12
Missouri	1,471	118,421	167.64	541	0		34,687	2,723,105	99.12
Montana		0			0			0	
Nebraska	9	4,778	542.95		0		6,199	428,298	79.21
Nevada	*	*	*	*	*	*	*	*	*
New Hampshire		0			0			0	
New Jersey		0			0			0	
New Mexico		0			0			0	
New York	2	2,000	869.57		0			0	
North Carolina	42	1,978	139.30		0		8,612	528,206	87.47
North Dakota		0			0			0	
Ohio	106	23,856	343.25	39	473	105.11	15,755	1,534,741	109.84
Oklahoma	443	63,174	152.04	80	0		1,063	45,067	58.26
Oregon		0			0			0	
Pennsylvania		0			0		26	3,900	150.00
Puerto Rico		0			0			0	
South Carolina	8	379	48.01		0		5,441	307,952	66.18
South Dakota		0		17	0		1,473	91,050	67.98
Tennessee	3,002	248,409	118.30	1	0		5,277	402,888	86.05
Texas	381	53,558	140.54		0		7,557	652,982	130.86
Utah		0			0			0	
Vermont		0			0			0	
Virginia		0			0		1,638	127,362	97.39
Washington		0			0		399	25,782	109.34
West Virginia		0			0			0	
Wisconsin		0		945	71	16.14	290	53,386	184.28
Wyoming		0			0			0	
United States	130,456	9,341,018	134.87	9,075	81,014	147.14	257,160	23,717,545	110.59

* Data withheld to avoid disclosure of individual operators.
[1] Not including acres which receive no cost share.
Total acres treated may not add due to rounding.
FSA, Conservation and Environmental Programs Division, (530) 792-5594.

Table 12-12.—Conservation Reserve Program (CRP): Enrollment by practice, under contract, August 2015

(CP34, CP36, and CP37)

State	CP34 Flood control structure			CP36 Longleaf pine			CP37 Duck nesting habitat		
	Total acres treated	Total cost share	Avg cost share per acre treated [1]	Total acres treated	Total cost share	Avg cost share per acre treated [1]	Total acres treated	Total cost share	Avg cost share per acre treated [1]
Alabama		0		13,550	2,460,974	206.72		0	
Alaska		0			0			0	
Arkansas		0			0			0	
California		0			0			0	
Colorado		0			0			0	
Delaware		0			0			0	
Florida		0		1,847	453,663	250.60		0	
Georgia		0		104,928	19,468,997	294.01		0	
Hawaii		0			0			0	
Idaho		0			0			0	
Illinois		0		2	247	117.76		0	
Indiana		0			0			0	
Iowa		0			0		2,476	219,005	129.31
Kansas		0			0			0	
Kentucky		0			0			0	
Louisiana		0		59	7,030	119.97		0	
Maine		0			0			0	
Maryland		0			0			0	
Massachusetts		0			0			0	
Michigan		0			0			0	
Minnesota	69	5,623	112.91		0		12,740	1,073,534	93.95
Mississippi		0		418	32,207	97.27		0	
Missouri		0			0			0	
Montana		0			0		1,978	23,211	18.39
Nebraska		0			0			0	
Nevada	*	*	*	*	*	*	*	0	*
New Hampshire		0			0			0	
New Jersey		0			0			0	
New Mexico		0			0			0	
New York		0			0			0	
North Carolina		0		6,715	1,072,047	220.96		0	
North Dakota		0			0		137,523	3,927,184	47.82
Ohio		0			0			0	
Oklahoma		0			0			0	
Oregon		0			0			0	
Pennsylvania		0			0			0	
Puerto Rico		0			0			0	
South Carolina		0		11,208	1,295,700	140.73		0	
South Dakota		0			0		145,698	4,847,260	41.201
Tennessee		0			0			0	
Texas		0			0			0	
Utah		0			0			0	
Vermont		0			0			0	
Virginia		0		527	62,573	127.31		0	
Washington		0			0			0	
West Virginia		0			0			0	
Wisconsin		0			0			0	
Wyoming		0			0			0	
United States	69	5,623	112.91	139,254	24,853,439	261.95	300,415	10,090,194	47.12

* Data withheld to avoid disclosure of individual operators.
[1] Not including acres which receive no cost share.
Total acres treated may not add due to rounding.
FSA, Conservation and Environmental Programs Division, (530) 792-5594.

Table 12-13.—Conservation Reserve Program (CRP): Enrollment by practice, under contract, August 2015
(CP38 and CP39)

State	CP38 State acres for wildlife enhancement			CP39 Constructed wetlands		
	Total acres treated	Total cost share	Avg cost share per acre treated [1]	Total acres treated	Total cost share	Avg cost share per acre treated [1]
Alabama	2,114	126,966	116.01	0	0	
Alaska	0	0		0	0	
Arkansas	15,404	791,285	171.06	0	0	
California	0	0		0	0	
Colorado	29,267	1,631,956	72.29	0	0	
Delaware	0	0		0	0	
Florida	0	0		0	0	
Georgia	10,449	748,469	108.59	0	0	
Hawaii	0	0		0	0	
Idaho	108,395	7,840,264	76.90	0	0	
Illinois	20,540	3,432,062	228.00	17	0	
Indiana	27,988	3,470,261	144.68	0	0	
Iowa	80,626	10,885,131	154.82	251	60,210	284.55
Kansas	82,585	2,344,175	42.82	4	1,150	261.36
Kentucky	10,765	956,730	208.41	0	0	
Louisiana	215	21,000	150.00	0	0	
Maine	2,135	384,722	419.21	0	0	
Maryland	0	0		0	0	
Massachusetts	0	0		0	0	
Michigan	6,314	1,465,682	357.40	0	0	
Minnesota	50,484	4,477,436	107.15	71	18,278	258.53
Mississippi	8,418	1,696,133	384.88	0	0	
Missouri	30,423	4,193,165	225.43	0	0	
Montana	34,667	3,188,025	191.32	0	0	
Nebraska	57,532	4,114,563	86.25	0	0	
Nevada	*	*	*	*	*	*
New Hampshire	0	0		0	0	
New Jersey	705	138,645	235.24	0	0	
New Mexico	2,749	0		0	0	
New York	1,252	129,204	220.45	0	0	
North Carolina	880	109,686	219.50	0	0	
North Dakota	116,448	2,569,478	44.94	0	0	
Ohio	17,664	1,448,431	132.52	38	54,829	2,383.89
Oklahoma	7,038	2,749,021	498.80	0	0	
Oregon	1,046	49,637	55.35	0	0	
Pennsylvania	0	0		0	0	
Puerto Rico	0	0		0	0	
South Carolina	1,105	189,901	214.32	0	0	
South Dakota	101,763	2,665,017	41.89	0	0	
Tennessee	6,047	766,608	138.63	0	0	
Texas	88,057	7,358,065	93.49	0	0	
Utah	0	0		0	0	
Vermont	0	0		0	0	
Virginia	308	37,305	121.05	0	0	
Washington	73,032	9,549,456	137.32	0	0	
West Virginia	0	0		0	0	
Wisconsin	10,219	1,465,836	178.16	0	0	
Wyoming	9,688	852,240	87.97.	0	0	
United States	1,016,323	81,846,552	108.72	381	134,467	434.19

* Data withheld to avoid disclosure of individual operators.
[1] Not including acres which receive no cost share.
Total acres treated may not add due to rounding.
FSA, Conservation and Environmental Programs Division, (530) 792-5594.

Table 12-14.—Conservation Reserve Program (CRP): Enrollment by practice, under contract, August 2015

(CP40, CP41, and CP42)

State	CP40 Aquaculture wetlands			CP41 Flooded prairie wetlands			CP42 Pollinator Habitat[2]		
	Total acres treated	Total cost share	Avg cost share per acre treated[1]	Total acres treated	Total cost share	Avg cost share per acre treated[1]	Total acres treated	Total cost share	Avg cost share per acre treated[1]
Alabama	37	504	36.52	0	0	....	21	3,800	200.00
Alaska	0	0	....	0	0	....	0	0	....
Arkansas	2,191	573,522	314.28	0	0	....	32	6,553	412.15
California	0	0	....	0	0	....	1,821	0	....
Colorado	446	29,853	67.01	0	0	....	13,289	1,073,401	104.59
Delaware	0	0	....	0	0	....	4	2,800	700.00
Florida	0	0	....	0	0	....	0	0	....
Georgia	0	0	....	0	0	....	0	0	....
Hawaii	0	0	....	0	0	....	0	0	....
Idaho	0	0	....	0	0	....	7,828	767,212	103.54
Illinois	0	0	....	0	0	....	12,887	5,688,018	453.60
Indiana	0	0	....	0	0	....	1,910	244,482	137.52
Iowa	0	0	....	125	23,200	185.30	15,304	7,804,973	517.15
Kansas	0	0	....	0	0	....	1,763	90,400	78.57
Kentucky	0	0	....	0	0	....	302	57,898	201.30
Louisiana	3,324	609,562	183.36	0	0	....	17	1,413	83.60
Maine	0	0	....	0	0	....	0	0	....
Maryland	0	0	....	0	0	....	19	0	....
Massachusetts	0	0	....	0	0	....	0	0	....
Michigan	0	0	....	0	0	....	280	45,858	175.95
Minnesota	0	0	....	700	37,490	63.95	879	149,019	200.76
Mississippi	10,816	1,645,072	180.91	0	0	....	0	0	....
Missouri	206	0	....	0	0	....	3,829	448,401	152.03
Montana	0	0	....	0	0	....	4,900	188,718	48.91
Nebraska	88	0	....	0	0	....	5,734	392,920	85.11
Nevada	*	*	*	*	*	*	*	*	*
New Hampshire	0	0	....	0	0	....	0	0	....
New Jersey	0	0	....	0	0	....	1	244	407.00
New Mexico	0	0	....	0	0	....	742	84,804	120.00
New York	0	0	....	0	0	....	23	9,825	514.37
North Carolina	58	692	197.71	0	0	....	4	3,510	900.00
North Dakota	0	0	....	32,803	929,661	40.40	1,791	286,724	174.27
Ohio	0	0	....	0	0	....	1,323	72,080	106.71
Oklahoma	0	0	....	0	0	....	142	7,088	105.33
Oregon	0	0	....	0	0	....	352	13,728	85.85
Pennsylvania	0	0	....	0	0	....	0	0	....
Puerto Rico	0	0	....	0	0	....	0	0	....
South Carolina	0	0	....	0	0	....	0	0	....
South Dakota	0	0	....	10,121	413,700	44.22	1,131	164,800	160.91
Tennessee	0	0	....	0	0	....	417	35,062	178.77
Texas	0	0	....	0	0	....	12,131	975,905	126.98
Utah	0	0	....	0	0	....	766	64,790	84.58
Vermont	0	0	....	0	0	....	0	0	....
Virginia	0	0	....	0	0	....	61	3,576	236.82
Washington	0	0	....	0	0	....	3,005	325,008	123.73
West Virginia	0	0	....	0	0	....	0	0	....
Wisconsin	0	0	....	0	0	....	215	49,375	304.92
Wyoming	0	0	....	0	0	....	15	2,310	150.00
United States	17,166	2,859,205	194.43	43,750	1,404,051	42.45	92,953	19,067,488	248.31

* Data withheld to avoid disclosure of individual operations.   [1] Not including acres which receive no cost share.   [2] Does not include about 40,000 acres from Signup 39 before CP42 was implemented.
Total acres treated may not add due to rounding.
FSA, Conservation and Environmental Programs Division, (530) 792-5594.

Table 12-15.—Emergency Conservation Program: Assistance, fiscal years 2005–2014 [1]

Year	Emergency Conservation Program
2005	56,376
2006	58,973
2007	30,754
2008	27,845
2009	73,028
2010	76,735
2011	32,262
2012	56,114
2013	41,206
2014	22,548

[1] Totals are from unrounded data.
FSA, Conservation and Environmental Protection Division, (202) 720-0048.

Table 12-16.—Conservation Reserve Program (CRP): Enrollment by State, August 2015

State [1]	Number of contracts	Number of farms	Acres	Annual rent ($1,000)	Avg Payments [2] ($/acre)
Alabama	6,723	4,855	278,058	12,762	45.90
Alaska	37	28	17,518	619	35.32
Arkansas	5,744	3,212	234,443	16,310	69.57
California	298	226	81,722	3,661	44.80
Colorado	10,807	5,579	1,894,647	66,041	34.86
Connecticut	6	5	52	4	69.51
Delaware	570	305	6,166	729	118.17
Florida	899	736	36,299	1,686	46.46
Georgia	7,950	5,412	260,424	14,540	55.83
Hawaii	19	15	1,000	50	49.64
Idaho	4,299	2,451	589,817	30,624	51.92
Illinois	77,953	43,036	894,671	135,752	151.73
Indiana	35,327	19,775	231,719	32,593	140.66
Iowa	100,779	51,030	1,484,096	243,576	164.12
Kansas	40,428	23,304	2,192,408	92,340	42.12
Kentucky	14,349	7,622	262,607	35,902	136.72
Louisiana	4,787	3,086	295,392	23,390	79.18
Maine	342	243	8,315	385	46.31
Maryland	5,931	3,286	67,588	10,948	161.98
Massachusetts	3	3	10	2	207.20
Michigan	12,727	7,145	166,681	17,888	107.32
Minnesota	55,576	29,669	1,151,977	102,505	88.98
Mississippi	17,821	11,191	744,311	44,861	60.27
Missouri	31,207	18,201	1,014,671	102,486	101.00
Montana	8,696	3,823	1,504,134	44,810	29.79
Nebraska	22,694	13,167	794,829	56,375	70.93
Nevada	*	*	*	*	*
New Hampshire	3	3	13	1	69.04
New Jersey	314	208	2,026	173	85.20
New Mexico	2,000	1,282	432,391	16,141	37.33
New York	2,526	1,824	42,896	3,332	77.68
North Carolina	6,578	4,309	89,498	6,524	72.89
North Dakota	24,246	12,526	1,525,887	67,790	44.43
Ohio	36,172	20,163	267,555	37,994	142.01
Oklahoma	6,424	4,324	738,215	25,374	34.37
Oregon	4,365	2,327	541,450	32,489	60.00
Pennsylvania	9,926	6,566	164,432	18,760	114.09
Puerto Rico	18	18	1,387	104	74.91
Rhode Island	*	*	*	*	*
South Carolina	4,856	2,699	95,714	3,844	40.16
South Dakota	29,679	13,769	922,585	72,001	78.04
Tennessee	5,837	3,861	142,790	12,142	85.04
Texas	19,768	14,449	3,041,690	111,817	36.76
Utah	869	535	171,887	5,851	34.04
Vermont	413	300	2,876	297	103.25
Virginia	5,437	4,149	54,694	3,442	62.94
Washington	11,640	5,174	1,309,759	76,152	58.14
West Virginia	514	421	6,534	513	78.59
Wisconsin	17,015	10,692	244,083	25,281	103.58
Wyoming	810	537	192,839	5,129	26.60
United States	655,384	367,543	24,204,930	1,615,993	66.77

* Data withheld to avoid disclosure of individual operations. [1] State in which land is located. [2] Payments scheduled to be made October 2015.
FSA, Conservation and Environmental Programs Division, (530) 792-5594.

Table 12-17.—Forest land: Total forest land and area and ownership of timberland, by regions, Jan. 1, 2012[1]

Region	Total forest land[2]	Timberland[3]							
		All ownerships	Federal			State, county, and municipal	Private		
			Total	National forest	Other		Total	Forest industry	Farmer and other private[4]
	1,000 acres	1,000 acres	1,000 acres	1,000 acres	1,000 acres	1,000 acres	1,000 acres	1,000 acres	1,000 acres
Northeast	84,846	79,822	3,122	2,443	679	10,454	66,246	21,290	44,956
North Central	90,730	87,556	9,049	7,902	1148	16,706	61,802	7,602	54,199
North	175,575	167,378	12,172	10,345	1,827	27,159	128,047	28,892	99,155
Southeast	89,844	86,755	7,763	5,230	2,533	5,561	73,430	26,499	46,931
South Central	154,872	123,292	10,304	7,466	2,838	3,788	109,200	34,712	74,488
South	244,716	210,048	18,067	12,696	5,371	9,351	182,631	61,212	121,419
Great Plains	6,724	6,179	1,272	1,071	201	232	4,675	136	4,539
Intermountain	124,614	64,844	46,790	43,870	2,920	2,906	15,148	3,772	11,376
Rocky Mountains	131,338	71,023	48,062	44,941	3,121	3,138	19,823	3,908	15,915
Alaska	128,577	12,817	4,699	3,677	1,022	4,822	3,297	2,776	521
Pacific Northwest	52,222	42,197	19,860	17,512	2,348	3,700	18,638	10,428	8,210
Pacific Southwest[5] ..	33,805	17,690	9,451	9,137	313	500	7,740	4,064	3,677
Pacific Coast	214,604	72,705	34,010	30,326	3,684	9,020	29,675	17,268	12,407
All regions	766,234	521,154	112,310	98,308	14,002	48,668	360,175	111,279	248,896

[1] Forest Resources of the United States, 2012. Data may not add to totals because of rounding.   [2] Forest land is land at least 10 percent stocked by forest trees of any size, including land that formerly had such tree cover and that will be naturally or artificially regenerated. Forest land includes transition zones, such as areas between heavily forested and nonforested lands that are at least 10 percent stocked with forest trees, and forest areas. The minimum area for classification of forest land is 1 acre. Roadside, streamside, and shelterbelt strips of timber must have a crown width at least 120 feet wide to qualify as forest land. Unimproved roads and trails, streams, and clearings in forest areas are classified as forest if less than 120 feet in width.   [3] Timberland is forest land that is producing or is capable of producing crops of industrial wood and that is not withdrawn from timber utilization by statute or administrative regulation. Areas qualifying as timberland have the capability of producing more than 20 cubic feet per acre per year of industrial wood in natural stands. Currently inaccessible and inoperable areas are included.   [4] Includes Indian lands.   [5] Includes Hawaii.
FS, Economics & Statistics Research, RWU-4851, (608) 231–9376.

Table 12-18.—Timber volume: Net volume of growing stock and sawtimber on timberland, by softwoods and hardwoods, and regions, 2012 [1]

Region	Growing stock [2]			Sawtimber [3]		
	All species	Softwoods	Hardwoods	All species	Softwoods	Hardwoods
	Million cubic feet	Million cubic feet	Million cubic feet	Million board feet	Million board feet	Million board feet
Northeast	152,096	36,463	115,634	169,834	40,537	129,296
North Central	115,707	22,298	93,409	138,625	25,419	113,207
North	267,803	58,761	209,043	308,459	65,956	242,503
Southeast	134,872	62,061	72,812	157,981	65,724	92,257
South Central	171,751	66,895	104,855	200,586	70,556	130,031
South	306,623	128,956	177,667	385,567	136,280	222,288
Great Plains	4,760	2,021	2,739	8,285	2,548	5,737
Intermountain	131,145	122,068	9,077	150,347	139,341	11,006
Rocky Mountains	135,905	124,089	11,816	158,632	141,889	16,743
Alaska	35,762	32,453	3,309	37,458	33,967	3,491
Pacific Northwest	158,206	145,473	12,732	166,312	153,099	13,213
Pacific Southwest [4]	68,096	57,887	10,209	72,125	61,377	10,748
Pacific Coast	262,063	235,813	26,250	275,895	248,444	27,451
All regions	972,395	547,619	424,776	1,101,554	592,568	508,985

[1] Forest Resources of the United States, 2012. Data may not add to totals because of rounding. [2] Live trees of commercial species meeting specified standards of quality or vigor. Cull trees are excluded. Includes only trees 5.0-inches diameter or larger at 4½ feet above ground. [3] Live trees of commercial species containing at least one 12-foot sawlog or two noncontiguous 8-foot logs, and meeting regional specifications for freedom from defect. Softwood trees must be at least 9.0-inches diameter and hardwood trees must be at least 11.0-inches diameter at 4½ feet above ground. [4] Includes Hawaii.
FS, Economics & Statistics Research, RWU-4851, (608) 231–9376.

Table 12-19.—Timber removals: Roundwood product output, logging residues and other removals from growing stock and other sources, by softwoods and hardwoods, 2011 [1]

Roundwood products, logging residues, and other removals	All sources			Growing stock [2]			Other sources [3]		
	All species	Soft-woods	Hard-woods	All species	Soft-woods	Hard-woods	All species	Soft-woods	Hard-woods
	Million cubic feet	Million cubic feet	Million cubic feet	Million cubic feet	Million cubic feet	Million cubic feet	Million cubic feet	Million cubic feet	Million cubic feet
Roundwood products:									
Sawlogs	4,988	3,647	1,241	4,752	3,494	1,257	236	152	84
Pulpwood	5,033	3,353	1,680	4,483	3,064	1,419	550	289	261
Veneer logs	700	624	76	680	608	73	19	16	3
Other products [4]	283	223	60	209	159	50	74	64	10
Fuelwood [5]	1,804	515	1,289	633	135	498	1,170	379	791
Total	12,808	8,362	4,346	10,757	7,460	3,297	2,049	900	1,149
Logging residues [6]	3,730	1,941	1,790	1,031	465	566	2,700	1,475	1,224
Other removals [7]	1,633	526	1,106	1,065	393	672	568	133	435
Total	5,363	2,467	2,896	2,096	858	1,238	3,268	1,608	1,659

[1] Forest Resources of the United States, 2012. Data may not add to totals because of rounding.   [2] Includes live trees of commercial species meeting specified standards of quality or vigor. Cull trees are excluded. Includes only trees 5.0-inches diameter or larger at 4½ feet above ground.   [3] Includes salvable dead trees, rough and rotten trees, trees of noncommercial species, trees less than 5.0-inches diameter at 4½ feet above ground, tops, and roundwood harvested from nonforest land (for example, fence rows).   [4] Includes such items as cooperage, pilings, poles, posts, shakes, shingles, board mills, charcoal and export logs.   [5] Downed and dead wood volume left on the ground after trees have been cut on timberland.   [6] Net of wet rot or limbs.   [7] Unutilized wood volume from cut or otherwise killed growing stock, from non-growing stock sources on timberland (for example, precommercial thinnings), or from timberland clearing. Does not include volume removed from inventory through reclassification of timberland to reserved timberland.
FS, Economics & Statistics Research, RWU-4851, (608) 231–9376.

Table 12-20.—Timber growth, removals and mortality: Net annual growth, removals, and mortality of growing stock on timberland by softwoods and hardwoods and regions, 2011 [1]

Region	Growth [2]			Removals [3]			Mortality [4]		
	All species	Soft-woods	Hard-woods	All species	Soft-woods	Hard-woods	All species	Soft-woods	Hard-woods
	Million cubic feet	Million cubic feet	Million cubic feet	Million cubic feet	Million cubic feet	Million cubic feet	Million cubic feet	Million cubic feet	Million cubic feet
Northeast	3,548	896	2,652	1,119	376	743	1,235	347	888
North Central	2,968	616	2,352	1,240	264	977	1,246	217	1,029
North	6,516	1,512	5,004	2,360	640	1,720	2,481	564	1,917
Southeast	6,135	3,908	2,227	3,844	2,658	1,186	1,145	627	518
South Central	7,674	4,899	2,775	4,204	2,676	1,528	1,710	621	1,089
South	13,809	8,808	5,002	8,048	5,335	2,714	2,855	1,248	1,607
Great Plains	103	21	82	36	28	8	58	29	29
Intermountain	818	717	101	391	382	9	2,018	1,919	99
Rocky Mountains	921	739	183	427	410	17	2,076	1,947	128
Alaska	253	133	121	66	59	7	303	275	27
Pacific Northwest	3,331	2,040	291	1,599	1,521	78	933	822	111
Pacific Southwest [5]	1,582	1,433	149	355	355	0	367	305	62
Pacific Coast	5,167	4,605	561	1,020	1,935	85	1,603	1,402	201
All regions	26,413	15,663	10,750	12,854	8,319	4,535	9,015	5,162	3,853

[1] Forest Resources of the United States, 2012. Data may not add to totals because of rounding.   [2] The net increase in the volume of trees during a specified year. Components include the increment in net volume of trees at the beginning of the specific year surviving to its end, plus the net volume of trees reaching the minimum size class during the year, minus the volume of trees that died during the year, and minus the net volume of trees that became cull trees during the year.   [3] The net volume of trees removed from the inventory during a specified year by harvesting, cultural operations such as timber stand improvement, or land clearing.   [4] The volume of sound wood in trees that died from natural causes during a specified year.   [5] Includes Hawaii.
FS, Economics & Statistics Research, RWU-4851, (608) 231–9376.

Table 12-21.—Timber volume: Net volume of sawtimber on timberland in the West, by regions and species, 2012 [1]

Species	Total West	Inter-mountain	Alaska	Pacific Northwest	Pacific South-west [2]	Great Plains
	Million board feet	Million board feet	Million board feet	Million board feet	Million board feet	Million board feet
Softwoods:						
Douglas-fir	127,075	30,129	0	77,829	19,117	0
Ponderosa and Jeffrey pines	41,424	16,938	0	12,221	10,390	1,875
True fir	53,662	22,885	19	16,531	14,227	0
Western hemlock	35,190	1,397	12,785	20,885	123	0
Sugar pine	3,678	0	0	704	2,974	0
Western white pine	1,085	400	0	380	305	0
Redwood	5,366	0	0	19	5,347	0
Sitka spruce	11,995	0	10,372	1,422	201	0
Engelmann and other spruces	25,174	19,283	4,282	1,601	8	0
Western larch	6,126	4,013	3	2,110	0	0
Incense cedar	4,206	0	0	810	3,396	0
Lodgepole pine	24,083	19,569	86	3,390	1,038	0
Other	20,839	7,453	4,906	7,573	762	145
Total	359,902	122,068	32,453	145,473	57,887	2,021
Hardwoods:						
Cottonwood and aspen	11,761	8,844	1,079	776	106	956
Red alder	6,866	20	130	6,348	368	0
Oak	5,439	0	0	605	4,351	483
Other	13,998	213	2,100	5,002	5,383	1,300
Total	38,066	9,077	3,309	12,732	10,209	2,739
All species	397,969	131,145	35,762	158,206	68,096	4,760

[1] Forest Resources of the United States, 2012. International ¼-inch rule. Data may not add to totals because of rounding.　[2] Includes Hawaii.
FS, Economics & Statistics Research, RWU-4851, (608) 231–9376.

Table 12-22.—Timber volume: Net volume of sawtimber on timberland in the East, by regions and species, 2012 [1]

Species	Total East	North			South		
		Total	Northeast	North Central	Total	Southeast	South Central
	Million board feet	Million board feet	Million board feet	Million board feet	Million board feet	Million board feet	Million board feet
Softwoods:							
Longleaf and slash pines	17,129	0	0	0	17,129	12,729	4,400
Loblolly and shortleaf pines	93,937	1,837	813	1,024	92,100	37,343	54,757
Other yellow pines	8,744	2,199	1,827	372	6,545	4,785	1,760
White and red pines	24,196	21,349	12,567	8,782	2,847	2,284	563
Jack pine	988	988	8	980	0	0	0
Spruce and balsam fir	13,221	13,148	8,904	4,244	73	73	0
Eastern hemlock	11,200	19,244	8,919	1,325	956	460	496
Cypress	7,286	23	14	9	7,263	3,917	3,346
Other	11,016	8,972	3,410	5,562	2,044	470	1,574
Total	187,717	58,761	36,463	22,298	128,956	62,061	66,895
Hardwoods:							
Select white oaks	35,696	16,125	5,962	10,163	19,571	7,417	12,154
Select red oaks	27,986	18,502	11,592	6,911	9,484	3,362	6,122
Other white oaks	23,356	7,683	5,461	2,222	15,673	5,704	9,969
Other red oaks	46,157	14,286	6,051	8,235	31,871	11,792	20,079
Hickory	23,036	9,391	4,029	5,362	13,645	3,765	9,880
Yellow birch	4,748	4,659	3,910	749	89	81	8
Hard maple	26,749	24,127	14,029	10,098	2,622	504	2,118
Soft maple	41,869	33,267	22,220	11,047	8,602	5,148	3,454
Beech	9,069	6,445	5,376	1,069	2,624	888	1,736
Sweetgum	20,181	944	658	286	19,237	7,914	11,323
Tupelo and black gum	12,227	1,140	833	307	11,087	6,243	4,844
Ash	19,352	13,577	6,886	6,691	5,775	1,782	3,993
Basswood	6,178	5,407	1,975	3,432	771	333	438
Yellow-poplar	31,941	10,141	7,246	2,895	21,800	13,364	8,436
Cottonwood and aspen	15,607	14,933	3,881	11,052	674	113	561
Black walnut	3,064	2,304	459	1,845	760	213	547
Black cherry	10,198	8,997	6,635	2,362	1,201	429	772
Other	29,295	17,118	8,433	8,685	12,177	3,658	8,419
Total	386,710	209,043	115,634	93,409	177,667	72,812	104,855
All species	574,426	267,803	152,096	115,707	306,623	134,872	171,751

[1] Forest Resources of the United States, 2012. International ¼-inch rule. Data may not add to totals because of rounding.
FS, Economics & Statistics Research, RWU-4851, (608) 231–9376.

Table 12-23.—National Forest System: National Forest System lands and other lands in States and Territories, 2015

State or other area	Gross acreage	National Forest System acreage [1]	Other acreage [2]
	1,000 acres	*1,000 acres*	*1,000 acres*
Alabama	1,290	671	620
Alaska	23,977	22,167	1,809
Arizona	11,813	11,204	609
Arkansas	3,549	2,593	956
California	24,284	20,762	3,522
Colorado	15,960	14,483	1,477
Florida	1,423	1,197	226
Georgia	1,796	867	929
Idaho	21,714	20,444	1,270
Illinois	958	304	653
Indiana	647	204	443
Kansas	109	109	0
Kentucky	2,205	820	1,386
Louisiana	1,032	609	424
Maine	94	54	41
Michigan	4,887	2,874	2,013
Minnesota	5,488	2,844	2,643
Mississippi	2,375	1,192	1,183
Missouri	3,093	1,506	1,587
Montana	19,189	17,182	2,007
Nebraska	562	351	211
Nevada	6,293	5,760	533
New Hampshire	847	748	98
New Mexico	10,249	9,225	1,024
New York	17	16	0
North Carolina	3,027	1,255	1,771
North Dakota	1,103	1,103	0
Ohio	856	244	612
Oklahoma	759	399	359
Oregon	17,659	15,696	1,963
Pennsylvania	741	514	227
South Carolina	1,381	632	749
South Dakota	2,433	2,006	427
Tennessee	1,291	720	571
Texas	2,002	757	1,245
Utah	9,211	8,190	1,022
Vermont	837	410	427
Virginia	3,254	1,666	1,588
Washington	11,990	9,329	2,662
West Virginia	1,893	1,046	847
Wisconsin	2,003	1,524	479
Wyoming	9,726	9,215	511
Puerto Rico	56	29	27
Virgin Islands	0	0	0
Total	234,074	192,922	41,152

Reference: Land Areas of the National Forest System 2015 Report.

[1] *National Forest System acreage.*—A nationally significant system of Federally owned units of forest, range, and related land consisting of national forests, purchase units, national grasslands, land utilization project areas, experimental forest areas, experimental range areas, designated experimental areas, other land areas; water areas, and interests in lands that are administered by USDA Forest Service or designated for administration through the Forest Service.

National forests.—Units formally established and permanently set aside and reserved for national forest purposes.

Purchase units.—Units designated by the Secretary of Agriculture or previously approved by the National Forest Reservation Commission for purposes of Weeks Law Acquisition.

National grasslands.—Units designated by the Secretary of Agriculture and permanently held by the Department of Agriculture under Title III of the Bankhead-Jones Farm Tenant Act.

Land utilization projects.—Units designated by the Secretary of Agriculture for conservation and utilization under Title III of the Bankhead-Jones Farm Tenant Act.

Research and experimental areas.—Units reserved and dedicated by the Secretary of Agriculture for forest or range research and experimentation.

Other areas.—Units administered by the Forest Service that are not included in the above groups.

[2] *Other acreage.*—Lands within the unit boundaries in private, State, county, and municipal ownership and Federal lands over which the Forest Service has no jurisdiction. Areas of such lands which have been offered to the United States and have been approved for acquisition and subsequent Forest Service administration, but to which title had not yet been accepted by the United States.

FS, Economics & Statistics Research, RWU-4851, (608) 231–9376.

Table 12-24.—Forest products cut on National Forest System lands: Volume and value of timber cut and value of all products, United States, fiscal years 2006–2015

Year [1]	Timber cut [2]		Value of miscellaneous forest products [4]	Total value including free-use timber
	Volume	Value [3]		
	Million bd. ft.	*1,000 dollars*	*1,000 dollars*	*1,000 dollars*
2006	2,296	221,390	3,262	221,512
2007	1,960	166,641	3,262	169,992
2008	2,049	138,119	3,262	141,231
2009	1,954	94,963	3,262	78,050
2010	2,137	104,809	3,262	80,265
2011	2,440	129,301	3,262	108,374
2012	2,500	137,680	3,262	117,712
2013	2,408	140,533	3,262	(NA)
2014	2,437	146,361	3,262	(NA)
2015	2,543	162,651	3,262	(NA)

(NA) Not available. [1] U.S. Timber Production, Trade, Consumption, and Price Statistics. Fiscal years Oct. 1–Sept. 30. [2] Commercial and cost sales and land exchanges. [3] Includes collections for forest restoration or improvement under the Knutson-Vandenberg Act, 1930. [4] Includes materials not measurable in board feet, such as Christmas trees, tanbark, turpentine, seedlings, Spanish moss, etc.
FS, Economics & Statistics Research, RWU-4851, (608) 231–9376.

Table 12-25.—Lumber: Production, United States, 2006–2015

Year	Total	Softwoods	Hardwoods
	Billion bd. ft.	*Billion bd. ft.*	*Billion bd. ft.*
2006	49.740	38.726	11.014
2007	45.766	35.158	10.608
2008	35.964	29.177	6.787
2009	30.229	23.240	6.989
2010	30.461	24.802	5.659
2011	33.304	26.754	6.550
2012	34.791	28.256	6.535
2013	37.321	29.982	7.339
2014	39.810	31.490	8.320
2015	40.304	32.028	8.276

FS, Economics & Statistics Research, RWU-4851, (608) 231–9376. From data published by the American Forest and Paper Association.

Table 12-26.—Timber products: Production, imports, exports, and consumption, United States, 2006–2015 [1]

Year	Industrial roundwood used for—										
	Lumber				Plywood and veneer				Pulpwood chip		Other industrial products, production and consumption [4]
	Production	Imports	Exports	Consumption	Production	Imports	Exports	Consumption	Imports	Exports	
	Million cu. ft.³	Million cu. ft.³	Million cu. ft.³	Million cu. ft.³	Million cu. ft.³	Million cu. ft.³	Million cu. ft.³	Million cu. ft.³	Million cu. ft.³	Million cu. ft.³	Million cu. ft.³
2006	7,505	3,415	390	10,530	989	339	35	1,293	4	151	320
2007	6,921	2,743	359	9,305	898	264	40	1,122	3	205	325
2008	5,395	1,894	345	6,945	754	184	45	884	5	257	290
2009	4,576	1,347	288	5,636	616	146	34	728	9	196	294
2010	4,569	1,422	341	5,650	655	137	55	736	9	235	294
2011	5,005	1,403	387	6,022	651	139	52	738	9	235	294
2012	5,219	1,480	374	6,324	679	139	57	761	9	235	294
2013	5,607	1,675	406	6,911	703	147	55	796	9	234	294
2014	5,995	1,869	406	7,458	693	187	46	834	9	234	294
2015	6,065	2,100	371	7,793	683	222	38	867	9	235	294

Year	Industrial roundwood used for—									
	Pulpwood-based products				Logs		Total			
	Production	Imports [2]	Exports [2]	Consumption	Imports	Exports	Production	Imports	Exports	Consumption
	Million cu. ft.³	Million cu. ft.³	Million cu. ft.³	Million cu. ft.³	Million cu. ft.³	Million cu. ft.³	Million cu. ft.³	Million cu. ft.³	Million cu. ft.³	Million cu. ft.³
2006	5,456	1,460	693	6,223	94	339	14,760	5,312	1,607	18,464
2007	5,259	1,269	771	5,757	67	350	13,959	4,345	1,726	16,578
2008	4,899	1,068	818	5,149	35	313	11,900	3,187	1,779	13,308
2009	4,733	902	785	4,849	17	322	10,737	2,421	1,625	11,532
2010	4,779	911	832	4,857	20	407	10,938	2,498	1,870	11,566
2011	4,748	887	899	4,736	19	485	11,418	2,457	2,058	11,817
2012	5,805	883	875	5,815	28	432	12,663	2,539	1,972	13,231
2013	5,959	942	1,073	5,827	26	463	13,260	2,799	2,231	13,862
2014	6,562	1,205	1,340	6,427	22	444	14,222	3,292	2,470	15,044
2015	6,569	1,172	1,348	6,392	22	279	14,124	3,524	2,272	15,376

Year	Fuelwood production and consumption	Production, all products	Consumption, all products
	Million cu. ft.³	Million cu. ft.³	Million cu. ft.³
2006	1,555	16,315	20,019
2007	1,605	15,564	18,183
2008	1,510	13,410	14,818
2009	1,400	12,137	12,932
2010	1,380	12,318	12,946
2011	1,366	12,784	13,183
2012	1,440	14,103	14,671
2013	1,520	14,780	15,382
2014	1,566	15,788	16,610
2015	1,566	15,690	16,942

[1] U.S. Timber Production, Trade, Consumption, and Price Statistics. Data may not add to totals because of rounding. [2] Includes both pulpwood and the pulpwood equivalent of wood pulp, paper, and board. [3] Roundwood equivalent. [4] Includes cooperage logs, poles and piling, fence posts, hewn ties, round mine timbers, excelsior bolts, chemical wood, shingle bolts, and miscellaneous items.
FS, Economics and Statistics Research, RWU-4851, (608) 231–9376.

Table 12-27.—Timber products: Producer price indexes, selected products, United States, 2006–2015

[2009=100]

Year	Lumber	Softwood plywood	Woodpulp	Paper	Paperboard
2006	126.2	110.8	95.9	93.2	92.7
2007	116.9	115.0	107.5	94.3	97.3
2008	109.4	112.3	114.1	102.6	105.2
2009	100.0	100.0	100.0	100.0	100.0
2010	112.0	114.7	123.8	101.4	108.5
2011	111.5	108.1	129.7	106.5	111.2
2012	115.4	127.2	123.8	106.7	110.0
2013	133.0	137.7	120.3	106.2	117.6
2014	143.8	142.7	122.3	107.3	120.0
2015	133.4	136.4	120.9	105.7	117.3

FS, Economics & Statistics Research, RWU-4851, (608) 231–9376. Compiled from reports of the U.S. Department of Labor, Bureau of Labor Statistics.

Table 12-28.—Timber prices: Average stumpage prices for sawtimber sold from national forests, by selected species, 2006–2015

Year	Southern pine [1]	Ponderosa pine [2]	Western hemlock [3]	All eastern hardwoods [4]	Oak, white, red, and black [4]	Maple, sugar [5]
	Dollars per 1,000 bd. ft.	Dollars per 1,000 bd. ft.	Dollars per 1,000 bd. ft.	Dollars per 1,000 bd. ft.	Dollars per 1,000 bd. ft.	Dollars per 1,000 bd. ft.
2006	112.46	39.17	101.09	275.31	180.26	533.29
2007	176.38	60.90	54.56	276.62	220.35	361.56
2008	152.65	33.52	46.18	198.25	156.27	479.60
2009	(NA)	(NA)	77.50	171.43	119.53	274.98
2010	(NA)	(NA)	65.90	118.80	214.32	432.53
2011	(NA)	(NA)	83.30	105.30	370.81	504.30
2012	(NA)	(NA)	88.69	142.46	389.17	491.68
2013	(NA)	(NA)	58.20	108.27	216.81	447.78
2014	(NA)	(NA)	60.20	144.89	211.82	548.71
2015	(NA)	(NA)	69.04	238.10	308.72	521.51

(NA) Not available. [1] Southern region. [2] Pacific Southwest region. Includes Jeffrey pine. [3] Pacific Northwest region. [4] Eastern and Southern regions. [5] Eastern region.

Forest Service National Forest prices in this table are for timber sold on a Scribner Decimal C log rule basis, except in the Northeastern States where International ¼-inch log rule is used. Prices include KV payments; exclude timber sold by land exchanges and from land utilization project lands. Data are high bid prices which include specified road costs.

FS, Economics & Statistics Research, RWU-4851, (608) 231–9376. U.S. Timber Production, Trade, Consumption, and Price Statistics.

Table 12-29.—Timber products: Pulpwood consumption, woodpulp production, and paper and board production and consumption, United States, 2006–2015[1]

Year	Pulpwood consumption[2]	Woodpulp production[3]	Paper and board[4]		
			Production	Consumption or new supply[5]	Per capita consumption
	1,000 cords[6]	*1,000 tons*	*1,000 tons*	*1,000 tons*	*Pounds*
2006	85,883	60,568	91,800	102,439	707
2007	82,696	55,636	91,570	99,825	662
2008	80,341	56,745	87,619	93,640	615
2009	79,992	52,122	78,299	81,767	532
2010	80,307	55,343	82,968	85,331	551
2011	80,040	55,125	81,519	82,858	531
2012	81,422	55,475	80,916	82,321	524
2013	91,251	54,466	80,478	78,603	497
2014	92,186	53,367	79,488	78,046	490
2015	93,141	(NA)	79,024	77,444	481

(NA) Not available.   [1] Revised to match data from American Forest and Paper Association and American Pulpwood Association.   [2] Includes changes in stocks.   [3] Excludes defibrated and exploded woodpulp used for hard pressed board.   [4] Excludes hardboard. Includes wet machine board and construction grades.   [5] Production plus imports and minus exports (excludes products); changes in inventories not taken into account.   [6] One cord equals 128 cubic feet.
U.S. Timber Production, Trade, Consumption, and Price Statistics.FS, Economic & Statistics Research, RWU-4851, (608) 231–9376. Compiled from U.S. Department of Commerce and American Forest and Paper Association.

Table 12-30.—Lumber and competing engineered wood products: Production by type of product, 2006–2015

Year	Laminated veneer lumber[1]	Wood glulam[1]	Structural Panels[1]		Lumber[2]	
			Oriented strandboard	Softwood Plywood	Hardwood	Softwood
	Million cubic feet	*Million board feet*	*Million square feet (3/8-in. basis)*	*Million square feet (3/8-in. basis)*	*Billion board feet*	*Billion board feet*
2006	80	461	14,960	13,428	11.0	38.7
2007	68	358	14,763	12,243	10.6	35.2
2008	47	256	13,003	10,237	6.8	29.2
2009	30	167	9,598	8,608	7.0	23.2
2010	37	176	10,299	9,131	5.7	24.8
2011	38	184	10,039	8,980	6.6	26.8
2012	46	204	11,038	9,181	6.5	28.3
2013	54	230	12,492	9,346	7.3	30.0
2014	60	233	13,008	8,985	8.3	31.5
2015	61	252	13,283	8,751	8.3	32.0

[1] APA-The Engineered Wood Association.   [2] US Department of Commerce, Bureau of the Census; American Forest and Paper Association; Luppold and Dempsey. Hardwood Market Report: 2008-present.
FS, Economics & Statistics Research, RWU-4851, (608) 231–9376.

CHAPTER XIII

CONSUMPTION AND FAMILY LIVING

The statistics in this chapter deal with the consumption of food by both rural and urban people, retail price levels, and other aspects of family living of farm people. Data presented here on quantities of food available for consumption are based on material presented in the earlier commodity chapters, but they are shown here at the retail level, a form that is more useful for an analysis of the demand situation faced by the producer. Data on quantities of farm-produced food consumed directly by farm households are presented in the commodity chapters. Its value and the rental value of the farm home are given in the section on farm income.

Table 13-1.—Population: Number of people eating from civilian food supplies, United States, Jan. 1 and July 1, 2006-2015

Year	Jan. 1	July 1
	Millions	Millions
2006	296.0	297.4
2007	299.0	300.4
2008	301.9	303.2
2009	304.6	305.8
2010	307.2	308.1
2011	309.4	310.5
2012	311.8	312.9
2013	314.1	315.2
2014	316.5	317.7
2015	319.1	320.2

Monthly Population Estimates for the United States: April 1, 2010 to December 1, 2015. Source: U.S. Census Bureau, Population Division. Release Dates: Monthly, April 2010 to December 2015.
ERS, Rural Economy Branch (202) 694–5435.

Table 13-2.—Macronutrients: Quantities available for consumption per capita per day, United States, 2001–2010 [1]

Year	Food energy	Carbo-hydrate	Dietary fiber	Protein	Total fat	Saturated fatty acids	Monounsaturated fatty acids	Polyunsat-urated fatty acids	Choles-terol
	Kilo-calories	Grams	Grams	Grams	Grams	Grams	Grams	Grams	Milli-grams
2001	4,100	500	24	122	192	61	84	38	470
2002	4,200	495	24	123	199	63	87	40	480
2003	4,200	492	24	124	202	64	88	40	480
2004	4,200	491	25	125	202	64	88	40	490
2005	4,100	488	25	123	194	61	84	40	470
2006	4,100	485	25	124	196	62	83	41	480
2007	4,100	480	24	123	196	61	82	43	470
2008	4,100	478	24	121	194	60	80	45	470
2009	4,000	471	24	120	185	57	75	42	460
2010	4,000	474	25	120	190	59	77	44	460

[1] Data are based on Economic Research Service estimates of per capita quantities of food available for consumption. Center for Nutrition Policy and Promotion (CNPP), (703) 305–7600.

Table 13-3.—Vitamins: Quantities available for consumption per capita per day, United States, 2001–2010 [1]

Year	Vitamin A	Caro-tenes	Vitamin E	Vitamin C	Thiamin	Ribo-flavin	Niacin	Vitamin B₆	Folate	Folate, Dfe	Vitamin B₁₂
	Micro-grams RAE	Micro-grams RE	Milli-grams AT	Milli-grams	Milli-grams	Milli-grams	Milli-grams	Milli-grams	Micro-grams	Micro-grams	Micro-grams
2001	960	670	20.6	113	3	3.1	35	2.6	690	900	10
2002	940	630	21.7	112	3	3.1	35	2.6	682	888	10
2003	950	660	21.7	114	3	3.1	35	2.6	690	899	10
2004	960	670	21.7	113	3	3.1	35	2.7	685	893	10.1
2005	910	650	21.4	111	3	3.1	35	2.6	682	889	9.9
2006	930	650	21.4	109	3	3.1	35	2.6	684	892	9.9
2007	940	630	22.1	107	3	3.1	35	2.6	678	886	9.8
2008	930	630	22.4	106	2.9	3	34	2.5	673	881	9.6
2009	920	610	21.3	105	2.9	3	34	2.5	673	881	9.6
2010	920	630	21.9	105	2.9	3	34	2.5	681	889	9.3

[1] Data are based on USDA Economic Research Service estimates of per capita quantities of food available for consumption. Seafood estimates from the National Oceanic and Atmospheric Administration. Game harvest estimates from the individual States or from the Wildlife Management Institute. Imputed data for foods no longer reported or available. Values in the USDA Agricultural Research, National Nutrient Database for Standard Reference, Release 26. Components may not add to 100 because of rounding.
Calculated by USDA/Center for Nutrition Policy and Promotion, (703) 305–7600. Data last updated May 15, 2014.

Table 13-4.—Minerals: Quantities available for consumption per capita per day, United States, 2001–2010 [1]

Year	Calcium	Phosphorus	Magnesium	Iron	Zinc	Copper	Potassium	Sodium	Selenium
	Milligrams	Milligrams	Milligrams	Milli-grams	Milli-grams	Milli-grams	Milligrams	Milli-grams	Micro-grams
2001	1020	1850	410	24.8	18.1	2.1	3,960	1,270	191.8
2002	1020	1840	410	24.7	18.1	2.1	3,930	1,280	192.2
2003	1020	1870	410	24.9	18.2	2.1	3,970	1,340	195.8
2004	1030	1880	420	25	18.3	2.2	3,990	1,340	195.7
2005	1010	1840	410	24.9	18.1	2.2	3,890	1,250	191.2
2006	1020	1850	420	25.1	18.1	2.2	3,890	1,230	192.9
2007	1030	1850	410	24.9	18.2	2.2	3,870	1,240	193.7
2008	1030	1830	410	24.6	17.8	2.1	3,800	1,210	189.4
2009	1040	1820	400	24.4	17.7	2.1	3,770	1,220	188.8
2010	1030	1810	410	24.6	17.6	2.1	3,760	1,210	187.2

[1] Data are based on USDA Economic Research Service estimates of per capita quantities of food available for consumption. Seafood estimates from the National Oceanic and Atmospheric Administration. Game harvest estimates from the individual States or from the Wildlife Management Institute. Imputed data for foods no longer reported or available. Values in the USDA Agricultural Research, National Nutrient Database for Standard Reference, Release 26. Components may not add to 100 because of rounding.
Calculated by USDA/Center for Nutrition Policy and Promotion, (703) 305–7600. Data last updated May 15, 2014.

Table 13-5.—Food nutrients: Percentage of total contributed by major food groups, 2010 [1]

Nutrient	Meat, poultry, fish	Dairy products	Eggs	Legumes, nuts, soy	Grain products	Fruits		
						Citrus	Non-citrus	Total
	Percent	Percent	Percent	Percent	Percent	Percent	Percent	Percent
Kilocalories	16.1	10.0	1.2	3.4	23.4	0.7	2.1	2.8
Carbohydrate	0.1	6.0	0.1	2.4	41.5	1.4	4.4	5.8
Fiber	0.0	1.1	0.0	16.2	35.1	2.1	9.3	11.3
Protein	43.1	18.9	3.6	6.4	20.7	0.4	0.7	1.1
Total fat	24.8	11.5	1.8	4.1	2.3	0.0	0.5	0.5
Saturated fatty acids	28.9	23.1	1.8	2.4	1.6	0.0	0.3	0.3
Monounsaturated fatty acids	28.0	7.8	1.7	4.6	1.3	0.0	0.7	0.7
Polyunsaturated fatty acids	11.5	1.8	1.1	5.2	3.7	0.0	0.4	0.4
Cholesterol	47.7	15.7	31.9	0.0	0.0	0.0	0.0	0.0
Vitamin A (RAE)	21.0	21.5	5.2	0.0	6.5	0.5	2.0	2.5
Carotene	0.0	2.3	0.0	0.1	0.8	1.1	6.8	7.9
Vitamin E	4.1	2.1	1.8	5.6	4.2	0.6	2.1	2.7
Vitamin C	1.9	1.4	0.0	0.1	6.0	24.2	17.8	42.0
Thiamin	18.1	5.3	0.7	5.1	60.0	1.4	1.4	2.8
Riboflavin	17.7	27.5	5.4	2.1	38.2	0.4	1.6	1.9
Niacin	38.6	1.3	0.1	4.3	43.4	0.4	1.5	1.9
Vitamin B6	40.6	6.4	2.0	4.1	19.8	1.1	5.3	6.4
Folate (DFE)	3.2	2.8	1.8	8.2	70.7	2.5	1.6	4.1
Vitamin B12	72.2	22.8	4.8	0.0	0.1	0.0	0.0	0.0
Calcium	3.6	71.8	1.8	4.6	4.8	1.0	1.3	2.3
Phosphorus	26.7	31.6	3.7	6.7	18.9	0.5	1.2	1.8
Magnesium	13.9	13.9	1.0	14.7	24.1	1.5	4.1	5.6
Iron	17.1	1.8	2.6	7.8	51.2	0.3	1.7	2.1
Zinc	41.7	16.2	2.2	5.5	24.7	0.2	0.9	1.1
Copper	21.5	3.0	1.6	20.4	21.5	1.2	3.8	5.0
Potassium	19.8	17.8	1.2	10.1	9.9	2.9	6.9	9.8
Sodium	19.0	37.5	4.0	0.3	1.1	0.1	2.7	2.8
Selenium	30.0	12.5	5.9	7.1	40.4	0.1	0.3	0.3

Nutrient	Vegetables					Fats, oils	Sugars, sweeteners	Miscellaneous
	White potatoes	Dark green, deep yellow	Tomatoes	Other	Total			
	Percent	Percent	Percent	Percent	Percent	Percent	Percent	Percent
Kilocalories	1.9	0.4	0.5	1.1	3.8	22.5	15.8	0.9
Carbohydrate	3.6	0.8	1.0	2.0	7.4	0.0	35.3	1.5
Fiber	6.5	4.0	3.8	8.4	22.7	0.0	0.0	13.5
Protein	1.7	0.6	0.6	1.5	4.4	0.1	0.0	1.8
Total fat	0.1	0.1	0.1	0.1	0.3	53.8	0.0	0.9
Saturated fatty acids	0.1	0.0	0.0	0.1	0.2	40.6	0.0	1.1
Monounsaturated fatty acids	0.0	0.0	0.0	0.1	0.1	54.9	0.0	0.8
Polyunsaturated fatty acids	0.1	0.1	0.1	0.3	0.5	75.2	0.0	0.6
Cholesterol	0.0	0.0	0.0	0.0	0.0	4.6	0.0	0.0
Vitamin A (RAE)	0.0	25.9	1.9	5.5	33.3	8.2	0.0	1.7
Carotene	0.0	70.4	4.3	7.8	82.5	1.7	0.0	4.7
Vitamin E	0.3	1.4	2.7	1.3	5.7	72.8	0.0	0.4
Vitamin C	14.4	15.2	6.8	11.1	47.4	0.0	0.0	1.3
Thiamin	3.5	0.9	0.7	2.2	7.2	0.0	0.1	0.6
Riboflavin	0.8	1.2	0.9	2.0	4.9	0.1	0.6	1.5
Niacin	3.6	0.8	1.6	1.8	7.8	0.0	0.0	2.7
Vitamin B6	8.9	3.0	2.4	4.5	18.8	0.0	0.2	1.8
Folate (DFE)	1.4	1.9	1.0	3.7	8.0	0.2	0.0	1.2
Vitamin B12	0.0	0.0	0.0	0.0	0.0	0.2	0.0	0.0
Calcium	0.8	1.5	0.9	3.1	6.4	0.3	0.5	3.9
Phosphorus	2.2	1.0	0.9	2.4	6.5	0.1	0.1	3.9
Magnesium	4.1	2.1	1.9	4.1	12.1	0.1	0.6	14.1
Iron	2.7	1.1	1.6	2.7	8.2	0.1	0.6	8.7
Zinc	1.5	0.7	0.6	1.8	4.6	0.1	0.2	3.7
Copper	3.6	1.7	2.6	2.9	10.7	0.0	1.3	14.9
Potassium	9.1	3.5	4.6	5.4	22.5	0.1	0.5	8.3
Sodium	2.9	1.4	12.1	9.0	25.3	6.5	2.8	0.6
Selenium	0.6	0.2	0.2	0.6	1.5	0.0	0.9	1.4

[1] Data are based on USDA Economic Research Service estimates of per capita quantities of food available for consumption. Seafood estimates from the National Oceanic and Atmospheric Administration. Game harvest estimates from the individual States or from the Wildlife Management Institute. Imputed data for foods no longer reported or available. Values in the USDA Agricultural Research, National Nutrient Database for Standard Reference, Release 26. Components may not add to 100 because of rounding.

Calculated by USDA/Center for Nutrition Policy and Promotion, (703) 305–7600. Data last updated May 15, 2014.

Table 13-6.—Consumption: Per capita consumption of major food commodities, United States, 2010–2014 [1]

Commodity	2010	2011	2012	2013	2014
	Pounds	Pounds	Pounds	Pounds	Pounds
Red meats [2][3]	102.1	97.7	97.9	97.9	95.4
Beef	56.7	54.4	54.5	53.5	51.5
Veal	0.3	0.3	0.3	0.3	0.2
Lamb and mutton	0.7	0.6	0.6	0.7	0.7
Pork	44.4	42.3	42.5	43.4	43.1
Fish [2]	15.8	14.9	14.2	14.3	14.5
Canned	3.9	3.7	3.4	3.6	3.3
Fresh and frozen	11.5	10.9	10.5	10.4	10.8
Cured	0.3	0.3	0.3	0.3	0.3
Poultry [2][3]	70.6	71.0	69.2	70.2	71.1
Chicken	58.0	58.4	56.6	57.6	58.7
Turkey	12.6	12.6	12.6	12.5	12.4
Eggs	32.5	32.0	32.8	33.3	34.3
Dairy products [4]					
Total dairy products	603.0	603.3	613.1	605.3	614.3
Fluid milk and cream	201.8	196.7	194.6	190.4	178.2
Plain and flavored whole milk	49.2	47.5	46.5	46.2	45.9
Plain reduced fat and light milk (2%, 1%, and 0.5%)	86.3	85.7	84.6	82.5	79.6
Plain fat free milk (skim)	27.3	26.5	24.8	22.6	19.9
Flavored lower fat free milk	13.0	12.7	12.5	12.3	12.0
Buttermilk	1.5	1.5	1.5	1.6	1.5
Eggnog	0.5	0.5	0.4	0.4	0.4
Yogurt (excl. frozen)	13.4	13.6	14.0	14.9	14.8
Sour cream and dip	4.0	4.0	4.1	4.0	4.1
Cheese (excluding cottage) [5]	32.7	33.0	33.3	33.4	33.9
American	13.3	13.0	13.2	13.3	13.5
Cheddar	10.0	9.6	9.6	9.6	9.7
Italian	13.5	14.0	13.8	13.8	14.18
Mozzarella	10.6	10.8	10.7	10.7	11.2
Cottage cheese	2.3	2.3	2.3	2.1	2.1
Condensed and evaporated milk	7.2	7.2	7.3	7.2	6.8
Ice cream	13.5	12.8	12.8	12.7	12.3
Butter	4.9	5.4	5.5	5.5	5.5
Fruits and vegetables [3][6]	648.7	627.2	641.6	643.0	644.6
Fruits	252.2	244.8	250.2	261.4	259.2
Fresh	128.5	129.3	131.7	136.1	135.9
Citrus	21.6	22.8	23.5	23.9	23.3
Noncitrus	107.0	106.6	108.2	112.2	112.6
Processing	123.6	115.5	118.6	125.2	123.4
Citrus	57.0	54.2	61.0	65.2	65.3
Noncitrus	66.6	61.3	57.6	60.0	58.0
Vegetables	396.5	382.4	391.3	381.6	385.4
Fresh	190.0	185.7	188.8	184.1	185.7
Processing	206.5	196.7	202.5	197.6	199.7
Flour and cereal products [7]	194.2	172.5	174.1	174.7	174.4
Wheat flour [8]	134.8	132.5	134.3	135.0	134.7
Corn products	33.1	34.1	33.9	33.9	34.0
Oat products	4.7	4.8	4.7	4.5	4.5
Barley and rye products	1.1	1.1	1.1	1.2	1.7
Caloric sweeteners (dry weight basis) [3]	133.7	131.2	131.1	130.1	131.0
Sugar (refined)	65.9	65.8	66.6	67.9	68.3
Corn sweeteners [9]	66.0	63.6	62.8	60.4	60.7
Honey and edible syrups	1.7	1.7	1.7	1.8	2.0
Others					
Coffee (green bean equivalent)	9.2	9.6	9.7	9.9	10.0
Cocoa (chocolate liquor equivalent) [10]	4.4	4.4	4.2	4.2	4.1
Tea (dry leaf equivalent)	1.0	1.0	0.9	0.9	0.9
Peanuts (shelled)	7.0	7.0	6.7	6.9	7.0
Tree nuts (shelled)	3.9	3.8	4.2	4.0	4.2

[1] Quantity in pounds, retail weight unless otherwise shown. [2] Boneless, trimmed weight equivalent. [3] Total may not add due to rounding. [4] Total dairy products reported on a milk-equivalent, milkfat basis. All other dairy categories reported on a product weight basis. [5] Natural equivalent of cheese and cheese products. [6] Farm weight. [7] White, whole wheat, semolina, and durum flour. [7] Annual data and per capita estimates for rice were discontinued after 2010, reducing totals for flour and cereal products from 2011 forward. [8] White, whole wheat, semolina, and durum flour. [9] High fructose, glucose, and dextrose. [10] Chocolate liquor is what remains after cocoa beans have been roasted and hulled; it is sometimes called ground or bitter chocolate.

ERS, Food Economics Division, (202) 694-5400. Historical consumption and supply-disappearance data for food may be found at http://www.ers.usda.gov/data-products/food-availability-(per-capita)-data-system.aspx.

Table 13-7.—Food plans: Food cost at home, at four cost levels, for families and individuals in the United States, for week and month, December 2015 [1]

Age-gender groups	Weekly cost [2]				Monthy cost [2]			
	Thrifty plan	Low-cost plan	Moderate-cost plan	Liberal plan	Thrifty plan	Low-cost plan	Moderate-cost plan	Liberal plan
	Dollars	Dollars	Dollars	Dollars	Dollars	Dollars	Dollars	Dollars
Individuals: [3]								
Child:								
1 year	21.70	29.40	33.30	40.60	94.20	127.50	144.20	175.80
2-3 year	23.80	30.40	36.70	44.80	103.20	131.60	159.10	194.00
4-5 years	25.00	31.50	39.10	47.70	108.30	136.50	169.50	206.90
6-8 years	32.10	44.80	53.40	63.20	139.30	193.90	231.50	273.80
9-11 years	36.30	47.70	62.00	72.20	157.40	206.80	268.80	313.00
Male:								
12-13 years	38.90	55.00	69.10	81.00	168.50	238.20	299.30	351.20
14-18 years	39.90	55.80	71.20	81.90	172.90	241.90	308.60	355.00
19-50 years	43.20	55.80	69.90	86.10	187.00	241.70	302.90	373.20
51-70 years	39.30	52.50	65.30	78.90	170.30	227.30	282.90	341.90
71+ years	39.50	52.00	64.60	80.00	171.20	225.30	279.80	346.70
Female:								
12-13 years	38.90	47.30	57.20	70.10	168.30	205.00	247.80	303.50
14-18 years	38.10	47.60	57.70	71.10	165.20	206.40	250.20	308.20
19-50 years	38.10	48.40	59.80	76.20	165.20	209.50	259.10	330.10
51-70 years	37.70	46.90	58.40	70.40	163.40	203.30	253.20	305.20
71+ years	37.00	46.50	57.90	69.80	160.20	201.60	250.90	302.60
Families:								
Family of 2: [4]								
19-50 years	89.40	114.50	142.70	178.50	387.40	496.30	618.20	773.60
51-70 years	84.70	109.30	136.10	164.30	367.00	473.60	589.70	711.80
Family of 4:								
Couple, 19-50 years and children.								
2-3 and 4-5 years	130.10	166.00	205.50	254.80	563.70	719.30	890.60	1,104.20
6-8 and 9-11 years	149.70	196.60	245.10	297.70	648.80	851.90	1,062.20	1,290.10

[1] The Food Plans represent a nutritious diet at four different cost levels. The nutritional bases of the Food Plans are the 1997-2005 Dietary References Intakes, 2005 Dietary Guidelines for Americans, and 2005 MyPyramid food intake recommendations. In addition to cost, differences among plans are in specific foods and quantities of foods. Another basis of the Food Plans is that all meals and snacks are prepared at home. For specific foods and quantities of foods in the Food Plans, see *Thrifty Food Plan, 2006* and *The Low-Cost, Moderate-Cost, and Liberal Food Plans, 2007.* All four Food Plans are based on 2001-02 data and updated to current dollars by using the Consumer Price Index for specific food items.
[2] All costs are rounded to nearest 10 cents.
[3] The costs given are for individuals in 4–person families. For individuals in other size families, the following adjustments are suggested: 1 person-add 20 percent; 2 person-add 10 percent; 3 person-add 5 percent; 4 person-no adjustment; 5- or 6- person–subtract 5 percent; 7- (or more) person-subtract 10 percent. To calculate overall household food costs, (1) adjust food costs for each person in household and then (2) sum these adjusted food costs.
[4] Ten percent added for family size adjustment.
Center for Nutrition Policy and Promotion, (703) 305–7600. This file and the referenced Food Plans may be accessed on CNPP's home page at http://www.cnpp.usda.gov.

Table 13-8.—SNAP: Participation and federal costs, fiscal years 2006–2015

Fiscal year [1]	Average monthly participation [2]		Recipient benefits	Total cost [3]	Average monthly benefit [4]	
	Persons	Housholds			Per person	Per household
	1,000	*1,000*	*1,000 dollars*	*1,000 dollars*	*Dollars*	*Dollars*
2006	26,549	11,733	30,187,347	32,903,063	94.75	214.41
2007	26,316	11,788	30,373,271	33,173,525	96.18	214.72
2008	28,223	12,727	34,608,397	37,639,643	102.19	226.60
2009	33,490	15,232	50,359,919	53,620,011	125.31	275.51
2010	40,302	18,618	64,702,165	68,283,942	133.79	289.60
2011	44,709	21,072	71,810,924	75,687,179	133.85	283.99
2012	46,609	22,330	74,619,345	78,411,049	133.41	278.48
2013	47,636	23,052	76,066,319	79,929,150	133.07	274.98
2014	46,664	22,744	69,998,836	74,074,101	125.01	256.47
2015	45,767	22,522	69,655,429	73,975,324	126.83	257.73

SNAP is the Special Nutrition Assistance Program, formerly known as the Food Stamp Program.   [1] October 1 to September 30.   [2] Participation data are 12-month averages.   [3] Total cost includes matching funds for state administrative expenses (e.g., certification of households, quality control, anti-fraud activities; employment and training); and for other Federal costs (e.g., benefit redemption processing; computer support; electronic benefit transfer systems; retailer redemption and monitoring; certification of SSI recipients; nutrition education and program information).   [4] The sharprise in FY 2009 reflects April 2009 implementation of higher benefits mandated by the American Recovery Reinvestment Act.
FNS, Budget Division/Program Reports, Analysis and Monitoring Branch, (703) 305–2165.

Table 13-9.—Food and Nutrition Service Programs: Federal costs of the National School Lunch, School Breakfast, Child Care Food, Summer Food Service, WIC, Special Milk, and Food Distribution Programs, fiscal years 2006–2015 [1]

Fiscal year [2]	Child Nutrition					WIC [6]	Special Milk	Food Distribution Programs [7]
	Cash payments				Cost of food distribution programs [5]			
	School Lunch	School Breakfast	Child & Adult Care [3]	Summer Food [4]				
	1,000 dollars	*1,000 dollars*	*1,000 dollars*	*1,000 dollars*	*1,000 dollars*	*1,000 dollars*	*1,000 dollars*	*1,000 dollars*
2006	7,387,935	2,041,914	2,079,250	274,309	876,545	5,072,736	14,581	528,994
2007	7,706,080	2,163,478	2,160,391	288,631	1,112,559	5,409,580	13,619	487,968
2008	8,264,807	2,365,478	2,315,212	324,912	1,141,080	6,188,804	14,859	543,025
2009	8,874,484	2,582,645	2,438,419	345,119	1,216,491	6,471,624	14,112	873,582
2010	9,751,746	2,859,231	2,544,639	357,250	1,223,236	6,689,911	11,926	895,761
2011	10,105,030	3,034,160	2,621,363	371,662	1,298,930	7,178,702	12,296	829,794
2012	10,414,084	3,276,987	2,740,600	397,177	1,278,920	6,801,314	12,295	756,034
2013	11,057,484	3,513,920	2,869,431	426,049	1,288,881	6,490,974	10,722	998,071
2014	11,355,733	3,685,260	2,997,747	463,905	1,437,600	6,329,968	10,506	939,075
2015 [8]	11,698,118	3,892,673	3,159,664	487,327	1,457,012	6,170,008	10,500	840,143

[1] See table 13-8 for Special Nutrition Assistance Program costs.   [2] October 1–September 30.   [3] Includes sponsor administrative, audit, and startup costs.   [4] Includes sponsor administrative, State administrative and health clinic costs.   [5] Includes entitlement commodities, bonus commodities, and cash-in-lieu for the National School Lunch, School Breakfast, Child and Adult Care Food, and Summer Food Service Programs.   [6] Includes food costs, administrative costs, program evaluation funds, special grants, and Farmer's Market projects for the Special Supplemental Food Program for Women, Infants and Children.   [7] Includes entitlement and bonus commodities, cash-in-lieu of commodities, and administrative costs of the following programs: Food Distribution to Indian Reservations, Nutrition Services Incentive Program (formerly Nutrition Program for the Elderly), Commodity Supplemental Food, Charitable Institutions, Summer Camps, Emergency Food Assistance Program (TEFAP), Disaster Feeding, Bureau of Federal Prisons, Veteran Affairs Administration, and the Food Stamp Program Elderly Pilot Project.   [8] Preliminary.
FNS, Budget Division/Program Reports, Analysis and Monitoring Branch, (703) 305–2165.

Table 13-10.—Food and Nutrition Service program benefits: Cash payments made under the National School Lunch, School Breakfast, Child and Adult Care, Summer Food and Special Milk Programs and the value of food benefits provided under the SNAP, WIC, Commodity Distribution and The Emergency Feeding Food Assistance Programs, fiscal year 2015 [1]

State/Territory	Child Nutrition Program (cash payments only) [2]					Special Supple- mental Food (WIC)	SNAP (formerly Food Stamp Program) [4]	Emer- gency food assist- ance (TEFAP)	Com- modity distribu- tion [3]	Total [4]
	Child and Adult Care Food	Summer Food	Special Milk	National School Lunch	School Breakfast					
	1,000 dollars	1,000 dollars	1,000 dollars	1,000 dollars	1,000 dollars	1,000 dollars	1,000 dollars	1,000 dollars	1,000 dollars	1,000 dollars
Alabama	43,375	7,227	29	210,937	71,828	76,722	1,341,907	8,012	21,503	1,781,540
Alaska	8,848	1,670	2	34,604	11,274	12,539	168,054	1,074	3,582	241,648
Amer. Samoa [4]	0	0	0	0	0	5,418	0	0	0	5,418
Arizona	46,230	4,197	53	270,321	87,683	91,442	1,459,585	9,936	38,257	2,007,704
Arkansas	54,927	6,347	0	130,038	49,726	44,359	648,770	4,428	17,929	956,524
California	383,227	23,153	376	1,437,934	479,625	727,290	7,528,040	59,347	192,527	10,831,520
Colorado	23,386	4,173	145	131,240	47,482	45,144	771,960	6,502	17,122	1,047,154
Connecticut	15,403	3,504	185	96,057	28,208	29,818	715,335	4,781	10,448	903,738
Delaware	14,130	1,878	30	33,605	12,277	9,243	228,935	1,379	6,016	307,494
District of Columbia	8,079	2,522	8	26,157	10,437	7,801	224,104	1,134	3,664	283,906
Florida	223,682	35,403	25	745,670	225,105	245,028	5,688,712	25,462	101,890	7,290,974
Georgia	106,884	14,950	10	506,294	177,516	127,746	2,803,607	14,799	56,004	3,807,810
Guam	392	0	0	7,993	2,964	6,928	109,109	275	35	127,696
Hawaii	6,723	469	0	45,320	11,743	19,494	505,466	1,296	3,776	594,287
Idaho	7,028	3,745	138	52,431	18,394	17,149	273,758	1,971	7,318	381,933
Illinois	137,208	12,755	2,317	458,967	133,091	144,179	3,303,103	21,346	52,225	4,265,191
Indiana	51,401	8,111	181	255,828	73,693	71,452	1,244,188	11,150	39,125	1,755,130
Iowa	25,167	3,401	79	101,822	24,460	28,412	516,608	2,797	17,423	720,170
Kansas	29,624	3,669	64	105,594	30,061	27,458	374,433	3,325	16,145	590,375
Kentucky	33,387	5,992	29	203,759	78,151	59,649	1,112,381	6,049	29,256	1,528,652
Louisiana	85,216	9,678	34	216,194	75,053	74,332	1,298,446	6,042	39,124	1,804,118
Maine	8,786	1,944	20	33,114	11,643	11,959	282,016	1,563	5,055	356,100
Maryland	50,994	8,005	319	160,694	63,633	77,631	1,149,658	4,877	22,962	1,538,772
Massachusetts	58,239	7,399	294	175,868	50,094	52,285	1,202,313	8,593	25,064	1,580,148
Michigan	59,893	11,262	297	290,102	103,300	113,773	2,369,234	15,144	49,003	3,012,008
Minnesota	56,185	7,961	797	156,014	46,499	63,404	627,558	5,432	28,001	991,852
Mississippi	40,591	6,531	5	169,324	61,237	57,230	916,552	6,115	23,884	1,281,469
Missouri	47,148	11,701	502	207,179	73,080	61,808	1,258,536	8,503	32,919	1,701,376
Montana	9,595	1,588	17	26,473	8,048	9,220	171,414	1,159	8,855	236,368
Nebraska	30,325	2,526	62	67,471	16,399	20,062	242,093	1,852	18,408	399,198
Nevada	8,784	1,647	111	95,344	28,806	32,498	605,593	4,448	16,807	794,038
New Hampshire	4,357	791	182	22,914	5,127	5,832	132,498	1,465	5,288	178,454
New Jersey	69,868	7,465	414	246,426	85,040	104,298	1,291,436	11,866	36,239	1,853,050
New Mexico	29,431	5,829	0	95,353	45,878	24,666	685,207	3,506	16,703	906,572
New York	224,051	54,233	765	674,381	194,558	308,293	5,046,487	29,953	99,742	6,632,461
North Carolina	88,207	10,599	152	375,638	129,355	128,028	2,395,550	13,615	45,936	3,187,079
North Dakota	9,348	606	24	19,178	4,907	7,005	77,913	765	8,679	128,426
Northern Marianas [4]	0	0	0	0	0	3,099	0	0	0	3,099
Ohio	81,609	10,719	396	350,526	113,886	102,123	2,528,635	16,963	51,587	3,256,645
Oklahoma	53,140	4,639	15	160,296	57,366	52,487	864,951	5,201	43,469	1,241,564
Oregon	31,912	5,951	113	110,966	37,001	48,142	1,152,977	5,935	15,644	1,408,640
Pennsylvania	102,783	14,563	350	351,616	98,821	137,232	2,699,655	20,067	66,117	3,491,204
Puerto Rico [4]	27,129	11,818	0	114,696	31,332	171,873	0	6,554	19,372	382,774
Rhode Island	8,680	1,261	63	29,324	9,376	11,087	282,777	1,176	3,457	347,202
South Carolina	31,591	8,537	9	201,254	76,048	63,263	1,208,605	6,356	21,578	1,617,240
South Dakota	8,157	1,303	31	28,784	7,587	11,314	148,867	704	10,925	217,671
Tennessee	61,028	11,039	27	269,065	103,191	68,925	1,884,709	8,977	26,452	2,433,413
Texas	330,521	37,866	25	1,394,902	541,661	307,834	5,265,414	36,495	180,623	8,095,342
Utah	26,726	1,013	75	98,325	19,912	26,101	313,811	3,282	16,090	505,335
Vermont	5,645	1,233	75	15,411	5,569	8,807	124,409	716	2,854	164,718
Virginia	41,355	10,014	152	224,514	71,857	51,415	1,230,788	8,391	23,459	1,661,944
Virgin Islands	1,317	511	0	4,894	1,067	4,150	56,466	170	737	69,312
Washington	43,299	5,422	239	195,067	53,032	91,891	1,527,741	8,901	26,918	1,952,509
West Va	14,414	1,413	22	71,663	37,139	19,662	497,269	3,517	8,819	653,918
Wisconsin	36,052	7,593	1,212	167,915	47,944	57,070	1,051,154	7,472	31,233	1,407,646
Wyoming	4,592	773	34	14,273	3,510	4,075	46,448	760	3,025	77,491
DoD [5]	0	0	0	8,391	0	0	0	0	1,997	10,389
Total [4]	3,010,069	428,599	10,500	11,698,118	3,892,673	4,190,148	69,655,429	451,601	1,671,270	95,008,406

[1] Excludes all administrative and program evaluation costs.   [2] Excludes totals for Food Safety Education and $14.8 million for Team Nutrition.   [3] Includes distribution of bonus and entitlement commodities to the National School Lunch, Child and Adult Care, Summer Food Service, Charitable Institutions, Summer Camps, Food Distribution on Indian Reservations, Nutrition Services Incentive Program (NSIP, formerly Nutrition Program for the Elderly), Commodity Supplemental Food, and Disaster Feeding Programs. Also includes cash-in-lieu of commodities for the National School Lunch and the Child and Adult Care Food programs (NSIP cash grants were transferred to the Agency on Aging, DHHS, in FY 2003).   [4] Excludes Nutrition Assistance grants of $2,001 million for Puerto Rico, $22.5 million for the Northern Marianas, and $23.0 million for American Samoa.   [5] Dept. of Defense represents food service to children of armed forces personnel in overseas schools.
FNS, Budget Division/Program Reports, Analysis and Monitoring Branch (703) 305–2165. May 8, 2014.

Table 13-11.—Food and Nutrition Service Programs: Persons participating, fiscal years 2006–2015

Fiscal year	National School Lunch Program [1]	School Breakfast Program [1]	Child and Adult Care Program [2]	Summer Food Service [3]	WIC Program [4]
	Thousands	*Thousands*	*Thousands*	*Thousands*	*Thousands*
2006	30,128	9,760	3,112	1,912	8,088
2007	30,630	10,119	3,207	1,977	8,285
2008	31,016	10,608	3,254	2,130	8,705
2009	31,310	11,076	3,320	2,260	9,122
2010	31,753	11,669	3,411	2,304	9,175
2011	31,842	12,175	3,431	2,278	8,961
2012	31,652	12,868	3,547	2,348	8,908
2013	30,678	13,202	3,681	2,428	8,663
2014	30,459	13,635	3,891	2,663	8,258
2015	30,493	14,092	4,183	2,569	8,024

[1] Average monthly participation (excluding summer months). [2] Average daily attendance (data reported quarterly). [3] Average daily attendance for peak month (July). [4] Average monthly participation. WIC is an abbreviation for the Special Supplemental Food Program for Women, Infants and Children.
FNS, Budget Division/Program Reports, Analysis and Monitoring Branch, (703) 305–2165.

Table 13-12.—Consumers' prices: Index number of prices paid for goods and services, United States, 2006–2015 [1]

[1982–84=100]

Year	Food	Nonfood items					All items
		Apparel and upkeep	Housing		Transportation	Medical care	
			Total	Rent			
2006	195.2	119.5	203.2	241.9	180.9	336.2	201.6
2007	202.9	118.9	209.6	250.8	184.7	351.1	207.3
2008	214.1	118.9	216.3	257.2	195.5	364.1	215.3
2009	218.0	120.1	217.1	259.9	179.3	375.6	214.5
2010	219.6	119.5	216.3	258.8	193.4	388.4	218.1
2011	227.8	122.1	219.1	262.2	212.4	400.3	224.9
2012	233.8	126.3	222.7	267.8	217.3	414.9	229.6
2013	237.0	127.4	227.4	274.0	217.4	425.1	233.0
2014	242.7	127.5	233.2	281.8	215.9	435.3	236.7
2015	247.2	125.9	238.1	290.4	199.1	446.8	237.0

[1] Reflects retail prices of goods and services usually bought by average families in urban areas of the United States. This index is the official index released monthly by the U.S. Department of Labor. Data are for all urban consumers.
ERS, Food Markets Branch, (202) 694–5349. Compiled from data of the U.S. Department of Labor.

CHAPTER XIV

STATISTICS OF FERTILIZERS AND PESTICIDES

This chapter contains statistics on percentages of crop acres treated by various types of fertilizers and pesticides. Nitrogen, phosphate, potash, and sulfur are the most common fertilizers; herbicides, insecticides, fungicides, and other chemicals are the main categories of pesticides. Other chemicals include soil fumigants, vine killers, and desiccants. The tables show data for field crops for 2003–2015, fruits for 2015, and vegetables for 2014. NASS collects data for field crops on an annual basis; data collection for fruits and vegetables was suspended for fiscal year 2013. The surveyed States are generally the major producing States for each crop shown in the tables and represent 65–95 percent of the U.S. planted acres, depending on the selected crop. Application data for specific pesticide active ingredients and additional fertilizer data are available in the series of NASS "Agricultural Chemical Usage" reports and data sets.

Table 14-1.—Field crops: Fertilizer, and percent of area receiving applications, all States surveyed, 2006–2015 [1]

Crop	Nitrogen	Phosphate	Potash	Sulfur
	Percent	*Percent*	*Percent*	*Percent*
2006:				
Rice	97	67	54	18
Soybeans	18	23	25	3
Wheat, Durum	92	74	7	4
Wheat, Other Spring	95	85	27	13
Wheat, Winter	80	57	17	14
2007:				
All Cotton	92	67	52	42
2009:				
Wheat, Durum	99	85	11	9
Wheat, Other Spring	94	84	21	14
Wheat, Winter	83	54	16	16
2010:				
Corn	97	78	61	15
Cotton, Upland	90	62	52	42
Potatoes, Fall	99	96	90	73
2011:				
Barley	86	68	28	30
Sorghum	81	54	9	16
2012:				
Soybeans	27	37	37	7
Wheat, Durum	98	89	12	12
Wheat, Other Spring	97	87	27	20
Wheat, Winter	85	55	13	19
2013:				
Rice	97	75	54	23
Peanuts	40	42	42	17
2014:				
Corn	97	80	65	29
Potatoes, Fall	99	97	90	79
2015:				
Cotton	78	56	42	30
Oats	76	62	40	24
Soybeans	28	39	38	8
Wheat, Spring (Excl Durum)	97	89	40	30
Wheat, Spring Durum	98	95	32	31
Wheat, Winter	88	60	16	22

[1] Refers to percent of planted acres receiving one or more applications of a specific fertilizer ingredient.   See tables 14-2 through 14-21 for surveyed States.   Note: See planted acreage estimates in tables 1-56 for barley, 1-36 for corn, 2-2 for upland and all cotton, 1-48 for oats, 3-16 for peanuts, 1-27 for rice, 1-65 for sorghum, 3-32 for soybeans and 1-8 for wheat.
NASS, Environmental, Economics, and Demographics Branch, (202) 720–6146.

Table 14-2.—Barley: Pesticide usage, 2003 and 2011 [1]

State and Year	Percent treated and amount applied							
	Herbicide		Insecticide		Fungicide		Other Chemicals	
	Area applied	Pounds applied	Area applied	Pounds applied	Area applied	Pounds applied	Area applied	Pounds applied
	Percent	Thousands	Percent	Thousands	Percent	Thousands	Percent	Thousands
AZ:								
2011	37	16	*	*	....	....	....	....
CA:								
2003	67	32	....	....	....	....	....	....
2011	39	48	....	*	....	....	....	....
CO:								
2011	96	36	22	1	25	2	....	....
ID:								
2003	94	573	3	16	....	....	5	9
2011	91	302	11	1	18	11	14	27
MN:								
2003	89	88	8	3	39	9	....	....
2011	58	19	10	2	32	3	....	....
MT:								
2003	93	1,005	2	5	....	....	....	....
2011	92	1,065	*	*	26	16	*	*
ND:								
2003	98	1,067	4	12	11	20	....	....
2011	95	236	6	4	48	26	....	....
OR:								
2011	74	21	*	*	31	1	*	*
PA:								
2003	32	8	....	....	....	....	....	....
2011	39	7	*	*	3	**	....	....
SD:								
2003	86	34	....	....	....	....	....	....
UT:								
2003	75	17	....	....	....	....	....	....
VA:								
2011	59	29	20	**	15	1	*	*
WA:								
2003	94	358	....	....	....	....	....	....
2011	93	111	....	....	19	2	....	....
WI:								
2003	21	5	....	....	....	....	....	....
2011	25	4	....	....	....	....	....	....
WY:								
2003	83	57	10	....	....	....	....	....
2011	76	32	*	*	13	1	....	....

* Insufficient number of reports to publish data.   ** Amount applied is less than 500 lbs.   [1] Data not available for all States for all years.   Note: Planted acres are in table 1-56.
NASS, Environmental, Economics, and Demographics Branch, (202) 720–6146.

Table 14-3.—Barley: Fertilizer usage, 2003 and 2011 [1]

State	Nitrogen		Phosphate		Potash		Sulfur [2]	
	Area applied	Pounds applied	Area applied	Pounds applied	Area applied	Pounds applied	Area applied	Pounds applied
	Percent	Millions	Percent	Millions	Percent	Millions	Percent	Millions
AZ:								
2011	97	8.2	39	1.5	*	*	21	0.3
CA:								
2003	72	5.2	32	0.6	2			
2011	41	2.6	*	*			14	0.2
CO:								
2011	82	7.2	40	1.5	28	0.6	36	0.4
ID:								
2003	91	56.2	58	15.4	25	5.7		
2011	93	44.2	66	10.0	37	4.9	65	9.6
MN:								
2003	91	11.4	87	5.6	66	4.0		
2011	60	3.3	56	1.3	44	1.0	11	0.1
MT:								
2003	92	44.2	88	30.2	52	9.7		
2011	90	29.1	83	16.6	37	3.6	16	1.2
ND:								
2003	98	116.5	91	50.7	20	4.2		
2011	96	28.3	92	11.1	16	1.0	4	0.1
OR:								
2011	77	1.6	26	0.1	*	*	36	0.1
PA:								
2003	69	2.2	39	1.1	40	1.2		
2011	64	1.9	32	0.8	32	0.8	10	0.1
SD:								
2003	82	2.6	78	1.9	13	0.2		
UT:								
2003	58	2.1	14	0.3				
VA:								
2011	77	4.9	39	1.7	35	1.7	24	0.3
WA:								
2003	99	22.5	58	2.5	8	0.5		
2011	91	9.0	59	1.1	9	0.2	90	1.6
WI:								
2003	37	0.5	36	0.7	44	1.8		
2011	34	0.2	*	*	34	1.0	14	0.1
WY:								
2003	78	7.3	60	2.4	22	0.7		
2011	73	5.2	55	1.8	30	0.8	18	0.4

* Insufficient number of reports to publish data.   ** Area applied is less than 0.5 percent.   *** Amount applied is less than 50,000 lbs.   [1] Data not available for all States for all years.   [2] Estimates began in 2005.   Note: Planted acres are in table 1-56.
NASS, Environmental, Economics, and Demographics Branch, (202) 720–6146.

Table 14-4.—Corn: Pesticide usage, 2005–2014 [1] [2]

State	Herbicide		Insecticide [2]		Fungicide [2]		Other Chemicals	
	Area applied	Pounds applied	Area applied	Pounds applied	Area applied	Pounds applied	Area applied	Pounds applied
	Percent	Thousands	Percent	Thousands	Percent	Thousands	Percent	Thousands
CO:								
2005	90	1,494	24	252				
2010	95	3,176	12	81				
GA:								
2005	91	495	14	25				
2010	97	559	30	58	*	*		
IL:								
2005	99	30,967	52	1,426				
2010	99	29,354	28	399	23	339	*	*
2014	98	26,870	23	237	21	321	*	*
IN:								
2005	97	14,136	41	722				
2010	97	15,060	14	114	5	31		
2014	100	14,609	9	45	12	80	*	*
IA:								
2005	96	24,726	11	187				
2010	100	26,195	8	148	11	149		
2014	95	30,240	13	122	19	296	*	*
KS:								
2005	87	7,436	11	89				
2010	96	14,727	7	93	*	*	*	*
2014	97	11,432	10	244	3	13		
KY:								
2005	100	3,187	18	26				
2010	98	4,661	39	12	*	*	*	*
2014	94	4,102	29	15	11	19	*	*
MI:								
2005	99	5,145	14	153				
2010	98	4,520	*	*	*	*	*	*
2014	98	5,710	4	5	12	36		
MN:								
2005	100	10,361	12	214				
2010	95	11,619	7	108	*	*		
2014	98	13,596	2	26	*	*	*	*
MO:								
2005	96	7,707	11	41				
2010	94	8,304	17	7	*	*		
2014	98	8,517	24	43	14	61		
NE:								
2005	98	18,416	20	456				
2010	97	20,418	4	231	8	100	*	*
2014	95	21,320	16	213	10	118	*	*
NY:								
2005	96	2,325	21	146				
2010	98	2,721	13	23				
NC:								
2005	98	1,669	17	130				
2010	98	2,349	40	30	*	*		
ND:								
2005	99	1,094						
2010	100	3,761	*	*				
2014	97	4,489	*	*	4	16		
OH:								
2005	99	9,322	9	215				
2010	99	9,149	8	93				
2014	99	9,231	13	64	9	81	*	*
PA:								
2005	97	3,346	21	154				
2010	99	4,668	30	31				
2014	97	4,080	20	46	4	6	4	2
SD:								
2005	100	6,036	12	239				
2010	100	8,480	1	4				
2014	98	10,803	*	*	9	47		
TX:								
2005	94	3,344	24	236				
2010	97	3,755	13	133	*	*		
2014	89	4,338	14	528	23	27	*	*
WI:								
2005	97	6,369	22	134				
2010	97	8,676	11	64	*	*		
2014	90	6,952	8	22	*	*	*	*

* Insufficient number of reports to publish data. [1] Data not available for all States for all years. [2] Amount applied excludes Bt (bacillus thuringiensis) and other biologicals. Note: Planted acres are in table 1-36.
NASS, Environmental, Economics, and Demographics Branch, (202) 720–6146.

Table 14-5.—Corn: Fertilizer usage, 2005–2014 [1]

State and Year	Nitrogen		Phosphate		Potash		Sulfur [2]	
	Area applied	Pounds applied	Area applied	Pounds applied	Area applied	Pounds applied	Area applied	Pounds applied
	Percent	Millions	Percent	Millions	Percent	Millions	Percent	Millions
CO:								
2005	89	126.2	63	24.4	21	4.2	33	3.3
2010	98	165.1	52	19.3	8	1.6	19	3.4
GA:								
2005	98	38.7	86	16.1	87	24.5	53	2.5
2010	90	46.7	74	14.8	78	22.6	51	2.4
IL:								
2005	98	1,728.3	84	780.4	84	1,160.5	4	14.9
2010	98	2,061.5	85	988.1	81	1,080.0	9	20.5
2014	99	1,918.3	85	938.4	81	1,066.9	8	17.5
IN:								
2005	100	869.3	93	420.2	88	648.2	14	8.1
2010	99	1,041.0	90	366.5	87	613.5	10	10.4
2014	96	877.7	81	351.1	79	530.5	22	16.0
IA:								
2005	92	1,653.2	70	579.0	71	762.3	5	4.5
2010	95	1,806.6	72	620.3	68	734.7	8	11.5
2014	96	1,842.2	64	600.5	63	705.3	18	40.2
KS:								
2005	97	482.1	81	112.7	26	34.9	17	5.3
2010	99	629.7	81	146.7	37	72.1	29	20.3
2014	98	549.6	87	148.5	35	60.5	29	10.1
KY:								
2005	98	210.5	78	75.5	77	86.9	....	....
2010	96	210.6	88	116.8	88	124.6	*	*
2014	98	214.1	88	109.0	87	119.0	14	4.5
MI:								
2005	97	277.8	88	89.6	81	148.4	21	3.7
2010	99	288.8	93	72.1	83	186.9	35	4.5
2014	98	338.7	84	89.2	82	192.6	25	7.4
MN:								
2005	94	953.9	86	378.1	77	400.3	9	8.2
2010	87	835.4	72	277.3	68	327.7	9	8.5
2014	94	1,091.8	86	491.4	80	570.2	51	72.5
MO:								
2005	99	489.5	79	149.5	78	180.1	19	10.0
2010	99	392.5	89	177.3	61	109.9	15	4.0
2014	98	603.3	81	191.1	83	228.0	32	16.2
NE:								
2005	99	1,162.5	75	237.3	22	38.8	30	35.0
2010	99	1,270.1	69	256.4	20	46.1	23	28.4
2014	99	1,359.9	77	329.9	29	100.8	42	71.6
NY:								
2005	94	62.2	88	33.2	79	34.9	....	....
2010	86	52.2	75	28.3	59	23.2	*	*
NC:								
2005	97	90.5	74	25.5	86	53.1	18	1.1
2010	94	109.3	83	30.6	81	57.6	29	5.2
ND:								
2005	99	169.3	94	58.8	38	13.3	8	0.9
2010	100	326.7	94	85.5	53	35.1	8	2.2
2014	99	349.7	88	132.1	44	34.7	43	12.3
OH:								
2005	99	551.7	87	224.9	76	264.5	12	3.2
2010	100	481.8	90	198.8	83	257.8	18	11.2
2014	98	558.8	85	228.9	80	282.2	15	5.1
PA:								
2005	88	108.4	64	40.7	58	37.4	6	3.0
2010	94	109.2	39	25.0	46	32.2	28	5.9
2014	96	137.7	67	39.7	66	40.5	35	5.6
SD:								
2005	95	477.7	79	154.2	37	41.9	13	5.5
2010	99	580.8	85	196.0	35	46.6	19	11.7
2014	97	684.8	83	216.4	50	113.2	35	24.0
TX:								
2005	94	282.0	81	73.9	28	10.6	29	6.9
2010	99	292.0	72	59.5	32	12.2	32	8.8
2014	95	310.5	77	77.9	37	19.2	43	10.7
WI:								
2005	93	380.9	84	118.8	84	191.7	22	9.1
2010	93	330.9	72	121.1	77	159.1	20	7.3
2014	98	407.6	85	127.9	88	222.2	49	31.7

*Insufficient number of reports to publish data. [1] Data not available for all States for all years. [2] Estimates began in 2005. Note: Planted acres are in table 1-36.
NASS, Environmental, Economics, and Demographics Branch, (202) 720–6146.

Table 14-6.—Cotton, Upland: Pesticide usage, 2005–2015 [1][2]

State and Year	Herbicide		Insecticide [3]		Fungicide		Other Chemicals	
	Area applied	Pounds applied	Area applied	Pounds applied	Area applied	Pounds applied	Area applied	Pounds applied
	Percent	Thousands	Percent	Thousands	Percent	Thousands	Percent	Thousands
Alabama:								
2005	98	1,186	74	192	2	3	89	697
2007	98	941	55	88	*	*	75	423
2015	100	997	76	107	(D)	(D)	97	673
Arizona:								
2015	85	353	30	12	(D)	(D)	67	46
Arkansas:								
2005	95	2,997	84	2,669	6	18	87	1,910
2007	97	2,399	92	1,092	2	16	96	1,780
2010	92	1,587	97	655	*	*	98	1,473
2015	99	752	99	322	(D)	(D)	100	605
California:								
2005	92	551	96	574	4	2	96	1,570
2007	90	565	90	506	2	1	93	1,414
2015	93	295	89	97	(D)	(D)	78	266
Georgia:								
2005	99	2,958	88	1,145	**	1	95	2,539
2007	100	3,163	85	956	*	*	96	3,955
2010	100	4,098	75	1,121	*	*	95	4,300
2015	100	4,370	78	513	(D)	(D)	97	2,643
Louisiana:								
2005	98	1,897	94	1,358	3	7	99	888
2007	98	992	99	562	*	*	100	567
Mississippi:								
2005	100	3,947	92	1,917	6	28	98	1,880
2007	100	2,132	97	1,231	2	3	99	1,146
2010	100	1,457	92	634	....	....	100	853
2015	100	1,495	92	385	....	....	100	769
Missouri:								
2007	100	995	83	270	*	*	100	867
2010	100	1,191	92	279	*	*	99	556
2015	100	590	78	193	(D)	(D)	100	470
North Carolina:								
2005	99	2,181	82	597	7	41	92	1,642
2007	100	1,479	79	300	3	15	99	896
2010	100	1,581	83	315	*	*	95	1,103
2015	98	1,448	77	137	(D)	(D)	90	613
South Carolina:								
2007	100	535	92	85	13	13	86	291
2015	100	739	87	227	....	....	83	211
Tennessee:								
2005	99	1,339	87	253	11	23	94	1,030
2007	100	1,482	94	228	*	*	99	985
2010	98	1,291	94	146	*	*	97	790
2015	100	511	49	69	....	....	48	180
Texas:								
2005	93	8,677	53	5,946	....	....	47	3,075
2007	96	11,532	43	2,624	*	*	74	5,702
2010	99	13,111	35	2,891	*	*	80	5,425
2015	90	14,783	15	383	(D)	....	67	3,022

Planted acres are in table 2-2.
* Insufficient number of reports to publish data.   ** Area applied is less than 0.5 percent. (D) Withheld to avoid disclosing data for individual operations. [1] Data not available for all States for all years.   [2] 2007 data are for all cotton (pima and upland).   [3] Amount applied excludes Bt (bacillus thuringiensis) and other biologicals.
NASS, Environmental, Economics, and Demographics Branch, (202) 720–6146.

Table 14-7.—Cotton, Upland: Fertilizer usage, 2005–2015 [1] [2]

State and Year	Percent treated and amount applied							
	Nitrogen		Phosphate		Potash		Sulfur	
	Area applied	Pounds applied	Area applied	Pounds applied	Area applied	Pounds applied	Area applied	Pounds applied
	Percent	Millions	Percent	Millions	Percent	Millions	Percent	Millions
Alabama:								
2005	98	51.4	87	27.0	90	37.0	39	3.4
2007	97	34.2	87	17.0	90	23.3	46	2.1
2015	99	28.3	78	13.1	79	23.7	48	2.2
Arizona:								
2015	98	12.5	50	2.3	15	0.2	8	0.1
Arkansas:								
2005	96	112.8	73	33.3	82	71.2	33	8.5
2007	98	94.1	83	29.4	85	63.9	46	5.5
2010	98	57.6	83	19.3	86	34.7	27	1.6
2015	100	27.0	79	7.8	80	14.3	46	1.2
California:								
2005	96	79.8	32	10.2	22	8.3	4	0.2
2007	96	53.6	39	13.2	20	4.3	*	*
2015	94	14.9	28	1.4	33	2.6	5	0.1
Georgia:								
2005	97	112.6	88	63.8	90	103.7	56	11.7
2007	98	90.9	91	56.3	91	81.3	67	10.5
2010	97	122.7	86	66.9	91	127.0	53	8.1
2015	99	100.4	82	47.1	97	105.7	56	9.3
Louisiana:								
2005	99	47.5	47	12.3	49	23.3	35	1.3
2007	100	29.3	70	8.1	63	16.1	*	*
Mississippi:								
2005	99	144.5	35	22.6	58	82.7	17	2.8
2007	100	77.3	33	12.4	54	37.7	28	2.0
2010	95	40.0	32	6.4	53	20.2	29	1.4
2015	96	33.1	44	8.0	65	24.3	17	0.7
Missouri:								
2007	98	36.1	88	10.3	95	24.5	64	2.6
2010	100	40.2	81	12.2	93	25.2	61	3.7
2015	100	17.8	74	4.8	92	10.6	44	1.5
North Carolina:								
2005	95	57.9	74	25.7	95	79.0	40	7.1
2007	92	31.3	71	11.0	89	44.2	25	2.9
2010	98	35.7	68	17.2	87	45.6	38	2.6
2015	99	25.2	75	10.7	89	27.4	27	1.3
South Carolina:								
2007	99	16.2	79	6.5	94	16.0	33	0.8
2015	99	20.1	73	6.7	96	23.7	19	0.7
Tennessee:								
2005	100	60.6	90	31.1	99	58.3	42	2.1
2007	100	52.3	95	25.2	100	45.1	60	2.8
2010	100	36.9	96	19.1	99	33.4	75	3.2
2015	100	17.2	96	9.8	49	7.0	79	2.7
Texas:								
2005	77	310.9	64	144.9	32	35.4	40	32.3
2007	86	347.7	60	109.8	24	19.8	42	26.2
2010	85	296.6	52	92.0	30	25.0	38	26.8
2015	66	207.2	46	75.8	16	10.9	24	11.0

*Insufficient number of reports to publish data. [1] Data not available for all States for all years. [2] 2007 data are for all cotton (pima and upland). Note: Planted acres are in table 2-3.
NASS, Environmental, Economics, and Demographics Branch, (202) 720–6146.

Table 14-8.—Peanuts: Pesticide usage, 2004 and 2013 [1]

State and Year	Percent treated and amount applied							
	Herbicide		Insecticide		Fungicide		Other Chemicals	
	Area applied	Pounds applied	Area applied	Pounds applied	Area applied	Pounds applied	Area applied	Pounds applied
	Percent	*Thousands*	*Percent*	*Thousands*	*Percent*	*Thousands*	*Percent*	*Thousands*
Alabama:								
2004	100	277	81	200	100	896	....	....
2013	99	219	32	33	95	632	*	*
Florida:								
2004	100	298	88	199	100	835	....	....
2013	93	237	59	28	100	670	*	*
Georgia:								
2004	99	878	77	569	99	2,275	....	....
2013	100	853	48	177	97	1,692	*	*
North Carolina:								
2004	100	221	92	161	96	164	43	1,404
2013	99	190	71	54	87	144	*	*
South Carolina:								
2013	96	150	60	33	81	261	....	....
Texas:								
2004	94	258	3	2	67	154	....	....
2013	65	93	11	2.0	33	42	....	....

* Insufficient number of reports to publish data. [1] Data not available for all States for all years. Note: Planted acres are in table 3-16.
NASS, Environmental, Economics, and Demographics Branch, (202) 720–6146.

Table 14-9.—Peanuts: Fertilizer usage, 2004 and 2013 [1]

State and Year	Percent treated and amount applied							
	Nitrogen		Phosphate		Potash		Sulfur [2]	
	Area applied	Pounds applied	Area applied	Pounds applied	Area applied	Pounds applied	Area applied	Pounds applied
	Percent	*Millions*	*Percent*	*Millions*	*Percent*	*Millions*	*Percent*	*Millions*
Alabama:								
2004	70	4.3	79	8.6	75	12.4	....	....
2013	48	1.5	54	3.0	57	5.2	19	0.6
Florida:								
2004	71	3.3	80	5.4	94	12.7	....	....
2013	81	2.2	86	4.9	93	11.7	36	1.3
Georgia:								
2004	48	5.3	59	17.5	51	23.7	....	....
2013	27	2.1	31	6.5	26	6.5	*	*
North Carolina:								
2004	37	1.0	35	1.2	64	6.7	....	....
2013	21	0.5	22	0.7	51	4.1	23	0.5
South Carolina:								
2013	13	0.5	14	0.4	22	1.5	*	*
Texas:								
2004	86	14.4	77	10.6	62	9.3	....	....
2013	57	5.3	47	2.4	34	1.4	37	0.7

* Insufficient number of reports to publish data. [1] Data not available for all States for all years. [2] Estimates began in 2005. Note: Planted acres are in table 3-16.
NASS, Environmental, Economics, and Demographics Branch, (202) 720–6146.

Table 14-10.—Oats: Pesticide usage, 2005 and 2015 [1]

State and Year	Herbicide Area applied	Herbicide Pounds applied	Insecticide Area applied	Insecticide Pounds applied	Fungicide Area applied	Fungicide Pounds applied
	Percent	*Thousands*	*Percent*	*Thousands*		
California:						
2005	36	59				
Idaho:						
2005	26	17				
Illinois:						
2005	7	1				
2015	(D)	(D)			(D)	(D)
Iowa:						
2005	3	2				
2015	(D)	(D)			(D)	(D)
Kansas:						
2005	27	13				
2015	41	28			(D)	(D)
Michigan:						
2005	61	26				
2015	66	33	(D)	(D)	(D)	(D)
Minnesota:						
2005	21	26				
2015	43	66	(D)	(D)	13	4
Montana:						
2005	34	18				
Nebraska:						
2005	7	4				
2015	35	21			(D)	(D)
New York:						
2005	51	23				
2015	63	26	(D)	(D)	(D)	(D)
North Dakota:						
2005	54	167				
2015	82	186	(D)	(D)	9	2
Ohio:						
2015	59	22				
Pennsylvania:						
2005	58	46				
2015	60	43	(D)	(D)		
South Dakota:						
2005	37	52				
2015	81	130	(D)	(D)	21	9
Texas:						
2005	26	80	18	35		
2015	45	95	14	28	9	10
Wisconsin:						
2005	18	25				
2015	30	37	(D)	(D)	6	2

Planted acres are in table 1-48.
(D) Withheld to avoid disclosing data for individual operations.   [1] Data not available for all States for all years.
NASS, Environmental, Economics, and Demographics Branch, (202) 720–6146.

Table 14-11.—Oats: Fertilizer usage, 2005 and 2015 [1]

State and Year	Percent treated and amount applied							
	Nitrogen		Phosphate		Potash		Sulfur	
	Area applied	Pounds applied	Area applied	Pounds applied	Area applied	Pounds applied	Area applied	Pounds applied
	Percent	Millions	Percent	Millions	Percent	Millions	Percent	Millions
California:								
2005	26	4.4	....	....	....	....	....	....
Idaho:								
2005	42	1.6	22	1.4	5	0.1	12	0.2
Illinois:								
2005	15	0.4	12	0.4	26	1.7	....	....
2015	28	0.4	20	0.5	28	0.9	(D)	(D)
Iowa:								
2005	31	1.8	30	2.5	40	6.9	....	....
2015	27	1.4	26	2.1	28	2.7	11	0.3
Kansas:								
2005	84	4.4	39	1.4	17	0.8	....	....
2015	87	4.5	33	0.9	12	0.3	(D)	(D)
Michigan:								
2005	82	2.6	72	2.8	77	3.4	....	....
2015	81	2.0	75	1.7	78	2.0	6	(Z)
Minnesota:								
2005	28	4.2	22	2.4	28	5.9	5	0.2
2015	60	8.3	50	3.6	53	6.7	21	0.7
Montana:								
2005	53	2.0	35	1.0	14	0.4	9	0.1
Nebraska:								
2005	68	4.5	24	1.3	7	0.1	5	....
2015	77	5.1	65	2.6	36	0.8	29	0.4
New York:								
2005	75	1.9	72	2.7	72	2.8	....	....
2015	66	1.3	63	1.7	62	1.8	(D)	(D)
North Dakota:								
2005	71	15.8	49	5.7	9	0.7	5	0.1
2015	88	12.1	70	7.2	27	0.9	22	0.6
Ohio:								
2015	70	1.3	69	2.2	66	2.7	(D)	(D)
Pennsylvania:								
2005	90	4.5	81	4.9	82	5.1	2	0.1
2015	81	2.2	76	2.9	75	2.7	(D)	(D)
South Dakota:								
2005	64	11.8	46	5.6	17	1.7	....	....
2015	91	22.1	74	9.2	20	1.1	17	0.7
Texas:								
2005	79	45.4	56	12.7	39	4.9	25	1.7
2015	97	27.8	82	12.1	39	3.2	51	3.0
Wisconsin:								
2005	23	2.1	24	3.9	35	15.1	8	0.4
2015	51	3.9	36	2.9	45	8.4	21	1.0

Planted acres are in table 1-48.
(Z) Less than half the rounding unit.   [1] Data not available for all States for all years.
NASS, Environmental, Economics, and Demographics Branch, (202) 720–6146.

Table 14-12.—Potatoes, Fall: Pesticide usage, 2005–2014 [1]

State and Year	Percent treated and amount applied							
	Herbicide		Insecticide [2]		Fungicide		Other Chemicals	
	Area applied	Pounds applied	Area applied	Pounds applied	Area applied	Pounds applied	Area applied	Pounds applied
	Percent	Thousands	Percent	Thousands	Percent	Thousands	Percent	Thousands
CO:								
2005	78	101	57	10	78	87	34	9,678
2010	92	173	93	108	98	191	74	5,156
2014	78	83	50	37	96	120	74	5,696
ID:								
2005	90	694	65	331	81	813	49	37,732
2010	98	829	67	190	93	983	63	36,080
2014	89	717	89	287	95	1,618	57	36,437
ME:								
2005	100	35	91	18	100	607	12	46
2010	93	36	91	8	100	373	73	30
2014	97	30	84	9	90	329	78	37
MI:								
2005	98	68	97	20	98	391	2	55
2010	90	70	90	18	96	298	67	46
2014	99	86	96	32	99	442	73	35
MN:								
2005	97	33	97	10	98	578	8	7
2010	90	44	99	9	100	586	48	3,099
2014	91	34	93	13	91	612	52	4,067
ND:								
2005	89	57	76	11	96	854	7	15
2010	89	71	85	13	96	874	60	6,872
2014	90	71	89	10	99	801	30	1,653
WA:								
2005	96	328	97	517	99	1,394	70	17,171
2010	98	376	97	392	99	1,197	80	16,901
2014	92	566	98	489	99	1,109	87	16,635
WI:								
2005	99	78	97	62	99	810	49	3,327
2010	82	73	91	12	96	866	92	2,742
2014	95	96	99	16	100	987	96	4,913

[1] Data not available for all States for all years. [2] Amount applied excludes Bt (bacillus thuringiensis) and other biologicals.
NASS, Environmental, Economics, and Demographics Branch, (202) 720–6146.

Table 14-13.—Potatoes, Fall: Fertilizer usage, 2005–2014 [1]

State and Year	Percent treated and amount applied							
	Nitrogen		Phosphate		Potash		Sulfur [2]	
	Area applied	Pounds applied	Area applied	Pounds applied	Area applied	Pounds applied	Area applied	Pounds applied
	Percent	Millions	Percent	Millions	Percent	Millions	Percent	Millions
Colorado:								
2005	92	9.4	86	7.9	64	3.2	89	2.6
2010	97	10.0	92	10.4	66	3.9	93	5.2
2014	94	9.2	93	7.2	70	1.6	94	2.6
Idaho:								
2005	100	72.9	99	56.9	92	40.0	82	21.7
2010	100	75.0	96	53.2	91	35.9	81	30.0
2014	99	78.1	98	47.5	85	36.3	90	23.6
Maine:								
2005	100	10.2	100	10.1	100	11.9	....	....
2010	99	8.8	99	8.2	95	10.2	*	*
2014	98	8.3	98	7.1	98	8.7	12	0.1
Michigan:								
2005	99	9.2	94	4.9	100	10.2	58	1.4
2010	100	8.6	98	3.5	100	8.6	*	*
2014	100	7.9	99	3.8	100	9.6	52	1.7
Minnesota:								
2005	100	8.2	100	5.0	81	7.7	55	0.7
2010	98	9.2	100	4.6	100	14.6	80	1.3
2014	100	6.2	88	2.8	80	5.9	66	0.9
North Dakota:								
2005	100	14.7	100	8.4	96	13.7	54	1.3
2010	100	15.1	89	7.9	84	14.3	44	1.3
2014	100	12.3	97	8.9	100	8.7	79	1.7
Washington:								
2005	100	37.8	98	30.2	92	38.2	89	9.5
2010	100	32.6	99	24.9	94	36.0	90	9.7
2014	100	29.2	100	20.1	97	18.2	82	7.2
Wisconsin:								
2005	100	17.9	99	9.1	99	20.5	72	4.1
2010	100	16.5	100	5.3	93	17.7	91	3.9
2014	100	15.0	96	5.4	99	20.8	89	4.2

*Insufficient number of reports to publish data.   [1] Data not available for all States for all years.   [2] Estimates began in 2005.

NASS, Environmental, Economics, and Demographics Branch, (202) 720–6146.

Table 14-14.—Rice: Pesticide usage, 2006 and 2013 [1]

State and Year	Percent treated and amount applied							
	Herbicide		Insecticide		Fungicide		Other Chemicals	
	Area applied	Pounds applied	Area applied	Pounds applied	Area applied	Pounds applied	Area applied	Pounds applied
	Percent	Thousands	Percent	Thousands	Percent	Thousands	Percent	Thousands
Arkansas:								
2006	95	3,054	10	14	37	109	5	269
2013	97	2,646	30	10	45	103	5	251
California:								
2006	93	2,500	14	2	50	738	....	....
2013	96	2,986	28	6	45	553	*	*
Louisiana:								
2006	96	475	42	49	46	30	....	....
2013	99	680	17	13	73	82	*	*
Mississippi:								
2006	100	502	55	14	46	16	3	36
2013	100	364	51	2	35	9	6	37
Missouri:								
2006	100	454	....	....	25	12	....	....
2013	100	512	18	1	55	22	20	159
Texas:								
2006	97	496	77	83	55	21	....	....
2013	88	263	35	3	35	11	*	*

* Insufficient number of reports to publish data.　[1] Data not available for all States for all years.　Note: Planted acres are in table 1-27.
NASS, Environmental, Economics, and Demographics Branch, (202) 720–6146.

Table 14-15.—Rice: Fertilizer usage, 2006 and 2013 [1]

State and Year	Percent treated and amount applied							
	Nitrogen		Phosphate		Potash		Sulfur	
	Area applied	Pounds applied	Area applied	Pounds applied	Area applied	Pounds applied	Area applied	Pounds applied
	Percent	Millions	Percent	Millions	Percent	Millions	Percent	Millions
Arkansas:								
2006	97	281.2	68	54.7	60	64.9	9	6.0
2013	96	195.8	76	53.2	56	51.3	14	2.1
California:								
2006	94	61.4	75	18.2	40	7.2	31	4.0
2013	98	83.9	81	21.2	38	7.2	41	4.4
Louisiana:								
2006	99	52.8	78	14.6	75	16.2	4	0.3
2013	100	63.9	87	16.0	79	23.0	14	1.3
Mississippi:								
2006	99	35.8	29	2.5	4	0.5	42	1.5
2013	100	23.3	13	0.8	*	*	50	1.2
Missouri:								
2006	100	45.2	47	5.5	42	5.7	29	0.7
2013	100	32.7	56	4.7	*	*	29	0.6
Texas:								
2006	97	29.2	92	5.8	89	6.0	30	0.6
2013	90	22.3	82	5.5	79	5.4	11	0.1

* Insufficient number of reports to publish data.　[1] Data not available for all States for all years.　Note: Planted acres are in table 1-27.
NASS, Environmental, Economics, and Demographics Branch, (202) 720–6146.

Table 14-16.—Sorghum: Pesticide usage, 2003 and 2011 [1]

State and Year	Percent treated and amount applied							
	Herbicide		Insecticide		Fungicide		Other chemicals	
	Area applied	Pounds applied	Area applied	Pounds applied	Area applied	Pounds applied	Area applied	Pounds applied
	Percent	Thousands	Percent	Thousands				
Colorado:								
2003	52	132	....	....	....	....	....	....
2011	75	285	....	....	....	....	....	....
Kansas:								
2003	90	9,014	....	....	....	....	....	....
2011	96	9,411	4	3	....	....	....	....
Missouri:								
2003	98	571	6	4	....	....	....	....
Nebraska:								
2003	98	2,030	4	29	....	....	....	....
2011	84	362	*	*	....	....	....	....
Oklahoma:								
2003	84	329	*	*	....	....	....	....
2011	74	505	*	*	....	....	....	....
South Dakota:								
2003	87	430	....	....	....	....	....	....
2011	89	289	*	*	....	....	....	....
Texas:								
2003	78	2,881	20	208	....	....	....	....
2011	72	1,901	10	9	*	*	....	....

Planted acres are in table 1-65.
* Insufficient number of reports to publish data.   [1] Data not available for all States for all years.
NASS, Environmental, Economics, and Demographics Branch, (202) 720–6146.

Table 14-17.—Sorghum: Fertilizer usage, 2003 and 2011 [1]

State	Percent treated and amount applied							
	Nitrogen		Phosphate		Potash		Sulfur [2]	
	Area applied	Pounds applied	Area applied	Pounds applied	Area applied	Pounds applied	Area applied	Pounds applied
	Percent	Millions	Percent	Millions	Percent	Millions	Percent	Millions
Colorado:								
2003	61	7.8	39	5.5	....	....	....	....
2011	75	7.4	41	1.5	*	*	*	*
Kansas:								
2003	97	261.8	55	57.5	4	4.7	....	....
2011	83	135.7	51	32.8	8	2.9	9	1.5
Missouri:								
2003	100	25.0	75	9.1	72	10.8	....	....
Nebraska:								
2003	99	56.7	40	6.1	1	0.1	....	....
2011	88	11.0	41	1.9	15	0.2	22	0.4
Oklahoma:								
2003	69	15.5	36	3.6	11	0.8	....	....
2011	81	11.9	58	3.8	13	0.2	13	0.3
South Dakota:								
2003	84	13.0	54	4.4	3	0.1	....	....
2011	88	8.9	45	1.9	*	*	*	*
Texas:								
2003	63	182.8	43	45.5	14	5.5	....	....
2011	77	94.1	63	27.2	11	3.7	29	5.1

Planted acres are in table 1-65.
* Insufficient number of reports to publish data.   [1] Data not available for all States for all years.   [2] Estimates began in 2005.
NASS, Environmental, Economics, and Demographics Branch, (202) 720–6146.

Table 14-18.—Soybeans: Pesticide usage, 2012 and 2015 [1]

State and Year	Percent treated and amount applied							
	Herbicide		Insecticide [2]		Fungicide		Other Chemicals [2]	
	Area applied	Pounds applied	Area applied	Pounds applied	Area applied	Pounds applied	Area applied	Pounds applied
	Percent	Thousands	Percent	Thousands	Percent	Thousands	Percent	Thousands
Arkansas:								
2012	97	7,531	36	238	27	132	....	....
2015	89	6,565	23	129	10	57	....	....
Illinois:								
2012	98	15,478	16	384	11	105	*	*
2015	99	19,119	7	89	12	173	3	37
Indiana:								
2012	99	8,730	16	362	13	95	*	*
2015	96	11,216	11	118	8	60	....	....
Iowa:								
2012	99	16,093	21	486	16	173	*	*
2015	93	17,021	25	425	18	235	(D)	(D)
Kansas:								
2012	96	7,337	*	*	*	*	....	....
2015	98	9,017	9	62	(D)	(D)	(D)	(D)
Kentucky:								
2012	98	2,553	13	6	7	9	....	....
2015	97	3,199	23	55	20	74	(D)	(D)
Louisiana:								
2012	99	3,224	84	604	71	130	*	*
2015	97	4,186	63	....	43	107	(D)	(D)
Michigan:								
2012	97	3,650	25	194	*	*	....	....
2015	....	3,041	(D)	(D)	9	63	(D)	(D)
Minnesota:								
2012	100	11,017	26	493	8	54	*	*
2015	96	11,723	56	656	11	118	(D)	(D)
Mississippi:								
2012	100	5,448	38	364	33	60	*	*
2015	97	7,493	38	123	25	103	7	718
Missouri:								
2012	91	9,287	15	120	4	18	....	....
2015	90	8,460	6	19	(D)	(D)	....	....
Nebraska:								
2012	98	8,947	*	*	5	23	....	....
2015	98	9,967	9	25	7	44	....	....
North Carolina								
2012	90	2,852	31	40	14	29	....	....
2015	97	3,620	19	54	13	64	(D)	(D)
North Dakota								
2012	99	8,719	10	125	3	17	....	....
2015	99	9,737	37	367	7	79	(D)	(D)
Ohio:								
2012	98	7,941	6	45	8	40	....	....
2015	96	8,377	7	13	10	67	(D)	(D)
South Dakota								
2012	97	7,771	17	363	5	77	....	....
2015	93	9,025	29	291	5	33	....	....
Tennessee:								
2012	95	2,861	12	6	23	33	....	....
2015	100	4,427	22	14	15	40	(D)	(D)
Virginia:								
2012	99	1,138	37	29	26	17	....	....
2015	96	1,224	37	25	16	19	(D)	(D)
Wisconsin:								
2012	100	2,404	29	177	5	8	....	....
2015	96	2,829	(D)	(D)	11	30	(D)	(D)

Planted acres are in table 3-32.
* Insufficient number of reports to publish data. (D) Withheld to avoid disclosing data for individual operations. [1] Data not available for all States for all years. [2] Amount applied excludes Bt (bacillus thuringiensis) and other biologicals.
NASS, Environmental, Economics, and Demographics Branch, (202) 720–6146.

Table 14-19.—Soybeans: Fertilizer usage, 2012 and 2015 [1]

State and Year	Percent treated and amount applied							
	Nitrogen		Phosphate		Potash		Sulfur	
	Area applied	Pounds applied	Area applied	Pounds applied	Area applied	Pounds applied	Area applied	Pounds applied
	Percent	*Millions*	*Percent*	*Millions*	*Percent*	*Millions*	*Percent*	*Millions*
Arkansas:								
2012	7	8.5	49	84.2	53	149.2	*	*
2015	(D)	(D)	59	113.6	64	168.2	(D)	(D)
Illinois:								
2012	19	44.7	23	148.8	39	387.5	4	8.6
2015	21	53.2	28	190.3	44	486.0	3	4.8
Indiana:								
2012	18	13.8	30	83.9	47	224.5	....	....
2015	35	35.9	39	115.9	50	317.3	6	2.2
Iowa:								
2012	13	19.2	26	134.5	36	270.0	7	8.8
2015	15	32.1	30	156.8	32	264.3	5	7.3
Kansas:								
2012	29	14.7	31	44.0	14	24.3	3	1.1
2015	33	18.9	36	45.3	20	23.5	12	2.2
Kentucky:								
2012	24	5.1	44	34.6	54	64.9	*	*
2015	37	17.8	54	59.6	58	85.0	4	0.8
Louisiana:								
2012	6	0.9	22	10.3	24	17.8	*	*
2015	(D)	(D)	29	22.0	29	27.9	7	1.8
Michigan:								
2012	44	11.2	43	30.4	69	107.2	*	*
2015	36	10.2	33	27.5	49	85.1	11	1.8
Minnesota:								
2012	27	24.0	30	80.2	27	101.0	6	4.4
2015	19	19.4	27	102.7	29	131.2	6	4.6
Mississippi:								
2012	14	6.2	13	12.0	26	43.7	*	*
2015	8	2.7	26	37.6	33	68.3	(D)	(D)
Missouri:								
2012	24	23.3	45	130.4	46	190.9	8	3.3
2015	20	16.8	41	103.1	46	159.9	7	4.9
Nebraska:								
2012	42	27.4	52	129.6	14	24.6	21	10.1
2015	36	30.5	45	115.8	15	24.3	25	19.1
North Carolina								
2012	50	15.8	44	31.6	61	80.3	16	2.3
2015	53	21.8	49	42.0	64	96.8	12	3.2
North Dakota								
2012	51	27.3	61	106.0	14	12.2	3	0.9
2015	48	28.0	51	114.0	14	23.3	13	7.2
Ohio:								
2012	25	17.2	31	74.1	58	267.4	4	1.8
2015	32	25.9	34	82.3	49	232.2	4	3.1
South Dakota								
2012	43	25.4	49	110.5	22	33.4	2	1.2
2015	41	22.3	49	102.5	27	36.4	4	1.6
Tennessee:								
2012	48	15.7	68	45.4	72	73.6	13	3.2
2015	48	22.1	66	78.1	70	116.5	17	2.9
Virginia:								
2012	53	7.3	50	13.3	56	27.3	9	1.5
2015	58	9.2	64	19.6	79	40.6	25	2.0
Wisconsin:								
2012	45	13.4	42	25.5	75	114.9	28	11.6
2015	39	13.1	44	34.4	71	116.7	29	6.9

Planted acres are in table 3-32.
* Insufficient number of reports to publish data.   (D) Withheld to avoid disclosing data for individual operations.   [1] Data not available for all States for all years.
NASS, Environmental, Economics, and Demographics Branch, (202) 720–6146.

Table 14-20.—Wheat: Pesticide usage, 2012 and 2015 [1]

State and Year	Herbicide		Insecticide		Fungicide [2]		Other Chemicals [2]	
	Area applied	Pounds applied	Area applied	Pounds applied	Area applied	Pounds applied	Area applied	Pounds applied
	Percent	Thousands	Percent	Thousands	Percent	Thousands	Percent	Thousands
Winter								
Colorado:								
2012	81	3,875	*	*	*	*	....	....
2015	75	2,816	(D)	(D)	36	127	....	....
Idaho:								
2012	95	444	*	*	46	45	*	*
2015	90	....	(D)	(D)	56	50	(D)	(D)
Illinois:								
2012	50	59	15	4	36	32	....	....
2015	45	56	9	1	25	21	(D)	(D)
Kansas:								
2012	67	4,567	*	*	21	227	....	....
2015	58	3,043	(D)	(D)	17	174	....	....
Missouri:								
2012	36	54	21	5	30	33	....	....
2015	40	98	22	5	27	38	....	....
Montana:								
2012	95	2,365	....	....	7	16	....	....
2015	97	3,299	(D)	(D)	18	42	....	....
Nebraska:								
2012	61	588	*	*	17	30	....	....
2015	69	607	(D)	(D)	(D)	(D)	(D)	(D)
North Carolina:								
2015	78	127	41	7	29	21	....	....
Ohio:								
2012	39	37	9	1	23	18	....	....
2015	35	50	7	5	23	19	(D)	(D)
Oklahoma:								
2012	41	1,089	*	*	*	*	....	....
2015	44	1,539	(D)	(D)	(D)	(D)	....	....
Oregon:								
2012	98	886	*	*	39	44	*	*
2015	100	865	(D)	(D)	15	20	(D)	(D)
South Dakota:								
2012	75	799	6	1	56	61	....	....
2015	86	959	(D)	(D)	32	57	....	....
Texas:								
2012	27	997	4	81	8	68	....	....
2015	41	2,528	12	234	11	102	....	....
Washington:								
2012	99	1,181	*	*	62	166	....	....
2015	95	1,569	(D)	(D)	44	98	(D)	(D)
Durum								
Montana:								
2012	96	604	....	....	9	3	....	....
2015	98	678	(D)	(D)	26	12	....	....
North Dakota:								
2012	100	1,146	5	8	51	88	....	....
2015	100	673	(D)	(D)	72	116	....	....
Other Spring								
Minnesota:								
2012	97	664	41	37	76	144	....	....
2015	97	676	41	47	76	241	....	....
Montana:								
2012	93	2,642	*	*	10	25	....	....
2015	98	3,197	(D)	(D)	16	38	....	....
North Dakota:								
2012	99	3,855	10	81	64	460	....	....
2015	96	4,288	(D)	(D)	62	597	....	....
South Dakota:								
2012	93	519	*	*	39	43	....	....
2015	90	612	9	5	38	74	....	....

Planted acres are in table 1-8.   * Insufficient number of reports to publish data. (D) Withheld to avoid disclosing data for individual operations.   [1] Data not available for all States for all years.   [2] Amount applied excludes biologicals.
NASS, Environmental, Economics, and Demographics Branch, (202) 720–6146.

Table 14-21.—Wheat: Fertilizer usage, 2012 and 2015 [1]

State and Year	Percent treated and amount applied							
	Nitrogen		Phosphate		Potash		Sulfur [2]	
	Area applied	Pounds applied	Area applied	Pounds applied	Area applied	Pounds applied	Area applied	Pounds applied
	Percent	Millions	Percent	Millions	Percent	Millions	Percent	Millions
Winter Wheat								
Colorado:								
2012	72	59.2	40	21.3	3	1.4	11	1.2
2015	79	75.3	52	25.5	6	0.9	16	1.6
Idaho:								
2012	94	91.9	61	14.0	34	6.2	58	10.4
2015	96	88.2	65	13.5	31	4.9	80	14.4
Illinois:								
2012	88	63.3	78	41.2	81	54.2	10	1.3
2015	90	38.8	72	27.5	58	33.1	10	1.2
Kansas:								
2012	94	456.2	65	183.1	8	24.4	9	4.4
2015	94	518.7	63	174.4	9	33.1	11	4.8
Missouri:								
2012	99	72.3	85	37.6	82	47.0	33	3.3
2015	92	55.4	80	33.4	80	42.5	19	1.6
Montana:								
2012	91	135.3	88	57.4	21	5.2	23	3.9
2015	99	154.4	93	69.5	32	7.9	31	5.6
Nebraska:								
2012	83	63.0	64	30.7	7	1.3	23	3.1
2015	83	65.6	58	30.2	3	0.8	23	3.2
North Carolina:								
2015	80	39.7	43	11.9	61	32.7	24	3.6
Ohio:								
2012	89	39.6	83	22.5	80	25.1	25	1.6
2015	88	35.4	74	20.9	70	26.0	22	1.8
Oklahoma:								
2012	85	288.8	43	72.1	8	5.7	5	2.6
2015	94	264.3	65	95.0	12	10.2	5	1.6
Oregon:								
2012	100	64.2	22	3.7	11	3.4	50	7.5
2015	97	33.3	15	1.4	13	1.4	60	3.8
South Dakota:								
2012	84	81.5	75	29.7	9	1.3	34	3.3
2015	99	118.2	79	46.3	13	2.2	33	3.3
Texas:								
2012	65	203.5	32	55.6	7	6.6	21	14.5
2015	67	209.2	42	63.4	10	14.7	22	10.7
Washington:								
2012	97	133.3	40	13.4	8	2.6	74	15.7
2015	99	129.6	59	16.6	14	2.2	86	18.3
Durum								
Montana:								
2012	98	32.2	90	12.7	12	0.6	15	0.5
2015	98	39.9	97	16.8	23	1.8	23	1.2
North Dakota:								
2012	99	95.0	89	31.4	12	1.7	11	0.6
2015	99	88.7	94	37.8	36	6.1	35	3.2
Other Spring								
Minnesota:								
2012	98	151.8	88	48.8	58	23.1	13	2.2
2015	95	147.2	87	49.0	69	34.5	30	4.6
Montana:								
2012	93	146.4	84	70.4	17	7.3	28	8.7
2015	94	158.6	81	55.5	28	11.7	36	7.7
North Dakota:								
2012	98	515.0	90	195.3	29	45.3	18	8.9
2015	99	633.9	95	244.8	41	49.5	30	13.8
South Dakota:								
2012	97	84.8	78	32.9	10	3.8	24	3.0
2015	98	116.6	80	43.4	21	8.2	24	3.2

Planted acres are in table 1-8.   * Insufficient number of reports to publish data.   [1] Data not available for all States for all years.
NASS, Environmental, Economics, and Demographics Branch, (202) 720–6146.

Table 14-22.—Fruits, Pesticides: Percent of bearing acres receiving applications, for surveyed States, 2015 [1]

Crop	Herbicide	Insecticide	Fungicide	Other
	Percent			
Apples	37	91	81	80
Apricots	55	77	70	58
Avocados	35	40	(D)	42
Blackberries	95	83	76	24
Blueberries	66	85	86	20
Cherries, Sweet	36	83	83	72
Cherries, Tart	54	90	95	68
Dates	23	8		
Figs	60	36	36	26
Grapefruit	70	94	84	85
Grapes, All	55	53	83	45
Grapes, Raisin	51	71	91	61
Grapes, Table	74	60	96	80
Grapes, Wine	53	48	81	35
Kiwifruit	50	12	(D)	59
Lemons	60	74	34	66
Nectarines	49	55	55	54
Olives	39	26	43	10
Oranges	72	90	72	76
Peaches	43	78	82	57
Pears	45	92	90	95
Plums	47	70	51	66
Prunes	57	63	63	40
Raspberries	91	94	92	25
Tangelos	75	97	96	90
Tangerines	76	88	72	71

* Insufficient number of reports to publish data.   [1] Refers to acres receiving one or more applications of a specific agricultural chemical.
NASS, Environmental, Economics, and Demographics Branch, (202) 720–6146.

Table 14-23.—Fruit, Fertilizers: Percent of acres receiving applications, for surveyed States, 2015 [1]

Crop	Nitrogen	Phosphate	Potash	Sulfur
	Percent			
Apples	60	42	48	21
Apricots	78	16	34	29
Avocados	81	58	63	25
Blackberries	54	71	79	26
Blueberries	72	78	81	30
Cherries, Sweet	80	57	59	15
Cherries, Tart	81	30	64	22
Dates	74	83	83	(D)
Figs	72	(D)	39	
Grapefruit	95	86	89	35
Grapes, All	77	51	67	24
Grapes, Raisin	64	62	64	8
Grapes, Table	142	28	88	76
Grapes, Wine	75	54	67	25
Kiwifruit	84	58	71	61
Lemons	75	56	73	26
Nectarines	73	63	79	29
Olives	53	28	34	5
Oranges	50	62	70	35
Peaches	79	54	61	11
Pears	66	42	37	33
Plums	74	50	79	48
Prunes	63	9	51	20
Raspberries	71	98	97	13
Tangelos	91	82	95	(D)
Tangerines	89	72	78	41

(D) Withheld to avoid disclosing data for individual operations.   [1] Refers to acres receiving one or more applications of a specific agricultural chemical.
NASS, Environmental, Economics, and Demographics Branch, (202) 720–6146.

Table 14-24.—Vegetables, Pesticides: Percent of acres receiving applications, for surveyed States, 2014 [1]

Crop	Herbicide	Insecticide	Fungicide	Other
		Percent		
Asparagus	76	87	54	
Beans, Snap, Fresh	(D)	(D)	(D)	(D)
Beans, Snap, Proc	92	73	57	
Broccoli	65	89	38	(D)
Cabbage, Fresh				
Cantatoupes	18	63	52	(D)
Carrots, Fresh	(D)	(D)	(D)	
Carrots, Proc	46	38	58	18
Cauliflower	53	90	13	(D)
Celery	73	86	77	(D)
Sweet Corn, Fresh	65	75	18	1
Sweet Corn, Proc	79	64	33	(D)
Cucumbers, Fresh	46	79	88	11
Cucumbers, Pickles	(D)	(D)	(D)	
Eggplant				
Garlic	97	59	95	
Honeydews	(D)	(D)	(D)	(D)
Head Lettuce	50	93	88	(D)
Other Lettuce	48	88	83	(D)
Onions	87	86	82	28
Green Peas, Proc	90	42	10	(D)
Bell Peppers	(D)	(D)	(D)	(D)
Pumpkins	65	77	74	1
Spinach	31	95		(D)
Squash	37	57	63	2
Strawberries	10	93	97	65
Tomatoes, Fresh	(D)	(D)	(D)	(D)
Tomatoes, Proc	59	66	71	23
Watermelons	49	74	90	13

(D) Withheld to avoid disclosing data for individual operations.   [1] Refers to acres receiving one or more applications of a specific agricultural chemical.
NASS, Environmental, Economics, and Demographics Branch, (202) 720–6146.

Table 14-25.—Vegetables, Fertilizers: Percent of acres receiving applications, for surveyed States, 2014 [1]

Crop	Nitrogen	Phosphate	Potash	Sulfur
		Percent		
Asparagus	99	35	69	38
Beans, Snap, Fresh	(D)	(D)	(D)	(D)
Beans, Snap, Proc	88	84	84	58
Broccoli	97	83	45	40
Cabbage, Fresh	97	94	92	44
Cantatoupes	94	82	73	(D)
Carrots, Fresh	(D)	(D)	(D)	(D)
Carrots, Proc	99	77	62	42
Cauliflower	98	87	66	64
Celery	96	91	82	18
Sweet Corn, Fresh	98	87	85	22
Sweet Corn, Proc	88	73	68	49
Cucumbers, Fresh	97	94	96	37
Cucumbers, Pickles	(D)	(D)	(D)	(D)
Eggplant				
Garlic	98	73	9	(D)
Honeydews	(D)	(D)	(D)	(D)
Head Lettuce	97	92	(D)	(D)
Other Lettuce	99	97	(D)	(D)
Onions	98	91	68	51
Green Peas, Proc	78	58	56	40
Bell Peppers	(D)	(D)	(D)	(D)
Potatoes, fall	99	97	90	79
Pumpkins	93	71	89	26
Spinach	95	91	29	(D)
Squash	88	76	81	21
Strawberries	96	94	96	74
Tomatoes, Fresh	(D)	(D)	(D)	(D)
Tomatoes, Proc	98	64	58	13
Watermelons	94	85	94	34

(D) Withheld to avoid disclosing data for individual operations.   [1] Refers to acres receiving one or more applications of a specific agricultural chemical.
NASS, Environmental, Economics, and Demographics Branch, (202) 720–6146.

CHAPTER XV
MISCELLANEOUS AGRICULTURAL STATISTICS

This chapter contains miscellaneous data which do not fit into the preceding chapters. Included here are summary tables on foreign trade in agricultural products; statistics on fishery products; tables on refrigerated warehouses; and statistics on crops in Alaska.

Foreign Agricultural Trade Statistics

Agricultural products, sometimes referred to as food and fiber products, cover a broad range of goods from unprocessed bulk commodities like soybeans, feed corn and wheat to highly-processed, high-value foods and beverages like sausages, bakery goods, ice cream, or beer sold in retail stores and restaurants. All of the products found in Chapters 1-24 (except for fishery products in Chapter 3) of the U.S. Harmonized Tariff Schedule are considered agricultural products. These products generally fall into the following categories: grains, animal feeds, and grain products (like bread and pasta); oilseeds and oilseed products (like canola oil); livestock, poultry and dairy products including live animals, meats, eggs, and feathers; horticultural products including all fresh and processed fruits, vegetables, tree nuts, as well as nursery products and beer and wine; unmanufactured tobacco; and tropical products like sugar, cocoa, and coffee. Certain other products are considered "agricultural," the most significant of which are essential oils (Chapter 33), protein isolates and modified starches (Chapter 35), raw rubber (Chapter 40), raw animal hides and skins (Chapter 41), and wool and cotton (Chapters 51-52). Manufactured products derived from plants or animals, but which are not considered "agricultural" by USDA's definition are cotton yarn, textiles and clothing; leather and leather articles of apparel; and cigarettes and spirits. The World Trade Organization's definition of agricultural products varies in that it includes some products like spirits and tobacco products.

U.S. foreign agricultural trade statistics are based on documents filed by exporters and importers and compiled by the Bureau of the Census. Puerto Rico is a Customs district within the U.S. Customs territory, and its trade with foreign countries is included in U.S. export and import statistics. U.S. export and import statistics include merchandise trade between the U.S. Virgin Islands and foreign countries even though the Virgin Islands of the United States are not officially a part of the U.S. Customs territory.

The export value, the value at the port of exportation, is based on the selling price and includes inland freight, insurance, and other charges to the port. The country of destination is the country of ultimate destination or where the commodities are consumed or further processed.

Agricultural products, like manufactured goods, are often transhipped from the one country to another. Shippers are asked to identify the ultimate destination of a shipment. However, transhipment points are often recorded as the ultimate destination even though the actual point of consumption may be in a neighboring state. Thus, exports to countries which act as transhipment points are generally overstated, while exports to neighboring countries are often understated. Major world transhipment points include the Netherlands, Hong Kong, and Singapore. In such cases, exports are over reported for the Netherlands, but under reported for Germany, Belgium and the United Kingdom. They are overstated to Hong Kong, but under reported to China, and they are overstated to Singapore, but understated to Malaysia and Indonesia.

Imports for consumption are a combination of entries for immediate consumption and withdrawals from bonded warehouses for consumption. The import value, defined generally as the market value in the foreign country, excludes import duties, ocean freight, and marine insurance. The country of origin is defined as the country where the commodities were grown or processed. Where the country of origin is not known, the imports are credited to the country of shipment.

Table 15-1.—Foreign trade: Value of total agricultural exports and imports, United States, fiscal years 2006–2015

Fiscal year ending Sep. 30 [1]	U.S. total domestic exports			U.S. total imports for consumption, customs value			Surplus agricultural exports over agricultural imports
	Total merchandise exports	Agricultural exports [2]	Agricultural exports share of total exports	Total merchandise imports	Agricultural imports	Agricultural imports share of total imports	
	Million dollars	Million dollars	Percent	Million dollars	Million dollars	Percent	Million dollars
2006	993,203	68,593	7	1,834,339	64,026	3	4,567
2007	1,110,804	82,220	7	1,906,928	70,063	4	12,157
2008	1,297,574	114,911	9	2,152,782	79,320	4	35,591
2009 [1] ...	1,058,869	96,296	9	1,594,328	73,404	5	22,892
2010	1,224,652	108,529	9	1,844,486	78,963	4	29,566
2011	1,446,591	137,465	10	2,147,138	94,511	4	42,955
2012	1,534,887	135,907	9	2,272,484	103,371	5	32,536
2013	1,563,016	141,144	9	2,262,244	103,872	5	37,272
2014	1,617,435	152,312	9	2,334,187	109,221	5	43,091
2015	1,545,861	139,741	9	2,289,605	114,026	5	25,715

[1] Fiscal years Oct. 1–Sept. 30 revised. [2] Includes food exported for relief or charity by individuals and private agencies.
ERS, Market and Trade Economics Division, (202) 694–5257.

Table 15-2.—Foreign trade: Value and quantity of bulk commodity exports, United States, fiscal years 2006–2015 [1]

Fiscal year	Wheat, unmilled	Rice, milled	Feed grains [2]	Oilseeds [3]	Tobacco unmanufactured	Cotton and linters	Bulk commodities
	Value						
	Million dollars	Million dollars	Million dollars	Million dollars	Million dollars	Million dollars	Million dollars
2006	4,289	1,291	6,808	7,161	1,058	4,678	25,286
2007	6,579	1,273	9,783	9,339	1,143	4,305	32,423
2008	12,332	2,010	15,750	15,580	1,280	4,762	51,714
2009	5,997	2,241	9,982	14,790	1,199	3,561	37,771
2010	5,840	2,296	9,806	17,951	1,223	4,836	41,951
2011	11,504	2,117	13,969	21,571	1,116	8,982	59,260
2012	8,376	1,963	11,719	21,101	1,051	6,551	50,761
2013	10,126	2,212	6,198	22,456	1,193	5,649	47,834
2014	8,257	2,002	12,486	25,753	1,114	4,614	54,226
2015	5,846	2,109	11,046	23,179	1,250	4,155	47,585
	Quantity						
	1,000 metric tons	1,000 metric tons	1,000 metric tons	1,000 metric tons	1,000 metric tons	1,000 metric tons	1,000 metric tons
2006	25,005	4,014	61,363	27,593	169	3,707	121,851
2007	29,636	3,306	59,051	31,592	180	3,128	126,893
2008	32,847	3,899	68,205	32,148	184	2,970	140,253
2009	22,545	3,388	51,442	35,713	168	2,805	116,062
2010	25,698	4,260	53,849	42,499	185	2,746	129,236
2011	34,583	3,920	49,170	41,365	174	3,113	132,325
2012	26,969	3,578	40,024	39,247	167	2,731	112,716
2013	31,110	3,848	20,460	35,862	165	2,921	94,366
2014	27,081	3,355	56,605	46,031	152	2,202	135,426
2015	21,145	4,129	56,236	51,224	177	2,540	135,451

[1] Fiscal years, Oct. 1–Sept. 30. [2] Corn, barley, sorghum, rye, and oats. [3] Soybeans, peanuts, rapeseed, cottonseed, sunflowerseed, safflowerseed, and others.
ERS, Market and Trade Economics Division, (202) 694–5257.

Table 15-3.—Agricultural exports: Value to top 50 countries of destination, United States, fiscal years 2013–2015 [1]

Country	2013	2014	2015
	Million dollars	Million dollars	Million dollars
China	23,360.9	25,699.4	22,544.9
Canada	21,458.3	21,786.9	21,337.2
Mexico	17,938.5	19,498.2	18,004.9
European Union-28	11,548.6	12,709.9	12,311.4
Japan	12,419.4	13,360.0	11,702.5
Korea, South	5,203.4	6,867.5	6,428.3
Hong Kong	3,630.1	4,052.4	3,934.5
Taiwan	3,178.9	3,491.3	3,314.9
Colombia	1,418.0	2,310.8	2,594.7
Indonesia	2,619.5	2,964.7	2,443.0
Philippines	2,428.7	2,774.3	2,420.7
Vietnam	2,066.1	2,231.9	2,413.4
Thailand	1,445.9	1,613.5	1,719.2
Turkey	2,176.9	2,066.3	1,619.7
Australia(*)	1,345.0	1,536.4	1,449.7
United Arab Emirates	1,173.8	1,259.1	1,317.1
Saudi Arabia	1,128.4	1,405.0	1,268.1
Peru	682.4	958.3	1,245.7
Dominican Republic	1,132.3	1,281.2	1,178.9
India	971.6	987.8	1,113.0
Guatemala	939.9	1,101.4	1,111.9
Egypt	1,661.1	1,851.4	1,044.1
Venezuela	1,600.4	1,322.9	897.8
Malaysia	1,024.5	1,010.1	864.9
Chile	902.2	856.7	803.4
Brazil	1,556.7	1,642.3	788.1
Singapore	734.4	801.2	733.9
Costa Rica	482.9	600.8	687.1
Nigeria	1,110.7	901.2	674.9
Panama	571.4	668.2	653.9
Honduras	588.7	619.0	585.4
Israel(*)	625.6	715.9	524.7
El Salvador	466.4	525.3	517.5
Switzerland(*)	459.3	389.6	482.1
Pakistan	360.7	271.2	452.0
New Zealand(*)	394.8	424.0	449.5
Bangladesh	248.6	289.7	414.2
Russia	1,316.1	1,165.2	405.9
Ecuador	466.0	437.9	381.5
Trinidad and Tobago	380.4	398.1	381.0
Jamaica	447.0	418.1	371.9
Haiti	408.1	363.9	352.4
Morocco	455.5	444.6	322.9
Bahamas, The	243.9	240.2	254.0
Jordan	238.9	237.7	234.3
Nicaragua	216.1	212.9	230.4
Kuwait	234.9	237.0	227.8
South Africa	279.8	281.3	213.1
Iraq	113.2	213.2	212.4
Angola	282.8	276.3	199.2
Other Countries	5,006.2	4,539.4	3,906.4
World Total [2]	141,143.7	152,311.6	139,740.8

[1] Fiscal years Oct. 1–Sept. 30. [2] Totals may not add due to rounding.
ERS, Market and Trade Economics Divison, (202) 694–5257.

Table 15-4.—Foreign trade in agricultural products: Value of exports by principal commodity groups, United States, fiscal years 2012–2015 [1]

Commodity	2012	2013	2014	2015
	1,000 dollars	1,000 dollars	1,000 dollars	1,000 dollars
Total Merchandise Exports	1,534,887,2130	1,563,016,139	1,617,435,259	1,545,861,228
Nonagricultural U.S. Exports (Na)	1,398,980,520	1,421,872,425	1,465,123,648	1,406,120,388
Total Agricultural exports	135,906,693	141,143,714	152,311,611	139,740,840
Animals and animal products	28,951,218	30,563,844	32,967,495	28,550,460
Animals, live	792,824	826,574	660,926	522,954
Cattle	375,719	365,913	194,661	108,569
Horses,mules,burros-live	379,217	423,082	442,306	391,860
Swine, live	32,920	30,833	17,899	17,259
Sheep, live	3,605	5,621	4,930	3,649
Other live Animals	1,364	1,125	1,129	1,618
Red meat and products	12,426,938	12,629,569	14,054,351	13,096,435
Beef and Veal	4,770,766	5,217,247	6,036,279	5,847,713
Beef and Veal, fresh/frozen	4,596,817	5,019,369	5,757,388	5,554,420
Beef prep/pres	173,949	197,879	278,891	293,292
Horsemeat, fresh/frozen	27	58	60	3
Lamb and Goat, fr/frozen	16,326	16,734	17,318	12,186
Pork	5,549,347	5,167,839	5,672,702	4,935,499
Pork, fresh/frozen	4,957,912	4,470,063	4,919,156	4,183,729
Pork, prep/pres	591,435	697,776	753,546	751,770
Variety meats	1,456,200	1,557,060	1,674,100	1,662,106
Beef variety meats	712,544	720,139	783,918	877,973
Pork variety meats	691,389	778,117	825,558	717,294
Other variety meats	52,268	58,805	64,624	66,839
Other meats, fresh/frozen	634,271	670,630	653,892	638,928
Poultry and products	6,153,107	6,489,572	6,406,657	5,487,967
Poultry, live	198,750	206,944	211,355	175,707
Baby chicks	187,288	193,219	196,389	162,777
Other live poultry	11,462	13,724	14,966	12,930
Poultry meats	4,918,546	5,190,337	5,045,749	4,266,077
Chickens, fresh/frozen	3,897,792	4,070,972	3,857,742	3,148,681
Turkeys, fresh/frozen	577,831	590,751	634,181	524,061
Other poultry, fresh/frozen	9,674	5,752	5,103	7,525
Poultry meats, prep	433,249	522,862	548,722	585,810
Poultry, misc	573,574	469,847	524,690	385,156
Eggs	462,236	622,445	624,864	661,027
Dairy products	5,274,236	6,212,803	7,499,843	5,659,138
Evaporated/condensed milk	47,817	92,844	67,067	41,904
Nonfat dry milk	1,418,128	1,875,632	2,435,597	1,554,962
Butter and milkfat	165,302	247,159	309,429	80,213
Cheese	1,104,752	1,242,158	1,703,060	1,444,477
Whey,fluid/dried	812,526	896,067	1,046,924	745,200
Other dairy products	1,725,711	1,858,944	1,937,767	1,792,381
Fats, oils, and greases	977,229	716,282	672,082	477,641
Lard	64,264	51,211	60,315	51,530
Tallow, inedible	534,282	371,983	373,629	242,000
Other animal fats	378,682	293,088	238,138	184,111
Hides and skins	2,770,110	3,058,069	2,989,633	2,627,816
Bovine hides, whole	1,123,174	1,289,832	1,435,506	1,340,334
Other cattle hides	68,544	90,244	53,050	10,051
Calf skins, whole	227,170	116,369	98,323	61,278
Horse hides, whole	431,316	445,581	404,630	328,360
Sheep and lamb skins	30,534	23,133	21,011	14,389
Other hides and Skin	361,408	487,213	478,489	381,903
Furskins	527,964	605,697	498,624	491,501
Mink pelts	480,372	541,561	418,697	454,336
Other furskins	47,592	64,136	79,926	37,164
Wool and mohair	19,309	22,948	21,811	22,111
Sausage casings	197,714	225,866	241,421	242,313
Bull semen	131,784	136,458	157,350	164,401
Misc animal products	207,967	245,703	263,420	249,685
Grains and feeds	33,678,863	31,694,568	36,681,392	31,885,661
Wheat,unmilled	8,375,731	10,126,210	8,257,060	5,846,267
Wheat flour	150,795	142,726	132,983	156,638
Other wheat products	170,846	180,993	186,792	194,047
Rice-paddy, milled	1,962,970	2,212,059	2,001,730	2,108,749
Feed grains and products	12,289,277	6,739,813	13,001,684	11,524,499
Feed grain	11,719,232	6,198,202	12,486,009	11,046,279
Barley	59,416	46,522	87,072	68,582
Corn	11,216,737	5,570,678	11,088,625	8,791,341
Grain sorghum	431,331	570,124	1,297,689	2,179,031
Oats	5,054	4,147	6,595	4,097
Rye	6,694	6,731	6,028	3,228

See footnote(s) at end of table.

Table 15-4.—Foreign trade in agricultural products: Value of exports by principal commodity groups, United States, fiscal years 2012–2015[1]—Continued

Commodity	2012	2013	2014	2015
	1,000 dollars	1,000 dollars	1,000 dollars	1,000 dollars
Feed grains and products--Continued				
Feed grain products	570,046	541,611	515,675	478,220
Popcorn	90,015	103,944	119,579	119,220
Blended food products	118,372	69,710	60,014	61,846
Other grain products	3,519,141	3,754,795	3,893,572	3,796,913
Feeds and fodders	7,001,716	8,364,318	9,027,980	8,077,483
Corn by-products	718,565	914,538	923,183	731,028
Alfalfa meal and cubes	58,516	59,124	63,557	64,141
Beef pulp	99,055	110,911	114,386	100,798
Citrus pulp pellets	30,740	36,643	19,938	3,468
Other feeds and fodders	6,094,841	7,243,102	7,906,917	7,178,047
Fruit and prep	6,430,836	6,603,903	6,752,461	6,550,101
Fruits, fresh	4,829,857	4,987,769	4,998,833	4,793,483
Citrus fruits	1,008,711	1,004,071	973,927	995,359
Grapefruit	158,852	144,746	125,390	121,753
Lemons and limes	126,292	140,851	219,313	216,891
Oranges and tangerines	709,795	704,456	621,251	638,134
Other citrus	13,773	14,017	7,972	18,581
Noncitrus Fruits	3,821,146	3,983,698	4,024,906	3,798,124
Apples	1,025,215	1,161,541	1,052,325	1,095,235
Berries	696,356	746,736	752,005	711,333
Cherries	502,294	412,312	469,986	431,144
Grapes	799,994	827,400	949,466	825,812
Melons,	161,513	159,238	163,124	151,077
Peaches	168,287	173,358	167,217	161,132
Pears	203,678	223,261	224,493	199,242
Plums	79,410	72,233	67,105	59,138
Other noncitrus	184,399	207,620	179,185	164,011
Fruits, dried	690,630	659,410	711,964	652,240
Raisin	386,351	364,571	432,677	341,347
Prunes	170,865	187,258	172,651	186,144
Other dried fruits	133,414	107,582	106,636	124,750
Fruits, canned	532,129	547,351	581,479	656,606
Fruits, frozen	129,127	147,750	158,842	129,411
Other fruits, prep	249,093	261,624	301,344	318,361
Fruits, juices	1,295,231	1,284,983	1,252,045	1,108,493
Apple juice	36,856	39,186	39,527	43,915
Grape juice	97,209	97,668	84,969	70,964
Grapefruit juice	56,574	56,303	46,886	40,771
Orange juice	457,338	461,506	465,846	388,706
Other fruit juices	647,254	630,319	614,817	564,136
Wine	1,285,765	1,474,169	1,447,068	1,527,295
Nuts and prep	6,435,595	7,767,580	8,646,761	9,452,299
Almonds	3,301,381	3,876,728	4,552,878	5,111,513
Filberts	86,576	93,185	113,445	114,335
Peanuts	324,399	603,484	513,114	544,411
Pistachios	916,303	1,131,652	1,236,453	1,065,621
Walnuts	1,051,572	1,258,140	1,421,835	1,649,159
Pecans	435,558	430,842	372,702	506,252
Other nuts	319,806	373,550	436,334	461,007
Vegetables and prep	6,107,950	6,587,469	7,028,460	6,940,940
Vegetables fresh	2,153,699	2,329,931	2,404,896	2,400,346
Aspargus	36,000	32,540	30,417	22,625
Broccoli	128,279	129,714	135,231	123,657
Carrots	121,957	122,427	116,165	111,271
Cabbage	28,483	42,182	35,589	37,176
Celery	74,251	96,514	81,309	83,498
Cauliflower	124,706	132,155	142,525	132,667
Corn sweet	52,561	52,422	46,327	43,869
Cucumbers	17,740	16,589	16,421	17,179
Garlic	11,452	13,055	15,957	14,514
Lettuce	455,254	493,796	473,155	502,358
Mushrooms	36,668	33,478	33,728	30,882
Onions and shallots	170,847	186,731	205,965	166,852
Peppers	82,141	86,815	91,828	82,645
Potatoes	208,230	213,332	201,937	184,111
Tomatoes	150,308	152,265	165,666	146,081
Other fresh vegetables	454,821	525,916	612,676	700,961
Vegetables, frozen	1,303,749	1,352,405	1,435,236	1,344,323
Corn, sweet	87,627	89,164	93,305	98,827
Potatoes	1,023,842	1,064,105	1,131,200	1,030,812
Other frozen vegetables	192,281	199,136	210,731	214,685
Vegetables, canned	545,417	651,418	764,985	731,612
Pulses	682,027	747,062	797,204	699,288
Dried Beans	361,126	361,821	358,772	321,290
Dried Peas	166,627	178,971	252,473	197,517
Dried Lentils	103,007	138,152	141,155	148,471
Dried chickpeas	51,267	68,117	44,804	32,011
Other vegetables, prep/pres	1,423,058	1,506,653	1,626,139	1,765,371

See footnote(s) at end of table.

Table 15-4.—Foreign trade in agricultural products: Value of exports by principal commodity groups, United States, fiscal years 2012–2015[1]—Continued

Commodity	2012	2013	2014	2015
	1,000 dollars	*1,000 dollars*	*1,000 dollars*	*1,000 dollars*
Oilseeds and products	28,813,150	31,992,749	34,881,752	31,559,548
Oilcake and meal	3,955,009	5,587,337	5,810,383	5,319,112
Bran and residues	32,117	29,152	19,163	16,270
Corn oilcake and meal	22,358	8,543	47,558	28,483
Soybean meal	3,844,954	5,472,503	5,687,203	5,238,591
Other oilcake and meal	55,579	77,139	56,459	35,768
Oilseeds	21,100,708	22,455,538	25,753,288	23,179,231
Rapeseed	150,566	105,708	110,359	94,614
Safflower seeds	143	1,433	1,438	1,032
Soybeans	19,856,690	20,886,753	24,099,072	21,636,938
Sunflowerseeds	110,978	156,398	138,988	119,796
Peanuts, oilstock	68,698	215,846	174,413	120,292
Other oilseeds	90,572	121,978	113,385	112,836
Protein substances	823,060	967,424	1,115,633	1,093,722
Vegetable oils	3,757,434	3,949,874	3,318,081	3,061,206
Soybean oil	829,936	1,147,763	809,767	771,004
Cottonseed oil	72,287	60,005	41,029	35,537
Sunflower oil	36,799	44,545	57,629	42,009
Corn oil	619,427	567,261	464,861	391,211
Peanut oil	8,237	13,840	10,438	14,941
Rapeseed oil	346,219	246,541	123,320	95,416
Safflower oil	28,650	32,303	19,799	29,750
Other vegetable oils/waxes	1,815,878	1,837,614	1,791,239	1,681,339
Tobacco, unmfg	1,051,037	1,192,610	1,113,706	1,250,428
Tobacco, light air	236,807	262,622	210,593	239,938
Tobacco, flue	553,811	705,626	685,739	745,162
Other tobacco, unmfg	260,419	224,362	217,374	265,328
Cotton, excluding linters	6,533,609	5,604,211	4,598,448	4,141,498
Cotton linters	17,692	44,764	15,464	13,023
Essential oils	1,582,468	1,688,297	1,711,321	1,812,248
Seeds, field and garden	1,444,510	1,573,728	1,692,488	1,539,863
Sugar and tropical products	5,181,578	5,258,495	5,267,221	5,068,761
Sugar and related products	1,977,000	2,017,478	1,857,763	1,766,139
Sugar, cane or beet	169,688	162,624	151,250	93,865
Related sugar products	1,807,312	1,854,854	1,706,513	1,672,274
Coffee	1,079,304	983,628	959,746	911,753
Cocoa	220,200	218,457	218,178	214,692
Chocolate and prep	1,284,533	1,424,038	1,557,370	1,545,615
Tea, including herbal	377,424	391,511	423,930	375,541
Spices	148,332	153,931	175,170	189,176
Rubber, crude	91,957	66,312	71,635	61,827
Fibers, excluding cotton	2,827	3,140	3,428	4,018
Other hort products	5,208,711	5,710,532	5,998,229	5,982,499
Hops, including extract	205,155	206,611	204,794	222,979
Starches, not wheat/corn	113,843	126,163	141,556	148,717
Yeasts	95,745	93,135	99,988	97,949
Misc hort products	4,793,968	5,284,622	5,551,892	5,512,853
Nursery & greenhouse	376,962	390,306	399,269	397,034
Beverages, excluding juice	1,511,519	1,711,507	1,858,031	1,960,688

[1] Fiscal years, Oct. 1–Sept. 30. Totals may not add due to rounding.
ERS, Market and Trade Economics Division, (202) 694–5257.
Compiled from reports of the U.S. Department of Commerce.

Table 15-5.—Foreign trade in agricultural products: Value of imports by principal groups, United States, fiscal years 2012–2015 [1]

Product	2012	2013	2014	2015
	1,000 dollars	1,000 dollars	1,000 dollars	1,000 dollars
Total merchandise imports	2,272,483,785	2,262,243,815	2,334,187,363	2,289,604,720
Non-agricultural U.S. imports	2,168,786,323	2,158,179,951	2,224,900,709	2,175,430,132
Total agricultural imports	103,697,462	104,063,864	109,286,654	114,174,588
Animals & prods.	13,382,155	13,923,417	16,179,418	19,602,243
Animals - live, excluding poultry	2,356,798	2,466,606	2,909,811	3,483,476
Cattle and calves	1,676,472	1,784,224	2,128,025	2,640,598
Horses, mules, burros	302,206	327,427	333,505	409,168
Swine	347,527	329,321	410,090	373,763
Sheep, Live	396	273	482	212
Other live animals	30,198	25,361	37,709	59,735
Red meat & products	5,864,029	6,009,557	7,484,347	10,060,525
Beef & veal	3,620,414	3,672,803	4,727,315	7,202,838
Beef & veal - fresh or frozen	3,278,320	3,363,740	4,407,196	6,775,427
Beef & veal - prep. or pres.	342,094	309,063	320,120	427,411
Pork	1,294,011	1,381,341	1,654,334	1,660,765
Pork - fr. or froz.	1,003,072	1,092,292	1,343,292	1,345,521
Pork - prep. or pres.	290,939	289,049	311,042	315,244
Mutton, goat & lamb	617,886	623,144	726,861	765,178
Horsemeat - fr. or froz.	1,769	1,482	1,799	2,173
Variety meats - fr. or froz.	208,170	210,433	262,604	304,839
Other meats - fr. or froz.	32,117	41,185	45,469	48,819
Other meats & prods.	89,662	79,169	65,964	75,913
Poultry and prods.	612,979	650,271	699,909	814,164
Poultry - live	41,267	45,456	42,887	44,921
Poultry meat	306,244	360,170	376,987	467,496
Eggs	42,653	40,955	57,218	114,440
Poultry, misc.	222,814	203,690	222,817	187,307
Dairy products	3,143,088	3,209,875	3,443,858	3,632,137
Milk & cream, fresh or dried	109,197	118,586	113,604	119,596
Butter & butterfat mixtures	84,122	80,436	74,959	173,424
Cheese	1,060,712	1,150,129	1,235,336	1,318,070
Casein & mixtures	690,676	621,857	702,819	678,599
Other dairy prods.	1,198,382	1,238,867	1,317,140	1,342,448
Fats, oils, & greases	192,847	190,861	181,344	178,667
Hides & skins	219,281	282,069	293,240	255,093
Sheep & lamb skins	868	760	867	390
Other hides & skins	49,117	51,344	58,266	56,743
Furskins	169,296	229,966	234,107	197,960
Wool - unmfg.	34,067	24,209	21,718	21,092
Apparel grade wool	19,692	14,650	11,844	11,365
Carpet grade wool	14,375	9,559	9,874	9,727
Sausage casings	140,509	166,277	192,720	208,335
Bull semen	28,500	31,827	41,741	39,196
Misc. animal prods	789,727	891,579	910,338	909,319
Silk, raw	328	286	392	239
Grains & feeds	9,676,173	11,452,596	11,002,476	11,036,345
Wheat, ex. seed	832,264	1,003,536	1,086,258	774,376
Corn, unmilled	186,685	1,176,437	150,532	227,345
Oats, unmilled	410,820	361,101	516,729	443,907
Barley, unmilled	139,386	135,746	151,536	122,046
Rice	659,303	712,872	772,406	744,932
Biscuits & wafers	2,820,351	2,996,483	3,102,149	3,356,290
Pasta & noodles	549,151	566,351	613,894	616,285
Other grains & preps.	2,492,348	2,790,714	2,894,802	3,004,296
Feeds & fodders, excluding oilcake	1,585,865	1,709,356	1,714,171	1,746,869
Fruits & preps.	10,428,752	11,415,575	12,590,631	13,734,903
Fruits - fr. or froz.	8,215,293	8,977,579	10,129,662	11,042,893
Apples, fresh	160,670	219,653	237,193	210,962
Avocados	846,883	1,003,422	1,480,924	1,625,484
Berries, excl. strawberries	1,154,714	1,231,636	1,415,928	1,734,695
Bananas & plantains - fresh or frozen	2,056,919	2,141,374	2,185,461	2,229,495
Citrus, fresh	515,774	588,558	782,151	821,698
Grapes, fresh	1,089,336	1,190,056	1,195,201	1,292,997
Kiwifruit, fresh	68,615	85,324	100,522	117,972
Mangoes	361,332	450,661	467,753	525,256
Melons	450,002	539,482	560,572	612,336
Peaches	53,654	48,490	40,388	65,808
Pears	78,480	106,961	119,466	134,883
Pineapples - fr. or froz.	526,363	555,252	655,172	643,485
Plums	34,077	42,912	30,259	49,891
Strawberries - fr. or froz.	494,337	446,980	503,931	580,603
Other fruits - fr. or froz.	324,135	326,817	354,741	397,328
Fruits - prep. or pres.	2,213,459	2,437,996	2,460,969	2,692,010
Bananas & plantains - prep. or pres.	111,499	116,561	125,050	136,833
Pineapples - canned or prep.	311,425	328,340	303,818	378,974
Other fruits - prep. or pres.	1,790,536	1,993,094	2,032,101	2,176,204
Fruit juices	1,861,827	1,858,973	1,797,764	1,789,947
Apple juice	678,064	680,988	597,078	496,512
Grape juice	133,202	169,039	129,266	79,877
Grapefruit juice	888	1,391	652	734
Lemon juice	78,971	53,351	88,132	128,298
Lime juice	24,963	22,090	28,870	27,149
Orange juice	481,409	537,597	547,389	573,349
Pineapple juice	110,168	114,290	113,826	127,537
Other fruit juice	354,162	280,228	292,551	356,490

See footnote(s) at end of table.

Table 15-5.—Foreign trade in agricultural products: Value of imports by principal groups, United States, fiscal years 2012-2015 [1]—Continued

Product	2012	2013	2014	2015
	1,000 dollars	*1,000 dollars*	*1,000 dollars*	*1,000 dollars*
Nuts & preps	1,981,130	1,922,273	2,188,130	2,693,985
Brazil nuts	53,091	52,616	60,683	74,463
Cashew nuts	900,198	933,191	986,209	1,224,423
Chestnuts	13,824	13,014	13,733	13,008
Coconut meat	175,891	151,125	196,109	251,449
Filberts	31,793	46,702	53,090	39,542
Macadamia nuts	85,058	111,438	116,480	151,110
Pecans	318,029	243,447	298,087	390,564
Pistachio nuts	7,650	8,304	6,068	10,749
Other nuts	395,596	362,437	457,671	538,678
Vegetables & preps.	10,426,336	11,196,554	11,458,161	11,801,276
Vegetables, fresh	5,837,387	6,551,970	6,660,236	6,882,279
Tomatoes	1,885,713	1,935,648	1,936,263	1,948,745
Asparagus	443,125	578,374	497,581	551,238
Beans	80,115	96,128	96,076	104,997
Cabbage	20,344	32,179	33,716	36,294
Carrots	66,435	73,577	74,816	87,950
Cauliflower & broccoli, fresh	128,651	157,531	176,819	215,024
Celery,	15,954	25,845	22,509	25,921
Cucumbers	482,080	556,948	627,885	610,724
Eggplant	64,262	64,267	58,927	60,364
Endive	4,715	4,650	4,648	3,880
Garlic	126,612	191,795	131,122	157,272
Lettuce	164,738	190,079	202,298	240,480
Okra	17,114	19,730	27,099	25,002
Onions	282,383	338,304	354,206	402,048
Peas	66,490	78,598	72,164	74,481
Peppers	1,030,965	1,183,174	1,273,512	1,205,496
Potatoes	149,070	126,043	172,820	144,589
Radishes	20,068	19,713	19,263	21,140
Squash	278,197	305,618	303,461	325,726
Other fresh vegetables	510,356	573,771	575,050	640,909
Vegetables - prep. or pres.	2,713,695	2,721,914	2,816,103	2,944,989
Bamboo shoots, preserved	38,212	34,353	28,094	26,722
Cucumbers, preserved	58,572	52,123	43,131	57,946
Garlic, dried	49,949	63,663	63,219	65,145
Olives - prep. or pres.	385,985	384,976	428,967	411,261
Mushrooms, canned	127,261	102,205	110,666	110,537
Mushrooms, dried	27,724	26,891	32,679	34,621
Onions, preserved	33,975	38,386	38,924	44,205
Artichokes - prep.	174,742	137,098	137,862	137,646
Asparagus- prep.	32,271	41,569	24,450	14,864
Tomatoes, incl. paste & sauce	168,864	174,204	181,166	193,315
Waterchestnuts	25,854	30,652	30,854	32,849
Peppers & pimentos, prep.	85,033	99,393	95,212	103,835
Veg Starches, excluding wheat & corn	142,144	137,821	147,075	166,480
Soups & sauces	326,437	361,177	396,971	431,472
Other vegetables - prep. or pres.	1,036,672	1,037,405	1,056,834	1,114,093
Vegetables, frozen	1,633,527	1,703,433	1,717,253	1,710,190
Tomatoes	5,861	5,127	6,395	5,059
Asparagus	18,699	13,970	15,667	10,550
Beans	74,139	84,876	78,954	72,400
Carrots	4,555	4,601	4,839	4,975
Cauliflower & broccoli	321,108	300,041	322,089	342,429
Okra	8,765	7,487	10,184	15,076
Peas	36,603	39,350	42,872	41,305
Potatoes	709,313	795,935	753,521	712,264
Other frozen vegetables	454,485	452,046	482,733	506,131
Pulses	241,728	219,236	264,569	263,818
Dried peas	35,915	53,662	67,733	48,506
Dried beans	162,879	124,154	140,944	160,733
Dried lentils	23,695	19,552	30,915	29,850
Dried chickpeas	19,239	21,867	24,977	24,728
Sugar & related prods.	4,986,705	4,326,443	4,482,086	4,637,381
Sugar - cane & beet	2,555,825	1,698,728	1,652,323	1,762,164
Molasses	168,505	184,901	144,233	154,290
Confectionery prods.	1,454,815	1,535,424	1,656,979	1,630,573
Other sugar & related prods.	807,561	907,389	1,028,552	1,090,354
Cocoa & products	4,175,589	4,090,083	4,745,435	4,755,464
Coffee & products	7,783,814	6,058,130	6,012,099	6,393,574
Tea	661,211	704,577	694,965	734,683
Spices & herbs	1,291,465	1,420,618	1,581,192	1,812,124
Pepper	707,717	764,402	816,148	979,267
Other spices & herbs	583,747	656,217	765,044	832,857

See footnote(s) at end of table.

Table 15-5.—Foreign trade in agricultural products: Value of imports by principal groups, United States, fiscal years 2012–2015[1]—Continued

Product	2012	2013	2014	2015
	1,000 dollars	*1,000 dollars*	*1,000 dollars*	*1,000 dollars*
Drugs, crude & natural	1,476,942	1,562,378	1,640,086	1,702,235
Essential oils	2,166,621	2,378,411	2,594,070	2,722,880
Fibers, excl. cotton	90,759	78,289	89,314	108,199
Rubber & gums	3,791,086	2,672,791	2,138,519	1,577,932
Tobacco - unmfg.	947,371	931,868	838,154	784,025
Tobacco - filler	917,356	901,137	804,240	759,506
Tobacco - scrap	24,274	22,527	22,480	15,410
Other tobacco	5,740	8,203	11,434	9,109
Beverages, ex. fruit juice	11,253,367	11,652,510	12,451,031	13,150,154
Wine	5,096,414	5,364,895	5,496,241	5,485,185
Malt beverages	3,759,652	3,619,872	4,128,393	4,391,117
Other beverages	2,397,300	2,667,744	2,826,397	3,273,852
Oilseeds & prods.	8,301,758	8,551,107	9,658,703	8,515,757
Oilseeds & oilnuts	892,520	1,266,536	1,983,636	1,129,539
Flaxseed	130,408	118,234	126,670	116,081
Rapeseed	299,758	259,468	468,482	262,927
Soybeans	266,620	684,079	1,098,150	493,587
Sunflower seeds	35,617	42,735	46,586	54,299
Other oilseeds & oilnuts	160,117	162,021	243,748	202,644
Oils & waxes - vegetable	6,434,906	5,970,181	6,112,530	6,086,371
Castor oil	86,171	79,197	80,634	87,106
Coconut oil	767,542	562,498	735,495	860,973
Cottonseed oil	4,532	9,337	13,498	6,737
Olive oil	960,345	1,077,865	1,150,460	1,184,700
Palm oil	1,123,090	1,142,199	1,029,905	842,545
Palm kernel oil	439,969	282,433	312,751	328,023
Peanut oil	23,784	21,585	27,752	15,714
Rapeseed oil	1,906,072	1,619,207	1,598,346	1,487,438
Soybean oil	83,729	114,754	79,344	97,732
Sesame oil	60,412	65,584	70,119	81,103
Other vegetable oils	979,260	995,522	1,014,226	1,094,301
Oilcake & meal	974,332	1,314,390	1,562,537	1,299,847
Cotton, excl. linters	13,080	6,679	9,143	5,055
Cotton, linters	2,177	-	488	489
Seeds - field & garden	1,283,962	1,644,131	1,379,586	990,493
Cut flowers	962,416	1,000,903	1,013,218	1,016,472
Nursery stock, bulbs, etc.	663,081	669,545	705,680	712,765
Other hort products	6,089,685	4,546,012	4,036,306	3,896,206
Hops, including extract	33,356	41,383	43,828	50,374
Starches, ex wheat/corn	100,487	105,421	117,403	114,388
Yeasts	254,318	278,839	292,893	299,217
Misc hort products	5,701,525	4,120,369	3,582,182	3,432,228

[1] Fiscal years, Oct. 1–Sept. 30.
ERS, Market and Trade Economics Division, (202) 694–5257. Compiled from reports of the U.S. Department of Commerce.

Table 15-6.—Agricultural exports: Value of U.S. exports to the top market, China, by commodity, fiscal years 2013–2015 [1]

Commodity	Value		
	2013	2014	2015
	1,000 dollars	*1,000 dollars*	*1,000 dollars*
Total agricultural exports	23,360,863	25,699,412	22,544,943
Animals and animal products	3,703,483	3,381,282	2,626,492
Animals Live, ex Poultry	17,240	9,878	4,564
Cattle, live	-	-	-
Horses, live	490	3,652	2,085
Swine, Live	16,750	6,226	2,479
Other live animals	-	-	-
Red meat and Products	789,101	569,331	353,850
Beef and Veal	-	121	3
Beef and Veal, fresh or frozen	-	121	3
Beef, prep or pres	-	-	-
Lamb, goat, fr-ch-froz	-	-	-
Pork	345,563	287,046	169,750
Pork-fresh or frozen	320,058	279,180	169,683
Pork-prep or pres	25,505	7,866	67
Variety meats, Ed Offals	340,673	205,351	141,372
Beef variety meats	159	150	-
Pork variety meats	340,114	204,284	139,930
Other variety meats	400	917	1,441
Other meats-fr or froz	102,865	76,813	42,726
Poultry and poultry products	529,405	379,918	170,168
Poultry-Live	49,552	31,212	14,899
Baby chicks	47,069	28,436	14,887
Other live poultry	2,482	2,775	12
Poultry meats	247,506	210,425	97,839
Chickens, fresh or frozen	164,945	144,451	53,904
Turkeys, fresh or frozen	72,344	52,591	16,690
Other poultry, fresh or frozen	371	191	-
Poultry meats, prep or pres.	9,847	13,191	27,245
Poultry misc.	229,925	136,667	55,861
Eggs	2,422	1,614	1,569
Dairy prods	607,814	767,078	486,499
Evap and condensed milk	4,774	4,924	5,288
Nonfat dry milk	165,929	245,078	73,450
Butter and Milkfat	846	1,633	170
Cheese	39,537	58,954	52,263
Whey, fluid or dried	220,542	267,593	172,714
Other dairy products	176,185	188,896	182,614
Fats, oils and greases	3,305	2,937	2,676
Lard	2,182	2,088	1,873
Other animal fats	1,123	849	803
Hides and skins include furs	1,617,190	1,507,167	1,398,046
Bovine hides, whole	797,668	919,772	892,224
Other cattle hides, pieces	51,166	34,686	2,705
Calf skins, whole	53,480	31,587	28,633
Horse hides whole	294,130	245,501	207,370
Sheep and lamb skins	16,349	15,230	11,393
Other hides and skins, ex.furs	222,718	186,826	127,564
Furskins	181,680	73,564	128,157
Mink pelts	162,576	66,121	123,597
Other furskins	19,103	7,443	4,560
Wool and Mohair	10,247	6,929	10,800
Sausage casings	94,107	96,855	132,775
Bull semen	8,288	15,500	18,016
Misc animal products-other	26,787	25,690	49,099
Grains and feeds	3,296,030	4,587,554	4,542,618
Wheat, unmilled	1,038,466	439,795	145,158
Wheat flour	407	145	547
Other wheat products	1,831	2,349	1,158
Rice-paddy,milled parb	353	253	-
Feed grains and products	709,365	1,670,566	2,207,616
Feed grains	705,663	1,665,355	2,201,362
Corn	678,214	571,144	142,944
Grain sorghums	27,433	1,094,194	2,058,414
Oats	16	18	4
Rye	-	-	-
Feed grain products	3,702	5,211	6,254
Popcorn	10,878	17,422	24,068
Blended food products	1,765	1,290	855
Other grain prods	71,821	84,395	93,743
Feeds and fodders, ex.oilcakes	1,461,145	2,371,339	2,069,473
Corn by-products	1,520	1,779	1,081
Alfalfa meal and cubes	1,614	14,836	21,660
Citrus pulp pellet	32	-	-
Other feeds and fodder	1,457,980	2,354,723	2,046,732

See footnote(s) at end of table.

Table 15-6.—Agricultural exports: Value of U.S. exports to the top market, China, by commodity, fiscalyears 2013–2015 [1]—Continued

Commodity	Value		
	2013	2014	2015
	1,000 dollars	1,000 dollars	1,000 dollars
Fruits and prep. ex.juice	164,950	171,942	243,265
Fruits, fresh	104,424	103,842	151,915
Citrus fruits	33,382	13,374	36,439
Grapefruit	523	411	541
Lemons and limes	5,105	7,446	9,444
Oranges and tangerines	26,927	5,518	25,935
Other citrus fruits	826	-	520
Noncitrus fruits	71,042	90,467	115,476
Apple	1,011	933	20,406
Berries	692	157	96
Cherries	33,399	49,894	65,822
Grapes	28,829	32,752	19,908
Melons	22	-	-
Peaches	289	116	101
Pears	897	2,851	4,718
Plums	4,881	2,518	4,191
Other noncitrus	1,022	1,246	233
Fruits, dried	35,099	32,499	38,542
Raisins	20,823	22,953	24,812
Prunes	5,921	2,099	2,269
Other dried fruits	8,354	7,447	11,461
Fruits-canned, excluding juice	22,986	31,257	45,116
Fruits-frozen, excluding juice	634	881	1,441
Other fruits, prep. or pres	1,806	3,464	6,251
Fruit juices, including frozen	44,392	51,221	32,974
Apple juice	185	314	356
Grape juice	2,362	3,108	2,137
Grapefruit juice	3,033	2,344	2,390
Orange juice	4,648	4,741	3,769
Other fruit juices	34,163	40,714	24,323
Wine	71,052	74,999	61,537
Nuts and prep	386,380	308,702	229,875
Almonds	142,158	104,828	87,950
Filberts	4,880	2,870	1,620
Peanuts, shelled or prep	2,130	3,031	3,579
Pistachios	62,385	84,681	32,381
Walnuts	164,252	102,508	82,492
Pecans	7,420	2,364	7,704
Other nuts	3,156	8,420	14,150
Vegetables and prep	174,983	191,949	174,431
Vegetables, fresh	3,283	2,536	5,104
Asparagus	-	-	-
Broccoli	193	-	-
Carrots	20	-	-
Celery	211	442	320
Cauliflower	-	-	-
Corn, sweet	236	180	349
Cucumbers	-	-	-
Garlic	-	139	-
Lettuce	-	-	-
Mushrooms	-	-	-
Onion and Shallots	48	3	69
Peppers	66	79	208
Potatoes	48	11	151
Tomatoes	314	713	953
Other fresh vegetables	2,148	970	3,054
Vegetables, frozen	107,010	112,034	111,414
Corn, sweet	7,219	10,237	11,634
Potatoes	98,032	100,714	97,444
Other frozen vegetables	1,759	1,084	2,335
Vegetables, canned	19,292	25,560	2,235
Pulses	25,155	27,277	28,713
Dried beans	222	524	969
Dried peas	24,798	25,917	27,041
Dried lentils	16	836	702
Dried chick peas	118	-	-
Other vegetables, prep or pres	20,243	24,542	26,965

See footnote(s) at end of table.

Table 15-6.—Agricultural exports: Value of U.S. exports to the top market, China, by commodity, fiscal years 2013–2015 [1]—Continued

Commodity	Value		
	2013	2014	2015
	1,000 dollars	*1,000 dollars*	*1,000 dollars*
Oilseeds and products	12,505,524	14,846,034	12,858,095
Oilcake and meal	1,652	37,986	32,728
Bran and residues	15	-	-
Corn oilcake and meal	-	31,449	19,375
Soybean meal	1,607	6,501	13,333
Other oilcake and meal	30	36	19
Oilseeds	12,211,483	14,592,965	12,775,164
Rapeseed	0	0	329
Soybeans	12,191,031	14,556,765	12,737,629
Sunflowerseeds	1,104	2,134	1,010
Peanuts, including oilstock	2,718	9,851	13,958
Other oilseeds	1,793	4,916	1,390
Protein substances	14,836	19,299	20,848
Vegetable oils	292,390	215,083	50,203
Soybean oil	232,329	164,943	88
Cottonseed oil	-	-	-
Sunflower oil	359	570	164
Corn oil	173	225	184
Peanut oil	430	1,360	1,822
Rapeseed oil	6,835	312	381
Safflower oil	1,128	139	226
Other Vegetable oils and Waxes	51,136	47,534	47,339
Tobacco-unmanufactured	172,492	213,475	197,977
Tobacco-light air cured	-	460	-
Tobacco-flue cured	171,340	205,465	197,846
Other tobacco-unmanufactured	1,153	7,549	131
Cotton, excluding linters	2,236,056	1,249,616	1,024,266
Cotton linters	25,617	4,001	5,736
Essential oils	137,766	129,013	130,431
Seeds-field and garden	94,846	118,070	106,979
Sugar and tropical products	115,942	103,128	81,886
Sugar and related products	15,137	18,464	19,708
Sugar, cane or beet	367	1,157	249
Related sugar product	14,770	17,307	19,458
Coffee	21,327	15,007	10,219
Cocoa	43,777	17,058	12,933
Chocolate and prep	13,532	23,552	20,594
Tea, including herbal	19,620	25,790	14,498
Spices	1,655	1,855	1,967
Rubber, crude	771	1,390	1,941
Fibers, excluding cotton	122	11	25
Other hort products	202,575	238,224	197,377
Hops, including extract	11,645	15,332	16,448
Starches, not wheat/corn	671	549	508
Yeasts	8,878	5,780	4,792
Misc hort products	181,381	216,563	175,629
Nursery and greenhouse	3,547	4,914	6,332
Beverages, excluding juice	25,229	25,288	24,670

[1] Fiscal years Oct. 1–Sept. 30.
ERS, Market and Trade Economics Division, (202) 694–5257.

Table 15-7.—Agricultural imports for consumption: Value of Top 50 countries of origin, United States, fiscal years 2013–2015 [1]

Country	2013	2014	2015
	1,000 dollars	*1,000 dollars*	*1,000 dollars*
Canada	21,555,694	22,819,837	22,267,541
Mexico	17,218,508	18,863,749	20,632,874
European Union-28	17,329,270	18,744,268	19,691,000
Australia	2,720,569	3,343,874	4,652,453
China	4,459,021	4,343,127	4,330,323
Brazil	3,759,965	3,674,507	3,534,224
New Zealand	2,177,524	2,460,561	2,997,541
India	3,696,291	2,971,621	2,951,609
Indonesia	3,127,109	3,093,719	2,913,731
Chile	2,842,554	2,742,661	2,807,212
Colombia	2,142,268	2,358,266	2,471,774
Thailand	2,217,847	2,241,561	2,294,891
Guatemala	1,867,379	1,825,719	1,944,287
Vietnam	1,450,047	1,658,565	1,877,825
Peru	1,239,315	1,556,807	1,591,073
Costa Rica	1,469,319	1,552,097	1,501,925
Argentina	1,821,288	1,566,335	1,455,672
Switzerland	1,045,945	1,095,311	1,277,267
Malaysia	1,689,724	1,501,893	1,246,150
Ecuador	939,721	1,074,279	1,201,249
Philippines	1,002,441	1,098,989	1,174,146
Cote d'Ivoire	889,883	1,136,841	920,650
Turkey	708,800	784,324	736,006
Honduras	574,532	600,516	665,809
Nicaragua	431,199	503,699	607,412
Japan	577,819	555,840	549,959
Korea, South	413,919	441,767	468,928
Dominican Republic	355,690	367,590	424,663
Uruguay	273,569	264,720	389,800
Taiwan	321,416	331,940	345,998
Israel	323,314	344,422	337,590
South Africa	266,624	260,667	273,091
El Salvador	276,130	206,209	248,434
Ghana	212,410	210,975	216,778
Tunisia	132,657	78,115	190,737
Morocco	117,212	164,830	181,127
Sri Lanka	95,492	110,538	174,303
Ukraine	37,477	47,232	160,387
Bolivia	142,169	206,940	150,065
Ethiopia	113,591	114,115	146,939
Kenya	100,433	99,251	144,188
Madagascar	60,002	79,618	126,875
Pakistan	121,753	112,621	120,162
Singapore	106,371	101,801	117,299
Paraguay	185,551	170,688	111,301
Egypt	92,859	96,294	109,304
Jamaica	69,683	81,562	95,571
Hong Kong	91,036	84,589	95,249
Norway	79,074	84,817	84,007
Papua New Guinea	65,519	80,817	82,418
Other countries	863,734	909,516	936,320
Total U.S. Agricultural Imports [2]	103,871,717	109,220,600	114,026,137

[1] Fiscal years Oct. 1–Sept. 30. [2] Totals may not add due to rounding.
ERS, Market and Trade Economics Division, (202) 694–5257. Compiled from reports of the U.S. Department of Commerce.

Table 15-8.—European Union: Value of agricultural imports by origin, 2006–2015

Year [1]	United States	EU countries [2]	Other countries	Total
	Million dollars	Million dollars	Million dollars	Million dollars
2006	8,889	238,982	84,774	332,645
2007	10,279	285,430	106,527	402,236
2008	12,206	335,188	129,056	476,450
2009	8,653	295,743	107,350	411,746
2010	9,999	304,902	114,064	428,965
2011	11,916	361,756	143,326	516,998
2012	11,151	350,066	134,045	495,262
2013	13,699	382,260	134,698	530,657
2014	14,341	382,750	136,635	533,726
2015	13,555	330,931	123,088	467,574

[1] Data on calendar year basis. Users should use cautious interpretation on reports that include summarized reporter groupings. These groupings will only include the members that have reported data for any particular year.   [2] EU-28. Based on import data from the United Nations.

ERS, Market and Trade Economics Division, (202) 694–5257. Data Source: United National Commodity Trade Statistics, United Nations Statistics Division.

Table 15-9.—Fisheries: U.S. Commercial landings and value of principal species: 2007–2014 [1]

Species	Landings							
	2007	2008	2009	2010	2011	2012	2013	2014
	Million pounds	Million pounds	Million pounds	Million pounds	Million pounds	Million pounds	Million pounds	Million pounds
Fish:								
Cod, Atlantic	17	19	20	18	18	11	5	5
Flounder	483	663	575	625	707	703	717	713
Haddock	8	14	13	22	13	4	4	10
Halibut	70	67	60	56	43	34	30	23
Herring, sea	232	259	313	254	276	270	298	309
Jack mackerel	1	1	0	0	0	*	2	4
Menhaden	1,482	1,341	1,567	1,472	1,875	1,771	1,466	1,256
Ocean perch, Atlantic	2	3	3	4	4	8	7	10
Pollock	3,085	2,298	1,883	1,959	2,827	2,872	3,003	3,145
Salmon, Pacific	885	658	705	788	780	636	1,069	720
Tuna	51	48	49	48	50	60	56	59
Whiting (silver hake)	14	14	17	18	17	16	14	16
Shellfish:								
Clams (meats)	116	108	101	89	86	91	91	91
Crabs	294	325	326	350	369	367	332	295
Lobsters, American	81	82	96	115	126	150	149	148
Oysters (meats)	38	30	36	28	29	33	35	34
Scallops (meats)	59	54	58	57	59	57	41	34
Shrimp	281	257	301	259	313	303	283	295
	Value							
	Million dollars	Million dollars	Million dollars	Million dollars	Million dollars	Million dollars	Million dollars	Million dollars
Fish:								
Cod, Atlantic	27	31	25	28	33	22	10	9
Flounder	154	184	153	146	160	177	198	175
Haddock	12	16	14	22	16	8	6	11
Halibut	227	218	139	139	213	152	117	115
Herring, sea	35	45	56	44	38	49	49	42
Jack mackerel	93	91	98	107	144	128	129	117
Menhaden	(2)	(2)	(2)	(2)	(2)	(2)	(2)	(2)
Ocean perch, Atlantic	1	1	2	2	3	6	4	6
Pollock	306	334	281	282	363	343	406	400
Salmon, Pacific	381	395	370	555	618	489	757	617
Tuna	94	107	96	108	136	164	146	135
Whiting (silver hake)	8	8	9	11	11	10	9	11
Shellfish:								
Clams (meats)	194	187	191	201	187	193	209	215
Crabs	472	562	485	573	650	681	714	686
Lobsters, American	376	306	300	397	424	429	460	567
Oysters (meats)	140	132	137	118	132	155	193	240
Scallops (meats)	387	372	382	457	587	561	470	428
Shrimp	433	442	370	414	518	490	565	681

*Total is less than the weight threshhold.   [1] Landings are reported in round (live) weight for all items except univalve and bivalve mullusks such as clams, oysters, and scallops, which are reported in weight of meats (excluding the shell). Landings for Mississippi River drainage are not available.   [2] Less than $500,000.

2014 data are preliminary. Totals may not add due to rounding. Data do not include landings by U.S.-flag vessels at Puerto Rico or other ports outside the 50 States. Data do not include aquaculture products, except oysters and clams.

U.S. Department of Commerce, NOAA, NMFS, Fisheries Statistics Division, (301) 427–8103.

Table 15-10.—Fresh and frozen fishery products: Production and value, 2007–2014

Product	Production							
	2007	2008	2009	2010	2011	2012	2013[2]	2014
	Million pounds	Million pounds	Million pounds	Million pounds	Million pounds	Million pounds	Million pounds	Million pounds
Fish fillets [1]	632	656	511	585	775	692	783	786
Cod	32	39	36	49	66	64	72	77
Flounder	21	21	18	32	18	16	14	17
Haddock	11	9	14	23	26	11	11	14
Ocean perch, Atlantic	1	1	1	1	1	1	1	2
Rockfish	2	2	3	2	2	2	2	2
Pollock, Atlantic	2	3	3	2	2	2	2	2
Pollock, Alaska	401	364	277	290	461	415	473	479
Other	162	217	159	186	199	181	181	193
	Value							
	Million dollars	Million dollars	Million dollars	Million dollars	Million dollars	Million dollars	Million dollars	Million dollars
Fish fillets [1]	1,304	1,392	1,223	1,486	1,773	1,844	2,230	2,085
Cod	102	112	102	131	173	216	245	253
Flounder	69	69	57	53	55	51	62	62
Haddock	59	44	60	89	101	54	56	68
Ocean perch, Atlantic	3	3	3	3	5	4	5	5
Rockfish	6	4	6	6	6	8	5	7
Pollock, Atlantic	5	8	8	7	7	8	8	7
Pollock, Alaska	494	450	341	368	568	643	715	711
Other	566	702	646	829	858	861	1,134	972

[1] Fresh and frozen. [2] Data for 2013 has been revised.
U.S. Department of Commerce, NOAA, NMFS, Fisheries Statistics Division, (301) 427–8103.

Table 15-11.—Canned fishery products: Production and value, 2007–2014

Product	Production							
	2007	2008	2009	2010	2011	2012	2013	2014
	Million pounds	Million pounds	Million pounds	Million pounds	Million pounds	Million pounds	Million pounds	Million pounds
Tuna [1]	436	474	369	395	385	387	384	391
Salmon	142	124	142	146	148	120	203	89
Clam products [2]	110	105	100	110	106	72	73	77
Sardines, Maine	([4])	([4])	([4])	([4])	([4])	([4])	([4])	([4])
Shrimp [3]	([5])	([4])	([4])	([4])	([4])	([4])	([4])	([5])
Crabs	([5])	([5])	([5])	1	1	([5])	([5])	([5])
Oysters	([5])	([5])	([5])	([4])	([4])	([4])	([4])	([4])
Total [6]	1,070	1,316	934	956	947	881	964	733
	Value							
	Million dollars	Million dollars	Million dollars	Million dollars	Million dollars	Million dollars	Million dollars	Million dollars
Tuna [1]	702	845	756	724	769	886	852	783
Salmon	274	225	322	356	377	410	572	354
Clam products [2]	89	95	89	98	100	66	90	74
Sardines, Maine	([4])	([4])	([4])	([4])	([4])	([4])	([4])	([4])
Shrimp [4]	1	([4])	([4])	([4])	([4])	([4])	([4])	([5])
Crabs	([5])	([5])	([5])	9	([5])	2	([5])	([5])
Oysters	([5])	([5])	([5])	([4])	([4])	([4])	([4])	([4])
Total [6]	1,324	1,422	1,408	1,414	1,476	1,615	1,780	1,375

[1] Flakes included with chunk. [2] "Cut out" or "drained" weight of can contents are given for whole or minced clams, and net contents for other clam products. [3] Drained weight. [4] Confidential data. [5] Less than 500,000 pounds or $500,000. [6] Includes other products not shown separately.
U.S. Dept. of Commerce, NOAA, NMFS, Fisheries Statistics Division, (301) 427–8103.

Table 15-12.—Fisheries: Fishermen and craft, 1977, and catch, 2013–2014 by area

Area	1977 [1]			2013		2014	
	Fishermen	Fishing vessels	Fishing boats [2]	Total catch	Value	Total catch	Value
	1,000	Number	1,000	Million pounds	Million dollars	Million pounds	Million dollars
New England States	31.7	929	15.4	636	1,162	643	1,199
Middle Atlantic States	17.3	573	11.3	583	435	601	471
South Atlantic States	11.6	1,463	6.7	92	160	104	185
Gulf States	29.3	5,328	11.0	1,457	905	1,205	989
Pacific Coast States	54.0	7,643	15.4	7,051	2,672	6,884	2,481
Great Lakes States	1.2	217	0.5	19	23	16	21
Hawaii	2.7	101	1.3	32	108	33	101
United States	182.1	17,545	89.2	9,870	5,466	9,486	5,448

[1] Exclusive of duplication among regions. Computation of area amounts will not equal U.S. total. Mississippi River data included with total.   [2] Refers to craft having capacity of less than 5 net tons.   Note: Table may not add due to rounding.
U.S. Department of Commerce, NOAA, NMFS, Fisheries Statistics Division, (301) 427–8103.

Table 15-13.—Fisheries: Quantity and value of domestic catch, 2005–2014

Year	Quantity [1]			Ex-vessel value	Average price per lb.
	Total	For human food	For industrial products [2]		
	Million pounds	Million pounds	Million pounds	Million dollars	Cents
2005	9,707	7,997	1,710	3,942	40.6
2006	9,483	7,842	1,641	4,024	42.4
2007	9,309	7,490	1,819	4,192	45.0
2008	8,325	6,633	1,692	4,383	52.6
2009	8,031	6,198	1,833	3,891	48.4
2010	8,231	6,526	1,705	4,520	54.9
2011	9,858	7,909	1,949	5,289	53.7
2012	9,634	7,477	2,157	5,103	53.0
2013	9,870	8,043	1,827	5,466	55.4
2014	9,486	7,828	1,658	5,448	57.4

[1] Live weight.   [2] Meals, oil, fish solubles, homogenized condensed fish, shell products, bait, and animal food.
U.S. Department of Commerce, NOAA, NMFS, Fisheries Statistics Division, (301) 427–8103.

Table 15-14.—Fisheries: Disposition of domestic catch, 2010–2014 [1]

Disposition	2010	2011	2012	2013	2014
	Million pounds	Million pounds	Million pounds	Million pounds	Million pounds
Fresh and frozen	6,515	7,817	7,541	8,009	7,916
Canned	373	371	299	365	196
Cured	102	52	82	45	63
Reduced to meal, oil, etc	1,241	1,618	1,712	1,451	1,311
Total	8,231	9,858	9,634	9,870	9,486

[1] Live weight catch. In addition to whole fish, a large portion of waste (400–500 mil. lb.) derived from canning, filleting, and dressing fish and shellfish is utilized in production of fish meal and oil in each year shown.
U.S. Department of Commerce, NOAA, NMFS, Fisheries Statistics Division, (301) 427–8103.

Table 15-15.—Fishery products: Supply, [1] 2010–2014 [2]

Item	2010	2011	2012	2013	2014
	Million pounds	Million pounds	Milion pounds	Million pounds	Million pounds
For human food	17,560	18,732	18,066	18,572	18,733
Finfish	12,504	13,644	13,159	13,787	13,680
Shellfish [3]	5,056	5,088	4,906	4,786	5,053
For industrial use	2,188	2,374	2,692	2,416	2,317
Domestic catch	8,231	9,858	9,634	9,870	9,486
Percent of total	41.7	46.7	46.4	47	45
For human food	6,526	7,909	7,478	8,043	7,828
Finfish	5,216	6,540	6,163	6,777	6,587
Shellfish [3]	1,310	1,369	1,314	1,266	1,240
For industrial use	1,705	1,949	2,157	1,827	1,658
Imports [4]	11,517	11,248	11,123	11,118	11,564
Percent of total	58.3	53.3	53.6	53	55
For human food	11,034	10,823	10,588	10,529	10,905
Finfish	7,288	7,104	6,996	7,009	7,092
Shellfish [3]	3,746	3,719	3,592	3,520	3,813
For industrial use [5]	483	425	535	589	659
Total	19,748	21,106	20,757	20,988	21,050

[1] Supply totals are the domestic catch and imports without accounting for exports taken out. [2] Live weight, except percent. May not add due to rounding. [3] For univalve and bivalves mollusks (conchs, clams, oysters, scallops, etc.), the weight of meats, excluding the shell is reported. [4] Excluding imports of edible fishery products consumed in Puerto Rico; includes landings of tuna caught by foreign vessels in American Samoa. [5] Fish meal and sea herring.
U.S. Department of Commerce, NOAA, NMFS, Fisheries Statistics Division, (301) 427–8103.

Table 15-16.—Processed fishery products: Production and value, 2011–2014 [1]

Item	Production				Value			
	2011	2012	2013	2014	2011	2012	2013	2014
	Million pounds	Million pounds	Million pounds	Million pounds	Million pounds	Million pounds	Million pounds	Million pounds
Fresh and frozen:.								
Fillets	767	687	775	780	1,738	1,818	2,186	2,054
Steaks	8	5	8	6	35	26	44	31
Fish sticks	80	58	58	66	105	87	87	101
Fish portions	172	152	147	132	346	260	256	252
Breaded shrimp	92	80	109	105	241	194	311	312
Canned products [2]	947	880	964	733	1,476	1,615	1,780	1,375
Fish and shellfish	641	582	662	562	1,251	1,373	1,534	1,225
Animal feed	306	299	302	171	225	242	246	150
Industrial products	(X)	(X)	(X)	(X)	435	498	479	590
Meal and scrap	621	586	508	515	239	280	242	300
Oil (body and liver)	143	115	176	139	63	55	57	85
Other	(X)	(X)	(X)	(X)	134	162	180	206

(X) Not applicable.   [1] Includes cured fish.   [2] Includes salmon eggs for baits.
U.S. Department of Commerce, NOAA, NMFS, Fisheries Statistics Division, (301) 427–8103.

Table 15-17.—Selected fishery products: Imports and exports, 2010–2014 [1]

Product	Quantity				
	2010	2011	2012	2013	2014
	Million pounds	Million pounds	Million pounds	Million pounds	Million pounds
Imports					
Edible	5,447	5,349	5,383	5,513	5,562
Fresh and frozen	4,526	4,451	4,525	4,652	4,697
Salmon [2]	228	291	242	225	214
Tuna	426	303	316	412	317
Groundfish fillets, blocks [3]	312	331	318	328	318
Other fillets and steaks	1,112	1,135	1,236	1,293	1,340
Scallops (meats)	50	55	34	60	59
Lobster, American and spiny	100	102	106	111	119
Shrimp and prawn	1,228	1,265	1,173	1,108	1,244
Canned	770	752	685	682	688
Sardines, in oil	21	20	25	26	27
Sardines and herring, not in oil	46	51	46	38	44
Tuna	442	413	354	347	342
Oysters	12	15	9	10	9
Pickled or salted	51	51	51	58	52
Cod, haddock, hake, pollock, cusk	6	4	7	8	8
Nonedible scrap and meal	86	76	96	105	118
Exports					
Canned salmon	91	112	91	100	95
Fish oil, nonedible	175	149	93	152	177

Product	Value				
	2010	2011	2012	2013	2014
	Million dollars	Million dollars	Million dollars	Million dollars	Million dollars
Imports					
Edible	14,811	16,618	16,690	18,102	20,246
Fresh and frozen	12,820	14,411	14,225	15,723	17,810
Salmon [2]	652	633	627	712	699
Tuna	680	578	749	841	709
Groundfish fillets, blocks [3]	560	648	581	657	1,711
Other fillets and steaks	3,107	3,466	3,637	3,932	4,440
Scallops (meats)	233	294	220	366	389
Lobster, American and spiny	871	900	895	928	1,079
Shrimp and prawn	4,272	5,148	4,442	5,240	6,656
Canned	1,581	1,781	1,923	1,838	1,890
Sardines, in oil	37	39	61	69	68
Sardines and herring, not in oil	62	70	63	56	64
Tuna	660	720	762	762	667
Oysters	30	44	28	30	26
Pickled or salted	86	87	95	106	95
Cod, haddock, hake, pollock, cusk	14	11	19	22	19
Nonedible scrap and meal	56	48	56	73	87
Exports					
Canned salmon	179	224	222	229	208
Fish oil, nonedible	96	103	100	147	166

[1] Includes Puerto Rico.   [2] Excludes fillets.   [3] Includes cod, cusk, haddock, hake, pollock, ocean perch, and whiting.
U.S. Dept. of Commerce, NOAA, NMFS, Fisheries Statistics Division (301) 427–8103.

Table 15-18.—Fishery products: Imports and exports, 2005–2014 [1]

Year	Imports [2]				Exports			
	Total value	Edible products		Non-edible, value	Total value	Edible products		Non-edible, value
		Quantity	Value			Quantity	Value	
	Million dollars	*Million pounds*	*Million dollars*	*Million dollars*	*Million dollars*	*Million pounds*	*Million dollars*	*Million dollars*
2005	25,120	5,115	12,099	13,021	15,431	2,929	4,074	11,357
2006	27,712	5,401	13,355	14,357	17,760	2,967	4,238	13,522
2007	28,777	5,346	13,696	15,081	20,054	2,869	4,269	15,785
2008	28,457	5,226	14,171	14,286	23,367	2,650	4,257	19,110
2009	23,554	5,161	13,124	10,430	19,636	2,546	3,980	15,656
2010	27,388	5,447	14,811	12,580	22,386	2,733	4,389	17,997
2011	30,943	5,349	16,618	14,326	26,183	3,265	5,442	20,742
2012	31,108	5,384	16,691	14,417	27,388	3,254	5,470	21,917
2013	33,254	5,514	18,102	15,151	29,116	3,324	5,584	23,533
2014	35,882	5,562	20,247	15,635	29,970	3,402	5,753	24,217

[1] Includes Puerto Rico. [2] Includes landings of tuna by foreign vessels in American Samoa.
U.S. Department of Commerce, NOAA, NMFS, Fisheries Statistics Division, (301) 427–8103.

Table 15-19.—Fish trips: Estimated number of fishing trips taken by marine recreational fishermen by subregion and year, Atlantic, Gulf, and Pacific Coasts, 2011–2014

Subregion	2011	2012	2013	2014
	Thousands	*Thousands*	*Thousands*	*Thousands*
Atlantic and Gulf:				
North Atlantic	6,059	6,164	6,286	6,651
Mid-Atlantic	15,974	14,434	14,217	14,347
South Atlantic [1]	17,676	17,793	16,618	17,646
Gulf	23,701	24,331	26,384	19,897
Total	63,410	62,722	63,505	58,541

Subregion	2011	2012	2013	2014
	Thousands	*Thousands*	*Thousands*	*Thousands*
Pacific: [2]				
Southern California	2,011	3,857	3,784	3,724
Northern California	1,691	1,579	1,592	1,515
Oregon	148	173	196	140
Washington	104	113	109	65
Hawaii	1,382	1,520	1,513	1,375
Alaska	491	473	595	583
Total	5,827	7,715	7,789	7,402

[1] Does not include trips from headboats (party boats) in the South Atlantic or Gulf of Mexico. [2] Pacific state estimates do not include salmon data collected by recreational surveys.
U.S. Department of Commerce, NOAA, NMFS, Fisheries Statistics Division, (301) 427–8103.

Table 15-20.—Fish harvested: Estimated number of fish harvested by marine recreational anglers by subregion and year, Atlantic, Gulf Coasts, and Pacific Coasts, 2011–2014

Subregion	2011	2012	2013	2014
	Thousands	Thousands	Thousands	Thousands
Atlantic and Gulf:				
North Atlantic	11,262	11,099	13,064	13,088
Mid-Atlantic	21,451	22,015	25,851	25,262
South Atlantic [1]	32,888	29,632	36,211	35,309
Gulf	62,012	65,015	82,018	61,488
Total	127,613	127,761	157,144	135,147
Pacific: [2]				
Southern California	4,540	5,678	5,120	5,188
Northern California	3,795	2,468	2,961	3,144
Oregon	367	455	490	386
Washington	294	302	310	209
Hawaii	2,503	2,763	3,651	3,720
Alaska	1,314	1,203	1,572	1,471
Total	12,813	12,869	14,104	14,118

[1] Does not include trips from headboats (party boats) in the South Atlantic or Gulf of Mexico.   [2] Data do not include recreational trips in Hawaii or Alaska.   Pacific state estimates do not include salmon data collected by recreational surveys.   Note: "Harvested" includes dead discards and fish used for bait but does not include fish released alive; totals may not match due to rounding.
U.S. Department of Commerce, NOAA, NMFS, Fisheries Statistics Division, (301) 427–8103.

Table 15-21.—Fish harvested: Estimated number of fish harvested by marine recreational anglers by mode and year, Atlantic, Gulf Coasts, and Pacific Coasts, 2011–2014

Mode .	2011	2012	2013	2014
	Thousands	Thousands	Thousands	Thousands
Atlantic and Gulf: [1]				
Shore	37,916	40,258	49,377	49,747
Party/charter [2]	9,789	10,921	12,213	10,582
Private/rental	79,908	76,582	95,554	74,818
Total	127,613	127,761	157,144	135,147
Pacific: [2]				
Shore	6,373	5,426	5,794	5,077
Party/charter	2,924	3,267	3,487	4,051
Private/rental	2,202	2,973	3,251	3,519
Total	11,499	11,666	12,532	12,647

[1] Does not include trips from headboats (party boats) in the South Atlantic or Gulf of Mexico.   [2] Data do not include recreational trips in Alaska.   Pacific state estimates do not include salmon data collected by recreational surveys.
Note: "Harvested" includes dead discards and fish used for bait but does not include fish released alive; totals may not match due to rounding.
U.S. Department of Commerce, NOAA, NMFS, Fisheries Statistics Division, (301) 427–8103.

Table 15-22.—Fish harvested: Estimated number of fish harvested by marine recreational anglers by species group and year, Atlantic and Gulf coasts, 2011–2014 [1]

Species group	2011	2012	2013	2014
	Thousands	*Thousands*	*Thousands*	*Thousands*
Atlantic Cod	580	338	391	282
Atlantic Croaker	7,315	7,193	9,438	8,663
Atlantic Mackerel	5,336	3,282	3,717	3,273
Black Drum	1,238	1,126	1,431	617
Black Sea Bass	1,513	2,408	1,716	2,626
Blue Runner	1,268	998	2,955	3,190
Conger Eels	4	1	22	1
Crevalle Jack	175	319	756	714
Cunner	45	23	99	71
Epinephelus Groupers	123	322	480	294
Florida Pompano	308	385	858	395
Freshwater Catfishes	504	708	461	623
Gray Snapper	731	1,297	2,230	2,558
Greater Amberjack	62	90	93	91
Gulf Flounder	229	335	364	327
King Mackerel	342	443	400	506
Kingfishes	5,543	5,978	7,120	6,741
Lane Snapper	64	206	368	372
Little Tunny/Atlantic Bonito	261	374	319	307
Moray Eels	-	-	-	-
Mycteroperca Groupers	140	188	278	156
Pigfish	840	876	697	716
Pinfishes	4,546	5,320	5,064	6,193
Pollock	410	208	568	376
Red Drum	4,385	3,518	4,896	1,473
Red Hake	224	76	103	187
Red Porgy	308	274	537	479
Red Snapper	557	633	1,308	519
Saltwater Catfishes	557	1,034	843	309
Sand Seatrout	6,224	5,191	3,344	2,134
Scup	3,058	3,670	5,040	4,400
Sheepshead	2,937	2,126	1,971	1,874
Silver Perch	195	504	147	270
Skates/Rays	69	44	72	71
Southern Flounder	1,331	1,249	1,503	655
Spanish Mackerel	2,454	2,672	4,472	2,592
Spot	6,004	4,759	8,156	8,712
Spotted Seatrout	15,827	15,413	13,883	3,515
Striped Bass	2,278	1,492	2,194	1,778
Striped Mullet	4,203	4,882	3,545	2,686
Summer Flounder	1,844	2,277	2,530	2,456
Tautog	430	495	539	1,040
Vermilion Snapper	664	365	835	940
Weakfish	36	236	137	84
White Grunt	1,509	1,825	2,184	2,371
White Perch	2,124	1,907	2,582	1,267
Winter Flounder	192	98	51	135
Yellowtail Snapper	282	439	818	820
Other Barracudas	59	93	84	134
Other Bluefish	5,219	5,640	6,021	6,095
Other Cods/Hakes	234	233	312	267
Other Dolphins	1,271	1,132	1,151	985
Other Drum	162	60	394	245
Other Eels	7	38	11	7
Other Flounders	180	108	123	151
Other Grunts	250	360	960	851
Other Herrings	20,698	22,165	31,202	32,315
Other Jacks	781	1,404	2,021	1,768
Other Mullets	4,185	4,175	6,419	4,788
Other Porgies	214	313	344	408
Other Puffers	1,197	708	494	123
Other Sculpins	-	-	3	3
Other Sea Basses	146	132	141	344
Other Searobins	110	120	354	137
Other Sharks	193	158	269	218
Other Snappers	75	124	142	159
Other Temperate Basses	-	-	-	-
Other Toadfishes	8	17	42	37
Other Triggerfishes/Filefishes	267	266	334	324
Other Tunas/Mackerels	383	545	595	485
Other Wrasses	69	203	174	193
Other fishes	2,636	2,170	4,039	5,251
Total [2]	127,613	127,761	157,144	135,147

- Represents zero.   [1] Data do not include headboats (party boats) in the South Atlantic and the Gulf of Mexico.   [2] Totals may not add due to rounding.   Note: "Harvested" includes dead discards and fish used for bait but does not include fish released alive.

U.S. Department of Commerce, NOAA, NMFS, Fisheries Statistics Division (301) 427–8103.

Table 15-23.—Fish harvested: Estimated number of fish harvested by marine recreational anglers by species group and year, Pacific coast, 2011–2014 [1]

Species group	2011	2012	2013	2014
	Thousands	*Thousands*	*Thousands*	*Thousands*
Albacore	-	-	-	1
Barred Sand Bass	238	152	64	69
Barred Surfperch	340	544	368	566
Bigeye Scad	656	347	810	731
Bigeye Trevally	1	2	3	5
Bigscale Soldierfish	2	8	11	24
Black Perch	66	42	33	24
Black Rockfish	621	718	1,022	771
Blackspot Sergeant	8	29	12	13
Blacktail Snapper	29	31	23	16
Blue Rockfish	177	161	272	328
Bluefin Trevally	76	60	88	107
Bluestripe Snapper	17	22	21	87
Bocaccio	164	209	188	185
Brown Rockfish	146	132	138	220
Cabezon	30	32	29	31
California Corbina	-	9	6	6
California Halibut	25	38	24	23
California Sheephead	31	31	48	40
Canary Rockfish	41	42	37	45
Chilipepper Rockfish	23	38	30	54
Chub Mackerel	1,115	848	576	1,040
Conger Eels	-	6	3	3
Convict Tang	138	141	109	65
Copper Rockfish	89	118	158	153
Giant Trevally	20	36	34	28
Goldring Surgeonfish	54	136	95	123
Gopher Rockfish	180	133	95	129
Greater Amberjack	-	3	2	3
Green Jobfish	18	30	8	19
Greenspotted Rockfish	49	52	32	29
Halfmoon	25	26	37	21
Hawaiian Flagtail	67	105	143	110
Hawaiian Hogfish	2	6	4	8
Highfin Rudderfish	6	21	7	6
Island Jack	1	8	10	9
Jacksmelt	366	239	248	200
Kawakawa	2	6	5	46
Kelp Bass	131	132	55	126
Kelp Greenling	55	35	35	25
Lingcod	192	229	279	295
Mackerel Scad	6	260	79	166
Manybar Goatfish	15	41	23	43
Moray Eels	-	6	8	3
Northern Anchovy	207	54	356	179
Olive Rockfish	39	67	49	77
Opaleye	11	41	31	40
Pacific Barracuda	46	50	18	27
Pacific Bonito	2	-	9	165
Pacific Cod	48	42	38	61
Pacific Hake	-	-	-	-
Pacific Herring	49	184	128	39
Pacific Tomcod	-	-	-	-
Pile Perch	5	10	8	5
Pink Snapper	26	47	46	42
Queenfish	40	66	34	22
Quillback Rockfish	11	15	7	4
Razorfishes	14	94	63	32
Redtail Surfperch	43	50	39	45
Rock Sole	-	1	-	-
Sanddabs	537	441	603	890
Shiner Perch	92	72	59	113
Silver Surfperch	28	16	21	30
Skates/Rays	2	6	8	5
Skipjack Tuna	125	197	380	199
Smallmouth Bonefish	14	27	23	30
Spiny Dogfish	2	-	1	-
Spotted Sand Bass	10	23	5	3
Squirrel Fishes	1	3	1	3
Starry Flounder	-	-	-	1
Striped Bass	30	14	7	29
Striped Mullet	13	16	34	27

See footnote(s) at end of table.

Table 15-23.—Fish harvested: Estimated number of fish harvested by marine recreational anglers by species group and year, Pacific coast, 2011–2014[1]—Continued

Species group	2011	2012	2013	2014
	Thousands	Thousands	Thousands	Thousands
Striped Seaperch	36	25	34	36
Surf Smelt	1,278	4	-	6
Unicornfishes	183	10	5	12
Wahoo	16	31	36	43
Walleye Surfperch	90	149	144	70
White Croaker	48	86	73	78
White Seaperch	10	17	7	12
Whitemouth Trevally	-	-	-	-
Whitesaddle Goatfish	7	11	4	7
Whitetip Soldierfish	6	3	3	-
Widow Rockfish	2	10	40	36
Yellowfin Tuna	141	182	150	219
Yellowstripe Goatfish	112	96	791	379
Yellowtail	-	13	16	159
Yellowtail Rockfish	173	168	171	182
Other Anchovies	135	61	19	1
Other Barracudas	5	9	2	6
Other Cods/Hakes	-	-	-	-
Other Damselfishes	5	27	20	16
Other Dolphins	63	163	94	92
Other Drum	43	140	92	58
Other Flounders	409	409	477	419
Other Goatfishes	38	11	54	109
Other Greenlings	1	14	1	8
Other Groupers	4	10	10	16
Other Herrings	470	855	634	89
Other Jacks	76	40	38	56
Other Mullets	13	9	39	48
Other Rockfishes	1,119	1,409	1,559	1,521
Other Sablefishes	10	18	18	12
Other Scorpionfishes	197	261	245	273
Other Sculpins	39	13	14	3
Other Sea Basses	-	-	-	1
Other Sea Chubs	5	16	33	39
Other Sharks	6	8	16	6
Other Silversides	74	195	207	221
Other Smelts	43	94	50	17
Other Snappers	24	67	55	58
Other Soldierfishes	3	2	45	54
Other Sturgeons	1	-	-	-
Other Surfperches	107	98	89	96
Other Surgeonfishes	28	68	63	77
Other Tunas/Mackerels	48	122	98	145
Other Wrasses	6	29	23	16
Other fishes	1,122	1,116	1,395	1,358
Total[2]	12,813	12,869	14,104	14,118

- Represents zero.　[1] Pacific estimates do not include salmon data collected by state recreational surveys.　[2] Totals may not add exactly due to rounding.

Note: "Harvested" includes dead discards and fish used for bait but does not include fish released alive.

U.S. Department of Commerce, NOAA, NMFS, Fisheries Statistics Division, (301) 427–8103.

Table 15-24.—Fish harvested: Estimated number of fish harvested by marine recreational anglers, by area of fishing and year, Atlantic and Gulf and Pacific Coast, 2011–2014

Area	2011	2012	2013	2014
	Thousands	*Thousands*	*Thousands*	*Thousands*
Atlantic and Gulf: [1]				
Inland	85,798	84,359	95,700	81,158
State Territorial Sea [2]	31,917	32,318	46,282	39,338
Federal Exclusive Ecomomic Zone [3]	9,898	11,084	15,162	14,651
Total	127,613	127,761	157,144	135,147
Pacific: [4]				
Inland	756	1,135	1,065	1,207
State Territorial Sea [2]	9,167	8,500	9,460	9,155
Federal Exclusive Ecomomic Zone [3]	1,576	2,031	2,007	2,285
Total	11,499	11,666	12,532	12,647

[1] Data does not include trips from headboats (party boats) in the South Atlantic or the Gulf of Mexico. [2] Open Ocean extending 0 to 3 miles from shore, except West Florida (10 miles). [3] Open ocean extending to 200 miles offshore from the outer edge of the State Territorial Seas. [4] Data does not include recreational catch from Alaska. Pacific estimates do not include salmon data collected by recreational surveys. Note: "Harvested" includes dead discards and fish used for bait but does not include fish released alive.
U.S. Department of Commerce, NOAA, NMFS, Fisheries Statistics Division, (301) 427–8103.

Table 15-25.—Catfish production: Water surface acre usage by State and United States, 2015–2016

State	Acres intended for utilization during January 1-June 30					Acres taken out of production during Jul 1-Dec 31 prev. year
	Foodsize	Fingerlings	Broodfish	Currently under or scheduled for:		
				Renovation	New construction	
	Acres	*Acres*	*Acres*	*Acres*	*Acres*	*Acres*
2015						
Alabama	14,900	250	85	480	65	240
Arkansas	4,200	1,200	195	95	(D)	(D)
California	1,100	185	100	40	(D)	(D)
Mississippi	28,800	7,000	1,500	3,400	(D)	960
North Carolina	1,200	135	35	70	(D)	(D)
Texas	1,800	100	70	105	30	500
Other States [1]	1,510	550	170	80	80	575
United States	53,510	9,420	2,155	4,270	175	2,275
2016						
Alabama	14,500	260	30	260	(D)	310
Arkansas	3,300	1,400	230	270	20	175
California	1,100	190	60	135	-	(D)
Mississippi	26,700	6,200	1,300	2,300	65	710
North Carolina	1,100	70	(D)	65	(D)	(D)
Texas	1,800	80	100	140	(D)	690
Other States [1]	935	475	190	100	115	215
United States	49,435	8,675	1,910	3,270	200	2,100

- Represents zero.   (D) Withheld to avoid disclosing data for individual operations.   [1] Other States include State estimates not shown and States suppressed due to disclosure.
NASS, Livestock Branch, (202) 720–3570.

Table 15-26.—Catfish: Number of operations and water surface acres used for production, 2015–2016, and total sales, 2014–2015, by State and United States

State	Water surface acres used for production during Jan 1 - July 30		Total sales	
	2015	2016	2014	2015
	Acres	Acres	1,000 dollars	1,000 dollars
Alabama	15,300	15,000	104,468	108,866
Arkansas	5,800	5,000	21,295	18,456
California	1,400	1,400	8,108	5,774
Mississippi	41,000	36,100	189,540	201,450
North Carolina	1,500	1,300	3,985	3,878
Texas	2,000	2,000	16,656	18,123
Other States [1]	2,910	1,740	7,888	4,911
United States	69,910	62,540	351,940	361,458

[1] Other States include State estimates not shown and States suppressed due to disclosure.
NASS, Livestock Branch, (202) 720–3570.

Table 15-27.—Catfish: Sales by size category, by State and United States, 2014–2015

Size category and State	Number of fish		Live weight		Sales			
					Total		Average price per pound	
	2014	2015	2014	2015	2014	2015	2014	2015
	1,000	1,000	1,000 pounds	1,000 pounds	1,000 dollars	1,000 dollars	Dollars	Dollars
Foodsize:								
Alabama	63,400	63,000	105,300	107,500	104,247	108,575	0.99	1.01
Arkansas	9,980	9,560	17,200	14,500	18,232	15,225	1.06	1.05
California	1,340	1,050	2,550	2,200	7,982	5,632	3.13	2.56
Mississippi	97,900	110,000	161,500	171,700	176,035	190,587	1.09	1.11
North Carolina	1,990	2,120	3,400	3,250	3,978	3,803	1.17	1.17
Texas	7,050	7,880	14,300	15,600	16,445	17,940	1.15	1.15
Other States [1]	2,940	1,590	3,248	1,944	5,463	3,271	1.68	1.68
United States	184,600	195,200	307,498	316,694	332,382	345,033	1.08	1.09
Stockers:								
Alabama	(D)	(D)	(D)	(D)	(D)	(D)	(D)	(D)
Arkansas	(D)	(D)	(D)	(D)	(D)	(D)	(D)	(D)
California	-	(D)	-	(D)	-	(D)	-	(D)
Mississippi	47,300	33,400	4,800	3,650	5,808	4,709	1.21	1.29
North Carolina	(D)	(D)	(D)	(D)	(D)	(D)	(D)	(D)
Texas	(D)	(D)	(D)	(D)	(D)	(D)	(D)	(D)
Other States [1]	23,380	25,575	2,354	2,614	3,114	3,489	1.32	1.33
United States	70,680	58,975	7,154	6,264	8,922	8,198	1.25	1.31
Fingerlings and fry:								
Alabama	2,450	690	41	15	221	115	5.40	7.65
Arkansas	8,100	9,870	278	262	798	383	2.87	1.46
California	860	670	10	9	76	63	7.60	6.95
Mississippi	141,000	101,000	4,800	3,600	7,440	5,724	1.55	1.59
North Carolina	(D)	(D)	(D)	(D)	(D)	(D)	(D)	(D)
Texas	(D)	(D)	(D)	(D)	(D)	(D)	(D)	(D)
Other States [1]	22,620	16,940	677	603	1,740	1,407	2.57	2.33
United States	175,030	129,170	5,806	4,489	10,275	7,692	1.77	1.71
Broodfish:								
Alabama	(D)	(D)	(D)	(D)	(D)	(D)	(D)	(D)
Arkansas	(D)	(D)	(D)	(D)	(D)	(D)	(D)	(D)
California	(D)	(D)	(D)	(D)	(D)	(D)	(D)	(D)
Mississippi	(D)	(D)	(D)	(D)	(D)	(D)	(D)	(D)
North Carolina	-	-	-	-	-	-	-	-
Texas	-	(D)	-	(D)	-	(D)	-	(D)
Other States [1]	50	65	283	342	361	535	1.28	1.56
United States	50	65	283	342	361	535	1.28	1.56

- Represents zero.　(D) Withheld to avoid disclosing data for individual operations.　[1] Other States include State estimates not shown and States suppressed due to disclosure.
NASS, Livestock Branch, (202) 720–0585.

Table 15-28.—Trout: Value of fish sold and distributed, by State (excluding eggs), and United States (including and excluding eggs), 2014–2015

State	Total value of fish sold		Total value of distributed fish	
	2014	2015	2014	2015
	1,000 dollars	*1,000 dollars*	*1,000 dollars*	*1,000 dollars*
Arkansas	-	-	3,122	2,612
California	4,990	5,390	12,794	7,761
Colorado	1,666	2,045	7,618	8,918
Georgia	455	483	1,259	1,075
Idaho	53,118	49,362	8,972	10,198
Michigan	619	1,347	966	992
Missouri	2,250	2,518	2,365	3,342
New York	631	636	3,149	2,835
North Carolina	7,888	8,469	(D)	(D)
Oregon	(D)	(D)	5,336	9,476
Pennsylvania	5,571	6,122	6,124	10,317
Utah	604	630	7,550	7,706
Virginia	1,481	1,610	2,087	1,464
Washington	(D)	(D)	20,735	19,453
West Virginia	1,233	1,052	(D)	(D)
Wisconsin	1,537	1,462	2,007	2,837
Other States [1]	21,173	23,267	28,155	28,558
United States				
Value excluding eggs	103,216	104,393	112,239	117,544
Value including eggs	111,990	113,099	(NA)	(NA)

- Represents zero. (D) Withheld to avoid disclosing data for individual operations. (NA) Not available. [1] Other States include State estimates not listed and States supressed due to disclosure.
NASS, Livestock Branch, (202) 720–3570.

Table 15-29.—Trout: Egg Sales, United States, 2012–2015

Year	Number of Eggs	Average Price per 1,000 Eggs	Total Sales [1]
	1,000	*Dollars*	*1,000 dollars*
2012	434,090	19.50	8,469
2013	445,805	19.40	8,639
2014	427,125	20.50	8,774
2015	431,475	20.20	8,706

[1] Total sales may not add due to rounding.
NASS, Livestock Branch, (202) 720-3570.

Table 15-30.—Trout: Sales by size category, by State and United States, 2014–2015

Size category and State	Number of fish		Live weight		Sales			
			Total[1]		Total[2]		Average price per pound	
	2014	2015	2014	2015	2014	2015	2014	2015
	1,000	1,000	1,000 pounds	1,000 pounds	1,000 dollars	1,000 dollars	Dollars	Dollars
12 inch or longer:								
Arkansas	-	-	-	-	-	-	-	-
California	1,310	1,350	1,550	1,600	4,697	5,056	3.03	3.16
Colorado	210	310	404	405	1,495	1,770	3.70	4.37
Georgia	(D)	(D)	(D)	(D)	(D)	(D)	(D)	(D)
Idaho	36,100	32,700	42,200	39,100	52,750	48,875	1.25	1.25
Michigan	105	(D)	119	(D)	458	(D)	3.85	(D)
Missouri	780	730	688	767	1,686	1,818	2.45	2.37
New York	(D)	(D)	(D)	(D)	(D)	(D)	(D)	(D)
North Carolina	3,310	3,220	4,050	3,700	6,885	7,807	1.70	2.11
Oregon	(D)	(D)	(D)	(D)	(D)	(D)	(D)	(D)
Pennsylvania	880	910	1,030	1,050	4,408	5,156	4.28	4.91
Utah	130	90	161	113	531	444	3.30	3.93
Virginia	480	590	489	567	1,213	1,435	2.48	2.53
Washington	(D)	(D)	(D)	(D)	(D)	(D)	(D)	(D)
West Virginia	420	550	518	464	1,166	984	2.25	2.12
Wisconsin	410	440	403	414	1,394	1,399	3.46	3.38
Other States[3]	4,150	4,460	9,121	9,767	19,058	21,693	2.09	2.22
United States	48,285	45,350	60,733	57,947	95,741	96,437	1.58	1.66
6 inch-12 inch:								
Arkansas	-	-	-	-	-	-	-	-
California	(D)	(D)	(D)	(D)	(D)	(D)	(D)	(D)
Colorado	(D)	(D)	(D)	(D)	(D)	(D)	(D)	(D)
Georgia	(D)	(D)	(D)	(D)	(D)	(D)	(D)	(D)
Idaho	(D)	(D)	(D)	(D)	(D)	(D)	(D)	(D)
Michigan	(D)	(D)	(D)	(D)	(D)	(D)	(D)	(D)
Missouri	460	(D)	177	(D)	460	(D)	2.60	(D)
New York	80	80	37	37	170	194	4.60	5.25
North Carolina	(D)	(D)	(D)	(D)	(D)	(D)	(D)	(D)
Oregon	(D)	240	(D)	91	(D)	376	(D)	4.13
Pennsylvania	510	420	190	176	1,045	906	5.50	5.15
Utah	(D)	(D)	(D)	(D)	(D)	(D)	(D)	(D)
Virginia	(D)	(D)	(D)	(D)	(D)	(D)	(D)	(D)
Washington	1,890	1,630	470	605	1,208	1,779	2.57	2.94
West Virginia	(D)	(D)	(D)	(D)	(D)	(D)	(D)	(D)
Wisconsin	70	(D)	36	(D)	138	(D)	3.84	(D)
Other States[3]	2,155	2,530	871	924	2,837	3,167	3.26	3.43
United States	5,165	4,900	1,781	1,833	5,858	6,422	3.29	3.50

Size category and State	Number of fish		Live weight		Sales			
			Total[1]		Total[2]		Average price per 1,000 fish	
	2014	2015	2014	2015	2014	2015	2014	2015
	1,000	1,000	1,000 pounds	1,000 pounds	1,000 dollars	1,000 dollars	Dollars	Dollars
1 inch-6 inch:								
Arkansas	-	-	-	-	-	-	-	-
California	(D)	(D)	(D)	(D)	(D)	(D)	(D)	(D)
Colorado	(D)	(D)	(D)	(D)	(D)	(D)	(D)	(D)
Georgia	(D)	(D)	(D)	(D)	(D)	(D)	(D)	(D)
Idaho	(D)	(D)	(D)	(D)	(D)	(D)	(D)	(D)
Michigan	(D)	(D)	(D)	(D)	(D)	(D)	(D)	(D)
Missouri	260	(D)	16	(D)	104	(D)	400	(D)
New York	(D)	(D)	(D)	(D)	(D)	(D)	(D)	(D)
North Carolina	(D)	(D)	(D)	(D)	(D)	(D)	(D)	(D)
Oregon	40	(D)	2	(D)	17	(D)	421	(D)
Pennsylvania	350	145	9	6	118	60	336	416
Utah	(D)	(D)	(D)	(D)	(D)	(D)	(D)	(D)
Virginia	(D)	(D)	(D)	(D)	(D)	(D)	(D)	(D)
Washington	(D)	(D)	(D)	(D)	(D)	(D)	(D)	(D)
West Virginia	(D)	(D)	(D)	(D)	(D)	(D)	(D)	(D)
Wisconsin	45	(D)	1	(D)	5	(D)	110	(D)
Other States[3]	7,655	8,325	178	218	1,373	1,474	179	177
United States	8,350	8,470	206	224	1,617	1,534	194	181

- Represents zero.　(D) Withheld to avoid disclosing data for individual operations.　[1] Due to rounding, total number of fish multiplied by the average pounds per unit may not exactly equal total live weight.　[2] Due to rounding, total number or live weight multiplied by average value per unit may not exactly equal total sales.　[3] Other States include State estimates not listed and States supressed due to disclosure.

NASS, Livestock Branch, (202) 720-3570.

Table 15-31.—General storages: Gross and usable cooler and freezer space, by State and United States, October 1, 2015 [1]

State	Cooler		Refrigerated		Total	
	Gross	Usable	Gross	Usable	Gross	Usable
	1,000 Cubic Feet					
Alabama	2,591	2,307	33,948	28,647	36,539	30,954
Alaska	945	777	2,700	2,318	3,645	3,095
Arizona	3,735	2,895	15,007	12,268	18,742	15,164
Arkansas	(D)	(D)	(D)	(D)	92,854	80,114
California	222,017	177,718	347,919	290,112	569,936	467,830
Colorado	(D)	(D)	(D)	(D)	28,251	23,284
Connecticut	(D)	(D)	(D)	(D)	6,018	5,099
Delaware	(D)	(D)	(D)	(D)	30,249	23,164
Florida	108,106	82,011	177,493	148,112	285,598	230,123
Georgia	61,467	49,664	188,738	156,185	250,205	205,849
Hawaii	(D)	(D)	(D)	(D)	(D)	(D)
Idaho	(D)	(D)	(D)	(D)	60,728	49,231
Illinois	27,944	23,833	172,350	137,973	200,294	161,806
Indiana	(D)	(D)	(D)	(D)	113,995	95,711
Iowa	22,094	16,519	69,155	57,681	91,249	74,199
Kansas	7,588	4,913	39,972	29,451	47,560	34,364
Kentucky	1,710	1,534	22,694	19,389	24,404	20,923
Louisiana	943	812	14,741	10,642	15,684	11,454
Maine	(D)	(D)	(D)	(D)	9,729	7,182
Maryland	4,758	4,198	34,770	29,256	39,528	33,454
Massachusetts	12,432	10,147	82,595	71,039	95,028	81,185
Michigan	12,355	9,900	89,543	72,943	101,898	82,842
Minnesota	23,898	16,407	69,665	57,082	93,563	73,489
Mississippi	(D)	(D)	(D)	(D)	15,630	12,123
Missouri	24,185	20,467	87,913	76,713	112,098	97,180
Montana	507	408	724	578	1,231	986
Nebraska	5,219	2,808	49,028	38,650	54,247	41,459
Nevada	(D)	(D)	(D)	(D)	(D)	8,835
New Hampshire	(D)	(D)	(D)	(D)	10,552	8,432
New Jersey	58,794	51,734	108,728	94,475	167,522	146,209
New Mexico	(D)	(D)	(D)	(D)	5,336	3,711
New York	34,917	28,290	67,772	57,653	102,689	85,943
North Carolina	4,303	3,236	59,729	42,954	64,032	46,190
North Dakota	(D)	(D)	(D)	(D)	10,325	7,800
Ohio	6,067	5,054	73,514	61,265	79,581	66,318
Oklahoma	(D)	(D)	(D)	(D)	14,498	12,103
Oregon	14,713	12,253	119,656	98,365	134,369	110,618
Pennsylvania	43,848	35,149	198,097	174,525	241,945	209,674
Rhode Island	(D)	(D)	(D)	(D)	(D)	(D)
South Carolina	(D)	(D)	(D)	(D)	27,738	21,628
South Dakota	(D)	(D)	(D)	(D)	11,673	7,125
Tennessee	(D)	(D)	(D)	(D)	67,732	60,156
Texas	62,338	53,156	190,371	152,460	252,709	205,616
Utah	12,038	10,675	34,492	27,400	46,530	38,075
Vermont	(D)	(D)	(D)	(D)	3,683	2,214
Virginia	17,752	13,993	60,655	53,071	78,406	67,064
Washington	23,474	18,905	187,735	151,115	211,209	170,020
West Virginia	(D)	(D)	(D)	(D)	(D)	(D)
Wisconsin	91,682	71,506	130,005	109,866	221,687	181,372
Wyoming	-	-	-	-	-	-
United States	973,079	779,301	3,195,842	2,644,645	4,168,921	3,423,944

- Represents zero.   (D) Withheld to avoid disclosing data for individual operations.   [1] Totals may not add due to rounding.

NASS, Livestock Branch, (202) 720–4751.

Table 15-32.—Refrigerated warehouses: Gross refrigerated space by type of warehouse, United States, biennially, October 2007–2015 [1][2]

Type	2007	2009	2011	2013	2015
	1,000 Cubic Feet				
General:					
Public	2,498,198	2,900,511	3,028,243	3,076,959	3,138,463
Private and Semiprivate	821,998	894,463	931,117	978,426	1,030,460
Total	3,320,194	3,794,974	3,959,354	4,055,385	4,168,921
Apple [3]:					
Public	8,170	23,474	....	....	....
Private and Semiprivate	683,798	613,118	....	....	....
Total	691,968	636,593	....	....	....
Total, all	4,012,162	4,431,567	....	....	....

[1] Warehouse space is defined as all space artificially cooled to temperatures of 50 degrees F. or less, in which food commodities are normally held for 30 days or longer. [2] Totals may not add due to rounding. [3] Apple discontinued in 2011. NASS, Livestock Branch, (202) 720–4751.

Table 15-33.—Alaska crops: Acreage harvested, volume harvested, and value of production, 2006–2015

Year	Oats for grain	Barley for grain	All hay	Potatoes	All vegetables [1]
	Acreage harvested				
	Acres	Acres	Acres	Acres	Acres
2006	800	4,200	20,000	840	341
2007	1,000	3,900	23,000	870	326
2008	500	3,400	18,000	780	347
2009	900	4,400	20,000	740	336
2010	800	4,200	20,000	750	327
2011	1,000	4,800	19,000	720	(NA)
2012	900	4,300	22,000	650	(NA)
2013	400	3,300	20,000	620	(NA)
2014	1,000	5,100	18,000	620	(NA)
2015	1,000	4,300	18,000	540	(NA)
	Volume harvested				
	Bushels	Bushels	Tons	Cwt.	Cwt.
2006	28,000	157,000	22,000	186,000	55,573
2007	47,000	158,000	31,000	176,000	47,340
2008	13,000	99,000	20,000	135,000	40,197
2009	37,000	183,000	23,000	137,000	43,420
2010	48,000	185,000	24,000	150,000	45,740
2011	80,000	175,000	22,000	134,000	(NA)
2012	59,000	207,000	27,000	140,000	(NA)
2013	15,000	110,000	15,000	130,000	(NA)
2014	57,000	217,000	25,000	155,000	(NA)
2015	47,000	146,000	20,000	135,000	(NA)
	Value of production				
	Dollars	Dollars	Dollars	Dollars	Dollars
2006	69,000	557,000	5,500,000	3,757,000	3,302,000
2007	132,000	577,000	8,370,000	3,538,000	3,072,000
2008	39,000	446,000	6,300,000	3,348,000	2,954,000
2009	113,000	814,000	7,130,000	3,315,000	3,155,000
2010	161,000	814,000	7,320,000	3,570,000	3,470,000
2011	276,000	788,000	6,600,000	3,176,000	(NA)
2012	212,000	1,107,000	8,505,000	3,864,000	(NA)
2013	56,000	594,000	5,925,000	3,055,000	(NA)
2014	211,000	1,183,000	9,625,000	3,395,000	(NA)
2015	179,000	796,000	7,400,000	2,903,000	(NA)

(NA) Not available. [1] Excludes greenhouse-grown vegetables. NASS, Crops Branch, (202) 720–2127.

Table 15-34.—Crop ranking: Major field crops, rank by production, major States, 2015

Rank	State	Corn, grain	State	Soybeans	State	All wheat
		1,000 bushels		*1,000 bushels*		*1,000 bushels*
1	IA	2,505,600	IA	553,700	ND	370,023
2	IL	2,012,500	IL	544,320	KS	321,900
3	NE	1,692,750	MN	377,500	MT	185,415
4	MN	1,428,800	NE	305,660	WA	111,540
5	IN	822,000	IN	275,000	TX	106,500
6	SD	799,770	OH	237,000	SD	103,406
7	KS	580,160	SD	235,520	OK	98,800
8	WI	498,780	ND	185,900	ID	88,294
9	MO	492,000	MO	181,035	MN	87,850
10	MI	437,360	AR	155,330	CO	79,635
	US	13,601,198	US	3,926,339	US	2,051,752

Rank	State	Winter wheat	State	Durum wheat	State	Other spring wheat
		1,000 bushels		*1,000 bushels*		*1,000 bushels*
1	KS	321,900	ND	42,463	ND	319,200
2	TX	106,500	MT	18,755	MN	85,800
3	OK	98,800	AZ	14,140	MT	75,640
4	MT	91,020	CA	6,180	SD	60,480
5	WA	89,040	ID	700	ID	29,750
6	CO	79,180	SD	246	WA	22,500
7	ID	57,400			OR	4,650
8	NE	45,980			UT	495
9	SD	42,680			CO	455
10	MI	38,475			NV	110
	US	1,370,188	US	82,484	US	599,080

Rank	State	Sorghum, grain	State	Barley	State	Oats
		1,000 bushels		*1,000 bushels*		*1,000 bushels*
1	KS	281,600	ND	67,200	WI	14,040
2	TX	149,450	ID	56,260	SD	12,615
3	AR	43,120	MT	44,720	MN	12,480
4	NE	23,040	MN	9,240	ND	10,360
5	CO	22,000	CO	8,190	IA	4,161
6	OK	21,320	WY	8,170	PA	3,575
7	SD	18,260	WA	5,040	MI	3,350
8	MO	13,160	PA	2,600	NE	2,680
9	MS	9,085	MD	2,415	TX	2,640
10	LA	6,290	OR	1,924	KS	2,600
	US	596,751	US	218,187	US	89,535

Rank	State	All cotton	State	Peanuts	State	Rice
		1,000 bales		*1,000 pounds*		*1,000 cwt.*
1	TX	5,778	GA	3,473,190	AR	94,341
2	GA	2,300	AL	659,950	CA	37,441
3	MS	670	FL	657,000	LA	28,791
4	AL	550	TX	588,000	MO	12,212
5	CA	525	NC	299,200	MS	10,594
6	CA	526	SC	262,400	TX	8,964
7	AR	471	MS	151,200		
8	MO	400	VA	73,150		
9	OK	374	OK	31,500		
10	AZ	308	NM	15,000		
	US	12,888	US	6,210,590	US	192,343

Rank	State	All hay, baled	State	Alfalfa hay, baled	State	Other hay, baled
		1,000 tons		*1,000 tons*		*1,000 tons*
1	TX	9,720	CA	5,451	TX	9,200
2	CA	6,777	ID	4,200	MO	5,670
3	SD	6,580	SD	4,180	OK	5,320
4	MO	6,398	MT	3,400	KY	5,060
5	NE	6,360	NE	3,400	TN	3,850
6	OK	5,914	WI	3,360	KS	3,420
7	KS	5,890	IA	3,003	NE	2,960
8	KY	5,689	CO	2,870	VA	2,420
9	ND	4,975	ND	2,850	SD	2,400
10	ID	4,860	MN	2,835	AR	2,240
	US	134,388	US	58,974	US	75,414

Rank	State	All tobacco	State	Dry edible beans	State	Potatoes
		1,000 pounds		*1,000 cwt.*		*1,000 cwt.*
1	NC	380,250	ND	8,901	ID	130,400
2	KY	149,830	MI	5,533	WA	100,300
3	VA	52,430	MN	3,896	WI	27,813
4	TN	48,770	NE	3,117	ND	27,600
5	GA	32,400	ID	2,141	CO	22,575
6	SC	26,000	WA	1,582	OR	21,784
7	PA	18,090	CA	1,029	MI	17,550
8	Other	4,566	CO	846	MN	16,200
9	OH	3,610	WY	713	ME	16,160
10			MT	634	CA	13,808
	US	715,946	US	30,121	US	441,205

NASS, Crops Branch, (202) 720–2127.

Table 15-35.—U.S. crop progress: 2015 crop and 5-year average

[In percent]

Week-ending date	Winter wheat							
	Planted		Emerged		Headed		Harvested	
	2014	Avg	2014	Avg	2015	Avg	2015	Avg
2014:								
Sep 7	3	4						
Sep 14	12	11						
Sep 21	25	22						
Sep 28	43	36	14	12				
Oct 5	56	53	28	24				
Oct 12	68	67	43	37				
Oct 19	76	77	56	50				
Oct 26	84	84	67	62				
Nov 2	90	89	77	72				
Nov 9	93	93	83	79				
Nov 16	95	97	87	84				
Nov 23	....	....	92	89				
2015:								
Apr 12					6	8		
Apr 19					16	15		
Apr 26					28	24		
May 3					43	34		
May 10 ...					56	45		
May 17 ...					68	56		
May 24 ...					77	67		
May 31 ...					84	77		
Jun 7					91	84	4	12
Jun 14					96	89	11	20
Jun 21					....	....	19	31
Jun 28					....	....	38	46
Jul 5					....	....	55	59
Jul 12					....	....	65	68
Jul 19					....	....	75	74
Jul 26					....	....	85	80
Aug 2					....	....	93	85
Aug 9					....	....	97	90

Week-ending date	Spring wheat							
	Planted		Emerged		Headed		Harvested	
	2015	Avg	2015	Avg	2015	Avg	2015	Avg
2015:								
Apr 12	17	11						
Apr 19	36	19						
Apr 26	55	29	11	9				
May 3	75	40	19	16				
May 10 ...	87	51	29	25				
May 17 ...	94	65	40	38				
May 24 ...	96	79	51	54				
May 31 ...	....	....	65	69				
Jun 7	....	....	79	80				
Jun 14			....	....				
Jun 21					23	15		
Jun 28					49	29		
Jul 5					76	47		
Jul 12					91	66		
Jul 19					96	83		
Jul 26					....	....	2	5
Aug 2					....	....	8	11
Aug 9					....	....	28	20
Aug 16					....	....	53	31
Aug 23					....	....	75	47
Aug 30					....	....	88	62
Sep 6					....	....	94	76
Sep 13					....	....	97	86

See footnote(s) at end of table.

Table 15-35.—U.S. crop progress: 2015 crop and 5-year average—Continued

[In percent]

Week-ending date	Rice							
	Planted		Emerged		Headed		Harvested	
	2015	Avg	2015	Avg	2015	Avg	2015	Avg
2015:								
Apr 5	14	18	3	4				
Apr 12	26	30	8	11				
Apr 19	32	42	17	23				
Apr 26	39	54	26	34				
May 3	61	62	37	45				
May 10	83	72	53	56				
May 17	89	82	70	66				
May 24	93	92	82	77				
May 31	96	98	90	87				
Jun 7			95	92				
Jun 14								
Jun 21					6	5		
Jun 28					16	9		
Jul 5					25	15		
Jul 12					30	22		
Jul 19					40	33		
Jul 26					51	45		
Aug 2					63	59		
Aug 9					81	71		
Aug 16					88	82	13	10
Aug 23					94	90	18	16
Aug 30					97	95	26	25
Sep 6							35	34
Sep 13							44	44
Sep 20							55	54
Sep 27							69	63
Oct 4							78	71
Oct 11							88	80
Oct 18							95	87

Week-ending date	Corn													
	Planted		Emerged		Silking		Dough		Dented		Mature		Harvested	
	2015	Avg	2015	Avg	2015	Avg	2015	Avg	2015	Avg	2015	Avg	2015	Avg
2015:														
Apr 12	2	5												
Apr 19	9	13												
Apr 26	19	25	2	6										
May 3	55	38	9	12										
May 10	75	57	29	24										
May 17	85	75	56	40										
May 24	92	88	74	62										
May 31	95	94	84	79										
Jun 7			91	90										
Jun 14			97	95										
Jun 28					4	8								
Jul 5					12	18								
Jul 12					27	34								
Jul 19					55	56								
Jul 26					78	77	14	17						
Aug 2					90	89	29	31						
Aug 9					96	96	50	49	9	15				
Aug 16							71	66	21	28				
Aug 23							85	81	39	43				
Aug 30							92	90	60	60	9	15		
Sep 6							96	95	76	75	20	26		
Sep 13									87	86	35	40	5	9
Sep 20									94	93	53	56	10	15
Sep 27									97	97	71	72	18	23
Oct 4											86	83	27	32
Oct 11											94	91	42	43
Oct 18											98	96	59	54
Oct 25													75	68
Nov 1													85	79
Nov 8													93	88
Nov 15													96	94

See footnote(s) at end of table.

Table 15-35.—U.S. crop progress: 2015 crop and 5-year average—Continued

[In percent]

Week-ending date	Sorghum									
	Planted		Headed		Coloring		Mature		Harvested	
	2015	Avg	2015	Avg	2015	Avg	2015	Avg	2015	Avg
2015:										
Apr 5	9	15								
Apr 12	16	20								
Apr 19	19	22								
Apr 26	24	25								
May 3	29	28								
May 10	32	33								
May 17	38	38								
May 24	41	46								
May 31	43	55								
Jun 7	56	68								
Jun 14	71	80								
Jun 21	85	89	18	21						
Jun 28	93	95	21	23						
Jul 5	97	98	24	25						
Jul 12			28	28	17	21				
Jul 19			33	35	20	24				
Jul 26			45	43	23	27				
Aug 2			57	53	29	30				
Aug 9			72	64	32	33				
Aug 16			83	75	39	38	24	26		
Aug 23			90	84	48	45	27	28		
Aug 30			95	90	58	54	29	30	20	23
Sep 6					71	65	33	34	23	24
Sep 13					83	75	43	39	27	26
Sep 20					90	83	52	45	31	28
Sep 27					96	89	65	55	36	32
Oct 4							77	65	43	37
Oct 11							85	75	51	44
Oct 18							91	84	61	52
Oct 25							95	92	71	61
Nov 1									79	72
Nov 8									85	82
Nov 15									91	89
Nov 22									94	93
Nov 29									98	97

Week-ending date	Soybeans											
	Planted		Emerged		Blooming		Setting Pods		Dropping Leaves		Harvested	
	2015	Avg	2015	Avg	2015	Avg	2015	Avg	2015	Avg	2015	Avg
2015:												
Apr 26	2	4										
May 3	13	9										
May 10	31	20										
May 17	45	36	13	12								
May 24	61	55	32	25								
May 31	71	70	49	45								
Jun 7	79	81	64	63								
Jun 14	87	90	75	77								
Jun 21	90	95	84	87								
Jun 28	94	97	89	94	12	9						
Jul 5	96	100	93	97	25	21						
Jul 12			96	100	45	37	6	7				
Jul 19					63	56	17	17				
Jul 26					77	72	34	31				
Aug 2					87	83	54	49				
Aug 9					94	91	69	66				
Aug 16					97	95	79	79				
Aug 23					100	98	87	88				
Aug 30							93	95	9	7		
Sep 6							96	99	18	16		
Sep 13									35	31		
Sep 20									56	50	7	7
Sep 27									74	70	21	16
Oct 4									85	83	42	32
Oct 11									92	91	62	54
Oct 18									96	96	77	68
Oct 25											87	80
Nov 1											92	88
Nov 8											95	93

See footnote(s) at end of table.

Table 15-35.—U.S. crop progress: 2015 crop and 5-year average—Continued

[In percent]

Week-ending date	Cotton, Upland									
	Planted		Squaring		Setting Bolls		Bolls Opening		Harvested	
	2015	Avg	2015	Avg	2015	Avg	2015	Avg	2015	Avg
2015:										
Apr 5	2	6								
Apr 12	4	8								
Apr 19	8	11								
Apr 26	10	16								
May 3	17	22								
May 10	26	32								
May 17	35	46								
May 24	47	61								
May 31	61	78	3	6						
Jun 7	81	89	7	10						
Jun 14	91	96	13	16						
Jun 21	94	100	22	26						
Jun 28	98	100	35	40	5	8				
Jul 5			48	55	10	14				
Jul 12			61	70	18	24				
Jul 19			76	81	33	36				
Jul 26			85	88	44	49				
Aug 2			92	94	57	64				
Aug 9			96	97	68	79	7	8		
Aug 16					73	88	10	12		
Aug 23					83	92	14	18		
Aug 30					94	96	22	27		
Sep 6					95	100	31	38		
Sep 13							46	51	4	7
Sep 20							57	61	7	9
Sep 27							69	70	11	12
Oct 4							77	78	16	18
Oct 11							89	84	22	25
Oct 18							94	89	31	32
Oct 25							95	94	42	43
Nov 1									50	54
Nov 8									58	65
Nov 15									64	74
Nov 22									70	82
Nov 29									80	88

Week-ending date	Barley								Oats							
	Planted		Emerged		Headed		Harvested		Planted		Emerged		Headed		Harvested	
	2015	Avg	2015	Avg	2015	Avg	2015	Avg	2015	Avg	2015	Avg	2015	Avg	2015	Avg
2015:																
Apr 5	....	......	....	......	....	......	....	......	32	37	26	30				
Apr 12	27	15	....	......	....	......	....	......	43	45	28	33				
Apr 19	43	24	....	......	....	......	....	......	59	53	32	37				
Apr 26	56	35	18	9	....	......	....	......	71	60	43	43				
May 3	75	47	39	17	....	......	....	......	85	67	57	50				
May 10	88	58	59	28	....	......	....	......	93	76	72	59				
May 17	95	70	72	40	....	......	....	......	96	84	83	69				
May 24	....	......	86	55	....	......	....	......	....	......	91	79	26	30		
May 31	....	......	95	70	....	......	....	......	....	......	95	88	30	33		
Jun 7	....	......	....	......	....	......	....	......	....	......	....	......	38	39		
Jun 14	....	......	....	......	....	......	....	......	....	......	....	......	51	49		
Jun 21	....	......	....	......	38	14	....	......	....	......	....	......	67	60		
Jun 28	....	......	....	......	62	26	....	......	....	......	....	......	83	71		
Jul 5	....	......	....	......	84	47	....	......	....	......	....	......	92	82		
Jul 12	....	......	....	......	95	69	....	......	....	......	....	......	96	90	11	15
Jul 19	....	......	....	......	....	......	....	......	....	......	....	......	....	......	16	23
Jul 26	....	......	....	......	....	......	5	3	....	......	....	......	....	......	27	34
Aug 2	....	......	....	......	....	......	17	8	....	......	....	......	....	......	43	48
Aug 9	....	......	....	......	....	......	42	18	....	......	....	......	....	......	62	62
Aug 16	....	......	....	......	....	......	66	32	....	......	....	......	....	......	80	75
Aug 23	....	......	....	......	....	......	86	50	....	......	....	......	....	......	90	85
Aug 30	....	......	....	......	....	......	93	67	....	......	....	......	....	......	95	91
Sep 6	....	......	....	......	....	......	95	82	....	......	....	......	....	......	....	......

See footnote(s) at end of table.

Table 15-35.—U.S. crop progress: 2015 crop and 5-year average—Continued

[In percent]

Week-ending date	Sunflower				Sugarbeets			
	Planted		Harvested		Planted		Harvested	
	2015	Avg	2015	Avg	2015	Avg	2015	Avg
2015:								
Apr 5	....	....	....	....	5	7		
Apr 12	....	....	....	....	15	14		
Apr 19	....	....	....	....	57	25		
Apr 26	....	....	....	....	78	42		
May 3	....	....	....	....	96	51		
May 24	16	15						
May 31	32	29						
Jun 7	49	47						
Jun 14	68	66						
Jun 21	80	81						
Jun 28	89	91						
Jul 5	98	96						
Sep 13	....	....	....	....	....	....	11	5
Sep 20	....	....	....	....	....	....	14	9
Sep 27	....	....	....	....	....	....	17	13
Oct 4	....	....	....	....	....	....	44	27
Oct 11	....	....	10	16	....	....	70	47
Oct 18	....	....	33	28	....	....	79	67
Oct 25	....	....	54	44	....	....	86	81
Nov 1	....	....	69	59	....	....	91	90
Nov 8	....	....	80	74	....	....	96	97
Nov 15	....	....	88	84				
Nov 22	....	....	95	91				

Week-ending date	Peanuts					
	Planted		Pegging		Harvested	
	2015	Avg	2015	Avg	2015	Avg
2015:						
Apr 26	5	6				
May 3	10	14				
May 10	26	27				
May 17	47	46				
May 24	68	67				
May 31	83	83				
Jun 7	92	91				
Jun 14	96	96	2	5		
Jun 21	....	....	16	12		
Jun 28	....	....	32	24		
Jul 5	....	....	45	39		
Jul 12	....	....	59	55		
Jul 19	....	....	73	67		
Jul 26	....	....	82	78		
Aug 2	....	....	88	87		
Aug 9	....	....	94	92		
Aug 16	....	....	97	96		
Sep 13	....	....	....	....	4	3
Sep 20	....	....	....	....	9	7
Sep 27	....	....	....	....	18	14
Oct 4	....	....	....	....	23	24
Oct 11	....	....	....	....	32	37
Oct 18	....	....	....	....	45	53
Oct 25	....	....	....	....	58	67
Nov 1	....	....	....	....	72	79
Nov 8	....	....	....	....	77	88
Nov 15	....	....	....	....	82	94
Nov 22	....	....	....	....	87	97
Nov 29	....	....	....	....	93	99

NASS, Crops Branch, (202) 720–2127.

Appendix I
Telephone Contact List

Appreciation is expressed to the following agencies for their help in this publication. The information offices are listed to provide help to those users who require additional information about specific tables in this publication.

Agricultural Marketing Service:
USDA/AMS
Room 2619 South Bldg.
Washington, DC 20250
202–720–8998

Agricultural Research Service:
USDA/ARS
5601 Sunnyside Ave
Bldg 1, Rm 2251
Beltsville, MD 20705–5128
301–504–1636

Animal and Plant Health Inspection Service:
USDA/APHIS
4700 River Rd
Riverdale, MD 20737
301–851–4100

Center for Nutrition Policy and Promotion:
USDA/CNPP
3101 Park Center Drive
Alexandria, VA 22302
703–305–7600

Economic Research Service:
USDA/ERS
355 E Street, S.W.
Washington, DC 20024-3221
202–694–5050

Farm Credit Administration:
FCA
1501 Farm Credit Dr.
McLean, VA 22102
703–883–4056

Farm Service Agency:
USDA/FSA
Room 4078 South Bldg.
Washington, DC 20250
202–720–7163

Food and Nutrition Service:
USDA/FNS
3101 Park Center Drive, Room 914
Alexandria, VA 22302
703–305–2062

Foreign Agricultural Service:
USDA/FAS
Room 5076 South Bldg.
Washington, DC 20250
202–720–7115

Forest Service:
USDA/FS
2nd Floor Central Wing, Yates Bldg.
Washington, DC 20250
202–205–8333

National Agricultural Statistics Service:
USDA/NASS
Room 5038 South Bldg.
Washington, DC 20250
202–720–3878

National Marine Fisheries Service:
USDC/NOAA/NMFS
1315 East/West Highway,
SSMC III - Room 12405
Silver Spring, MD 20910–3282
301–427–8103

Natural Resources Conservation Service:
USDA/NRCS
Room 6121 South Bldg.
Washington, DC 20250
202–720–2182

Risk Management Agency:
USDA/RMA
Room 403
Kansas City, MO 64133-4675
816–926–1805

Rural Business-Cooperatives Service:
USDA/RECD/RBS
Room 4801 South Bldg.
Washington, DC 20250
202–720–1019

Rural Utilities Service:
USDA/RD/RUS
Stop 1510, Room 5135
1400 Independence Avenue, SW
Washington, DC 20250
202–720–9540

INDEX